Production and Operations Analysis

THE MCGRAW-HILL/IRWIN SERIES
Operations and Decision Sciences

Production and Operations Analysis

Steven Nahmias
Santa Clara University

McGraw-Hill Irwin

Boston   Burr Ridge, IL   Dubuque, IA   Madison, WI   New York   San Francisco   St. Louis
Bangkok   Bogotá   Caracas   Kuala Lumpur   Lisbon   London   Madrid   Mexico City
Milan   Montreal   New Delhi   Santiago   Seoul   Singapore   Sydney   Taipei   Toronto

McGraw-Hill Higher Education

*A Division of The **McGraw-Hill** Companies*

PRODUCTION AND OPERATIONS ANALYSIS

Published by McGraw-Hill/Irwin, an imprint of The McGraw-Hill Companies, Inc. 1221 Avenue of the Americas, New York, NY, 10020. Copyright © 2001, 1997, 1993, 1989, by The McGraw-Hill Companies, Inc. All rights reserved. No part of this publication may be reproduced or distributed in any form or by any means, or stored in a data base or retrieval system, without the prior written consent of The McGraw-Hill Companies, Inc., including, but not limited to, in any network or other electronic storage or transmission, or broadcast for distance learning. Some ancillaries, including electronic and print components, may not be available to customers outside the United States.

This book is printed on acid-free paper.

1 2 3 4 5 6 7 8 9 0 CSI/CSI 0 9 8 7 6 5 4 3 2 1 0

ISBN   0-07-231265-3

Publisher: *Jeffrey Shelstad*
Sponsoring editor: *Scott Isenberg*
Developmental editor: *Christina Sanders*
Marketing manager: *Zina Craft*
Project manager: *Kimberly D. Hooker*
Production supervisor: *Michael McCormick*
Freelance design coordinator: *Gino Cieslik*
Supplement coordinator: *Cathy Tepper*
Media technology producer: *Ed Przyzycki*
Cover and interior design: *Kiera Cunningham*
Compositor: *Interactive Composition Corporation*
Typeface: *10.5/12 Times Roman*
Printer: *Courier Stoughton, Inc.*

Library of Congress Cataloging-in-Publication Data

Nahmias, Steven.
    Production and operations analysis / Steven Nahmias.—4th ed.
      p. cm. (The McGraw-Hill/Irwin series in operations and decision sciences)
    Includes index.
    ISBN 0-07-231265-3 (alk. paper)
    1. Production management. I. Title. II. McGraw-Hill/Irwin series. Operations and decision sciences.
    TS155.N112   2001
    658.5—dc21   00-060068

www.mhhe.com

In memory of my parents
Liza and Morris

Preface to the Fourth Edition

I am very pleased that this book has made it to a fourth edition, and am grateful to all of you who have adopted it over the past 12 years. This book is rare in that it spans two distinct markets: business and engineering. A previous survey revealed that adoptions are split almost evenly between the two areas, with about two-thirds of the units in schools of business. Keeping in mind the findings of this survey, and responding to both general market demands as well as specific reviewer suggestions, I have made several key changes to the fourth edition of *Production and Operations Analysis* that I believe have improved the book considerably. These changes include:

- The addition of a new chapter on supply chain management
- Enhanced international examples
- A stronger and simpler strategic framework, outlined in Chapter 1
- Increased and enhanced use and integration of Excel spreadsheets
- Selective updating and addition of "Snapshot Applications" boxes

While the fourth edition has 12 chapters, they are not the same 12 chapters that were in the first three editions. The fourth edition is the first to have a new chapter and also the first to have a chapter eliminated. Chapter 6, "Supply Chain Management," is new. It follows the style of other chapters in being a mix of quantitative and qualitative material. The chapter does include some material from the third edition that appeared in Chapters 1, 7, and 12, but is mostly new. Chapter 12 from previous editions, "Recent Advances in Operations Planning and Control," has been eliminated. What were recent advances in 1989 are no longer so, and I was concerned that much of the material in this chapter was being ignored because it was separated from the core topics. I indicate below where some of the material formerly in Chapter 12 now appears.

A more coherent treatment of operations strategy is provided in Chapter 1. Section 1.2, "A Framework for Operations Strategy," is new. The Snapshot Application featuring Wal-Mart has been moved to the new supply chain chapter, as was the other supply chain material that was formerly part of Chapter 1. A new Snapshot Application on Read-Rite's strategic alliance with Sumitomo Metals now appears in this chapter in the section on global operations. Most of the remainder of the chapter has remained intact. Chapters 2–5 are largely unchanged. The linear programming supplement after Chapter 3 has been updated to reflect the fact that Excel's Solver is now the standard for spreadsheet-based linear programming.

Chapter 6 is the new chapter on supply chain management. This chapter was placed here because of its conceptual tie with inventory management, the topic of the two previous chapters. Most of this chapter is new and it provides, I believe, one of the most comprehensive treatments of supply chain management in operations texts. The chapter starts with the Wal-Mart Snapshot Application, followed by a comprehensive treatment of the transportation and trans-shipment problems. This section includes a new Snapshot Application on the successful multiechelon inventory model developed for IBM's spare parts management. Section 5 on DRP is new. Section 6 on vehicle routing formerly appeared in Chapter 7. Section 7 on product design incorporates some material from the third edition and some new material. The Snapshot Application on Dell Computer featured in this section is new. Section 8 on the role of information in the supply chain is mostly drawn from the third edition. Sections 9 (multilevel distribution systems) and 10 (global issues) are new. Section 10 also features a new Snapshot Application on the work done for DEC in the 1990s (prior to its acquisition by Compaq).

Chapters 7–12 are the former Chapters 6–11 respectively. The material from the old Chapter 12 on flexible manufacturing systems is now part of Chapter 10, and the material from the old Chapter 12 on design for manufacturability is now in Chapter 11. The chapter on reliability and maintenance is now the final chapter, and it remains essentially unchanged from the last edition.

There are many people to thank for their help with this edition. First, I am grateful to the reviewers retained by the publisher who provided valuable feedback and suggestions for improvements. They include:

Chung Lee
Texas A&M University

Ted Lloyd
Eastern Kentucky University

Joe Milner
Washington University

Izzet Sahin
University of Wisconsin at Milwaukee

William Sherrard
San Diego State University

Sridhar Tayur
Carnegie Mellon University

David Yao
Columbia University

I continue to be indebted to colleagues at Santa Clara University—Chuck Feinstein, Rhonda Righter, Steve Smith, and Andy Tsay—and to the reviewers of previous editions, whose comments and suggestions helped immeasurably in the early stages of this book's development. They include Nicholas G. Hall (Ohio State University), Arie Harel, John Hebert (University of Akron), Michael Kim (University of California at Riverside), Hau Lee (Stanford University), Binshan Lin (Louisiana State University at Shreveport), John Loucks (Bowling Green State University) Sanjay Mehrotra (Northwestern University), Kamran Moizadeh (University of Washington), Eddy Patuwo (Kent State University), Jayant Rjgopal (University of Pittsburgh), Larry Robinson, Meir Rosenblatt, Linus

Schrage (University of Chicago), David Simchi-Levi (Massachusetts Institute of Technology), and Candace Yano (University of California at Berkeley).

I would also like to thank the team at McGraw-Hill/Irwin for their help throughout this project. They include Sponsoring Editor, Scott Isenberg; Developmental Editor, Christina Sanders; Marketing Manager, Zina Craft; Project Manager, Kimberly Hooker; Designer, Kiera Cunningham; Production Manager, Michael McCormick; Supplements Coordinator, Cathy Tepper; and New Media Producer, Ed Przyzycki. I am especially grateful to you, the users, who have continued to make this book a success. Many of you made suggestions, and all of them have been carefully considered in this revision. Please feel free to let me know via email (snahmias@scu.edu) your opinion of the fourth edition. Finally, special thanks to my wife Vivian for her continued encouragement and support.

Introduction

1 The Production and Operations Management Function

This book concerns production in the broadest sense of the word: that is, production of both *goods* and *services*. The term *production* has its roots in studies of factory automation. The term *operations management* has become popular in recent years to reflect broader interests. This book is about the analytical methods used to support the production and operations management function. Production and operations management is the process of managing people and resources in order to create a product or a service. Operations involve the logistics required to support the production function.

Marketing, finance, and production are the three major functional areas of the firm. Marketing acts as a buffer to the outside world. Marketing chooses the manner in which products are presented to the consumer and discovers the directions of consumer preferences. Finance is responsible for finding sources of outside funding and managing the capital within the firm. Production is responsible for converting raw materials into products or for providing a specific set of services. The activities of the three functional areas of the firm must be carefully coordinated for the firm to operate efficiently.

Bowman and Fetter (1967) list the following as the major economic problems of production management:

1. Inventory.
2. Production scheduling and control.
3. Equipment selection and replacement.
4. Maintenance.
5. Size and location of plants.
6. Plant layout and structure.
7. Quality control and inspection.
8. Traffic and materials handling.
9. Methods.

Although this list was compiled over 30 years ago, it still accurately represents the most important economic decisions of production management. This text treats each of these areas except for the last two. Although work methods and materials handling are both important topics, they are not areas in which analytical methods have had a significant impact.

2 History of Production Management

According to Skinner (1985), there are five periods of industrial history that stand out in the development of manufacturing management:

1780–1850	Manufacturing leaders as technology capitalists.
1850–1890	Manufacturing leaders as architects of mass production.
1890–1920	Manufacturing management moves down in the organization.
1920–1960	Manufacturing management refines its skills in controlling and stabilizing.
1960–1980	Shaking the foundations of industrial management.

During the early years of the Industrial Revolution, production began to shift from low-volume activity to larger-scale operations. Although the scale of these early operations was large, the machinery was not particularly complex and production operations were rigid. The management of these operations remained essentially in the hands of top management with the aid of overseers. Working conditions during this period were often abysmal.

The major thrust of the Industrial Revolution took place in the second 40-year period from 1850–1890. During this period, the concepts of mass production and the assembly line were born. Since coal could be efficiently transported, plants could be located in a larger variety of locations. The plant foreman had enormous power and influence during this period.

According to Skinner, the job of production manager actually came into being in the period 1890–1920. Manufacturing processes became too complex to be handled by top-management personnel any longer. With this complexity came the need for scientific management techniques. Frederick Taylor (often called the father of industrial engineering) is generally credited with being the originator of the concept of scientific management. Most of the scientific management techniques introduced around the turn of the century involved merely breaking a task down into its various components. These techniques are probably less scientific than just orderly. With the new levels of complexity, the single plant foreman could no longer coordinate the demands of producing a varied product line and changing production schedules. The age of the specialist was upon us.

The enormous worldwide depression that took place in the 1930s notwithstanding, in many ways the period 1920–1960 can be considered a golden age for the development of industry in the United States. By 1960, the United States was the preeminent economic power in the world. With the growth of the labor movement, working conditions had improved enormously. True scientific methods started finding their way into the factory. Mathematical models for learning, inventory control, quality control, production scheduling, and project management gained acceptance by the user community. Top management often came through the ranks of production professionals during this period.

Since 1960, many American companies have relinquished their domination of certain markets. Products that were traditionally produced in the United States are now imported from Germany, Japan, and the Far East. Many products are produced more cheaply and with higher quality overseas. Furthermore, management–employee relations are often better in foreign companies. Quality circles, introduced in Japan, allowed employees to input opinions about product development and production procedures. Far more sophisticated scientific production methods have been adopted in Japan than in other countries. For example, there are many more robots and modern flexible manufacturing systems in Japan than in the United States.

Production managers will require better training and preparation than ever before. They will have to be well versed in many new technologies and be able to deal with complex information systems and decision support systems. Computers will continue to improve. As powerful computing capability is brought directly to the factory floor, sophisticated mathematical models will play an ever-increasing role in modern production and operations management. The aim of this text is to provide the prospective production professional with the technical tools necessary to face the coming century.

3 Overview of the Text

Albert Einstein said that physics should be explained as simply as possible, but no simpler. This reflects the philosophy behind this book.

Features of this book include

- Breadth.
- Depth.
- Clarity.
- Modularity.
- Example.
- Reinforcement.
- Historical notes and summaries.
- Spreadsheet problems.
- Snapshot applications.

Almost every chapter includes material rarely found in survey texts, and some include topics only treated in published articles. The writing is economical, but all concepts are carefully explained. The chapters are designed to stand alone. Many sections can be skipped without loss of continuity, allowing the instructor to tailor the sequencing and the depth of coverage to his or her needs. I make extensive use of examples. Some are case-type examples spanning most of a chapter, while others are short examples used to illustrate a particular concept or technique. I have tried to make the examples realistic enough to show the student how a method would actually be used in practice.

I have chosen to include problems at the end of each section as well as at the end of most chapters. Section problems have two advantages. They give the student immediate reinforcement of the material discussed in that particular section. Also, it is easier for an instructor only covering portions of chapters to assign appropriate problems. There are more than 550 problems in the book.

Each chapter, except the first, includes a section titled "Historical Notes." As a student, I always found textbooks to be somewhat sterile, conveying little of the history of the developments discussed. I think it is interesting to know when and how things were developed. I realize that I am taking a big risk by including these sections, since I may have incorrectly identified some original sources. I encourage knowledgeable readers to let me know if this is the case. The summaries at the end of each chapter will provide the reader with an overview of the material treated in the chapters.

Problems designed to be solved with the aid of a computer (in particular, with a spreadsheet program) are included. These problems are marked with a graphic to identify them and almost always appear in the problem set at the end of the chapters.

How topics should be sequenced is a personal issue. Undoubtedly, many of you would sequence the material differently. My rationale is as follows. The first chapter is primarily expository and provides a nontechnical introduction to the important strategic issues facing firms today. The remainder of the book covers three general areas: inventory (broadly defined), scheduling and facilities design, and quality. The inventory sections include Chapters 2–6, covering forecasting, aggregate planning, deterministic and stochastic single-item systems, and supply chains respectively. MRP and JIT, treated in Chapter 7, overlaps with both inventory and scheduling, and provides the link to the next three chapters. These are Chapters 8–10 and include job shop scheduling, project scheduling, and layout and location. The final two chapters treat quality, broadly defined. Chapter 11 is a very comprehensive treatment of statistical quality control, acceptance sampling, and several additional quality topics. Chapter 12 is a chapter rarely, if ever, found in operations texts. It is my feeling that a comprehensive treatment of quality should include reliability and maintenance as well.

Previous editions of the book have had over 100 adopters from both business and engineering schools. While the book has not lost its quantitative character, revisions have been aimed at widening the audience by adding more qualitative material. The new supply chain chapter does include several quantitative sections but is, by and large, focused on strategic concerns. Because of the book's modular design, it can be adapted easily to the needs of the instructor. The first six chapters can serve as a reasonable one-quarter course on inventory and related topics; Chapters 7–10 provide a reasonable treatment of scheduling including MRP and JIT; and the last two chapters could form the basis for a course in quality and reliability. Alternatively, a survey course could be comprised of a few sections from each chapter. I believe there is sufficient material in the text for a two-quarter or two-semester sequence as well.

4 Overview of Each Chapter

Chapter 1. Strategy and Competition

Chapter 1 provides a brief overview of the significant issues in competition and operations strategy. The introduction positions manufacturing strategy as part of the overall business strategy of the firm. Section 1 considers the issue that we are evolving into a service economy and the concerns that this should engender if true. Section 2 provides an overview for the general framework for strategic issues. Section 3 considers the classic view of operations strategy developed by B. F. Skinner of Harvard. Section 4 considers issues arising from globalization of manufacturing and includes a new Snapshot Application on Read-Rite Corporation's strategic alliance with a Japanese metals company. Sections 5 to 8 consider several modern strategic initiatives undertaken in recent years. These include business process reengineering, just-in-time, the emphasis on quality, and time-based competition. Section 9 considers the product life cycle and how product and process life cycles need to be matched.

Section 10 treats learning and experience curves. These are certainly of strategic importance to anyone interested in projecting costs and efficiencies. The final section of Chapter 1 treats models for planning for capacity growth. This section also includes a qualitative discussion of some of the issues that arise in plant location decisions (mathematical models used for making location decisions are discussed in depth in Chapter 10).

Chapter 2. Forecasting

This chapter provides an in-depth treatment of this important topic. Care has been taken to provide both a rigorous treatment of the topic and one that is accessible to a wide audience. Issues such as the periods from and to which one forecasts are considered.

The first four sections of the chapter include an expository discussion of several important forecasting issues. In Sections 5 and 6 the notation used throughout the chapter is introduced and measures of performance are defined. Section 7 is devoted to a discussion of the most popular methods for predicting stationary series including simple moving averages and simple exponential smoothing. The difference between one-step-ahead and multiple-step forecasts is made clear. Also a careful comparison between exponential smoothing and moving averages is made. We discuss consistent values of N for moving averages and α for exponential smoothing in Appendix A and prove that consistent values of these parameters result from equating the variance of forecast errors. The Snapshot Application at the end of Section 7 discusses Sport Obermeyer's sophisticated forecasting system.

Section 8 considers methods for forecasting series in which a linear trend is present. We consider simple linear regression and Holt's method. Undoubtedly, there are many other methods that we could have considered, but these seemed to be the most representative. Holt's method is presented rather than R. G. Brown's double exponential smoothing since, in the opinion of this writer, Holt's method is more intuitive.

Section 9 considers methods for seasonal series. We present a popular method for deseasonalizing a series using moving averages and a careful and detailed exposition of Winter's exponential smoothing method for seasonal series. Few texts at this level consider Winter's method. Section 10 considers several practical issues when implementing forecasting methods. Section 11 provides a brief overview of advanced methods including forecasting in the presence of lost sales. Section 12 discusses the distinction between variance of demand and variance of forecast error. This issue is also treated in Chapter 5.

Chapter 3. Aggregate Planning

This chapter presents a comprehensive treatment of aggregate planning and considers several issues generally overlooked. One is the meaning of an aggregate unit of production. Students are often confused about the difference between normal units and aggregate units and few authors explain this difference. Prior to introducing the aggregate planning methodology, we discuss the need for aggregate planning and the relevant costs, and introduce a case-type problem used throughout the chapter.

Section 4 considers the two simplest aggregate plans: constant workforce plan and the zero inventory plan. The application of linear programming to aggregate planning is carefully developed. Since aggregate planning is an integer programming problem, rounding considerations cannot be ignored. We present a detailed example with computer outputs to illustrate the proper steps in rounding.

Sections 7 and 8 consider the HMMS Linear Decision Rule and Bowman's management coefficients. The material on disaggregating aggregate plans in Section 9 is rarely discussed in most texts. Section 10 considers global issues in aggregate planning and includes a Snapshot Application.

Supplement 1. Supplement on Linear Programming

This supplement differs from the chapters only in that there are no problems. The purpose of this supplement is to bring the student up to speed on linear programming for those who may not have been exposed to this important area. As more business programs drop the management science requirement, POM texts will have to include material traditionally taught in these courses. This supplement is more detailed than most and is designed to give the student sufficient background on the *application* (as opposed to the theory) of LP to be able to apply it. The major change in this section is the elimination of the discussion of LINDO. A comprehensive discussion of Solver is included, in recognition that Solver is now the de facto standard.

Chapter 4. Inventory Control Subject to Known Demand

Chapters 4 and 5 treat inventory models in depth. Sections 1, 2, and 3 of Chapter 4 present an expository description of types of inventories, motivation for holding inventories, and characteristics of inventory control systems. Section 4 presents a detailed discussion of the forms of the cost functions on which the models for Chapters 4 and 5 are based. The treatment of the simple EOQ model in Section 5 includes discussions of sensitivity analysis and order lead times. Section 6 treats the extension of the simple EOQ to the case of a finite production rate.

Section 7 concerns quantity discount models including both all units and incremental discounts. The material of Section 8, resource-constrained multiple product systems, is rarely found in OM texts. In this section we show under what circumstances such problems can be solved easily and under what circumstances Lagrange multipliers are required. Section 9 on EOQ models for production planning considers the interface between production scheduling and classical inventory models. Section 10 provides a brief overview of software for inventory control. The Snapshot Application highlights Mervyn's successful inventory control system.

Chapter 5. Inventory Control Subject to Uncertain Demand

This chapter presents a detailed and comprehensive treatment of stochastic inventory models. It begins with a general discussion of randomness in the context of inventory control and presents a case example that is used to illustrate methodologies presented in the remainder of the chapter. A rigorous derivation of the newsboy model is included as well as a careful discussion of the interpretation of costs. In this presentation I have chosen the overage and underage cost formulation instead of the formulation with order, holding, and penalty costs that one finds in many OR texts. I have found this particular formulation easier for the student to understand and apply. A unique feature of our coverage is the extension of the newsboy model to multiple planning periods.

Section 4 is a comprehensive treatment of lot size–reorder point models. Models of this type form the basis for many commercial inventory control systems. The section begins with definitions and descriptions of variables and costs. Section 5 treats service levels and the difference between setting the probability of not stocking out in the lead time and the fill rate, an important distinction. Other issues discussed in this section include scaling of lead time demand, imputed shortage cost, linking of inventory and forecasting, and random lead times.

Although a formal derivation of (s, S) policies is beyond the scope of this text, (s, S) policies are mentioned along with a brief discussion of service levels in periodic review systems. Section 7 considers ABC analysis and exchange curves for multiproduct inventory systems. The chapter concludes with a summary of several advanced topics in inventory modeling and a brief discussion of EDI (which is also discussed in more detail in Chapter 12).

There are five appendices to Chapter 5. I recommend that instructors be aware of the material contained there as they may wish to include it in their lectures. Appendix 5–A is a basic probability review. Appendix 5–B adds considerable technical depth to the discussion of the newsboy model, including the correct interpretation of overage and underage costs for infinite horizon versions of the backorder and the lost sales cases. Appendix 5–C provides details of the derivation of the (Q, R) policy. Appendix 5–D extends the models treated in the chapter to the Poisson and Laplace distributions.

Chapter 6. Supply Chain Management

This chapter, new with the fourth edition, provides a comprehensive overview of the key topics in this important new area of operations management. The chapter starts with definitions and a Snapshot Application featuring Wal-Mart. Sections 1–4 of the chapter consider the transportation problem, generalizations, and more general network formulations of transshipment problems. Using Excel to solve these problems is also treated in this section. Section 4 concludes with a Snapshot Application of IBM's spare parts management system.

Section 5 treats the DRP approach to resource planning. Section 6 on vehicle routing formerly appeared in Chapter 7 on Operations Scheduling. Section 7 on product design for supply chains includes some material that formerly appeared in Chapter 12 of the third edition and some new material. Dell Computer is featured in the Snapshot Application in this section.

Section 8 on the role of information in supply chains includes some material from the third edition and some new material as well. This section includes a discussion of the bullwhip effect. Sections 9 on multilevel distribution systems and 10 on global supply chains are new. In Section 9, I discuss some of the multiechelon inventory models and how they relate to modern supply chain issues. In Section 10 I discuss some of the important global issues that must be taken into account when designing a supply chain. The chapter concludes with another Snapshot Application featuring DEC's global supply chain design.

Chapter 7. Push and Pull Production Control Systems: MRP and JIT

We treat MRP and JIT as two philosophies for manufacturing control. The first half of the chapter treats MRP and lot sizing. This chapter has several unique features. One is that dynamic lot sizing is included here rather than in Chapter 4 or in a separate chapter on production planning. In this way the importance of lot sizing in MRP can be made clear. How lot sizing at one level affects the pattern of requirements at lower levels is often overlooked in presentations of this material.

A case study is presented to show how the explosion calculus works. The example is carried through to three levels of the product structure. Optimal lot sizing using the Wagner–Whitin algorithm has been relegated to Appendix A and can be skipped without loss of continuity. We consider Silver–Meal, least unit cost, and part period balancing heuristics as well. (The period order quantity heuristic is also discussed in the problems.) By presenting these lot sizing algorithms here, we can show how lot sizing decisions impact on the materials requirements plan at the various levels of the product structure. Also, we present an original heuristic (similar to others that have appeared in the literature) for constrained lot sizing.

Section 7 provides an in-depth treatment of JIT. Some of this material appeared in Chapter 12 of earlier editions, but there are several new topics as well, including SMED and CONWIP. MRP and JIT philosophies are compared and contrasted. The chapter concludes with a discussion of the growing market for MRP software.

We discuss problems that arise when implementing an MRP system in Section 8 of this chapter. These include uncertainty, capacity planning, rolling horizons, and system nervousness, to name a few.

Chapter 8. Operations Scheduling

This chapter presents an overview of the most important results for sequence scheduling. The effect of sequencing rules on several measures of performance is made clear by presenting a detailed example showing the effect of the four sequencing rules: first-come first-served, shortest processing time first, earliest due date, and critical ratio. Several algorithms for single machine sequencing are presented, including Moore's algorithm for minimizing the number of tardy jobs and Lawler's algorithm for precedence constraints.

The next portion of the chapter considers sequencing algorithms for multiple machines. We present Johnson's algorithm for both two and three machine scheduling and

Aaker's graphical procedure for two jobs on m machines. This is probably the only text also to consider results for stochastic scheduling. We present several of the more elegant results in this area and include some proofs to enhance the student's understanding of the material. This edition also includes a discussion of stochastic scheduling in a dynamic environment. This section is an application of queuing theory to sequence scheduling.

The last major topic considered in Chapter 8 is assembly line balancing. We present only one heuristic for solving the problem: the ranked positional weight technique. The section discusses experimenting with different values of the cycle time in order to achieve an optimal balance.

A brief review of post-MRP scheduling software is provided here, as well as an overview of using simulation for scheduling. The Snapshot Application provides descriptions of several successful scheduling initiatives in the real world.

Supplement 2. Supplement on Queuing Theory

This material supports the section on dynamic stochastic scheduling in Chapter 8 by providing the student with a good overview of queuing theory. It includes a derivation of the steady-state probability for the $M/M/1$ queue and results without proof for more complex models. As with the LP Supplement, it should be helpful to students who have not been exposed to queuing theory elsewhere.

Chapter 9. Project Scheduling

This chapter begins with a description of the project scheduling problem and how projects are represented as networks. A case study is introduced to illustrate the methodologies discussed throughout the remainder of the chapter. We present the details of CPM including the calculation of earliest and latest start and finish times as well as identification of the critical path. Time costing methods also are considered in Section 4.

The PERT method for handling random activity times is treated in Section 5. The traditional PERT approach, which assumes that the length of the critical path has the normal distribution and is the path with the longest *expected* activity time, is presented. An alternative approximation is to assume that the time required to complete two or more paths are independent random variables. In some cases, the assumption of path independence could give more accurate results than the traditional PERT approach, which assumes that the path with the longest expected value is the longest path.

This chapter also includes material rarely covered in production texts. We show by example how one incorporates resource constraints into the standard project planning framework and how one constructs resource loading profiles. This chapter also includes a section showing how one would apply linear programming to solve project scheduling problems. In addition to linear programming, one also could use special-purpose project scheduling software, reviewed here. Organizational structures to facilitate project management also are discussed, as well as PC project management software.

Chapter 10. Facilities Layout and Location

Chapter 10 is a comprehensive treatment of the major developments in the area of layout and location. It makes liberal use of examples to illustrate complex methodology. We discuss the traditional charts for assisting with layout decisions, including the from-to chart and the rel chart. In Section 4 we show how one uses the simple assignment model for solving location problems in which there is no interaction among facilities, and only a small set of alternative locations. The quadratic assignment model for solving more complex problems also is presented, although the details of implementing the algorithm are beyond the scope of this text.

Section 6 presents a detailed description of CRAFT. We show how centroids (used by CRAFT) are computed in Appendix A of the chapter and apply the method to a specific example. Other computerized layout methods are considered in less detail: these include COFAD, ALDEP, CORELAP, and PLANET. This section concludes with an interesting debate from the literature about whether computers or humans find better layouts.

In Sections 8 through 11 of Chapter 10 we treat the problem of locating new facilities. We consider qualitative location issues treated in detail in Chapter 1 as part of our discussion of strategy, so that discussion is not repeated here. In Section 9 we discuss the method for locating one new facility subject to a rectilinear distance measure. Other topics considered in this section include contour lines and a minimax objective rather than a weighted objective. Squared Euclidean and Euclidean distance measures are considered in Section 10. In Section 11 we include brief discussions of several other extensions including models for locating multiple facilities, location-allocation problems, discrete location models, and network location models.

Chapter 11. Quality and Assurance

This chapter is one of the longer ones in the text and presents a detailed coverage of the important techniques and issues in the quality area. The first part of the chapter is devoted to control charts. A brief review of the relevant results from probability theory is presented first. As with most of the other techniques discussed in the text, we start the section with a case-type example. The case is used to illustrate the construction of both $\overline{X}$ and R charts. A point discussed in Section 2 often overlooked in other presentations of this material is the interpretation of Type 1 and Type 2 errors in the context of control charts. Both p and c charts are discussed as well.

Section 6 includes material rarely found in survey texts, namely the economic design of $\overline{X}$ charts. Assuming costs for sampling, searching for an assignable cause, and operating out of control are known, we show how to find the minimum expected cost design of the $\overline{X}$ chart. The theory is illustrated with one of the chapter's case examples.

The second part of the chapter is devoted to acceptance sampling. Again a case example is introduced to illustrate the methodology. We discuss single sampling plans, double sampling plans, and sequential sampling plans. Also included in this section is a detailed treatment of average outgoing quality with derivations of the various formulas. The chapter concludes with a discussion of several popular management methods for improving quality including quality circles, total quality control, and zero defects.

Section 13 provides a discussion of total quality management including definitions, competition based on quality, organizing for quality, benchmarking quality, and a discussion of the Deming Prize and the Baldrige Award. Also included here are sections on listening to the customer and quality function deployment. Section 14 examines the interface between quality and design and the application of Taguchi methods. It also includes the material on design for manufacturability that was in Chapter 12 of the third edition.

Chapter 12. Reliability and Maintainability

Although reliability and maintainability are considered to be part of the body of information comprising production and operations management, few survey texts include this material. Part of the reason is that reliability is a mathematically difficult subject. However, reliability and quality are closely tied together: part of quality control is understanding how parts fail. Furthermore, few would argue that maintenance is an important part of production and operations analysis, and maintenance models are based on reliability theory.

Because of the nature of the material, this chapter is perhaps the most mathematically sophisticated in the text. Every attempt has been made to make this material as palatable as possible without trivializing it. Examples are used throughout to give the reader intuition about difficult concepts such as the failure rate function, and the pure randomness associated with the exponential failure law.

The Poisson process is an important part of this chapter. We introduce it when discussing series systems of components that fail completely at random, but make use of it later in the chapter as well. The reliability portion of the chapter concludes with a discussion of failure laws for groups of components.

Sections 5 through 7 of Chapter 12 are devoted to maintenance models. We treat age replacement models for both deterministic and random failure processes. We develop a general deterministic model using exponential functions to model maintenance costs and depreciation. Planned replacement under uncertainty for one item and for a block of items is the subject of Section 7.

In Section 8 we discuss warranties. This topic is rarely mentioned in textbooks. The section is based on results from the literature. Real-life examples are used to illustrate the practical value of this theory. The chapter concludes with a brief overview of key issues in software reliability.

References

Bowman, E. H., and R. B. Fetter. *Analysis for Production and Operations Management.* 3rd ed. New York: McGraw-Hill/Irwin, 1967.

Skinner, W. "The Taming of Lions: How Manufacturing Leadership Evolved, 1780–1984." Chapter 2 in *The Uneasy Alliance: Managing the Productivity-Technology Dilemma,* ed. K. B. Clark, R. H. Hayes, and C. Lorenz. Boston, MA: Harvard Business School Press, 1985.

Contents in Brief

Contents

Production and Operations Analysis

CHAPTER 1

STRATEGY AND COMPETITION

Success requires a vision, and visions must be articulated so all of the firm's employees can share in that vision. The formal articulation of the vision is known as the company mission statement. A good mission statement should provide a clear statement of the goals of the firm and is the first step toward formulating a coherent business strategy. Poor mission statements tend to be wordy and full of generalities. Good mission statements are direct, clear, and concise. In their book, Jones and Kahaner (1995) list the 50 corporate mission statements that they perceive as the best. One example is the Gillette Corporation. Their mission statement is: "Our Mission is to achieve or enhance clear leadership, worldwide, in the existing or new core consumer product categories in which we choose to compete." They then go on to list exactly which areas they perceive as their core competencies. Intel defines their mission as: "Do a great job for our customers, employees, and stockholders by being the preeminent building block supplier to the computing industry." The statement then provides details on "Values and Objectives." In many cases, their objectives are quite specific (e.g., "Lead in LAN products and Smart Network Services"). Certainly, the award for conciseness has to go to the General Electric Corporation, whose mission statement is three words: "Boundaryless . . . Speed . . . Stretch." The commentary following the statement provides an explanation of exactly what these words mean in the context of the corporation.

Once having articulated a vision, the next step is to plan a strategy for achieving that vision. This is the firm's business strategy. The overall business strategy includes defining

1. The market in which the enterprise competes.
2. The level of investment.
3. The means of allocating resources to and the integrating of separate business units.
4. Functional area strategies, including
   - The marketing strategy.
   - The financial strategy.
   - The operations strategy.

Broadly defined, the operations strategy is the means by which the firm deploys its resources to achieve its competitive goals. For manufacturing firms, it is the sum total of all decisions concerning the production, storage, and distribution of goods. Important operations strategy decisions include where to locate new manufacturing facilities, how large these facilities should be, what processes to use for manufacturing and moving goods through the system, and what workers to employ. Service firms also require an operations strategy. The United States continues to be a leader in financial services, which must be supported by effective and reliable operations departments in order to remain competitive. The Disney theme park's continuing record of success is due in part to its careful attention to detail in every phase of its operations.

Does the American culture place too much emphasis on marketing (selling the product) and finance (leveraged buyouts, mergers, stock prices) and too little on operations (making and delivering the product)? Years ago, this was certainly the case. However, we are quick learners. The enormous success of the Japanese auto industry, for example, provided strong motivation for the American big three to close their inefficient plants and change the way things are done. The dramatic differences that were brought to light by Womack, Jones, and Roos (1990) have largely been eliminated. Today, the best American auto plants rival their Japanese counterparts for quality and efficiency.

Still, a coherent operations strategy is essential. When the Apple Macintosh was introduced, the product was extremely successful. However, the company was plagued with backorders and failed to keep up with consumer demand. According to Debbi Coleman, Apple's former director of worldwide manufacturing:

> Manufacturing lacked an overall strategy which created problems that took nine months to solve . . . we had extremely poor forecasting. Incoming materials weren't inspected for defects and we didn't have a mechanism for telling suppliers what was wrong, except angry phone calls. Forty percent of Mac materials were coming from overseas and no one from Apple was inspecting them before they were shipped. . . . One of the biggest tasks that high-tech manufacturers face is designing a manufacturing strategy that allows a company to be flexible so it can ride with the highs and lows of consumer and business buying cycles. (Fallon, 1985)

Although it is easy to be critical of American management style, we must be aware of the factors motivating American managers and those motivating managers from other cultures. For example, the Japanese have not achieved their dramatic successes without cost. Sixteen-hour work days and a high rate of nervous breakdowns among management are common in Japan.

Measuring a firm's success by the performance of its share price can result in shortsighted management practices. Boards of directors are more concerned with the next quarterly report than with funding major long-term projects. In fact, Hayes and Wheelwright (1984) make a compelling argument that such factors led to a myopic management style in the United States, characterized by the following:

1. Managers' performance is measured on the basis of **return on investment** (ROI), which is simply the ratio of the profit realized by a particular operation or project over the investment made in that operation or project.
2. Performance is measured over short time horizons. There is little motivation for a manager to invest in a project that is not likely to bear fruit until after he or she has moved on to another position.

In order to improve ROI, a manager must either increase the numerator (profits) or decrease the denominator (investment). In the short term, decreasing the denominator by cutting back on the investment in new technologies or new facilities is easier than trying to increase profits by improving efficiency, the quality of the product, or the productivity of the operating unit. The long-term effects of decreasing investment are devastating. At some point, the capital costs required to modernize old factories become more than the firm can bear, and the firm loses its competitive position in the market-place.

Measuring individual performance over the short term is a philosophy that seems to pervade American life. Politicians are elected for two-, four-, or six-year terms. There is a strong incentive for them to show results in time for the next election. Even university professors are evaluated yearly on their professional performance in many institutions, even though most serious academic projects extend over many years.

1.1 Manufacturing Matters

A question that is being debated and has been debated by economists for the past 20 years is the importance of a strong manufacturing base. The decline of manufacturing domestically has led to a shift in jobs from the manufacturing sector to the service sector. Because there are major disparities in labor costs in different parts of the world, there are strong incentives for American firms to locate volume manufacturing facilities overseas to reduce labor costs. Is a strong manufacturing base important for the health of the economy?

There is no question that a larger portion of manufactured goods come from overseas. Dertouzes et al. (1989), using Department of Commerce data, show that from the period 1972 to 1986 imports in eight major industries increased significantly as a percentage of the total U.S. market and exports in these same industries decreased significantly. (The industries are automobiles, chemicals, commercial aircraft, consumer electronics, machine tools, semiconductors (including computers and office equipment), steel, and textiles.) This shows that manufacturing in these industries moved offshore at a rapid pace. Is this a cause for concern?

An argument put forth by several scholars (e.g., Daniel Bell, 1976) is that we are simply evolving from an industrial to a service economy. In this view, the three stages of economic evolution are (1) agrarian, (2) industrial, and (3) service. In the early years of our country, we were primarily an agrarian economy. With the industrial revolution, a large portion of the labor force shifted from agriculture to manufacturing. In recent years it seems that there is less interest in manufacturing. These scholars would argue that we are merely entering the third stage of the evolutionary process: moving from an industrial economy to a service economy.

It is comforting to think that the American economy is healthy and simply evolving from an industrial to a service economy. One might even argue that manufacturing is not important for economic well-being. According to economist Gary S. Becker (1986), "Strong modern economies do not seem to require a dominant manufacturing sector."

It is far from clear, however, that we evolved from an agrarian economy to an industrial economy. Although fewer American workers are employed in the agricultural sector of the economy, agricultural production has *not* declined. Based on U.S. Department of Commerce data, Cohen and Zysman (1987) state that "agriculture has sustained, over the long term, the highest rate of productivity increase of any sector." By utilizing new technologies, agriculture has been able to sustain growth while consuming fewer labor hours. Hence, the figures simply do not bear out the argument that our economy has shifted from an agricultural one to an industrial one.

The argument that the economy is undergoing natural stages of evolution is simply not borne out by the facts. I believe that all sectors of the economy—agricultural, manufacturing, and service—are important and that domestic economic well-being depends upon properly linking the activities of these sectors.

The return on innovations will be lost if new products are abandoned after development. The payoff for R&D can come only when the product is produced and sold. If manufacturing is taken offshore, then the "rent on innovation" cannot be recaptured. Furthermore, manufacturing naturally leads to innovation. It will be difficult for the United States to retain its position as a leader in innovation if it loses its position as a leader in manufacturing.

That manufacturing naturally leads to innovation is perhaps best illustrated by the Japanese experience in the video market. After Japan had captured the lion's share of the world market for televisions, the next major innovation in consumer video technology, the videocassette recorder, was developed in Japan, not the United States. Virtually all VCRs sold today are manufactured in Asia.

It is difficult to support the argument that we can shift easily from an industrial economy to a service economy. Many services exist to support manufacturing. If manufacturing activities shift to foreign soil, it is likely that the services that complement manufacturing will suffer. According to Cohen and Zysman (1987):

> Were America to lose mastery and control of manufacturing, vast numbers of service jobs would be relocated after a few short rounds of product and process innovation, largely to destinations outside the United States, and real wages in all service activities would fall, impoverishing the nation.

Still, there is some positive news. Unit labor costs in manufacturing in the United States have risen only modestly in recent years (an average annual rate of about 1 percent only in the period 1990 to 1995), while in Japan these costs rose more than an average of 4 percent annually over the same period. Manufacturing labor productivity in the United States has been increasing at an average annual rate of about 3 percent during this period and has been almost completely flat in Japan. The United States still owns the majority of the world's manufacturing. The balance of payments in the United States is still dismal, however. The balance on the account in 1995 is projected to be a negative $167.5 billion—the highest since 1987 (*World Economic Outlook,* October 1994).

If manufacturing jobs are moving offshore, does that mean that there will be few opportunities for new careers in manufacturing in coming years? According to the *Occupational Outlook Handbook for 1994–1995* published by the Department of Commerce, manufacturing employment in the United States is expected to decline over the next several years. This is a trend that we do not see reversing in the near future. However, over that same period the composition of manufacturing employment will undergo a shift from production jobs to professional and management positions. Hence, there will be an even greater demand for professionals with the proper training to manage the manufacturing function. Manufacturing management positions and jobs relating to the development of new manufacturing technologies will continue to grow.

Innovation is a means by which some American firms have kept ahead of the competition. Notable examples are Sun Microsystems in California and 3M Company in Minnesota. Sun subcontracts much of its manufacturing and has maintained an impressive share of the lucrative workstation market by innovative and timely new product introductions. 3M operates on the philosophy that 70 percent of its sales in five years will come from products that don't exist today (Drucker, 1991).

However, there are disturbing trends suggesting that the Japanese, in particular, may be surpassing the United States in new innovations. In 1980 the 10 top patent winners in the United States (in order) were General Electric, Bayer, RCA, U.S. Navy, AT&T, IBM, Hitachi, Westinghouse Electric, Siemens, and General Motors. In 1990, the sequence was Hitachi, Toshiba, Canon, Mitsubishi, General Electric, Fuji Foto, Eastman Kodak, Philips, IBM, and Siemens. Not only are half the firms on this list Japanese, but they account for five of the top six patent winners (Stewart, 1991).

1.2 A Framework for Operations Strategy

Classical literature on competitiveness claims that firms position themselves strategically in the marketplace along one of two dimensions: lower cost or product differentiation (Porter, 1990).

Often new entrants to a market position themselves as the low-cost providers. Firms that have adopted this approach include the Korean automakers (Hyundai, Daewoo, Kia), discount outlets such as Costco, and retailers such as Wal-Mart. While being the low-cost provider can be successful over the near term, it is a risky strategy. Consumers ultimately will abandon products that they perceive as poor quality regardless of cost. For example, many manufacturers of low-cost PC clones popular in the 1980s are long gone.

Most firms that have a long record of success in the marketplace have differentiated themselves from their competitors. By providing uniqueness to buyers, they are able to sustain high profit margins over time. One example is BMW, one of the most profitable auto firms in the world. BMW continues to produce high-performance, well-made cars that are often substantially more expensive than those of competitors in their class. Product differentiation within a firm has also been a successful strategy. Consider the success of General Motors in the early years compared to Ford. GM was able to successfully capture different market segments at the same time by forming five distinct divisions, while Henry Ford's insistence on providing only a single model almost led the company to bankruptcy (Womack et al., 1990).

Strategic Dimensions

However, cost and product differentiation are not the only two dimensions along which firms distinguish themselves. The following additional factors relate directly to the operations function:

- Quality
- Delivery speed
- Delivery reliability
- Flexibility

What does *quality* mean? It is a word often bandied about, but one that means different things in different contexts. Consider the following hypothetical remarks.

1. "That hairdryer was a real disappointment. It really didn't dry my hair as well as I expected."
2. "I was thrilled with my last car. I sold it with 150,000 miles and hardly had any repairs."
3. "I love buying from that catalogue. I always get what I order within two days."
4. "The refrigerator works fine, but I think the shelves could have been laid out better."

5. "That park had great rides, but the lines were a mess."
6. "Our quality is great. We've got less than six defectives per one million parts produced."

In each case, the speaker is referring to a different aspect of quality. In the first case, the product simply didn't perform the task it was designed to do. That is, its function was substandard. The repair record of an automobile is really an issue of reliability rather than quality, per se. In the third case, it is delivery speed that translates to quality service for that customer. The fourth case refers to a product that does what it is supposed to do, but the consumer is disappointed with the product design. The product quality (the rides) at the amusement park were fine, but the logistics of the park management were a disappointment. The final case refers to the statistical aspects of quality control.

Hence the word *quality* means different things in different contexts. A Geo Prism is a quality product and so is a Ferrari Testarosa. Consumers buying these products are both looking for quality cars but have fundamentally different objectives. The fact is that *everyone* competes on quality. For this reason, Terry Hill (1993) would classify quality as an order qualifier rather than an order winner. An option is immediately eliminated from consideration if it does not meet minimum quality standards. It is the particular aspect of quality on which one chooses to focus that determines the nature of the competitive strategy and the positioning of the firm.

Delivery speed can be an important competitive weapon in some contexts. Some firms base their primary competitive position on delivery speed, such as UPS and Federal Express. Mail-order and web-based retailers also must be able to deliver products reliably and quickly to remain competitive. Building contractors that complete projects on time will have an edge.

Delivery reliability means being able to deliver products or services when promised. Online brokerages that execute trades reliably and quickly will retain customers. Contract manufacturers are measured on several dimensions, one being whether they can deliver on time. As third-party sourcing of manufacturing continues to grow, the successful contract manufacturers will be the ones that put customers first and maintain a record of delivering high-quality products in a reliable fashion.

Flexibility means offering a wide range of products and being able to adjust to unexpected changes in the demand of the product mix offered. Successful manufacturers in the 21st century will be those that can respond the fastest to unpredictable changes in customer tastes. This writer was fortunate enough to tour Toyota's Motomachi Plant located in Toyoda City, Japan. What was particularly impressive was the ability to produce several different models in the same plant. In fact, each successive car on the assembly line was a different model. A right-hand drive Crown sedan, for the domestic market, was followed by a left-hand drive Lexus coupe, designated for shipment to the United States. Each car carried unique sets of instructions that could be read by both robot welders and human assemblers. This flexibility allowed Toyota to adjust the product mix on a real-time basis and to embark on a system in which customers could order custom-configured cars directly from terminals located in dealer showrooms (Port, 1999).

Hence, one way to think of operations strategy is the strategic positioning the firm chooses along one of the dimensions of cost, quality, delivery speed, delivery reliability, and flexibility. Operations management is concerned with implementing the strategy to achieve leadership along one of these dimensions.

1.3 The Classical View of Operations Strategy

The traditional view of manufacturing strategy was put forward by Wickham Skinner of the Harvard Business School and enhanced by several researchers, mostly from Harvard as well. The traditional view treats most strategic issues in the context of a single plant rather than the entire firm. The broad issues discussed in Section 1.2 relate to operations strategy on the firm level. While the classical view might be considered a bit old-fashioned, the issues are still important and relevant. Classical operations strategy thinking relates to the following issues:

1. Time horizon
2. Focus
3. Evaluation
4. Consistency

Time Horizon

Time horizon refers to the length of time required for the strategy to have an effect. A natural classification here is to separate strategy decisions into short-, medium-, and long-range decisions. Short-term operations decisions may have an impact that can be measured in days or even hours. These include decisions regarding purchasing, production and personnel scheduling, policies for control of quality and maintenance functions, short-term inventory control issues, production schedules, and so forth. It is short-term decisions of this type that concern the operations manager, and they are the primary focus of this text.

Medium-range operations decisions are those whose impact can be measured in terms of weeks and months. They include demand and requirements forecasting, employment-planning decisions including determining the workforce size and mix, decisions concerning the distribution of goods in existing distribution channels, and setting of company targets for inventory and service levels. Changing the levels that signal an out-of-control situation for a control chart is a short-term quality decision, whereas restructuring the production line and seeking more reliable vendors are issues that would affect product quality farther into the future.

Strategy is generally associated with long-term decisions. Choosing the timing, the location, and the scale of construction of new manufacturing facilities are typical long-term manufacturing and operations strategy decisions. Making these decisions requires information about the forecast for new and existing products, the changing characteristics of the marketplace, and the changes in the costs and availability of resources. Manufacturing strategy must address the groundwork for building the proper channels for sales and distribution as well as facilities design and development. Figure 1–1 shows a breakdown of the production and operations decisions in terms of the timing of the consequences of these decisions.

Time horizons affect the impact of decisions, the uncertainties surrounding those decisions, and the penalty for wrong decisions. Short time horizons involve many decisions, each of whose impact may be small, but cumulatively can make a difference. For example, managers at The Gap stores restock shelves every day. The manager phones orders into the DC (distribution center) that generally are delivered the following day (assuming the DC is in stock on requested items). The manager relies on sales data and personal judgment to select the mix of items to reorder. Mistakes mean that

FIGURE 1–1

Decision horizons of manufacturing strategy

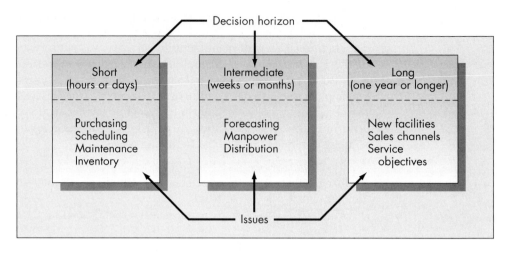

items may go out of stock, resulting in lost sales and irate customers, or not sell, resulting in wasted shelf space.

Buyers in the San Francisco headquarters of the firm decide on what lines of clothes to stock for the coming season. These may be considered intermediate-term decisions. The buyers have less information available to make their decisions than does the store manager, and the decisions they make have greater impact. The buyers have to judge fashion trends and color preferences. A line that doesn't sell must be marked down and sold at a loss.

Top management must make the long-term decisions. For The Gap, some long-term decisions might include (1) the number, location, and size of distribution centers; (2) terms and conditions of long-term contracts with suppliers; (3) arrangements for firmwide supply chain logistics (in-house versus outsourced); and (4) selection of management personnel. There is more uncertainty involved with these decisions. Demographics change, which could make prior decisions on the location and sizing for DCs wrong. A long-term contract with a plant overseas, for example, in China, could backfire. Quotas can be imposed by the U.S. government, or tariffs could be imposed by the Chinese government. The importance and sensitivity of long-term decisions are probably among the reasons top management personnel are paid so well in the United States.

Focus

The notion of **focus in manufacturing strategy** was first considered by Skinner (1974). He defined the five key characteristics of the focused factory as follows:

1. *Process technologies.* A natural means of focusing the operations of one plant or factory is by the process employed. Management should limit new unproven process technologies to one per factory and keep the number of different mature process technologies to a level that the plant manager can oversee efficiently.

2. *Market demands.* The marketplace often determines the focus of a product or line of products produced at a factory. Typical areas in which the market dictates plant focus are

   • *Price.* There is evidence that the American consumer is more price conscious than many overseas consumers. Price has always been a key product differentiation factor in the United States.

- *Lead time.* Products not protected by patents must be produced and distributed quickly in order to reach the marketplace before those of competitors.
- *Reliability.* Reliability specifications differ by market segment, often for identical products. As an example, consider a company producing a line of integrated circuits that are sold to a variety of different customers. One uses them in refrigerators and another in heart-lung machines. The reliability specifications are likely to be far greater for the heart-lung machine.

3. *Product volumes.* The production volumes within a single plant should be similar so that plant tooling, materials-handling systems, and production lines are neither under- nor overutilized.

4. *Quality level.* The level of quality of products produced in a single plant should be similar so that the firm can establish a consistent quality control standard. Quality standards are the result of several factors: the statistical control techniques used, the monitoring procedures, and the workers' training, procedures, and attitudes.

5. *Manufacturing tasks.* The productivity of a plant producing a broad line of different products will suffer from disruptions that result from frequent setups and re-structuring of the production lines. When management limits the number of distinct manufacturing tasks at one location, workers can concentrate on perfecting existing processes.

There is evidence that focused companies are more successful. Hayes and Wheelwright (1984) relate the operating profit margin for 11 companies to the number of major product types these companies produce. Based on these data, they show that firms that produce fewer products tend to be more profitable. It is interesting to note that for the 11 firms surveyed, focus, as measured in this fashion, was a better predic-tor of profitability than size of sales.

It should be noted that many firms successfully serve several diverse markets simul-taneously. One example (and there are certainly many others in both the United States and Japan) is Yamaha. Yamaha has established itself as a high-quality manufacturer of products as diverse as stereo equipment, musical instruments (including band instru-ments, pianos, and digital synthesizers), and sports equipment. However, focused fac-tories are certainly preferred even for firms that have a broad product line.

Evaluation

There are several dimensions along which one can evaluate production/operations strategy. Here are the most significant:

1. *Cost.* Where pricing is a key to market differentiation and competitiveness, a major means of strategy evaluation is the cost of products delivered to the customer. Direct costs of production include costs of materials, equipment, and labor. Costs of distribution involve inventory carrying costs, particularly inventory in the pipeline, transportation, and distribution costs. Cost of plant and process overhead also must be factored into the cost calculation. Overhead costs of new processes may be the most difficult to evaluate because of the uncertainties of both the process reliability and the useful lifetime.

2. *Quality.* In markets where product quality is a major determinant of product suc-cess in the marketplace, or high reliability is required to meet product specifications,

strategy should be evaluated along the quality dimension. For example, product quality is the primary means of evaluating manufacturing performance in Japan.

3. *Profitability.* Ultimately, it is the profitability of a product line that determines the success of the strategy undertaken to produce and to sell it. However, as noted above, short-term profit maximization could be a poor strategy for the firm if it entails reductions in the investment in new capacity and technology. If the time horizon associated with the evaluation of any particular strategy is not correct, top management could be making poor decisions. Strategies that achieve short-term profitability may not necessarily be in the best interests of the firm in the long term.

4. *Customer satisfaction.* Successful firms have come to realize that ultimate success is achieved only by maintaining a satisfied and loyal customer base. This means that the customer must not only be satisfied with a product when purchased, but must have confidence that the firm will stand behind its guarantees by supplying efficient and cost-effective service after a sale is made. *Market-driven quality* is a term we often hear these days. Its recent emphasis shows that companies are becoming more aware that customer service must be made an explicit part of the product delivery process.

Consistency

It is often the case that rather than having a single consistent strategy for manufacturing, the "strategy" is simply the composite of all company policies that affect manufacturing. Personnel and wage policies should be designed to encourage efficiency and improve productivity. Inventory control, scheduling, and production plans are geared toward minimizing production costs and improving measures such as worker idle time, work-in-process inventory, and production lead times. Process design is geared toward producing high-quality products. The problem is that each of these individual goals may be optimizing a different objective. The result is a complex plant structure in which management and labor assume adversarial positions.

Skinner (1974) cites a number of causes for the common inconsistencies that are observed in most firms. These include the following:

1. *Professionalism in the plant.* The number of different job titles and job functions has increased while the scope of many of these jobs has narrowed. Professionals in different areas have a stake in making themselves look good, and hence seek to maximize their personal contributions. Unfortunately, these professionals are not necessarily working toward the same goals. Some are attempting to minimize costs, others to improve the quality of the work environment, and others to optimize the cash flow within the firm.

2. *Product proliferation.* As noted above, those firms that produce a smaller number of products tend to be more profitable. As the number of distinct products grows, retaining a consistent set of goals within the plant is more difficult. Why do firms increase the size of the product line produced at one plant? Recall the discussion in the first part of this chapter concerning the consequences of measuring performance based on the return on investment. One sure way of decreasing the investment in new capacity is simply to produce new products in old plants. From management's point of view, this is a means of reducing overhead, but it could also result in poor product quality and inefficient production control.

FIGURE 1–2

The elements of production and operations strategy

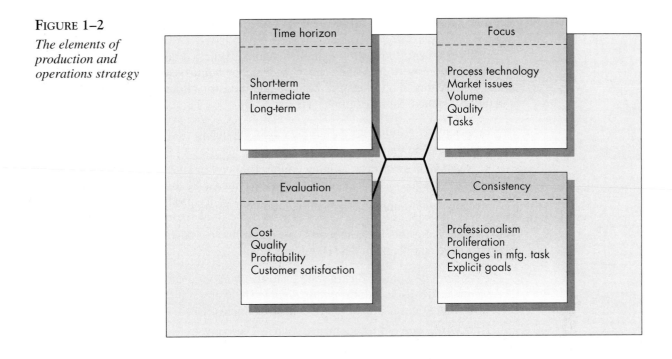

3. *Changes in the manufacturing task.* Management decides to add productive capacity in response to the needs dictated by the marketplace. Markets change, however, and as a result the function of the plant may change as well. The goals and the means for evaluating those goals may not change along with the demands placed upon the plant, however. It may happen that the objectives that made sense when the plant was constructed no longer make sense after the plant's function has changed.

4. *The manufacturing task was never made explicit.* Management must understand how the goals of manufacturing fit in with the overall corporate strategy in order for a consistent policy to emerge. If there is no clear corporate strategy, there is little hope that the functional areas of the firm—finance, marketing, and production—can develop a meaningful and consistent strategy independently.

The essential elements of the discussion of this section are summarized in Figure 1–2.

1.4 Competing in the Global Marketplace

International competitiveness has become a national obsession. Americans are concerned that their standard of living is eroding while it seems to improve elsewhere. Evidence exists that there is some truth to this perception. Our balance of trade with Japan has been in the red for decades, with no evidence of a reversal. American firms once held a dominant position worldwide in industries that have nearly disappeared domestically. Consumer electronics, steel, and machine tools are some examples. All the news is not bad, however. The American economy is strong and continues to grow. American firms still have the lion's share of the world market in many industries.

In his excellent study of international competitiveness, Porter (1990) poses the following question: Why does one country become the home base for successful international competitors in an industry? That certain industries flourish in certain countries cannot be disputed. Some examples are

1. Germany: printing presses, luxury cars, chemicals.
2. Switzerland: pharmaceuticals, chocolate.
3. Sweden: heavy trucks, mining equipment.
4. United States: personal computers, software, films.
5. Japan: automobiles, consumer electronics, robotics.

What accounts for this phenomenon? One can offer several compelling explanations, but most have counterexamples. Here are a few:

1. *Historical.* Some industries are historically strong in some countries and are not easily displaced. *Counterexample:* The demise of the steel industry in the United States is one of many counterexamples.

2. *Tax structure.* Some countries, such as Germany, have no capital gains tax, thus providing a more fertile environment for industry. *Counterexample:* However, there is no reason that favorable tax treatment should favor certain industries over others.

3. *National character.* Many believe that workers from other countries, particularly from Pacific Rim countries, are better trained and more dedicated than American workers. *Counterexample:* If this is true, why then do American firms dominate in some industry segments? How does one explain the enormous success Japanese-based corporations have had running plants in the United States with an American workforce?

4. *Natural resources.* There is no question that some industries are highly resource dependent and these industries have a distinct advantage in some countries. One example is the forest products industry in the United States and Canada. *Counterexample:* Many industry sectors are essentially resource independent but still seem to flourish in certain countries.

5. *Government policies.* Some governments provide direct assistance to fledgling industries, such as MITI in Japan. The role of the U.S. government is primarily regulatory. For example, environmental standards in the United States are probably more stringent than almost anywhere else. *Counterexample:* This does not explain why some industries dominate in countries with strict environmental and regulatory standards.

6. *Advantageous macroeconomic factors.* Exchange rates, interest rates, and government debt are some of the macroeconomic factors that provide nations with competitive advantage. For example, in the 1980s when interest rates were much higher in the United States than they were in Japan, it was much easier for Japanese firms to borrow for new projects. *Counterexample:* These factors do not explain why many nations have a rising standard of living despite rising deficits (Japan, Italy, and Korea are some examples).

7. *Cheap, abundant labor.* Although cheap labor can attract new industry, most countries with cheap labor are very poor. On the other hand, many countries (Germany, Switzerland, and Sweden are examples) have a high standard of living, high wage rates, and shortages of qualified labor.

8. *Management practices.* There is evidence that Japanese management practices are more effective in general than Western-style practices. *Counter-example:* If American management practices are so ineffective, why do we continue to dominate certain industries, such as personal computers, software development, and pharmaceuticals?

Talking about competitiveness is easier than measuring it. What are the appropriate ways to measure one country's success over another? Some possibilities are

- Balance of trade.
- Share of world exports.
- Creation of jobs.
- Low labor costs.

Arguments can be made against every one of these as a measure of international competitiveness. Switzerland and Italy have trade deficits, and at the same time have experienced rising standards of living. Similar arguments can be made for countries that import more than they export. The number of jobs created by an economy is a poor gauge of the health of that economy. More important is the quality of the jobs created. Finally, low labor costs correlate with low standard of living. These counterexamples show that it is no easy task to develop an effective measure of international competitiveness.

Porter (1990) argues that the appropriate measure to compare national performance is the rate of productivity growth. Productivity is the value of output per unit of input of labor or capital. Porter argues that productivity growth in some industries appears to be stronger in certain countries, and that there are reasons for this. In some cases we can find the reasons in domestic factor advantages. The factor theory says all countries have access to the same technology (an assumption that is not strictly true) and that national advantages accrue from endowments of production factors such as land, labor, natural resources, and capital.

There are some excellent examples of factor theory. Korea has relatively low labor costs, so it exports labor-intensive goods such as apparel and electronic assemblies. Sweden's iron ore is low in impurities, which contributes to a strong Swedish steel industry. As compelling as it is, there are counterexamples to the factor endowment theory as well. For example, after the Korean War, South Korea developed and excelled in several highly capital-intensive industries such as steel and shipbuilding even though the country was cash poor. Also, many countries have similar factor endowments, but some seem to excel in certain industries, nonetheless. These examples suggest that factor endowments do not explain all cases of nations with dominant industry segments.

Porter (1990) suggests the following four determinants of national advantage:

1. *Factor conditions* (discussed above).

2. *Demand conditions.* If domestic consumers are sophisticated and demanding, they apply pressure on local industry to innovate faster, which gives firms an edge internationally. Consumers of electronics in Japan are very demanding, thus positioning this industry competitively in the international marketplace.

3. *Related and supporting industries.* Having world-class suppliers nearby is a strong advantage. For example, the Italian footwear industry is supported by a strong leather industry and a strong design industry.

4. *Firm strategy, structure, and rivalry.* The manner in which firms are organized and managed contributes to their international competitiveness. Japanese management style is distinctly different from American. In Germany, many senior executives possess technical background, producing a strong inclination to product and process improvement. In Italy, there are many small family-owned companies, which encourages individualism.

Even though Porter makes a very convincing argument for national competitive advantage in some industries, there is a debate among economists as to whether the notion of international competitiveness makes any sense at all. Companies compete, not countries. This is the point of view taken by Paul Krugman (1994). According to Krugman, the United States and Japan are simply not competitors in the same way that Ford and Toyota are. The standard of living in a country depends on its own domestic economic performance and not on how it performs relative to other countries.

Krugman argues that too much emphasis on international competitiveness can lead to misguided strategies. Trade wars are much more likely in this case. This was the case in mid-1995 when the Clinton administration was planning to impose high tariffs on makers of Japanese luxury cars. Most economists agree that trade wars and their consequences, such as tariffs, benefit no one in the long run. Another problem that arises from national competitive pride is it can lead to poorly conceived government expenditures. France has spent billions propping up its failing computer industry. (Certainly, not all government investments in domestic industry can be considered a mistake. The Japanese government, for example, played a major role in nurturing the flat-panel display industry. Japanese-based firms now dominate this multibillion dollar industry.)

Another point supporting Krugman's position is that the lion's share of U.S. GDP is consumed in the United States, thus making a firm's success in our domestic market more important than its success in the world market. Krugman agrees that productivity growth is a valid concern. He argues, however, that we should be more productive in order to produce more, not to better our international competitors.

The debate over competitive advantage will continue. Policy makers need to be aware of all points of view, and weigh each carefully in formulating policy. Although Krugman makes several telling points, there is no question that globalization is a trend that shows no sign of reversing. We cannot stick our heads in the sand and say that foreign markets are not important to us. Economic borders are coming down all across the globe.

Problems for Sections 1–4

1. Why is it undesirable for the United States to evolve into a service economy?

2. What disadvantages do you see if the CEO is primarily concerned with short-term ROI?

3. Can you think of companies that have gone out of business because they focused only on cost and were not able to achieve a minimum quality standard?

SNAPSHOT APPLICATION

Read-Rite Wins with International Strategic Alliance

Read-Rite Corporation, founded in 1983 and based in Milpitas, California, is one of the largest American-based independent manufacturers of heads for hard drives. The head is the integrated circuit that makes reading and writing data onto the magnetic media possible. Head technology is complex and volatile; the head must be redesigned for each new generation of disk drives.

Building new fabrication facilities is extremely expensive. During the 1980s, the company struggled to keep up with rapidly changing hard drive technology. By 1990 the firm was once again profitable, but several problems remained. One was that they had only a single customer (Conner Peripherals). Two other problems were finding cash for expansion and keeping ahead of the rapidly advancing Japanese competition.

To stem the advancing tide of the Japanese into this arena, it would be important to compete with the Japanese on their home turf, reasoned CEO Cyril Yansouni. Strong home demand (as noted by Porter) is an important determinant of national competitive advantage. Access to Japanese markets is always difficult for outsiders, so Read-Rite sought a strategic alliance with a Japan-based partner.

The partner that Read-Rite chose was Sumitomo Metal Industries. On the surface, this sounds like a strange choice. However, although Sumitomo had no presence in technology at that time, the company was making several investments in technology-based businesses. Also, as the third-largest steel company in Japan, Sumitomo had plenty of cash. As part of the agreement, Sumitomo agreed to pay $30 million in exchange for an equity stake in Read-Rite.

The biggest risk faced by Read-Rite was the loss of their technology to a potential competitor. In order to protect themselves, according to Yansouni, "we're going to make divorce very painful for both sides." Neither side would be allowed to sell heads in Japan for many years after a separation, at which time any acquired technology would surely be obsolete.

How did things turn out? With the infusion of cash from Sumitomo, Read-Rite was able to attract new customers in the United States, including Western Digital, Quantum, and Maxtor. The following year the firm went public and, with additional cash from the IPO, had enough to start building assembly plants in Asia. The alliance (known as Read-Rite SMI) became very successful in Japan, capturing the number two spot in Japanese sales of heads after TDK. The success of the joint venture has helped to smooth out some of Read-Rite's domestic revenue swings and has been an important part of the firm's success in the 1990s. Almost a decade later, the joint venture continues to thrive in Japan, despite economic problems there.

Source: "Pacific Overtures," *The Institutional Investor,* January 1996.

4. What are the different quality standards referred to in the example comparing the GEO and the Ferrari?
5. List the four elements of operations strategy and discuss the role of each in strategic planning.
6. What are the advantages and disadvantages of producing new products in existing facilities?
7. Consider the following fictitious industrial scenarios. In each case, identify the possible errors in strategy.
    a. A manufacturer of Winchester disk drives located in northern California is expanding the product line to include optical disks. Both products will be manufactured at the same facility even though the optical disks require a different process technology.

    b. A conflict has developed between the head of quality control and the vice president for operations at a tool and die manufacturer in the Midwest. The quality control manager has requested new die-casting equipment costing $400,000 that has an excellent reliability record, while the operations vice president has decided to purchase a similar product that is less reliable from another manufacturer for $300,000.

    c. The chief financial officer of a firm has established the policy of no new investments in facilities for the next year in order to improve the firm's current profitability profile.

    d. A microprocessor designed in the United States and produced in a plant in Mexico is used to monitor the pressure of fluids. It is sold both to an automobile manufacturer and to a maker of life-sustaining equipment.

8. What are the four determinants of national advantage suggested by Porter? Give examples of companies that have thrived as a result of each of these factors.

9. What factor advantage favors the aluminum industry in the United States over Japan and makes aluminum much cheaper to produce here? (Hint: aluminum production is very energy intensive. In what part of the country is an inexpensive energy source available?)

10. Paul Krugman argues that because most of our domestic product is consumed domestically, we should not dwell on international competition. What industries in the United States have been hardest hit by foreign competition? What are the potential threats to the United States if these industries fail altogether?

11. Krugman points out some misguided government programs that have resulted from too much emphasis on international competitiveness. What risks are there from too *little* emphasis on international competitiveness?

1.5 Strategic Initiatives: Reengineering the Business Process

Seemingly on schedule, every few years a hot new production control method or management technique comes along, almost always described by a three-word acronym. While it is easy to be skeptical, by and large, the methods are sound and can have substantial value to corporations when implemented intelligently. *Business process reengineering* (BPR) caught on after the publication of the book by Hammer and Champy (1993). BPR is not a specific technique, such as MRP, or a production-planning concept like JIT. Rather, it is the idea that entrenched business processes can be changed and can be improved. The process is one of questioning why things are done a certain way, and not accepting the answer, "because that's the way we do it."

Hammer and Champy, who define BPR as "starting over," provide several examples of successful reengineering efforts. The first is the IBM Credit Corporation, a wholly owned subsidiary of IBM, that, if independent, would rank among the *Fortune* 100 service companies. This arm of IBM is responsible for advancing credit to new customers purchasing IBM equipment. The traditional credit approval process followed five steps:

1. An IBM salesperson would call in with a request for financing and the request would be logged on a piece of paper.

2. Someone carried the paper upstairs to the credit department, where someone else would enter the information into a computer system and check the potential borrower's credit rating. The specialist wrote the results of the credit check on a piece of paper, which was sent off to the business practices department.

3. A third person, in the business practices department, modified the standard loan document in response to the customer request. These modifications, which were done on yet another computer system, were then attached to the original request form and the credit department specialist's report.

4. Next the request went to a pricer, who keyed the information into a spreadsheet to determine the appropriate interest rate to charge the customer. The pricer's recommendation was written on a piece of paper and delivered (with the other papers) to the clerical group.

5. The information was turned into a quote letter that would be delivered to the field salesperson by Federal Express.

This process required an average of six days and sometimes as long as two weeks. Sales reps logged endless complaints about this delay: during this time, the customer could find other financing or another vendor. In an effort to see if this process could be streamlined, two senior managers decided to walk a new request through all five steps, asking personnel to put aside what they were doing and process it as they normally would. They found that the entire five-step process required an average of only 90 minutes of work! The rest of the time, requests were either in transit from one department to another or queueing up on somebody's desk waiting to be processed. Clearly the problem did not lie with the efficiency of the personnel but with the design of the credit approval process itself.

The solution was simple: the four specialists handling each loan request were replaced by a single loan generalist who handled each request from beginning to end. Up-to-date software was designed to support the generalist, who had no trouble dealing with most requests. The credit approval process was designed assuming that each request was sufficiently complex to require someone with special knowledge in each area. In truth, most requests were routine, and specialists generally did little more than a simple table look-up to determine the appropriate figure.

What was the result of this change? The six-day turnaround for loan requests was slashed to only four hours! And this was accomplished with fewer personnel and with a hundredfold increase in the number of deals handled.

While each reengineering effort requires careful thought and no two solutions will be exactly alike, Hammer and Champy (1993) suggest that reengineering efforts utilize the following general principles:

1. *Several jobs are combined into one.* Few examples of BPR are as dramatic as that of IBM Credit, but there are other success stories in the literature as well. Many of the successful cases have a common thread: the reduction of a complex process requiring many steps to a simpler one requiring fewer steps. In the case of IBM Credit, a five-step process was reduced to only a single step. This suggests a general principle. The IBM Credit process was a natural evolution of the concept of division of labor. The economist Adam Smith espoused this principle as far back as the 18th century (see the quote from *The Wealth of Nations* at the beginning of Section 1.10 of this chapter). However, a good thing can be carried too far. If one divides a process into too many steps, one eventually reaches the point of diminishing returns. BPR's most dramatic

successes have come from complex processes that were simplified by reducing the number of steps required.

2. *Workers make decisions.* One goal is to reduce the number of levels of reporting by allowing workers to make decisions that were previously reserved for management. In the case of IBM Credit, most decisions once reserved for specialists are now done by a single generalist. Giving workers greater decision-making power may pose a threat to management, who might see such a step as encroaching on their prerogatives.

3. *The steps in the process are performed in a natural order.* Process steps should not be performed necessarily in rigid linear sequence, but in an order that makes sense in the context of the problem being solved. In particular, in many cases, some tasks can be done simultaneously rather than in sequence. (These ideas, of course, are well known and form the basis for the concepts of project management in Chapter 8.)

4. *Processes should have multiple versions.* One should allow for contingencies, not by designing multiple independent processes, but by designing one flexible process that can react to different circumstances. In the case of IBM Credit, for example, the final credit issuance process had three versions: one for straightforward cases (handled by computer), one for cases of medium difficulty (handled by the deal structurer), and one for difficult cases (performed by the deal structurer with help from specialist advisers).

5. *Work is performed where it makes the most sense.* One of the basic principles of reengineering is not to carry the idea of division of labor too far. Another is not to carry the idea of centralization too far. For example, in most companies, purchasing is done centrally. This means that every purchase request is subject to the same minimum overhead in time and paperwork. A consequence might be that the cost of processing a request exceeds the cost of the item being purchased! A great deal can be saved in this case by allowing individual departments to handle their own purchasing for low-cost items. (Hammer and Champy discuss such a case.)

The authors list several other basic principles, involving minimizing checks and reconciliations, having a single point of contact, and being able to employ hybrid centralized/decentralized operations.

It is easier to list the steps one might consider in a reengineering effort than to actually implement one. In the real world, political realities cannot be ignored. For many of the success stories in the literature, not only are the processes simplified, but the headcount of personnel is reduced as well. It is certainly understandable for employees to see BPR as a thinly veiled excuse for downsizing (euphemistically called "right-sizing"). This was exactly the case in one financial services company. When word got out that management was planning a reengineering effort, most assumed that there would be major layoffs. Some even thought the company was on the verge of bankruptcy. In another instance, union leadership saw reengineering as a means for management to throw away the job categories and work rules they had won in hard-fought negotiations over the years, and persuaded the members to strike. In a third case, a senior manager was unhappy with the potential loss of authority that might accompany a reengineering effort. He resigned to start his own company. (These examples are related in a follow-up book by Hammer and Stanton, 1995.)

These stories show that starting a reengineering effort is not without risks. Rarely is the process as simple as IBM's. Reengineering has been described by Ronald Compton, the CEO of Aetna Life and Casualty, as "agonizingly, heartbreakingly, tough." There needs to be some cost–benefit analysis done up front to be sure the potential gains compensate for the risks.

Process optimization is not new. In its early years, the field of industrial engineering dealt with optimal design of processes, setting standards using time and motion studies, and flowcharting for understanding the sequence of events and flow of material in a factory. Why is BPR different? For one, BPR is concerned with business process flows rather than manufacturing process flows. Second, the concept is not one of optimizing an existing process, but one of rethinking how things should be done from scratch. As such, it is more revolutionary than evolutionary. It is likely to be more disruptive but could have larger payoffs. To make BPR work, employees at every level have to buy into the approach, and top management must champion it. Otherwise, the reengineering effort could be a costly failure.

1.6 Strategic Initiatives: Just-in-Time

Just-in-time (JIT) is a manufacturing process on one hand and a broad-based operations strategy on the other. The process elements of JIT will be discussed in detail in Chapter 7 as part of a complete analysis of push and pull inventory systems. However, JIT (or lean production, as it is also known) is a philosophy that includes treatment of inventory in the plant, relationships with suppliers, and distribution strategies. The core of the philosophy is to eliminate waste. This is accomplished by efficient scheduling of incoming orders, work-in-process inventories, and finished goods inventories.

JIT is an outgrowth of the **kanban system** introduced by Toyota. Kanban is a Japanese word meaning card or ticket. Originally, kanban cards were the only means of implementing JIT. The kanban system was introduced by Toyota to reduce excess work-in-process (WIP) inventories. Today, JIT is more ambitious. Both quality control systems and relationships with suppliers are part of an integrated JIT system. JIT systems can be implemented in ways other than using kanban cards. Integrating JIT philosophies with sophisticated information systems makes information transfer faster. The speed with which information can be transferred from one part of the firm to another is an important factor in the success of the JIT system.

JIT is a philosophy of operating a company that includes establishing understandings and working relationships with suppliers, providing for careful monitoring of quality and work flow, and ensuring that products are produced only as they are needed. Although JIT can be used simply as it was originally designed by Toyota, namely as a means of moving work-in-process from one work center to another, proponents of the method recommend much more. They would have a firm integrate the JIT philosophy into its overall business strategy.

Inventory and material flow systems are classified as either **push** or **pull systems.** A push system is one in which decisions concerning how material will flow through the system are made centrally. Based on these decisions, material is produced and "pushed" to the next level of the system. A typical push system is MRP (materials requirements planning), which is discussed in detail in Chapter 7. In MRP, appropriate production amounts for all levels of the production hierarchy are computed all at once based on forecasts of end-product demand and the relationship between components and end items. In JIT, production is initiated at one level as a result of a request from a higher level. Units are then "pulled" through the system.

JIT has many advantages over conventional systems. Eliminating WIP inventories results in reduced holding costs. Less inventory means less money tied up in inventory. JIT also allows quick detection of quality problems. Since units are produced only as they are needed, the situation in which large amounts of defective WIP inventory are produced before a quality problem is detected should never occur in a properly running JIT system. JIT also means that relationships with suppliers must be tightened up. Suppliers must be willing to absorb some uncertainty and adjust delivery quantities and the timing of deliveries to match the rates of product flows.

Part of what made the kanban system so effective for Toyota was its success in reducing setup times for critical operations. The most dramatic example of setup time reduction is the so-called SMED, or single-minute exchange of dies. Each time a major change in body style is initiated, it is necessary to change the dies used in the process. The die-changing operation typically took from four to six hours. During the die-changing operation the production line was closed down. Toyota management heard that Mercedes Benz was able to reduce its die-changing operation to less than one hour. Realizing that even more dramatic reductions were possible, Toyota set about focusing on the reduction of the time required for die changing. In a series of dramatic improvements, Toyota eventually reduced this critical operation to only several minutes. The essential idea behind SMED is to make as many changes as possible off-line, while the production process continues.

An important part of JIT is forming relationships with suppliers. What separates JIT purchasing from conventional purchasing practices? Freeland (1991) gives a list of characteristics contrasting the conventional and JIT purchasing behavior. Some of these include

Conventional Purchasing	*JIT Purchasing*
1. Large, infrequent deliveries.	1. Small, frequent deliveries.
2. Multiple suppliers for each part.	2. Few suppliers; single sourcing.
3. Short-term purchasing agreements.	3. Long-term agreements.
4. Minimal exchange of information.	4. Frequent information exchange.
5. Prices established by suppliers.	5. Prices negotiated.
6. Geographical proximity unimportant.	6. Geographical proximity important.

In his study, Freeland notes that the industries that seemed to benefit most from JIT purchasing were those that typically had large inventories. Companies without JIT purchasing tended to be more job-shop oriented or make-to-order oriented. Vendors that entered into JIT purchasing agreements tended to carry more safety stock, suggesting manufacturers are reducing inventories at the expense of the vendors. The JIT deliveries were somewhat more frequent, but the differences were not as large as one might expect. Geographical separation of vendors and purchasers was a serious impediment to successful implementation of JIT purchasing. The automotive industry was one that reported substantial benefit from JIT purchasing arrangements. In other industries, such as computers, the responses were mixed; some companies reported substantial benefits and some reported few benefits.

Although reducing excess work-in-process inventory can have many benefits, JIT is not necessarily the answer for all manufacturing situations. According to Stasey and McNair (1990),

> Inventory in a typical plant is like insurance, insurance that a problem in one area of a plant won't affect work performed in another. When problems creating the need for insurance are solved, then inventories disappear from the plant floor.

The implication is that we merely eliminate all sources of uncertainty in the plant and the need for inventories disappears. The problem is that there are some sources of variation that can never be eliminated. One is variation in consumer demand. JIT is effective only if final demand is regular. Another may be sources of variation inherent in the production process or in the equipment. Can one simply legislate away all sources of uncertainty in the manufacturing environment? Of course not. Hence, although the underlying principles of JIT are sound, it is not a cure-all and will not necessarily be the right method for every production situation.

1.7 Strategic Initiatives: Time-Based Competition

Professor Terry Hill of the London School of Business has proposed an interesting way to look at competitive factors. He classifies them into two types: "qualifiers" and "order winners." A product not possessing a qualifying factor is eliminated from consideration. The order winner is the factor that determines who gets the sale among the field of qualifiers.

Two factors about which we hear a great deal are quality and **time to market.** In the past decade, the Japanese and Germans gained a loyal following among U.S. consumers by producing quality products. American firms are catching up on the quality dimension. From the discussion in the previous section, we see that successful U.S.–based companies have been able to produce products that match the defect rates of foreign competitors. If this trend continues, product quality will be assumed by the consumer. Quality may become an order qualifier rather than an order winner.

If that is the case, what factors will determine order winners in years to come? Japanese automobile companies provided and continue to provide high-quality automobiles. In recent years, however, the major automobile producers in Japan have begun to focus on aesthetics and consumer tastes. They have branched out from the stolid small cars of the 1970s and 1980s to new markets with cars such as the Toyota-made Lexus luxury line and Mazda's innovative and successful Miata.

The timely introduction of new features and innovative design will determine the order winners in the automobile industry. In the computer industry, Compaq built its reputation partly on its ability to be the first to market with new technology. Time-based competition is a term that we will hear more and more frequently in coming years.

What is **time-based competition?** It is not the time and motion studies popular in the 1930s that formed the basis of the industrial engineering discipline. Rather, according to Blackburn (1991),

> Time-based competitors focus on the bigger picture, on the entire value-delivery system. They attempt to transform an entire organization into one focused on the total time required to deliver a product or service. Their goal is not to devise the best way to perform a task, but to either eliminate the task altogether or perform it in parallel with other tasks so that over-all system response time is reduced. Becoming a time-based competitor requires making revolutionary changes in the ways that processes are organized.

Successful retailers understand time-based competition. The success of the fashion chains The Gap and The Limited is due largely to their ability to deliver the latest fashions to the customer in a timely manner. Part of the success of the enormously successful Wal-Mart chain is its time-management strategy. Each stock item in a Wal-Mart store is replenished twice a week, while the industry average is once every two weeks. This allows Wal-Mart to achieve better inventory turnover rates than its competition and respond more quickly to changes in customer demand. Wal-Mart's strategies have enabled it to become the industry leader, with a growth rate three times the industry average and profits two times the industry average (Blackburn, 1991, Chapter 3).

Time-based management is a more complex issue for manufacturers, and in some industries it is clearly the key factor leading to success or failure. The industry leaders in the DRAM (dynamic random access memory) industry changed four times between 1978 and 1987. In each case, the firm that was first to market with the next-generation DRAM dominated that market. The DRAM experience is summarized in the following table (Davis, 1989):

Product	Firm	Year Introduced	First Year of Volume Production	Market Leaders in First Year of Volume Production
16K	Mostek	1976	1978	Mostek (25%) NEC (20%)
64K	Hitachi	1979	1982	Hitachi (19%) NEC (15%)
256K	NEC	1982	1984	NEC (27%) Hitachi (24%)
1 MB	Toshiba	1985	1987	Toshiba (47%) Mitsubishi (16%)

I am aware of no other example that shows so clearly and so predictably the value of getting to the market first.

1.8 Strategic Initiatives: Competing on Quality

What competitive factors do American managers believe will be important in the next decade? Based on a survey of 217 industry participants, the following factors were deemed as the most important for gaining a competitive edge in the coming years; they are listed in the order of importance.

1. Conformance quality.
2. On-time delivery performance.
3. Quality.
4. Product flexibility.
5. After-sale service.
6. Price.
7. Broad line (features).
8. Distribution.
9. Volume flexibility.
10. Promotion.

In this list we see some important themes. **Quality** and **time management** emerge as leading factors. Quality control has been given much press in recent years with the establishment of the prestigious Malcolm Baldrige Award (modeled after the Japanese Deming Prize, which has been around a lot longer). Quality means different things in different contexts, so it is important to understand how it is used in the context of manufactured goods. A high-quality product is one that performs as it was designed to perform. Products will perform as they are designed to perform if there is little variation in the manufacturing process. With this definition of quality, it is possible for a product with a poor design to be of high quality, just as it is possible for a well-designed product to be of poor quality. Even granting this somewhat narrow definition of quality, what is the best measure? Defect rates are a typical barometer. However, a more appropriate measure might be reliability of the product after manufacture. This measure is typically used to monitor quality of products such as automobiles and consumer electronics.

There has been an enormous ground swell of interest in the quality issue in the United States in recent years. With the onslaught of Japanese competition, many American industries are fighting for their lives. The business of selling quality is at an all-time high. Consulting companies that specialize in providing quality programs to industry, such as the Juran Institute and Philip Crosby Associates, are doing a booming business. The question is whether American firms are merely paying lip service to quality or are seriously trying to change the way they do business. There is evidence that, in some cases at least, the latter is true.

For example, in a comparison of American and Japanese auto companies, quality as measured by defects reported in the first three months of ownership declined significantly from 1987 to 1990 for U.S. companies, narrowing the gap with Japan significantly. The Buick Division of General Motors, a winner of the Baldrige Award, has made dramatic improvements along these lines. Between 1987 and 1990 Buick decreased this defect rate by about 70 percent, equaling the rate for Hondas in 1990 (*Business Week,* October 22, 1990).

There are many success stories in U.S. manufacturing. Ford Motors achieved dramatic success with the Taurus. Ford improved both quality and innovation, providing buyers with reliable and technologically advanced cars. In 1980 James Harbour reported that Japanese automakers could produce a car for $1,500 less than their American counterparts. That gap has been narrowed by Ford to within a few hundred dollars. Part of Ford's success lies in former CEO Donald Petersen's decision not to invest billions in new plants incorporating the latest technology as GM did in the mid-1980s. This is only part of the story, however. According to Faye Wills (1990),

> If you are looking for surprise answers to Ford's ascendancy, for hidden secrets, forget it. Good solid everyday management has turned the trick—textbook planning and execution, common-sense plant layouts and procedures, intelligent designs that not only sell cars, but also cut costs and bolster profit margins. It's that simple.

We can learn from our successes. The machine tool industry was one in which the Japanese made dramatic inroads in the 1980s. Many American firms fell to the onslaught of Asian competition, but not the Stanley Works of New Britain, Connecticut. In 1982, the firm's president was considering whether Stanley should remain in the hardware business as Asian firms flooded the U.S. market with low-priced hammers, screwdrivers, and other tools. Stanley decided to fight back. It modernized its plants and introduced new quality-control systems. Between 1982 and 1988 scrap rates dropped from 15 percent to only 3 percent at New Britain. Stanley not only met the competition head-on here at home, but also competed successfully in Asia. Stanley now runs a profitable operation selling its distinctive yellow tape measures in Asia.

Where are most PC clones made? Taiwan? Korea? Guess again. The answer may surprise you: Texas. Two Texas firms have been extremely successful in this marketplace. One is Compaq Computer, which entered the market in the early 1980s with the first portable PC. It continued to build well-designed and high-quality products, and today it commands 20 percent of the world's PC market. Compaq has established itself as a market leader. The other successful PC maker is from Texas Dell Computer. The sudden rise of Dell is an interesting story. Michael Dell, a former University of Texas student, started reselling IBM PCs in the early 1980s. He later formed PC's Limited, which marketed one of the first mail-order PC clones. Dell is now a serious competitor in the PC marketplace, offering a combination of state-of-the-art designs, high-quality products, and excellent service.

Another American firm that has made a serious commitment to quality is Motorola. Motorola, winner of the Baldrige Award in 1987, has been steadily driving down the rate of defects in its manufactured products. Defects were reported to be near 40 parts per million at the end of 1991, down from 6,000 parts per million in 1986. Motorola has announced that its goal is to reach 6-sigma (meaning six standard deviations away from the mean of a normal distribution), which translates to 3.4 parts per million. Motorola feels that the process of applying for the Baldrige Award was so valuable that it now requires all of its suppliers to apply for the award as well.

Success stories like these show that the United States can compete successfully with Japan and other overseas rivals on the quality dimension. However, total quality management must become ingrained into our culture if we are going to be truly world class. The fundamentals must be there. The systems must be in place to monitor the traditional quality measures: conformance to specifications and defect-free products. However, quality management must expand beyond statistical measures. Quality must pervade the way we do business, from quality in design, quality in manufacture, and quality in building working systems with vendors, to quality in customer service and satisfaction.

Problems for Sections 5–8

12. What is an operational definition of quality? Is it possible for a 13-inch TV selling for $100 to be of superior quality to a 35-inch console selling for $1,800?

13. Studies have shown that the defect rates for many Japanese products are much lower than for their American-made counterparts. Speculate on the reasons for these differences.

14. What does "time-based competition" mean? Give an example of a product that you purchased that was introduced to the marketplace ahead of its competitors.

15. Consider the old maxim, "Build a better mousetrap and the world will beat a path to your door." Discuss the meaning of this phrase in the context of time-based competition. In particular, is getting to the market first the only factor in a product's eventual success?

16. What general features would you look for in a business process that would make that process a candidate for reengineering? Discuss a situation from your own experience in which it was clear that the business process could have been improved.

17. In what ways might the following techniques be useful as part of a reengineering effort?
    • Computer-based simulation.
    • Flowcharting.

- Project management techniques.
- Mathematical modeling.
- Cross-functional teams.

18. What problems can you foresee arising in the following situations?
    a. Top management is interested in reengineering to cut costs, but the employees are skeptical.
    b. Line workers would like to see a reengineering effort undertaken to give them more say-so in what goes on, but management is uninterested.

19. Just-in-time has been characterized as a system whose primary goal is to eliminate waste. Discuss how waste can be introduced in (*a*) relationships with vendors, (*b*) receipt of material into the plant, and (*c*) movement of material through the plant. How do JIT methods cut down on these forms of waste?

20. In what ways can JIT systems improve product quality?

1.9 Matching Process and Product Life Cycles

The Product Life Cycle

The demand for new products typically undergoes cycles that can be identified and mapped over time. Understanding the nature of this evolution helps to identify appropriate strategies for production and operations at the various stages of the product cycle. A typical **product life cycle** is pictured in Figure 1–3. The product life cycle consists of four major segments:

1. Start-up.
2. Rapid growth.
3. Maturation.
4. Stabilization or decline.

FIGURE 1–3

The product life-cycle curve

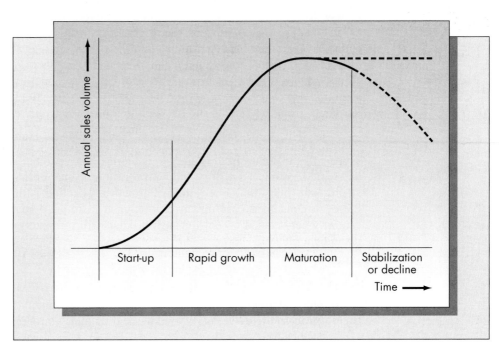

During the start-up phase, the market for the product is developing, production and distribution costs are high, and competition is generally not a problem. During this phase the primary strategy concern is to apply the experiences of the marketplace and of manufacturing to improve the production and marketing functions. At this time, serious design flaws should be revealed and corrected.

The period of rapid growth sees the beginning of competition. The primary strategic goal during this period is to establish the product as firmly as possible in the marketplace. To do this, management should consider alternative pricing patterns that suit the various customer classes and should reinforce brand preference among suppliers and customers. The manufacturing process should be undergoing improvements and standardization as product volume increases. Flexibility and modularization of the manufacturing function are highly desirable at this stage.

During the maturation phase of the product life cycle, the objective should be to maintain and improve the brand loyalty that the firm cultivated in the growth phase. Management should seek to increase market share through competitive pricing. Cost savings should be realized through improved production control and product distribution. During this phase the firm must listen to the messages of the marketplace. Most problems with product design and quality should have been corrected during the start-up and growth phases, but additional improvements should also be considered during this phase.

The appropriate shape of the life-cycle curve in the final stage depends on the nature of the product. Many products will continue to sell, with the potential for annual growth continuing almost indefinitely. Examples of such products are commodities such as household goods, processed food, automobiles, and so on. For such products the company's primary goals in this phase would be essentially the same as those described above for the maturation phase. Other products will experience a natural decline in sales volume as the market for the product becomes saturated or as the product becomes obsolete. If this is the case, the company should adopt a strategy of squeezing out the most from the product or product line while minimizing investment in new manufacturing technology and media advertising.

Although a useful concept, the product life-cycle curve is not accurate in all circumstances. Marketing departments that base their strategies on the life-cycle curve may make poor decisions. Dhalla and Yuspeh (1976) report an example of a firm that shifted advertising dollars from a successful stable product to a new product. The assumption was that the new product was entering the growth phase of its life cycle and the stable product was entering the declining phase of its life cycle. However, the new product never gained consumer acceptance, and because of a drop in the advertising budget, the sales of the stable product went into a decline and never recovered. They suggest that in some circumstances it is more effective to build a model that is consistent with the product's history and with consumer behavior than to blindly assume that all products follow the same pattern of growth and decline. Although we believe that the life-cycle concept is a useful way of looking at customer demand patterns in general, a carefully constructed model for each product will ultimately be a far more effective planning tool.

The Process Life Cycle

Abernathy and Townsend (1975) have classified three major stages of the **manufacturing process life cycle:** early, middle, and mature. These phases do not necessarily coincide exactly with the stages of the product life cycle, but they do provide a conceptual framework for planning improvements in the manufacturing process as the product matures.

In the first phase of the process life cycle, the manufacturing function has the characteristics of a job shop. It must cope with a varied mix of relatively low-volume orders and be responsive to changes in the product design. The types and quality of the inputs may vary considerably, and the firm has little control over suppliers.

In the middle phase of the process life cycle, automation begins to play a greater role. The firm should be able to exert more control over suppliers as the volume of production increases. Unit production costs decline as a result of learning effects. The production process may involve batch processing and some transfer lines (assembly lines).

In the last phase of the process life cycle, most of the major operations are automated, the production process is standardized, and few manufacturing innovations are introduced. The production process may assume the characteristics of a continuous flow operation.

This particular evolutionary scenario is not appropriate for all new manufacturing ventures. Companies that thrive on small one-of-a-kind orders will maintain the characteristics of a job shop, for example. The process life-cycle concept applies to new products that eventually mature into high-volume items. The issue of matching the characteristics of the product with the characteristics of the process is discussed below.

Experience curves show that unit production costs decline as the cumulative number of units produced increases. One may think of the experience curve in terms of the process life cycle shown in Figure 1–4. An accurate understanding of the relationship between the experience curve and the process life cycle can be very valuable. By matching the decline in unit cost with the various stages of the process life cycle, management can gain insight into the consequences of moving from one phase of the

FIGURE 1–4

The process life cycle and the experience curve

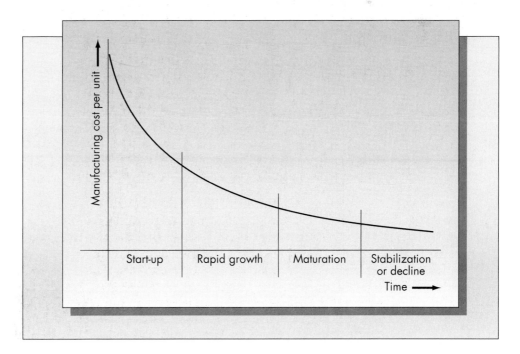

process life cycle into another. This insight will assist management in determining the proper timing of improvements in the manufacturing process.

The Product-Process Matrix

Hayes and Wheelwright (1979) consider linking the product and process life cycles using the **product-process matrix** pictured in Figure 1–5. The matrix is based on four phases in the evolution of the manufacturing process: (1) jumbled flow, (2) disconnected line flow, (3) connected line flow, and (4) continuous flow. This matrix may be viewed in two ways. One is to match the appropriate industry in its mature phase with the appropriate process. This point of view recognizes that not all industries necessarily follow the process evolution described in the previous section. Certain companies or certain products could remain in an early phase of the process life cycle indefinitely. However, even firms that do not evolve to a position in the lower right-hand corner of the matrix should, in most cases, be located somewhere on the diagonal of the matrix.

FIGURE 1–5

The product-process matrix

Process structure Process life-cycle stage	I Low volume, low standardization, one of a kind	II Multiple products, low volume	III Few major products, higher volume	IV High volume, high standardization, commodity products
I Jumbled flow (job shop)	Commercial printer			Void
II Disconnected line flow (batch)		Heavy equipment		
III Connected line flow (assembly line)			Auto assembly	
IV Continuous flow	Void			Sugar refinery

Source: Robert H. Hayes and Steven C. Wheelwright, "Link Manufacturing Process and Product Life Cycles" in the *Harvard Business Review* (January–February 1979). © 1979 by the President and Fellows of Harvard College; all rights reserved. Reprinted by permission.

Located in the upper left-hand corner of this matrix are companies that specialize in "one of a kind" jobs in which the manufacturing function has the characteristics of a jumbled flow shop. A commercial printer is an example of a jumbled flow shop. Production is in relatively small lots, and the shop is organized for maximum flexibility.

Farther down the diagonal are firms that still require a great deal of flexibility but produce a limited line of standardized items. Manufacturers of heavy equipment would fall into this category because they would produce in somewhat higher volumes. A disconnected line would provide enough flexibility to meet custom orders while still retaining economies of limited standardization.

The third category down the diagonal includes firms that produce a line of standard products for a large-volume market. Typical examples are producers of home appliances or electronic equipment, and automobile manufacturers. The assembly line or transfer line would be an appropriate process technology in this case.

Finally, the lower right-hand portion of the matrix would be appropriate for products involving continuous flow. Chemical processing, gasoline and oil refining, and sugar refining are examples. Such processes are characterized by low unit costs, standardization of the product, high sales volume, and extreme inflexibility of the production process.

What is the point of this particular classification scheme? It provides a means of assessing whether a firm is operating in the proper portion of the matrix; that is, if the process is properly matched with the product structure. Firms choosing to operate off the diagonal should have a clear understanding of the reasons for doing so. One example of a successful firm that operates off the diagonal is Rolls-Royce. Another is a company producing handmade furniture. The manufacturing process in these cases would have the characteristics of a jumbled flow shop, but competitors might typically be located in the second or third position on the diagonal.

There is another way to look at the product-process matrix. It can be used to identify the proper match of the production process with the phases of the product life cycle. In the start-up phase of product development, the firm would typically be positioned in the upper left-hand corner of the matrix. As the market for the product matures, the firm would move down the diagonal to achieve economies of scale. Finally, the firm would settle at the position on the matrix that would be appropriate based on the characteristics of the product.

Problems for Section 9

21. *a.* What are the four phases of the manufacturing process that appear in the product-process matrix?
    b. Discuss the disadvantages of operating off the diagonal of the matrix.

22. Give an example of a product that has undergone the four phases of the product life cycle and has achieved stability.

23. Discuss the following: "All firms should evolve along the diagonal of the product-process matrix."

24. Locate the following operations in the appropriate position on the product-process matrix.
    a. A small shop that repairs musical instruments.
    b. An oil refinery.
    c. A manufacturer of office furniture.

    d. A manufacturer of major household appliances such as washers, dryers, refrigerators, and so on.

    e. A manufacturing firm in the start-up phase.

1.10 Learning and Experience Curves

As experience is gained with the production of a particular product, either by a single worker or by an industry as a whole, the production process becomes more efficient. As noted by the economist Adam Smith as far back as the 18th century in his landmark work, *The Wealth of Nations:*

> The division of labor, by reducing every man's business to some one simple operation, and by making this operation the sole employment of his life, necessarily increases very much the dexterity of the worker.

By quantifying the relationship that describes the gain in efficiency as the cumulative number of units produced increases, management can accurately predict the eventual capacity of existing facilities and the unit costs of production. Today we recognize that many other factors besides the improving skill of the individual worker contribute to this effect. Some of these factors include the following:

- Improvements in production methods.
- Improvements in the reliability and efficiency of the tools and machines used.
- Better product design.
- Improved production scheduling and inventory control.
- Better organization of the workplace.

Studies of the aircraft industry undertaken during the 1920s showed that the direct-labor hours required to produce a unit of output declined as the cumulative number of units produced increased. The term **learning curve** was adopted to explain this phenomenon. Similarly, it has been observed in many industries that marginal production costs also decline as the cumulative number of units produced increases. The term **experience curve** has been used to describe this second phenomenon.

Learning Curves

As workers gain more experience with the requirements of a particular process, or as the process is improved over time, the number of hours required to produce an additional unit declines. The learning curve, which models this relationship, is also a means of describing dynamic economies of scale. Experience has shown that these curves are accurately represented by an exponential relationship. Let $Y(u)$ be the number of labor hours required to produce the uth unit. Then the learning curve is of the form

$$Y(u) = au^{-b},$$

where a is the number of hours required to produce the first unit and b measures the rate at which the marginal production hours decline as the cumulative number of units produced increases. Traditionally, learning curves are described by the percentage decline of the labor hours required to produce item $2n$ compared to the labor hours required to produce item n, and it is assumed that this percentage is independent of n. That is, an 80 percent learning curve means that the time required to produce unit $2n$ is 80 percent of the time required to produce unit n for any value of n. For an 80 percent

FIGURE 1–6

An 80 percent learning curve

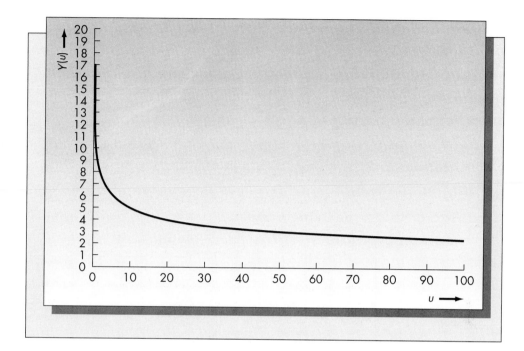

learning curve

$$\frac{Y(2u)}{Y(u)} = \frac{a(2u)^{-b}}{au^{-b}} = 2^{-b} = .80.$$

It follows that

$$-b\ln(2) = \ln(.8)$$

$$\text{or } b = -\ln(.8)/\ln(2) = .3219. \text{ (ln is the natural logarithm.)}$$

More generally, if the learning curve is a $100L$ percent learning curve, then

$$b = -\ln(L)/\ln(2).$$

Figure 1–6 shows an 80 percent learning curve. When graphed on double-log paper, the learning curve should be a straight line if the exponential relationship we have assumed is accurate. If logarithms of both sides of the expression for $Y(u)$ are taken, a linear relationship results, since

$$\ln(Y(u)) = \ln(a) - b\ln(u).$$

Linear regression is used to fit the values of a and b to actual data after the logarithm transformation has been made. (General equations for finding least squares estimators in linear regression appear in Appendix 2–B.)

Example 1.1

XYZ has kept careful records of the average number of labor hours required to produce one of its new products, a pressure transducer used in automobile fuel systems. These records are represented in the table on the following page.

According to the theory, there should be a straight-line relationship between the logarithm of the cumulative number of units produced and the logarithm of the hours required for the last unit of production. The graph of the logarithms of these quantities for the data above appears in Figure 1–7. The figure suggests that the

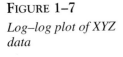

FIGURE 1–7

Log–log plot of XYZ data

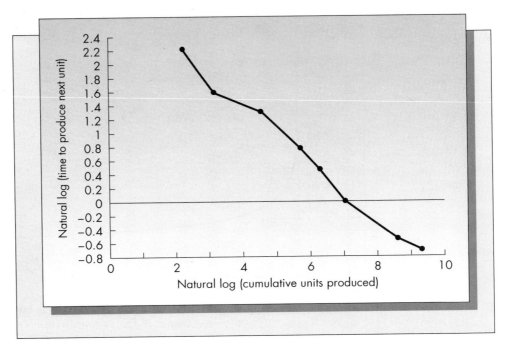

exponential learning curve is fairly accurate in this case. Using the methods outlined in Appendix 2–B, we have obtained estimators for the slope and the intercept of the least squares fit of the data in Figure 1–7. The values of the least squares estimators are

$$\text{Intercept} = 3.1301,$$

$$\text{Slope} = -.42276.$$

Cumulative Number of Units Produced (A)	Ln (Column A)	Hours Required for Next Unit (B)	Ln (Column B)
10.00	2.30	9.22	2.22
25.00	3.22	4.85	1.58
100.00	4.61	3.80	1.34
250.00	5.52	2.44	0.89
500.00	6.21	1.70	0.53
1,000.00	6.91	1.03	0.53
5,000.00	8.52	0.60	−0.51
10,000.00	9.21	0.50	−0.69

Since the intercept is $\ln(a)$, the value of a is $\exp(3.1301) = 22.88$. Hence, it should have taken about 23 hours to produce the first unit. The slope term is the constant $-b$. From above we have that

$$\ln(L) = -b \ln(2) = (-.42276)(.6931) = -.293.$$

It follows that $L = \exp(-.293) = .746$.

Hence, these data show that the learning effect for the production of the transducers can be accurately described by a 75 percent learning curve. This curve can be used to predict the number of labor hours that will be required for continued production of these particular transducers. For example, substituting $u = 50{,}000$ into the relationship

$$Y(u) = 22.88u^{-.42276}$$

gives a value of $Y(50{,}000) = .236$ hour. One must interpret such results with caution, however. A learning curve relationship may not be valid indefinitely. Eventually the product will reach the end of its natural life cycle, which could occur before 50,000 units have been produced in this example. Alternatively, there could be some absolute limit on the number of labor hours required to produce one unit that, because of the nature of the manufacturing process, can never be improved. Even with these limitations in mind, learning curves can be a valuable planning tool when properly used. ■

Experience Curves

Learning curves are a means of calibrating the decline in marginal labor hours as workers become more familiar with a particular task or as greater efficiency is introduced into the production process. Experience curves measure the effect that accumulated experience with production of a product or family of products has on overall cost and price. Experience curves are most valuable in industries that are undergoing major changes, such as the microelectronics industry, rather than very mature industries in which most radical changes have already been made, such as the automobile industry. The steady decline in the prices of integrated circuits (ICs) is a classic example of an experience curve. Figure 1–8 (Noyce, 1977) shows the average price per unit as a function of the industry's accumulated experience, in millions of units of production, during the period 1964 to 1972. This graph is shown on log–log scale and the points fall very close to a straight line. This case represents a 72 percent experience curve. That is, the average price per unit declines to about 72 percent of its previous value for each doubling of the cumulative production of ICs throughout the industry.

FIGURE 1–8

Prices of integrated circuits during the period 1964–1972

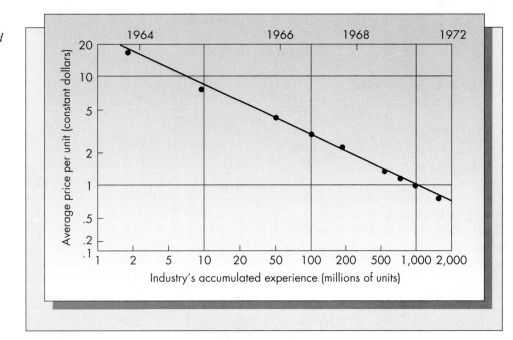

Source: Robert N. Noyce, "Microelectronics," in *Scientific American,* September 1977. © 1977 by Scientific American, Inc. All rights reserved. Reprinted with permission of the publisher.

FIGURE 1–9

*"Kinked" experi-
ence curve due to
umbrella pricing*

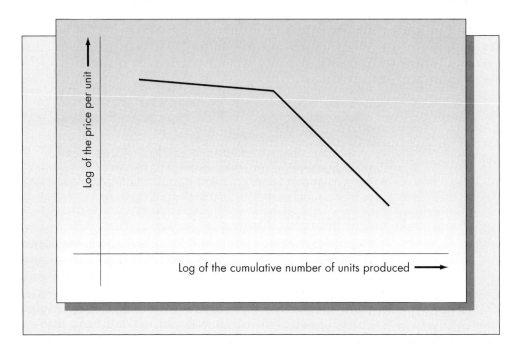

Experience curves are generally measured in terms of cost per unit of production. In most circumstances, the price of a product or family of products closely tracks the cost of production. However, in some cases umbrella pricing occurs. That is, prices remain fairly stable during a period in which production costs decline. Later, as competitive pressures of the marketplace take hold, prices decline more rapidly than costs until they catch up. This can cause a kink in the experience curve when price rather than cost is measured against cumulative volume. This type of phenomenon is pictured in Figure 1–9. Hayes and Wheelwright (1984, p. 243) give two examples of umbrella pricing and its effect on the experience curve. The experience curves for the pricing of free-standing gas ranges and polyvinyl chloride are examples of this phenomenon.

Learning curves have been the subject of criticism in the literature recently on a number of grounds: (*a*) they lack theoretical justification; (*b*) they confuse the effects of learning, economies of scale, and other technological improvements; and (*c*) they focus on cost rather than profit (Devinney, 1987). However, it is clear that such curves are accurate descriptors of the way that marginal labor hours and costs decline as a function of the cumulative experience gained by the firm or industry.

Learning and Experience Curves and Manufacturing Strategy

We define a **learning curve strategy** as one in which the primary goal is to reduce costs of production along the lines predicted by the learning curve. Ford Motors adopted a learning curve strategy in seeking cost reductions in the Model T during the period 1909 to 1923. Abernathy and Wayne (1974) showed that the selling price of the Model T during this period closely followed an 85 percent experience curve. Ford's strategy during this time was clearly aimed at cost cutting; the firm acquired or built new facilities including blast furnaces, logging operations and saw mills, a railroad, weaving mills, coke ovens, a paper mill, a glass plant, and a cement plant. This allowed Ford to vertically integrate operations, resulting in reduced throughput time and inventory levels—a strategy similar in spirit to the just-in-time philosophy discussed earlier in this chapter.

A learning curve strategy may not necessarily be the best choice over long planning horizons. Abernathy and Wayne (1974) make the argument that when manufacturing strategy is based on cost reduction, innovation is stifled. As consumer tastes changed in the 1920s, Ford's attention to cost cutting and standardization of the Model T manufacturing process resulted in its being slow to adapt to changing patterns of customer preferences. Ford's loss was General Motors's gain. GM was quick to respond to customer needs, recognizing that the open car design of the Model T would soon become obsolete. Ford thus found itself fighting for survival in the 1930s after having enjoyed almost complete domination of the market. Survival meant a break from the earlier rigid learning curve strategy to one based on innovation.

Another example of a firm that suffered from a learning curve strategy is Douglas Aircraft. The learning curve concept was deeply rooted in the airframe industry. Douglas made several commitments in the 1960s for delivery of jet aircraft based on extrapolation of costs down the learning curve. However, because of unforeseen changes in the product design, the costs were higher than anticipated, and commitments for delivery times could not be met. Douglas was forced into a merger with McDonnell Company as a result of the financial problems it experienced.

We are not implying by these examples that a learning curve strategy is wrong. Standardization and cost reduction based on volume production have been the keys to success for many companies. Failure to achieve quick time-to-volume can spell disaster in a highly competitive marketplace. What we are saying is that the learning curve strategy must be balanced with sufficient flexibility to respond to changes in the marketplace. Standardization must not stifle innovation and flexibility.

Problems for Section 10

25. What are the factors that contribute to the learning curve/experience curve phenomenon?

26. What is a "learning curve strategy"? Describe how this strategy led to Ford's success up until the mid-1920s and Ford's problems after that time.

27. What are some of the pitfalls that can occur when using learning curves and experience curves to predict costs? Refer to the experience of Douglas Aircraft.

28. Consider the example of XYZ Corporation presented in this section. If the learning curve remains accurate, how long will it take to produce the 100,000th unit?

29. A start-up firm has kept careful records of the time required to manufacture its product, a shutoff valve used in gasoline pipelines.

Cumulative Number of Units Produced	Number of Hours Required for Next Unit
50	3.3
100	2.2
400	1.0
600	.8
1,000	.5
10,000	.2

FIGURE 1–10
(*for problem 31*)

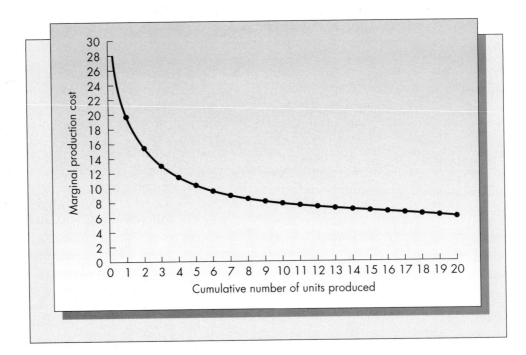

a. Compute the logarithms of the numbers in each column. (Use natural logs.)
b. Graph the ln(hours) against the ln(cumulative units) and eyeball a straightline fit of the data. Using your approximate fit, estimate *a* and *b*.
c. Using the results of part (*b*), estimate the time required to produce the first unit and the appropriate percentage learning curve that fits these data.
d. Repeat parts (*b*) and (*c*), but use an exact least squares fit of the logarithms computed in part (*a*).

30. Consider the learning curve derived in the previous problem. How much time will be required to produce the 100,000th unit, assuming the learning curve remains accurate?

31. Consider the experience curve plotted in Figure 1–10. What percentage experience curve does this represent?

32. Discuss the limitations of learning and experience curves.

33. An analyst predicts that an 80 percent experience curve should be an accurate predictor of the cost of producing a new product. Suppose that the cost of the first unit is $1,000. What would he predict is the cost of producing the
a. 100th unit?
b. 10,000th unit?

1.11 Capacity Growth Planning: A Long-Term Strategic Problem

The capacity of a plant is the number of units that the plant can produce in a given time. Capacity policy plays a key role in determining the firm's competitive position in the marketplace. A capacity strategy must take into account a variety of factors, including

• Predicted patterns of demand.
• Costs of constructing and operating new facilities.

- New process technology.
- Competitors' strategy.

Capacity planning is an extremely complex issue. Each time a company considers expanding existing productive capacity, it must sift through a myriad of possibilities. First, the decision must be made whether to increase capacity by modifying existing facilities. From an overhead point of view, this is an attractive alternative. It is cheaper to effect major changes in existing processes and plants than to construct new facilities. However, such a strategy ultimately could be penny wise and pound foolish. There is substantial evidence that plants that have focus are the most productive. Diminishing returns quickly set in if the firm tries to push the productive capacity of a single location beyond its optimal value.

Given the decision to go ahead with construction of a new plant, many issues remain to be resolved. These include

1. *When.* The timing of construction of new facilities is an important consideration. Lead times for construction and changing patterns of demand are two factors that affect timing.

2. *Where.* Locating new facilities is a complex issue. Consideration of the logistics of material flows suggests that new facilities be located near suppliers of raw materials and market outlets. If labor costs were the key issue, overseas locations might be preferred. Tax incentives are sometimes given by states and municipalities trying to attract new industry. Cost of living and geographical desirability are factors that would affect the company's ability to hire and keep qualified employees.

3. *How much.* Once management has decided when and where to add new capacity, it must decide on the size of the new facility. Adding too much capacity means that the capacity will be underutilized. This is an especially serious problem when capital is scarce. On the other hand, adding too little capacity means that the firm will soon be faced with the problem of increasing capacity again.

Economies of Scale and Economies of Scope

Economies of scale are generally considered the primary advantages of expanding existing capacity. Panzer and Willig (1981) introduced the concept of **economies of scope,** which they defined as the cost savings realized from combining the production of two or more product lines at a single location. The idea is that the manufacturing processes for these product lines may share some of the same equipment and personnel so that the cost of production at one location could be less than at two or more different locations.

The notion of economies of scope extends beyond the direct cost savings that the firm can realize by combining the production of two or more products at a single location. It is often necessary to duplicate a variety of support functions at different locations. These functions include information storage and retrieval systems and clerical and support staff. Such activities are easier to coordinate if they reside at the same location. The firm also can realize economies of scope by locating different facilities in the same geographic region. In this way employees can, if necessary, call upon the talents of key personnel at a nearby location.

Goldhar and Jelinek (1983) argue that considerations of economies of scope support investment in new manufacturing technology. Flexible manufacturing systems and computer-integrated manufacturing result in "efficiencies wrought by variety, not volume." These types of systems, argue the authors, allow the firm to produce multiple products in small lot sizes more cheaply using the same multipurpose equipment. (Flexible manufacturing systems are discussed in greater detail in Chapter 10.)

Management must weigh the benefits that the firm might realize by combining product lines at a single location against the disadvantages of lack of focus discussed above. Too many product lines produced at the same facility could cause the various manufacturing operations to interfere with each other. The proper sizing and diversity of the functions of a single plant must be balanced so that the firm can realize economies of scope without allowing the plant to lose its essential focus.

Make or Buy: A Prototype Capacity Expansion Problem

A classic problem faced by the firm is known as the **make-or-buy decision.** The firm can purchase the product from an outside source for c_1 per unit, but can produce it internally for a lower unit price, $c_2 < c_1$. However, in order to produce the product internally, the company must invest \$$K$ to expand production capacity. Which strategy should the firm adopt?

The make-or-buy problem contains many of the elements of the general capacity expansion problem. It clarifies the essential trade-off of investment and economies of scale. The total cost of the firm to produce x units is $K + c_2x$. This is equivalent to $K/x + c_2$ per unit. As x increases, the cost per unit of production decreases, since K/x is a decreasing function of x. The cost to purchase outside is c_1 per unit, independent of the quantity ordered. By graphing the total costs of both internal production and external purchasing, we can find the point at which the costs are equal. This is known as the break-even quantity. The break-even curves are pictured in Figure 1–11.

The break-even quantity solves

$$K + c_2x = c_1x,$$

giving $x = K/(c_1 - c_2)$.

FIGURE 1–11

Break-even curves

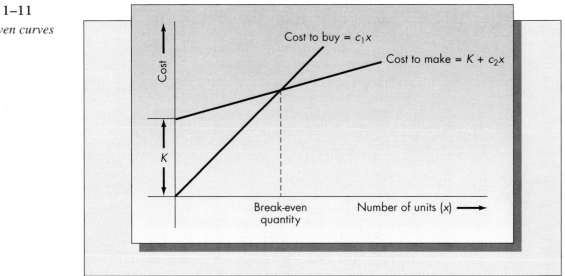

Example 1.2

A large international computer manufacturer is designing a new model of personal computer and must decide whether to produce the keyboards internally or to purchase them from an outside supplier. The supplier is willing to sell the keyboards for $50 each, but the manufacturer estimates that his firm can produce the keyboards for $35 each. Management estimates that expanding the current plant and purchasing the necessary equipment to make the keyboards would cost $8 million. Should they undertake the expansion?

The break-even quantity is

$$x = 8,000,000/(50 - 35) = 533,333.$$

Hence, the firm would have to sell at least 533,333 keyboards in order to justify the $8 million investment required for the expansion. ■

Break-even curves such as this are useful for getting a quick ballpark estimate of the desirability of a capacity addition. Their primary limitation is that they are static. They do not consider the dynamic aspects of the capacity problem, which cannot be ignored in most cases. These include changes in the anticipated pattern of demand and considerations of the time value of money. Even as static models, break-even curves are only rough approximations. They ignore the learning effects of production; that is, the marginal production cost should decrease as the number of units produced increases. (Learning curves are discussed in detail in Section 10 of this chapter.) Depending on the structure of the production function, it may be economical to produce some units internally and purchase some units outside. Manne (1967) discusses the implications of some of these issues.

Dynamic Capacity Expansion Policy

Capacity decisions must be made in a dynamic environment. In particular, the dynamics of the changing demand pattern determine when the firm should invest in new capacity. Two competing objectives in capacity planning are

1. Maximizing market share.
2. Maximizing capacity utilization.

A firm that bases its long-term strategy on maximization of capacity utilization runs the risk of incurring shortages in periods of higher-than-anticipated demand. An alternative strategy to increasing productive capacity is to produce to inventory and let the inventory absorb demand fluctuations. However, this can be very risky. Inventories can become obsolete, and holding costs can become a financial burden.

Alternatively, a firm may assume the strategy of maintaining a "capacity cushion." This capacity cushion is excess capacity that the firm can use to respond to sudden demand surges; it puts the firm in a position to capture a larger portion of the marketplace if the opportunity arises.

Consider the case where the demand exhibits an increasing linear trend. Two policies, (a) and (b), are represented in Figure 1–12. In both cases the firm is acquiring new capacity at equally spaced intervals x, $2x$, $3x$, . . . , and increasing the capacity by the same amount at each of these times. However, in case (a) capacity leads demand, meaning that the firm maintains excess capacity at all times; whereas in case (b), the capacity lags the demand, meaning that the existing capacity is fully utilized at all

FIGURE 1–12

*Capacity planning
strategies*

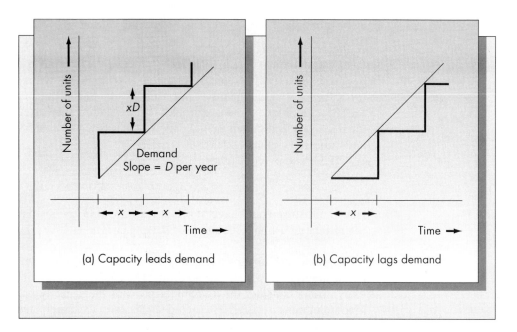

FIGURE 1–12

*Capacity planning
strategies*

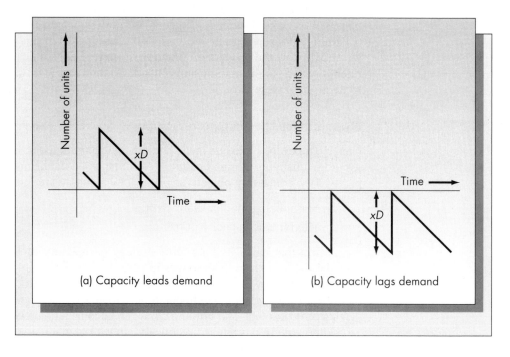

FIGURE 1–13

*Time path of excess
or deficient capacity*

times. Following policy (*a*) or (*b*) results in the time path of excess capacity (or capacity shortfall, if appropriate) given in Figure 1–13.

Consider the following specific model that appears in Manne (1967). Define

D = Annual increase in demand.
x = Time interval between introduction of successive plants.
r = Annual discount rate, compounded continuously.
$f(y)$ = Cost of opening a plant of capacity y.

From Figure 1–11*a* (which is the strategy assumed in the model), we see that if the time interval for plant replacement is x, it must be true that the plant size at each replacement is xD. Furthermore, the present value of a cost of \$1 incurred t years into the future is given by e^{-rt}. (A discussion of discounting and the time value of money appears in Appendix 1–A.)

Define $C(x)$ as the sum of discounted costs for an infinite horizon given a plant opening at time zero. It follows that

$$C(x) = f(xD) + e^{-rx}f(xD) + e^{-2rx}f(xD) + \cdots$$
$$= f(xD)[1 + e^{-rx} + (e^{-rx})^2 + (e^{-rx})^3 + \cdots]$$
$$= \frac{f(xD)}{1 - e^{-rx}}.$$

Experience has shown that a representation of $f(y)$ that explains the economies of scale for plants in a variety of industries is

$$f(y) = ky^a,$$

where k is a constant of proportionality. The exponent a measures the ratio of the incremental to the average costs of a unit of plant capacity. A value of 0.6 seems to be common (known as the six-tenths rule). As long as $a < 1$, there are economies of scale in plant construction, since a doubling of the plant size will result in less than a doubling of the construction costs. To see this, consider the ratio

$$\frac{f(2y)}{f(y)} = \frac{k(2y)^a}{k(y)^a} = 2^a.$$

Substituting $a = 0.6$, we obtain $2^a = 1.516$. This means that if $a = 0.6$ is accurate, the plant capacity can be doubled by increasing the dollar investment by about 52 percent. Henceforth, we assume that $0 < a < 1$ so that there are economies of scale in the plant sizing.

Given a specific form for $f(y)$, we can solve for the optimal timing of plant additions and hence the optimal sizing of new plants. If $f(y) = ky^a$, then

$$C(x) = \frac{k(xD)^a}{1 - e^{-rx}}.$$

Consider the logarithm of $C(x)$:

$$\log[C(x)] = \log[k(xD)^a] - \log[1 - e^{-rx}]$$
$$= \log(k) + a\log(xD) - \log[1 - e^{-rx}].$$

It can be shown that the cost function $C(x)$ has a unique minimum with respect to x and furthermore that the value of x for which the derivative of $\log[C(x)]$ is zero is the value of x that minimizes $C(x)$. It is easy to show[1] that the optimal solution satisfies

$$\frac{rx}{e^{rx} - 1} = a.$$

---

[1] $\dfrac{d\log[C(x)]}{dx} = \dfrac{aD}{xD} - \dfrac{(-e^{-rx})(-r)}{1 - e^{-rx}} = \dfrac{a}{x} - \dfrac{r}{e^{rx} - 1} = 0,$

which gives $\dfrac{rx}{e^{rx} - 1} = a.$

FIGURE 1–14

The function
$u/(e^u - 1)$

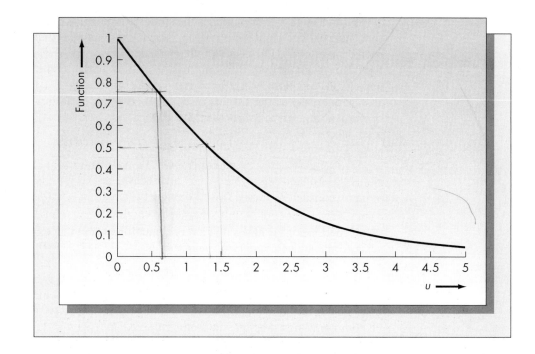

The function $f(u) = u/(e^u - 1)$ appears in Figure 1–14, where $u = rx$. By locating the value of a on the ordinate axis, one can find the optimal value of u on the abscissa axis.

Example 1.3

A chemicals firm is planning for an increase of production capacity. The firm has estimated that the cost of adding new capacity obeys the law

$$f(y) = .0107y^{.62},$$

where cost is measured in millions of dollars and capacity is measured in tons per year. For example, substituting $y = 20,000$ tons gives $f(y) = \$4.97$ million plant cost. Furthermore, suppose that the demand is growing at a constant rate of 5,000 tons per year and future costs are discounted using a 16 percent interest rate. From Figure 1–14 we see that, if $a = .62$, the value of u is approximately 0.9. Solving for x, we obtain the optimal timing of new plant openings:

$$x = u/r = .9/.16 = 5.625 \text{ years.}$$

The optimal value of the plant capacity should be $xD = (5.625)(5,000) = 28,125$ tons. Substituting $y = 28,125$ into the equation for $f(y)$ gives the cost of each plant at the optimal solution as \$6.135 million. ∎

Much of the research into the capacity expansion problem consists of extensions of models of this type. This particular model could be helpful in some circumstances but ignores a number of fundamental features one would expect to find in the real world:

1. *Finite plant lifetime.* The assumption of the model is that, once constructed, a plant has an infinite lifetime. However, companies close plants for a variety of reasons:

Equipment becomes obsolete or unreliable and cannot be replaced easily. Labor costs or requirements may dictate moving either to less expensive locations domestically or to overseas locations. Major changes in the process technology may not be easily adaptable to existing facilities.

2. *Demand patterns.* We have assumed that demand grows at a constant rate per year. Models have been proposed to account for more complex growth patterns of demand. In truth, demand uncertainty is a key factor. In many industries, foreign competition has made significant inroads into established markets, thus forcing rethinking of earlier strategies.

3. *Technological developments.* The model assumes that the capacity of all new plants constructed remains constant and that the cost of building a plant of given size remains constant as well. This is obviously unreasonable. Major changes in process technology occur on a regular basis, changing both the maximum size of new plants and the costs associated with a fixed plant size.

4. *Government regulation.* Environmental and safety restrictions may limit choices of plant location and scale.

5. *Overhead costs.* Most capacity expansion and location models do not explicitly account for the costs of overhead. During the energy crunch in the late 1970s, costs of energy overhead soared, wreaking havoc with plant overhead budgets.

6. *Tax incentives.* The financial implications of the sizing and location of new facilities must be considered in the context of tax planning. Tax incentives are offered by local or state municipalities to major corporations considering sites for the construction of new facilities.

An interesting question is whether models of this type really capture the way that companies have made capacity expansion decisions in the past. There is some preliminary evidence that they do not. Lieberman (1987) attempted to assess the factors that motivated firms to construct new chemical plants during the period 1957 to 1982. He found that the size of new plants increased by about 8 percent per year independent of market conditions. In periods of high demand, firms constructed more plants. This preliminary study indicates that the rational thinking that leads to models such as the one above does not accurately reflect the way that companies make plant-sizing decisions. His results suggest that firms build the largest plants possible with the existing technology. (Given sufficient economies of scale, however, this policy may theoretically be optimal.)

Issues in Plant Location

This section has been concerned with capacity expansion decisions, specifically, determining the amount and timing of new capacity additions. A related issue is the **location of the new facility.** Deciding where to locate a plant is a complex problem. Many factors must be carefully considered by management before making the final choice.

The following information about the plant itself is relevant to the location decision:

1. *Size of the plant.* This includes the required acreage, the number of square feet of space needed for the building structure, and constraints that might arise as a result of special needs.
2. *Product lines to be produced.*
3. *Process technology to be used.*

4. *Labor force requirements.* These include both the number of workers required and the specification of the particular skills needed.

5. *Transportation needs.* Depending on the nature of the product produced and the requirements for raw materials, the plant may have to be located near major interstate highways or rail lines.

6. *Utilities requirements.* These include special needs for power, water, sewage, or fossil fuels such as natural gas. Plants that have unusual power needs should be located in areas where energy is less expensive or near sources of hydroelectric power.

7. *Environmental issues.* Because of government regulations, there will be few allowable locations if the plant produces significant waste products.

8. *Interaction with other plants.* If the plant is a satellite of existing facilities, it is likely that management would want to locate the new plant near the others.

9. *International considerations.* Whether to locate a new facility domestically or overseas is a very sensitive issue. Although labor costs may be lower in some locations, such as the Far East, tariffs, import quotas, inventory pipeline costs, and market responsiveness also must be considered.

10. *Tax treatment.* Tax consideration is an important variable in the location decision. Favorable tax treatment is given by some countries, such as Ireland, to encourage new industry. There are also significant differences in state tax laws designed to attract domestic manufacturers.

Mathematical models are useful for assisting with many operational decisions. However, they generally are of only limited value for determining a suitable location for a new plant. Because so many factors and constraints enter into the decision process, such decisions are generally made based on the inputs of one or more of the company's divisions, and the decision process can span several years. (Mathematical techniques for making location decisions will be explored in Chapter 9, on facilities layout and location.) Schmenner (1982) has examined the decision process at a number of the *Fortune* 500 companies. His results showed that in most firms, the decision of where to locate a new facility was made either by the corporate staff or by the chief executive officer (CEO), even though the request for new facilities might have originated at the division level. The degree of decision-making autonomy enjoyed at the division level depended on the firm's management style.

Based on a sample survey, Schmenner reported that the major factors that influenced new facilities location decisions were the following:

1. *Labor costs.* This was a primary concern for industries such as apparel, leather, furniture, and consumer electronics. It was less of a concern for capital-intensive industries.

2. *Unionization.* A motivating factor for a firm considering expanding an existing facility, as opposed to one considering building a new facility, is the potential for eliminating union influence in the new facility. A fresh labor force may be more difficult to organize.

3. *Proximity to markets.* When transportation costs account for a major portion of the cost of goods sold, locating new plants near existing markets is essential.

4. *Proximity to supplies and resources.* The decision about where to locate plants in certain industries is based on the location of resources. For example,

firms producing wood or paper products must be located near forests, firms producing processed food near farms, and so forth.

5. *Proximity to other facilities.* Many companies tend to place manufacturing divisions and corporate facilities in the same geographic area. For example, IBM originated in Westchester County in New York, and located many of its divisions in that state. By locating key personnel near each other, the firm has been able to realize economies of scope.

6. *Quality of life in the region.* When other issues do not dictate the choice of a location, choosing a site that will be attractive to employees may help in recruiting key personnel. This is especially true in high-tech industries that must compete for workers with particular skills.

Problems for Section 11

34. A start-up company, Macrotech, plans to produce a device to translate Morse code to a written message on a home computer and to send written messages in Morse code over the airwaves. The device is primarily of interest to ham radio enthusiasts. The president, Ron Lodel, estimates that it would require a $30,000 initial investment. Each unit costs him $20 to produce and each sells for $85.
    a. How many units must be sold in order for the firm to recover its initial investment?
    b. What is the total revenue at the break-even volume?
    c. If the price were increased to $100 each, find the break-even volume.

35. For the previous problem, suppose that sales are expected to be 100 units in the first year and increase at a rate of 40 percent per year. How many years will it take to recoup the $30,000 initial investment? Assume that each unit sells for $85.

36. A domestic producer of baby carriages, Pramble, buys the wheels from a company in the north of England. Currently the wheels cost $4 each, but for a number of reasons the price will double. In order to produce the wheels themselves, Pramble would have to add to existing facilities at a cost of $800,000. It estimates that its unit cost of production would be $3.50. At the current time, the company sells 10,000 carriages annually. (Assume that there are four wheels per carriage.)
    a. At the current sales rate, how long would it take to pay back the investment required for the expansion?
    b. If sales are expected to increase at a rate of 15 percent per year, how long will it take to pay back the expansion?

37. Based on past experience, a chemicals firm estimates that the cost of new capacity additions obeys the law

$$f(y) = .0205 y^{.58}$$

where y is measured in tons per year and $f(y)$ in millions of dollars. Demand is growing at the rate of 3,000 tons per year and the accounting department recommends a rate of 12 percent per year for discounting future costs.
    a. Determine the optimal timing of plant additions and the optimal size of each addition.
    b. What is the cost of each addition?
    c. What is the present value of the cost of the next four additions? Assume an addition has just been made for the purposes of your calculation. (Refer to Appendix 1–A for a discussion of cost discounting.)

38. A major oil company is considering the optimal timing for the construction of new refineries. From past experience, each doubling of the size of a refinery at a single location results in an increase in the construction costs of about 68 percent. Furthermore, a plant size of 10,000 barrels per day costs $6 million. Assume that the demand for the oil is increasing at a constant rate of two million barrels yearly and the discount rate for future costs is 15 percent.
   a. Find the values of k and a assuming a relationship of the form $f(y) = ky^a$. Assume that y is in units of barrels per day.
   b. Determine the optimal timing of plant additions and the optimal size of each plant.
   c. Suppose that the largest single refinery that can be built with current technology is 15,000 barrels per day. Determine the optimal timing of plant additions and the optimal size of each plant in this case. (Assume 365 days per year for your calculations.)
39. Discuss the pros and cons of locating manufacturing facilities in a foreign country.

1.12 Summary

This chapter discussed the importance of **operations strategy** and its relationship to the overall business strategy of the firm. Operations continues to grow in importance in the firm. While a larger portion of direct manufacturing continues to move off-shore, the importance of the manufacturing function should not be underestimated. The success of the operations strategy can be measured along several dimensions. These include the obvious measures of cost and product characteristics, but also include quality, delivery speed, delivery reliability, and flexibility.

The classical view of manufacturing strategy, due primarily to Wickham Skinner, considers the following four dimensions of strategy: **time horizon, focus, evaluation, and consistency.** Different types of decisions relate to different time frames. A plant should be designed with a specific focus in mind, whether it be to minimize unit cost or to maximize product quality. Several evaluation criteria may be applied to analyze the effectiveness of a strategy.

We hear more and more frequently that we are part of a global community. When buying products today, we are less concerned with the country of origin than with the characteristics of the product. How many consumers of cell phones are even aware that Nokia is headquartered in Finland, Ericsson in Sweden, and Motorola in the United States? An interesting question explored by Michael Porter is: Why do some industries seem to thrive in some countries? While the answer is complex, Porter suggests that the following four factors are most important: *factor conditions; demand conditions; related and supporting industries;* and *firm strategy, structure, and rivalry.*

Changing the way that one does things can be difficult. Even more difficult is changing the way that a company does things. For that reason, **business process engineering (BPR)** is a painful process, even when it works. The most dramatic successes of BPR have come in service functions, but the concept can be applied to any environment. It is the process of rethinking how and why things are done in a certain way. Intelligently done, BPR can lead to dramatic improvements. However, it can also be a time-consuming and costly process.

Just-in-time (JIT) is a philosophy that grew from the kanban system developed by Toyota. At the heart of the approach is the elimination of waste. Systems are put in place to reduce material flows to small batches to avoid large build-ups of work-in-process inventories. While JIT developed on the factory floor, it is a concept that has been applied to the purchasing function as well. Successful application of JIT purchasing requires the development of long-term relationships, and usually requires close proximity to suppliers. The mechanics of JIT are discussed in more detail in Chapter 7.

Being able to get to the market quickly with products that people want in the volumes that the marketplace requires is crucial if one wants to be a market leader. **Time-based competition** means that the time from product conception to its appearance in the marketplace must be reduced. To do so, one performs as many tasks concurrently as possible. In many instances, time to market is less important than time to volume. Being the first to the market may not mean much if one cannot meet product demand.

The dramatic successes of the Japanese during the 1970s and 1980s were to a large extent due to the outstanding **quality** of their manufactured products. Two Americans, Deming and Juran, visited Japan in the early 1950s and played an important role in making the Japanese aware of the importance of producing quality products. The quality movement in the United States has resulted in a much greater awareness of the importance of quality, recognition for outstanding achievement in this arena vis the Malcolm Baldrige Award, and initiation of important programs such as quality circles and the six sigma program at Motorola. (Both the statistical and the organizational issues concerning quality are discussed in detail in Chapter 11.)

It is important to understand both **product and process life cycles.** Both go through the four cycles of start-up, rapid growth, maturation, and stabilization or decline. It is also important to understand which types of processes are appropriate for which types of products and industries. To this end, Hayes and Wheelwright have developed the concept of the product-process matrix.

Learning and experience curves are useful in modeling the decline in labor hours or the decline in product costs as experience is gained in the production of an item or family of items. These curves have been shown to obey an exponential law, and can be useful predictors of the cost or time required for production. (Moore's Law, due to Gordon Moore, a founder of Intel, predicted the doubling of chip performance every 18 months. This is an example of an experience curve, and the prediction has continued to be accurate to the present day.)

We discussed two methods for assisting with **capacity expansion** decisions. Break-even curves provide a means of determining the sales volume necessary to justify investing in new or existing facilities. A simple model for a dynamic expansion policy is presented that gives the optimal timing and sizing of new facilities assuming constant demand growth and discounting of future costs. We also discussed issues that arise in trying to decide where to locate new facilities. This problem is very complex in that there are many factors that relate to the decision of where to locate production, design, and management facilities.

Additional Problems for Chapter 1

40. What is a production and operations strategy? Discuss the elements in common with marketing and financial strategies and the elements that are different.

41. What is the difference between the product life cycle and the process life cycle? In what way are these concepts related?

42. Suppose that the Mendenhall Corporation, a producer of women's handbags, has determined that a 73 percent experience curve accurately describes the evolution of its production costs for a new line. If the first unit costs $100 to produce, what should the 10,000th unit cost based on the experience curve?

43. Delon's Department Store sells several of its own brands of clothes and several well-known designer brands as well. Delon's is considering building a plant in Malaysia to produce silk ties. The plant will cost the firm $5.5 million. The plant will be able to produce the ties for $1.20 each. On the other hand, Delon's can subcontract to have the ties produced and pay $3.00 each. How many ties will Delon's have to sell worldwide to break even on its investment in the new plant?

44. A Japanese steel manufacturer is considering expanding operations. From experience, it estimates that new capacity additions obey the law

$$f(y) = .00345y^{.51},$$

where the cost, $f(y)$, is measured in millions of dollars and y is measured in tons of steel produced. If the demand for steel is assumed to grow at the constant rate of 8,000 tons per year and future costs are discounted using a 10 percent discount rate, what is the optimal number of years between new plant openings?

The following problems are designed to be solved by spreadsheet.

45. Consider the following break-even problem: the cost of producing Q units, $c(Q)$, is described by the curve

$$c(Q) = 48Q(1 - \exp(-.08Q)),$$

where Q is in hundreds of units of items produced and $c(Q)$ is in thousands of dollars.
   a. Graph the function $c(Q)$. What is its shape? What economic phenomenon gives rise to a cumulative cost curve of this shape?
   b. At what production level does the cumulative production cost equal $1,000,000?
   c. Suppose that these units can be purchased from an outside supplier at a cost of $800 each, but the firm must invest $850,000 to build a facility that would be able to produce these units at a cost $c(Q)$. At what cumulative volume of production does it make sense to invest in the facility?

46. Maintenance costs for a new facility are expected to be $112,000 for the first year of operation. It is anticipated that these costs will increase at a rate of 8 percent per year. Assuming a rate of return of 10 percent, what is the present value of the stream of maintenance costs over the next 30 years?

47. Suppose the supplier of keyboards described in Example 1.2 is willing to offer the following incremental quantity discount schedule:

Cost per Keyboard	Order Quantity
$50	$Q \le 100{,}000$
$45	$100{,}000 < Q \le 500{,}000$
$40	$500{,}000 < Q$

Determine the cost to the firm for order quantities in increments of 20,000 for $Q = 200{,}000$ to $Q = 1{,}000{,}000$, and compare that to the cost to the firm of producing internally for these same values of Q. What is the break-even order quantity?

APPENDIX 1–A   PRESENT WORTH CALCULATIONS

Including the time value of money in the decision process is common when considering alternative investment strategies. The idea is that a dollar received today has greater value than one received a year from now. For example, if a dollar were placed in a simple passbook account paying 5 percent, it would be worth $1.05 in one year. More generally, if it were invested at a rate of return of r (expressed as a decimal), it would be worth $1 + r$ in a year, $(1 + r)^2$ in two years, and so on.

In the same way, a cost of $1 incurred in a year has a present value of less than $1 today. For example, at 5 percent interest, how much would one need to place in an account today so that the total principal plus interest would equal $1 in a year? The answer is $1/(1.05) = 0.9524$. Similarly, the present value of a $1 cost incurred in two years at 5 percent is $1/(1.05)^2 = 0.9070$. In general, the present value of a cost of $1 incurred in t years assuming a rate of return r is $(1 + r)^{-t}$.

These calculations assume that there is no compounding. Compounding means that one earns interest on the interest, so to speak. For example, 5 percent compounded semiannually means that one earns 2.5 percent on $1 after six months and 2.5 percent on the original $1 plus interest earned in the first six months. Hence, the total return is

$$(1.025)(1.025) = \$1.050625$$

after one year, or slightly more than 5 percent. If the interest were compounded quarterly, the dollar would be worth

$$(1 + .05/4)^4 = 1.0509453$$

at the end of the year. The logical extension of this idea is continuous compounding. One dollar invested at 5 percent compounded continuously would be worth

$$\lim_{n \to \infty} (1 + .05/n)^n = e^{.05} = 1.05127$$

at the end of a year. The number $e = 2.7172818 \ldots$ is defined as

$$e = \lim_{n \to \infty} (1 + 1/n)^n.$$

Notice that continuous compounding only increases the effective simple interest rate from 5 percent to 5.127 percent.

More generally, C invested at a rate of r for t years compounded continuously is worth Ce^{rt} at the end of t years.

Reversing the argument, the present value of a cost of C incurred in t years assuming continuous compounding at a discount rate r is Ce^{-rt}. A stream of costs $C_1, C_2, \ldots, C_n$ incurred at times $t_1, t_2, \ldots, t_n$ has present value

$$\sum_{i=1}^{n} C_i e^{-rt_i}.$$

A comprehensive treatment of discounting and its relationship to the capacity expansion problem can be found in Freidenfelds (1981).

Bibliography

Abernathy, W. J., and P. L. Townsend. "Technology, Productivity, and Process Change." *Technological Forecasting and Social Change* 7 (1975), pp. 379–96.

Abernathy, W. J., and K. Wayne. "Limits of the Learning Curve." *Harvard Business Review* 52 (September–October 1974), pp. 109–19.

Becker, G. S. Quoted in *Business Week,* January 27, 1986, p. 12.

Bell, D. *The Coming of the Post-Industrial Society: A Venture in Social Forecasting.* New York: Basic Books, 1976.

Blackburn, J. D. *Time-Based Competition: The Next Battleground in American Manufacturing.* New York: McGraw-Hill/Irwin, 1991.

Buffa, E. S., and R. K. Sarin. *Modern Production/Operations Management.* 8th ed. New York: John Wiley & Sons, 1987.

Business Week, October 22, 1990, pp. 94–95.

Business Week, "The Quality Imperative," October 25, 1991.

Cohen, S. S., and J. Zysman. *Manufacturing Matters: The Myth of the Post-Industrial Economy.* New York: Basic Books, 1987.

Davis, D. "Beating the Clock." *Electronic Business,* May 1989.

Dertouzes, M. L.; R. K. Lester; and R. M. Solow. *Made in America: Regaining the Productive Edge.* Cambridge, MA: MIT Press, 1989.

Devinney, T. M. "Entry and Learning." *Management Science* 33 (1987), pp. 706–24.

Dhalla, N. K., and S. Yuspeh. "Forget about the Product Life Cycle Concept." *Harvard Business Review* 54 (January–February 1976), pp. 102–12.

Drucker, P. F. "Japan: New Strategies for a New Reality." *The Wall Street Journal,* October 2, 1991.

Fallon, M. *San Jose Mercury News,* September 30, 1985, p. 11D.

Fine, C. H., and A. C. Hax. "Manufacturing Strategy: A Methodology and an Illustration." *Interfaces* 15 (1985), pp. 27–47.

Freeland, J. R. "A Survey of Just-in-Time Purchasing Practices in the United States." *Production and Inventory Management Journal,* Second Quarter 1991, pp. 43–50.

Freidenfelds, J. *Capacity Expansion: Analysis of Simple Models with Applications.* New York: Elsevier North Holland, 1981.

Goldhar, J. P., and M. Jelinek. "Plan for Economies of Scope." *Harvard Business Review* 61 (1983), pp. 141–48.

Hammer, M. S., and J. Champy. *Reengineering the Corporation: A Manifesto for Business Revolution.* New York: Harper Business, 1993.

Hammer, M. S., and S. A. Stanton. *The Reengineering Revolution.* New York: Harper Business, 1995.

Hayes, R. H., and S. Wheelwright. "Link Manufacturing Process and Product Life Cycles." *Harvard Business Review* 57 (January–February 1979), pp. 133–40.

Hayes, R. H., and S. Wheelwright. *Restoring Our Competitive Edge: Competing through Manufacturing.* New York: John Wiley & Sons, 1984.

Hill, T. J. *Manufacturing Strategy—Text and Cases.* New York: McGraw-Hill/Irwin, 1993.

Jones, P., and L. Kahaner. *Say It and Live It. The Fifty Corporate Mission Statements That Hit the Mark.* New York: Currency Doubleday, 1995.

Krugman, P. *Peddling Prosperity: Economic Sense and Nonsense in the Age of Diminished Expectations.* New York: W. W. Norton and Company, 1994.

Lieberman, M. "Scale Economies, Factory Focus, and Optimal Plant Size." Paper presented at Stanford Graduate School of Business, July 22, 1987.

Manne, A. S., ed. *Investments for Capacity Expansion: Size, Location, and Time Phasing.* Cambridge, MA: MIT Press, 1967.

Noyce, R. N. "Microelectronics." *Scientific American,* September 1977.

Panzer, J. C., and R. O. Willig. "Economies of Scope." *American Economic Review,* 1981, pp. 268–72.

Port, O. "Customers Move into the Driver's Seat." *Business Week,* October 4, 1999, pp. 103–06.

Porter, M. E. *Competitive Strategy: Techniques for Analyzing Industries and Competitors.* New York: The Free Press, 1980.

Porter, M. E. *The Competitive Advantage of Nations.* New York: The Free Press, 1990.

Prestowitz, C. V., Jr. *Trading Places.* New York: Basic Books, 1989.

Schmenner, R. W. *Making Business Location Decisions.* Englewood Cliffs, NJ: Prentice Hall, 1982.

Skinner, W. "The Focused Factory." *Harvard Business Review* 52 (May–June 1974), pp. 113–21.

Skinner, W. *Manufacturing in the Corporate Strategy.* New York: John Wiley & Sons, 1978.

Stasey, R., and C. J. McNair. *Crossroads: A JIT Success Story.* New York: McGraw-Hill/Irwin, 1990.

Stewart, T. A. "The New American Century: Where We Stand." *Fortune,* Spring–Summer 1991.

Wills, F. "Bully for Taurus." *Business Month,* February 1990, p. 13.

Womack, J. P.; D. T. Jones; and D. Ross. *The Machine That Changed the World.* New York: Harper Perennial, 1990.

World Economic Outlook, published by the International Monetary Fund. Washington, DC: October 1994.

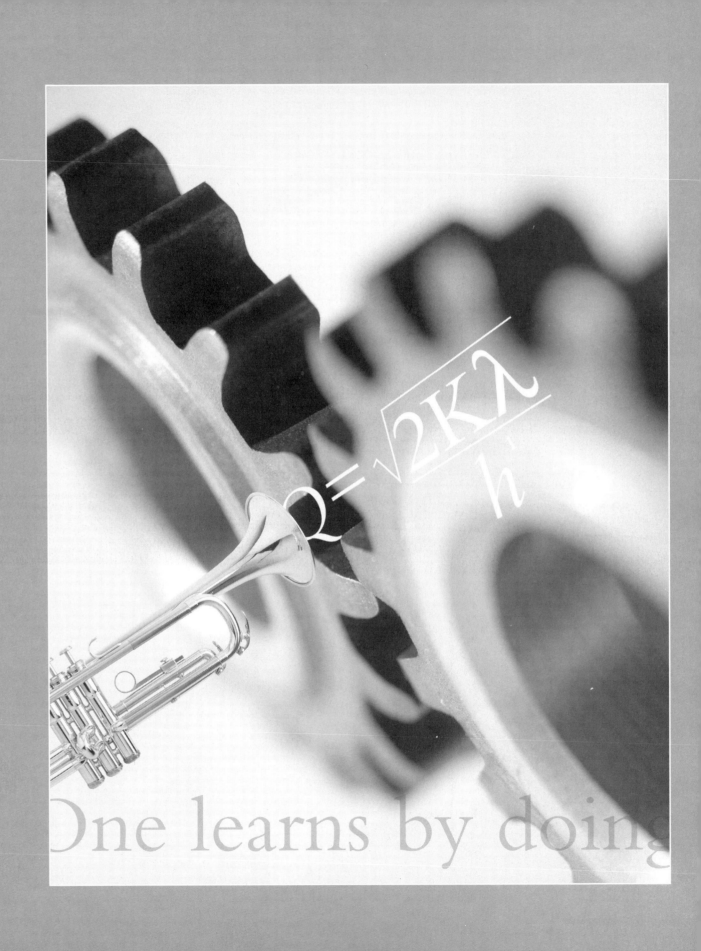

$Q = \sqrt{2K\lambda}h$

One learns by doing

CHAPTER 2

FORECASTING

As was so eloquently stated by Charles F. Kettering, "My concern is with the future since I plan to spend the rest of my life there." Forecasting is the process of predicting the future. All business planning is based to some extent on a forecast. Sales of existing products, customer demand patterns for new products, needs and availabilities of raw materials, changing skills of workers, interest rates, capacity requirements, and international politics are examples of factors likely to affect the future success of a firm.

The two functional areas of the firm that make the most use of forecasting methods are marketing and production. Marketing typically forecasts sales for both new and existing product lines. Sales forecasts are used by the production department for operations planning. In some cases the forecasts prepared by marketing may not meet the needs of production. For example, in order to determine suitable spare parts inventory levels, one must know maintenance schedules and forecast machine breakdowns. Also, marketing may be providing forecasts on product groups or families, but production may require forecasts for individual SKUs (stockkeeping units) for its planning.

We have seen firms benefit from good forecasting and pay the price for poor forecasting. During the 1960s, consumer tastes in automobiles slowly shifted from large, heavy gas guzzlers to smaller, more fuel efficient automobiles. Detroit, slow to respond to this change, suffered when the OPEC oil embargo hit in the late 1970s and tastes shifted more dramatically to smaller cars. Compaq Computer became a market leader in the early 1980s by properly forecasting consumer demand for a portable version of the IBM PC, which gained a popularity that far exceeded expectations. Forecasting played a role in Ford Motors's early success and later demise. Henry Ford saw that the consumer wanted a simpler, less expensive car that was easier to maintain than most manufacturers were offering in the early 1900s. His Model T dominated the industry. However, Ford did not see that consumers would tire of the open Model T design. Ford's failure to forecast consumer desires for other designs nearly resulted in the end of a firm that had monopolized the industry only a few years before.

Can all events be accurately forecasted? The answer is clearly no. Consider the experiment of tossing a coin. Assuming that it is a fair coin and the act of tossing does not introduce bias, the best you can say is that the probability of getting heads is 50 percent on any single toss. No one has been able to consistently top the 50 percent prediction rate for such an experiment over a long period of time. Many real phenomena are accurately described by a type of coin-flipping experiment. Games of chance played at casinos are random. By tipping the probabilities in its favor, the house is always guaranteed to win over the long term. There is evidence that daily prices of stocks follow a purely random process, much like a coin-flipping experiment. Studies have shown that professional money managers rarely outperform stock portfolios generated purely at random.

In production and operations management, we are primarily interested in forecasting product demand. Because demand is likely to be random in most circumstances, can forecasting methods provide any value? In most cases, the answer is yes. Although some portions of the demand process may be unpredictable, other portions may be predictable. Trends, cycles, and seasonal variation may be present, all of which give us an advantage over trying to predict the outcome of a coin toss. In this chapter we consider methods for predicting future values of a series based on past observations.

2.1 The Time Horizon in Forecasting

We may classify forecasting problems along several dimensions. One is the time horizon. Figure 2–1 is a schematic showing the three time horizons associated with forecasting and typical forecasting problems encountered in operations planning associated with each. Short-term forecasting is crucial for day-to-day planning. Short-term forecasts, typically measured in days or weeks, are required for inventory management, production plans that may be derived from a materials requirements planning system (to be discussed in detail in Chapter 7), and resource requirements planning. Shift scheduling may require forecasts of workers' availabilities and preferences.

The intermediate term is measured in weeks or months. Sales patterns for product families, requirements and availabilities of workers, and resource requirements are typical intermediate-term forecasting problems encountered in operations management.

Long-term production and manufacturing decisions, discussed in Chapter 1, are part of the overall firm's manufacturing strategy. One example is long-term planning

FIGURE 2–1

Forecast horizons in operation planning

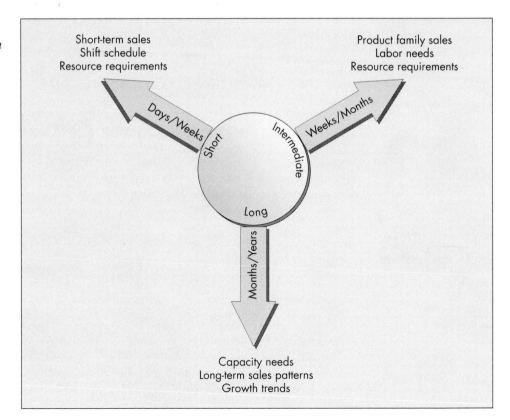

of capacity needs. When demands are expected to increase, the firm must plan for the construction of new facilities and/or the retrofitting of existing facilities with new technologies. Capacity planning decisions may require downsizing in some circumstances. For example, General Motors Corporation historically commanded about 45 percent of the domestic car market. However, in the 1990s that percentage dropped to 35 percent. As a result, GM was forced to significantly curtail its manufacturing operations to remain profitable.

2.2 Characteristics of Forecasts

1. *They are usually wrong.* As strange as it may sound, this is probably the most ignored and most significant property of almost all forecasting methods. Forecasts, once determined, are often treated as known information. Resource requirements and production schedules may require modifications if the forecast of demand proves to be inaccurate. The planning system should be sufficiently robust to be able to react to unanticipated forecast errors.

2. *A good forecast is more than a single number.* Given that forecasts are generally wrong, a good forecast also includes some measure of the anticipated forecast error. This could be in the form of a range, or an error measure such as the variance of the distribution of the forecast error.

3. *Aggregate forecasts are more accurate.* Recall from statistics that the variance of the average of a collection of independent identically distributed random variables is lower than the variance of each of the random variables; that is, the variance of the sample mean is smaller than the population variance. This same phenomenon is true in forecasting as well. On a percentage basis, the error made in forecasting sales for an entire product line is generally less than the error made in forecasting sales for an individual item.

4. *The longer the forecast horizon, the less accurate the forecast will be.* This property is quite intuitive. One can predict tomorrow's value of the Dow Jones Industrial Average more accurately than next year's value.

5. *Forecasts should not be used to the exclusion of known information.* A particular technique may result in reasonably accurate forecasts in most circumstances. However, there may be information available concerning the future demand that is not presented in the past history of the series. For example, the company may be planning a special promotional sale for a particular item so that the demand will probably be higher than normal. This information must be manually factored into the forecast.

2.3 Subjective Forecasting Methods

We classify forecasting methods as either **subjective** or **objective.** A subjective forecasting method is based on human judgment. There are several techniques for soliciting opinions for forecasting purposes:

1. *Sales force composites.* In forecasting product demand, a good source of subjective information is the company sales force. The sales force has direct contact with consumers and is therefore in a good position to see changes in their preferences. To develop a sales force composite forecast, members of the sales force submit sales estimates of the products they will sell in the coming year. These estimates might be

individual numbers or several numbers, such as pessimistic, most likely, and optimistic estimates. Sales managers would then be responsible for aggregating individual estimates to arrive at overall forecasts for each geographic region or product group. Sales force composites may be inaccurate when compensation of sales personnel is based on meeting a quota. In that case, there is clearly an incentive for the sales force to lowball its estimates.

2. *Customer surveys.* Customer surveys can signal future trends and shifting preference patterns. To be effective, however, surveys and sampling plans must be carefully designed to guarantee that the resulting data are statistically unbiased and representative of the customer base. Poorly designed questionnaires or an invalid sampling scheme may result in the wrong conclusions.

3. *Jury of executive opinion.* When there is no past history, as with new products, expert opinion may be the only source of information for preparing forecasts. The approach here is to systematically combine the opinions of experts to derive a forecast. For new product planning, opinions of personnel in the functional areas of marketing, finance, and production should be solicited. Combining individual forecasts may be done in several ways. One is to have the individual responsible for preparing the forecast interview the executives directly and develop a forecast from the results of the interviews. Another is to require the executives to meet as a group and come to a consensus.

4. *The Delphi method.* The Delphi method, like the jury of executive opinion method, is based on soliciting the opinions of experts. The difference lies in the manner in which individual opinions are combined. (The method is named for the Delphic oracle of ancient Greece, who purportedly had the power to predict the future.) The Delphi method attempts to eliminate some of the inherent shortcomings of group dynamics, in which the personalities of some group members overshadow those of other members. The method requires a group of experts to express their opinions, preferably by individual sample survey. The opinions are then compiled and a summary of the results is returned to the experts, with special attention to those opinions that are significantly different from the group averages. The experts are asked if they wish to reconsider their original opinions in light of the group response. The process is repeated until (ideally) an overall group consensus is reached.

As with any particular technique, the Delphi method has advantages and disadvantages. Its primary advantage is that it provides a means of assessing individual opinion without the usual concerns of personal interactions. On the negative side, the method is highly sensitive to the care in the formulation of the questionnaire. Because discussions are intentionally excluded from the process, the experts have no mechanism for resolving ambiguous questions. Furthermore, it is not necessarily true that a group consensus will ever be reached. An interesting case study of a successful application of the Delphi method can be found in Helmer and Rescher (1959).

2.4 Objective Forecasting Methods

Objective forecasting methods are those in which the forecast is derived from an analysis of data. A **time series** method is one that uses only past values of the phenomenon we are predicting. **Causal models** are ones that use data from sources other than the series being predicted; that is, there may be other variables with values that are *linked* in some way to what is being forecasted. We discuss these first.

Causal Models

Let Y represent the phenomenon we wish to forecast and $X_1, X_2, \ldots, X_n$ be n variables that we believe to be related to Y. Then a causal model is one in which the forecast for Y is some function of these variables, say,

$$Y = f(X_1, X_2, \ldots, X_n).$$

Econometric models are special causal models in which the relationship between Y and $(X_1, X_2, \ldots, X_n)$ is linear. That is,

$$Y = \alpha_0 + \alpha_1 X_1 + \alpha_2 X_2 + \cdots + \alpha_n X_n$$

for some constants $(\alpha_1, \ldots, \alpha_n)$. The method of least squares is most commonly used to find estimators for the constants. (We discuss the method in Appendix 2–B for the case of one independent variable.)

Let us consider a simple example of a causal forecasting model. A realtor is trying to estimate his income for the succeeding year. In the past he has found that his income is close to being proportional to the total number of housing sales in his territory. He also has noticed that there has typically been a close relationship between housing sales and interest rates for home mortgages. He might construct a model of the form

$$Y_t = \alpha_0 + \alpha_1 X_{t-1},$$

where Y_t is the number of sales in year t and X_{t-1} is the interest rate in year $t - 1$. Based on past data he would then determine the least squares estimators for the constants α_0 and α_1. Suppose that the values of these estimators are currently $\alpha_0 = 385.7$ and $\alpha_1 = -1,878$. Hence, the estimated relationship between home sales and mortgage rates is

$$Y_t = 385.7 - 1,878 X_{t-1},$$

where X_{t-1}, the previous year's interest rate, is expressed as a decimal. Then if the current mortgage interest rate is 10 percent, the model would predict that the number of sales the following year in his territory would be $385.7 - 187.8 = 197.9$, or about 198 houses sold.

Causal models of this type are common for predicting economic phenomena such as the GNP (gross national product) and the GDP (gross domestic product). Both MIT and the Wharton School of Business at the University of Pennsylvania have developed large-scale econometric models for making these predictions. Econometric prediction models are typically used by the economics and finance arms of the firm to forecast values of macroeconomic variables. Time series methods are more commonly used for operations planning applications.

Time Series Methods

Time series methods are often called naive methods, as they require no information other than the past values of the variable being predicted. *Time series* is just a fancy term for a collection of observations of some economic or physical phenomenon drawn at discrete points in time, usually equally spaced. The idea is that information can be inferred from the pattern of past observations and can be used to forecast future values of the series.

SNAPSHOT APPLICATION

Advanced Forecasting Inc. Serves the Semiconductor Industry

Advanced Forecasting, Inc. (AFI), is a Cupertino-based firm that specializes in providing forecasts for semiconductor sales and sales of related industries, such as semiconductor equipment and suppliers. The firm has had a history of accurately predicting the turning points in the sales patterns of semiconductors for more than a decade. Forecasts are determined from quantitative models (such as the ones discussed in this chapter). Although the actual models used are proprietary, forecasts are based on basic economic factors related to the semiconductor industry. According to the firm's founder Dr. Moshe Handelsman, the problem with most forecaster's predictions is that they are based on sub-

jective opinions and qualitative data. AFI uses a mathematical model to derive its forecasts, which are not second guessed. While the firm is only a small player in the semiconductor forecasting arena, their success has been dramatic. They have consistently been able to predict major shifts in the market for semiconductors, which is a fundamental need for management. According to Jean-Philippe Dauvin, vice president and chief economist for SGS-Thomson Microelectronics: "Our top management pays more attention to Advanced Forecasting's predictions than to any other industry source." Accurate forecasts allow management to deal with important strategic issues such as when production capacity should be expanded, what personnel needs will be, and what the demands will be on marketing and sales. The success of AFI demonstrates that quantitative-based forecasting can provide consistently accurate forecasts and, over the long term, are far more reliable than subjective methods.

Source: Advanced Forecasting, Inc., website, http://www.adv-forecast.com/afi/

In time series analysis we attempt to isolate the patterns that arise most often. These include the following:

1. *Trend.* Trend refers to the tendency of a time series to exhibit a stable pattern of growth or decline. We distinguish between linear trend (the pattern described by a straight line) and nonlinear trend (the pattern described by a nonlinear function, such as a quadratic or exponential curve). When the pattern of trend is not specified, it is generally understood to be linear.

2. *Seasonality.* A seasonal pattern is one that repeats at fixed intervals. In time series we generally think of the pattern repeating every year, although daily, weekly, and monthly seasonal patterns are common as well. Fashion wear, ice cream, and heating oil exhibit a yearly seasonal pattern. Consumption of electricity exhibits a strong daily seasonal pattern.

3. *Cycles.* Cyclic variation is similar to seasonality, except that the length and the magnitude of the cycle may vary. One associates cycles with long-term economic variations (that is, business cycles) that may be present in addition to seasonal fluctuations.

4. *Randomness.* A pure random series is one in which there is no recognizable pattern to the data. One can generate patterns purely at random that often appear to have structure. An example of this is the methodology of stock market chartists who impose forms on random patterns of stock market price data. On the other side of the coin, data that appear to be random could have a very definite structure. Truly random data that fluctuate around a fixed mean form what is called a horizontal pattern.

Examples of time series exhibiting some of these patterns are given in Figure 2–2.

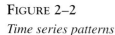
FIGURE 2–2
Time series patterns

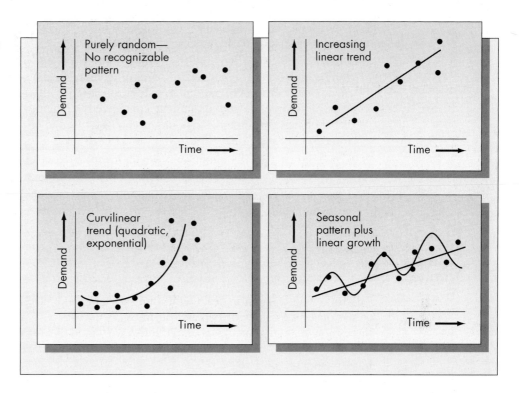

Problems for Sections 1–4

1. Name the four components of time series (i.e., the four distinct patterns exhibited by time series).

2. What distinguishes seasonality from cycles in time series analysis?

3. What is the appropriate type of forecasting method to use in each of the following scenarios:
   a. Holiday Inn, Inc., is attempting to predict the demand next year for motel rooms, based on a history of demand observations.
   b. Standard Brands has developed a new type of outdoor paint. The company wishes to forecast sales based on new housing starts.
   c. IBM is trying to ascertain the cost of a stock-out of a critical tape drive component. They do so by sample survey of managers at various national spare parts centers. The surveys are sent back to the managers for a reassessment and the process is repeated until a consensus is reached.

4. Discuss the role of forecasting for the following functions of the firm:
   a. Marketing.
   b. Accounting.
   c. Finance.
   d. Production.

5. Distinguish between the following types of forecasts:
   a. Aggregate versus single item.
   b. Short-term versus long-term.
   c. Causal versus naive.

6. What is the advantage of the Delphi method over the jury of executive opinion method? What do these methods have in common?

7. Consider the problem of choosing an appropriate college to attend when you were a high school senior. What forecasting concerns did you have when you made that decision? In particular, list the short-term, intermediate-term, and long-term forecasts you might have considered in making your final decision. What objective sources of data might you have used to provide you with better forecasts in each case?

8. Discuss the following quotation from an inventory control manager: "It's not my fault we ran out of those parts. The demand forecast was wrong."

9. Discuss the following statement: "Economists are predicting that interest rates will continue to be under 10 percent for at least 15 years."

2.5 Notation Conventions

The following discussion deals with time series methods. Define $D_1, D_2, \ldots, D_t, \ldots$ as the observed values of demand during periods $1, 2, \ldots, t, \ldots$. We will assume throughout that $\{D_t, t \geq 1\}$ is the time series we would like to predict. Furthermore, we will assume that if we are forecasting in period t, then we have observed $D_t, D_{t-1}, \ldots$ but have not observed D_{t+1}.

Define F_t as the forecast made *for* period t in period $t-1$. That is, it is the forecast made at the end of period $t-1$ after having observed $D_{t-1}, D_{t-2}, \ldots$, but before observing D_t. Hence, for now, we are assuming that all forecasts are one-step-ahead forecasts; that is, they are made for the demand in the next period. Multiple-step-ahead forecasts will be discussed in a later section.

Finally, note that a time series forecast it obtained by applying some set of weights to past data. That is,

$$F_t = \sum_{n=0}^{\infty} a_n D_{t-n} \qquad \text{for some set of weights } a_0, a_1, \ldots.$$

The time series methods discussed in this chapter are distinguished only by the choice of weights.

2.6 Evaluating Forecasts

Define the forecast error in period t, e_t, as the difference between the forecast value for that period and the actual demand for that period. For multiple-step-ahead forecasts,

$$e_t = F_{t-\tau, t} - D_t,$$

and for one-step-ahead forecasts,

$$e_t = F_t - D_t.$$

Let $e_1, e_2, \ldots, e_n$ be the forecast errors observed over n periods. Two common measures of forecast accuracy during these n periods are the mean absolute deviation

(MAD) and the mean squared error (MSE), given by the following formulas:

$$\text{MAD} = (1/n) \sum_{i=1}^{n} |e_i|$$

$$\text{MSE} = (1/n) \sum_{i=1}^{n} e_i^2$$

Note that the MSE is similar to the variance of a random sample. The MAD is often the preferred method of measuring the forecast error because it does not require squaring. Furthermore, when forecast errors are normally distributed, as is generally assumed, an estimate of the standard deviation of the forecast error, σ_e, is given by 1.25 times the MAD.

Although the MAD and the MSE are the two most common measures of forecast accuracy, other measures are used as well. One that is not dependent on the magnitude of the values of demand is known as the mean absolute percentage error (MAPE) and is given by the formula

$$\text{MAPE} = \left[(1/n) \sum_{i=1}^{n} |e_i/D_i| \right] \times 100.$$

Example 2.1

Artel, a manufacturer of SRAMs (static random access memories), has production plants in Austin, Texas, and Sacramento, California. The managers of these plants are asked to forecast production yields (measured in percent) one week ahead for their plants. Based on six weekly forecasts, the firm's management wishes to determine which manager is more successful at predicting his plant's yields. The results of their predictions are given in the following table.

Week	P1	O1	|E1|	|E1/O1|	P2	O2	|E2|	|E2/O2|
1	92	88	4	.0455	96	91	5	.0549
2	87	88	1	.0114	89	89	0	.0000
3	95	97	2	.0206	92	90	2	.0222
4	90	83	7	.0843	93	90	3	.0333
5	88	91	3	.0330	90	86	4	.0465
6	93	93	0	.0000	85	89	4	.0449

Interpret P1 as the forecast made by the manager of plant 1 at the beginning of each week, O1 as the yield observed at the end of each week in plant 1, and E1 as the difference between the predicted and the observed yields. The same definitions apply to plant 2.

Let us compare the performance of these managers using the three measures MAD, MSE, and MAPE defined above. To compute the MAD we simply average the observed absolute errors:

$$\text{MAD}_1 = 17/6 = 2.83$$

$$\text{MAD}_2 = 18/6 = 3.00.$$

Based on the MADs, the first manager has a slight edge. To compute the MSE in each case, square the observed errors and average the results to obtain

$$MSE_1 = 79/6 = 13.17$$

$$MSE_2 = 70/6 = 11.67.$$

The second manager's forecasts have a lower MSE than the first, even though the MADs go the other way. Why the switch? The reason that the first manager now looks worse is that the MSE is more sensitive to one large error than is the MAD. Notice that the largest observed error of 7 was incurred by manager 1.

Let us now compare their performances based on the MAPE. To compute the MAPE we average the ratios of the errors and the observed yields:

$$MAPE_1 = .0325$$

$$MAPE_2 = .0336.$$

Using the MAD or the MAPE, the first manager has a very slight edge, but using the MSE, the second manager looks better. The forecasting abilities of the two managers would seem to be very similar. Who is declared the "winner" depends on which method of evaluation management chooses. ■

A desirable property of forecasts is that they should be unbiased. Mathematically, that means that $E(e_i) = 0$. One way of tracking a forecast method is to graph the values of the forecast error e_i over time. If the method is unbiased, forecast errors should fluctuate randomly above and below zero. An example is presented in Figure 2–3.

FIGURE 2–3

Forecast errors over time

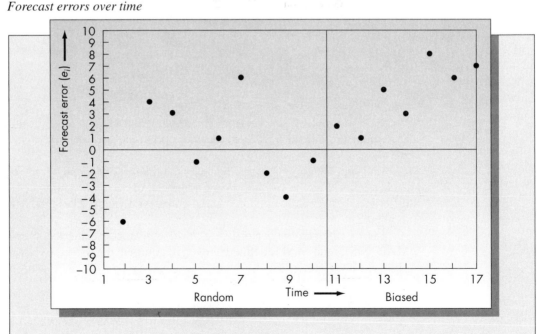

An alternative to a graphical method is to compute the cumulative sum of the forecast errors, $\Sigma\, e_i$. If the value of this sum deviates too far from zero (either above or below), it is an indication that the forecasting method is biased. At the end of this chapter we discuss a smoothing technique that also can be used to signal a bias in the forecasts. Statistical control charts also are used to identify unusually large values of the forecast error. (Statistical control charts are discussed in Chapter 8.)

Problems for Section 6

10. In *Dave Pelz's Short Game Bible,* the author attempts to characterize the skill of a golfer from a specific distance in terms of the ratio of the error in the shot and the intended shot distance. For example, a five iron that is hit 175 yards and is 20 yards off target would have an accuracy rating of 20/175 = .114, while a sand wedge hit 60 yards that is 10 yards off target would have an accuracy rating of 10/60 = .1667 (the lower the rating, the better). To what evaluation method discussed in this section in this most similar? Why does this evaluation method make more sense in golf than absolute or squared errors?

11. A forecasting method used to predict can opener sales applies the following set of weights to the last five periods of data: .1, .1, .2, .2, .4 (with .4 being applied to the most recent observation). Observed values of can opener sales are

Period:	1	2	3	4	5	6	7	8
Observation:	23	28	33	26	21	38	32	41

Determine the following:
a. The one-step-ahead forecast for period 9.
b. The one-step-ahead forecast that was made for period 6.

12. A simple forecasting method for weekly sales of floppy disk drives used by a local computer dealer is to form the average of the two most recent sales figures. Suppose sales for the drives for the past 12 weeks were

Week:	1	2	3	4	5	6	7	8	9	10	11	12
Sales:	86	75	72	83	132	65	110	90	67	92	98	73

a. Determine the one-step-ahead forecasts made for periods 3 through 12 using this method.
b. Determine the forecast errors for these periods.
c. Compute the MAD, the MSE, and the MAPE based on the forecast errors computed in part (b).

13. Two forecasting methods have been used to evaluate the same economic time series. The results are

Forecast from Method 1	Forecast from Method 2	Realized Value of the Series
223	210	256
289	320	340
430	390	375
134	112	110
190	150	225
550	490	525

Compare the effectiveness of these methods by computing the MSE, the MAD, and the MAPE. Do each of the measures of forecasting accuracy indicate that the same forecasting technique is best? If not, why?

14. What does the term *biased* mean in reference to a particular forecasting technique?

15. What is the estimate of the standard deviation of forecast error obtained from the data in Problem 12?

2.7 Methods for Forecasting Stationary Series

In this section we will discuss two popular techniques, moving averages and exponential smoothing, for forecasting stationary time series. A stationary time series is one in which each observation can be represented by a constant plus a random fluctuation. In symbols,

$$D_t = \mu + \epsilon_t,$$

where μ is an unknown constant corresponding to the mean of the series and ϵ_t is a random error with mean zero and variance σ^2.

The methods we consider in this section are more precisely known as single or simple exponential smoothing and single or simple moving averages. In addition, single moving averages also include weighted moving averages, which we do not discuss. For convenience, we will not use the modifiers *single* and *simple* in what follows. The meaning of the terms will be clear from the context.

Moving Averages

A simple but popular forecasting method is the method of moving averages. A moving average of order N is simply the arithmetic average of the most recent N observations. For the time being we restrict attention to one-step-ahead forecasts. Then F_t, the forecast made in period $t - 1$ for period t, is given by

$$F_t = (1/N) \sum_{i=t-N}^{t-1} D_i = (1/N) (D_{t-1} + D_{t-2} + \cdots + D_{t-N}).$$

In words, this says that the mean of the N most recent observations is used as the forecast for the next period. We will use the notation MA(N) for N-period moving averages.

Example 2.2

Quarterly data for the failures of certain aircraft engines at a local military base during the last two years are 200, 250, 175, 186, 225, 285, 305, 190. Both three-quarter and six-quarter moving averages are used to forecast the numbers of engine failures. Determine the one-step-ahead forecasts for periods 4 through 8 using three-period moving averages, and the one-step-ahead forecasts for periods 7 and 8 using six-period moving averages.

Solution

The three-period moving-average forecast for period 4 is obtained by averaging the first three data points.

$$F_4 = (1/3)(200 + 250 + 175) = 208.$$

The three-period moving-average forecast for period 5 is

$$F_5 = (1/3)(250 + 175 + 186) = 204.$$

The six-period moving-average forecast for period 7 is

$$F_7 = (1/6)(200 + 250 + 175 + 186 + 225 + 285) = 220.$$

Other forecasts are computed in a similar fashion. Arranging the forecasts and the associated forecast errors in a table, we obtain

Quarter	Engine Failures	MA(3)	Error	MA(6)	Error
1	200				
2	250				
3	175				
4	186	208	22		
5	225	204	−21		
6	285	195	−90		
7	305	232	−73	220	−85
8	190	272	82	238	48

An interesting question is, how does one obtain multiple-step-ahead forecasts? For example, suppose in Example 2.2 that we are interested in using MA(3) in period 3 to forecast for period 6. Because the moving-average method is based on the assumption that the demand series is stationary, the forecast made in period 3 for *any* future period will be the same. That is, the multiple-step-ahead and the one-step-ahead forecasts are identical (although the one-step-ahead forecast will generally be more accurate). Hence, the MA(3) forecast made in period 3 for period 6 is 208. In fact, the MA(3) forecast made in period 3 for any period beyond period 3 is 208 as well.

An apparent disadvantage of the moving-average technique is that one must recompute the average of the last N observations each time a new demand observation becomes available. For large N this could be tedious. However, recalculation of the full N-period average is not necessary every period, since

$$F_{t+1} = (1/N) \sum_{i=t-N+1}^{t} D_i = (1/N)\left[D_t + \sum_{i=t-N}^{t-1} D_i - D_{t-N}\right]$$

$$= F_t + (1/N)[D_t - D_{t-N}]$$

This means that for one-step-ahead forecasting, we need only compute the difference between the most recent demand and the demand N periods old in order to update the forecast. However, we still need to keep track of all N past observations. Why?

Moving Average Lags behind the Trend. Consider a demand process in which there is a definite trend. For example, suppose that the observed demand is 2, 4, 6, 8, 10, 12, 14, 16, 18, 20, 22, 24. Consider the one-step-ahead MA(3) and MA(6) forecasts for this series.

Period	Demand	MA(3)	MA(6)
1	2		
2	4		
3	6		
4	8	4	
5	10	6	
6	12	8	
7	14	10	7
8	16	12	9
9	18	14	11
10	20	16	13
11	22	18	15
12	24	20	17

The demand and the forecasts for the respective periods are pictured in Figure 2–4. Notice that both the MA(3) and the MA(6) forecasts lag behind the trend. Furthermore,

FIGURE 2–4

Moving-average forecasts lag behind a trend

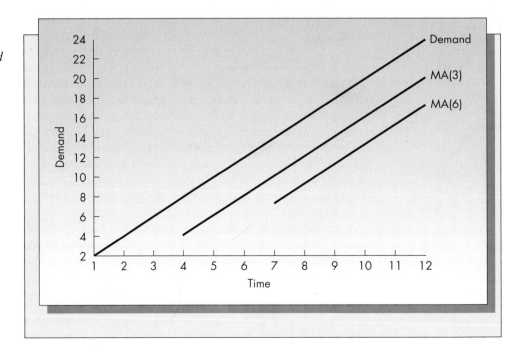

MA(6) has a greater lag. This implies that the use of simple moving averages is not an appropriate forecasting method when there is a trend in the series.

Problems on Moving Averages

Problems 16 through 21 are based on the following data. Observations of the demand for a certain part stocked at a parts supply depot during the calendar year 1999 were

Month	Demand	Month	Demand
January	89	July	223
February	57	August	286
March	144	September	212
April	221	October	275
May	177	November	188
June	280	December	312

16. Determine the one-step-ahead forecasts for the demand for January 1986 using 3-, 6-, and 12-month moving averages.

17. Using a four-month moving average, determine the one-step-ahead forecasts for July through December 1999.

18. Using a four-month moving average, determine the two-step-ahead forecast for July through December 1999. (Hint: The two-step-ahead forecast for July is based on the observed demands in February through May.)

19. Compute the MAD for the forecasts obtained in Problems 17 and 18 above. Which method gave better results? Based on forecasting theory, which method should have given better results?

20. Compute the one-step-ahead three-month and six-month moving-average forecasts for July through December. What effect does increasing N from 3 to 6 have on the forecasts?

21. What would an MA(1) forecasting method mean? Compare the accuracy of MA(1) and MA(4) forecasts for July through December 1999.

Exponential Smoothing

Another very popular forecasting method for stationary time series is exponential smoothing. The current forecast is the weighted average of the last forecast and the current value of demand. That is,

$$\text{New forecast} = \alpha(\text{Current observation of demand})$$
$$+ (1 - \alpha)(\text{Last forecast}).$$

In symbols,

$$F_t = \alpha D_{t-1} + (1 - \alpha)F_{t-1},$$

where $0 < \alpha \leq 1$ is the smoothing constant, which determines the relative weight placed on the current observation of demand. Interpret $(1 - \alpha)$ as the weight placed on past observations of demand. By a simple rearrangement of terms, the exponential smoothing equation for F_t can be written

$$F_t = F_{t-1} - \alpha(F_{t-1} - D_{t-1})$$
$$= F_{t-1} - \alpha e_{t-1}.$$

Written this way, we see that exponential smoothing can be interpreted as follows: the forecast in any period t is the forecast in period $t - 1$ minus some fraction of the observed forecast error in period $t - 1$. Notice that if we forecast high in period $t - 1$, e_{t-1} is positive and the adjustment is to decrease the forecast. Similarly, if we forecast low in period $t - 1$, the error is negative, and the adjustment is to increase the current forecast.

As before, F_t is the one-step-ahead forecast for period t made in period $t - 1$. Notice that since

$$F_{t-1} = \alpha D_{t-2} + (1 - \alpha)F_{t-2},$$

we can substitute above to obtain

$$F_t = \alpha D_{t-1} + \alpha(1 - \alpha)D_{t-2} + (1 - \alpha)^2 F_{t-2}.$$

We can now substitute for F_{t-2} in the same fashion. If we continue in this way we obtain the infinite expansion for F_t,

$$F_t = \sum_{i=0}^{\infty} \alpha(1 - \alpha)^i D_{t-i-1} = \sum_{i=0}^{\infty} a_i D_{t-i-1},$$

where the weights are $a_0 > a_1 > a_2 > \cdots > a_i = \alpha(1 - \alpha)^i$, and

$$\sum_{i=0}^{\infty} a_i = \sum_{i=0}^{\infty} \alpha(1 - \alpha)^i = \alpha \sum_{i=0}^{\infty} (1 - \alpha)^i = \alpha \times 1/[1 - (1 - \alpha)] = 1.$$

Hence, exponential smoothing applies a declining set of weights to all past data. The weights are graphed as a function of i in Figure 2–5.

In fact, we could fit the continuous exponential curve $g(i) = \alpha \exp(-\alpha i)$ to these weights, which is why the method is called exponential smoothing. The smoothing constant α plays essentially the same role here as the value of N does in moving averages. If α is large, more weight is placed on the current observation of demand and less weight on past observations, which results in forecasts that will react quickly to changes in the demand pattern but may have much greater variation from period to period. If α is small, then more weight is placed on past data and the forecasts are more stable.

When using an automatic forecasting technique to predict demand for a production application, stable forecasts (that is, forecasts that do not vary a great deal from period to period) are very desirable. Demand forecasts are used as the starting point for production planning and scheduling. Substantial revision in these forecasts can wreak havoc with employee work schedules, component bills of materials, and external purchase orders. For this reason, a value of α between .1 and .2 is generally recommended for production applications. (See, for example, Brown, 1962.)

Notice that the exponential smoothing equation also can be written in the form

$$F_t = F_{t-1} - \alpha(F_{t-1} - D_{t-1}) = F_{t-1} - \alpha e_{t-1}.$$

In this form we see that exponential smoothing adjusts the previous forecast by a fraction of the previous forecast error to obtain the current forecast.

Multiple-step-ahead forecasts are handled the same way for simple exponential smoothing as for moving averages; that is, the one-step-ahead and the multiple-step-ahead forecasts are the same.

FIGURE 2–5

Weights in exponential smoothing

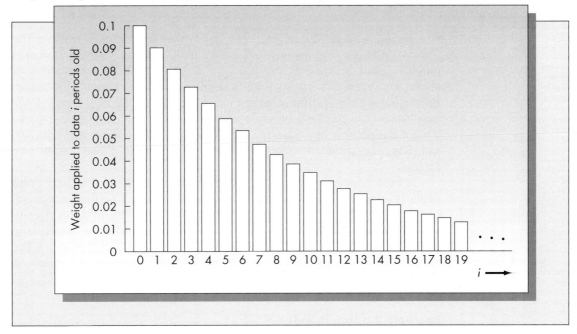

Example 2.3

Consider Example 2.2 (on page 66), in which moving averages were used to predict aircraft engine failures. The observed numbers of failures over a two-year period were 200, 250, 175, 186, 225, 285, 305, 190. We will now forecast using exponential smoothing. In order to get the method started, let us assume that the forecast for period 1 was 200. Suppose that $\alpha = 0.1$. The one-step-ahead forecast for period 2 is

$$F_2 = \alpha D_1 + (1 - \alpha)F_1 = (.1)(200) + (.9)(200) = 200.$$

Similarly,

$$F_3 = \alpha D_2 + (1 - \alpha)F_2 = (.1)(250) + (.9)(200) = 205.$$

Other one-step-ahead forecasts are computed in the same fashion. The observed numbers of failures and the one-step-ahead forecasts for each quarter are the following:

Quarter	Failures	Forecast
1	200	200 (by assumption)
2	250	200
3	175	205
4	186	202
5	225	201
6	285	203
7	305	211
8	190	220

Notice the effect of the smoothing constant. Although the original series shows high variance, the forecasts are quite stable. Repeat the calculations with a value of $\alpha = .4$. There will be much greater variation in the forecasts. ■

Because exponential smoothing requires that at each stage we have the previous forecast, it is not obvious how to get the method started. We could assume that the initial forecast is equal to the initial value of demand, as we did in Example 2.3 above. However, this approach has a serious drawback. Exponential smoothing puts substantial weight on past observations, so the initial value of demand will have an unreasonably large effect on early forecasts. This problem can be overcome by allowing the process to evolve for a reasonable number of periods (say 10 or more) and using the arithmetic average of the demand during those periods as the initial forecast.

In order to appreciate the effect of the smoothing constant, we have graphed a particularly erratic series in Figure 2–6, along with the resulting forecasts using values of $\alpha = .1$ and $\alpha = .4$. Notice that for $\alpha = .1$ the predicted value of demand results in a relatively smooth pattern, whereas for $\alpha = .4$, the predicted value exhibits significantly greater variation. Although smoothing with the larger value of α does a better job of tracking the series, the stability afforded by a smaller smoothing constant is very desirable for planning purposes.

Example 2.3 (continued)

Consider again the problem of forecasting aircraft engine failures. Suppose that we were interested in comparing the performance of MA(3) with the exponential smoothing forecasts obtained above (ES(.1)). The first period for which we have a

FIGURE 2–6

Exponential smoothing for different values of alpha

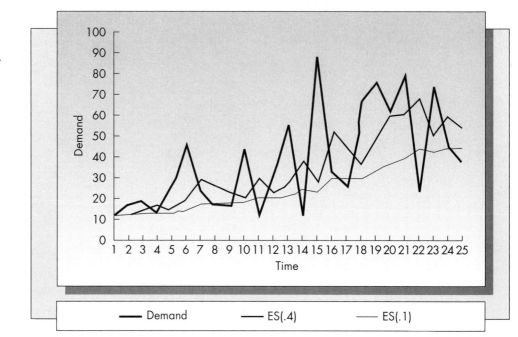

forecast using MA(3) is period 4, so we will make the comparison for periods 4 through 8 only.

Quarter	Failures	MA(3)	|Error|	ES(.1)	|Error|
4	186	208	22	202	16
5	225	204	21	201	24
6	285	195	90	203	82
7	305	232	73	211	94
8	190	272	82	220	30

The arithmetic average of the absolute errors, the MAD, is 57.6 for the three-period moving average and 49.2 for exponential smoothing. The respective values of the MSE are 4,215.6 and 3,458.4. Based on this comparison only, one might conclude that exponential smoothing is a superior method for this series. This is not necessarily true, however. ■

In Example 2.3 we compared exponential smoothing with $\alpha = 0.1$ and a moving average with $N = 3$. How do we know that those parameter settings are consistent? The MA(3) forecasts exhibit much greater variability than the ES(.1) forecasts do, suggesting that $\alpha = .1$ and $N = 3$ are not consistent values of these parameters.

Determining consistent values of α and N can be done in two ways. One is to equate the average age of the data used in making the forecast. A moving-average forecast consists of equal weights of $1/N$ applied to the last N observations. Multiplying the weight placed on each observation by its "age," we get the average age of data for moving averages as

$$\text{Average age} = (1/N)(1 + 2 + 3 + \cdots + N) = (1/N)(N)(N + 1)/2$$

$$= (N + 1)/2.$$

For exponential smoothing, the weight applied to data i periods old is $\alpha(1 - \alpha)^{i-1}$. Assume that we have an infinite history of past observations of demand. Hence, the average age of data in an exponential smoothing forecast is

$$\text{Average age} = \sum_{i=1}^{\infty} i\alpha(1 - \alpha)^{i-1} = 1/\alpha.$$

We omit the details of this calculation.

Equating the average age of data for the two methods, we obtain

$$\frac{N + 1}{2} = \frac{1}{\alpha},$$

which is equivalent to

$$\alpha = 2/(N + 1) \qquad \text{or} \qquad N = \frac{2 - \alpha}{\alpha}.$$

Hence, we see that we would have needed a value of $N = 19$ for $\alpha = .1$ or a value of $\alpha = .5$ for $N = 3$ in order for the methods to be consistent in the sense of average age of data.

In Appendix 2–A of this chapter, we derive the mean and variance of the forecast error for both moving averages and exponential smoothing in terms of the variance of each individual observation, assuming that the underlying demand process is stationary. We show that both methods are unbiased; that is, the expected value of the forecast error is zero. Furthermore, by equating the expressions for the variances of the forecast error, one obtains the same relationship between α and N as by equating the average age of data. This means that if both exponential smoothing and moving averages are used to predict the same stationary demand pattern, forecast errors are normally distributed, and $\alpha = 2/(N + 1)$, then both methods will have exactly the same distribution of forecast errors. (However, this does *not* mean that the forecasts obtained by the two methods are the same.)

Multiple-Step-Ahead Forecasts

Thus far, we have talked only about one-step-ahead forecasts. That is, we have assumed that a forecast in period t is for the demand in period $t + 1$. However, there are cases where we are interested in making a forecast for more than one step ahead. For example, a retailer planning for the Christmas season might need to make a forecast for December sales in June in order to have enough time to prepare. Since the underlying model assumed for both moving averages and exponential smoothing is stationary (i.e., not changing in time), the one-step-ahead and multiple-step-ahead forecasts for moving averages and exponential smoothing are the same. That is, a forecast made in June for July sales is the same as a forecast made in June for December sales. (In the case of the retailer, the assumption of stationarity would probably be wrong, since December sales would likely by greater than a typical month's sales. That would suggest that these methods would *not* be appropriate in this case.)

Comparison of Exponential Smoothing and Moving Averages

There are several similarities and several differences between exponential smoothing and moving averages.

Similarities

1. Both methods are derived with the assumption that the underlying demand process is stationary (that is, can be represented by a constant plus a random fluctuation with zero mean). However, we should keep in mind that although the methods are appropriate for stationary time series, we don't necessarily believe that the series are stationary forever. By adjusting the values of N and α we can make the two methods more or less responsive to shifts in the underlying pattern of the data.

2. Both methods depend on the specification of a single parameter. For moving averages the parameter is N, the number of periods in the moving average, and for exponential smoothing the parameter is α, the smoothing constant. Small values of N or large values of α result in forecasts that put greater weight on current data, and large values of N and small values of α put greater weight on past data. Small N and large α may be more responsive to changes in the demand process, but will result in forecast errors with higher variance.

3. Both methods will lag behind a trend if one exists.

4. When $\alpha = 2/(N + 1)$, both methods have the same distribution of forecast error. This means that they should have roughly the same level of accuracy, but it does *not* mean that they will give the same forecasts.

Differences

1. The exponential smoothing forecast is a weighted average of *all* past data points (as long as the smoothing constant is strictly less than 1). The moving-average forecast is a weighted average of only the last N periods of data. This can be an important advantage for moving averages. An outlier (an observation that is not representative of the sample population) is washed out of the moving-average forecast after N periods, but remains forever in the exponential smoothing forecast.

2. In order to use moving averages, one must save all N past data points. In order to use exponential smoothing, one need only save the last forecast. This is the most significant advantage of the exponential smoothing method and one reason for its popularity in practice. In order to appreciate the consequence of this difference, consider a system in which the demand for 300,000 inventory items is forecasted each month using a 12-month moving average. The forecasting module alone requires saving $300,000 \times 12 = 3,600,000$ pieces of information. If exponential smoothing were used, only 300,000 pieces of information need to be saved. This issue is less important today than it has been, as the cost of information storage has decreased enormously in recent years. However, it is still easier to manage a system that requires less data. It is primarily for this reason that exponential smoothing appears to be more popular than moving averages for production-planning applications.

Problems for Section 7

22. Handy, Inc., produces a solar-powered electronic calculator that has experienced the following monthly sales history for the first four months of the year, in thousands of units:

January	23.3	March	30.3
February	72.3	April	15.5

   a. If the forecast for January was 25, determine the one-step-ahead forecasts for February through May using exponential smoothing with a smoothing constant of $\alpha = .15$.

   b. Repeat the calculation in part (*a*) for a value of $\alpha = .40$. What difference in the forecasts do you observe?

   c. Compute the MSEs for the forecasts you obtained in parts (*a*) and (*b*) for February through April. Which value of α gave more accurate forecasts, based on the MSE?

23. Compare and contrast exponential smoothing when α is small (near zero) and when α is large (near 1).

24. Observed weekly sales of ball peen hammers at the town hardware store over an eight-week period have been 14, 9, 30, 22, 34, 12, 19, 23.

   a. Suppose that three-week moving averages are used to forecast sales. Determine the one-step-ahead forecasts for weeks 4 through 8.

   b. Suppose that exponential smoothing is used with a smoothing constant of $\alpha = .15$. Find the exponential smoothing forecasts for weeks 4 through 8. (To get the method started, use the same forecast for week 4 as you used in part (*a*).)

SNAPSHOT APPLICATION

Sport Obermeyer Slashes Costs with Improved Forecasting[1]

Sport Obermeyer is a leading supplier in the U.S. fashion ski apparel market. The firm, founded in 1950 by engineer/ski instructor Klaus Obermeyer, enjoys about a 45 percent market share in the rapidly expanding children's market and an 11 percent share of the adult market. Virtually all of the firm's offerings are redesigned annually to incorporate changes in style, fabrics, and colors. For more than 30 years, the firm was able to successfully meet demands by producing during the summer months after receiving firm orders from customers.

During the past decade, things have changed and problems have developed. First, volumes increased. There was insufficient capacity among suppliers to produce the required volume in the summer. Second, the firm developed a complex global supply chain strategy (see Section 10 of Chapter 6) to reduce costs. A parka sold in the United States might be sewn in China from fabrics and parts from Japan, South Korea, and Germany. Together these changes lengthened the production lead time, thus requiring the firm to commit to production before orders were placed by customers.

The firm undertook several "quick response" initiatives to reduce lead times. These included encouraging some customers to place orders earlier, locating raw materials near the Far East production facility, and instituting an air freight system to expedite delivery from the Far East to its Denver distribution center. Even with these changes in place, the problem of stockouts and markdowns due to oversupply were not solved. The company still had to commit about half the production based on forecasts. In the fashion industry, there is often no statistical history on which to base forecasts, and forecast errors can be huge. Products that outsell original forecasts by a factor of 2 or undersell original forecasts by a factor of 10 are common.

Sport Obermeyer needed some help with forecasting to avoid expensive miscalculations. The customary procedure was to base the forecasts on a consensus of

members of the buying committee. The problem with consensus forecasting is that the dominant personalities in a group carry more weight. A forecast obtained in this way might represent only the opinion of one person. To overcome this problem, the research team (Fisher et al., 1994) recommended that members of the committee supply *individual* forecasts.

The dispersion among individual forecasts turned out to be a reliable indicator of forecast accuracy. When committee members' forecasts were close, forecasts were more accurate. This provided a mechanism for signaling the products whose sales were likely to be poorly forecast. This did not solve the problem of poorly forecast items, but it allowed the firm to commit first to production of items whose forecasts were likely to be accurate. By the time production had to begin on the problem items, information on early sales patterns would be available.

The team noticed that retailers were remarkably similar. That meant that even if only the first 20 percent of orders for a product were in, that information could dramatically improve forecasts. Production plans for these "trouble" items could now be committed with greater confidence. In this way, the firm could separate products into two categories: reactive and nonreactive. The nonreactive items are those for which the forecast is likely to be accurate. These are produced early in the season. The reactive items are those whose forecasts are updated later in the season from early sales figures. The firm's experience was that stockout and markdown costs were reduced from 10.2 percent of sales to 1.8 percent of sales on items that could be produced reactively. Sport Obermeyer was able to produce 30 percent of its season's volume reactively and experienced a cost reduction of about 5 percent of sales.

What are the lessons here? One is that even in cases where there is no statistical history, statistical methodology can be successfully applied to improve forecasting accuracy. Another is not to assume that things should be done a certain way. Sport Obermeyer assumed that consensus forecasting was the best approach. In fact, by requiring the buying committee to reach a consensus, valuable information was being ignored. The *differences* among individual forecasts proved to be important.

---

[1] This application is based on the work of a team from the Wharton School and the Harvard Business School. The results are reported in Fisher et al. (1994).

    c. Based on the MAD, which method did better?
    d. What is the exponential smoothing forecast made at the end of week 6 for the sales in week 12?

25. Determine the following:
    a. The value of α consistent with $N = 6$ in moving averages.
    b. The value of N consistent with $\alpha = .05$.
    c. The value of α that results in a variance of forecast error, σ_e^2, 10 percent higher than the variance of each observation, σ^2 (refer to the formulas derived in Appendix 2–A).

26. Referring to the data in Problem 22, what is the exponential smoothing forecast made at the end of March for the sales in July? Assume $\alpha = .15$.

27. For the data for Problems 16 through 21, use the arithmetic average of the first six months of data as a baseline to initialize the exponential smoothing.
    a. Determine the one-step-ahead exponential smoothing forecasts for August through December, assuming $\alpha = .20$.
    b. Compare the accuracy of the forecasts obtained in part (*a*) with the one-step-ahead six-month moving-average forecasts determined in Problem 20.
    c. Comment on the reasons for the result you obtained in part (*b*).

2.8 Trend-Based Methods

Both exponential smoothing and moving-average forecasts will lag behind a trend if one exists. We will consider two forecasting methods that specifically account for a trend in the data: regression analysis and Holt's method. Regression analysis is a method that fits a straight line to a set of data. Holt's method is a type of double exponential smoothing that allows for simultaneous smoothing on the series and on the trend.

Regression Analysis

Let $(x_1, y_1), (x_2, y_2), \ldots, (x_n, y_n)$ be n paired data points for the two variables X and Y. Assume that y_i is the observed value of Y when x_i is the observed value of X. Refer to Y as the dependent variable and X as the independent variable. We believe that a relationship exists between X and Y that can be represented by the straight line

$$\hat{Y} = a + bX.$$

    Interpret $\hat{Y}$ as the predicted value of Y. The goal is to find the values of a and b so that the line $\hat{Y} = a + bX$ gives the best fit of the data. The values of a and b are chosen so that the sum of the squared distances between the regression line and the data points is minimized (see Figure 2–7). In Appendix 2–B we derive the optimal values of a and b in terms of the given data.

    When applying regression analysis to the forecasting problem, the independent variable often corresponds to time and the dependent variable to the series to be forecast. Assume that $D_1, D_2, \ldots, D_n$ are the values of the demand at times $1, 2, \ldots, n$. Then it is shown in Appendix 2–B that the optimal values of a and b are given by

$$b = \frac{S_{xy}}{S_{xx}}$$

and

$$a = \overline{D} - b(n + 1)/2,$$

FIGURE 2–7

An example of a regression line

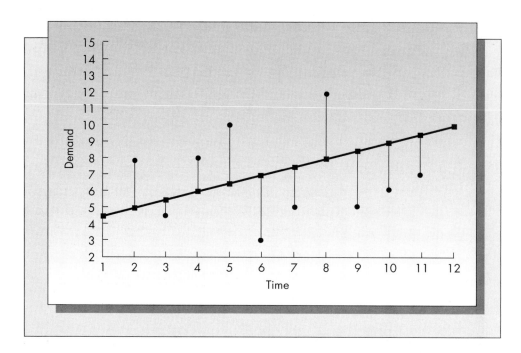

where

$$S_{xy} = n \sum_{i=1}^{n} iD_i - \frac{n(n+1)}{2} \sum_{i=1}^{n} D_i,$$

$$S_{xx} = \frac{n^2(n+1)(2n+1)}{6} - \frac{n^2(n+1)^2}{4},$$

and $\overline{D}$ is the arithmetic average of the observed demands during periods $1, 2, \ldots, n$.

Example 2.4

We will apply regression analysis to the problem, treated in Examples 2.2 and 2.3, of predicting aircraft engine failures. Recall that the demand for aircraft engines during the last eight quarters was 200, 250, 175, 186, 225, 285, 305, 190. Suppose that we use the first five periods as a baseline in order to estimate our regression parameters. Then

$$S_{xy} = 5[200 + (250)(2) + (175)(3) + (186)(4) + (225)(5)]$$
$$-[(5)(6)/2][200 + 250 + 175 + 186 + 225]$$
$$= -70,$$
$$S_{xy} = 5(25)(6)(11)/6 - (25)(36)/4 = 50.$$

Then

$$b = S_{xy}/S_{xx} = -70/50 = -7/5,$$
$$a = 207.2 - (-27/5)(3) = 211.4.$$

It follows that the regression equation based on five periods of data is

$$\hat{D}_t = 211.4 - (7/5)t.$$

$\hat{D}_t$ is the predicted value of demand at time t. We would use this regression equation to forecast from period 5 to any period beyond period 5. For example, the forecast made in period 5 for period 8 would be obtained by substituting $t = 8$ in the above equation, which would result in the forecast $211.4 - (7/5)(8) = 200.2$. Note that if we were interested in forecasting in period 7 for period 8, then this regression equation would not be appropriate. We would have to repeat the entire calculation using the data from periods 1 through 7. In fact, one of the serious drawbacks of using regression for forecasting is that updating forecasts as new data become available is very cumbersome. ■

Problems for Section 8

28. Shoreline Park in Mountain View, California, has kept close tabs on the number of patrons using the park since its opening in January 1993. For the first six months of operation, the following figures were recorded:

Month	Number of Patrons	Month	Number of Patrons
January	133	April	640
February	183	May	1,876
March	285	June	2,550

   a. Draw a graph of these six data points. Assume that January = period 1, February = period 2, and so on. Using a ruler, "eyeball" the best straight-line fit of the data. Estimate the slope and intercept from your graph.
   b. Compute the exact values of the intercept a and the slope b from the regression equations.
   c. What are the forecasts obtained for July through December 1993 from the regression equation determined in part (b)?
   d. Comment on the results you obtained in part (c). Specifically, how confident would you be about the accuracy of the forecasts that you obtained?

29. The Mountain View Department of Parks and Recreation must project the total use of Shoreline Park for calendar year 1994.
   a. Determine the forecast for the total number of people using the park in 1994 based on the regression equation.
   b. Determine the forecast for the same quantity using a six-month moving average.
   c. Draw a graph of the most likely shape of the curve describing the park's usage by month during the calendar year, and predict the same quantity using your graph. Is your prediction closer to the answer you obtained for part (a) or part (b)? Discuss your results.

Double Exponential Smoothing Using Holt's Method

Holt's method is a type of double exponential smoothing designed to track time series with linear trend. The method requires the specification of two smoothing constants, α and β, and uses two smoothing equations: one for the value of the series (the intercept) and one for the trend (the slope). The equations are

$$S_t = \alpha D_t + (1 - \alpha)(S_{t-1} + G_{t-1}),$$
$$G_t = \beta(S_t - S_{t-1}) + (1 - \beta)G_{t-1}.$$

Interpret S_t as the value of the intercept at time t and G_t as the value of the slope at time t. The first equation is very similar to that used for simple exponential smoothing. When the most current observation of demand, D_t, becomes available, it is averaged with the prior forecast of the current demand, which is the previous intercept, S_{t-1}, plus 1 times the previous slope, G_{t-1}. The second equation can be explained as follows. Our new estimate of the intercept, S_t, causes us to revise our estimate of the slope to $S_t - S_{t-1}$. This value is then averaged with the previous estimate of the slope, G_{t-1}. The smoothing constants may be the same, but for most applications more stability is given to the slope estimate (implying $\beta \leq \alpha$).

The τ-step-ahead forecast made in period t, which is denoted by $F_{t, t+\tau}$, is given by

$$F_{t, t+\tau} = S_t + \tau G_t.$$

Example 2.5

Let us apply Holt's method to the problem of developing one-step-ahead forecasts for the aircraft engine failure data. Recall that the original series was 200, 250, 175, 186, 225, 285, 305, 190. Assume that both α and β are equal to 0.1. In order to get the method started, we need estimates of both the intercept and the slope at time zero. Suppose that these are $S_0 = 200$ and $G_0 = 10$. Then we obtain

$$S_1 = (.1)(200) + (.9)(200 + 10) = 209.0$$
$$G_1 = (.1)(209 - 200) + (.9)(10) = 9.9$$
$$S_2 = (.1)(250) + (.9)(209 + 9.9) = 222.0$$
$$G_2 = (.1)(222 - 209) + (.9)(9.9) = 10.2$$
$$S_3 = (.1)(175) + (.9)(222 + 10.2) = 226.5$$
$$G_3 = (.1)(226.5 - 222) + (.9)(10.2) = 9.6$$

and so on.

Comparing the one-step-ahead forecasts to the actual numbers of failures for periods 4 through 8, we obtain the following:

Period	Actual	Forecast	$|Error|$
4	186	236.1	50.1
5	225	240.3	15.3
6	285	247.7	37.3
7	305	260.8	44.2
8	190	275.0	85.0

Averaging the numbers in the final column, we obtain a MAD of 46.4. Notice that this is lower than that for simple exponential smoothing or moving averages. Holt's method does better for this series because it is explicitly designed to track the trend in the data, whereas simple exponential smoothing and moving averages are not. Note that the forecasts in the table above are one-step-ahead forecasts. Suppose you needed to forecast the demand in period 2 for period 5. This forecast is $F_{2, 5} = S_2 + (3)G_2 = 222 + (3)(10.2) = 252.6$. ∎

The initialization problem also arises in getting Holt's method started. The best approach is to establish some set of initial periods as a baseline and use regression analysis to determine estimates of the slope and intercept using the baseline data.

Both Holt's method and regression are designed to handle series that exhibit trend. However, with Holt's method it is far easier to update forecasts as new observations become available.

More Problems for Section 8

30. For the data in Problem 28, use the results of the regression equation to estimate the slope and intercept of the series at the end of June. Use these numbers as the initial values of slope and intercept required in Holt's method. Assume that $\alpha = .15, \beta = .10$ for all calculations.

    a. Suppose that the actual number of visitors using the park in July was 2,150 and the number in August was 2,660. Use Holt's method to update the estimates of the slope and intercept based on these observations.

    b. What are the one-step-ahead and two-step-ahead forecasts that Holt's method gives for the number of park visitors in September and October?

    c. What is the forecast made at the end of July for the number of park attendees in December?

31. Due to serious flooding, the park was closed for most of December 1993. During that time only 53 people visited. Comment on the effect this observation would have on predictions of future use of the park. If you were in charge of forecasting the park's usage, how would you deal with this data point?

32. Discuss some of the problems that could arise when using either regression analysis or Holt's method for obtaining multiple-step-ahead forecasts.

2.9 Methods for Seasonal Series

The last topic of this chapter is forecasting methods for seasonal problems. A seasonal series is one that has a pattern that repeats every N periods for some value of N (which is at least 3). A typical seasonal series is pictured in Figure 2–8.

We refer to the number of periods before the pattern begins to repeat as the length of the season (N in the picture). Note that this is different from the popular usage of the word *season* as a time of year. In order to use a seasonal model, one must be able to specify the length of the season.

There are several ways to represent seasonality. The most common is to assume that there exists a set of multipliers c_t, for $1 \leq t \leq N$, with the property that $\sum c_t = N$. The multiplier c_t represents the average amount that the demand in the tth period of the season is above or below the overall average. For example, if $c_3 = 1.25$ and $c_5 = .60$, then, on average, the demand in the third period of the season is 25 percent above the average demand and the demand in the fifth period of the season is 40 percent below the average demand. These multipliers are known as seasonal factors.

Seasonal Factors for Stationary Series

In this section we present a simple method of computing seasonal factors for a time series with seasonal variation and no trend. In the next section we consider a method that is likely to be more accurate when there is also trend. Both methods require a minimum of two seasons of data.

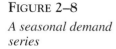

FIGURE 2–8

A seasonal demand series

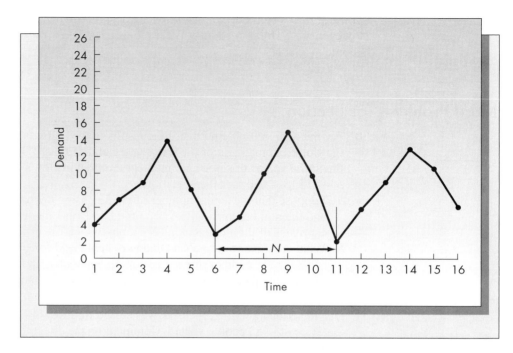

The method is as follows:

1. Compute the sample mean of all the data.
2. Divide each observation by the sample mean. This gives seasonal factors for each period of observed data.
3. Average the factors for like periods within each season. That is, average all the factors corresponding to the first period of a season, all the factors corresponding to the second period of a season, and so on. The resulting averages are the N seasonal factors. They will always add to exactly N.

Example 2.6

The County Transportation Department needs to determine usage factors by day of the week for a popular toll bridge connecting different parts of the city. In the current study, they are considering only working days. Suppose that the numbers of cars using the bridge each working day over the past four weeks were (in thousands of cars)

	Week 1	*Week 2*	*Week 3*	*Week 4*
Monday	16.2	17.3	14.6	16.1
Tuesday	12.2	11.5	13.1	11.8
Wednesday	14.2	15.0	13.0	12.9
Thursday	17.3	17.6	16.9	16.6
Friday	22.5	23.5	21.9	24.3

Find the seasonal factors corresponding to daily usage of the bridge.

Solution

1. First we compute the average of all the observations. You should satisfy yourself that the mean is 16.425.

2. Next, we divide each observation by the mean value. Doing so gives the following:

0.986	1.053	0.889	0.980
0.743	0.700	0.798	0.718
0.865	0.913	0.791	0.785
1.053	1.072	1.029	1.011
1.370	1.431	1.333	1.479

3. Finally, we average factors corresponding to the same period of the season. That is, average all factors for Mondays, all factors for Tuesdays, and so on. The resulting five seasonal factors for bridge usage by day of the week are

Monday:	0.98
Tuesday:	0.74
Wednesday:	0.84
Thursday:	1.04
Friday:	1.40

You should satisfy yourself that these factors do sum to 5 as required. Interpret the results in the following way: Mondays are an average day because the factor for Mondays is very close to 1. The usage is lowest on Tuesdays, about 26 percent below the average, and highest on Fridays, about 40 percent above the average.

Forecasts for the numbers of cars using the bridge on any day of the week would be obtained by multiplying the sample mean of 16.425 by the appropriate seasonal factor. Hence, usage forecasts by day of the week (in thousands of cars) would be

Monday:	16.1
Tuesday:	12.1
Wednesday:	13.8
Thursday:	17.1
Friday:	23.0

∎

Seasonal Decomposition Using Moving Averages

A slightly more complex method for estimating seasonal factors requires the computation of N-period moving averages, where N is the length of the season. The method is best illustrated by example.

Example 2.7

Suppose that the original demand history of a certain item for the past eight quarters is given by 10, 20, 26, 17, 12, 23, 30, 22. The graph of this demand series is given in Figure 2–9.

The picture suggests that these data represent two seasons, with each season being four periods long. The next step is to compute all four-period moving averages. This gives

Period	Demand	MA(4)
1	10	
2	20	
3	26	
4	17	18.25
5	12	18.75
6	23	19.50
7	30	20.50
8	22	21.75

Next, the moving averages must be "centered." The moving average 18.25 is computed by averaging the demand for periods 1, 2, 3, and 4. The center of these numbers is 2.5. The next average of 18.75 corresponds to periods 2, 3, 4, and 5. The center of these numbers is 3.5. We repeat this process for the remaining averages and

FIGURE 2–9

Demand history for Example 2.7

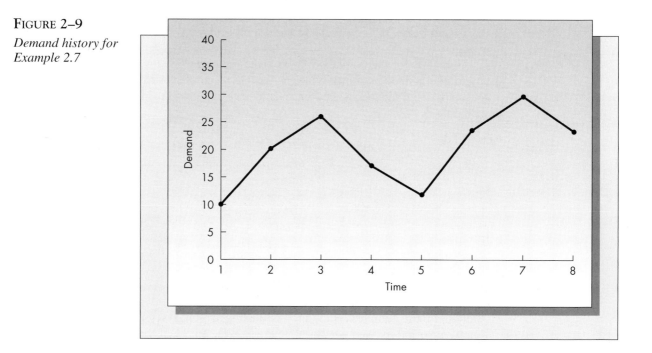

locate them in the centered positions in the following way:

Period	Demand	MA(4)	Centered
1	10		
2	20		18.25
3	26		18.75
4	17	18.25	19.50
5	12	18.75	20.50
6	23	19.50	21.75
7	30	20.50	
8	22	21.75	

Had N been odd instead of even, the centered values would have occurred on periods instead of between periods. (For example, for $N = 5$ the first MA(5) value is centered at period 3.)

Next, we must get these centered values back "on" periods. To do so, we average adjacent values. Averaging 18.25 and 18.75 gives 18.5, which corresponds to the average of periods 2.5 and 3.5, which is 3. Repeating this process for the other centered values gives centered moving averages for periods 4, 5, and 6 as well.

The next step is to obtain values for periods 1, 2, 7, and 8. In order to obtain values for periods 1 and 2, we average the values for periods 3 and 4, and to obtain values for periods 7 and 8, we average the values for periods 5 and 6. In this way, we obtain a centered moving average for each period in which we have an observation of demand.

(A) Period	(B) Demand	(C) Centered MA	(B/C) Ratio
1	10	18.81	.532
2	20	18.81	1.063
3	26	18.50	1.405
4	17	19.125	.888
5	12	20.00	.600
6	23	21.125	1.089
7	30	20.56	1.463
8	22	20.56	1.070

Once we have obtained the centered moving average for each period, we form the ratio of the demand for that period over the centered MA. These values, reported in the fourth column above, are estimates of the seasonal factors for each period. The next step is to form the average of the factors that correspond to the same periods of each season. As we have exactly two seasons of data, we average $(.532 + .600)/2 = .566$, $(1.063 + 1.089)/2 = 1.076$, $(1.405 + 1.463)/2 = 1.434$, and $(.888 + 1.070)/2 = .979$. In this case, this procedure yields exactly four seasonal factors. In general, it will result in exactly N seasonal factors.

Finally, we must be sure that the sum of the seasonal factors is exactly N, or exactly 4 for this example. We find $.566 + 1.076 + 1.434 + .979 = 4.055$. In order to make the sum equal to exactly 4, we multiply each factor by 4/4.055 = .9864. The final seasonal factors are

Period	Factor
1	.558
2	1.061
3	1.415
4	.966

The sum of these final seasonal factors is exactly 4. Based on the given data, these factors tell us that, on average, the first quarter of each year results in sales that are about 45 percent below the yearly average, the second quarter in sales that are about 6 percent above the yearly average, and so on. The next step in the process is to *divide* each observation by the appropriate seasonal factor in order to obtain the deseasonalized demand.

(A) Demand	(B) Factor	Deseasonalized Demand (A/B)
10	.558	17.92
20	1.061	18.85
26	1.415	18.39
17	.966	17.60
12	.558	21.50
23	1.061	21.68
30	1.415	21.22
22	.966	22.77

The deseasonalized series will still contain all components of the signal of the original series except for seasonality. A forecast can now be made based on the deseasonalized demand, which must be "reseasonalized" by multiplying by the appropriate seasonal factor.

As an example, suppose that we were using a six-month moving average to forecast the deseasonalized series. The average of the last six observations is 20.52. This number is then multiplied by the appropriate seasonal factor to obtain forecasts of future demand. The forecast for period 9 is (20.52)(.5580) = 11.45, for period 10 is (20.52)(1.061) = 21.77, and so on.

However, because the deseasonalized series exhibits a trend, it would be more appropriate to apply a trend-based method such as Holt's method or regression analysis to the deseasonalized series. If we use regression analysis, we obtain the least squares fit of the data as $\hat{D}_t = 16.8 + .7092t$. Suppose that we are interested in forecasting at the end of period 8 for periods 9 through 12. Substituting t equals 9 through 12 into the regression equation gives the forecasts of the deseasonalized series for the following year as 23.18, 23.89, 24.60, and 25.31. Each of these forecasts is then reseasonalized by multiplying by the appropriate seasonal factor. The forecasts obtained by this method for the following year (periods 9 through 12) are respectively 12.93, 25.35, 34.81, and 24.45. ■

Problems for Section 9

33. Sales of walking shorts at Hugo's Department Store in downtown Rolla appear to exhibit a seasonal pattern. The proprietor of the store, Wally Hugo, has kept careful records of the sales of several of his popular items, including walking shorts. During the past two years the monthly sales of the shorts have been

	Year 1	Year 2		Year 1	Year 2
Jan.	12	16	July	112	130
Feb.	18	14	Aug.	90	83
March	36	46	Sep.	66	52
April	53	48	Oct.	45	49
May	79	88	Nov.	23	14
June	134	160	Dec.	21	26

Assuming no trend in shorts sales over the two years, obtain estimates for the monthly seasonal factors.

34. A popular brand of tennis shoe has had the following demand history by quarters over a three-year period.

1990	Demand	1991	Demand	1992	Demand
1	12	1	16	1	14
2	25	2	32	2	45
3	76	3	71	3	84
4	52	4	62	4	47

  a. Determine the seasonal factors for each quarter by the method of centered moving averages.
  b. Based on the result of (a), determine the deseasonalized demand series.
  c. Predict the demand for the first quarter of 1993 for the deseasonalized series from a six-quarter moving average.
  d. Using the results from parts (a) and (c), predict the demand for the shoes for the first quarter of 1993.
  e. Compare the result from part (a) with seasonal factors obtained by the "quick and dirty" method described at the beginning of Section 9.

Winters's Method for Seasonal Problems

The moving-average method described above can be used to predict a seasonal series with or without a trend. However, as new data become available, the method requires that all seasonal factors be recalculated from scratch. Winters's method is a type of triple exponential smoothing, and this has the important advantage of being easy to update as new data become available.

We assume a model of the form

$$D_t = (\mu + G_t) c_t + \epsilon_t.$$

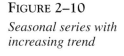

FIGURE 2–10

Seasonal series with increasing trend

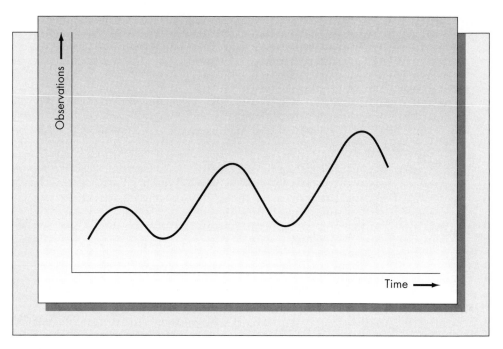

Interpret μ as the base signal or intercept at time $t = 0$ excluding seasonality, G as the trend or slope component, c_t as the multiplicative seasonal component in period $t,$ and finally ϵ_t as the error term. Because the seasonal factor multiplies both the base level and the trend term, we are assuming that the underlying series has a form similar to that pictured in Figure 2–10.

Again assume that the length of the season is exactly N periods and that the seasonal factors are the same each season and have the property that $\Sigma\, c_t = N.$ Three exponential smoothing equations are used each period to update estimates of deseasonalized series, the seasonal factors, and the trend. These equations may have different smoothing constants, which we will label $\alpha, \beta,$ and $\gamma.$

1. *The series.* The current level of the deseasonalized series, $S_t,$ is given by

$$S_t = \alpha(D_t/c_{t-N}) + (1 - \alpha)(S_{t-1} + G_{t-1}).$$

Notice what this equation does. By dividing by the appropriate seasonal factor, we are deseasonalizing the newest demand observation. This is then averaged with the current forecast for the deseasonalized series, as in Holt's method.

2. *The trend.* The trend is updated in a fashion similar to Holt's method.

$$G_t = \beta[S_t - S_{t-1}] + (1 - \beta)G_{t-1}.$$

3. *The seasonal factors.*

$$c_t = \gamma(D_t/S_t) + (1 - \gamma)c_{t-N}.$$

The ratio of the most recent demand observation over the current estimate of the deseasonalized demand gives the current estimate of the seasonal factor. This is then averaged with the previous best estimate of the seasonal factor, $c_{t-N}.$ Each time that a seasonal factor is updated, it is necessary to norm the most recent N factors to add to $N.$

Finally, the forecast made in period t for any future period $t + \tau$ is given by

$$F_{t, t+\tau} = (S_t + \tau G_t)c_{t+\tau-N}.$$

Note that this forecasting equation assumes that $t \leq N$. If $N < \tau \leq 2N$, the appropriate seasonal factor would be $c_{t+\tau-2N}$; if $2N < \tau \leq 3N$, the appropriate seasonal factor would be $c_{t+\tau-3N}$; and so on.

Initialization Procedure. In order to get the method started, we need to obtain initial estimates for the series, the slope, and the seasonal factors. Winters suggests that a minimum of two seasons of data be available for initialization. Let us assume that exactly two seasons of data are available; that is, $2N$ data points. Suppose that the current period is $t = 0$, so that the past observations are labeled $D_{-2N+1}, D_{-2N+2}, \ldots, D_0$.

1. Calculate the sample means for the two separate seasons of data.

$$V_1 = \frac{1}{N} \sum_{j=-2N+1}^{-N} D_j$$

$$V_2 = \frac{1}{N} \sum_{j=-N+1}^{0} D_j$$

2. Define $G_0 = (V_2 - V_1)/N$ as the initial slope estimate. If $m > 2$ seasons of data are available for initialization, then compute $V_1, \ldots, V_m$ as above and define $G_0 = (V_m - V_1)/[(m - 1)N]$. If we locate V_1 at the center of the first season of data (at period $(-3N + 1)/2$) and V_2 at the center of the second season of data (at period $(-N + 1)/2$), then G_0 is simply the slope of the line connecting V_1 and V_2 (refer to Figure 2–11).

3. Set $S_0 = V_2 + G_0[(N - 1)/2]$. This estimates the value of the series at time $t = 0$. Note that S_0 is the value assumed by the line connecting V_1 and V_2 at $t = 0$ (see Figure 2–11).

FIGURE 2–11

Initialization for Winters's method

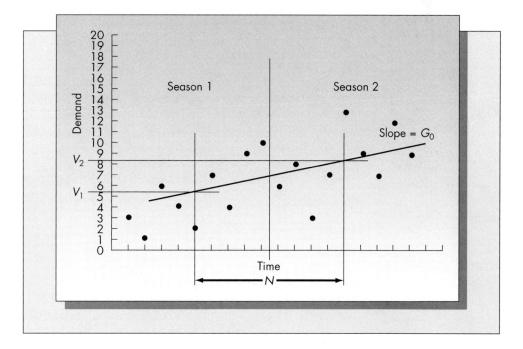

4. *a.* The initial seasonal factors are computed for each period in which data are available and then averaged to obtain one set of seasonal factors. The initial seasonal factors are obtained by dividing each of the initial observations by the corresponding point along the line connecting V_1 and V_2. This can be done graphically or by using the following formula:

$$c_t = \frac{D_t}{V_i - [(N+1)/2 - j]G_0} \qquad \text{for } -2N + 1 \le t \le 0,$$

where $i = 1$ for the first season, $i = 2$ for the second season, and j is the period of the season. That is, $j = 1$ for $t = -2N + 1$ and $t = -N + 1$; $j = 2$ for $t = -2N + 2$ and $t = -N + 2$, and so on.

    b. Average the seasonal factors. Assuming exactly two seasons of initial data, we obtain

$$c_{-N+1} = \frac{c_{-2N+1} + c_{-N+1}}{2}, \ldots, c_0 = \frac{c_{-N} + c_0}{2}.$$

    c. Normalize the seasonal factors.

$$c_j = \left[\frac{c_j}{\sum\limits_{i=0}^{-N+1} c_i} \right] \cdot N \qquad \text{for } -N + 1 \le j \le 0.$$

The initialization procedure discussed above is the one suggested by Winters. It is not the only means of initializing the system. Seasonal factors could be determined by the method of moving averages discussed in the previous section. Another alternative would be to fit a linear regression to the baseline data and use the resulting slope and intercept values, as was done in Holt's method, to obtain S_0 and G_0. The seasonal factors would be obtained by dividing each demand observation in the baseline period by the corresponding value on the regression line, averaging like periods, and norming. The actual values of the initial estimates of the intercept, the slope, and the seasonal factors will be similar no matter which initialization scheme is employed.

Example 2.8

Assume that the initial data set is the same as that of Example 2.7, in which centered moving averages were used to find the seasonal factors. Recall that we have two seasons of data: 10, 20, 26, 17, 12, 23, 30, 22. Then

$$V_1 = (10 + 20 + 26 + 17)/4 = 18.25,$$
$$V_2 = (12 + 23 + 30 + 22)/4 = 21.75,$$
$$G_0 = (21.75 - 18.25)/4 = .875,$$
$$S_0 = 21.75 + (.875)(1.5) = 23.06.$$

The initial seasonal factors are computed as follows:

$$c_{-7} = \frac{10}{18.25 - (5/2 - 1)(.875)} = .5904,$$

$$c_{-6} = \frac{20}{18.25 - (5/2 - 1)(.875)} = 1.123.$$

The other factors are computed in a similar fashion. They are

$$c_{-5} = 1.391, \qquad c_{-4} = .869, \qquad c_{-3} = .5872,$$
$$c_{-2} = 1.079, \qquad c_{-1} = 1.352, \qquad c_0 = .9539.$$

We then average c_{-7} and c_{-3}, c_{-6} and c_{-2}, and so on, to obtain the four seasonal factors:

$$c_{-3} = .5888, \qquad c_{-2} = 1.1010, \qquad c_{-1} = 1.3720, \qquad c_0 = .9115.$$

Finally, norming the factors to ensure that the sum is 4 results in

$$c_{-3} = .5900, \qquad c_{-2} = 1.1100, \qquad c_{-1} = 1.3800, \qquad c_0 = .9200.$$

Notice how closely these factors agree with those obtained from the moving-average method.

Suppose that we wish to forecast the following year's demand at time $t = 0$. The forecasting equation is

$$F_{t, t+\tau} = (S_t + \tau G_t)c_{t+\tau -N},$$

which results in

$$F_{0,1} = (S_0 + G_0)c_{-3} = (23.06 + .875)(.59) = 14.12,$$
$$F_{0,2} = (S_0 + 2G_0)c_{-2} = [23.06 + (2)(.875)](1.11) = 27.54,$$
$$F_{0,3} = 35.44,$$
$$F_{0,4} = 24.38.$$

Now, suppose that at time $t = 1$ we observe a demand of $D_1 = 16$. We now need to update our equations. Assume that $\alpha = .2, \beta = .1$, and $\gamma = .1$. Then

$$S_1 = \alpha(D_1/c_{-3}) + (1 - \alpha)(S_0 + G_0)$$
$$= (.2)(16/.59) + (.8)(23.06 + .875) = 24.57,$$

$$G_1 = \beta(S_1 - S_0) + (1 - \beta)G_0$$
$$= (.1)(24.57 - 23.06) + (.9)(.875) = .9385,$$

$$c_1 = \gamma(D_1/S_1) + (1 - \gamma)c_{-3}$$
$$= (.1)(16/24.57) + (.9)(.59) = .5961.$$

At this point, we would renorm c_{-2}, c_{-1}, c_0, and the new value of c_1 to add to 4. Because the sum is 4.0061, it is close enough (the norming would result in rounding c_1 down to .59 once again).

Forecasting from period 1, we obtain

$$F_{1,2} = (S_1 + G_1)c_{-2} = (24.57 + .9385)(1.11) = 28.3144,$$
$$F_{1,3} = (S_1 + 2G_1)c_{-1} = [24.57 + (2)(.9385)](1.38) = 36.4969,$$

and so on.

Now suppose that we have observed one full year of demand, given by $D_1 = 16$, $D_2 = 33$, $D_3 = 34$, and $D_4 = 26$. Each time a new observation becomes available, the intercept, slope, and most current seasonal factor estimates are updated. One obtains

$$S_2 = 26.35, \qquad S_3 = 26.83, \qquad S_4 = 27.89,$$
$$G_2 = 1.0227, \qquad G_3 = 0.9678, \qquad G_4 = 0.9770,$$
$$c_2 = 1.124, \qquad c_3 = 1.369, \qquad c_4 = 0.9212.$$

As c_1, c_2, c_3, and c_4 sum to 4.01, normalization is not necessary. Suppose that we were interested in the forecast made in period 4 for period 10. The forecasting equation is

$$F_{t, t+\tau} = (S_t + \tau G_t)c_{t+\tau-2N},$$

which results in

$$F_{4,10} = (S_4 + 6G_4)c_2 = [27.89 + 6(.9770)](1.124) = 37.94.$$

An important consideration is the choice of the smoothing constants α, β, and γ to be used in Winters's method. The issues here are the same as those discussed for simple exponential smoothing and Holt's method. Large values of the smoothing constants will result in more responsive but less stable forecasts. One method for setting α, β, and γ is to experiment with various values of the parameters that retrospectively give the best fit of previous forecasts to the observed history of the series. Because one must test many combinations of the three constants, the calculations are tedious. Furthermore, there is no guarantee that the best values of the smoothing constants based on past data will be the best values for future forecasts. The most conservative approach is to guarantee stable forecasts by choosing the smoothing constants to be between 0.1 and 0.2. ■

More Problems for Section 9

35. Consider the data for Problem 34.
    a. Using the data from 1991 and 1992, determine initial values of the intercept, slope, and seasonal factors for Winters's method.
    b. Assume that the observed demand for the first quarter of 1993 was 18. Using $\alpha = .2$, $\beta = .15$, and $\gamma = .10$, update the estimates of the series, the slope, and the seasonal factors.
    c. What are the forecasts made at the end of the first quarter of 1993 for the remaining three quarters of 1993?

36. Suppose the observed quarterly demand for 1993 was 18, 51, 86, 66. Compare the accuracy of the forecasts obtained for the last three quarters of 1993 in Problems 34(d) and 35(c) by computing both the MAD and the MSE.

37. Determine updated estimates of the slope, the intercept, and the seasonal factors for the end of 1993 based on the observations given in Problem 36. Using these updated estimates, determine the forecasts that Winters's method gives for all of 1995 made at the end of 1993. Use the values of the smoothing constants given in Problem 35(b).

2.10 Practical Considerations

Model Identification and Monitoring

Determining the proper model depends both on the characteristics of the history of observations and on the context in which the forecasts are required. When historical data are available, they should be examined carefully in order to determine if obvious patterns exist, such as trend or seasonal fluctuations. Usually, these patterns can be spotted by graphing the data. Statistical tests, such as significance of regression, can be used to verify the existence of a trend, for example. Identifying complex relationships requires more sophisticated methods. The *sample autocorrelation function* can reveal intricate relationships that simple graphical methods cannot. The Box-Jenkins (1970) methodology is based on identifying the appropriate model from an examination of the autocorrelation structure. Box-Jenkins methods require substantial past history (at least 72 data points are recommended) and separate model identification and optimization for each series considered. Judging from the literature, such methods are rarely used in production applications.

Once a model has been chosen, forecasts should be monitored regularly to see if the model is appropriate or if some unforeseen change has occurred in the series. As we indicated, a forecasting method should not be biased. That is, the expected value of the forecast error should be zero. In addition to the methods mentioned in Section 6, one means of monitoring the bias is the *tracking signal* developed by Trigg (1964). Following earlier notation, let e_t be the observed error in period t and $|e_t|$ the absolute value of the observed error. The smoothed values of the error and the absolute error are given by

$$E_t = \beta e_t + (1 - \beta)E_{t-1},$$
$$M_t = \beta|e_t| + (1 - \beta)M_{t-1}.$$

The tracking signal is the ratio

$$T_t = \left|\frac{E_t}{M_t}\right|.$$

If forecasts are unbiased, the smoothed error E_t should be small compared to the smoothed absolute error M_t. Hence, a large value of the tracking signal indicates biased forecasts, which suggest that the forecasting model is inappropriate. The value of T_t that signals a significant bias depends on the smoothing constant β. For example, Trigg (1964) claims that a value of T_t exceeding 0.51 indicates nonrandom errors for a β of 0.1. The tracking signal also can be used directly as a variable smoothing constant. This is considered in Problem 48.

Simple versus Complex Time Series Methods

The literature on forecasting is voluminous. In this chapter we have touched only on a number of fairly simple techniques. The reader is undoubtedly asking himself or herself, do these methods actually work? The results from the literature suggest that the simplest methods are often as accurate as sophisticated ones. Armstrong (1984) reviews 25 years of forecasting case studies with the goal of ascertaining whether or not sophisticated methods work better. In comparing the results of 39 case studies, he found that in 20 cases sophisticated methods performed about as well as simple ones, in 11 cases they outperformed simple methods, and in 7 cases they performed significantly worse.

A more sophisticated forecasting method is one that requires the estimation of a larger number of parameters from the data. Trouble can arise when these parameters are estimated incorrectly. To give some idea of the nature of this problem, consider a comparison of simple moving averages and regression analysis for the following series: 7, 12, 9, 23, 27. Suppose that we are interested in forecasting at the end of period 5 for the demand in period 15 (that is, we require $F_{5,15}$). The five-period moving-average forecast made at the end of period 5 is 15.6, and this would be the forecast for period 15. The least squares fit of the data is $\hat{D}_t = 0.3 + 5.1t$. Substituting $t = 15$, we obtain the regression forecast of 76.8. In Figure 2–12 we picture the realization of the demand through period 15. Notice what has happened. The apparent trend that existed

FIGURE 2–12

*The difficulty with
long-term forecasts*

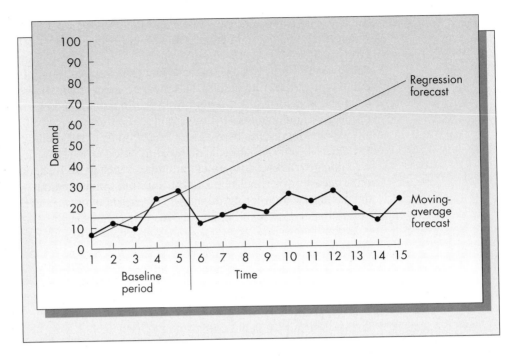

in the first five periods was extrapolated to period 15 by the regression equation. However, there really was no significant trend in this particular case. The more complex model gave *significantly* poorer results for the long-term forecast.

There is some evidence that the arithmetic average of forecasts obtained from different methods is more accurate than a single method (see Makridakis and Winkler, 1983). This is perhaps because often a single method is unable to capture the underlying signal in the data and different models capture different aspects of the signal. (See a discussion of this phenomenon following Armstrong, 1984.)

What do these observations tell us about the application of forecasting techniques to production planning? At the aggregate level of planning, forecast accuracy is extremely important and multiple-step-ahead forecasts play an integral role in the planning of workforce and production levels. For that reason, blind reliance on time series methods is not advised at this level. At a lower level in the system, such as routine inventory management for spare parts, the use of simple time series methods such as moving averages or exponential smoothing makes a great deal of sense. At the individual item level, short-term forecasts for a large number of items are required, and monitoring the forecast for each item is impractical at best. The risk of severe errors is minimized if simple methods are used.

2.11 Overview of Advanced Topics in Forecasting

Box-Jenkins Methods

This chapter has provided a description of the most popular forecasting methods for production planning applications. The bulk of the discussion has centered on time series methods. These are automatic methods that base forecasts on a history of observations. A popular advanced time series technique is the so-called *Box-Jenkins method* (Box and Jenkins, 1970). This method exploits possible dependencies among values of the series from period to period. (The other methods we reviewed in this chapter assume that successive observations are independent.) Accounting for these dependencies can often substantially improve forecasts.

For example, suppose a company has instituted a sales incentive system that provides a bonus for the employee with the best improvement in bookings from one month to the next. With such an incentive in place, a month of poor sales is often followed by a month of good sales. Similarly, a month of good sales would usually be followed by a lull. This means that sales in consecutive months tend to be negatively correlated. This information can be used to improve sales forecasts.

Autocorrelation is the correlation (i.e., degree of dependency) among values of observed data separated by a fixed number of periods. In the example above, we would say that the series has a negative autocorrelation of order one. If high values tend to be followed by high values, the autocorrelation of order one is positive. We can also measure autocorrelations of order greater than one. For example, if low values tend to be followed by low values two periods later, then the series exhibits a positive autocorrelation of order two. To use these methods, one determines sample autocorrelations from the data for one period of separation, two periods of separation, and so on.

The most general Box-Jenkins model is known as the *ARIMA* (autoregressive integrated moving average) model. Autoregressive processes and moving-average processes (not to be confused with simple moving averages discussed earlier in the chapter) are two models of time series with autocorrelation. The term *integrated* refers to differencing. First-order differencing means deriving a new series that is equal to the first differences of successive values of the original series. Differencing is a means of eliminating trend and polynomial growth.

Interested readers should refer to both Box and Jenkins (1970) and Nelson (1973) for a complete discussion. ARIMA models are not as straightforward to construct as the other methods discussed in this chapter. Developing the proper form of the model requires a deep understanding of these processes. One must employ sophisticated mathematical techniques to estimate the model's parameters. Furthermore, to obtain a reasonable estimate for the autocorrelation function, one must have a substantial history of observations. At least 72 data points are recommended. The payoff is that under the right circumstances, these methods can significantly outperform simpler ones.

Simulation as a Forecasting Tool

Computer simulation is a powerful technique for tackling complex problems. A computer simulation is a description of a problem reduced to a computer program. The program is designed to recreate the key aspects of the dynamics of a real situation. When a problem is too complex to model mathematically, simulation is a popular alternative. By rerunning the program under different starting conditions and/or different scenarios, one can, by a kind of trial-and-error process, discover the best strategy for managing a system.

Simulation is a common tool for modeling manufacturing planning problems such as complex material flow problems in the plant. It is less commonly used as a forecasting tool. Compaq Computer, a successful producer of personal computers based in Houston, Texas, has experimented with a powerful simulation-based forecasting tool for assisting with the process of new product introductions (McWilliams, 1995). The program recommends the optimal timing and pricing of new product introductions by incorporating forecasts of component availability and price changes, fluctuating demand for a given feature or price, and the impact of rival models.

Using this tool, Compaq decided to delay announcement of several new Pentium-based models in late 1994. This strategy "went against everything the company believed." Compaq's basic strategy had always been to be a technology leader, but their forecasting tool suggested that corporate customers were not quite ready to switch to Pentium-based machines at the end of 1994. The strategy proved to be very profitable for Compaq, which subsequently posted record earnings.

Forecasting Demand in the Presence of Lost Sales

Retailers rely heavily on forecasting. Basic items (items that don't change appreciably from season to season, such as men's dress shirts) generally have substantial sales history, arguing for the use of time series methods to forecast demand. However, there is an important difference between what is observed and what one wants to forecast. The goal is to forecast *demand,* but one only observes *sales*. What's the difference? Suppose a customer wants to buy a blouse in a certain size and color and finds it's not available on the shelf? What will she do? Perhaps she will place a special order with a salesperson, but, more likely, she will just leave the store and try to find the product somewhere else. This is known as a lost sale. The difficulty is that most retailers have no way to track lost sales. Thus, they observe sales but need to estimate demand.

As an example, consider an item that is restocked to 10 units at the beginning of each week. Suppose that over the past 15 weeks the sales history for the item was 7, 5, 10, 10, 8, 3, 6, 10, 10, 9, 5, 0, 10, 10, 4. Consider those weeks in which sales were 10 units. What were the demands in those weeks? The answer is that we don't know. We only know that it was *at least* 10. If you computed the sample mean and sample variance of these numbers, they would underestimate the true mean and variance of demand.

How does one go about forecasting demand in this situation? In the parlance of classical statistics, this is known as a censored sample. That means that we know the values of demand for only a portion of the sample. For the other portion of the sample, we know only a lower bound on the demand. Special statistical methods that incorporate censoring give significantly improved estimates of the population mean and variance in this case. These methods can be imbedded into sequential forecasting schemes, such as exponential smoothing, to provide significantly improved forecasts.

Nahmias (1994) considered the problem of forecasting in the presence of lost sales when the true demand distribution was normal. He compared the method of maximum likelihood for censored samples and a new method, either of which could be incorporated into exponential smoothing routines. He also showed that both of these methods would result in substantially improved forecasts of both the mean and the variation of the demand.

To see how dramatic this difference can be, consider a situation in which the true weekly demand for a product is a normal random variable with mean 100 and standard deviation 30. Suppose that items are stocked up to 110 units at the start of each week. Exponential smoothing is used to obtain two sets of forecasts: the first accounts for lost sales (includes censoring) and the second does not (does not including censoring). Figures 2–13 and 2–14 show the estimators for the mean and the standard deviation with and without censoring. Notice the severe low bias when lost sales are ignored in both cases. That means that by not correctly accounting for the difference between sales and demand, one underestimates both the mean and the variance of the demand.

FIGURE 2–13

Tracking the mean when lost sales are present

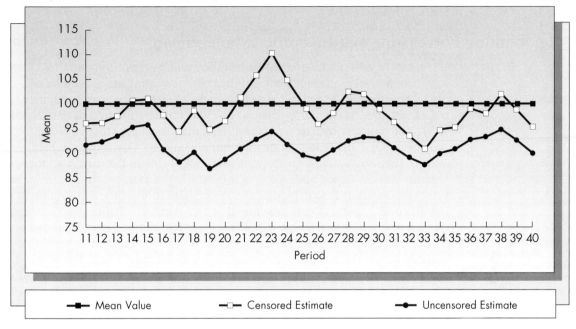

FIGURE 2–14

Tracking the standard deviation when lost sales are present

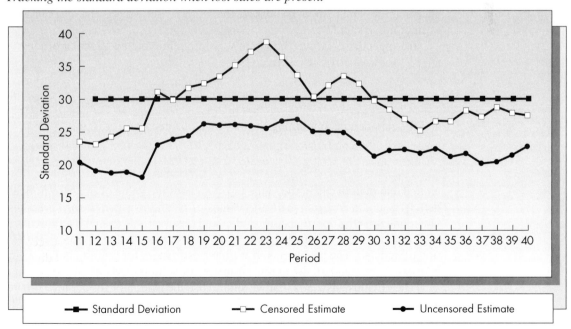

Since both the mean and the variance of demand are inputs for determining optimal stocking levels, these levels could be severely underestimated.

2.12 Linking Forecasting and Inventory Management

Inventory control under demand uncertainty will be treated in detail in Chapter 5. In practice, inventory management and demand forecasting are closely linked. The forecasting method could be any one of the methods discussed in this chapter. One of the inputs required for inventory control models is the distribution of the demand over a period or over an order replenishment lead time. Is there a link between the distribution of forecast errors and the distribution of demand? The answer is that there is such a link. The forecast error distribution plays a key role in the correct application of inventory models in practice.

The first issue is to decide on the appropriate form of the distribution of demand. Most commercial systems assume that the demand distribution is normal. That means we need only estimate the mean μ and the standard deviation σ to specify the entire distribution. (Statistical methods, known as goodness-of-fit techniques, can be applied to test the accuracy of the normality assumption.) Whether or not the inventory system is linked to a forecasting system, we must have a history of observations of demand to obtain statistical estimates of the mean and variance. (When no history of demand exists, personal judgment must be substituted for statistical estimation. Methods for aggregating subjective judgment are discussed in Section 3 of this chapter.)

In the beginning of Chapter 5, we discuss how one would estimate the mean and the variance of demand directly from a history of demand observations. In practice, however, we don't generally use a long history of observations, because we believe that the underlying demand distribution does not remain constant indefinitely. That is why we adjust the N for moving averages or the α for exponential smoothing to balance stability and responsiveness. If that is the case, what is the appropriate variance estimator one should use?

In Appendix 2–A, we show that for simple moving averages of order N, the variance of forecast error, σ_e^2, is given by

$$\sigma_e^2 = \sigma^2 \left(\frac{N + 1}{N} \right)$$

and for exponential smoothing the variance of forecast error is

$$\sigma_e^2 = \sigma^2 \left(\frac{2}{2 - \alpha} \right).$$

Notice that in both cases, the value of σ_e^2 *exceeds* the value of σ^2. Also notice that as N gets large and as α gets small, the values of σ_e and σ grow close. These cases occur when one uses the entire history of demand observations to make a forecast. In Chapter 5 one of the inputs needed to determine safety stocks for inventory is the distribution of demand. The problem is, if we have estimators for both σ_e and σ, which should be used as the standard deviation estimator for setting safety stocks?

The obvious answer is that we should use the estimator for σ, since this represents the standard deviation of demand. The correct answer, however, is that we should use σ_e. The reason is that the process of forecasting introduces sampling error into the estimation process, and this sampling error is accounted for in the value of σ_e. The forecasting error variance is higher than the demand variance because the forecast is based on only a limited portion of the demand history.

We also can provide an intuitive explanation. If a forecast is used to estimate the mean demand, we keep safety stocks in order to protect against the error in this forecast. Hence, the distribution of forecast errors is more relevant than the distribution of demands. This is an important practical point that has been the source of much confusion in the literature. We will return to this issue in Chapter 5 and discuss its relevance in the context of safety stock calculations.

Most inventory control systems use the method suggested by R. G. Brown (1959 and 1962) to estimate the value of σ_e. (In fact, it appears that Brown was the first to recognize the importance of the distribution of forecast errors in inventory management applications.) The method requires estimating the MAD of forecast errors using exponential smoothing. This is accomplished using the smoothing equation

$$\text{MAD}_t = \alpha \, \text{MAD}_{t-1} + (1 - \alpha)|F_t - D_t|.$$

The MAD is converted to an estimate of the standard deviation of forecast error by multiplying by 1.25. That is, the estimator for σ_e obtained at time t is

$$\hat{\sigma}_e = 1.25 \, \text{MAD}_t.$$

A small value of α, generally between 0.1 and 0.2, is used to assure stability in the MAD estimator. This approach to estimating the MAD works for any of the forecasting methods discussed in this chapter. Safety stocks are then computed using this estimator for σ_e.

2.13 Historical Notes and Additional Topics

Forecasting is a rich area of research. Its importance in business applications cannot be overstated. The simple time series methods discussed in this chapter have their roots in basic statistics and probability theory. The method of exponential smoothing is generally credited to R. G. Brown (1959 and 1962), who worked as a consultant to A. D. Little at the time. Although not grounded in basic theory, exponential smoothing is probably one of the most popular forecasting methods used today. Brown was also the first to recognize the importance of the forecast error distribution and its implications for inventory management. However, interest in using statistical methods for forecasting goes back to the turn of the century or before. (See, for example, Yule, 1926.)

Not discussed in this chapter is forecasting of time series using spectral analysis and state space methods. These methods are highly sophisticated and require substantial structure to exist in the data. They often rely on the use of the autocorrelation function, and in that sense are similar conceptually to Box-Jenkins methods discussed briefly in Section 11. The groundbreaking work in this area is due to Norbert Wiener (1949) and Rudolph Kalman (1960). However, there have been few applications of these methods to forecasting economic time series. Most applications have been in the area of signal processing in electrical engineering. (Davenport and Root [1958] provide a good summary of the fundamental concepts in this area.)

The *Kalman filter* is a type of exponential smoothing technique in which the value of the smoothing constant changes with time and is chosen in some sort of optimal fashion. The idea of adjusting the value of the smoothing constant based on some measure of prior performance of the model has been used in a number of ad hoc ways. A typical approach is the one suggested by Trigg and Leach (1967), which requires the calculation of the tracking signal. The tracking signal is then used as the value of the smoothing constant for the next forecast. The idea is that when the tracking signal is large, it suggests that the time series has undergone a shift; a larger value of the smoothing constant should be more responsive to a sudden shift in the underlying signal. Other methods have also been suggested. We have intentionally not included an explicit discussion of adaptive response rate methods for one reason: there is little evidence that they work in the context of predicting economic time series. Most studies that compare the effectiveness of different forecasting methods for many different series show no advantage for adaptive response rate models. (See, for example, Armstrong, 1984.) The method of Trigg and Leach is discussed in more detail in Problem 48.

A good starting point for further reading is the text by Makridakis, Wheelwright, and McGhee (1983), which gives a more detailed exposition of time series methods at a level similar to ours. Business applications are discussed in Wilson and Keating (1990).

2.14 Summary

This chapter provided an introduction to a number of the more popular time series forecasting techniques as well as a brief discussion of other methods, including the Delphi method and causal models. A *moving-average* forecast is obtained by computing the arithmetic average of the N most recent observations of demand. An *exponential smoothing* forecast is obtained by computing the weighted average of the current observation of demand and the most recent forecast of that demand. The weight applied to the current observation is α and the weight applied to the last forecast (that is, the past observations) is $1 - \alpha$. Small N and large α result in responsive forecasts, and large N and small α result in stable forecasts. Although the two methods have similar properties, exponential smoothing is generally preferred because it requires storing only the previous forecast, whereas moving averages requires storing the last N demand observations.

When there is a trend in the series, both moving averages and exponential smoothing lag behind the trend. We discussed two time series techniques that are designed to track the trend. One is *regression analysis,* which uses least squares to fit a straight line to the data, and the other is *Holt's method,* which is a type of double exponential smoothing. Holt's method has the advantage that forecasts are easier to update as new demand observations become available.

We also discussed techniques for seasonal series. We employed *classical decomposition* to show how simple moving averages could be used to estimate seasonal factors and obtain the deseasonalized series when there is a trend and showed how seasonal factors could be estimated quickly when there is no trend. The extension of Holt's method to deal with seasonal problems, called *Winters's method,* is a type of triple exponential smoothing technique.

Section 2.11 provided a brief overview of advanced methods that are beyond the scope of this text. The final section discussed the relationship between forecasting and inventory control. The key point, which will be elaborated on in Chapter 5, is that the standard deviation of forecast error is the appropriate measure of variation for computing safety stocks.

Additional Problems on Forecasting

38. John Kittle, an independent insurance agent, uses a five-year moving average to forecast the number of claims made in a single year for one of the large insurance companies he sells for. He has just discovered that a clerk in his employ incorrectly entered the number of claims made four years ago as 1,400 when it should have been 1,200.

    a. What adjustment should Mr. Kittle make in next year's forecast to take into account the corrected value of the number of claims four years ago?

    b. Suppose that Mr. Kittle used simple exponential smoothing with $\alpha = 0.2$ instead of moving averages to determine his forecast. What adjustment is now required in next year's forecast? (Note that you do need to know the value of the forecast for next year in order to solve this problem.)

39. A method of estimating the MAD discussed in Section 12 recomputes it each time a new demand is observed according to the following formula:

$$\text{MAD}_t = \alpha \, |e_t| + (1 - \alpha) \, \text{MAD}_{t-1}.$$

    Consider the one-step-ahead forecasts for aircraft engine failures for quarters 2 through 8 obtained in Example 2.3. Assume an initial value of the MAD = 50 in period 1. Using the same value of α, what values of the MAD does this method give for periods 2 through 8? Discuss the advantages and disadvantages of this approach vis-à-vis direct computation of the MAD.

40. Herman Hahn is attempting to set up an integrated forecasting and inventory control system for his hardware store, Hahn's Hardware. When Herman indicates that outdoor lights are a seasonal item on the computer, he is prompted by the program to input the seasonal factors by quarter.

    Unfortunately, Herman has not kept any historical data, but he estimates that first-quarter demand for the lights is about 30 percent below average, the second-quarter demand about 20 percent below average, third-quarter demand about average, and fourth-quarter demand about 50 percent above average. What should he input for the seasonal factors?

41. Irwin Richards, a publisher of business textbooks, publishes in the areas of management, marketing, accounting, production, finance, and economics. The president of the firm is interested in getting a relative measure of the sizes of books in the various fields. Over the past three years, the average numbers of pages of books published in each area were

	Year 1	Year 2	Year 3
Management	835	956	774
Marketing	620	540	575
Accounting	440	490	525
Production	695	680	624
Finance	380	425	410
Economics	1,220	1,040	1,312

Using the quick and dirty methods discussed in Section 9, find multiplicative factors for each area giving the percentage above or below the mean number of pages.

42. Over a two-year period, the Topper Company sold the following numbers of lawn mowers:

Month:	1	2	3	4	5	6	7	8	9	10	11	12
Sales:	238	220	195	245	345	380	270	220	280	120	110	85
Month:	13	14	15	16	17	18	19	20	21	22	23	24
Sales:	135	145	185	219	240	420	520	410	380	320	290	240

   a. In column A input the numbers 1 to 24 representing the months and in column B the observed monthly sales. Compute the three-month moving-average forecast and place this in the third column. Be sure to align your forecast with the period *for* which you are forecasting (the average of sales for months 1, 2, and 3 should be placed in row 4; the average of sales for months 2, 3, and 4 in row 5; and so on.) In the fourth column, compute the forecast error for each month in which you have obtained a forecast.
   b. In columns 5, 6, and 7 compute the absolute error, the squared error, and the absolute percentage error. Using these results, find the MAD, MSE, and MAPE for the MA(3) forecasts for months 4 through 24.
   c. Repeat parts (*a*) and (*b*) for six-month moving averages. (These calculations should appear in columns 7 through 11.) Which method, MA(3) or MA(6), was more accurate for these data?

43. Repeat the calculations in Problem 42 using simple exponential smoothing, and allow the smoothing constant α to be a variable. That is, the smoothing constant should be a cell location. By experimenting with different values of α, determine the value that appears to minimize the
   a. MAD.
   b. MSE.
   c. MAPE.
   Assume that the forecast for month 1 is 225.

44. Baby It's You, a maker of baby foods, has found a high correlation between the aggregate company sales (in $100,000) and the number of births nationally the *preceding* year. Suppose that the sales and the birth figures during the past eight years are

	Year							
	1	*2*	*3*	*4*	*5*	*6*	*7*	*8*
Sales (in $100,000)	6.1	6.4	8.3	8.8	5.1	9.2	7.3	12.5
U.S. births (in millions)	2.9	3.4	3.5	3.1	3.8	2.8	4.2	3.7

   a. Assuming that U.S. births represent the independent variable and sales the dependent variable, determine a regression equation for predicting sales based on births. Use years 2 through 8 as your baseline. (Hint: You will require the general regression formulas appearing in Appendix 2–B to solve this problem.)

b. Suppose that births are forecasted to be 3.3 million in year 9. What forecast for sales revenue in year 10 do you obtain using the results of (a)?

c. Suppose that simple exponential smoothing with $\alpha = .15$ is used to predict the number of births. Use the average of years 1 to 4 as your initial forecast for period 5, and determine an exponentially smoothed forecast for U.S. births in year 9.

d. Combine the results in (a), (b), and (c) to obtain a forecast for the sum of total aggregate sales in years 9 and 10.

45. Hy and Murray are planning to set up an ice cream stand in Shoreline Park, described in Problem 28. After six months of operation, the observed sales of ice cream (in dollars) and the number of park attendees are

	Month					
	1	*2*	*3*	*4*	*5*	*6*
Ice cream sales	325	335	172	645	770	950
Park attendees	880	976	440	1,823	1,885	2,436

a. Determine a regression equation treating ice cream sales as the dependent variable and time as the independent variable. Based on this regression equation, what should the dollar sales of ice cream be in two years (month 30)? How confident are you about this forecast? Explain your answer.

b. Determine a regression equation treating ice cream sales as the dependent variable and park attendees as the independent variable. (Hint: You will require the general regression equations in Appendix 2–B in order to solve this part.)

c. Suppose that park attendance is expected to follow a logistic curve (see Figure 2–15).

The park department expects the attendance to peak out at about 6,000 attendees per month. Plot the data of park attendees by month and "eyeball" a logistics curve fit of the data using 6,000 as your maximum value. Based on your curve and the regression equation determined in part (b), predict ice cream sales for months 12 through 18.

46. A suggested method for determining the "right" value of the smoothing constant α in exponential smoothing is to retrospectively determine the α value that results in the minimum forecast error for some set of historical data. Comment on the appropriateness of this method and some of the potential problems that could result.

47. Lakeroad, a manufacturer of hard disks for personal computers, was founded in 1981 and has sold the following numbers of disks:

Year	Number Sold (in 000s)	Year	Number Sold (in 000s)
1981	.2	1985	34.5
1982	4.3	1986	68.2
1983	8.8	1987	85.0
1984	18.6	1988	58.0

FIGURE 2–15

Logistic curve (for Problem 45)

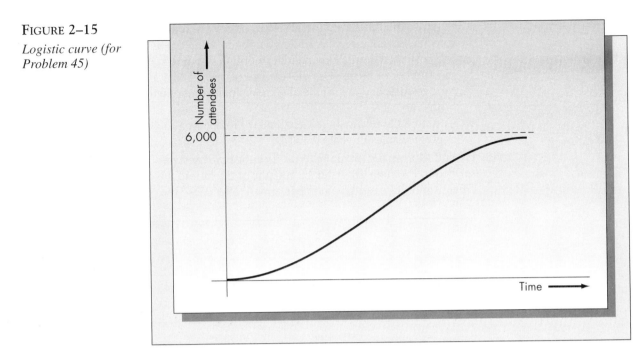

a. Suppose the firm uses Holt's method for forecasting sales. Assume $S_0 = 0$ and $G_0 = 8$. Using $\alpha = .2$ and $\beta = .2$, find one-step-ahead forecasts for 1982 through 1989 and compute the MAD and MSE for the forecasts during this period. What is the sales forecast for the year 2000 made at the end of 1988? Based on the results of 1988, why might this forecast be very inaccurate?

b. By experimenting with various values of α and β, determine the values of the smoothing constants that appear to give the most accurate forecasts.

48. Trigg and Leach (1967) suggest the following adaptive response-rate exponential smoothing method. Along with smoothing the original series, also smooth the error e_t and the absolute error $|e_t|$ according to the equations

$$E_t = \beta e_t + (1 - \beta)E_{t-1},$$
$$M_t = \beta|e_t| + (1 - \beta)M_{t-1}$$

and define the smoothing constant to be used in forecasting the series in period t as

$$\alpha_t = \left| \frac{E_t}{M_t} \right|.$$

The forecast made in period t for period $t + 1$ is obtained by the usual exponential smoothing equation, using α_t as the smoothing constant. That is,

$$F_{t+1} = \alpha_t D_t + (1 - \alpha_t)F_t.$$

The idea behind the approach is that when E_t is close in magnitude to M_t, it suggests that the forecasts are biased. In that case, a larger value of the smoothing constant results that makes the exponential smoothing more responsive to sudden changes in the series.

a. Apply the Trigg-Leach method to the data in Problem 22. Using the MAD, compare the accuracy of these forecasts with the simple exponential smoothing forecasts obtained in Problem 22. Assume that $E_1 = e_1$ (the observed error in January) and $M_1 = |e_1|$. Use $\beta = .1$.

b. For what types of time series will the Trigg-Leach adaptive response rate give more accurate forecasts, and under what circumstances will it give less accurate forecasts? Comment on the advisability of using such a method for situations in which forecasts are not closely monitored.

49. The owner of a small brewery in Milwaukee, Wisconsin, is using Winters's method to forecast his quarterly beer sales. He has been using smoothing constants of $\alpha = .2$, $\beta = .2$, and $\gamma = .2$. He has currently obtained the following values of the various slope, intercept, and seasonal factors: $S_{10} = 120$, $G_{10} = 14$, $c_{10} = 1.2$, $c_9 = 1.1$, $c_8 = .8$, and $c_7 = .9$.

a. Determine the forecast for beer sales in quarter 11.

b. Suppose that the actual sales turn out to be 128 in quarter 11. Find S_{11} and G_{11}, and find the updated values of the seasonal factors. Also determine the forecast made at the end of quarter 11 for quarter 13.

50. The U.S. gross national product (GNP) in billions of dollars during the period 1964 to 1984 was as follows:

Year	GNP	Year	GNP
1964	649.8	1975	1,598.4
1965	705.1	1976	1,782.8
1966	772.0	1977	1,990.5
1967	816.4	1978	2,249.7
1968	892.7	1979	2,508.2
1969	963.9	1980	2,732.0
1970	1,015.5	1981	3,052.6
1971	1,102.7	1982	3,166.0
1972	1,212.8	1983	3,401.6
1973	1,359.3	1984	3,774.7
1974	1,472.8		

Source: *Economic Report of the President,* February 1986.

a. Use Holt's method to predict the GNP. Determine a regression fit of the data for the period 1964 to 1974 to estimate the initial values of the slope and intercept. (Hint: If you are doing the regression by hand, transform the years by subtracting 1963 from each value to make the calculations less cumbersome.) Using Holt's method, determine forecasts for 1975 to 1984. Assume that $\alpha = .2$ and $\beta = .1$. Compute the MAD and the MSE of the one-step-ahead forecasts for the period 1975 to 1984.

b. Determine the percentage increase in GNP from 1964 to 1984 and graph the resulting series. Use a six-month moving average and simple exponential smoothing with $\alpha = .2$ to obtain one-step-ahead forecasts of this series for the period 1975 to 1984. (Use the arithmetic average of the observations from 1964 to 1974 to initialize the exponential smoothing.)

In both cases (i.e., MA and ES forecasts), convert your forecasts of the percentage increase for the following year to a forecast of the GNP itself and

compute the MAD and the MSE of the resulting forecasts. Compare the
accuracy of these methods with that of part (a).

c. Discuss the problem of predicting GNP. What methods other than the ones
used in parts (a) and (b) might give better predictions of this series?

APPENDIX 2–A   FORECAST ERRORS FOR MOVING AVERAGES AND
EXPONENTIAL SMOOTHING

The forecast error e_t is the difference between the forecast for period t and the actual
demand for that period. In this section we will derive the distribution of the forecast
error for both moving averages and exponential smoothing.

The demand is assumed to be generated by the process

$$D_t = \mu + \epsilon_t,$$

where ϵ_t is normal with mean zero and variance σ^2.

Case 1. Moving Averages

Consider first the case in which forecasts are generated by moving averages. Then the
forecast error is $e_t = F_t - D_t$, where F_t is given by

$$F_t = \frac{1}{N} \sum_{i=t-N}^{t-1} D_i.$$

It follows that

$$E(F_t - D_t) = (1/N) \sum_{i=t-N}^{t-1} E(D_i) - E(D_t) = (1/N)(N\mu) - \mu = 0.$$

This proves that when demand is stationary, moving-average forecasts are unbiased.
Also,

$$\text{Var}(F_t - D_t) = \text{Var}(F_t) + \text{Var}(D_t) = (1/N^2) \sum_{i=t-N}^{t-1} \text{Var}(D_i) + \text{Var}(D_t)$$

$$= (1/N^2)(N\sigma^2) + \sigma^2$$

$$= \sigma^2(1 + 1/N) = \sigma^2[(N + 1)/N].$$

It follows that the standard deviation of the forecast error, σ_e, is

$$\sigma_e = \sigma \sqrt{\frac{N + 1}{N}}.$$

This is the standard deviation of the forecast error for simple moving averages in terms
of the standard deviation of each observation.

Having derived the mean and the variance of the forecast error, we still need to
specify the *form* of the forecast error distribution. By assumption, the values of D_t form
a sequence of independent, identically distributed, normal random variables. Since F_t

is a linear combination of $D_{t-1}, D_{t-2}, \ldots D_{t-N}$, it follows that F_t is normally distributed and independent of D_t. It now follows that e_t is normal as well. Hence, the distribution of e_t is completely specified by its mean and variance.

As the expected value of the forecast error is zero, we say the method is unbiased. Notice that this is a result of the assumption that the demand process is stationary. Consider the variance of the forecast error. The value of N that minimizes σ_e is $N = +\infty$. This means that the variance is minimized if the forecast is the average of all of the past data. However, our intuition tells us that we can do better if we use more recent data to make our forecast. The discrepancy arises because we really do not believe our assumption that the demand process is stationary for all time. A smaller value of N will allow the moving-average method to react more quickly to unforeseen changes in the demand process.

Case 2. Exponential Smoothing

Now consider the case in which forecasts are generated by exponential smoothing. In this case F_t may be represented by the weighted infinite sum of past values of demand.

$$F_t = \alpha D_{t-1} + \alpha(1 - \alpha)D_{t-2} + \alpha(1 - \alpha)^2 D_{t-3} + \cdots,$$

$$E(F_t) = \mu[\alpha + \alpha(1 - \alpha) + \alpha(1 - \alpha)^2 + \cdots] = \mu.$$

Notice that this means that $E(e_t) = 0$, so that both exponential smoothing and moving averages are unbiased forecasting methods when the underlying demand process is a constant plus a random term.

$$\text{Var}(F_t) = \alpha^2\sigma^2 + (1 - \alpha)^2\alpha^2\sigma^2 + \cdots$$

$$= \sigma^2\alpha^2 \sum_{n=0}^{\infty} (1 - \alpha)^{2n}.$$

It can be shown that

$$\sum_{n=0}^{\infty} (1 - \alpha)^{2n} = \frac{1}{1 - (1 - \alpha)^2}$$

so that

$$\text{Var}(F_t) = \frac{\sigma^2\alpha^2}{1 - (1 - \alpha)^2} = \frac{\sigma^2\alpha}{2 - \alpha}.$$

Since

$$\text{Var}(e_t) = \text{Var}(F_t) + \text{Var}(D_t),$$

$$\text{Var}(e_t) = \sigma^2[\alpha/(2 - \alpha) + 1] = \sigma^2[2/(2 - \alpha)],$$

or

$$\sigma_e = \sigma \sqrt{\frac{2}{2 - \alpha}}.$$

This is the standard deviation of the forecast error for simple exponential smoothing in terms of the standard deviation of each observation. The distribution of the forecast error for exponential smoothing is normal for essentially the same reasons as stated above for moving averages.

Notice that if we equate the variances of the forecast error for exponential smoothing and moving averages, we obtain

$$2/(2 - \alpha) = (N + 1)/N$$

or $\alpha = 2/(N + 1)$, which is exactly the same result as we obtained by equating the average age of data for the two methods.

APPENDIX 2–B DERIVATION OF THE EQUATIONS FOR THE SLOPE AND INTERCEPT FOR REGRESSION ANALYSIS

In this appendix we derive the equations for the optimal values of a and b for the regression model. Assume that the data are $(x_1, y_1), (x_2, y_2), \ldots, (x_n, y_n)$, and the regression model to be fitted is $Y = a + bX$. Define

$$g(a, b) = \sum_{i=1}^{n} [y_i - (a + bx_i)]^2.$$

Interpret $g(a, b)$ as the sum of the squares of the distances from the line $a + bx$ to the data points y_i. The object of the analysis is to choose a and b to minimize $g(a, b)$. This is accomplished where

$$\frac{\partial g}{\partial a} = \frac{\partial g}{\partial b} = 0.$$

That is,

$$\frac{\partial g}{\partial a} = -\sum_{i=1}^{n} 2[y_i - (a + bx_i)] = 0,$$

$$\frac{\partial g}{\partial b} = -\sum_{i=1}^{n} 2x_i[y_i - (a + bx_i)] = 0,$$

which results in the two equations

$$an + b \sum_{i=1}^{n} x_i = \sum_{i=1}^{n} y_i, \tag{1}$$

$$a \sum_{i=1}^{n} x_i + b \sum_{i=1}^{n} x_i^2 = \sum_{i=1}^{n} x_i y_i. \tag{2}$$

These are two linear equations in the unknowns a and b. Multiplying Equation (1) by Σx_i and Equation (2) by n gives

$$an \sum_{i=1}^{n} x_i + b \left(\sum_{i=1}^{n} x_i \right)^2 = \left(\sum_{i=1}^{n} x_i \right) \left(\sum_{i=1}^{n} y_i \right), \tag{3}$$

$$an \sum_{i=1}^{n} x_i + bn \sum_{i=1}^{n} x_i^2 = n \sum_{i=1}^{n} x_i y_i. \tag{4}$$

Subtracting Equation (3) from Equation (4) results in

$$b \left[n \sum_{i=1}^{n} x_i^2 - \left(\sum_{i=1}^{n} x_i \right)^2 \right] = n \sum_{i=1}^{n} x_i y_i - \left(\sum_{i=1}^{n} x_i \right) \left(\sum_{i=1}^{n} y_i \right). \tag{5}$$

Define $S_{xy} = n \sum x_i y_i - (\sum x_i)(\sum y_i)$ and $S_{xx} = n\sum x_i^2 - (\sum x_i)^2$. It follows that Equation (5) may be written $bS_{xx} = S_{xy}$, which gives

$$b = \frac{S_{xy}}{S_{xx}}. \tag{6}$$

From Equation (1) we have

$$an = \sum_{i=1}^{n} y_i - b \sum_{i=1}^{n} x_i$$

or

$$a = \bar{y} - b\bar{x}, \tag{7}$$

where $\bar{y} = (1/n)\sum y_i$ and $\bar{x} = (1/n)\sum x_i$.

These formulas can be specialized to the forecasting problem when the independent variable is assumed to be time. In that case, the data are of the form $(1, D_1), (2, D_2), \ldots,$ (n, D_n), and the forecasting equation is of the form $\hat{D}_t = a + bt$. The various formulas can be simplified as follows:

$$\sum x_i = 1 + 2 + 3 + \cdots + n = \frac{n(n + 1)}{2},$$

$$\sum x_i^2 = 1 + 4 + 9 + \cdots + n^2 = \frac{n(n + 1)(2n + 1)}{6}.$$

Hence, we can write

$$S_{xy} = n \sum_{i=1}^{n} iD_i - n(n + 1)/2 \sum_{i=1}^{n} D_i,$$

$$S_{xx} = \frac{n^2(n + 1)(2n + 1)}{6} - \frac{n^2(n + 1)^2}{4}.$$

APPENDIX 2–C   GLOSSARY OF NOTATION FOR CHAPTER 2

a = Estimate of the intercept in regression analysis.

α = Smoothing constant used for single exponential smoothing. One of the smoothing constants used for Holt's method or one of the smoothing constants used for Winters's method.

b = Estimate of the slope in regression analysis.

β = Second smoothing constant used for either Holt's method or Winters's method.

c_t = Seasonal factor for the tth period of a season.

γ = Third smoothing constant used for Winters's method.

D_t = Demand in period t. Refers to the series whose values are to be forecasted.

$e_t = F_t - D_t$ = (Observed) forecasting error in period t.

ϵ_t = Random variable representing the random component of the demand.

F_t = One-step-ahead forecast made in period $t - 1$ for demand in period t.

$F_{t, t+\tau}$ = τ-step-ahead forecast made in period t for the demand in period $t + \tau$.

G_t = Smoothed value of the slope for Holt's and Winters's methods.

μ = Mean of the demand process.

MAD = Mean absolute deviation = $(1/n) \sum_{i=1}^{n} |e_i|$.

MAPE = Mean absolute percentage error = $(1/n) \sum_{i=1}^{n} |e_i/D_i| \times 100$.

MSE = Mean squared error = $(1/n) \sum_{i=1}^{n} e_i^2$.

S_t = Smoothed value of the series (intercept) for Holt's and Winters's methods.

σ^2 = Variance of the demand process.

T_t = Value of the tracking signal in period t (refer to Problem 48).

Bibliography

Armstrong, J. S. "Forecasting by Extrapolation: Conclusions from Twenty-Five Years of Research." *Interfaces* 14 (1984), pp. 52–66.

Box, G. E. P., and G. M. Jenkins. *Time Series Analysis, Forecasting, and Control.* San Francisco: Holden Day, 1970.

Brown, R. G. *Statistical Forecasting for Inventory Control.* New York: McGraw-Hill, 1959.

Brown, R. G. *Smoothing, Forecasting, and Prediction of Discrete Time Series.* Englewood Cliffs, NJ: Prentice Hall, 1962.

Chambers, J. C.; S. K. Mullick; and D. D. Smith. "How to Choose the Right Forecasting Technique." *Harvard Business Review* 65 (1971), pp. 45–74.

Davenport, W. B., and W. L. Root. *An Introduction to the Theory of Random Signals and Noise.* New York: McGraw-Hill, 1958.

Draper, N. R., and H. Smith. *Applied Regression Analysis.* New York: John Wiley & Sons, 1968.

Fisher, M. L.; J. H. Hammond; W. R. Obermeyer; and A. Raman, "Making Supply Meet Demand in an Uncertain World." *Harvard Business Review,* May–June 1994, pp. 221–40.

Gross, C. W., and R. T. Peterson. *Business Forecasting.* 2nd ed. New York: John Wiley & Sons, 1983.

Helmer, O., and N. Rescher. "On the Epistemology of the Inexact Sciences." *Management Science* 6 (1959), pp. 25–52.

Holt, C. C. "Forecasting Seasonal and Trends by Exponentially Weighted Moving Averages." Office of Naval Research Memorandum No. 52, 1957.

Kalman, R. E. "A New Approach to Linear Filtering and Prediction Problems." *Journal of Basic Engineering,* Ser. D 82, 1960, pp. 35–44.

Makridakis, S.; S. C. Wheelwright; and V. E. McGhee. *Forecasting: Methods and Applications.* 2nd ed. New York: John Wiley & Sons, 1983.

Makridakis, S., and R. L. Winkler. "Averages of Forecasts." *Management Science* 29 (1983), pp. 987–96.

McWilliams, G. "At Compaq, a Desktop Crystal Ball." *Business Week,* March 20, 1995, pp. 96–97.

Montgomery, D. C., and L. A. Johnson. *Forecasting and Time Series Analysis.* New York: McGraw-Hill, 1976.

Muth, J. F. "Optimal Properties of Exponentially Weighted Forecasts of Time Series with Permanent and Transitory Components." *Journal of the American Statistical Association* 55 (1960), pp. 299–306.

Nahmias, S. "Demand Estimation in Lost Sales Inventory Systems." *Naval Research Logistics* 41 (1994), pp. 739–57.

Nelson, C. R. *Applied Time Series Analysis for Managerial Forecasting.* San Francisco: Holden Day, 1973.

Trigg, D. W. "Monitoring a Forecasting System." *Operational Research Quarterly* 15 (1964), pp. 271–74.

Trigg, D. W., and A. G. Leach. "Exponential Smoothing with Adaptive Response Rate." *Operational Research Quarterly* 18 (1967), pp. 53–59.

Wiener, N. *Extrapolation, Interpolation, and Smoothing of Stationary Time Series*. Cambridge, MA: MIT Press, 1949.

Wilson, J. H., and B. Keating. *Business Forecasting*. Homewood, IL: Richard D. Irwin, 1990.

Winters, P. R. "Forecasting Sales by Exponentially Weighted Moving Averages." *Management Science* 6 (1960), pp. 324–42.

Yule, G. U. "Why Do We Sometimes Get Nonsense Correlations between Time Series? A Study of Sampling and the Nature of Time Series." *Journal of the Royal Statistical Society* 89 (1926), pp. 1–64.

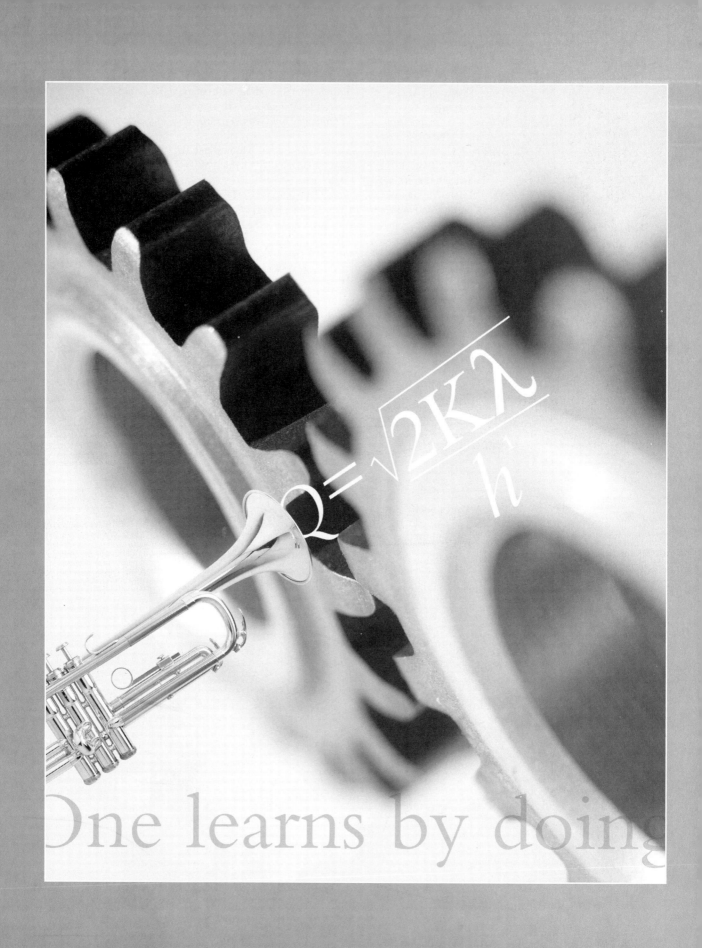

$$Q = \sqrt{\frac{2K\lambda}{h}}$$

One learns by doing

CHAPTER 3

AGGREGATE PLANNING

Aggregate planning, which might also be called macro production planning, addresses the problem of deciding how many employees the firm should retain and, for a manufacturing firm, the quantity and the mix of products to be produced. Macro planning is not limited to manufacturing firms. Service organizations must determine employee staffing needs as well. For example, airlines must plan staffing levels for flight attendants and pilots, and hospitals must plan staffing levels for nurses. Macro planning strategies are a fundamental part of the firm's overall business strategy. Some firms operate on the philosophy that costs can be controlled only by making frequent changes in the size and/or composition of the workforce. The aerospace industry in California in the 1970s adopted this strategy. As government contracts shifted from one producer to another, so did the technical workforce. Other firms have a reputation for retaining employees, even in bad times. Until recently, IBM and AT&T were two well-known examples.

Whether a firm provides a service or produces a product, macro planning begins with the forecast of demand. Techniques for demand forecasting were presented in Chapter 2. How responsive the firm can be to anticipated changes in the demand depends on several factors. These factors include the general strategy the firm may have regarding retaining workers and its commitments to existing employees.

As noted in Chapter 2, demand forecasts are generally wrong because there is almost always a random component of the demand that cannot be predicted exactly in advance. The aggregate planning methodology discussed in this chapter requires the assumption that demand is deterministic, or known in advance. This assumption is made to simplify the analysis and allow us to focus on the systematic or predictable changes in the demand pattern, rather than on the unsystematic or random changes. Inventory management subject to randomness is treated in detail in Chapter 5.

Traditionally, most manufacturers have chosen to retain primary production in house. Some components might be purchased from outside suppliers (see the discussion of the make-or-buy problem in Chapter 1), but the primary product is traditionally produced by the firm. Henry Ford was one of the first American manufacturers to design a completely vertically integrated business. Ford even owned a stand of rubber trees so it would not have to purchase rubber for tires.

That philosophy is undergoing a dramatic change, however. In dynamic environments, firms are finding that they can be more flexible if the manufacturing is outsourced; that is, if it is done on a subcontract basis. One example is Sun Microsystems, a California-based producer of computer workstations. Sun, a market leader, adopted the strategy of focusing on product innovation and design rather than on manufacturing. They have developed close ties to contract manufacturers such as San Jose–based Solectron Corporation, winner of the Baldrige Award for Quality. Subcontracting its primary manufacturing function has allowed Sun to be more flexible and to focus on innovation in a rapidly changing market.

Aggregate planning involves competing objectives. One objective is to react quickly to anticipated changes in demand, which would require making frequent and potentially large changes in the size of the labor force. Such a strategy has been called a chase strategy. This may be cost effective, but could be a poor long-run business strategy. Workers who are laid off may not be available when business turns around. For this reason, the firm may wish to adopt the objective of retaining a stable workforce. However, this strategy often results in large buildups of inventory during periods of low demand. Service firms may incur substantial debt to meet payrolls in slow periods. A third objective is to develop a production plan for the firm that maximizes profit over the planning horizon subject to constraints on capacity. When profit maximization is the primary objective, explicit costs of making changes must be factored into the decision process.

Aggregate planning methodology is designed to translate demand forecasts into a blueprint for planning staffing and production levels for the firm over a predetermined planning horizon. Aggregate planning methodology is not limited to top-level planning. Although generally considered to be a macro planning tool for determining overall workforce and production levels, large companies may find aggregate planning useful at the plant level as well. Production planning may be viewed as a hierarchical process in which purchasing, production, and staffing decisions must be made at several levels in the firm. Aggregate planning methods may be applied at almost any level, although the concept is one of managing groups of items rather than single items. The inventory planning methods discussed in Chapters 4 and 5 are geared toward single-item control.

This chapter reviews several techniques for determining aggregate plans. Some of these are heuristic (i.e., approximate) and some are optimal. We hope to convey to the reader an understanding of the issues involved in aggregate planning, a knowledge of the basic tools available for providing solutions, and an appreciation of the difficulties associated with implementing aggregate plans in the real world.

3.1 Aggregate Units of Production

The aggregate planning approach is predicated on the existence of an aggregate unit of production. When the types of items produced are similar, an aggregate production unit can correspond to an "average" item, but if many different types of items are produced, it would be more appropriate to consider aggregate units in terms of weight (tons of steel), volume (gallons of gasoline), amount of work required (worker-years of programming time), or dollar value (value of inventory in dollars). What the appropriate aggregating scheme should be is not always obvious. It depends on the context of the particular planning problem and the level of aggregation required.

Example 3.1

A plant manager working for a large national appliance firm is considering implementing an aggregate planning system to determine the workforce and production levels in his plant. This particular plant produces six models of washing machines. The characteristics of the machines are

Model Number	Number of Worker-Hours Required to Produce	Selling Price
A5532	4.2	$285
K4242	4.9	345
L9898	5.1	395
L3800	5.2	425
M2624	5.4	525
M3880	5.8	725

The plant manager must decide on the particular aggregation scheme to use. One possibility is to define an aggregate unit as one dollar of output. Unfortunately, the selling prices of the various models of washing machines are not consistent with the number of worker-hours required to produce them. The ratio of the selling price divided by the worker-hours is $67.86 for A5532 and $125.00 for M3880. (The company bases its pricing on the fact that the less expensive models have a higher sales volume.) The manager notices that the percentages of the total number of sales for these six models have been fairly constant, with values of 32 percent for A5532, 21 percent for K4242, 17 percent for L9898, 14 percent for L3800, 10 percent for M2624, and 6 percent for M3880. He decides to define an aggregate unit of production as a fictitious washing machine requiring $(.32)(4.2) + (.21)(4.9) + (.17)(5.1) + (.14)(5.2) + (.10)(5.4) + (.06)(5.8) = 4.856$ hours of labor time. He can obtain sales forecasts for aggregate production units in essentially the same way by multiplying the appropriate fractions by the forecasts for unit sales of each type of machine. ∎

The approach used by the plant manager in Example 3.1 was possible because of the relative similarity of the products produced. However, defining an aggregate unit of production at a higher level of the firm is more difficult. In cases in which the firm produces a large variety of products, a natural aggregate unit is sales dollars. Although, as we saw in the example, this will not necessarily translate to the same number of units of production for each item, it will generally provide a good approximation for planning at the highest level of a firm that produces a diverse product line.

Aggregate planning (and the associated problem of disaggregating the aggregate plans or converting them into detailed master schedules) is closely related to hierarchical production planning (HPP) championed by Hax and Meal (1975). HPP considers workforce sizes and production rates at a variety of levels of the firm as opposed to simply the top level, as in aggregate planning. For aggregate planning purposes, Hax and Meal recommend the following hierarchy:

1. *Items.* These are the final products to be delivered to the customer. An item is often referred to as an SKU (for stockkeeping unit) and represents the finest level of detail in the product structure.

2. *Families.* These are defined as a group of items that share a common manufacturing setup cost.

3. *Types.* Types are groups of families with production quantities that are determined by a single aggregate production plan.

FIGURE 3–1

The hierarchy of production planning decisions.

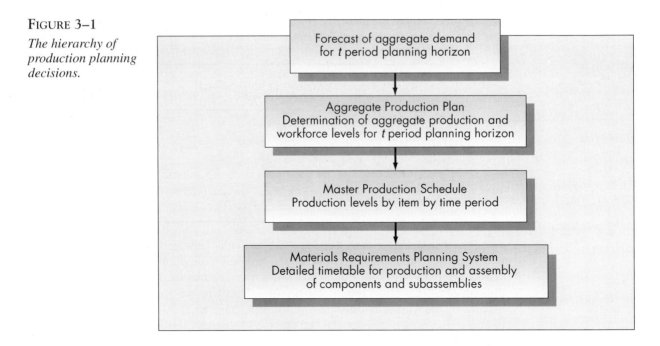

In the example above, items would correspond to individual models of washing machines. A family might be all washing machines, and a type might be large appliances. The Hax-Meal aggregation scheme will not necessarily work in every situation. In general, the aggregation method should be consistent with the firm's organizational structure and product line.

In Figure 3–1 we present a schematic of the aggregate planning function and its place in the hierarchy of production planning decisions.

3.2 Overview of the Aggregate Planning Problem

Having now defined the appropriate aggregate unit for the level of the firm for which an aggregate plan is to be determined, we assume that there exists a forecast of the demand for a specified planning horizon, expressed in terms of aggregate production units. Let $D_1, D_2, \ldots, D_T$ be the demand forecasts for the next T planning periods. In most applications, a planning period is a month, although aggregate plans can be developed for other periods of time, such as weeks, quarters, or years. An important feature of aggregate planning is that the demands are treated as known constants (that is, the forecast error is assumed to be zero). The reasons for making this assumption will be discussed below.

The goal of aggregate planning is to determine aggregate production quantities and the levels of resources required to achieve these production goals. In practice, this translates to finding the number of workers that should be employed and the number of aggregate units to be produced in each of the planning periods 1, 2, . . . , T. The objective of aggregate planning is to balance the advantages of producing to meet demand as closely as possible against the disruptions caused by changing the levels of production and/or the workforce levels.

The primary issues related to the aggregate planning problem include

1. *Smoothing.* Smoothing refers to costs that result from changing production and workforce levels from one period to the next. Two of the key components of smoothing costs are the costs that result from hiring and firing workers. Aggregate planning methodology requires the specification of these costs, which may be difficult to estimate. Firing workers could have far-reaching consequences and costs that may be difficult to evaluate. Firms that hire and fire frequently develop a poor public image. This could adversely affect sales and discourage potential employees from joining the company. Furthermore, workers that are laid off might not simply wait around for business to pick up. Firing workers can have a detrimental effect on the future size of the labor force if those workers obtain employment in other industries. Finally, most companies are simply not at liberty to hire and fire at will. Labor agreements restrict the freedom of management to freely alter workforce levels. However, it is still valuable for management to be aware of the cost trade-offs associated with varying workforce levels and the attendant savings in inventory costs.

2. *Bottleneck problems.* We use the term *bottleneck* to refer to the inability of the system to respond to sudden changes in demand as a result of capacity restrictions. For example, a bottleneck could arise when the forecast for demand in one month is unusually high, and the plant does not have sufficient capacity to meet that demand. A breakdown of a vital piece of equipment also could result in a bottleneck.

3. *Planning horizon.* The number of periods for which the demand is to be forecasted, and hence the number of periods for which workforce and inventory levels are to be determined, must be specified in advance. The choice of the value of the forecast horizon, T, can be significant in determining the usefulness of the aggregate plan. If T is too small, then current production levels might not be adequate for meeting the demand beyond the horizon length. If T is too large, it is likely that the forecasts far into the future will prove inaccurate. If future demands turn out to be very different from the forecasts, then current decisions indicated by the aggregate plan could be incorrect. Another issue involving the planning horizon is the *end-of-horizon* effect. For example, the aggregate plan might recommend that the inventory at the end of the horizon be drawn to zero in order to minimize holding costs. This could be a poor strategy, especially if demand increases at that time. (However, this particular problem can be avoided by adding a constraint specifying minimum ending inventory levels.)

In practice, rolling schedules are almost always used. This means that at the time of the next decision, a new forecast of demand is appended to the former forecasts and old forecasts might be revised to reflect new information. The new aggregate plan may recommend different production and workforce levels for the current period than were recommended one period ago. When only the decisions for the current planning period need to be implemented immediately, the schedule should be viewed as dynamic rather than static.

Although rolling schedules are common, it is possible that because of production lead times, the schedule must be frozen for a certain number of planning periods. This means that decisions over some collection of future periods cannot be altered. The most direct means of dealing with frozen horizons is simply to label as period 1 the first period in which decisions are not frozen.

4. *Treatment of demand.* As noted above, aggregate planning methodology requires the assumption that demand is known with certainty. This is simultaneously a weakness and a strength of the approach. It is a weakness because it ignores the

possibility (and, in fact, likelihood) of forecast errors. As noted in the discussion of forecasting techniques in Chapter 2, it is virtually a certainty that demand forecasts are wrong. Aggregate planning does not provide any buffer against unanticipated forecast errors. However, most inventory models that allow for random demand require that the average demand be constant over time. Aggregate planning allows the manager to focus on the systematic changes that are generally not present in models that assume random demand. By assuming deterministic demand, the effects of seasonal fluctuations and business cycles can be incorporated into the planning function.

3.3 Costs in Aggregate Planning

As with most of the optimization problems considered in production management, the goal of the analysis is to choose the aggregate plan that minimizes cost. It is important to identify and measure those specific costs that are affected by the planning decision.

1. *Smoothing costs.* Smoothing costs are those costs that accrue as a result of changing the production levels from one period to the next. In the aggregate planning context, the most salient smoothing cost is the cost of changing the size of the workforce. Increasing the size of the workforce requires time and expense to advertise positions, interview prospective employees, and train new hires. Decreasing the size of the workforce means that workers must be laid off. Severance pay is thus one cost of decreasing the size of the workforce. Other costs, somewhat harder to measure, are (*a*) the costs of a decline in worker morale that may result and (*b*) the potential for decreasing the size of the labor pool in the future, as workers who are laid off acquire jobs with other firms or in other industries.

Most of the models that we consider assume that the costs of increasing and decreasing the size of the workforce are linear functions of the number of employees that are hired or fired. That is, there is a constant dollar amount charged for each employee hired or fired. The assumption of linearity is probably reasonable up to a point. As the supply of labor becomes scarce, there may be additional costs required to hire more workers, and the costs of laying off workers may go up substantially if the number of workers laid off is too large. A typical cost function for changing the size of the workforce appears in Figure 3–2.

2. *Holding costs.* Holding costs are the costs that accrue as a result of having capital tied up in inventory. If the firm can decrease its inventory, the money saved could be invested elsewhere with a return that will vary with the industry and with the specific company. (A more complete discussion of holding costs is deferred to the next chapter.) Holding costs are almost always assumed to be linear in the number of units being held at a particular point in time. We will assume for the purposes of the aggregate planning analysis that the holding cost is expressed in terms of dollars per unit held per planning period. We also will assume that holding costs are charged against the inventory remaining on hand at the *end* of the planning period. This assumption is made for convenience only. Holding costs could be charged against starting inventory or average inventory as well.

3. *Shortage costs.* Holding costs are charged against the aggregate inventory as long as it is positive. In some situations it may be necessary to incur shortages, which are represented by a negative level of inventory. Shortages can occur when forecasted demand exceeds the capacity of the production facility or when demands are higher than anticipated. For the purposes of aggregate planning, it is generally assumed that

FIGURE 3–2

Cost of changing the size of the workforce.

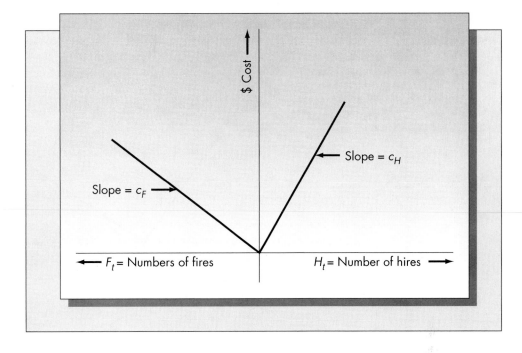

excess demand is backlogged and filled in a future period. In a highly competitive situation, however, it is possible that excess demand is lost and the customer goes elsewhere. This case, which is known as lost sales, is more appropriate in the management of single items and is more common in a retail than in a manufacturing context.

As with holding costs, shortage costs are generally assumed to be linear. Convex functions also can accurately describe shortage costs, but linear functions seem to be the most common. Figure 3–3 shows a typical holding/shortage cost function.

4. *Regular time costs.* These costs involve the cost of producing one unit of output during regular working hours. Included in this category are the actual payroll costs of regular employees working on regular time, the direct and indirect costs of materials, and other manufacturing expenses. When all production is carried out on regular time, regular payroll costs become a "sunk cost," because the number of units produced must equal the number of units demanded over any planning horizon of sufficient length. If there is no overtime or worker idle time, regular payroll costs do not have to be included in the evaluation of different strategies.

5. *Overtime and subcontracting costs.* Overtime and subcontracting costs are the costs of production of units not produced on regular time. Overtime refers to production by regular-time employees beyond the normal work day, and subcontracting refers to the production of items by an outside supplier. Again, it is generally assumed that both of these costs are linear.

6. *Idle time costs.* The complete formulation of the aggregate planning problem also includes a cost for underutilization of the workforce, or idle time. In most contexts, the idle time cost is zero, as the direct costs of idle time would be taken into account in labor costs and lower production levels. However, idle time could have other consequences for the firm. For example, if the aggregate units are input to another process, idle time on the line could result in higher costs to the subsequent process. In such cases, one would explicitly include a positive idle cost.

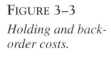

FIGURE 3–3

Holding and back-order costs.

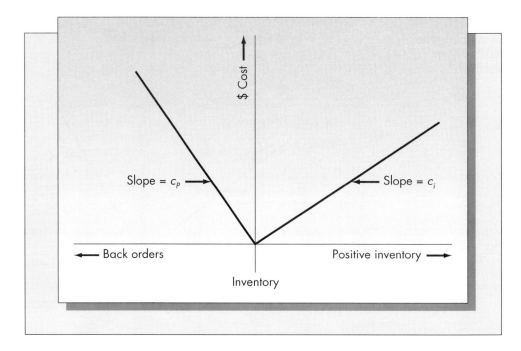

When planning is done at a relatively high level of the firm, the effects of intangible factors are more pronounced. Any solution to the aggregate planning problem obtained from a cost-based model must be considered carefully in the context of company policy. An optimal solution to a mathematical model might result in a policy that requires frequent hiring and firing of personnel. Such a policy may be infeasible because of prior contract agreements, or undesirable because of the potential negative effects on the firm's public image.

Problems for Sections 1–3

1. What does the term *aggregate unit of production* mean? Do aggregate production units always correspond to actual items? Do they ever? Discuss.
2. What is the aggregation scheme recommended by Hax and Meal?
3. Discuss the following terms and their relationship to the aggregate planning problem:
   a. Smoothing.
   b. Bottlenecks.
   c. Capacity.
   d. Planning horizon.
4. A local machine shop employs 60 workers who have a variety of skills. The shop accepts one-time orders and also maintains a number of regular clients. Discuss some of the difficulties with using the aggregate planning methodology in this context.
5. A large manufacturer of household consumer goods is considering integrating an aggregate planning model into its manufacturing strategy. Two of the company vice presidents disagree strongly as to the value of the approach. What arguments might each of the vice presidents use to support his or her point of view?

6. Describe the following costs and discuss the difficulties that arise in attempting to measure them in a real operating environment.
   a. Smoothing costs.
   b. Holding costs.
   c. Payroll costs.
7. Discuss the following statement: "Since we use a rolling production schedule, I really don't need to know the demand beyond next month."
8. St. Clair County Hospital is attempting to assess its needs for nurses over the coming four months (January to April). The need for nurses depends on both the numbers and the types of patients in the hospital. Based on a study conducted by consultants, the hospital has determined that the following ratios of nurses to patients are required:

| | | Patient Forecasts | | | |
Patient Type	Numbers of Nurses Required per Patient	Jan.	Feb.	Mar.	Apr.
Major surgery	.4	28	21	16	18
Minor surgery	.1	12	25	45	32
Maternity	.5	22	43	90	26
Critical care	.6	75	45	60	30
Other	.3	80	94	73	77

   a. How many nurses should be working each month to most closely match patient forecasts?
   b. Suppose the hospital does not want to change its policy of not increasing the nursing staff size by more than 10 percent in any month. Suggest a schedule of nurse staffing over the four months that meets this requirement and also meets the need for nurses each month.

3.4 A Prototype Problem

One can obtain adequate solutions for many aggregate planning problems by hand or by using relatively straightforward graphical techniques. Linear programming is a means of obtaining (nearly) optimal solutions. We illustrate the different solution techniques with the following example.

Example 3.2

Densepack is to plan workforce and production levels for the six-month period January to June. The firm produces a line of disk drives for mainframe computers that are plug compatible with several computers produced by major manufacturers. Forecast demands over the next six months for a particular line of drives produced in the Milpitas, California, plant are 1,280, 640, 900, 1,200, 2,000, and 1,400. There are currently (end of December) 300 workers employed in the Milpitas plant. Ending inventory in December is expected to be 500 units, and the firm would like to have 600 units on hand at the end of June.

There are several ways to incorporate the starting and the ending inventory constraints into the formulation. The most convenient is simply to modify the values of the predicted demand. Define net predicted demand in period 1 as the predicted demand minus initial inventory. If there is a minimum ending inventory constraint, then this amount should be added to the demand in period T. Minimum buffer inventories also can be handled by modifying the predicted demand. If there is a minimum buffer inventory in every period, this amount should be added to the first period's demand. If there is a minimum buffer inventory in only one period, this amount should be added to that period's demand and subtracted from the next period's demand. Actual ending inventories should be computed using the original demand pattern, however.

Returning to Example 3.2, we define the net predicted demand for January as 780 (1,280 − 500) and the net predicted demand for June as 2,000 (1,400 + 600). By considering net demand, we may make the simplifying assumption that starting and ending inventory are both zero. The net predicted demand and the net cumulative demand for the six months January to June are as follows:

Month	Net Predicted Demand	Net Cumulative Demand
January	780	780
February	640	1,420
March	900	2,320
April	1,200	3,520
May	2,000	5,520
June	2,000	7,520

The cumulative net demand is pictured in Figure 3–4. A production plan is the specification of the production levels for each month. If shortages are not permitted,

FIGURE 3–4

A feasible aggregate plan for Densepack

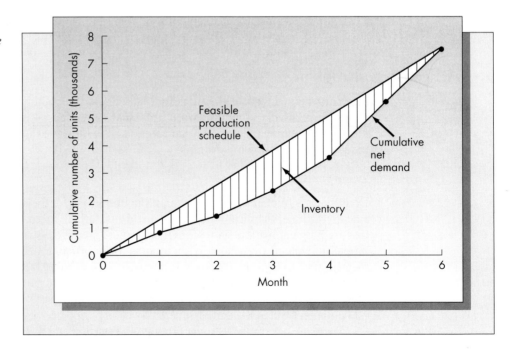

then cumulative production must be at least as great as cumulative demand each period. In addition to the cumulative net demand, Figure 3–4 also shows one feasible production plan.

In order to illustrate the cost trade-offs of various production plans, we will assume in the example that there are only three costs to be considered: cost of hiring workers, cost of firing workers, and cost of holding inventory. Define

C_H = Cost of hiring one worker = \$500,
C_F = Cost of firing one worker = \$1,000,
C_I = Cost of holding one unit of inventory for one month = \$80.

We require a means of translating aggregate production in units to workforce levels. Because not all months have an equal number of working days, we will use a day as an indivisible unit of measure and define

K = Number of aggregate units produced by one worker in one day.

In the past, the plant manager observed that over 22 working days, with the workforce level constant at 76 workers, the firm produced 245 disk drives. That means that on average the production rate was $245/22 = 11.1364$ drives per day when there were 76 workers employed at the plant. It follows that one worker produced an average of $11.1364/76 = .14653$ drive in one day. Hence, $K = .14653$ for this example. ∎

We will evaluate two alternative plans for managing the workforce that represent two essentially opposite management strategies. Plan 1 is to change the workforce each month in order to produce enough units to most closely match the demand pattern. This is known as a *zero inventory plan*. Plan 2 is to maintain the minimum constant workforce necessary to satisfy the net demand. This is known as the *constant workforce plan*.

Evaluation of a Chase Strategy (Zero Inventory Plan)

In this section we will develop a production plan for Densepack that minimizes the levels of inventory the firm must hold during the six-month planning horizon. Table 3–1 summarizes the input information for the calculations and shows the minimum number of workers required in each month.

TABLE 3–1   **Initial Calculations for Zero Inventory Plan for Densepack**

A	B	C	D	E
Month	Number of Working Days	Number of Units Produced per Worker (B × .14653)	Forecast Net Demand	Minimum Number of Workers Required (D/C rounded up)
January	20	2.931	780	267
February	24	3.517	640	182
March	18	2.638	900	342
April	26	3.810	1,200	315
May	22	3.224	2,000	621
June	15	2.198	2,000	910

TABLE 3–2   **Zero Inventory Aggregate Plan for Densepack**

A	B	C	D	E	F	G	H	I
Month	Number of Workers	Number Hired	Number Fired	Number of Units per Worker	Number of Units Produced (B × E)	Cumulative Production	Cumulative Demand	Ending Inventory (G − H)
January	267		33	2.931	783	783	780	3
February	182		85	3.517	640	1,423	1,420	3
March	342	160		2.638	902	2,325	2,320	5
April	315		27	3.810	1,200	3,525	3,520	5
May	621	306		3.224	2,002	5,527	5,520	7
June	910	289		2.198	2,000	7,527	7,520	7
Totals		755	145					30

One obtains the entries in the final column of Table 3–1, the minimum number of workers required each month, by dividing the forecasted net demand by the number of units produced per worker. The value of this ratio is then rounded *upward* to the next higher integer. We must round upward to guarantee that shortages do not occur. As an example, consider the month of January. Forming the ratio 780/2.931 gives 266.12, which is rounded up to 267 workers. The number of working days each month depends upon a variety of factors, such as paid holidays and worker schedules. The reduced number of days in June is due to a planned shutdown of the plant in the last week of June.

Recall that the number of workers employed at the end of December is 300. Hiring and firing workers each month to match forecast demand as closely as possible results in the aggregate plan given in Table 3–2.

The number of units produced each month (column F in Table 3–2) is obtained by the following formula:

$$\text{Number of units produced} = \text{Number of workers} \times \begin{array}{c}\text{Average number of}\\ \text{aggregate units produced in}\\ \text{a month by a single worker}\end{array}$$

rounded to the nearest integer.

The total cost of this production plan is obtained by multiplying the totals at the bottom of Table 3–2 by the appropriate costs. For this example, the total cost of hiring, firing, and holding is (755)(500) + (145)(1,000) + (30)(80) = $524,900. This cost must now be adjusted to include the cost of holding for the ending inventory of 600 units, which was netted out of the demand for June. Hence, the total cost of this plan is 524,900 + (600)(80) = $572,900. Note that the initial inventory of 500 units does not enter into the calculations because it will be netted out during the month of January.

It is usually impossible to achieve zero inventory at the end of each planning period because it is not possible to employ a fractional number of workers. For this reason, there will almost always be some inventory remaining at the end of each period in addition to the inventory required to be on hand at the end of the planning horizon.

It is possible that ending inventory in one or more periods could build up to a point where the size of the workforce could be reduced by one or more workers. In this example there is sufficient inventory on hand to reduce the workforce by one worker in the months of both March and May. Check that the resulting plan hires a total of 753 workers and fires a total of 144 workers and has a total of only 13 units of inventory. The cost of this modified plan comes to $569,540.

Evaluation of the Constant Workforce Plan

Now assume that the goal is to eliminate completely the need for hiring and firing during the planning horizon. In order to guarantee that shortages do not occur in any period, it is necessary to compute the minimum workforce required for *every* month in the planning horizon. For January, the net cumulative demand is 780 and there are 2.931 units produced per worker, resulting in a minimum workforce of 267 in January. There are exactly $2.931 + 3.517 = 6.448$ units produced per worker in January and February combined, which have a cumulative demand of 1,420. Hence, $1,420/6.448 = 220.22 \approx 221$ workers are required to cover both January and February. Continuing to form the ratios of the cumulative net demand and the cumulative number of units produced per worker for each month in the horizon results in Table 3–3.

The minimum number of workers required for the entire six-month planning period is the maximum entry in column D in Table 3–3, which is 411 workers. It is only a coincidence that the maximum ratio occurred in the final period.

Because there are 300 workers employed at the end of December, the constant workforce plan requires hiring 111 workers at the beginning of January. No further hiring and firing of workers are required. The inventory levels that result from a constant workforce of 411 workers appear in Table 3–4. The monthly production levels in column C of the table are obtained by multiplying the number of units produced per worker each month by the fixed workforce size of 411 workers. The total of the ending inventory levels is $5,962 + 600 = 6,562$. (Recall the 600 units that were netted out

---

TABLE 3–3 **Computation of the Minimum Workforce Required by Densepack**

A	B	C	D
Month	Cumulative Net Demand	Cumulative Number of Units Produced per Worker	Ratio B/C (rounded up)
January	780	2.931	267
February	1,420	6.448	221
March	2,320	9.086	256
April	3,520	12.896	273
May	5,520	16.120	343
June	7,520	18.318	411

TABLE 3–4   **Inventory Levels for Constant Workforce Schedule**

A	B	C	D	E	F
Month	Number of Units Produced per Worker	Monthly Production (B × 411)	Cumulative Production	Cumulative Net Demand	Ending Inventory (D − E)
January	2.931	1,205	1,205	780	425
February	3.517	1,445	2,650	1,420	1,230
March	2.638	1,084	3,734	2,320	1,414
April	3.810	1,566	5,300	3,520	1,780
May	3.224	1,325	6,625	5,520	1,105
June	2.198	903	7,528	7,520	8
Total					5,962

of the demand for June.) Hence the total inventory cost of this plan is $(6,562)(80) =$ $524,960. To this we add the cost of increasing the workforce from 300 to 411 in January, which is $(111)(500) = $55,500$, giving a total cost of this plan of $580,460. This is somewhat higher than the cost of the zero inventory plan, which was $569,540. However, because costs of the two plans are close, it is likely that the company would prefer the constant workforce plan in order to avoid any unaccounted for costs of making frequent changes in the workforce.

Mixed Strategies and Additional Constraints

The zero inventory plan and constant workforce strategies treated above are pure strategies: they are designed to achieve one objective. With more flexibility, small modifications can result in dramatically lower costs. One might question the interest in manual calculations considering that aggregate planning problems can be formulated and solved optimally by linear programming.

Manual calculations enhance intuition and understanding. Computers are dumb. It is easy to overlook a critical constraint or objective when using a computer. It's important to have a feel for the right solution before solving a problem on a computer, so that glaring mistakes are obvious. An important skill, largely ignored these days, is being able to do a ballpark calculation in one's head before pulling out the calculator or computer.

Figure 3–4 shows the constant workforce strategy for Densepack. The hatched area represents the inventory carried in each month. Suppose we allow a single change in the production rate during the six months. Can you identify a strategy from the figure that substantially reduces inventory without permitting shortages?

Graphically, the problem is to cover the cumulative net demand curve with two straight lines, rather than one straight line. This can be accomplished by driving the net inventory to zero at the end of period 4 (April). To do so, we need to produce enough in each of the months January through April to meet the cumulative net demand each month. That means we need to produce $3,520/4 = 880$ units in each of the first four months. In Figure 3–4, the line connecting the origin to the cumulative net demand in April lies wholly above the cumulative net demand curve for the prior months. If the graph is accurate, that means that there should be no shortages occurring in these months. The May and June production is then set to 2,000, exactly matching the net demand in these months. With this policy we obtain

Month	Cumulative Net Demand	Cumulative Production
January	780	880
February	1,420	1,760
March	2,320	2,640
April	3,520	3,520
May	5,520	5,520
June	7,520	7,520

As we will see in the next section, this policy turns out to be optimal for the Densepack problem.

The graphical solution method also can be used when additional constraints are present. For example, suppose that the production capacity of the plant is only 1,800 units per month. Then the policy above is infeasible in May and June. In this case the constraint means that the slope of the cumulative production curve is bounded by 1,800. One solution in this case would be to produce 980 in each of the first four months and 1,800 units in each of the last two months. Another constraint might be that the maximum change from one month to the next be no more than 750 units. Suggest a production plan to meet this constraint.

These are a few examples of constraints that might arise in using aggregate planning methodology. As the constraints become more complex, finding good solutions graphically becomes more difficult. Fortunately, most constraints of this nature can be incorporated easily into the linear programming formulations of aggregate planning problems.

Problems for Section 4

9. Harold Grey owns a small farm in the Salinas Valley that grows apricots. The apricots are dried on the premises and sold to a number of large supermarket chains. Based on past experience and committed contracts, he estimates that sales over the next five years in thousands of packages will be as follows:

Year	Forecasted Demand (thousands of packages)
1	300
2	120
3	200
4	110
5	135

Assume that each worker stays on the job for at least one year, and that Grey currently has three workers on the payroll. He estimates that he will have 20,000 packages on hand at the end of the current year. Assume that, on the average, each worker is paid $25,000 per year and is responsible for producing 30,000 packages. Inventory costs have been estimated to be 4 cents per package per year, and shortages are not allowed.

Based on the effort of interviewing and training new workers, Farmer Grey estimates that it costs $500 for each worker hired. Severance pay amounts to $1,000 per worker.

   a. Assuming that shortages are not allowed, determine the minimum constant workforce that he will need over the next five years.

   b. Evaluate the cost of the plan found in part (*a*).

10. For the data given in Problem 9, graph the cumulative net demand.

   a. Graphically determine a production plan that changes the production rate exactly once during the five years, and evaluate the cost of that plan.

   b. Graphically determine a production plan that changes the production rate exactly twice during the five years, and evaluate the cost of that plan.

11. An implicit assumption made in Problem 9 was that dried apricots unsold at the end of a year could be sold in subsequent years. Suppose that apricots unsold at the end of any year must be discarded. Assume a disposal cost of $0.20 per package. Resolve Problem 9 under these conditions.

12. The personnel department of the A & M Corporation wants to know how many workers will be needed each month for the next six-month production period. The following is a monthly demand forecast for the six-month period.

Month	*Forecasted Demand*
July	1,250
August	1,100
September	950
October	900
November	1,000
December	1,150

The inventory on hand at the end of June was 500 units. The company wants to maintain a minimum inventory of 300 units each month and would like to have 400 units on hand at the end of December. Each unit requires 5 employee-hours to produce, there are 20 working days each month, and each employee works an 8-hour day. The workforce at the end of June was 35 workers.

   a. Determine a minimum inventory production plan (i.e., one that allows arbitrary hiring and firing).

   b. Determine the production plan that meets demand but does not hire or fire workers during the six-month period.

13. Mr. Meadows Cookie Company makes a variety of chocolate chip cookies in the plant in Albion, Michigan. Based on orders received and forecasts of buying habits, it is estimated that the demand for the next four months is 850, 1,260, 510, and 980, expressed in thousands of cookies. During a 46-day period when there were 120 workers, the company produced 1.7 million cookies. Assume that the number of workdays over the four months are respectively 26, 24, 20, and 16. There are currently 100 workers employed, and there is no starting inventory of cookies.

    a. What is the minimum constant workforce required to meet demand over the next four months?

    b. Assume that c_I = 10 cents per cookie per month, c_H = $100, and c_F = $200. Evaluate the cost of the plan derived in (*a*).

14. A local semiconductor firm, Superchip, is planning its workforce and production levels over the next year. The firm makes a variety of microprocessors and uses sales dollars as its aggregate production measure. Based on orders received and sales forecasts provided by the marketing department, the estimate of dollar sales for the next year by month is as follows:

Month	Production Days	Predicted Demand (in $10,000)
January	22	340
February	16	380
March	21	220
April	19	100
May	23	490
June	20	625
July	24	375
August	12	310
September	19	175
October	22	145
November	20	120
December	16	165

    Inventory holding costs are based on a 25 percent annual interest charge. It is anticipated that there will be 675 workers on the payroll at the end of the current year and inventories will amount to $120,000. The firm would like to have at least $100,000 of inventory at the end of December next year. It is estimated that each worker accounts for an average of $60,000 of production per year (assume that one year consists of 250 working days). The cost of hiring a new worker is $200, and the cost of laying off a worker is $400.

    a. Determine the minimum constant workforce that will meet the predicted demand for the coming year.

    b. Evaluate the cost of the plan determined in part (*a*).

15. For the data in Problem 14, determine the cost of the plan that changes the workforce size each period to most closely match the demand.

16. Graph the cumulative net demand for the Superchip data of Problem 14. Graphically determine a production plan that changes production levels no more than three times, and determine the total cost of that plan.

3.5 Solution of Aggregate Planning Problems by Linear Programming

Linear programming is a term used to describe a general class of optimization problems. The objective is to determine values of *n* nonnegative real variables in order to maximize or minimize a linear function of these variables that is subject to *m* linear constraints of these variables.[1] The primary advantage in formulating a problem as a linear program is that optimal solutions can be found very efficiently by the simplex method.

When all cost functions are linear, there is a linear programming formulation of the general aggregate planning problem. Because of the efficiency of commercial linear programming codes, this means that (essentially) optimal solutions can be obtained for very large problems.[2]

Cost Parameters and Given Information

The following values are assumed to be known:

c_H = Cost of hiring one worker,
c_F = Cost of firing one worker,
c_I = Cost of holding one unit of stock for one period,
c_R = Cost of producing one unit on regular time,
c_O = Incremental cost of producing one unit on overtime,
c_U = Idle cost per unit of production,
c_S = Cost to subcontract one unit of production,
n_t = Number of production days in period t,
K = Number of aggregate units produced by one worker in one day,
I_0 = Initial inventory on hand at the start of the planning horizon,
W_0 = Initial workforce at the start of the planning horizon,
D_t = Forecast of demand in period t.

The cost parameters also may be time dependent; that is, they may change with t. Time-dependent cost parameters could be useful for modeling changes in the costs of hiring or firing due, for example, to shortages in the labor pool, or changes in the costs of production and/or storage due to shortages in the supply of resources, or changes in interest rates.

Problem Variables

The following are the problem variables:

W_t = Workforce level in period t,
P_t = Production level in period t,
I_t = Inventory level in period t,
H_t = Number of workers hired in period t,
F_t = Number of workers fired in period t,
O_t = Overtime production in units,
U_t = Worker idle time in units ("undertime"),
S_t = Number of units subcontracted from outside.

The overtime and idle time variables are determined in the following way. The term Kn_t represents the number of units produced by one worker in period t, so that Kn_tW_t would be the number of units produced by the entire workforce in period t.

---

[1] An overview of linear programming can be found in the supplement that follows this chapter.

[2] The qualifier is included because rounding may give suboptimal solutions. There will be more about this point later.

However, we do not require that $Kn_tW_t = P_t$. If $P_t > Kn_tW_t$, then the number of units produced exceeds what the workforce can produce on regular time. This means that the difference is being produced on overtime, so that the number of units produced on overtime is exactly $O_t = P_t - Kn_tW_t$. If $P_t < Kn_tW_t$, then the workforce is producing less than it should be on regular time, which means that there is worker idle time. The idle time is measured in units of production rather than in time, and is given by $U_t = Kn_tW_t - P_t$.

Problem Constraints

Three sets of constraints are required for the linear programming formulation. They are included to ensure that conservation of labor and conservation of units are satisfied.

1. Conservation of workforce constraints.

$$
\begin{array}{ccccccc}
W_t & = & W_{t-1} & + & H_t & - & F_t \\
\text{Number} & = & \text{Number} & + & \text{Number} & - & \text{Number} \\
\text{of workers} & & \text{of workers} & & \text{hired} & & \text{fired} \\
\text{in } t & & \text{in } t-1 & & \text{in } t & & \text{in } t
\end{array}
\qquad \text{for } 1 \le t \le T.
$$

2. Conservation of units constraints.

$$
\begin{array}{ccccccccc}
I_t & = & I_{t-1} & + & P_t & + & S_t & - & D_t \\
\text{Inventory} & = & \text{Inventory} & + & \text{Number} & + & \text{Number} & - & \text{Demand} \\
\text{in } t & & \text{in } t-1 & & \text{of units} & & \text{of units} & & \text{in } t \\
& & & & \text{produced} & & \text{subcontracted} \\
& & & & \text{in } t & & \text{in } t
\end{array}
\quad \text{for } 1 \le t \le T.
$$

3. Constraints relating production levels to workforce levels.

$$
\begin{array}{ccccccc}
Pt & = & Kn_tW_t & + & O_t & - & U_t \\
\text{Number} & = & \text{Number} & + & \text{Number} & - & \text{Number} \\
\text{of units} & & \text{of units} & & \text{of units} & & \text{of units} \\
\text{produced} & & \text{produced} & & \text{produced} & & \text{of idle} \\
\text{in } t & & \text{by regular} & & \text{on over-} & & \text{production} \\
& & \text{workforce} & & \text{time in } t & & \text{in } t. \\
& & \text{in } t
\end{array}
\qquad \text{for } 1 \le t \le T.
$$

In addition to these constraints, linear programming requires that all problem variables be nonnegative. These constraints and the nonnegativity constraints are the minimum that must be present in any formulation. Notice that (1), (2), and (3) constitute $3T$ constraints, rather than 3 constraints, where T is the length of the forecast horizon.

The formulation also requires specification of the initial inventory, I_0, and the initial workforce, W_0, and may include specification of the ending inventory in the final period, I_T.

The objective function includes all of the costs defined above. The linear programming formulation is to choose values of the problem variables W_t, P_t, I_t, H_t, F_t, O_t, U_t, and S_t to

$$
\text{Minimize } \sum_{t=1}^{T} (c_H H_t + c_F F_t + c_I I_t + c_R P_t + c_O O_t + c_U U_t + c_S S_t)
$$

subject to

$$W_t = W_{t-1} + H_t - F_t \qquad \text{for } 1 \le t \le T$$
$$\text{(conservation of workforce)}, \tag{A}$$

$$P_t = Kn_tW_t + O_t - U_t \qquad \text{for } 1 \le t \le T$$
$$\text{(production and workforce)} \tag{B}$$

$$I_t = I_{t-1} + P_t + S_t - D_t \qquad \text{for } 1 \le t \le T \text{ (inventory balance)}, \tag{C}$$

$$H_t, F_t, I_t, O_t, U_t, S_t, W_t, P_t \ge 0 \quad \text{(nonnegativity)}, \tag{D}$$

plus any additional constraints that define the values of starting inventory, starting workforce, ending inventory, or any other variables with values that are fixed in advance.

Rounding the Variables

In general, the optimal values of the problem variables will not be integers. However, fractional values for many of the variables do not make sense. These variables include the size of the workforce, the number of workers hired each period, and the number of workers fired each period, and also may include the number of units produced each period. (It is possible that fractional numbers of units could be produced in some applications.) One way to deal with this problem is to require in advance that some or all of the problem variables assume only integer values. Unfortunately, this makes the solution algorithm considerably more complex. The resulting problem, known as an integer linear programming problem, requires much more computational effort to solve than does ordinary linear programming. For a moderate-sized problem, solving the problem as an integer linear program is certainly a reasonable alternative.

If an integer programming code is unavailable or if the problem is simply too large to solve by integer programming, linear programming still provides a workable solution. However, after the linear programming solution is obtained, some of the problem variables must be rounded to integer values. Simply rounding off each variable to the closest integer may lead to an infeasible solution and/or one in which production and workforce levels are inconsistent. It is not obvious what is the best way to round the variables. We recommend the following conservative approach: round the values of the numbers of workers in each period t to W_t, the next larger integer. Once the values of W_t are determined, the values of the other variables, H_t, F_t, and P_t, can be found along with the cost of the resulting plan.

Conservative rounding will always result in a feasible solution, but will rarely give the optimal solution. The conservative solution generally can be improved by trial-and-error experimentation.

There is no guarantee that if a problem can be formulated as a linear program, the final solution makes sense in the context of the problem. In the aggregate planning problem, it does not make sense that there should be both overtime production and idle time in the same period, and it does not make sense that workers should be hired and fired in the same period. This means that either one or both of the variables O_t and U_t must be zero, and either one or both of the variables H_t and F_t must be zero for each t, $1 \le t \le T$. This requirement can be included explicitly in the problem formulation by adding the constraints

$$O_tU_t = 0 \qquad \text{for } 1 \le t \le T,$$
$$H_tF_t = 0 \qquad \text{for } 1 \le t \le T,$$

since if the product of two variables is zero it means that at least one must be zero. Unfortunately, these constraints are not linear, as they involve a product of problem variables. However, it turns out that it is not necessary to explicitly include these constraints, because the optimal solution to a linear programming problem always occurs at an extreme point of the feasible region. It can be shown that every extreme point solution automatically has this property. If this were not the case, the linear programming solution would be meaningless.

Extensions

Linear programming also can be used to solve somewhat more general versions of the aggregate planning problem. Uncertainty of demand can be accounted for indirectly by assuming that there is a minimum buffer inventory B_t each period. In that case we would include the constraints

$$I_t \geq B_t \qquad \text{for } 1 \leq t \leq T.$$

The constants B_t would have to be specified in advance. Upper bounds on the number of workers hired and the number of workers fired each period could be included in a similar way. Capacity constraints on the amount of production each period could easily be represented by the set of constraints:

$$P_t \leq C_t \qquad \text{for } 1 \leq t \leq T.$$

The linear programming formulation above assumed that inventory levels would never go negative. However, in some cases it might be desirable or even necessary to allow demand to exceed supply; for example, if forecast demand exceeded production capacity over some set of planning periods. In order to treat backlogging of excess demand, the inventory level I_t must be expressed as the difference between two nonnegative variables, say I_t^+ and I_t^-, satisfying

$$I_t = I_t^+ - I_t^-,$$
$$I_t^+ \geq 0, \qquad I_t^- \geq 0.$$

The holding cost would now be charged against I_t^+ and the penalty cost for back orders (say c_P) against I_t^-. However, notice that for the solution to be sensible, it must be true that I_t^+ and I_t^- are not both positive in the same period t. As with the overtime and idle time and the hiring and firing variables, the properties of linear programming will guarantee that this holds without having to explicitly include the constraint $I_t^+ I_t^- = 0$ in the formulation.

In the development of the linear programming model, we stated the requirement that all of the cost functions must be linear. This is not strictly correct. Linear programming also can be used when the cost functions are *convex piecewise-linear functions*.

A convex function is one with an increasing slope. A piecewise-linear function is one that is composed of straight-line segments. Hence, a convex piecewise-linear function is a function composed of straight lines that have increasing slopes. A typical example is presented in Figure 3–5.

In practice, it is likely that some or all of the cost functions for aggregate planning are convex. For example, if Figure 3–5 represents the cost of hiring workers, then the marginal cost of hiring one additional worker increases with the number of workers that have already been hired. This is probably more accurate than assuming

FIGURE 3–5

*A convex piecewise-
linear function*

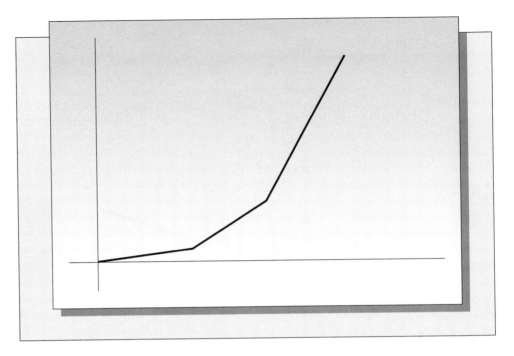

that the cost of hiring one additional worker is a constant independent of the number of workers previously hired. As more workers are hired, the available labor pool shrinks and more effort must be expended to hire the remaining available workers.

In order to see exactly how convex piecewise-linear functions would be incorporated into the linear programming formulation, we will consider a very simple case. Suppose that the cost of hiring new workers is represented by the function pictured in Figure 3–6. According to the figure, it costs c_{H1} to hire each worker until H^* workers are hired, and it costs c_{H2} for each worker hired beyond H^* workers, with $c_{H1} < c_{H2}$. The variable H_t, the number of workers hired in period t, must be expressed as the sum of two variables:

$$H_t = H_{1t} + H_{2t}.$$

Interpret H_{1t} as the number of workers hired up to H^* and H_{2t} as the number of workers hired beyond H^* in period t. The cost of hiring is now represented in the objective function as

$$\sum_{t=1}^{T} (c_{H1}H_{1t} + c_{H2}H_{2t}),$$

and the additional constraints

$$H_t = H_{1t} + H_{2t}$$

$$0 \leq H_{1t} \leq H^*$$

$$0 \leq H_{2t}$$

must also be included.

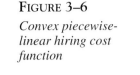

FIGURE 3–6

*Convex piecewise-
linear hiring cost
function*

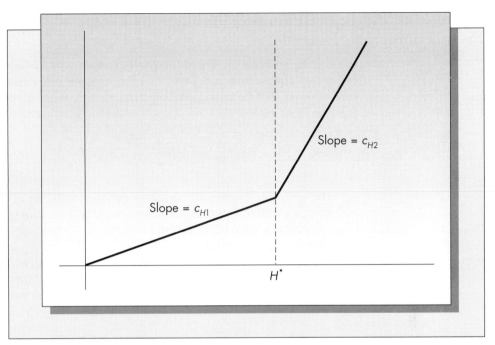

In order for the final solution to make sense, it can never be the case that $H_{1t} < H^*$ and $H_{2t} > 0$ for some t. (Why?) However, because linear programming searches for the minimum cost solution, it will force H_{1t} to its maximum value before allowing H_{2t} to become positive, since $c_{H1} < c_{H2}$. This is the reason that the cost functions must be convex. This approach can easily be extended to more than two linear segments and to any of the other cost functions present in the objective function. The technique is known as separable convex programming and is discussed in greater detail in Hillier and Lieberman (1990).

Other Solution Methods

Bowman (1956) suggested a transportation formulation of the aggregate planning problem when back orders are not permitted. Several authors have explored solving aggregate planning problems with specially tailored algorithms. These algorithms could be faster for solving large aggregate planning problems than a general linear programming code such as LINDO. In addition, some of these formulations allow for certain types of nonlinear costs. For example, if there are economies of scale in production, the production cost function is likely to be a concave function of the number of units produced. Concave cost functions are not as amenable to linear programming formulations as convex functions. We believe that a general-purpose linear programming package is sufficient for most reasonable-sized problems that one is likely to encounter in the real world. However, the reader should be aware that there are alternative solution techniques that could be more efficient for large problems and could provide solutions when some of the cost functions are nonlinear. A paper that details a transportation-type procedure more efficient than the simplex method for solving aggregate planning problems is Erenguc and Tufekci (1988).

3.6 Solving Aggregate Planning Problems by Linear Programming: An Example

We will demonstrate the use of linear programming by finding the optimal solution to the example presented in Section 4. As there is no subcontracting, overtime, or idle time allowed, and the cost coefficients are constant with respect to time, the objective function is simply

$$\text{Minimize} \left(500 \sum_{t=1}^{6} H_t + 1000 \sum_{t=1}^{6} F_t + 80 \sum_{t=1}^{6} I_t \right).$$

The boundary conditions comprise the specifications of the initial inventory of 500 units, the initial workforce of 300 workers, and the ending inventory of 600 units. These are best handled by including a separate additional constraint for each boundary condition.

The constraints are obtained by substituting $t = 1, \ldots, 6$ into Equations (A), (B), and (C). The full set of constraints expressed in standard linear programming format (with all problem variables on the left-hand side and nonnegative constants on the right-hand side) is as follows:

$$
\begin{aligned}
W_1 - W_0 - H_1 + F_1 &= 0, \\
W_2 - W_1 - H_2 + F_2 &= 0, \\
W_3 - W_2 - H_3 + F_3 &= 0, \\
W_4 - W_3 - H_4 + F_4 &= 0, \\
W_5 - W_4 - H_5 + F_5 &= 0, \\
W_6 - W_5 - H_6 + F_6 &= 0;
\end{aligned}
\tag{A}
$$

$$
\begin{aligned}
P_1 - I_1 + I_0 &= 1{,}280, \\
P_2 - I_2 + I_1 &= 640, \\
P_3 - I_3 + I_2 &= 900, \\
P_4 - I_4 + I_3 &= 1{,}200, \\
P_5 - I_5 + I_4 &= 2{,}000, \\
P_6 - I_6 + I_5 &= 1{,}400;
\end{aligned}
\tag{B}
$$

$$
\begin{aligned}
P_1 - 2.931 W_1 &= 0, \\
P_2 - 3.517 W_2 &= 0, \\
P_3 - 2.638 W_3 &= 0, \\
P_4 - 3.810 W_4 &= 0, \\
P_5 - 3.224 W_5 &= 0, \\
P_6 - 2.198 W_6 &= 0;
\end{aligned}
\tag{C}
$$

$$W_1, \ldots, W_6, P_1, \ldots, P_6, I_1, \ldots, I_6, F_1, \ldots, F_6, H_1, \ldots, H_6 \geq 0; \tag{D}$$

$$
\begin{aligned}
W_0 &= 300, \\
I_0 &= 500, \\
I_6 &= 600.
\end{aligned}
\tag{E}
$$

TABLE 3–5   Lenear Programming Solution of Densepack Problem

```
LP OPTIMUM FOUND AT STEP 24
        OBJECTIVE FUNCTION VALUE
 1)          379320.900
VARIABLE              VALUE           REDUCED COST
      H1             .000000          1500.000000
      H2             .000000           791.378100
      H3             .000000           222.439900
      H4             .000000             6.736023
      H5          464.782300             .000000
      H6             .000000           192.750300

      F1           27.047150             .000000
      F2             .000000           708.621900
      F3             .000000          1277.560000
      F4             .000000          1493.264000
      F5             .000000          1500.000000
      F6             .000000          1307.250000

      W0          300.000000             .000000
      W1          272.952900             .000000
      W2          272.952900             .000000
      W3          272.952900             .000000
      W4          272.952900             .000000
      W5          737.735200             .000000
      W6          737.735200             .000000

      I0          500.000000             .000000
      I1           20.024820             .000000
      I2          340.000000             .000000
      I3          160.049800             .000000
      I4             .000000            18.445950
      I5          378.458200             .000000
      I6          600.000000             .000000

      P1          800.024800             .000000
      P2          959.975200             .000000
      P3          720.049600             .000000
      P4         1039.950000             .000000
      P5         2378.458000             .000000
      P6         1621.542000             .000000
```

We have solved this linear program using the LINDO system developed by Schrage (1984). The output of the LINDO program is given in Table 3–5. In the supplement on linear programming, we also treat solving linear programs with Excel. The Excel Solver, of course, gives the same results. The value of the objective function at the optimal solution is $379,320.90, which is *considerably* less than that achieved with

TABLE 3–6   Aggregate Plan for Densepack Obtained from Rounding the Linear Programming Solution

A	B	C	D	E	F	G	H	I
Month	Number of Workers	Number Hired	Number Fired	Number of Units per Worker	Number of Units Produced (B × E)	Cumulative Production	Cumulative Net Demand	Ending Inventory (G − H)
January	273		27	2.931	800	800	780	20
February	273			3.517	960	1,760	1,420	340
March	273			2.638	720	2,480	2,320	160
April	273			3.810	1,040	3,520	3,520	0
May	738	465		3.224	2,379	5,899	5,520	379
June	738			2.198	1,622	7,521	7,520	1
Totals		465	27					900

either the zero inventory plan or the constant workforce plan. However, this cost is based on fractional values of the variables. The actual cost will be slightly higher after rounding.

Following the rounding procedure recommended earlier, we will round all the values of W_t to the next higher integer. That gives $W_1 = \cdots = W_4 = 273$ and $W_5 = W_6 = 738$. This determines the values of the other problem variables. This means that the firm should fire 27 workers in January and hire 465 workers in May. The complete solution is given in Table 3–6.

Again, because column H in Table 3–6 corresponds to net demand, we add the 600 units of ending inventory in June, giving a total inventory of $900 + 600 = 1,500$ units. Hence, the total cost of this plan is $(500)(465) + (1,000)(27) + (80)(1,500) = \$379,500$, which represents a substantial savings over both the zero inventory plan and the constant workforce plan.

The results of the linear programming analysis suggest another plan that might be more suitable for the company. Because the optimal strategy is to decrease the workforce in January and build it back up again in May, a reasonable alternative might be to not fire the 27 workers in January and to hire fewer workers in May. In this case, the most efficient method for finding the correct number of workers to hire in May is to simply re-solve the linear program, but without the variables $F_1, \ldots, F_6$, as no firing of workers means that these variables are forced to zero. (If you wish to avoid reentering the problem into the computer, simply append the old formulation with the constraints $F_1 = 0, F_2 = 0, \ldots, F_6 = 0$.) The optimal number of workers to hire in May turns out to be 374 if no workers are fired, and the cost of the plan is approximately $\$386,120$. This is only slightly more expensive than the optimal plan, and has the important advantage of not requiring the firing of any workers.

Problems for Sections 5 and 6

17. Formulate Problem 13 as a linear program. Be sure to define all variables and include all of the required constraints.

18. Problem 17 was solved on the computer. The linear programming output of the program is given below.

```
LP OPTIMUM FOUND   AT STEP      14

                OBJECTIVE FUNCTION VALUE

     1)           36185.0600

     VARIABLE              VALUE              REDUCED COST
         H1            6.143851                .000000
         H2           64.310690                .000000
         H3             .000000              300.000000
         H4          116.071400                .000000
         F1             .000000              300.000000
         F2             .000000              300.000000
         F3           87.662330                .000000
         F4             .000000              300.000000
         I1             .000000               59.415580
         I2             .000000              189.285700
         I3             .000000               31.006490
         W1          106.143900                .000000
         W2          170.454500                .000000
         W3           82.792210                .000000
         W4          198.863600                .000000
         P1          850.000000                .000000
         P2         1260.000000                .000000
         P3          510.000000                .000000
         P4          980.000000                .000000
         I4             .000000              120.292200
```

a. By rounding the values of the variables using the conservative approach described in this section, find a feasible solution to the problem. Determine the cost of the resulting plan.

b. Determine another plan that achieves a lower cost than that determined in part (a) while still retaining feasibility (that is, having a nonnegative amount of inventory on hand at the end of each period).

19. a. Formulate Problem 9 as a linear program.

b. Solve the problem using a linear programming code. Round the variables in the resulting solution and determine the cost of the plan you obtain.

*20. a. Formulate Problem 14 as a linear program.

b. Solve the problem using a linear programming code. Round off the solution and determine the cost of the resulting plan.

21. Consider Mr. Meadows Cookie Company, described in Problem 13. Suppose that the cost of hiring workers each period is $100 for each worker until 20 workers are hired, $400 for each worker when between 21 and 50 workers are hired, and $700 for each worker hired beyond 50.

    a. Write down the complete linear programming formulation of the revised problem.

    b. Solve the revised problem using a linear programming code. What difference does the new hiring cost function make in the solution?

*22. Leather-All produces a line of handmade leather products. At the present time, the company is producing only belts, handbags, and attaché cases. The predicted demand for these three types of items over a six-month planning horizon is as follows:

Month	Number of Working Days	Belts	Handbags	Attaché Cases
1	22	2,500	1,250	240
2	20	2,800	680	380
3	19	2,000	1,625	110
4	24	3,400	745	75
5	21	3,000	835	126
6	17	1,600	375	45

The belts require an average of two hours to produce, the handbags three hours, and the attaché cases six hours. All of the workers have the skill to work on any item. Leather-All has 46 employees who each has a share in the firm and cannot be fired. There are an additional 30 locals that are available and can be hired for short periods at higher cost. Regular employees earn $8.50 per hour on regular time and $14.00 per hour on overtime. Regular time comprises a seven-hour workday and the regular employees will work as much overtime as is available. The additional workers are hired for $11.00 per hour and are kept on the payroll for at least one full month. Costs of hiring and firing are negligible.

    Because of the competitive nature of the industry, Leather-All does not want to incur any demand back orders.

    a. Using worker hours as an aggregate measure of production, convert the forecasted demands to demands in terms of aggregate units.

    b. What would be the size of the workforce needed to satisfy the demand for the coming six months on regular time only? Would it be to the company's advantage to bring the permanent workforce up to this level? Why or why not?

    c. Determine a production plan that meets the demand using only regular-time employees and the total cost of that plan.

    d. Determine a production plan that utilizes only additional employees to absorb excess demand and the cost of that plan.

*23. *a.* Formulate the problem of optimizing Leather-All's hiring schedule as a linear program. Define all problem variables and include whatever constraints are necessary.

    b. Solve the problem formulated in part (*a*) using a linear programming code. Round all of the relevant variables and determine the cost of the resulting plan. Compare your results with those obtained in Problem 22(*c*) and (*d*).

3.7 The Linear Decision Rule

An interesting alternative approach to solving the aggregate planning problem has been suggested by Holt, Modigliani, Muth, and Simon (1960). They assume that all of the relevant costs, including inventory costs and the costs of changing production levels and numbers of workers, are represented by quadratic functions. That is, the total cost over the T-period planning horizon can be written in the form

$$\sum_{t=1}^{T} [c_1 W_t + c_2(W_t - W_{t-1})^2 + c_3(P_t - Kn_t W_t)^2 + c_4 P_t + c_5(I_t - c_6)^2]$$

subject to

$$I_t = I_{t-1} + P_t - D_t \qquad \text{for } 1 \le t \le T.$$

The values of the constants $c_1, c_2, \ldots, c_6$ must be determined for each particular application. Expressing the cost functions as quadratic functions rather than simple linear functions has some advantages over the linear programming approach. As quadratic functions are differentiable, the standard rules of calculus can be used to determine optimal solutions. The optimal solution will occur where the first partial derivatives with respect to each of the problem variables are zero. When quadratic functions are differentiated, they yield first-order equations (linear equations), which are easy to solve. Also, quadratic functions give a more accurate approximation to general nonlinear functions than do linear functions. (Any nonlinear function may be approximated by a Taylor series expansion in which the first two terms yield a quadratic approximation.)

However, the quadratic approach also has one serious disadvantage. It is that quadratic functions (that is, parabolas) are symmetric (see Figure 3–7). Hence, the cost of hiring a given number of workers must be the same as the cost of firing the same number

FIGURE 3–7

Quadratic cost functions used in deriving the linear decesion rule

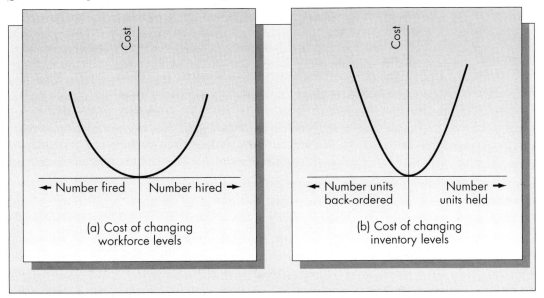

(a) Cost of changing workforce levels

(b) Cost of changing inventory levels

of workers, and the cost of producing a given number of units on overtime must be the same as the cost attached to the same number of units of worker idle time. The problem can be overcome somewhat by not setting the center of symmetry of the cost function at zero, but the basic problem of symmetry still prevails.

The most appealing feature of this approach is the simple form of the optimal policy. For example, the optimal production level in period t, P_t, has the form

$$P_t = \sum_{n=0}^{K} (a_n D_{t+n} + bW_{t-1} + cI_{t-1} + d).$$

The terms a_n, b, c, and d are constants that depend on the cost parameters. W_t has a similar form. The computational advantages of the linear decision rule are less important now than they were when this analysis was first done because of the widespread availability of computers today. Because linear programming provides a much more flexible framework for formulating aggregate planning problems, and reasonably large linear programs can be solved even on personal computers, one would have to conclude that linear programming is a preferable solution technique.

It should be recognized, however, that the text by Holt, Modigliani, Muth, and Simon represents a landmark work in the application of quantitive methods to production planning problems. The authors developed a solution method that results in a set of formulas that are easy to implement, and they actually undertook the implementation of the method. The work details the application of the approach to a large manufacturer of household paints in the Pittsburgh area. The analysis was actually implemented in the company, but a subsequent visit to the firm indicated that serious problems arose when the linear decision rule was followed, primarily because of the firm's policy of not firing workers when the model indicated that they should be fired. (See Vollman, Berry, and Whybark, 1992, p. 627.)

3.8 Modeling Management Behavior

Bowman (1963) developed an interesting technique for aggregate planning that should be applicable to other types of problems as well. His idea is to construct a sensible model for controlling production levels and fit the parameters of the model as closely as possible to the actual prior decisions made by management. In this way the model reflects the judgment and the experience of managers and avoids a number of the problems that arise when using more traditional modeling methods. One such problem is determining the accuracy of the assumptions required by the model. Another problem that is avoided is the need to determine values of input parameters that could be difficult to measure.

Let us consider the problem of producing a single product over T planning periods. Suppose that $D_1, \ldots, D_T$ are the forecasts of demand for the next T periods and that production levels $P_1, \ldots, P_T$ must be determined. The simplest reasonable decision rule is

$$P_t = D_t \qquad \text{for } 1 \leq t \leq T.$$

Trying to match production exactly to demand will generally result in a production schedule that is too erratic. Production smoothing can be accomplished using a decision rule of the form

$$P_t = D_t + \alpha(P_{t-1} - D_t),$$

where α is a smoothing factor for production, having the property $0 \leq \alpha \leq 1$. (The effect of smoothing here is the same as that of exponential smoothing, discussed in detail in Chapter 2.) When α equals zero, this rule is identical to the one above. When $\alpha = 1$, the production in period t is exactly the same as the production in period $t - 1$. The choice of α provides a means of placing a relative weight on matching production with demand versus holding production constant from period to period.

In addition to smoothing production, the firm also might be interested in keeping inventory levels near some target level, say I_N. A model that smooths both inventory and production simultaneously is

$$P_t = D_t + \alpha(P_{t-1} - D_t) + \beta(I_N - I_{t-1}),$$

where $0 \leq \beta \leq 1$ measures the relative weight placed on smoothing of inventory.

Finally, the model should incorporate demand forecasts. In this way, production levels can be increased or decreased in anticipation of a change in the pattern of demands. A rule that includes smoothing of production and inventory as well as forecasts of future demand is

$$P_t = \sum_{i=t}^{t+n} a_{t-i+1}D_i + \alpha(P_{t-1} - D_t) + \beta(I_N - I_{t-1}).$$

This decision rule is arrived at by a straightforward common-sense analysis of what a good production rule should be. It is interesting to note how similar this rule is to the linear decision rule discussed in the previous section, which was derived from a mathematical model. Undoubtedly, the previous work on the linear decision rule inspired the form of this rule. It is often the case that the most valuable feature of a model is indicating the best *form* of an optimal strategy, and not necessarily the best strategy itself.

This particular model for P_t requires determination of $a_1, \ldots, a_n, \alpha, \beta,$ and I_N. Bowman suggests that the values of these parameters be determined by retrospectively observing the system for a reasonable period of time, and fitting the parameters to the actual history of management actions during that time using a technique such as least squares. In this way the model becomes a reflection of past management behavior.

Bowman compares the actual experience of a number of companies with the experience that they would have had using his approach, and shows that in most cases there would have been a substantial cost reduction. The model merely emulates management behavior, so why should it outperform actual management experience? The theory is that a model derived in this manner is a reflection of management making rational decisions, as most of the time the system is stable. However, if a sudden unusual event occurs, such as a much-higher-than-anticipated demand or a sudden decline in productive capacity due, for example, to the breakdown of a machine or the loss of key personnel, it is likely that many managers would tend to overreact. However, the model would recommend production levels that are consistent with past decision making. Hence, the use of a simple but consistent model for making decisions would keep management from panicking in an unusual situation. Although the approach is conceptually appealing, there are few reports in the literature of successful implementations of such a technique.

Problems for Sections 7 and 8

24. The form of the optimal production rule derived by Holt, Modigliani, Muth, and Simon (1960) for the case of the paint company indicates that the production level in period t should be computed from the formula

$$P_t = .463D_t + .234D_{t+1} + .111D_{t+2} + .046D_{t+3} + .013D_{t+4}$$
$$- .022D_{t+5} - .008D_{t+6} - .010D_{t+7} - .009D_{t+8}$$
$$- .008D_{t+9} - .007D_{t+10} - .005D_{t+11}$$
$$+ .993W_{t-1} - .464I_{t-1} + 153,$$

and the workforce level in period t from the formula

$$W_t = .010D_t + .0088D_{t+1} + .0071D_{t+2} + .0054D_{t+3} + .0042D_{t+4}$$
$$+ .0031D_{t+5} + .0023D_{t+6} + .0016D_{t+7} + .0012D_{t+8}$$
$$+ .0009D_{t+9} + .0006D_{t+10} + .0005D_{t+11}$$
$$+ .743W_{t-1} - .01I_{t-1} - 2.09.$$

Suppose that forecast demands for the coming 12 months are 150, 164, 185, 193, 167, 143, 126, 93, 84, 72, 50, 66. The current number of employees is 180, and the current inventory level is 45 aggregate units.

a. Compute the values of the aggregate production level and the number of workers that the company should be using in the current period.

b. Repeat the calculation you performed in part (a) but ignore all terms with a coefficient that has an absolute value less than or equal to .01 when calculating P_t, and less than or equal to .005 when calculating W_t. How much difference do you observe in the answer?

c. Suppose that the current demand is realized to be 163 and the new forecast one year ahead is 72. Using the approximation suggested in part (b), determine the new values for the aggregate production level and the number of workers. (Hint: You must compute the new value I_{t-1} and assume that the new value of W_{t-1} is the value computed in part (a).)

25. Using the following values of management coefficients for Bowman's smoothed production model, determine the production level for which Harold Grey (described in Problem 9) should plan in the coming year. Assume that the current production level is 150,000 packages.

$$a_1 = .3475, \quad a_2 = .1211, \quad a_3 = .0556, \quad a_4 = .0663,$$
$$a_5 = .0023, \quad \alpha = .6, \quad \beta = .3, \quad I_N = 40,000.$$

26. a. Discuss the similarities and the differences between the linear decision rule of Holt, Modigliani, Muth, and Simon and Bowman's management coefficients approach.

b. Discuss the relative advantages and disadvantages of linear decision rules versus linear programming for solving aggregate planning problems.

3.9 Disaggregating Aggregate Plans

Aggregate planning can be done at several levels, but it is always the case that an aggregate plan corresponds to some grouping of items. These groupings can vary in size from several items, composing a product family, to the entire line of items produced by the firm. In either case, ensuring that the aggregate plan and the master production schedules for individual items are consistent could be an issue. When aggregate plans are the result of combining or aggregating individual item plans, disaggregation is straightforward. However, aggregate planning provides a top-down management perspective, and for that reason aggregate plans are not generally the result of a bottom-up aggregation of single-item master production schedules. A firm concerned with providing consistency between the aggregate plan and the individual item production schedules would need to develop a disaggregation scheme consistent with the definition of an aggregate unit and with the organizational structure of the firm.

It is not clear that coordinating the production schedules for individual items with the aggregate plan is a serious issue in all circumstances. Aggregate planning may be viewed primarily as a means of determining gross workforce levels over a specified planning horizon. Aggregate units may be a fictitious construct designed only to develop a meaningful representation of the overall sales activity of the firm. When this is the case, the detailed planning of individual item production plans can be treated as a separate problem that must be solved subject to the constraints determined by the size of the workforce.

However, if an aggregate unit of production does correspond to an actual item, or is determined using a hierarchical planning structure such as the one discussed earlier in the chapter (items, families, and types), it is important that the master production schedule and the aggregate plan be consistent. The problem of breaking down an aggregate plan into a master schedule by item has received little attention. Individual companies are probably dealing with the problem in ways that suit their specific production systems and organizational structures, but few general approaches to the problem have been developed.

A general scheme for dealing with the disaggregation issue that is consistent with the grouping of items into families and families into types has been championed by a group of researchers from MIT. The methods are discussed in detail in Hax and Candea (1984). We present a very brief overview of their approach.

Suppose that X^* represents the number of aggregate units of production of a given type indicated by the aggregate plan for the next immediate planning period. One computes the number of units of family j to be produced, Y_j, by solving the following mathematical programming problem:

$$\text{Minimize} \sum_{j=1}^{J} \frac{K_j \lambda_j}{Y_j}$$

subject to

$$\sum_{j=1}^{J} Y_j = X^*$$

and

$$a_j \leq Y_j \leq b_j \qquad \text{for } 1 \leq j \leq J,$$

where K_j is the setup cost for family j and λ_j is the annual demand for family j. The constants a_j and b_j are upper and lower bounds on the number of units of each family that can be produced. The term $K_j \lambda_j / Y_j$ is the average annual setup cost for family j.

SNAPSHOT APPLICATION

Welch's Uses Aggregate Planning for Production Scheduling

Welch's is a make-to-stock food manufacturer based in Concord, Massachusetts. They are probably best known for grape jelly and grape juices, but they produce a wide variety of processed foods. Allen and Schuster (1994) describe an aggregate planning model for Welch's primary production facility.

The characteristics of the production system for which their system was designed are

- Dynamic, uncertain demand, resulting in changing buffer stock requirements.
- Make-to-stock environment.
- Dedicated production lines.
- Production lines that each produces two or more families of products.
- Large setup times and setup costs for families, as opposed to low setup times and costs for individual items.

The two primary objectives of the production system as described by the authors are to smooth peak demands through time so as not to exceed production capacity and to allocate production requirements among the families to balance family holding and setup costs. The planning is done with a six-month time horizon for demand forecasting. The six-month period is divided into two portions: the next four weeks and the remaining five months. Detailed plans are developed for the near term, including regular and overtime production allocation.

The model has two primary components: a family planning model, which finds the optimal timing and sizing of family production runs, and a disaggregation planning model, which takes the results of family planning and determines lot sizes for individual items within families.

The authors also discuss several implementation issues specific to Welch's environment. Product run lengths must be tied to the existing eight-hour shift structure. To do so; they recommend that production run lengths be expressed as multiples of one-quarter shift (two hours).

The model was implemented on a personal computer. Computing times are very moderate. Solution techniques include a mixed integer mathematical program and a linear programming formulation with relaxation (that is, rounding of variables to integer values).

This case demonstrates that the concepts discussed in this chapter can be useful in a real production planning environment. Although the system described here is not based on any of the specific models discussed in this chapter, this application shows that aggregation and disaggregation are useful concepts. Hierarchical aggregation for production scheduling is a valuable planning tool.

(A detailed derivation of this term will be given in the next chapter, when we discuss models for individual item management.) Hence it follows that the objective function represents the total average annual setup cost for all of the families constituting this particular type. Holding cost is ignored at this level, because the objective function for the aggregate planning model that resulted in X^* is assumed to have included holding costs. The justification for using only the first-period results of the aggregate plan is that schedules are generated on a rolling horizon basis. As new information becomes available each period, the old schedules are revised and updated.

A number of feasibility issues need to be addressed before the family run sizes Y_j are further disaggregated into lots for individual items. The objective is to schedule the lots for individual items within a family so that they run out at the scheduled setup time for the family. In this way, items within the same family can be produced within the same production setup.

The concept of disaggregating the aggregate plan along organizational lines in a fashion that is consistent with the aggregation scheme is an appealing one. Whether or not the methods discussed in this section provide a workable link between aggregate plans and detailed item schedules remains to be seen.

Another approach to the disaggregation problem has been explored by Chung and Krajewski (1984). They develop a mathematical programming formulation of the problem. Inputs to the program include aggregate plans for each product family. This includes setup time, setup status, total production level for the family, inventory level, workforce level, overtime, and regular time availability. The goal of the analysis is to specify lot sizes and timing of production runs for each individual item, consistent with the aggregate information for the product family. Although such a formulation provides a potential link between the aggregate plan and the master production schedule, the resulting mathematical program requires many inputs and can result in a very large mixed integer problem that could be very time-consuming to solve. For these reasons, firms are unlikely to use such methods when there are large numbers of individual items in each product family.

Problems for Section 9

27. What does "disaggregation of aggregate plans" mean?

28. Discuss the following quotation made by a production manager: "Aggregate planning is useless to me because the results have nothing to do with my master production schedule."

3.10 Production Planning on a Global Scale

Globalization of manufacturing operations is commonplace. Many major corporations are now classified as multinationals; manufacturing and distribution activities routinely cross international borders. With the globalization of both sources of production and markets, firms must rethink production planning strategies. One issue explored in this chapter was smoothing of production plans over time; costs of increasing or decreasing workforce levels (and, hence, production levels) play a major role in the optimization of any aggregate plan. When formulating global production strategies, other smoothing issues arise. Exchange rates, costs of direct labor, and tax structure are just some of the differences among countries that must be factored into a global strategy.

Why the increased interest in global operations? In short, cost and competitiveness. According to McGrath and Bequillard (1989):

> The benefits of a properly executed international manufacturing strategy can be very substantial. A well developed strategy can have a direct impact on the financial performance and ultimately be reflected in increased profitability. In the electronic industry, there are examples of companies attributing 5% to 15% reduction in cost of goods sold, 10% to 20% increase in sales, 50% to 150% improvement in asset utilization, and 30% to 100% increase in inventory turnover to their internationalization of manufacturing.

Cohen et al. (1989) outline some of the issues that a firm must consider when planning production levels on a worldwide basis. These include

- In order to achieve the kinds of economies of scale required to be competitive today, multinational plants and vendors must be managed as a global system.
- Duties and tariffs are based on material flows. Their impact must be factored into decisions regarding shipments of raw material, intermediate product, and finished product across national boundaries.
- Exchange rates fluctuate randomly and affect production costs and pricing decisions in countries where the product is produced and sold.
- Corporate tax rates vary widely from one country to another.
- Global sourcing must take into account longer lead times, lower unit costs, and access to new technologies.
- Strategies for market penetration, local content rules, and quotas constrain product flow across borders.
- Product designs may vary by national market.
- Centralized control of multinational enterprises creates difficulties for several reasons, and decentralized control requires coordination.
- Cultural, language, and skill differences can be significant.

Determining optimal globalized manufacturing strategies is clearly a daunting problem for any multinational firm. One can formulate and solve mathematical models similar to the linear programming formulations of the aggregate planning models presented in this chapter, but the results of these models must always be balanced against judgment and experience. Cohen et al. (1989) consider such a model. They assume multiple products, plants, markets, raw materials, vendors, vendor supply contract alternatives, time periods, and countries. Their formulation is a large-scale mixed integer, nonlinear program.

One issue not treated in their model is that of exchange rate fluctuations and their effect on both pricing and production planning. Pricing, in particular, is traditionally done by adding a markup to unit costs in the home market. This completely ignores the issue of exchange rate fluctuations and can lead to unreasonable prices in some countries. For example, this issue has arisen at Caterpillar Tractor (Caterpillar Tractor Company, 1985). In this case, dealers all over the world were billed in U.S. dollars based on U.S. production costs. When the dollar was strong relative to other currencies, retail prices charged to overseas customers were not competitive in local markets. Caterpillar found themselves losing market share abroad as a result. In the early 1980s the firm switched to a locally competitive pricing strategy to counteract this problem.

The notion that manufacturing capacity can be used as a hedge against exchange rate fluctuations has been explored by several researchers. Kogut and Kulatilaka (1994), for example, develop a mathematical model for determining when it is optimal to switch production from one location to another. Since the cost of switching is assumed to be positive, there must be a sufficiently large difference in exchange rates before switching is recommended. As an example, they consider a situation where a firm can produce its product in either the United States or Germany. If production is currently being done in one location, the model provides a means of determining if it is economical to switch locations based on the relative strengths of the deutschmark and dollar. While such models are in the early stages of development, they provide a means of rationalizing international production planning strategies. Similar issues have been explored by Huchzermeier and Cohen (1996) as well.

3.11 Practical Considerations

Aggregate planning can be a valuable aid in planning production and manpower levels for a company, and provides a means of absorbing demand fluctuations by smoothing workforce and production levels. There are a number of advantages for planning on the aggregate level over the detail level. One is that the cost of preparing forecasts and determining productivity and cost parameters on an individual item basis can be prohibitive. The cost of data collection and input is probably a more significant drawback of large-scale mathematical programming formulations of detailed production plans than is the cost of actually performing the computations. A second advantage of aggregate planning is the relative improvement in forecast accuracy that can be achieved by aggregating items. As we saw in Chapter 2 on forecasting, aggregate forecasts are generally more accurate than individual forecasts.[3] Finally, an aggregate planning framework allows the manager to see the "big picture" and not be unduly influenced by specifics.

With all of these advantages, one would think that aggregate planning would have an important place in the planning of the production activities of most companies. However, this does not appear to be the case. There are a number of reasons for the lack of interest in aggregate planning methodology. First is the difficulty of properly defining an aggregate unit of production. There appears to be no simple way of specifying how individual items should be aggregated that will work in all situations. Second, once an aggregating scheme is developed, cost estimates and demand forecasts for aggregate units are required. It is difficult enough to obtain accurate cost and demand information for real units. Obtaining such information on an aggregate basis could be considerably more difficult, depending on the aggregation scheme that is used. Third, aggregate planning models rarely reflect the political and operational realities of the environment in which the company is operating. Assuming that workforce levels can be changed easily is probably not very realistic for most companies. Finally, as Silver and Peterson (1985) suggest, managers do not want to rely on a mathematical model for answers to the extremely sensitive and important issues that are addressed in this analysis.

Another issue is whether or not the goals of aggregate planning analysis can be achieved through methods other than those discussed in this chapter. Schwarz and Johnson (1978) claim that the cost savings achieved using a linear decision rule for aggregate planning could be obtained by improved management of the aggregate inventory alone. They substantiate their hypothesis using the data reported in Holt, Modigliani, Muth, and Simon (1960), and show that for the case of the paint company virtually all of the cost savings of the linear decision rule were a result of increasing the buffer inventory and not of making major changes in the size of the labor force. This

---

[3]More precisely, if the individual forecasts are not perfectly correlated, then the standard deviation of the forecast error for the aggregate will be less than the sum of the standard deviations of the individual forecast errors.

single comparison does not verify the authors' hypothesis in general, but suggests that better inventory management, using the methods to be discussed in Chapters 4 and 5, could return a large portion of the benefits that a company might realize with aggregate planning.

Even with this long list of disadvantages, the mathematical models discussed in this chapter can, and should, serve as an aid to production planners. Although optimal solutions to a mathematical model may not be true optimal solutions to the problem that they address, they do provide insight and could reveal alternatives that would not otherwise be evident.

3.12 Historical Notes

The aggregate planning problem was conceived in an important series of papers that appeared in the mid-1950s. The first, Holt, Modigliani, and Simon (1955), discussed the structure of the problem and introduced the quadratic cost approach, and the later study of Holt, Modigliani, and Muth (1956) concentrated on the computational aspects of the model. A complete description of the method and its application to production planning for a paint company is presented in Holt, Modigliani, Muth, and Simon (1960).

That production planning problems could be formulated as linear programs appears to have been known in the early 1950s. Bowman (1956) discussed the use of a transportation model for production planning. The particular linear programming formulation of the aggregate planning problem discussed in Section 5 is essentially the same as the one developed by Hansmann and Hess (1960). Other linear programming formulations of the production planning problem generally involve multiple products or more complex cost structures (see, for example, Newson, 1975a and 1975b).

More recent work on the aggregate planning problem has focused on aggregation and disaggregation issues (Axsater, 1981; Bitran and Hax, 1981; and Zoller, 1971), the incorporation of learning curves into linear decision rules (Ebert, 1976), extensions to allow for multiple products (Bergstrom and Smith, 1970), and inclusion of marketing and/or financial variables (Damon and Schramm, 1972, and Leitch, 1974).

Taubert (1968) considers a technique he refers to as the search decision rule. The method requires developing a computer simulation model of the system and searching the response surface using standard search techniques to obtain a (not necessarily optimal) solution. Taubert's approach, which was described in detail in Buffa and Taubert (1972), gives results that are comparable to those of Holt, Modigliani, Muth, and Simon (1960) for the case of the paint company.

Kamien and Li (1990) developed a mathematical model to examine the effects of subcontracting on aggregate planning decisions. The authors show that under certain circumstances it is preferred to producing in-house, and provides an additional means of smoothing production and workforce levels.

3.13 Summary

Determining optimal production levels for all products produced by a large firm can be an enormous undertaking. *Aggregate planning* addresses this problem by assuming that individual items can be grouped together. However, finding an effective aggregating scheme can be difficult. The aggregating scheme chosen must be consistent with the structure of the organization and the nature of the products produced. One particular aggregating scheme that has been suggested is *items, families,* and *types.* Items (or stockkeeping units), representing the finest level of detail, would be identified by separate part numbers. Families are groups of items that share a common manufacturing setup, and types are natural groupings of families. This particular aggregation scheme is fairly general, but there is no guarantee that it will work in every application.

In aggregate planning one assumes *deterministic demand.* Demand forecasts over a specified planning horizon are required input. This assumption is not made for the sake of realism, but to allow the analysis to focus on the changes in the demand that are systematic rather than random. The goal of the analysis is to determine the number of workers that should be employed each period and the number of aggregate units that should be produced each period.

The objective is to minimize costs of production, payroll, holding, and changing the size of the workforce. The costs of making changes are generally referred to as *smoothing costs.* Most of the aggregate planning models discussed in this chapter assume that all of the costs are *linear functions.* This means that the cost of hiring an additional worker is the same as the cost of hiring the previous worker, and the cost of holding an additional unit of inventory is the same as the cost of holding the previous unit of inventory. This assumption is probably a reasonable approximation for most real systems. It is unlikely that the primary problem with applying aggregate planning methodology to a real situation will be that the shape of the cost functions is incorrect; it is more likely that the primary difficulty will be in correctly estimating the costs and other required input information.

One can find approximate solutions to aggregate planning problems by graphical methods. A production schedule satisfies demand if the graph of cumulative production lies above the graph of cumulative demand. As long as all costs are linear, optimal solutions[4] can be found by *linear programming.* Because solutions of linear programming problems may be fractions, some or all of the variables must be rounded to integers. The best way to round the variables is not always obvious. Once an optimal solution is found, its appropriateness must be considered in the context of the planning problem, as there are intangible costs and constraints that might be difficult or impossible to include directly in the problem formulation.

Linear programming also can be used to determine optimal production levels at the individual item level, but the size of the resulting problem and the input data requirements might make such an approach unrealistic for large companies. However, detailed linear programming formulations could be used at the family level rather than the individual item level for large firms.

Early work by Holt, Modigliani, Muth, and Simon (1960) required that the cost functions be quadratic functions. With quadratic costs, the form of the resulting optimal policy is a simple linear function, known as the *linear decision rule.* However, the assumption of quadratic cost functions is probably too restrictive to be accurate for most real problems.

Bowman's *management coefficients* model fits a reasonable mathematical model of the system to the actual decisions made by management over a period of time.

---

[4]That is, optimal subject to rounding errors.

This approach has the advantage of incorporating management's intuition into the model, and encourages management personnel to be consistent in future decision making.

Aggregate plans will be of little use to the firm if they cannot be coordinated with detailed item schedules (i.e., the master production schedule). The problem of *disaggregating aggregate plans* is a difficult one, but one that must be addressed if the aggregate plans are to have value to the firm. There have been some mathematical programming formulations of the disaggregating problem suggested in the literature, but these disaggregation schemes have yet to be proven in practice.

Additional Problems on Aggregate Planning

29. An aggregate planning model is being considered for the following applications. Suggest an aggregate measure of production and discuss the difficulties of applying aggregate planning in each case.
   a. Planning for the size of the faculty in a university.
   b. Determining workforce requirements for a travel agency.
   c. Planning workforce and production levels in a fish-processing plant that produces canned sardines, anchovies, kippers, and smoked oysters.

30. A plant manager employed by a national producer of kitchen products is planning workforce levels for the next three months. An aggregate unit of production requires an average of four labor hours. Forecast demands for the three-month horizon are as follows:

Month	Workdays	Forecasted Demand (in aggregate units)
July	23	2,400
August	16	3,000
September	20	800

Assume that a normal workday is eight hours. Hiring costs are $500 per worker and firing costs are $1,000 per worker. Holding costs are $40.00 per aggregate unit held per month. Shortages are not permitted. Assume that the ending inventory for June was 600 and the manager wishes to have at least 800 units on hand at the end of September. June workforce level is 30 workers.
   a. Find the minimum constant workforce plan for the three months and the total hiring, firing, and holding costs of that plan.
   b. Draw a graph of the cumulative net demand and use that graph to suggest a more efficient plan that makes one change in workforce levels during the planning horizon. Find the cost of the plan and compare it to the plan considered in part (*a*). Assuming that this plan is more cost effective, comment on why.

31. A local firm manufactures children's toys. The projected demand over the next four months for one particular model of toy robot is

Month	Workdays	Forecasted Demand (in aggregate units)
July	23	3,825
August	16	7,245
September	20	2,770
October	22	4,440

   Assume that a normal workday is eight hours. Hiring costs are $350 per worker and firing costs (including severance pay) are $850 per worker. Holding costs are $4.00 per aggregate unit held per month. Assume that it requires an average of 1 hour and 40 minutes for one worker to assemble one toy. Shortages are not permitted. Assume that the ending inventory for June was 600 of these toys and the manager wishes to have at least 800 units on hand at the end of October. Assume that the current workforce level is 35 workers.
   a. Find the minimum constant workforce plan for the four months and the total hiring, firing, and holding costs of that plan.
   b. Determine the plan that changes the workforce level each month to most closely match the demand and the cost of that plan.

32. The Paris Paint Company is in the process of planning labor force requirements and production levels for the next four quarters. The marketing department has provided production with the following forecasts of demand for Paris Paint over the next year:

Quarter	Demand Forecast (in thousands of gallons)
1	380
2	630
3	220
4	160

   Assume that there are currently 280 employees with the company. Employees are hired for at least one full quarter. Hiring costs amount to $1,200 per employee and firing costs are $2,500 per employee. Inventory costs are $1 per gallon per quarter. It is estimated that one worker produces 1,000 gallons of paint each quarter.
   Assume that Paris currently has 80,000 gallons of paint in inventory and would like to end the year with an inventory of at least 20,000 gallons.
   a. Determine the minimum constant workforce plan for Paris Paint and the cost of the plan. Assume that stock-outs are not allowed.
   b. Determine the zero inventory plan that hires and fires workers each quarter to match demand as closely as possible and the cost of that plan.

   c. If Paris were able to back-order excess demand at a cost of $2 per gallon per quarter, determine a minimum constant workforce plan that holds less inventory than the plan you found in part (*a*), but incurs stock-outs in quarter 2. Determine the cost of the new plan.

33. Consider the problem of Paris Paint presented in Problem 32. Suppose that the plant has the capacity to employ a maximum of 370 workers. Suppose that regular-time employee costs are $12.50 per hour. Assume seven-hour days, five-day weeks, and four-week months. Overtime is paid on a time-and-a-half basis. Subcontracting is available at a cost of $7 per gallon of paint produced. Overtime is limited to three hours per employee per day, and no more than 100,000 gallons can be subcontracted in any quarter. Determine a policy that meets the demand, and the cost of that policy. (Assume stock-outs are not allowed.)

34. *a.* Formulate Problem 32 as a linear program. Assume that stock-outs are not allowed.
   b. Solve the linear program. Round the variables and determine the cost of the resulting plan.

35. *a.* Formulate Problem 33 as a linear program.
   b. Solve the linear program. Round the variables and determine the cost of the resulting plan.

36. The Mr. Meadows Cookie Company can obtain accurate forecasts for 12 months based on firm orders. These forecasts and the number of workdays per month are

Month	Demand Forecast (in thousands of cookies)	Workdays
1	850	26
2	1,260	24
3	510	20
4	980	18
5	770	22
6	850	23
7	1,050	14
8	1,550	21
9	1,350	23
10	1,000	24
11	970	21
12	680	13

    During a 46-day period when there were 120 workers, the firm produced 1,700,000 cookies. Assume that there are 100 workers employed at the beginning of month 1 and zero starting inventory.
   a. Find the minimum constant workforce needed to meet monthly demand.
   b. Assume $c_I = 0.10 per cookie per month, $c_H = 100, and $c_F = 200. Add columns that give the cumulative on-hand inventory and inventory cost. What is the total cost of the constant workforce plan?

c. Construct a plan that changes the level of the workforce to meet monthly demand as closely as possible. In designing the logic for your calculations, be sure that inventory does not go negative in any month. Determine the cost of this plan.

37. The Yeasty Brewing Company produces a popular local beer known as Iron Stomach. Beer sales are somewhat seasonal, and Yeasty is planning its production and manpower levels on March 31 for the next six months. The demand forecasts are

Month	Production Days	Forecasted Demand (in hundreds of cases)
April	11	85
May	22	93
June	20	122
July	23	176
August	16	140
September	20	63

As of March 31, Yeasty had 86 workers on the payroll. Over a period of 26 working days when there were 100 workers on the payroll, Yeasty produced 12,000 cases of beer. The cost to hire each worker is $125 and the cost of laying off each worker is $300. Holding costs amount to 75 cents per case per month.

As of March 31, Yeasty expects to have 4,500 cases of beer in stock, and it wants to maintain a minimum buffer inventory of 1,000 cases each month. It plans to start October with 3,000 cases on hand.

a. Based on this information, find the minimum constant workforce plan for Yeasty over the six months, and determine hiring, firing, and holding costs associated with that plan.

b. Suppose that it takes one month to train a new worker. How will that affect your solution?

c. Suppose that the maximum number of workers that the company can expect to be able to hire in one month is 10. How will that affect your solution to part (*a*)?

d. Determine the zero inventory plan and the cost of that plan (you may ignore the conditions in parts (*b*) and (*c*)).

38. *a.* Formulate the problem of planning Yeasty's production levels, as described in Problem 37, as a linear program. (You may ignore the conditions in parts (*b*) and (*c*).)

b. Solve the resulting linear program. Round the appropriate variables and determine the cost of your solution.

c. Suppose Yeasty does not wish to fire any workers. What is the optimal plan subject to this constraint?

39. A local canning company sells canned vegetables to a supermarket chain in the Minneapolis area. A typical case of canned vegetables requires an average of 0.2 day of labor to produce. The aggregate inventory on hand at the end of June is 800 cases. The demand for the vegetables can be accurately predicted for

about 18 months based on orders received by the firm. The predicted demands for the next 18 months are

Month	Forecasted Demand (hundreds of cases)	Workdays	Month	Forecasted Demand (hundreds of cases)	Workdays
July	23	21	April	29	20
August	28	14	May	33	22
September	42	20	June	31	21
October	26	23	July	20	18
November	29	18	August	16	14
December	58	10	September	33	20
January	19	20	October	35	23
February	17	14	November	28	18
March	25	20	December	50	10

The firm currently has 25 workers. The cost of hiring and training a new worker is $1,000, and the cost to lay off one worker is $1,500. The firm estimates a cost of $2.80 to store a case of vegetables for a month. They would like to have 1,500 cases in inventory at the end of the 18-month planning horizon.

a. Develop a spreadsheet to find the minimum constant workforce aggregate plan and determine the total cost of that plan.

b. Develop a spreadsheet to find a plan that hires and fires workers monthly in order to minimize inventory costs. Determine the total cost of that plan as well.

c. Graph the cumulative net demand and, using the graph, find a plan that changes the workforce level exactly once during the 18 months and has a lower cost than either of the plans considered in parts (a) and (b).

40. Consider the linear production rule of Holt, Modigliani, Muth, and Simon, appearing in Problem 24. Design a spreadsheet to compute P_t and W_t from demand forecasts. In particular, suppose that at the end of December 1995 the workforce was at 180 and the inventory was at 45 aggregate units.

a. Using the forecasts of demand for the months January through December 1996 given in Problem 24, find the recommended levels of aggregate production and workforce size for January.

*b. Design your spreadsheet so that it can efficiently update recommended levels as demands are realized. In particular, suppose that the demand for January was actually 175 and the forecast for January 1997 was 130. What are the recommended levels of aggregate production and workforce size for February? (Hint: Store the demand forecasts and coefficients in separate columns and modify the computing formula as new data are appended to the old.)

*c. The actual realizations of demand and new forecasts of demand for the remainder of 1996 are given below. Determine updated values for P_t and W_t each month.

Month	Actual Demand (1996)	New Forecast (1997)
February	148	160
March	200	175
April	190	120
May	180	190
June	130	230
July	95	280
August	120	210
September	68	110
October	88	95
November	75	92
December	60	77

APPENDIX 3–A   GLOSSARY OF NOTATION FOR CHAPTER 3

α = Smoothing constant for production and demand used in Bowman's model.

β = Smoothing constant for inventory used in Bowman's model.

c_F = Cost of firing one worker.

c_H = Cost of hiring one worker.

c_I = Cost of holding one unit of stock for one period.

c_O = Cost of producing one unit on overtime.

c_P = Penalty cost for back orders.

c_R = Cost of producing one unit on regular time.

c_S = Cost to subcontract one unit of production.

c_U = Idle cost per unit of production.

D_t = Forecast of demand in period t.

F_t = Number of workers fired in period t.

H_t = Number of workers hired in period t.

I_t = Inventory level in period t.

K = Number of aggregate units produced by one worker in one day.

λ_j = Annual demand for family j (refer to Section 9).

n_t = Number of production days in period t.

O_t = Overtime production in units.

P_t = Production level in period t.

S_t = Number of units subcontracted from outside.

T = Number of periods in the planning horizon.

U_t = Worker idle time in units ("undertime").

W_t = Workforce level in period t.

Bibliography

Allen, S. J., and E. W. Schuster. "Practical Production Scheduling with Capacity Constraints and Dynamic Demand: Family Planning and Disaggregation." *Production and Inventory Management Journal* 35, no. 4 (1994), pp. 15–20.

Axsater, S. "Aggregation of Product Data for Hierarchical Production Planning." *Operations Research* 29 (1981), pp. 744–56.

Bergstrom, G. L., and B. E. Smith. "Multi-Item Production Planning—An Extension of the HMMS Rules." *Management Science* 16 (1970), pp. 614–29.

Bitran, G. R., and A. Hax. "Disaggregation and Resource Allocation Using Convex Knapsack Problems with Bounded Variables." *Management Science* 27 (1981), pp. 431–41.

Bowman, E. H. "Production Scheduling by the Transportation Method of Linear Programming." *Operations Research* 4 (1956), pp. 100–103.

Bowman, E. H. "Consistency and Optimality in Managerial Decision Making." *Management Science* 9 (1963), pp. 310–21.

Buffa, E. S. *Modern Production/Operations Management.* 7th ed. New York: John Wiley & Sons, 1983.

Buffa, E. S., and W. H. Taubert. *Production-Inventory Systems: Planning and Control.* Rev. ed. New York: McGraw-Hill/Irwin, 1972.

"Caterpillar Tractor Company." Harvard Business School Case 385–276 (1985).

Chung, C., and L. J. Krajewski. "Planning Horizons for Master Production Scheduling." *Journal of Operations Management,* August 1984, pp. 389–406.

Cohen, M. A.; M. L. Fisher; and J. Jaikumar. "International Manufacturing and Distribution Networks." In *Managing International Manufacturing,* ed. K. Ferdows, pp. 67–93. Amsterdam: North Holland, 1989.

Damon, W. W., and R. Schramm. "A Simultaneous Decision Model for Production, Marketing, and Finance." *Management Science* 19 (1972), pp. 161–72.

Ebert, R. J. "Aggregate Planning with Learning Curve Productivity." *Management Science* 23 (1976), pp. 171–82.

Erenguc, S., and S. Tufekci. "A Transportation Type Aggregate Production Model with Bounds on Inventory and Backordering." *European Journal of Operations Research* 35 (1988), pp. 414–25.

Ferdows, K., ed. *Managing International Manufacturing.* Amsterdam: North Holland, 1989.

Hansmann, F., and S. W. Hess. "A Linear Programming Approach to Production and Employment Scheduling." *Management Technology* 1 (1960), pp. 46–51.

Hax, A. C., and D. Candea. *Production and Inventory Management.* Englewood Cliffs, NJ: Prentice Hall, 1984.

Hax, A. C., and H. C. Meal. "Hierarchical Integration of Production Planning and Scheduling." In *TIMS Studies in Management Science.* Volume 1, *Logistics,* ed. M. Geisler. New York: Elsevier, 1975.

Hillier, F. S., and G. J. Lieberman. *Introduction to Operations Research.* 5th ed. San Francisco: Holden Day, 1990.

Holt, C. C.; F. Modigliani; and J. F. Muth. "Derivation of a Linear Decision Rule for Production and Employment." *Management Science* 2 (1956), pp. 159–77.

Holt, C. C.; F. Modigliani; J. F. Muth; and H. A. Simon. *Planning Production, Inventories, and Workforce.* Englewood Cliffs, NJ: Prentice Hall, 1960.

Holt., C. C.; F. Modigliani; and H. A. Simon. "A Linear Decision Rule for Employment and Production Scheduling." *Management Science* 2 (1955), pp. 1–30.

Huchzermeier, A., and M. A. Cohen. "Valuing Operational Flexibility under Exchange Rate Risk." *Operations Research* 44 (1996), pp. 100–13.

Kamien, M. I., and L. Li. "Subcontracting, Coordination, Flexibility, and Production Smoothing in Aggregate Planning." *Management Science* 36 (1990), pp. 1352–63.

Kogut, B., and N. Kulatilaka. "Operating Flexibility, Global Manufacturing, and the Option Value of a Multinational Network." *Management Science* 40 (1994), pp. 123–39.

Leitch, R. A. "Marketing Strategy and Optimal Production Schedule." *Management Science* 20 (1974), pp. 903–11.

McGrath, M. E., and R. B. Bequillard. "International Manufacturing Strategies and Infrastructural Considerations in the Electronics Industry." In *Managing International Manufacturing,* ed. K. Ferdows, pp. 23–40. Amsterdam: North Holland, 1989.

Newson, E. F. P. "Multi-Item Lot Size Scheduling by Heuristic, Part 1: With Fixed Resources." *Management Science* 21 (1975a), pp. 1186–93.

Newson, E. F. P. "Multi-Item Lot Size Scheduling by Heuristic, Part 2: With Fixed Resources." *Management Science* 21 (1975b), pp. 1194–1205.

Schrage, L. *Linear, Integer, and Quadratic Programming with LINDO.* Palo Alto, CA: Scientific Press, 1984.

Schwarz, L. B., and R. E. Johnson. "An Appraisal of the Empirical Performance of the Linear Decision Rule for Aggregate Planning." *Management Science* 24 (1978), pp. 844–49.

Silver, E. A., and R. Peterson. *Decision Systems for Inventory Management and Production Planning.* 2nd ed. New York: John Wiley & Sons, 1985.

Taubert, W. H. "A Search Decision Rule for the Aggregate Scheduling Problem." *Management Science* 14 (1968), pp. B343–59.

Vollmann, T. E.; W. L. Berry; and D. C. Whybark. *Manufacturing, Planning, and Control Systems.* 3rd ed. New York: McGraw-Hill/Irwin, 1992.

Zoller, K. "Optimal Disaggregation of Aggregate Production Plans." *Management Science* 17 (1971), pp. B533–49.

$$Q = \sqrt{\frac{2K\lambda}{h}}$$

One learns by doing

SUPPLEMENT 1

LINEAR PROGRAMMING

1 Introduction

Linear programming is a mathematical technique for solving a broad class of optimization problems. These problems require maximizing or minimizing a linear function of n real variables subject to m constraints. One can formulate and solve a large number of real problems with linear programming. A partial list includes

1. Scheduling of personnel.
2. Several varieties of blending problems including cattle feed, sausages, ice cream, and steel making.
3. Inventory control and production planning.
4. Distribution and logistics problems.
5. Assignment problems.

Problems with thousands of variables and thousands of constraints are easily solvable on computers today. Linear programming was developed to solve logistics problems during World War II. George Dantzig, a mathematician employed by the RAND Corporation at the time, developed a solution procedure he labeled the Simplex Method. That the method turned out to be so efficient for solving large problems quickly was a surprise even to its developer. That fact, coupled with the simultaneous development of the electronic computer, established linear programming as an important tool in logistics management. The success of linear programming in industry spawned the development of the disciplines of operations research and management science. The Simplex Method has withstood the test of time. Only in recent years has another method been developed that potentially could be more efficient than the Simplex Method for solving very large, specially structured, linear programs. This method, known as Karmarkar's Algorithm, is named for the Bell Labs mathematician who conceived it.

In the next section we consider a typical manufacturing problem that we formulate and solve using linear programming. Later we explore how one solves small problems (that is, having exactly two decision variables) graphically, and how one solves large problems using a computer.

2 A Prototype Linear Programming Problem

Example S1.1

Sidneyville manufactures household and commercial furnishings. The Office Division produces two desks, rolltop and regular. Sidneyville constructs the desks in its plant outside Medford, Oregon, from a selection of woods. The woods are cut to a uniform thickness of 1 inch. For this reason, one measures the wood in units of square feet. One rolltop desk requires 10 square feet of pine, 4 square feet of cedar, and 15 square feet of maple. One regular desk requires 20 square feet of pine, 16 square feet of cedar, and 10 square feet of maple. The desks yield respectively $115 and $90 profit per sale. At the current time the firm has available 200 square feet of pine, 128 square feet of cedar, and 220 square feet of maple. They have backlogged orders for both desks and would like to produce the number of rolltop and regular desks that would maximize profit. How many of each should they produce?

Solution

The first step in formulating a problem as a linear program is to identify the decision variables. In this case there are two decisions required: the number of rolltop desks to produce and the number of regular desks to produce. We must assign symbol names to each of these decision variables.

Let

x_1 = Number of rolltop desks to be produced,
x_2 = Number of regular desks to be produced.

Now that we have identified the decision variables, the next step is to identify the objective function and the constraints. The objective function is the quantity we wish to maximize or minimize. The objective is to maximize the profits, so the objective function equals the total profit when producing x_1 rolltop desks and x_2 regular desks. Each rolltop desk contributes $115 to profit, so the total contribution to profit from all rolltop desks is $115x_1$. Similarly, the contribution to profit from all regular desks is $90x_2$. Hence, the total profit is $115x_1 + 90x_2$. This is known as the objective function.

The next step is to identify the constraints. The number of desks Sidneyville can produce is limited by the amount of wood available. The three types of wood constitute the critical resources. To obtain the constraints, we need to find expressions for the amount of each type of wood consumed by construction of x_1 rolltop desks and x_2 regular desks. Those expressions are then bounded by the amount of each type of wood available.

The number of square feet of pine used to make x_1 rolltop desks is $10x_1$. The number of square feet of pine used to make x_2 regular desks is $20x_2$. It follows that the total amount of pine consumed in square feet is $10x_1 + 20x_2$. This quantity cannot exceed the number of square feet of pine available, which is 200. Hence we obtain the first constraint:

$$10x_1 + 20x_2 \le 200.$$

The other two constraints are similar. The second constraint is to ensure that the firm does not exceed the available supply of cedar. Each rolltop desk requires 4 square feet of cedar, so x_1 rolltop desks require $4x_1$ square feet of cedar. Each regular desk requires 16 square feet of cedar, so x_2 regular desks require $16x_2$ square feet of cedar. It follows that the constraint on the supply of cedar is

$$4x_1 + 16x_2 \le 128.$$

In the same way, the final constraint ensuring that we do not exceed the supply of maple is

$$15x_1 + 10x_2 \le 220.$$

Because we cannot produce a negative number of desks, we also include nonnegativity constraints:

$$x_1 \ge 0,$$

$$x_2 \ge 0.$$

We have now constructed the complete linear programming formulation of the Sidneyville problem. The goal is to find values of x_1 and x_2 to maximize $115x_1 + 90x_2$, subject to the constraints.

$$10x_1 + 20x_2 \le 200,$$

$$4x_1 + 16x_2 \le 128,$$

$$15x_1 + 10x_2 \le 220,$$

$$x_1, x_2 \ge 0.$$

We next consider how such a problem is solved. We will briefly outline the theory behind the solution technique known as the Simplex Method. However, as linear programming problems are almost never solved by hand any longer, you will not need to understand the mechanics of the Simplex Method in order to use linear programming. You will need to know how to formulate problems as linear programs, enter the formulations into the computer, recognize special problems, and analyze the computer's output.

A typical computer output is given in Table S1–1.

This output tells us that at the optimal solution, the value of x_1 is 12 and the value of x_2 is 4. That is, Sidneyville should produce exactly 12 rolltop desks and 4 regular desks. The value of the objective function at the optimal solution is $1,740.00. That this is the (unique) optimal solution means the following: every other

---

TABLE S1–1   **Partial Computer Output for Sidneyville Problem**

```
LP OPTIMUM FOUND AT STEP 2

OBJECTIVE FUNCTION VALUE

1) 1740.000000

VARIABLE          VALUE          REDUCED COST
    X1          12.000000          .000000
    X2           4.000000          .000000
```

pair of values of x_1 and x_2 will result in either a lower profit, not meeting the constraints, or both.

Sidneyville's manager of production planning is very skeptical about mathematics. When presented with this solution, his response was. "There's only one problem with this production plan. We have a specialist make the rolltop portion of the rolltop desk. She can only do four desks a day and we want to be ready to ship out in two days. There is no way we can produce twelve of those desks in two days. I knew this math stuff was a lot of hooey!"

The manager was wrong. The trouble is not that the formulation is incorrect, but that it does not include all relevant constraints, as labor hours turned out to be a critical resource. The lesson here is that for the final solution to be meaningful, the model must include *all* relevant constraints. ■.

3 Statement of the General Problem

The Sidneyville problem is an example of a linear program in which there are two decision variables and three constraints. Linear programming problems may have any number of decision variables and any number of constraints. Suppose that there are n decision variables, labeled $x_1, x_2, \ldots, x_n$, subject to m resource constraints. Then we may write the problem of maximizing the objective subject to the constraints as

$$\text{Maximize } c_1x_1 + c_2x_2 + \cdots + c_nx_n,$$

$$\text{subject to } a_{11}x_1 + a_{12}x_2 + \cdots + a_{1n}x_n \leq b_1,$$

$$a_{21}x_1 + a_{22}x_2 + \cdots + a_{2n}x_n \leq b_2,$$

$$\vdots$$

$$a_{m1}x_1 + a_{m2}x_2 + \cdots + a_{mn}x_n \leq b_m,$$

$$x_1, x_2, \ldots, x_n \geq 0.$$

Interpret $c_1, c_2, \ldots, c_n$ as the profit coefficients per unit of output of the activities $x_1, x_2, \ldots, x_n$; a_{ij} as the amount of resource i consumed by one unit of activity j; and b_1 as the amount of resource i available, for $i = 1, \ldots, m$ and $j = 1, \ldots, n$. We require that the constants $b_1, \ldots, b_m$ be nonnegative. This particular formulation includes problems in which we want to maximize profit subject to constraints on the available resources. However, linear programming can be used to solve a much larger variety of problems. Other possible formulations will be discussed below.

Definitions of Commonly Used Terms

1. *Objective function.* This is the quantity we wish to maximize or minimize. In the formulation above, the objective function is the term $c_1x_1 + c_2x_2 + \cdots + c_nx_n$. In business applications, one typically minimizes cost or maximizes profit. We use the abbreviations "min" for a minimization problem and "max" for a maximization problem.

2. *Constraints.* Each constraint is a linear inequality or equation; that is, a linear combination of the problem variables followed by a relational operator ($\leq$ or $=$ or $\geq$) followed by a nonnegative constant. Although the formulation above shows all $\leq$ constraints, $\geq$ and $=$ type constraints are also common. For example, suppose there is a

contractual agreement that requires a minimum number of labor hours daily. This would result in a $\geq$ constraint.

3. *Right-hand side.* The right-hand side is the constant following the relational operator in a constraint. The constants b_1, b_2, . . . , b_m are the right-hand sides above. These constants are required to be nonnegative numbers. However, we do *not* require that the constants a_{ij} be nonnegative. This means that any constraint can be written with a nonnegative right-hand side by multiplying through by the constant -1 whenever the right-hand side is negative. Consider the following simple example. Suppose that when formulating a problem as a linear program we obtain the constraint

$$4x_1 - 2x_2 \leq -5.$$

Because the right-hand side is negative, this is not a legal constraint. However, if we multiply through by -1, this constraint becomes

$$-4x_1 + 2x_2 \geq 5,$$

which is acceptable.

4. *Feasible region.* The feasible region is the set of values of the decision variables, x_1, x_2, . . . , x_n, that satisfy the constraints. Because each of the constraints is generated by a linear equation or linear inequality, the feasible region has a particular structure. The technical term for this structure is *convex polytope*. In two dimensions, a convex polytope is a convex set with boundaries that are straight lines. In three dimensions, the boundaries are formed by planes. A convex set is characterized as follows: pick any two points in the set and connect them with a straight line; the line lies entirely within the set.

5. *Extreme points.* Because of the structure of the feasible region, there will be a finite number of feasible points, with the property that they cannot be expressed as a linear combination of any other set of feasible points. These points are known as extreme points or corner points, and they play an important role in linear programming. The concept of extreme points will become clearer when we discuss graphical solutions.

6. *Feasible solution.* A feasible solution is one particular set of values of the decision variables that satisfies the constraints. A feasible solution is also one point in the feasible region. It may be an extreme point or an interior point.

7. *Optimal solution.* The optimal solution is the feasible solution that maximizes or minimizes the objective function. In some cases the optimal solution may not be unique. When this is the case, there will be an infinite number of optimal solutions.

Features of Linear Programs

Linear programming is a very powerful tool. Many real problems have been successfully formulated and solved using this technique. However, to use the method correctly, one must be aware of its limitations. Two important features of linear programs are linearity and continuity. Many problems that may appear to be solvable by linear programming fail one or both of these two crucial tests.

Linearity. Optimization problems can be formulated as linear programs only when (*a*) the objective can be expressed as a linear function of the decision variables and (*b*) all constraints can be expressed as linear functions of the decision variables.

Linearity implies that quantities change in fixed proportions. For example, if it costs \$10 to produce one unit, then it costs \$20 to produce two units, and \$100 to produce ten units. If one ounce of orange juice supplies 30 mg of vitamin C, then three ounces must supply 90 mg. Linearity must hold in the objective function and the constraints. In the objective function, this means that the profit or cost per unit must be the same independent of the number of units. In the constraints, linearity means that the amount of each resource consumed is the same per unit whether one produces a single unit or many units.

However, one often observes nonlinear relationships in the real world. Economies of scale in production mean that the marginal cost of producing a unit decreases as the number of units produced increases. When this occurs, the cost of production is a nonlinear function of the number of units produced. An example of scale economies is a fixed setup cost for production. The formula for the EOQ discussed in Chapter 4 says that the lot size increases as the square root of the demand rate. Hence, EOQ is a nonlinear function of the demand. When either the objective function or a constraint is a nonlinear function of the decision variables, the problem is a nonlinear programming problem and cannot be solved by linear programming.[1]

Continuity.   This means that the decision variables should be continuous (that is, able to assume any nonnegative value) as opposed to discrete or integer valued. This can be a serious restriction. The solution to many problems makes sense only if the decision variables are integer valued. In particular, Example S1.1 is, strictly speaking, not a linear programming problem because the number of desks produced must be integer valued. (We were lucky that the optimal solution turned out to be integer valued in this case.) One might think that the optimal integer solution is equal to the continuous solution rounded off to the nearest integer. Unfortunately, this is not always the case. First, rounding may lead to infeasibility; that is, the rounded-off solution may lie outside the feasible region. Second, even if the rounded solution is feasible, it may not be optimal. It can happen that the optimal integer solution is in an entirely different portion of the feasible region than the rounded-off linear programming solution!

To give the reader some idea of the difficulties that can arise when the solution must be integer valued, consider Example S1.1. Suppose that the profit from selling rolltop desks was $150 rather than $115. Then the objective would be to maximize $150x_1 + 90x_2$ subject to the same set of constraints. The optimal linear programming solution is

$$x_1 = 14.666666\ldots.$$

$$x_2 = 0.$$

Rounding the solution to the nearest integer gives $x_1 = 15$ and $x_2 = 0$, which is infeasible. Substituting these values into the final constraint results in a requirement of 225 square feet of maple. However, only 220 square feet are available. Rounding x_1 down to 14 results in a feasible but suboptimal solution. At $x_1 = 14$ and $x_2 = 0$, there are 10 feet of maple still available, which is enough to make one regular desk. The optimal integer solution in this case is $x_1 = 14$ and $x_2 = 1$.

When the decision variables must be integer valued, we say that the problem is an integer linear programming problem. Finding optimal integer solutions to linear programs can be very time-consuming, even for modest-sized problems. However, Excel

---

[1]In some circumstances, convex programming problems can be solved by linear programming by approximating the objective function with a piecewise-linear function. An example of this approach appears at the end of Section 5 of Chapter 3.

does offer an option for defining some or all of the problem variables as integer valued. Excel does a fine job solving small integer linear programming problems. For larger problems, one would use a computer program designed to solve integer programming problems. In some cases, and especially when the values of the variables are relatively large, careful rounding of the linear programming solution should give acceptable results.

4 Solving Linear Programming Problems Graphically

Graphing Linear Inequalities

In this section we will show how to solve two-variable linear programming problems graphically. Although most real problems have more than two variables, understanding the procedure for solving two-variable linear programs will improve your grasp of the concepts underlying the Simplex Method.

The first step is to graph the linear inequalities represented by the constraints. A linear inequality corresponds to all points in the plane on one side of a straight line. There are thus two steps to graphing linear inequalities:

1. Draw the straight line representing the boundary of the region corresponding to the constraint expressed as an equation.
2. Determine which side of the line corresponds to the inequality.

To illustrate the method, consider the first constraint in Example S1.1.

$$10x_1 + 20x_2 \leq 200.$$

The boundary of the region represented by this inequality is the straight line

$$10x_1 + 20x_2 = 200.$$

The easiest way to graph a straight line is to determine the two intercepts. These are found by setting x_2 to zero and solving for x_1, and then setting x_1 to zero and solving for x_2. First setting x_2 to zero gives

$$10x_1 = 200 \qquad \text{or} \qquad x_1 = 20.$$

Similarly, setting x_1 to zero and solving for x_2 gives

$$20x_2 = 200 \qquad \text{or} \qquad x_2 = 10.$$

Hence, the line $10x_1 + 20x_2 = 200$ must pass through the points (20, 0) (the x_1 intercept) and (0, 10) (the x_2 intercept). A graph of this line is shown in Figure S1–1. Now that we have graphed the line defining the boundary of the half space, we must determine which side of the line corresponds to the inequality. To do so, we pick any point *not* on the line, substitute the values of x_1 and x_2, and see if the inequality is satisfied or not. If the inequality is satisfied, then that point belongs in the half space; if it is not, then that point does not belong in the half space. If the boundary line does not go through the origin (that is, the point (0, 0)) then the most straightforward approach is to use the origin as the point to be tested.

Substituting (0, 0) into the inequality, we obtain

$$(10)(0) + (20)(0) = 0 \leq 200.$$

FIGURE S1–1

Graphing a constraint boundary

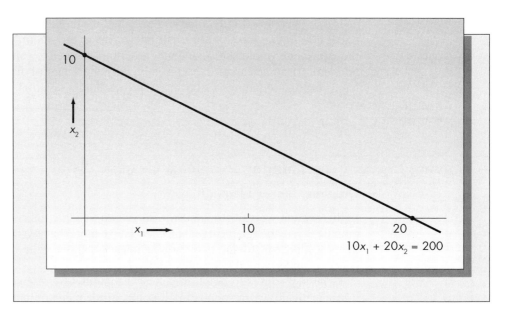

$$10x_1 + 20x_2 = 200$$

FIGURE S1–2

Half space representing the inequality $10x_1 + 20x_2 \leq 200$

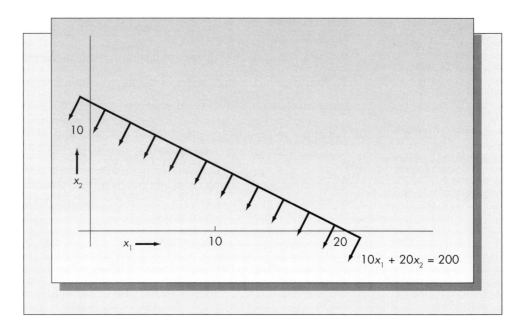

$$10x_1 + 20x_2 = 200$$

Because 0 is less than 200, the test is satisfied and the origin lies within the region represented by the inequality. This means that the graph of the inequality includes all the points below the boundary line in Figure S1–1, as shown in Figure S1–2.

The origin test to determine which side of the line is appropriate only works when the boundary line itself does not go through the origin. If it does, then some other point not on the line must be used for the test. For example, consider the constraint

$$4x_1 - 3x_2 \geq 0.$$

When we try to graph the line $4x_1 - 3x_2 = 0$, we see that substituting $x_1 = 0$ gives $x_2 = 0$ and substituting $x_2 = 0$ gives $x_1 = 0$. This means that the line passes through the origin.

FIGURE S1–3

Graphing a constraint boundary that passes through the origin

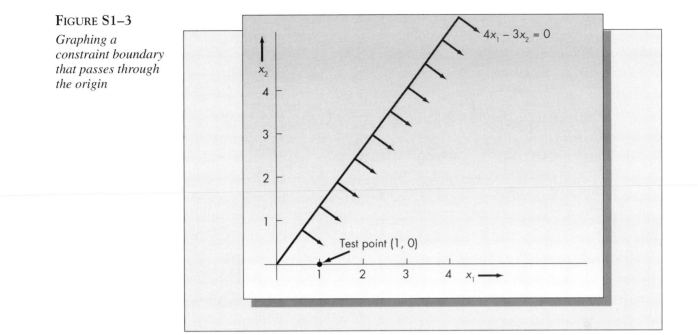

In order to graph the line, we must determine another point that lies on it. Just pick any value of x_1 and solve for the corresponding value of x_2. For example, substituting $x_1 = 3$ gives $x_2 = 4$, meaning that the point $(3, 4)$ lies on the line as well as the point $(0, 0)$. The boundary line is graphed by connecting these points, as shown in Figure S1–3.

Next, we determine which side of the inequality corresponds to the region of interest. As noted above, the origin test does not work when the line passes through the origin. We pick any point *not* on the line to do the test. In this case, one point that does not lie on the line is $x_1 = 1$ and $x_2 = 0$. Substituting these values into the inequality gives

$$(4)(1) - 0 = 4 > 0.$$

The inequality is satisfied, so the point $(1, 0)$ lies in the region. The inequality corresponds to the points below the line as pictured in Figure S1–3.

Graphing the Feasible Region

The graph of the feasible region is found by graphing the linear inequalities represented by the constraints and determining the region of intersection of the corresponding half spaces. We will determine a graph of the feasible region for the Sidneyville problem in this way.

We have graphed the feasible region corresponding to the first constraint in Figure S1–2. Consider the other two constraints:

$$4x_1 + 16x_2 \le 128,$$

$$15x_1 + 10x_2 \le 220.$$

The half spaces corresponding to these constraints are found in the same way. First we graph the straight lines corresponding to the region boundaries. In the first case the intercepts are $x_1 = 32$ and $x_2 = 8$, and in the second case they are $x_1 = 14.6667$ and

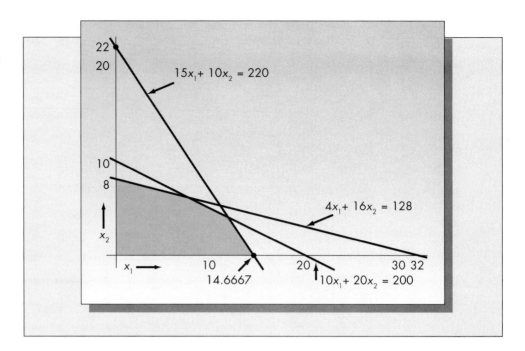

$x_2 = 22$. Using the origin test, we see that the appropriate half spaces are the points lying below both lines. In addition, we must also include the nonnegativity constraints, $x_1 \geq 0$ and $x_2 \geq 0$. The resulting feasible region is pictured in Figure S1–4.

Finding the Optimal Solution

The feasible region pictured in Figure S1–4 has several interesting properties that are common to all linear programming problems. Pick any two feasible solutions (that is, points in the region) and connect these points with a straight line. The resulting line lies completely in the region, meaning that the feasible region is a convex set. The region boundaries are straight lines. These lines intersect at points known as the extreme points. There are a total of five extreme points in the feasible region pictured in Figure S1–4.

An important property of linear programs is that the optimal solution always occurs at an extreme point of the feasible region. For the Sidneyville problem, this means that among the infinite number of solutions (i.e., points) in the feasible region, the optimal solution will be one of only five points![2] This is true no matter what objective function we assume. This means that we can find the optimal solution to the Sidneyville problem by identifying the five extreme points, substituting the (x_1, x_2) coordinates of these points into the objective function, and determining the point that results in the maximum profit. We will consider this method first, even though there is a more efficient graphical solution procedure.

From Figure S1–4 we see that one of the extreme points is the origin $(0, 0)$. Another is the x_2 intercept corresponding to the constraint $4x_1 + 16x_2 = 128$, which is $(0, 8)$. A third is the x_1 intercept corresponding to the constraint $15x_1 + 10x_2 = 220$, which is

---

[2]It can happen that two extreme points are optimal, in which case all the points on the line connecting them are optimal as well.

(14.6667, 0). The other two extreme points correspond to the intersections of pairs of boundary lines. They are found by simultaneously solving the equations corresponding to the boundaries.

First, we simultaneously solve the equations

$$4x_1 + 16x_2 = 128,$$
$$10x_1 + 20x_2 = 200.$$

These equations can be solved in several ways. Multiplying the first equation by 10 and the second by 4 gives

$$40x_1 + 160x_2 = 1,280,$$
$$40x_1 + 80x_2 = 800.$$

Subtracting the second equation from the first yields

$$80x_2 = 480$$
$$x_2 = 6.$$

The value of x_1 is found by substituting $x_2 = 6$ into either equation (both will yield the same value for x_1). Substituting into the first equation gives

$$4x_1 + (16)(6) = 128$$
$$4x_1 = 128 - 96 = 32$$
$$x_1 = 8.$$

Check that substituting $x_2 = 6$ into the equation $10x_1 + 20x_2 = 200$ gives the same result.

The last extreme point is found by solving

$$15x_1 + 10x_2 = 220,$$
$$10x_1 + 20x_2 = 200.$$

simultaneously. We will not present the details of this calculation. The reader should be able to show by the same methods as above that the simultaneous solution in this case is

$$x_1 = 12,$$
$$x_2 = 4.$$

We have now identified all five extreme points. The next step is to substitute the corresponding values of x_1 and x_2 into the objective function and see which gives the largest profit. The objective function is $115x_1 + 90x_2$.

Extreme Point	Value of Objective Function
(0, 0)	$(115)(0) + (90)(0) = 0$
(0, 8)	$(115)(0) + (90)(8) = 720$
(14.666 . . . ,0)	$(115)(14.666 \dots) + (90)(0) = 1,686.67$
(8, 6)	$(115)(8) + (90)(6) = 1,460$
(12, 4)	$(115)(12) + (90)(4) = 1,740$

The maximum value of the objective function is 1,740 and is achieved at the point (12, 4). This agrees with the computer output in Table S1–1.

Hence, we have shown in this section that one method of finding the optimal solution to a linear programming problem is to find all the extreme points, substitute their values into the objective function, and pick the one that gives the largest objective function value for maximization problems or the smallest objective function value for minimization problems. Next we show how one can quickly identify the optimal extreme point graphically.

Identifying the Optimal Solution Directly by Graphical Means

One identifies the optimal solution directly in the following way. The objective function is a linear combination of the decision variables. In our example the objective function is $115x_1 + 90x_2$. Consider the family of straight lines defined by the equation

$$Z = 115x_1 + 90x_2.$$

As Z is varied, one generates a family of parallel lines. The variable Z is the profit obtained when producing x_1 rolltop desks and x_2 regular desks such that (x_1, x_2) lies on the line $Z = 115x_1 + 90x_2$. As an example, consider $Z = 4,140$. Figure S1–5 shows the line $4,140 = 115x_1 + 90x_2$. Notice that it lies completely outside the feasible region, meaning that no feasible combination of x_1 and x_2 results in a profit of \$4,140. Reducing Z to 2,070 drops the Z line closer to the feasible region, as also pictured in Figure S1–5.

The graphical method of identifying the optimal solution is to pick one value of Z, such as $Z = 3,000$, that takes us beyond the feasible region, place a ruler on the Z line, and move the ruler parallel to the Z line toward the feasible region. The extreme point that is hit first is the optimal solution. Once one graphically determines which extreme point is optimal, the coordinates of that point are found by solving the appropriate equations as shown above. This approach avoids having to identify all the extreme points of the feasible region.

FIGURE S1–5

Approaching the feasible region with the Z line

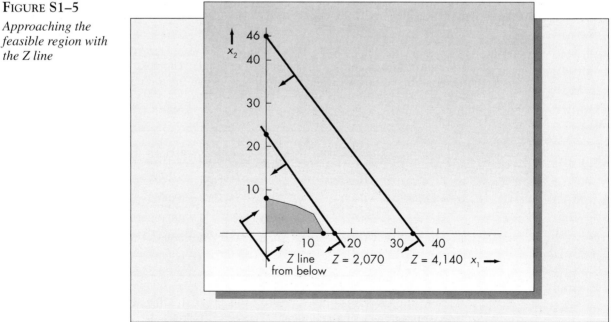

One problem can arise. If we pick a small starting value of Z and move toward the feasible region from below, this method will identify a *different* extreme point. In Figure S1–5, suppose that we chose $Z = -3,000$. Then the Z line would be located under the feasible region. As we moved the Z line toward the feasible region, the first extreme point encountered is the origin. This means that the origin solves Example S1.1 with a minimization objective rather than a maximization objective (i.e., if your goal is to *minimize* profit, the best strategy is not to produce any desks).

Hence, this approach identifies two extreme points. One corresponds to the maximum solution and one to the minimum solution. In this case it is obvious which is which. When it is not obvious, the (x_1, x_2) values for each extreme point should be found and substituted into the objective function to be certain which is the minimum solution and which is the maximum solution.

5 The Simplex Method: An Overview

The Simplex Method is an algorithm that moves sequentially from one extreme point to another until it reaches the optimal solution. If the origin (that is, all problem variables set to zero) is a feasible solution, it will serve as the initial extreme point.[3] At each iteration, the method considers all adjacent extreme points (those that can be reached by moving along an edge), and moves to the one that gives the best improvement in the objective function. The algorithm continues to move from one adjacent extreme point to another, finally terminating when the optimal solution is reached.

For the problem of Example S1.1 the origin is feasible, so that is the initial feasible solution. The two extreme points adjacent to the origin are $(x_1, x_2) = (14.6667, 0)$ and $(x_1, x_2) = (0, 8)$. The largest improvement in the objective function is obtained by moving to the point $(14.6667, 0)$. There are two extreme points adjacent to $(14.6667, 0)$. They are $(0, 0)$ and $(12, 4)$. Clearly, a greater improvement is obtained by moving to $(12, 4)$. At this point the method recognizes that the current solution is optimal, as a movement to another adjacent extreme point lowers the profit.

In the worst case, the Simplex Method could conceivably need to search all the extreme points of the feasible region before identifying the optimal solution. If this were common, the Simplex Method would not be a practical solution method for solving linear programming problems. Let us consider why.

Suppose that a linear program had 25 variables and 25 less than or equal to constraints. For each constraint, one adds a slack variable converting the problem to one with only equality constraints. This is known as standard form. Hence, in standard form we have a problem of 50 variables and 25 constraints. Each basic solution (extreme point) corresponds to setting 25 variables to zero and solving the resulting system of 25 linear equations in 25 unknowns. It follows that the number of such solutions (that is, extreme points) equals the number of combinations of 50 things taken 25 at a time. This turns out to be about 1.264×10^{14} (about 126 trillion). To give the reader some idea of the magnitude of this number, suppose that we had a computer program that could identify 100 extreme points every second. At that rate, it would take about 40,000 years to find all the extreme points for a problem of this size!

---

[3]When the origin is not feasible, there are techniques available for determining an initial feasible solution to get the method started.

It is indeed fortunate that the Simplex Method rarely needs to search all the extreme points of the feasible region to discover the optimal solution. In fact, for a problem of this size, on average it would need to evaluate only about 25 extreme points.[4] Hence, the Simplex Method turned out to be a very efficient solution procedure.

We will not explore the mechanics of the method or additional theory underpinning its concepts. With today's easy access to computing and the wide availability of excellent software, it is unlikely that anyone with a real problem would solve it manually. There are many excellent texts that delve more deeply into the theoretical and computational aspects of linear programming. A good starting point for the interested reader is Hillier and Lieberman (1990). A more detailed treatment of the theory can be found in Hadley (1962).

6 Solving Linear Programming Problems with Excel

Excel has become the standard for spreadsheet programs. One of the useful features of Excel is that it comes bundled with an add-in called Solver that solves both linear and nonlinear programming problems. While the reliability of the nonlinear and the integer portions of the program are suspect, the linear programming part of Solver is excellent.

Because Solver is part of a spreadsheet program, problems are not entered algebraically (as they are with a system known as LINDO discussed in previous editions of this book). Consider Example S1.1. The algebraic representation is

$$\text{Maximize } 115x_1 + 90x_2,$$

subject to

$$10x_1 + 20x_2 \le 200,$$
$$4x_1 + 16x_2 \le 128,$$
$$15x_1 + 10x_2 \le 220,$$
$$x_1, x_2 \ge 0.$$

It is convenient to write the problem in a matrix format before entering the information into the Excel spreadsheet. The matrix representation for this problem is

Variable names:	x_1	x_2	Operator	RHS
Objective function: subject to	115	90	max	
Constraint 1	10	20	$\le$	200
Constraint 2	4	16	$\le$	128
Constraint 3	15	10	$\le$	220

The spreadsheet will look very much like this. The only differences are that one must specify the locations of the variable values (which I recommend be a row located directly

---

[4]The reason for this was understood only very recently. The proof requires very sophisticated mathematics.

under the variable names) and the algebraic formulas for the objective function and the constraints. These will be located in a column between "Operator" and "RHS." We will label this column "Value" in the spreadsheet.

Note that the column labeled "Operator" is not required. It is a convenience for the user to help keep track of the direction of the objective function and constraints. Excel requires this information, and it is entered manually when Solver is invoked.

At this point the spreadsheet should look similar to Figure S1–6. Notice the additional row labeled "Var Values" and the additional column labeled "Value." It is in the column labeled "Value" that one enters the algebraic form of the linear programming problem. The locations of the variable values are cells C4 and D4. The algebraic form of the objective function will be entered in cell F5 and the constraints in cells F7, F8, and F9. The formula for cell F5 is =C5*$C$4+D5*$D$4. The formula can be typed in or entered using the mouse to point and click cell locations.

Notice that we have used absolute addressing for the variable values (C4 and D4). This allows us to copy the formula from cell F5 to cells F7, F8, and F9 without having to retype the algebraic form for the constraints. You may wish to assign name labels to these cells so that they can later be referred to by name rather than by cell location.

FIGURE S1–6

Excel spreadsheet for Sidneyville problem

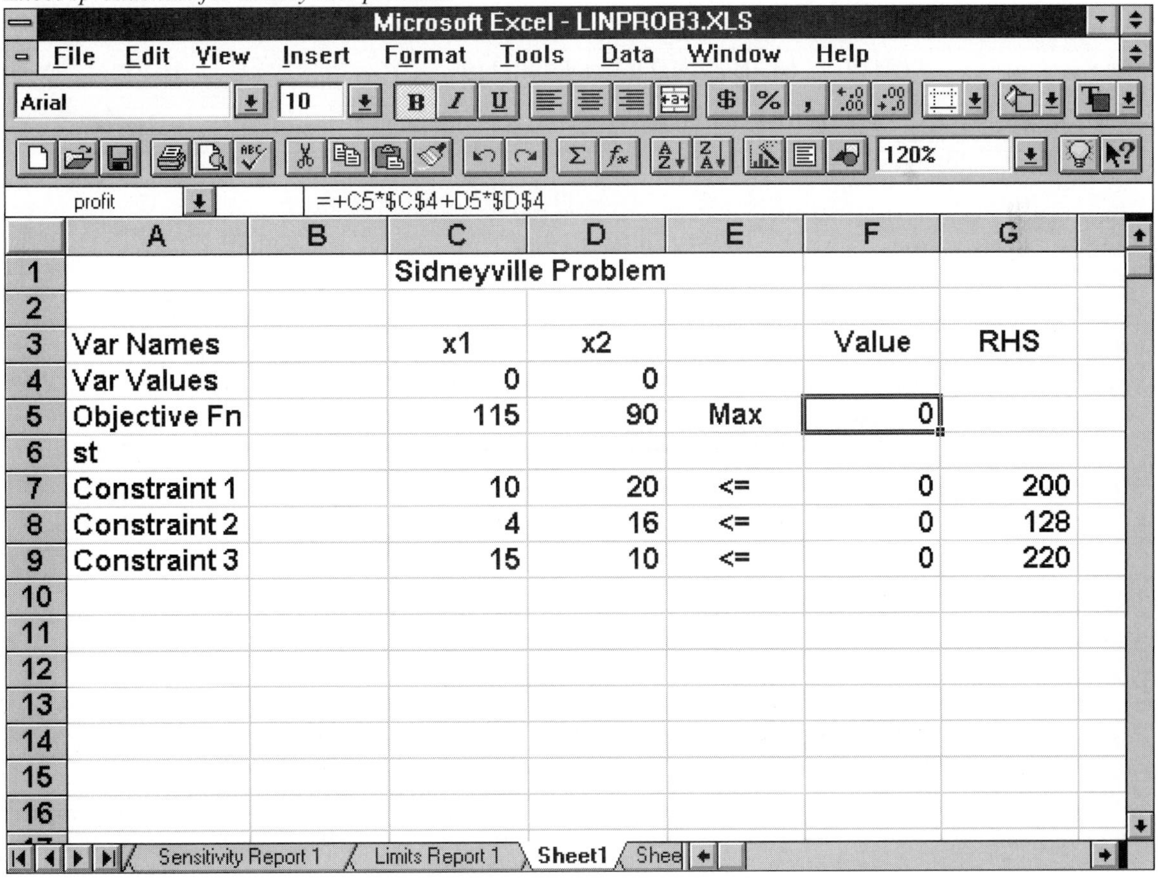

	A	B	C	D	E	F	G
1			Sidneyville Problem				
2							
3	Var Names		x1	x2		Value	RHS
4	Var Values		0	0			
5	Objective Fn		115	90	Max	0	
6	st						
7	Constraint 1		10	20	<=	0	200
8	Constraint 2		4	16	<=	0	128
9	Constraint 3		15	10	<=	0	220
10							
11							
12							
13							
14							
15							
16							

profit: =+C5*$C$4+D5*$D$4

Microsoft Excel - LINPROB3.XLS

(This is done most conveniently by invoking the formula bar, accessing the label area, and typing in a label of your choice. Note that the label "profit" appears just below the name of the current font. This was the name assigned to cell F5.)

Check that the algebraic form of the constraints is correct after copying cell F5 to cells F7, F8, and F9. For example, the formula corresponding to cell F7 should be: =C7*$C$4 + D7*$D$4.

Place 0's in the cells corresponding to the variable values (C4 and C5). After Solver has completed its search, the optimal values of these variables will appear in these cell locations.

The problem is now completely defined. Invoke Solver to obtain the solution. This is done by accessing Tools from the Menu Bar and choosing Solver. The Solver dialogue box will appear as in Figure S1–7. The first requirement is setting the location of the target cell. This corresponds to the cell location of the algebraic formula for the objective function—cell F5 in our spreadsheet. This can be accomplished by typing in F5 or clicking on the cell with the mouse. Notice that in this case I assigned the name "profit" to cell F5, so I simply type in "profit" for the target cell.

FIGURE S1–7

Spreadsheet with Solver dialog box

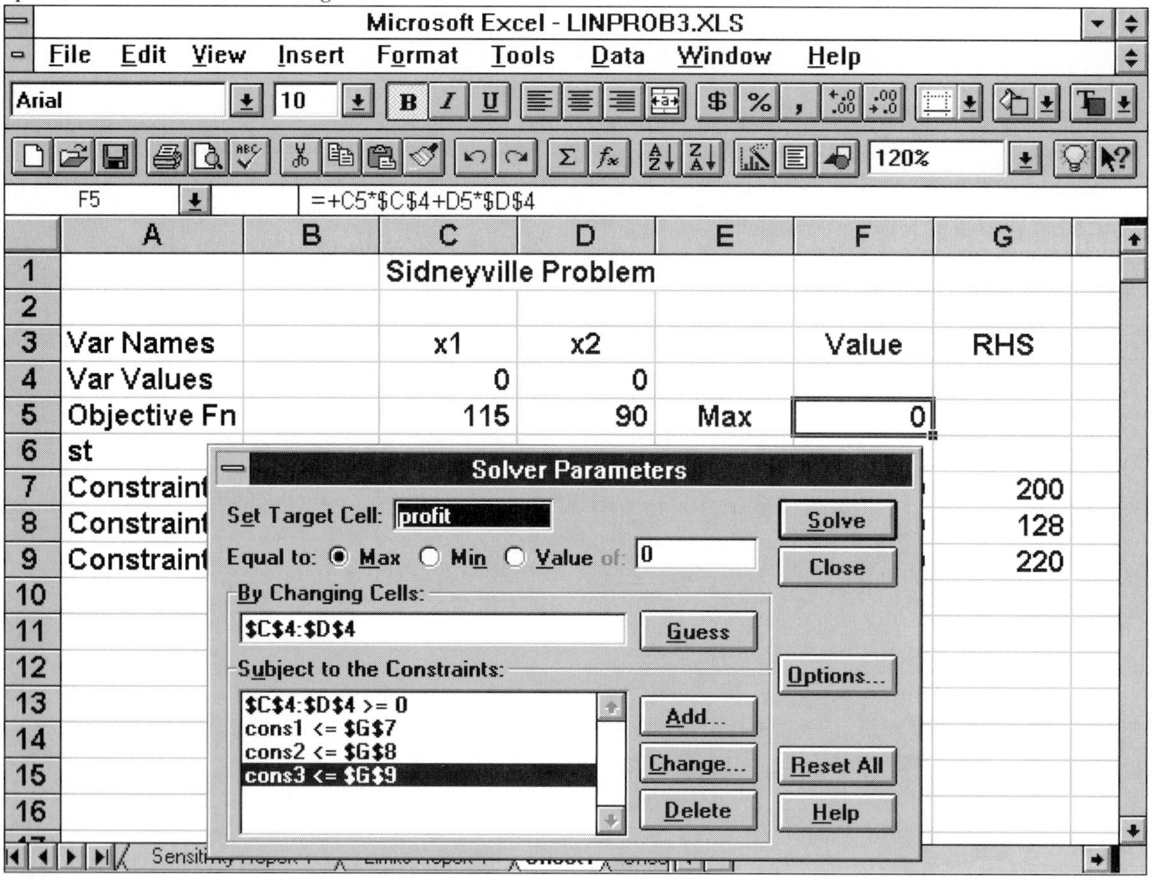

Next, specify if the problem is a min (minimum) or a max (maximum). Excel refers to the variables in the problem as "changing cells." You must tell Solver where to find these cells. In this spreadsheet they are C4 and D4. (These can be indicated by pointing and clicking with the mouse, manually typing in the cell locations, or entering preassigned cell names as we did for the objective function.)

Next, tell Excel where to find the algebraic definitions for the constraints. The constraints are entered into the system one at a time by clicking on the Add button. For each constraint you first tell the system where to find the algebraic form for the left-hand side of the constraint (F7, F8, and F9 in our case), the logical operator for the constraint ($\leq$, $=$, or $\geq$), and the location of the RHS value (G7, G8, and G9 in our case).

Because Solver is a general-purpose mathematical programming tool, two additional pieces of information must be included. Both of these can be done at the same time by clicking the Options key and choosing the two options "Assume linear model" and "Assume non-negative." In this way, the nonnegativity constraints do not need to be explicitly entered (note: The nonnegativity option was not available in Excel Version 5 or earlier. Nonnegativity constraints had to be entered into the problem explicitly.) Telling the program to assume a linear model assures that the problem is solved by the Simplex algorithm rather than the gradient method used for nonlinear problems. If your output includes values of Lagrange variables, you'll know that you forgot to specify this option.

At this point your dialog box should look like Figure S1–7. Notice that I have named some of the cells and the names appear in the dialog box rather than the cell locations. Using named cells that have some meaning relative to your problem will be very helpful later when you obtain the solution and sensitivity reports. In the dialog box, we have named the objective cell (F5) "profit" and the cells corresponding to the constraints "cons1," "cons2," and "cons3."

Check that all information is correctly entered and that you have specified a linear problem. Now simply click the mouse on the Solve button and Excel will whirl away and quickly produce a solution. After solving, the resulting spreadsheet should look like the one in Figure S1–8. Notice that the values of the variables in cells C4 and D4 now reflect the optimal solution of 12 and 4, respectively. The value in cell F5 is the value of the optimal profit of $1,740. The values in cells C7, C8, and C9 are the values of the left-hand sides of the constraints.

Although the optimal values of the variables and the optimal value of the objective function now appear in your spreadsheet, Excel has the option of printing several types of reports. The two that are relevant for linear programming are the Answer and Sensitivity reports. These reports appear on different sheets of the Excel workbook and are shown in Figures S1–9 and S1–10. Most of the information in the Answer report appears in the original spreadsheet. We also are told which constraints are binding. A variable that is labeled "not binding" is one that is nonzero.

Entering Large Problems Efficiently

The steps outlined above for solving linear programming problems on a spreadsheet are fine for small problems. Excel has several features that allow more efficient entry of larger problems, however.

One such feature is the SUMPRODUCT function. SUMPRODUCT is a vector or array product, which multiplies the elements of one array times another array term by term and adds the results. This means that the algebraic form for the objective function

FIGURE S1–8

The Excel spreadsheet displaying the optimal solution

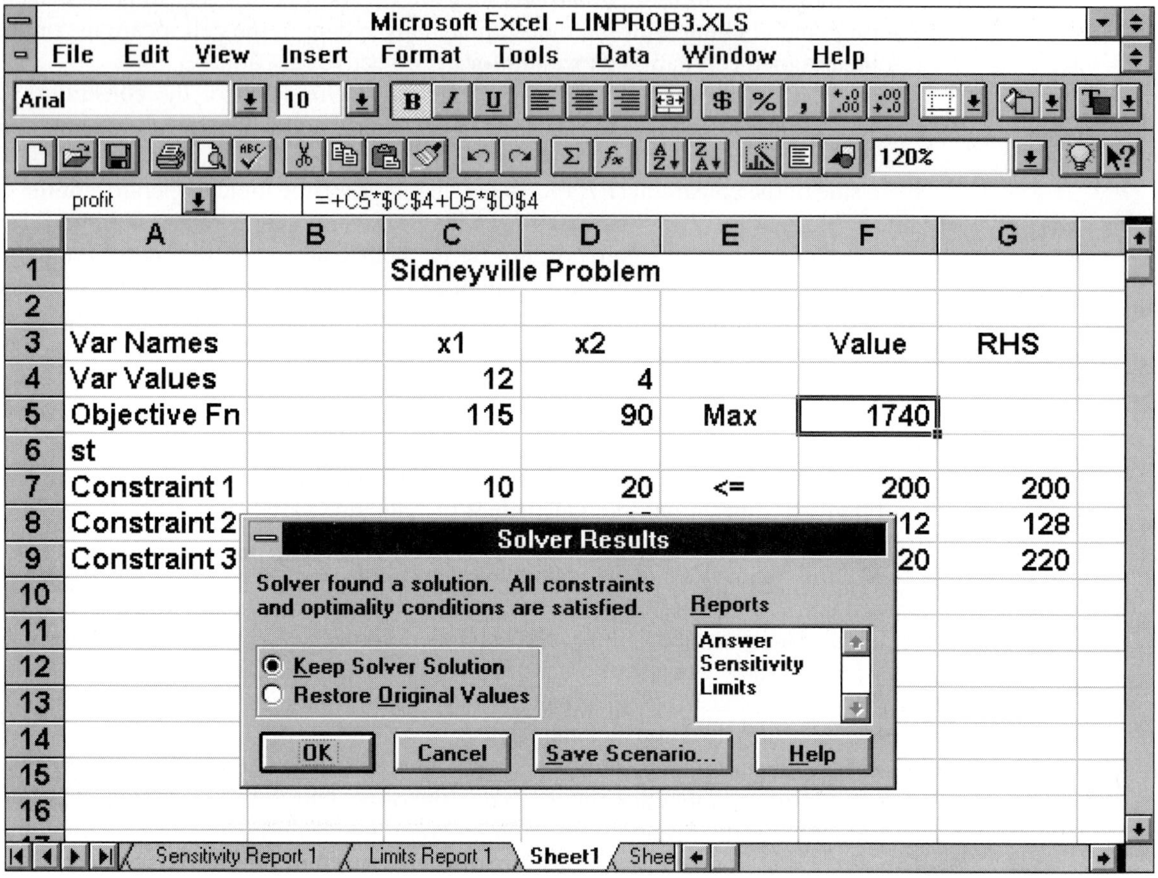

and for the constraints can be entered with this function. Recall that for cell F5 we used the formula =C5*$C$4+D5*$D$4. This formula also could have been entered as =SUMPRODUCT($C$4:$D$4, C5:D5). While this may not appear to be much of an improvement here, it saves a lot of typing when entering large problems.

The second shortcut for entering large problems is to group constraints by type and enter all constraints in a group at one time. In the case of the Sidneyville problem, there are three ≤ constraints. These constraints could be entered with one command, similar to the single command for entering the nonnegativity constraints. The appropriate formulas for the constraints appear in cells F7, F8, and F9. The single command for entering these three constraints into Solver is: $F$7:$F$9 ≤ $G$7:$G$9. This can be typed in directly or entered by pointing to the appropriate cells in the spreadsheet. The Solver dialog box for the Sidneyville problem using this approach is shown in Figure S1–11.

Using SUMPRODUCT and entering constraints in groups can save a lot of time for large problems. However, the Solver dialog box is not as informative, since only cell locations and not cell names appear. For problems of fewer than 10 constraints total, entering the constraints one at a time is fine.

FIGURE S1–9

Answer report for Sidneyville problem

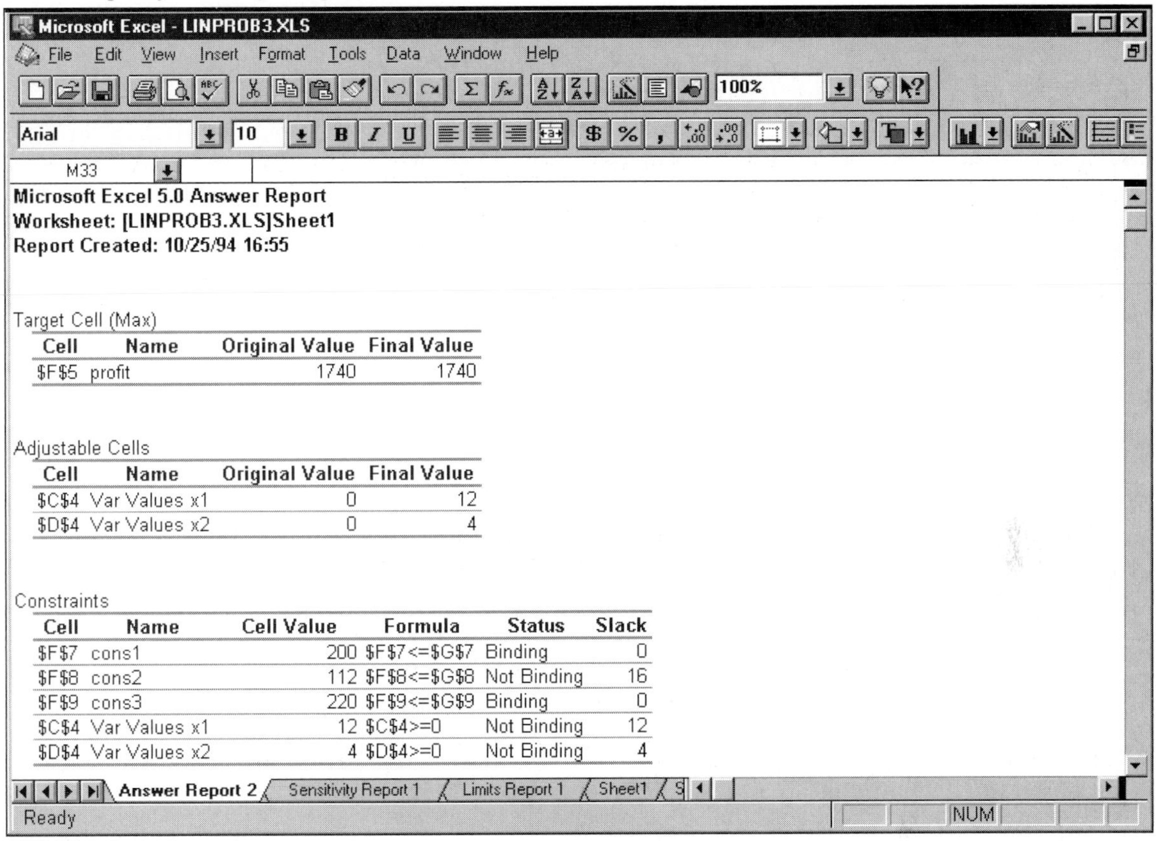

```
Microsoft Excel - LINPROB3.XLS

File   Edit   View   Insert   Format   Tools   Data   Window   Help

Arial           10    B  I  U

M33

Microsoft Excel 5.0 Answer Report
Worksheet: [LINPROB3.XLS]Sheet1
Report Created: 10/25/94 16:55


Target Cell (Max)
    Cell      Name        Original Value  Final Value
    $F$5  profit                    1740         1740


Adjustable Cells
    Cell      Name        Original Value  Final Value
    $C$4  Var Values x1                0           12
    $D$4  Var Values x2                0            4


Constraints
    Cell      Name        Cell Value   Formula        Status      Slack
    $F$7  cons1                  200  $F$7<=$G$7    Binding          0
    $F$8  cons2                  112  $F$8<=$G$8    Not Binding     16
    $F$9  cons3                  220  $F$9<=$G$9    Binding          0
    $C$4  Var Values x1           12  $C$4>=0       Not Binding     12
    $D$4  Var Values x2            4  $D$4>=0       Not Binding      4

Answer Report 2 / Sensitivity Report 1 / Limits Report 1 / Sheet1 / S

Ready                                                              NUM
```

We noted earlier that an advantage of the spreadsheet for solving linear programming is that one can construct a general-purpose template. A template might have up to 10 variables and 10 constraints. Variable names could be x1 through x10 and constraint names const1 through const10. The SUMPRODUCT functions for the objective function and the constraints would be programmed in advance. One would then simply enter the coefficients of the problem to be solved and save as a new file name. (This is much faster than typing in the full algebraic representation for every new problem as one must do with other linear programming systems.)

7 Interpreting the Sensitivity Report

Shadow Prices

An interesting issue is the value of additional resources in a linear programming problem. In Example S1.1, the resources are the three types of wood: pine, cedar, and maple. An increase in the level of any of these resources results in an increase in the value of the right-hand side of the appropriate constraint. The sensitivity report

FIGURE S1–10

Sensitivity report for Sidneyville problem

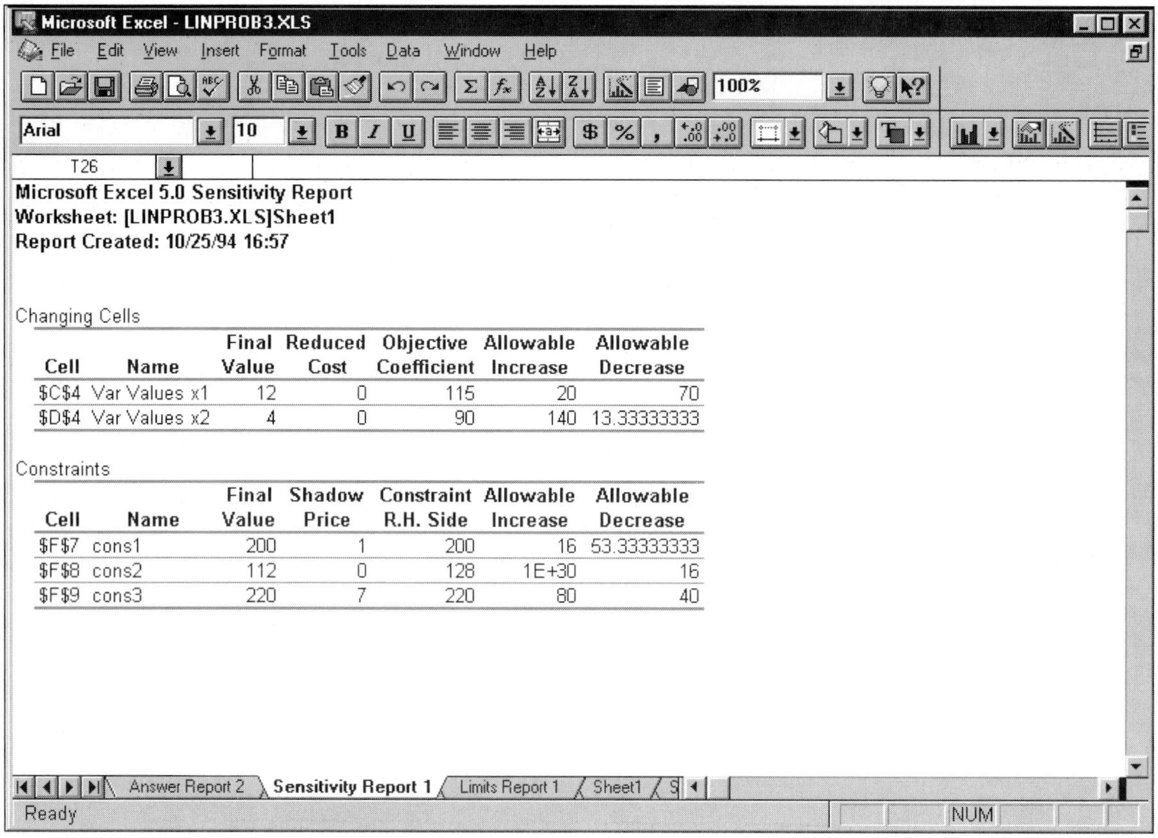

Microsoft Excel 5.0 Sensitivity Report
Worksheet: [LINPROB3.XLS]Sheet1
Report Created: 10/25/94 16:57

Changing Cells

Cell	Name	Final Value	Reduced Cost	Objective Coefficient	Allowable Increase	Allowable Decrease
$C$4	Var Values x1	12	0	115	20	70
$D$4	Var Values x2	4	0	90	140	13.33333333

Constraints

Cell	Name	Final Value	Shadow Price	Constraint R.H. Side	Allowable Increase	Allowable Decrease
$F$7	cons1	200	1	200	16	53.33333333
$F$8	cons2	112	0	128	1E+30	16
$F$9	cons3	220	7	220	80	40

(Figure S1–10) gives information about the value to the objective function of additional resources. This information is contained in the values of the shadow prices for the constraints.

The shadow price is defined as the improvement in the objective function realized by adding one additional unit of a resource. For Example S1.1, the first constraint refers to the amount of pine needed. Since this constraint is binding at the optimal solution (since the final value and the constraint right-hand-side values are the same), it is likely that if we had additional pine we could increase revenue. The shadow price tells us just how beneficial. Since the shadow price for this constraint is $1, it means that for each additional unit square foot of pine, the objective (profit) increases by $1. (This will hold only within a certain range as discussed below.) Consider the second constraint (cedar). Since the final value and the right-hand side for this constraint are different (112 versus 120), there is slack in this constraint. That means we are not consuming the entire quantity of cedar at the optimal solution, and additional cedar will not improve the profit. This is borne out by the fact that the shadow price for this constraint is zero. The final shadow price of 7 indicates that every additional square foot of maple contributes $7 to the profit.

FIGURE S1–11

Solver dialog box with efficient data entry

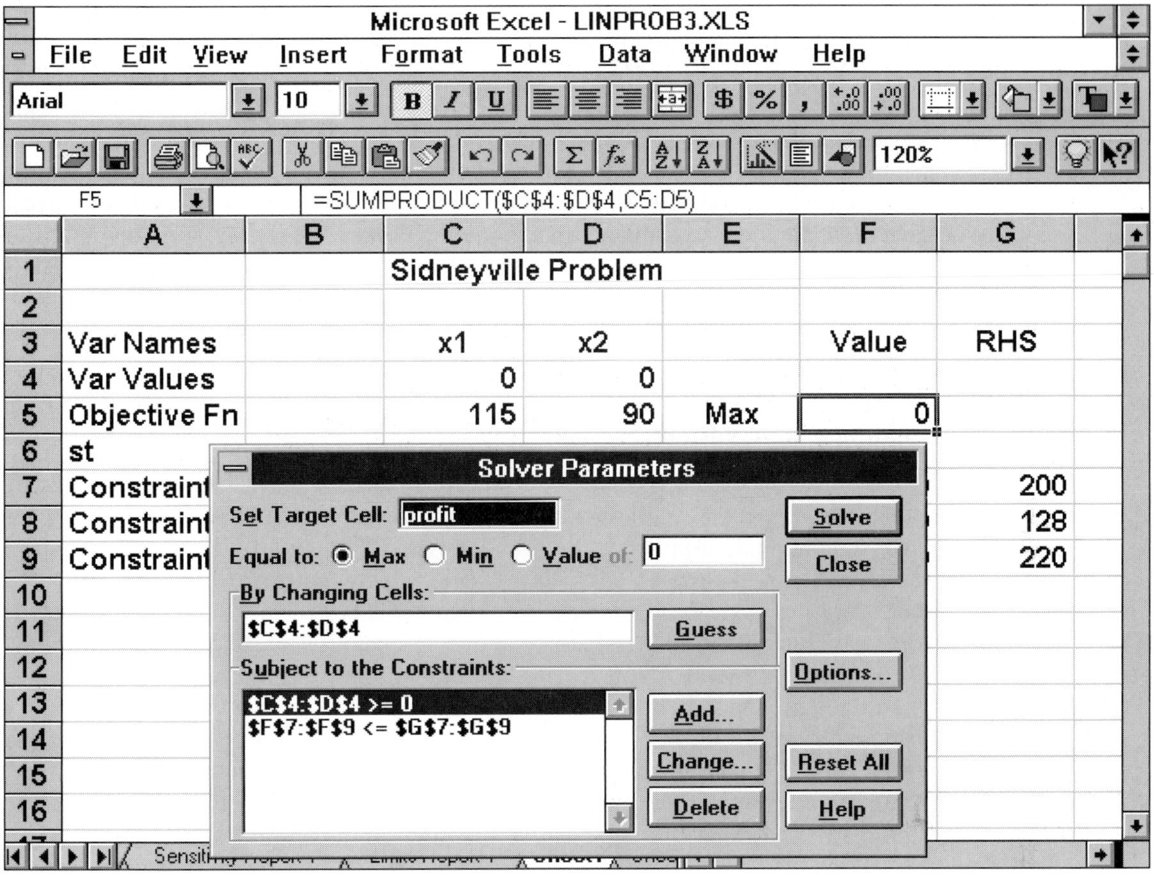

Objective Function Coefficients and Right-Hand Sides

The shadow prices remain valid as long as the optimal basis does not change. (The basis is the set of positive variables in the final solution.) The columns Allowable Increase and Allowable Decrease indicate over what range the shadow prices remain valid. This means that we can determine the effect on the objective function of changes in constraint right-hand sides without resolving the problem.

The first part of the sensitivity report (Figure S1–10) gives the values of the objective function coefficients for which the shadow prices remain valid. The current values of the profits are $115 and $90 respectively. The shadow prices are valid as long as the first objective function coefficient does not increase more than 20 or decrease more than 70 (i.e., the objective function coefficient for x_1 is between 45 and 135). Similarly, the allowable range for the objective function coefficient for x_2 is $76\frac{2}{3}$ to 230.

The second part of the sensitivity report reports the ranges on the right-hand sides of the constraints for which the shadow prices remain valid. Hence, the shadow price of $1 for pine is valid for an increase of 16 or less and a decrease of $53\frac{1}{3}$ or less in the right-hand side of the first constraint. That is, the right-hand side of the first constraint could be any value between $146\frac{2}{3}$ and 216. The shadow price of 0 for cedar is valid for any increase (1E+30 should be interpreted as infinity) and a decrease of 16 or less, and the shadow price for maple is valid for an increase of 80 or less and a decrease of 40 or less.

If either the objective function coefficients or the right-hand sides increase or decrease beyond the allowable ranges, the shadow prices no longer remain valid, and the problem would have to be resolved to determine the effect. Note that Excel uses the convention that a positive shadow price means an increase in the objective function per unit increase in the right-hand side of a constraint, and a negative shadow price means a decrease in the objective function per unit increase in the right-hand side, irrespective of whether the problem is a max or a min. (Other linear programming systems may have other conventions.) Also note that changes to objective function coefficients or right-hand sides can be only one at a time in these ranges. The rules for simultaneous changes in right-hand sides or objective function coefficients are much more complex and will not be discussed here.

Adding a New Variable

We can use the results of sensitivity analysis to determine whether it is profitable to add a new activity (variable) without re-solving the problem. For small problems, such as Example S1.1, simply inputting and solving the new problem on the computer is quick and easy. However, in real applications, the number of decision variables and constraints could be in the hundreds or even in the thousands. Reentering a problem of this magnitude is a major task, to be avoided whenever possible.

Suppose the firm is considering producing a third product, a vanity table, which would require the same woods used in making the desks. Each vanity table would contribute $75 to profit, but each would require 8 square feet of pine, 6 square feet of cedar, and 10 square feet of maple. We could determine if it would be worth producing vanity tables in addition to the desks by solving the problem with three activities and comparing the values of the objective functions at the optimal solutions.

There is a faster way. The dual prices in Figure S1–10 tell us the value of each unit of resource at the current solution. The decrease in profit resulting from reducing the supply of pine is $1 per square foot, which translates to $8.00 for 8 square feet of pine. There is no cost for decreasing the supply of cedar. The cost of decreasing the supply of maple by 10 square feet is (10)(7) = $70. Hence the total decrease in profit from the consumption of resources required to produce one vanity table is $78. The contribution to profit is only $75. We conclude that it is not optimal to produce vanity tables in addition to the desks with the current resources. Had we determined, however, that it was profitable to produce the vanity table, we would have had to re-solve the problem with three activities to find the optimal numbers of desks and vanity tables to produce.

Using Sensitivity Analysis

To cement your understanding of the information in Figure S1–10, consider the following questions:

Example S1.1 (continued)

a. Sidneyville's sales manager has renegotiated the contract for regular desks and now expects to make a profit of $125 on each. He excitedly conveys this information to the firm's production manager, expecting that the optimal mix of rolltop and regular desks will change as a result. Does it?

b. Suppose that the new contract also has a higher profit for the rolltop desks. If the new profit for the rolltop desks is $140, how will this change the optimal solution?

c. A logging company has offered to sell Sidneyville an additional 50 square feet of maple for $5.00 per square foot. Based on the original objective function, would you recommend that they accept the offer?

d. Assuming that Sidneyville purchases the 50 square feet of maple, how is the optimal solution affected?

e. The firm is considering a pine desk that would require 25 square feet of pine and no other wood. What profit for pine desks would be required to make its production worthwhile, assuming current levels of resources and original profits on regular and rolltop desks?

f. During inspection, the quality department discovered that 50 square feet of pine had water damage and could not be used. Will it be optimal to produce both desks under these circumstances? Will the product mix change?

Solution

a. According to Figure S1–10, the allowable increase in the coefficient of the objective function for variable x_2, the regular desks, is 140. Because the increase to $125 is still within the allowable range, the optimal mix of rolltop and regular desks will remain the same: namely, $x_1 = 12$ and $x_2 = 4$.

b. The allowable increase in the objective function for the rolltop desks (x_2) is 20, or to a maximum value of 135. As 140 is outside the allowable range, it is possible that the basis will change. However, the allowable ranges in Figure S1–10 are only valid if the profit for regular desks is $90. The allowable ranges will change when the profit for regular desks is changed to $125, even though the optimal solution does not. The output for part (*a*) (that is, with profits of $115 and $125) is

OBJ COEFFICIENT RANGES

VARIABLE	CURRENT COEF	ALLOWABLE INCREASE	ALLOWABLE DECREASE
X1	115.000000	72.500000	52.500000
X2	125.000000	105.000000	48.333330

This shows that the allowable increase in the coefficient for x_1 is now 72.5. Because 140 is within the allowable range, the solution for parts (*a*) and (*b*) will be the same, which is also the same as our original solution of $x_1 = 12$ and $x_2 = 4$.

c. Since the dual price for the third constraint corresponding to maple is 7, it is profitable to purchase the maple for $5.00 a square foot. The allowable increase of the right-hand side over which this dual price applies is 80, so that it is worth purchasing the full 50 additional square feet.

d. Because the increase of 50 is within the allowable right-hand-side range, we know that the basis will not change. That is, it still will be optimal to produce both the rolltop and the regular desks. However, if the right-hand side changes, the values of the basic variables *will* change. We must re-solve the problem with the new right-hand-side value to determine the updated solution. The solution is

---

```
                                    LP OPTIMUM FOUND AT STEP 1
OBJECTIVE FUNCTION VALUE
              1)                    2090.000000

        VARIABLE              VALUE            REDUCED COST
        X1                    17.000000             .000000
        X2                     1.500000             .000000
```

To retain feasibility we round x_2 to 1. (This is *not* the optimal integer solution, however. The optimal integer solution is $x_1 = 18$ and $x_2 = 0$ with a profit of \$2,070, which is obtained from Excel by identifying both x_1 and x_2 as integer variables. The suboptimal solution of $x_1 = 17$ and $x_2 = 1$ results in a profit of \$2,045.)

e. The dual price for pine is \$1 per square foot. As each desk consumes 25 square feet of pine, the profit for each pine desk must exceed \$25 for pine desks to be profitable to produce.

f. The right-hand side of the first constraint can decrease as much as 53.333330 and the current basis will remain optimal. That means that a decrease of 50 square feet will not change the basis; it will still be profitable to produce both desks. However, the production quantities will decrease. We must re-solve the problem to determine the correct levels of the new quantities. They are

---

```
    LP OPTIMUM  FOUND AT STEP 2
    OBJECTIVE FUNCTION VALUE
          1)         1690.000000
    VARIABLE           VALUE                    REDUCED COST
        X1           14.500000                      .000000
        X2             .250000                      .000000
```

Again, we need to round these variables. Rounding both x_1 and x_2 down guarantees feasibility. If we produce 14 rolltop desks and 0 regular desks, we require 140 square feet of pine (150 are available), 56 square feet of cedar (128 are available), and 210 square feet of maple (220 are available). There does not appear to be enough wood to produce an additional desk of either type, so we leave the solution at $x_1 = 14$ and $x_2 = 0$. (This is the optimal integer solution.) ■

---

8 Recognizing Special Problems

Several problems can occur when solving linear programming problems. In this section we will discuss the causes of these problems and how one recognizes them when using Excel.

Unbounded Solutions

The feasible region of a linear program is not necessarily bounded. The feasible region for the Sidneyville problem pictured in Figure S1–4 is bounded. However, consider the following linear programming problem:

Example S1.2

Maximize

$$2x_1 + 3x_2$$

subject to

$$x_1 + 4x_2 \geq 8$$
$$x_1 + x_2 \geq 5$$
$$2x_1 + x_2 \geq 7$$
$$x_1, x_2 \geq 0.$$

Figure S1–12 shows the feasible region. Notice that it is unbounded. Because we can make x_1 and x_2 as large as we like, there is no limit to the size of the objective function. When this occurs, the problem is unbounded and there is no optimal solution.

When this problem is inputted in Excel, Solver writes in very large values for the problem variables and displays a message that set target values do not converge. The Excel output for this problem appears in Figure S1–13. ■

FIGURE S1–12

Feasible region for Example S1.2

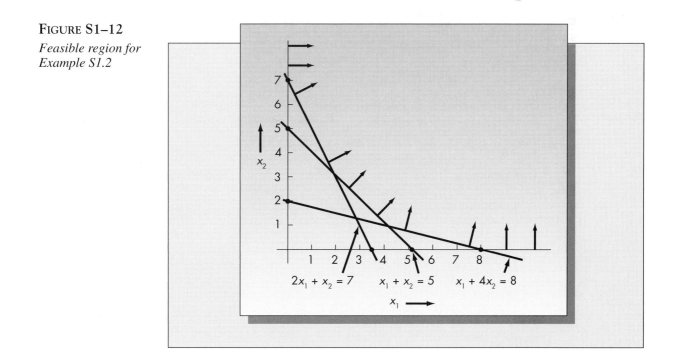

FIGURE S1–13

Excel output for Example S1.2

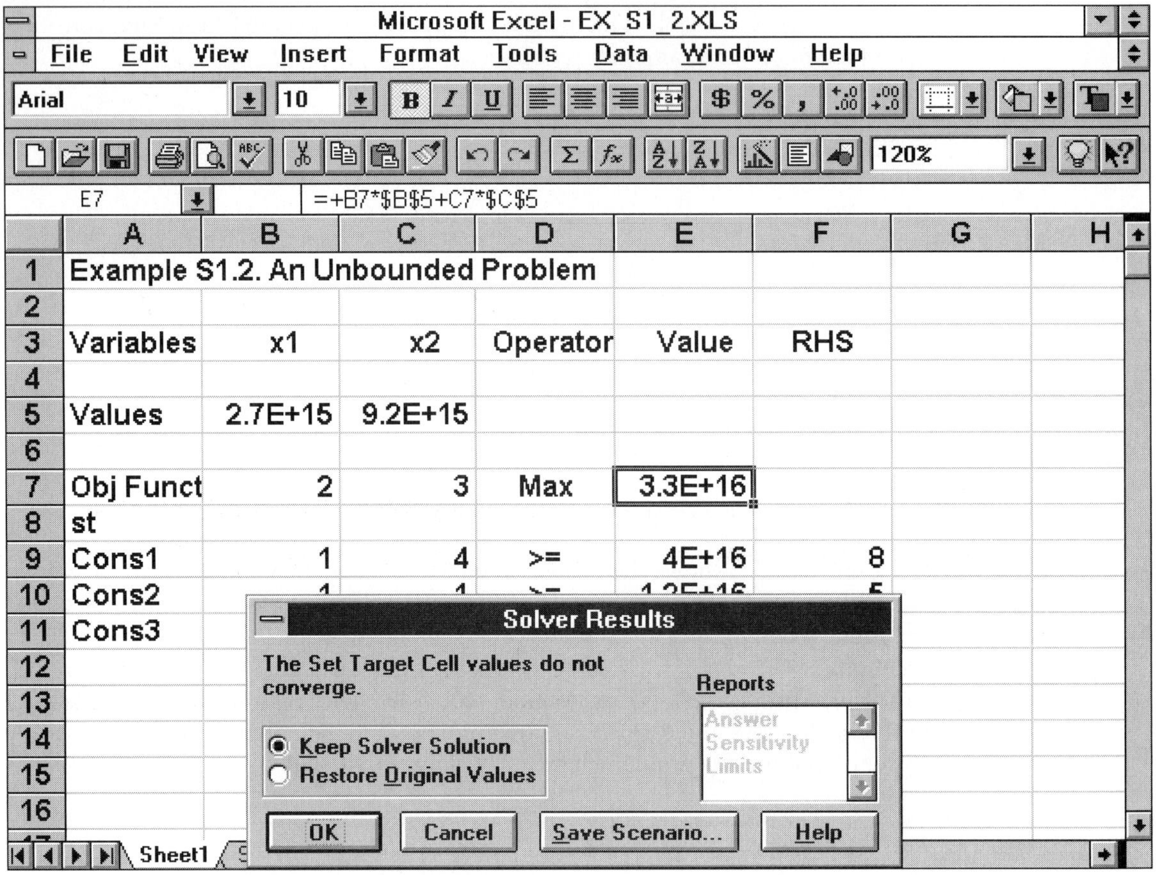

Empty Feasible Region

It is possible for two or more constraints to be inconsistent. When that occurs, there will be no feasible solution. Consider the following problem.

Example S1.3

Maximize

$$2x_1 + 3x_2$$

subject to

$$x_1 + 4x_2 \leq 8$$
$$x_1 + x_2 \geq 10$$
$$x_1, x_2 \geq 0.$$

The feasible region for this example appears in Figure S1–14. Notice that there is no intersection of the half spaces defined by the two constraints in the positive

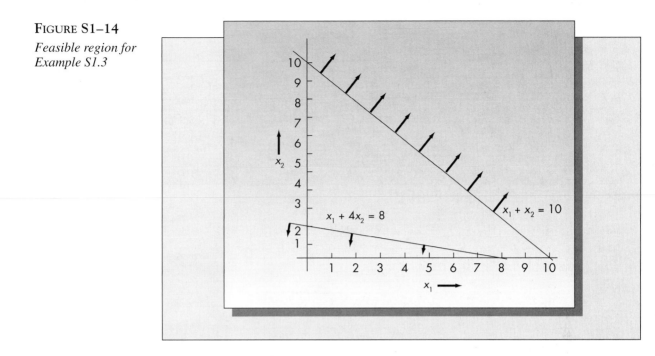

quadrant. In this case, the feasible region is empty and we say that the problem is infeasible. The Excel output appears in Figure S1–15. Note that the solution $x_1 = 8$ and $x_2 = 0$ shown is not feasible because it results in negative slack in the first constraint. ∎

Degeneracy

In linear programming there are two types of variables: basic variables and nonbasic variables. The number of basic variables equals the number of constraints, and basic variables may be either original variables or slack or surplus variables. What defines basic variables? Consider any linear program in standard form. By including slack and surplus variables, all constraints are expressed as equations. In standard form, there always will be more variables than constraints. Suppose that after a linear programming problem has been expressed in standard form, there are $n + m$ variables and n constraints. A basic solution is found by setting m variables to zero and solving the resulting n equations in n unknowns. In most cases the values of the n basic variables will be positive. A degenerate solution occurs when one or more basic variables are zero at the optimal solution. In Excel, a basic variable is one with zero reduced cost. Degeneracy occurs when the value of a variable is zero and its reduced cost or shadow price is also zero.

Why are we interested in degeneracy? It is possible that if degenerate solutions occur, the Simplex Method will cycle through some set of solutions and never recognize the optimal solution. The phenomenon of cycling has never been observed in practice, and most computer programs have means of guaranteeing that it never occurs. The bottom line is that degeneracy is an issue about which we need not worry, but one of which we should be aware.

FIGURE S1–15

Excel output for Example S1.3

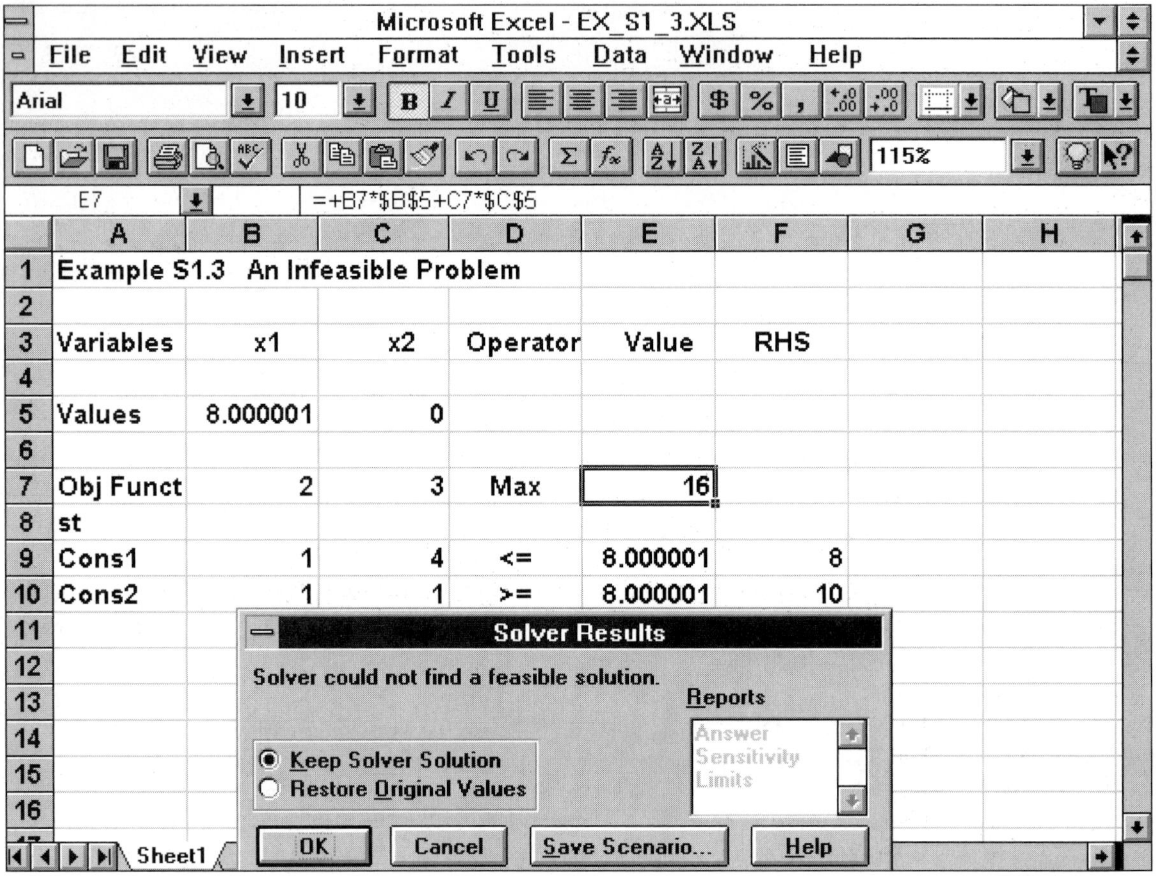

Multiple Optimal Solutions

The optimal solution to a linear program is not always unique. There are cases in which there are multiple optimal solutions. In Chapter 3 we saw that two-variable problems could be solved by graphical means by approaching the feasible region with the Z line. Assuming that we approach the feasible region from the correct side, the first feasible point with which the Z line comes into contact is the optimal solution.

However, suppose that the Z line is parallel to one of the constraints. In that case it does not contact a single point first, but an entire edge.

Example S1.4

Consider the feasible region pictured in Figure S1–12 corresponding to the constraints of Example S1.2. Suppose that the objective function is min $3x_1 + 3x_2$. Then the Z line has slope -1 and is parallel to the constraint boundary $x_1 + x_2 = 5$. As the Z line approaches the feasible region, it meets the edge defined by this constraint and both extreme points along this edge. This means that both extreme points *and* all points along the edge are optimal.

Unfortunately, Excel does not indicate that there are multiple optimal solutions to this problem. Our only clue is that the solution is degenerate; the surplus variable for the third constraint has both zero dual price and zero value. Our graphical solution tells us that both extreme points $(2, 3)$ and $(4, 1)$ are optimal, and so are all points along the edge connecting these extreme points. (The points along the edge can be written in the form

$$x_1 = \alpha(2) + (1 - \alpha)(4),$$
$$x_2 = \alpha(3) + (1 - \alpha)(1),$$

where α is a number between zero and one. This is known as a convex combination of these two extreme points.) ∎

Redundant Constraints

It is possible for one or more constraints to be redundant. That means that these constraints can be eliminated from the formulation without affecting the solution. In simple two-variable problems, redundant constraints can be recognized graphically because they lie outside the feasible region. Excel does not recognize or signal if one or more constraints are redundant. Sometimes redundant constraints can cause degeneracy, but degeneracy can result when constraints are not redundant as well.

Example S1.5

Consider Example S1.1. Suppose that we add the following additional constraint:

$$x_1 + x_2 \le 20.$$

Figure S1–16 shows the resulting feasible region. It is exactly the same as the feasible region pictured in Figure S1–4. The additional constraint has no effect, as it lies completely outside the feasible region defined by the first three constraints. The optimal solution will, of course, be exactly the same.

FIGURE S1–16

Feasible region for Example S1.5 showing a redundant constraint

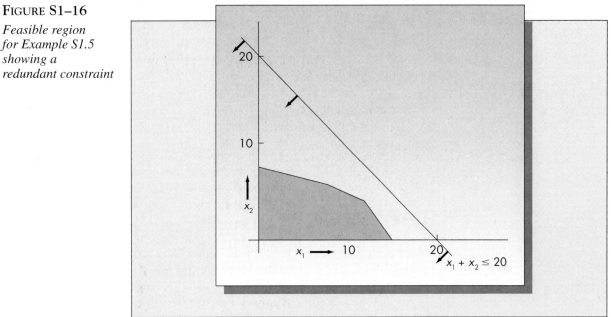

If we had originally formulated the problem with the four constraints

$$10x_1 + 20x_2 \leq 200,$$

$$4x_1 + 16x_2 \leq 128,$$

$$15x_1 + 10x_2 \leq 220,$$

$$x_1 + x_2 \leq 20,$$

and solved, the output indicates that the optimal solution is $x_1 = 12$ and $x_2 = 4$ as before, and gives us no clue that the last constraint is redundant. The only way to see that the final constraint is redundant is to graph the feasible region as we did in Figure S1–16. Graphing is possible, however, in two-variable problems only.

Does redundancy cause a problem? Not really. We would certainly like to be able to write our linear program as economically as possible, but if one or more constraints are redundant, the optimal solution is unaffected. ■

9 The Application of Linear Programming to Production and Operations Analysis

In Chapter 3 we showed how linear programming could be used to find optimal solutions (subject to rounding errors) for aggregate planning problems. Although this is the only explicit use of linear programming in this book, there have been successful linear programming applications for many operations management problems.[5] Scheduling and distribution are perhaps two areas in which applications are most common.

Fisher et al. (1982) describe an application of linear programming to the problem of providing a coordinated vehicle scheduling and routing system for delivery of consumable products to customers of the Du Pont Company. The primary issue was to determine delivery routes (loops) for the trucks to the company's clients in various regions of the country. A refrigerated truck drives a weekly loop that includes several dozen customers. The largest region considered, Chicago, had 16 loops and several hundred cities, whereas the Houston region, the smallest, had 4 loops and less than 80 cities.

The basic mathematical formulation of the problem was a generalized assignment problem. (The assignment problem is discussed in Chapter 9. It is a linear programming problem in which the decision variables are restricted to be zeros and ones.) The mathematical formulation used in this study is the following:

1. Given data
   d_{ik} = Cost of including customer i in loop k,
   a_i = Demand from customer i.

2. Problem variables

$$y_{ik} = \begin{cases} 1 & \text{if customer } i \text{ is assigned to loop } k, \\ 0 & \text{if customer } i \text{ is not assigned to loop } k. \end{cases}$$

3. Generalized assignment problem

$$\text{Min} \sum_{k=1}^{K} \sum_{i=1}^{n} d_{ik}\, y_{ik}$$

---

[5]In this section we interpret linear programming in the broad sense to include integer linear programming.

subject to

$$\sum_{k=1}^{K} y_{ik} = 1, \qquad \text{for } i = 1, \ldots, n,$$

$$\sum_{i=1}^{n} a_i y_{ik} \leq b_k, \qquad \text{for } k = 1, \ldots, K,$$

$$y_{ik} = 0 \text{ or } 1, \qquad \text{for all } i \text{ and } k,$$

where K is the total number of loops in the region and n is the number of customers.

The implementation of this model was reported to have saved Du Pont over $200 million. A more complex mathematical model solving a similar problem for Air Products Corporation was reported by Bell et al. (1983). This study won the Institute of Management Sciences Practice Award in 1983.

Linear programming (or, more generally, mathematical programming) has been an important tool for logistics planning for a wide variety of operations management problems. As computer technology improves and inexpensive desktop computing becomes more accessible to a large group of professionals, general-purpose solution techniques such as linear programming should play an important role in productions and operations planning in the next decade.

Bibliography

Bell, W. J.; L. M. Dalberto; M. L. Fisher; A. J. Greenfield; R. Jaikumar; P. Kedia; R. G. Mack; and P. J. Prvtzman. "Improving the Distribution of Industrial Gases with an On-Line Computerized Routing and Scheduling Optimizer." *Interfaces* 13 (1983), pp. 4–23.

Fisher, M.; A. J. Greenfield; R. Jaikumar; and J. T. Uster III. "A Computerized Vehicle Routing Application." *Interfaces* 12 (1982), pp. 42–52.

Hadley, G. *Linear Programming*. Reading, MA: Addison-Wesley, 1962.

Hillier, F. S., and G. J. Lieberman. *Introduction to Operations Research*. 5th ed. San Francisco: Holden Day, 1990.

Ragsdale, C. T. *Spreadsheet Modeling and Decision Analysis*. 2nd ed. Cincinnati: South-Western College Publishing,1998.

$$Q = \sqrt{\frac{2K\lambda}{h}}$$

One learns by doing

CHAPTER 4

INVENTORY CONTROL SUBJECT TO KNOWN DEMAND

The current investment in inventories in the United States is enormous. In the last quarter of 1999, the total dollar investment was estimated to be $1.37 trillion.[1] Figure 4–1 shows investment in inventories broken down by sectors of the economy. The inventory models we will be discussing in this chapter and in Chapter 5 can be applied to all the sectors of the economy shown in Figure 4–1, but are most applicable to the manufacturing, wholesale, and retail sectors, which compose approximately 82 percent of the total. The trillion-dollar investment in inventories accounts for between 20 and 25 percent of the total annual GNP. Clearly there is enormous potential for improving the efficiency of our economy by intelligently controlling inventories. Companies that use scientific inventory control methods have a significant competitive advantage in the marketplace.

A major portion of this text is devoted to presenting and analyzing several mathematical models that can assist with controlling the replenishment of inventories. Both Chapters 4 and 5 assume that the demand for the item is external to the system. In most cases, this means that the inventory is being acquired or produced to meet the needs of a customer. In a manufacturing environment, however, demands for certain parts are the result of production schedules for higher-level assemblies; the production-lot-sizing decisions at one level of the system result in the demand patterns at other levels. The interaction of components, subassemblies, and final products plays an important role in determining future demand. Systems of this type are referred to as *materials requirements planning* (MRP) systems or dependent demand systems. MRP is treated in detail in Chapter 7.

The fundamental problem of inventory management can be succinctly described by the two questions (1) When should an order be placed? and (2) How much should be ordered? The complexity of the resulting model depends upon the assumptions one makes about the various parameters of the system. The major distinction is between models that assume known demand (this chapter) and those that assume random demand (Chapter 5), although, as we will see, the form of the cost functions and the

---

[1] *Survey of Current Business* 79 (December 1999).

FIGURE 4–1

Breakdown of the total investment in inventories in the U.S. economy (1999)

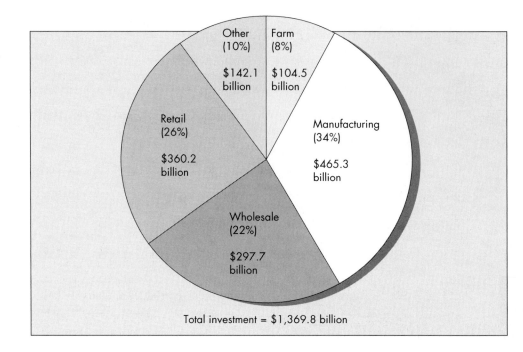

Other (10%)
$142.1 billion

Farm (8%)
$104.5 billion

Manufacturing (34%)
$465.3 billion

Retail (26%)
$360.2 billion

Wholesale (22%)
$297.7 billion

Total investment = $1,369.8 billion

assumptions one makes about physical characteristics of the system also play an important role in determining the complexity of the resulting model.

In general, the models that we discuss can be used interchangeably to describe either replenishment from an outside vendor or internal production. This means that from the point of view of the model, inventory control and production planning are often synonymous. For example, the lot-sizing methods treated in Chapter 7 could just as well have been included in this chapter. The issue is not the label that is placed on a technique, but whether it is being correctly applied to the problem being addressed.

4.1 Types of Inventories

When we consider inventories in the context of manufacturing and distribution, there is a natural classification scheme suggested by the value added from manufacturing or processing. (This certainly is not the only means of categorizing inventories, but it is the most natural one for manufacturing applications.)

1. *Raw materials.* These are the resources required in the production or processing activity of the firm.
2. *Components.* Components correspond to items that have not yet reached completion in the production process. Components are sometimes referred to as subassemblies.
3. *Work-in-process.* Work-in-process (WIP) is inventory either waiting in the system for processing or being processed. Work-in-process inventories include component inventories and may include some raw materials inventories as well. The level of work-in-process inventory is often used as a measure of the

efficiency of a production scheduling system. The just-in-time approach, discussed in detail in Chapter 7, is aimed at reducing WIP to a minimum.

4. *Finished goods.* Also known as end items, these are the final products of the production process. During production, value is added to the inventory at each level of the manufacturing operation, culminating with finished goods.

The appropriate label to place on inventory depends upon the context. For example, components for some operations might be the end products for others.

4.2 Motivation for Holding Inventories

1. *Economies of scale.* Consider a company that produces a line of similar items, such as air filters for automobiles. Each production run of a particular size of filter requires that the production line be reconfigured and the machines recalibrated. Because the company must invest substantial time and money in setting up to produce each filter size, enough filters should be produced at each setup to justify this cost. This means that it could be economical to produce a relatively large number of items in each production run and store them for future use. This allows the firm to amortize fixed setup costs over a larger number of units.[2]

2. *Uncertainties.* Uncertainty often plays a major role in motivating a firm to store inventories. Uncertainty of external demand is the most important. For example, a retailer stocks different items so that he or she can be responsive to consumer preferences. If a customer requests an item that is not available immediately, it is likely that the customer will go elsewhere. Worse, the customer may never return. Inventory provides a buffer against the uncertainty of demand.

Other uncertainties provide a motivation for holding inventories as well. One is the uncertainty of the lead time. Lead time is defined as the amount of time that elapses from the point that an order is placed until it arrives. In the production planning context, interpret the lead time as the time required to produce the item. Even when future demand can be predicted accurately, the company needs to hold buffer stocks to assure a smooth flow of production or continued sales when replenishment lead times are uncertain.

A third significant source of uncertainty is the supply. The OPEC oil embargo of the late 1970s is an example of the chaos that can result when supply lines are threatened. Two industries that relied (and continue to rely) heavily on oil and gasoline are the electric utilities and the airlines. Firms in these and other industries risked having to curtail operations because of fuel shortages.

Additional uncertainties that could motivate a firm to store inventory include the uncertainty in the supply of labor, the price of resources, and the cost of capital.

3. *Speculation.* If the value of an item or natural resource is expected to increase, it may be more economical to purchase large quantities at current prices and store the items for future use than to pay the higher prices at a future date. In the early 1970s, for example, the Westinghouse Corporation sustained severe losses on its contracts to build nuclear plants for several electric utility companies because it guaranteed to supply the uranium necessary to operate the plants at a fixed price. Unfortunately for Westinghouse, the price of the uranium skyrocketed between the time the contracts were signed and the time the plants were built.

---

[2]This argument assumes that the setup cost is a fixed constant. In some circumstances it can be reduced, thus justifying smaller lot sizes. This forms the basis of the just-in-time philosophy discussed in detail in Chapter 7.

Other industries require large quantities of costly commodities that have experienced considerable fluctuation in price. For example, silver is required for the production of photographic film. By correctly anticipating a major price increase in silver, a major producer of photographic film, such as Kodak, could purchase and store large quantities of silver in advance of the increase and realize substantial savings.

The speculative motive also can be a factor for a firm facing the possibility of a labor strike. The cost of production could increase significantly when there is a severe shortage of labor.

4. *Transportation.* In-transit or *pipeline* inventories exist because transportation times are positive. When transportation times are long, as is the case when transporting oil from the Middle East to the United States, the investment in pipeline inventories can be substantial. One of the disadvantages of producing overseas is the increased transportation time, and hence the increase in pipeline inventories. This factor has been instrumental in motivating some firms to establish production operations domestically.

5. *Smoothing.* Changes in the demand pattern for a product can be deterministic or random. Seasonality is an example of a deterministic variation, while unanticipated changes in economic conditions can result in random variation. Producing and storing inventory in anticipation of peak demand can help to alleviate the disruptions caused by changing production rates and workforce levels. Smoothing costs and planning for anticipated swings in the demand were considered in the aggregate planning models in Chapter 3.

6. *Logistics.* We use the term *logistics* to describe reasons for holding inventory different from those outlined above. Certain constraints can arise in the purchasing, production, or distribution of items that force the system to maintain inventory. One such case is an item that must be purchased in minimum quantities. Another is the logistics of manufacture; it is virtually impossible to reduce all inventories to zero and expect any continuity in a manufacturing process.

7. *Control costs.* An important issue, and one that often is overlooked, is the cost of maintaining the inventory control system. A system in which more inventory is carried does not require the same level of control as one in which inventory levels are kept to a bare minimum. It can be less costly to the firm in the long run to maintain large inventories of inexpensive items than to expend worker time to keep detailed records for these items. Even though control costs could be a major factor in determining the suitability of a particular technique or system, they are rarely factored into the types of inventory models we will be discussing.

4.3 Characteristics of Inventory Systems

1. *Demand.* The assumptions one makes about the pattern and characteristics of the demand often turn out to be the most significant in determining the complexity of the resulting control model.

   a. *Constant versus variable.* The simplest inventory models assume that the rate of demand is a constant. The EOQ (economic order quantity) model and its extensions are based on this assumption. Variable demand arises in a variety of contexts, including aggregate planning (Chapter 3) and materials requirements planning (Chapter 7).

   b. *Known versus random.* It is possible for demand to be constant in expectation but still be random. Synonyms for random are *uncertain* and *stochastic.*

Virtually all stochastic demand models assume that the average demand rate is constant. Random demand models are generally both more realistic and more complex than their deterministic counterparts.

2. *Lead time.* If items are ordered from the outside, the lead time is defined as the amount of time that elapses from the instant that an order is placed until it arrives. If items are produced internally, however, then interpret lead time as the amount of time required to produce a batch of items. We will use the Greek letter τ to represent lead time, which is expressed in the same units of time as demand. That is, if demand is expressed in units per year, then lead time should be expressed in years.

3. *Review time.* In some systems the current level of inventory is known at all times. This is an accurate assumption when demand transactions are recorded as they occur. One example of a system in which inventory levels are known at all times is a modern supermarket with a visual scanning device at the checkout stand that is linked to a storewide inventory database. As an item is passed through the scanner, the transaction is recorded in the database, and the inventory level is decreased by one unit. We will refer to this case as *continuous review.* In the other case, referred to as *periodic review,* inventory levels are known only at discrete points in time. An example of periodic review is a small grocery store in which physical stock-taking is required to determine the current levels of on-hand inventory.

4. *Excess demand.* Another important distinguishing characteristic is how the system reacts to excess demand (that is, demand that cannot be filled immediately from stock). The two most common assumptions are that excess demand is either back-ordered (held over to be satisfied at a future time) or lost (generally satisfied from outside the system). Other possibilities include partial back-ordering (part of the demand is back-ordered and part of the demand is lost) or customer impatience (if the customer's order is not filled within a fixed amount of time, he cancels). The vast majority of inventory models, especially the ones that are used in practice, assume full back-ordering of excess demand.

5. *Changing inventory.* In some cases the inventory undergoes changes over time that may affect its utility. Some items have a limited shelf life, such as food, and others may become obsolete, such as automotive spare parts. Mathematical models that incorporate the effects of perishability or obsolescence are generally quite complex and beyond the scope of this text. A brief discussion can be found in Section 8 of Chapter 5.

4.4 Relevant Costs

Because we are interested in optimizing the inventory system, we must determine an appropriate optimization or performance criterion. Virtually all inventory models use cost minimization as the optimization criterion. An alternative performance criterion might be profit maximization. However, cost minimization and profit maximization are essentially equivalent criteria for most inventory control problems. Although different systems have different characteristics, virtually all inventory costs can be placed into one of three categories: holding cost, order cost, or penalty cost. We discuss each in turn.

Holding Cost

The **holding cost,** also known as the carrying cost or the inventory cost, is the sum of all costs that are proportional to the amount of inventory physically on hand at any point in time. The components of the holding cost include a variety of seemingly unrelated items. Some of these are

- Cost of providing the physical space to store the items.
- Taxes and insurance.
- Breakage, spoilage, deterioration, and obsolescence.
- Opportunity cost of alternative investment.

The last item often turns out to be the most significant in computing holding costs for most applications. Inventory and cash are in some sense equivalent. Capital must be invested to either purchase or produce inventory, and decreasing inventory levels results in increased capital. This capital could be invested by the company either internally, in its own operation, or externally.

What is the interest rate that could be earned on this capital? You and I can place our money in a simple passbook account with an interest rate of 2 percent, or possibly a long-term certificate of deposit with a return of maybe 5 percent. We could earn somewhat more by investing in high-yield bond funds or buying short-term industrial paper or second deeds of trust.

In general, however, most companies must earn higher rates of return on their investments than do individuals in order to remain profitable. The value of the interest rate that corresponds to the opportunity cost of alternative investment is related to (but not the same as) a number of standard accounting measures, including the internal rate of return, the return on assets, and the hurdle rate (the minimum rate that would make an investment attractive to the firm). Finding the right interest rate for the opportunity cost of alternative investment is very difficult. Its value is estimated by the firm's accounting department and is usually an amalgam of the accounting measures listed above. For convenience, we will use the term *cost of capital* to refer to this component of the holding cost. We may think of the holding cost as an aggregated interest rate comprised of the four components listed above. For example,

$$
\begin{array}{rl}
28\% = & \text{Cost of capital} \\
2\% = & \text{Taxes and insurance} \\
6\% = & \text{Cost of storage} \\
\underline{1\%} = & \underline{\text{Breakage and spoilage}} \\
37\% = & \text{Total interest charge}
\end{array}
$$

This would be interpreted as follows: We would assess a charge of 37 cents for every dollar that we have invested in inventory during a one-year period. However, as we generally measure inventory in units rather than in dollars, it is convenient to express the holding cost in terms of dollars per unit per year rather than dollars per dollar per year. Let c be the dollar value of one unit of inventory, I be the annual interest rate, and h be the holding cost in terms of dollars per unit per year. Then we have the relationship

$$h = Ic.$$

Hence, in the example above, an item valued at $180 would have an annual holding cost of $h = (0.37)(\$180) = \$66.60$. If we held 300 of these items for five years, the total holding cost over the five years would be

$$(5)(300)(66.60) = \$99,900.$$

This example raises an interesting question. Suppose that during the five-year period the inventory level did not stay fixed at 300 but varied on a continuous basis.

FIGURE 4–2

*Inventory as a
function of time*

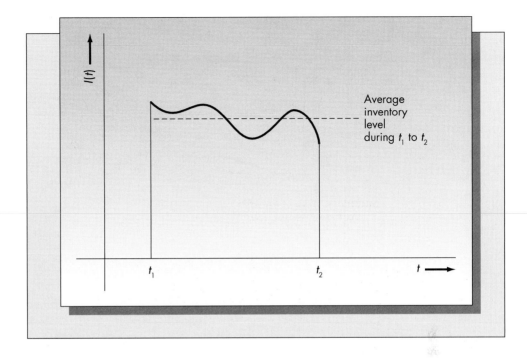

We would expect inventory levels to change over time. Inventory levels decrease when items are used to satisfy demand and increase when units are produced or new orders arrive. How would the holding cost be computed in such a case? In particular, suppose the inventory level $I(t)$ during some interval (t_1, t_2) behaves as in Figure 4–2.

The holding cost incurred at any point in time is proportional to the inventory level at that point in time. In general, the total holding cost incurred from a time t_1 to a time t_2 is h multiplied by the area under the curve described by $I(t)$. The *average* inventory level during the period (t_1, t_2) is the area under the curve divided by $t_2 - t_1$. For the cases considered in this chapter, simple geometry can be used to find the area under the inventory level curve. When $I(t)$ is described by a straight line, its average value is obvious. In cases such as that pictured in Figure 4–2, in which the curve of $I(t)$ is complex, the average inventory level would be determined by computing the integral of $I(t)$ over the interval (t_1, t_2) and dividing by $t_2 - t_1$.

Order Cost

The holding cost includes all of those costs that are proportional to the amount of inventory on hand, whereas the **order cost** depends on the amount of inventory that is ordered or produced.

In most applications, the order cost has two components: a fixed and a variable component. The fixed cost, K, is incurred independent of the size of the order as long as it is not zero. The variable cost, c, is incurred on a per-unit basis. We also refer to K as the setup cost and c as the proportional order cost. Define $C(x)$ as the cost of ordering (or producing) x units. It follows that

$$C(x) = \begin{cases} 0 & \text{if } x = 0, \\ K + cx & \text{if } x > 0. \end{cases}$$

The order cost function is pictured in Figure 4–3.

Figure 4–3

Order cost function

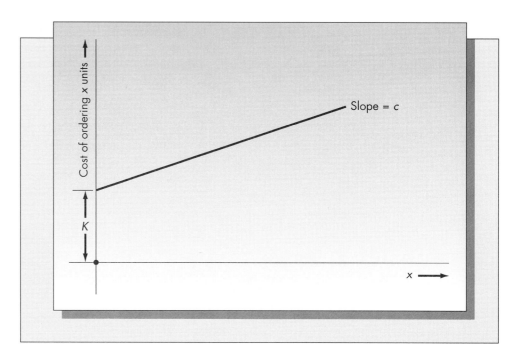

When estimating the setup cost, one should include *only* those costs that are relevant to the current ordering decision. For example, the cost of maintaining the purchasing department of the company is *not* relevant to daily ordering decisions and should not be factored into the estimation of the setup cost. This is an overhead cost that is independent of the decision of whether or not an order should be placed. The appropriate costs comprising K would be the bookkeeping expense associated with the order, the fixed costs independent of the size of the order that might be required by the vendor, costs of order generation and receiving, and handling costs.

Penalty Cost

The **penalty cost,** also known as the shortage cost or the stock-out cost, is the cost of not having sufficient stock on hand to satisfy a demand *when it occurs.* This cost has a different interpretation depending on whether excess demand is back-ordered (orders that cannot be filled immediately are held on the books until the next shipment arrives) or lost (known as lost sales). In the back-order case, the penalty cost includes whatever bookkeeping and/or delay costs might be involved. In the lost-sales case, it includes the lost profit that would have been made from the sale. In either case, it would also include the "loss-of-goodwill" cost, which is a measure of customer satisfaction. Estimating the loss-of-goodwill component of the penalty cost can be very difficult in practice.

We use the symbol p to denote penalty cost and assume that p is charged on a per-unit basis. That is, each time a demand occurs that cannot be satisfied immediately, a cost p is incurred independent of how long it takes to eventually fill the demand. An alternative means of accounting for shortages is to charge the penalty cost on a per-unit-per-unit-time basis (as we did with the holding cost). This approach is appropriate if the time that a back order stays on the books is important; for example, if a back order results in stopping a production line because of the unavailability of a part. The models considered in this chapter assume that penalty costs are charged on a per-unit basis only. Penalty cost models are not considered in this chapter. Penalty costs are included in Chapter 5, but models incorporating a time-weighted penalty cost are not.

We present those inventory models that have had the greatest impact in the user community. Many of the techniques discussed in both this chapter and in Chapter 5 form the basis for commercial inventory control systems or in-house systems. In most cases, the models are simple enough that optimal operating policies can be calculated by hand, but they are often complex enough to capture the essential trade-offs in inventory management.

Problems for Sections 1–4

1. What are the two questions that inventory control addresses?

2. Discuss the cost penalties incurred by a firm that holds too much inventory and one that holds too little inventory.

3. ABC, Inc., produces a line of touring bicycles. Specifically, what are the four types of inventories (raw materials, components, work-in-process, and finished goods) that would arise in the production of this item?

4. I Carry rents trucks for moving and hauling. Each truck costs the company an average of $8,000, and the inventory of trucks varies monthly depending on the number that are rented out. During the first eight months of last year, I Carry had the following ending inventory of trucks on hand:

Month	Number of Trucks	Month	Number of Trucks
January	26	May	13
February	38	June	9
March	31	July	16
April	22	August	5

I Carry uses a 20 percent annual interest rate to represent the cost of capital. Yearly costs of storage amount to 3 percent of the value of each truck, and the cost of liability insurance is 2 percent.

a. Determine the total handling cost incurred by I Carry during the period January to August. Assume for the purposes of your calculation that the holding cost incurred in a month is proportional to the inventory on hand at the end of the month.

b. Assuming that these eight months are representative, estimate the average annual cost of holding trucks.

5. Stationery Supplies is considering installing an inventory control system in its store in Provo, Utah. The store carries about 1,400 different inventory items and has annual gross sales of about $80,000. The inventory control system would cost $12,500 to install and about $2,000 per year in additional supplies, time, and maintenance. If the savings to the store from the system can be represented as a

fixed percentage of annual sales, what would that percentage have to be in order for the system to pay for itself in five years or less?

6. For Stationery Supplies, discussed in Problem 5, list and discuss all of the uncertainties that would motivate the store to maintain inventories of its 1,400 items.

7. Stationery Supplies orders plastic erasers from a company in Nürnberg, Germany. It takes six weeks to ship the erasers from Germany to Utah. Stationery Supplies maintains a standing order of 200 erasers every six months (shipped on the first of January and the first of July).

   a. Assuming the ordering policy the store is using does not result in large buildups of inventory or long-term stock-outs, what is the annual demand for erasers?

   b. Draw a graph of the pipeline inventory (that is, the inventory ordered but not received) of the erasers during one year. What is the average pipeline inventory of erasers during the year?

   c. Express the replenishment lead time in years and multiply the annual demand you obtained in part (a) by the lead time. What do you notice about the result that you obtain?

8. Penalty costs can be assessed only against the number of units of demand that cannot be satisfied, or against the number of units weighted by the amount of time that an order stays on the books. Consider the following history of supply and demand transactions for a particular part:

Month	Number of Items Received	Demand during Month
January	200	520
February	175	1,640
March	750	670
April	950	425
May	500	280
June	2,050	550

Assume that starting inventory at the beginning of January is 480 units.

   a. Determine the ending inventory each month. Assume that excess demands are back-ordered.

   b. Assume that each time a unit is demanded that cannot be supplied immediately, a one-time charge of $10 is made. Determine the stock-out cost incurred during the six months (1) if excess demand at the end of each month is lost, and (2) if excess demand at the end of each month is back-ordered.

   c. Suppose that each stock-out costs $10 per unit per month that the demand remains unfilled. If demands are filled on a first-come, first-served basis, what is the total stock-out cost incurred during the six months using this type of cost criterion? (Assume that the demand occurs at the beginning of the month for purposes of your calculation.) Notice that you must assume that excess demands are back-ordered for this case to make any sense.

   d. Discuss under what circumstances the cost criterion used in (b) might be appropriate and under what circumstances the cost criterion used in (c) might be appropriate.

9. HAL Ltd. produces a line of high-capacity disk drives for mainframe computers. The housings for the drives are produced in Hamilton, Ontario, and shipped to the main plant in Toronto. HAL uses the drive housings at a fairly steady rate of 720 per year. Suppose that the housings are shipped in trucks that can hold 40 housings at one time. It is estimated that the fixed cost of loading the housings onto the truck and unloading them on the other end is $300 for shipments of 120 or fewer housings (i.e., three or fewer truckloads). Each trip made by a single truck costs the company $160 in driver time, gasoline, oil, insurance, and wear and tear on the truck.

   a. Compute the annual costs of transportation and loading and unloading the housings for the following policies: (1) shipping one truck per week, (2) shipping one full truckload as often as needed, and (3) shipping three full truckloads as often as needed.

   b. For what reasons might the policy in (*a*) with the highest annual cost be more desirable from a systems point of view than the policy having the lowest annual cost?

4.5 The EOQ Model

The **EOQ model** (or *economic order quantity model*) is the simplest and most fundamental of all inventory models. It describes the important trade-off between fixed order costs and holding costs, and is the basis for the analysis of more complex systems.

The Basic Model

Assumptions:

1. The demand rate is known and is a constant λ units per unit time. (The unit of time may be days, weeks, months, and so on. In what follows we assume that the default unit of time is a year. However, the analysis is valid for other time units as long as all relevant variables are expressed in the same units.)

2. Shortages are not permitted.

3. There is no order lead time. (This assumption will be relaxed.)

4. The costs include
      a. Setup cost at K per positive order placed.
      b. Proportional order cost at c per unit ordered.
      c. Holding cost at h per unit held per unit time.

Assume with no loss in generality that the on-hand inventory at time zero is zero. Shortages are not allowed, so we must place an order at time zero. Let Q be the size of the order. It follows that the on-hand inventory level increases instantaneously from zero to Q at time $t = 0$.

Consider the next time an order is to be placed. At this time, either the inventory is positive or it is again zero. A little reflection shows that we can reduce the holding costs by waiting until the inventory level drops to zero before ordering again. At the instant that on-hand inventory equals zero, the situation looks exactly the same as it did at time $t = 0$. If it was optimal to place an order for Q units at that time, then it is still

FIGURE 4–4

*Inventory levels for
the EOQ model*

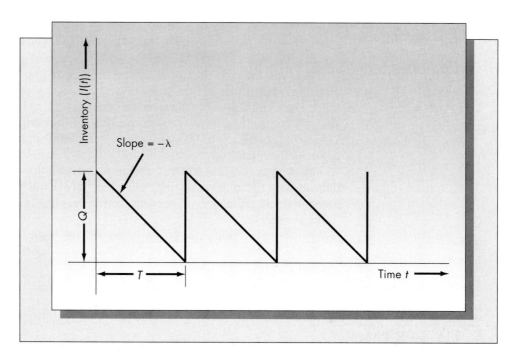

optimal to order Q units. It follows that the function that describes the changes in stock levels over time is the familiar sawtooth pattern of Figure 4–4.

The objective is to choose Q to minimize the average cost per unit time. Unless otherwise stated, we will assume that a unit of time is a year, so that we minimize the average annual cost. Other units of time, such as days, weeks, or months, are also acceptable, as long as all time-related variables are expressed in the same units. One might think that the appropriate optimization criterion would be to minimize the *total* cost in a cycle. However, this ignores the fact that the cycle length itself is a function of Q and must be explicitly included in the formulation.

Next, we derive an expression for the average annual cost as a function of the lot size Q. In each cycle, the total fixed plus proportional order cost is $C(Q) = K + cQ$. In order to obtain the order cost per unit time, we divide by the cycle length T. As Q units are consumed each cycle at a rate λ, it follows that $T = Q/\lambda$. This result also can be obtained by noting that the slope of the inventory curve pictured in Figure 4–4, $-\lambda$, equals the ratio $-Q/T$.

Consider the holding cost. Because the inventory level decreases linearly from Q to 0 each cycle, the average inventory level during one order cycle is $Q/2$. Because all cycles are identical, the average inventory level over a time horizon composed of many cycles is also $Q/2$. It follows that the average annual cost, say $G(Q)$, is given by

$$G(Q) = \frac{K + cQ}{T} + \frac{hQ}{2} = \frac{K + cQ}{Q/\lambda} + \frac{hQ}{2}$$

$$= \frac{K\lambda}{Q} + \lambda c + \frac{hQ}{2}.$$

The three terms composing $G(Q)$ are annual setup cost, annual purchase cost, and annual holding cost, respectively.

FIGURE 4–5

The average annual cost function G(Q)

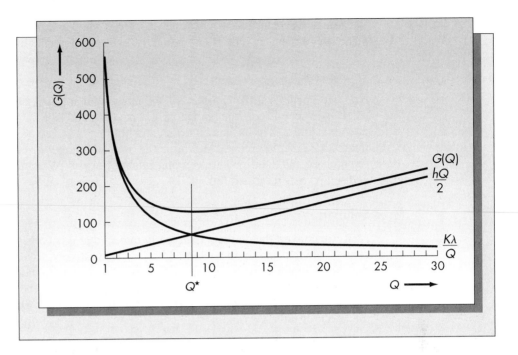

We now wish to find Q to minimize $G(Q)$. Consider the shape of the curve $G(Q)$. We have that

$$G'(Q) = -K\lambda/Q^2 + h/2$$

and

$$G''(Q) = 2K\lambda/Q^3 > 0 \qquad \text{for } Q > 0.$$

Since $G''(Q) > 0$, it follows that $G(Q)$ is a convex function of Q. Furthermore, since $G'(0) = -\infty$ and $G'(\infty) = h/2$, it follows that $G(Q)$ behaves as pictured in Figure 4–5.

The optimal value of Q occurs where $G'(Q) = 0$. This is true when $Q^2 = 2K\lambda/h$, which gives

$$Q^* = \sqrt{\frac{2K\lambda}{h}}.$$

Q^* is known as the economic order quantity (EOQ). There are a number of interesting points to note:

1. In Figure 4–5, the curves corresponding to the fixed order cost component $K\lambda/Q$ and the holding cost component $hQ/2$ also are included. Notice that Q^* is the value of Q where the two curves intersect. (If you equate $hQ/2$ and $K\lambda/Q$ and solve for Q, you will obtain the EOQ formula.) In general, the minimum of the sum of two functions will *not* occur at the intersection of the two functions. It is an interesting coincidence that it does in this case.

2. Notice that the proportional order cost component, c, does not appear explicitly in the expression for Q^*. This is because the term λc appearing in the definition of $G(Q)$ is independent of Q. As all feasible policies must replenish inventory at the rate of demand, the proportional order cost incurred per unit time is λc independent of Q.

Because λc is a constant, we generally ignore it when computing average costs. Notice that c *does* affect the value of Q^* indirectly, as h appears in the EOQ formula and $h = Ic$.

Example 4.1

Number 2 pencils at the campus bookstore are sold at a fairly steady rate of 60 per week. The pencils cost the bookstore 2 cents each and sell for 15 cents each. It costs the bookstore $12 to initiate an order, and holding costs are based on an annual interest rate of 25 percent. Determine the optimal number of pencils for the bookstore to purchase and the time between placement of orders. What are the yearly holding and setup costs for this item?

Solution

First, we convert the demand to a yearly rate so that it is consistent with the interest charge, which is given on an annual basis. (Alternatively, we could have converted the annual interest rate to a weekly interest rate.) The annual demand rate is $\lambda = (60)(52) = 3,120$. The holding cost h is the product of the annual interest rate and the variable cost of the item. Hence, $h = (0.25)(0.02) = 0.005$. Substituting into the EOQ formula, we obtain

$$Q^* = \sqrt{\frac{2K\lambda}{h}} = \sqrt{\frac{(2)(12)(3,120)}{0.005}} = 3,870.$$

The cycle time is $T = Q/\lambda = 3,870/3,120 = 1.24$ years. The average annual holding cost is $h(Q/2) = 0.005(3,870/2) = \$9.675$. The average annual setup cost is $K\lambda/Q$, which is also $\$9.675$. ■

Example 4.1 illustrates some of the problems that can arise when using simple models. The optimal solution calls for ordering almost 4,000 pencils every 15 months. Even though this value of Q minimizes the yearly holding and setup costs, it could be infeasible: the store may not have the space to store 4,000 pencils. Simple models cannot account for all of the constraints present in a real problem. For that reason, every solution must be considered in context and modified, if necessary, to fit the application.

Notice also that the optimal solution did not depend on the selling price of 15 cents. Even if each pencil sold for $2, we would recommend the same order quantity, because the pencils are assumed to sell at a rate of 60 per week no matter what their price. This is, of course, a simplification of reality. It is reasonable to assume that the demand is relatively stable for a range of prices. Inventory models explicitly incorporate selling price in the formulation only when setting the price is included as part of the optimization.

Inclusion of Order Lead Time

One of the assumptions made in our derivation of the EOQ model was that there was no order lead time. We now relax that assumption. Suppose in Example 4.1 that the pencils had to be ordered four months in advance. If we were to place the order exactly four months before the end of the cycle, the order would arrive at exactly the same point in time as in the zero lead time case. The optimal timing of order placement for Example 4.1 is shown in Figure 4–6.

Rather than say that an order should be placed so far in advance of the end of a cycle, it is more convenient to indicate reordering in terms of the on-hand inventory.

FIGURE 4–6

*Reorder point
calculation for
Example 4.1*

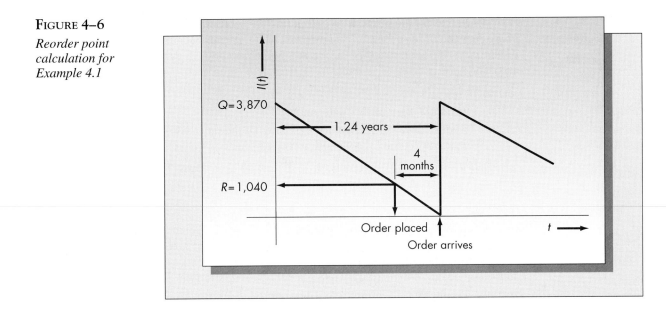

FIGURE 4–7

*Reorder point
calculation for lead
times exceeding one
cycle*

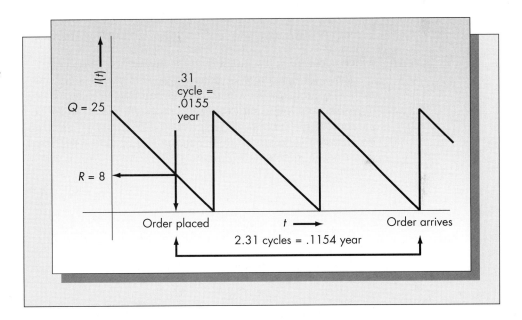

Define R, the reorder point, as the level of on-hand inventory at the instant an order should be placed. From Figure 4–6. We see that R is the product of the lead time and the demand rate ($R = \lambda\tau$). For the example, $R = (3,120)(.3333) = 1,040$. Notice that we converted the lead time to years before multiplying. *Always express all relevant variables in the same units of time.*

Determining the reorder point is more difficult when the lead time exceeds a cycle. Consider an item with an EOQ of 25, a demand rate of 500 units per year, and a lead time of six weeks. The cycle time is $T = 25/500 = 0.05$ year, or 2.6 weeks. Forming the ratio of τ/T, we obtain 2.31. This means that there are exactly 2.31 cycles in the lead time. Every order must be placed 2.31 cycles in advance (see Figure 4–7).

Notice that for the purpose of computing the reorder point, this is exactly the same as placing the order *0.31 cycle in advance*. This is true because the level of on-hand inventory is the same whether we are at a point 2.31 or 0.31 cycle before the arrival of an order. In this case, 0.31 cycle is 0.0155 year, thus giving a reorder point of $R = (0.0155)(500) = 7.75 \approx 8$. In general, when $\tau > T$ use the following procedure:

a. Form the ratio τ/T.

b. Consider only the fractional remainder of the ratio. Multiply this fractional remainder by the cycle length to convert back to years.

c. Multiply the result of step (*b*) by the demand rate to obtain the reorder point.

Sensitivity

In this section we examine the issue of how sensitive the annual cost function is to errors in the calculation of Q. Consider Example 4.1. Suppose that the bookstore orders pencils in batches of 1,000, rather than 3,870 as the optimal solution indicates. What additional cost is it incurring by using a suboptimal solution? To answer the question, we consider the average annual cost function $G(Q)$. By substituting $Q = 1,000$, we can find the average annual cost for this lot size and compare it to the optimal cost to determine the magnitude of the penalty. We have

$$G(Q) = K\lambda/Q + hQ/2 = (12)(3,120)/1,000 + (0.005)(1,000)/2 = \$39.94,$$

which is considerably larger than the optimal cost of $19.35.

One can find the cost penalty for suboptimal solutions in this manner for any particular problem. However, it is more instructive and more convenient to obtain a universal solution to the sensitivity problem. We do so by deriving an expression for the ratio of the suboptimal cost over the optimal cost as a function of the ratio of the optimal and suboptimal order quantities. Let G^* be the average annual holding and setup cost at the optimal solution. Then

$$G^* = K\lambda/Q^* + hQ^*/2 = \frac{K\lambda}{\sqrt{2K\lambda/h}} + \frac{h}{2}\sqrt{\frac{2K\lambda}{h}} = 2\sqrt{\frac{K\lambda h}{2}}$$

$$= \sqrt{2K\lambda h}.$$

It follows that for any Q,

$$\frac{G(Q)}{G^*} = \frac{K\lambda/Q + hQ/2}{\sqrt{2K\lambda h}}$$

$$= \frac{1}{2Q}\sqrt{\frac{2K\lambda}{h}} + \frac{Q}{2}\sqrt{\frac{h}{2K\lambda}}$$

$$= \frac{Q^*}{2Q} + \frac{Q}{2Q^*}$$

$$= \frac{1}{2}\left[\frac{Q^*}{Q} + \frac{Q}{Q^*}\right].$$

To see how one would use this result, consider using a suboptimal lot size in Example 4.1. The optimal solution was $Q^* = 3,870$, and we wished to evaluate the cost error of using $Q = 1,000$. Forming the ratio Q^*/Q gives 3.87. Hence, $G(Q)/G^* = (0.5)(3.87 + 1/3.87) = (0.5)(4.128) = 2.06$. This says that the average annual holding and setup cost with $Q = 1,000$ is 2.06 times the optimal average annual holding and setup cost.

In general, the cost function $G(Q)$ is relatively insensitive to errors in Q. For example, if Q is twice as large as it should be, Q/Q^* is 2 and G/G^* is 1.25. Hence, an error of 100 percent in the value of Q results in an error of only 25 percent in the annual holding and setup cost. Notice that you obtain the same result if Q is half Q^*, since $Q^*/Q = 2$. However, this does *not* imply that the average annual cost function is symmetric. In fact, suppose that the order quantity differed from the optimal by ΔQ units. A value of $Q = Q^* + \Delta Q$ would result in a *lower* average annual cost than a value of $Q = Q^* - \Delta Q$.

EOQ and JIT

Largely as a result of the success of Toyota's kanban system, a new philosophy is emerging about the role and importance of inventories in manufacturing environments. This philosophy, known as *just-in-time* (JIT), says that excess work-in-process inventories are not desirable, and inventories should be reduced to the bare essentials. (We discuss the just-in-time philosophy in more detail in Chapter 1 and the mechanics of kanban in Chapter 7.) EOQ is the result of traditional thinking about inventories and scale economies in economics. Are the EOQ and JIT approaches at odds with each other?

Proponents argue that an essential part of implementing JIT is reducing setup times, and hence setup costs. As setup costs decrease, traditional EOQ theory says that lot sizes should be reduced. In this sense, the two ways of thinking are compatible. However, there are times when they may not be. We believe that there is substantial value to the JIT approach that may not be incorporated easily into a mathematical model. Quality problems can be identified and rectified before inventories of defective parts accumulate. Plants can be more flexible if they are not burdened with excess in-process inventories. Certainly Toyota's success with the JIT approach, as evidenced by substantially lower inventory costs per car than are typical for U.S. auto manufacturers, is a testament to the value of JIT.

However, we believe that every new approach must be incorporated carefully into the firm's business and not adopted blindly without evaluating its consequences and appropriateness. JIT, although an exciting new development in material management, is not always the best approach. The principles underlying EOQ (and MRP, discussed in Chapter 7) are sound and should not be ignored. The following example illustrates this point.

Example 4.2

The Rahway, New Jersey, plant of Metalcase, a manufacturer of office furniture, produces metal desks at a rate of 200 per month. Each desk requires 40 Phillips head metal screws purchased from a supplier in North Carolina. The screws cost 3 cents each. Fixed delivery charges and costs of receiving and storing shipments of the screws amount to about $100 per shipment, independently of the size of the shipment. The firm uses a 25 percent interest rate to determine holding costs. Metalcase would like to establish a standing order with the supplier and is considering several alternatives. What standing order size should they use?

Solution

First we compute the EOQ. The annual demand for screws is

$$(200)(12)(40) = 96,000.$$

The annual holding cost per screw is $(0.25)(0.03) = 0.0075$. From the EOQ formula, the optimal lot size is

$$Q^* = \sqrt{\frac{(2)(100)(96,000)}{0.0075}} = 50,597.$$

Note that the cycle time is $T = Q/\lambda = 50,597/96,000 = 0.53$ year or about once every six months. Hence the optimal policy calls for replenishment of the screws about twice a year. A JIT approach would be to order the screws as frequently as possible to minimize the inventory held at the plant. Implementing such an approach might suggest a policy of weekly deliveries. Such a policy makes little sense in this context, however. This policy would require 52 deliveries per year, incurring setup costs of $5,200 annually. The EOQ solution gives a total annual cost of both setups and holding of less than $400. For a low-value item such as this with high fixed order costs, small lot sizes in accordance with JIT are inappropriate. The point is that no single approach should be blindly adopted for all situations. The success of a method in one context does not ensure its appropriateness in all other contexts. ■

Problems for Section 5

10. A specialty coffeehouse sells Colombian coffee at a fairly steady rate of 280 pounds annually. The beans are purchased from a local supplier for $2.40 per pound. The coffeehouse estimates that it costs $45 in paperwork and labor to place an order for the coffee, and holding costs are based on a 20 percent annual interest rate.
    a. Determine the optimal order quantity for Colombian coffee.
    b. What is the time between placement of orders?
    c. What is the average annual cost of holding and setup due to this item?
    d. If replenishment lead time is three weeks, determine the reorder level based on the on-hand inventory.

11. For the situation described in Problem 10, draw a graph of the amount of inventory on order. Using your graph, determine the average amount of inventory on order. Also compute the demand during the replenishment lead time. How do these two quantities differ?

12. A large automobile repair shop installs about 1,250 mufflers per year, 18 percent of which are for imported cars. All of the imported-car mufflers are purchased from a single local supplier at a cost of $18.50 each. The shop uses a holding cost based on a 25 percent annual interest rate. The setup cost for placing an order is estimated to be $28.
    a. Determine the optimal number of imported-car mufflers the shop should purchase each time an order is placed, and the time between placement of orders.
    b. If the replenishment lead time is six weeks, what is the reorder point based on the level of on-hand inventory?
    c. The current reorder policy is to buy imported-car mufflers only once a year. What are the additional holding and setup cost incurred by this policy?

13. Consider the coffeehouse discussed in Problem 10. Suppose that its setup cost for ordering was really only $15. Determine the error made in calculating the annual cost of holding and setup incurred as a result of its using the wrong value of K. (Note that this implies that its current order policy is suboptimal.)

14. A local machine shop buys hex nuts and molly screws from the same supplier. The hex nuts cost 15 cents each and the molly screws cost 38 cents each. A setup cost of $100 is assumed for all orders. This includes the cost of tracking and receiving the orders. Holding costs are based on a 25 percent annual interest rate. The shop uses an average of 20,000 hex nuts and 14,000 molly screws annually.
    a. Determine the optimal size of the orders of hex nuts and molly screws, and the optimal time between placement of orders of these two items.
    b. If both items are ordered and received simultaneously, the setup cost of $100 applies to the combined order. Compare the average annual cost of holding and setup if these items are ordered separately; if they are both ordered when the hex nuts would normally be ordered; and if they are both ordered when the molly screws would normally be ordered.

15. David's Delicatessen flies in Hebrew National salamis regularly to satisfy a growing demand for the salamis in Silicon Valley. The owner, David Gold, estimates that the demand for the salamis is pretty steady at 175 per month. The salamis cost Gold $1.85 each. The fixed cost of calling his brother in New York and having the salamis flown in is $200. It takes three weeks to receive an order. Gold's accountant, Irving Wu, recommends an annual cost of capital of 22 percent, a cost of shelf space of 3 percent of the value of the item, and a cost of 2 percent of the value for taxes and insurance.
    a. How many salamis should Gold have flown in and how often should he order them?
    b. How many salamis should Gold have on hand when he phones his brother to send another shipment?
    c. Suppose that the salamis sell for $3 each. Are these salamis a profitable item for Gold? If so, what annual profit can he expect to realize from this item? (Assume that he operates the system optimally.)
    d. If the salamis have a shelf life of only 4 weeks, what is the trouble with the policy that you derived in part (a)? What policy would he have to use in that case? Is the item still profitable?

16. In view of the results derived in the section on sensitivity analysis, discuss the following quotation of an inventory control manager: "If my lot sizes are going to be off the mark, I'd rather miss on the high side than on the low side."

4.6 Extension to a Finite Production Rate

An implicit assumption of the simple EOQ model is that the items are obtained from an outside supplier. When that is the case, it is reasonable to assume that the entire lot is delivered at the same time. However, if we wish to use the EOQ formula when the units are produced internally, then we are effectively assuming that the production rate is infinite. When the production rate is much larger than the demand rate, this assumption is probably satisfactory as an approximation. However, if the rate of production is comparable to the rate of demand, the simple EOQ formula will lead to incorrect results.

FIGURE 4–8

Inventory levels for finite production rate model

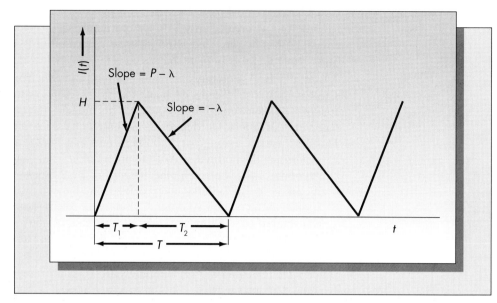

Assume that items are produced at a rate P during a production run. We require that $P > \lambda$ for feasibility. All other assumptions will be identical to those made in the derivation of the simple EOQ. When units are produced internally, the curve describing inventory levels as a function of time is slightly different from the sawtooth pattern of Figure 4–4. The change in the inventory level over time for the finite production rate case is shown in Figure 4–8.

Let Q be the size of each production run. Let T, the cycle length, be the time between successive production startups. Write $T = T_1 + T_2$, where T_1 is uptime (production time) and T_2 is downtime. Note that the maximum level of on-hand inventory during a cycle is *not Q*.

The number of units consumed each cycle is λT, which must be the same as the number of units produced each cycle, which is simply Q. It follows that $Q = \lambda T$, or $T = Q/\lambda$. Define H as the maximum level of on-hand inventory. As items are produced at a rate P for a time T_1, it follows that $Q = PT_1$, or $T_1 = Q/P$. From Figure 4–8 we see that $H/T_1 = P - \lambda$. This follows from the definition of the slope as the rise over the run. Substituting $T_1 = Q/P$ and solving for H gives $H = Q(1 - \lambda/P)$.

We now determine an expression for the average annual cost function. Because the average inventory level is $H/2$, it follows that

$$G(Q) = \frac{K}{T} + \frac{hH}{2} = \frac{K\lambda}{Q} + \frac{hQ}{2}\,(1 - \lambda/P).$$

Notice that if we define $h' = h(1 - \lambda/P)$, then this $G(Q)$ is identical to that of the infinite production rate case with h' substituted for h. It follows that

$$Q^* = \sqrt{\frac{2K\lambda}{h'}}.$$

Example 4.3

A local company produces a programmable EPROM for several industrial clients. It has experienced a relatively flat demand of 2,500 units per year for the product. The EPROM is produced at a rate of 10,000 units per year. The accounting department has estimated that it costs $50 to initiate a production run, each unit costs the company $2 to manufacture, and the cost of holding is based on a 30 percent annual interest rate. Determine the optimal size of a production run, the length of each production run, and the average annual cost of holding and setup. What is the maximum level of the on-hand inventory of the EPROMs?

Solution

First, we compute $h = (0.3)(2) = 0.6$ per unit per year. The modified holding cost is $h' = h(1 - \lambda/P) = (0.6)(1 - 2,500/10,000) = 0.45$. Substituting into the EOQ formula and using h' for h, we obtain $Q^* = 745$. Note that the simple EOQ equals 645, all out 14 percent less.

The time between production runs is $T = Q/\lambda = 745/2,500 = .298$ year. The uptime each cycle is $T_1 = Q/P = 745/10,000 = 0.0745$ year, and the downtime each cycle is $T_2 = T - T_1 = 0.2235$ year.

The average annual cost of holding and setup is

$$G(Q^*) = \frac{K\lambda}{Q^*} + \frac{h'Q^*}{2} = \frac{(50)(2,500)}{745} + \frac{(0.45)(745)}{2} = 335.41.$$

The maximum level of on-hand inventory is $H = Q^*(1 - \lambda/P) = 559$ units. ∎

Problems for Section 6

17. The Wod Chemical Company produces a chemical compound that is used as a lawn fertilizer. The compound can be produced at a rate of 10,000 pounds per day. Annual demand for the compound is 0.6 million pounds per year. The fixed cost of setting up for a production run of the chemical is $1,500, and the variable cost of production is $3.50 per pound. The company uses an interest rate of 22 percent to account for the cost of capital, and the costs of storage and handling of the chemical amount to 12 percent of the value. Assume that there are 250 working days in a year.

    a. What is the optimal size of the production run for this particular compound?

    b. What proportion of each production cycle consists of uptime and what proportion consists of downtime?

    c. What is the average annual cost of holding and setup attributed to this item? If the compound sells for $3.90 per pound, what is the annual profit the company is realizing from this item?

18. Determine the batch size that would result in Problem 17 if you assumed that the production rate was infinite. What is the additional average annual cost that would be incurred using this batch size rather than the one you found in Problem 17?

19. HAL Ltd., discussed in Problem 9, can produce the disk drive housings in the Hamilton, Ontario, plant at a rate of 150 housings per month. The housings cost HAL $85 each to produce and the setup cost for beginning a production run is $700. Assume an annual interest rate of 28 percent for determining the holding cost.

    a. What is the optimal number of housings for HAL to produce in each production run?

    b. Find the time between initiation of production runs, the time devoted to production, and the downtime each production cycle.

    c. What is the maximum dollar investment in housings that HAL has at any point in time?

20. Filter Systems produces air filters for domestic and foreign cars. One filter, part number JJ39877, is supplied on an exclusive contract basis to Oil Changers at a constant 200 units monthly. Filter Systems can produce this filter at a rate of 50 per hour. Setup time to change the settings on the equipment is 1.5 hours. Worker time (including overhead) is charged at the rate of $55 per hour and plant idle time during setups is estimated to cost the firm $100 per hour in lost profit.

    Filter Systems has established a 22 percent annual interest charge for determining holding cost. Each filter costs the company $2.50 to produce; they are sold for $5.50 each to Oil Changers. Assume 6-hour days, 20 working days per month, and 12 months per year for your calculations.

    a. How many JJ39877 filters should Filter Systems produce in each production run of this particular part to minimize annual holding and setup costs?

    b. Assuming that it produces the optimal number of filters in each run, what is the maximum level of on-hand inventory of these filters that the firm has at any point in time?

    c. What percentage of the working time does the company produce these particular filters, assuming that the policy in part (*a*) is used?

4.7 Quantity Discount Models

We have assumed up until this point that the cost *c* of each unit is independent of the size of the order. Often, however, the supplier is willing to charge less per unit for larger orders. The purpose of the discount is to encourage the customer to buy the product in larger batches. Such quantity discounts are common for many consumer goods.

Although many different types of discount schedules exist, there are two that seem to be the most popular: all-units and incremental. In each case we assume that there are one or more breakpoints defining changes in the unit cost. However, there are two possibilities: either the discount is applied to all of the units in an order (all-units), or it is applied only to the additional units beyond the breakpoint (incremental). The all-units case is more common.

Example 4.4

The Weighty Trash Bag Company has the following price schedule for its large trash can liners. For orders of less than 500 bags, the company charges 30 cents per bag; for orders of 500 or more but fewer than 1,000 bags, it charges 29 cents per bag; and for orders of 1,000 or more, it charges 28 cents per bag. In this case the breakpoints occur at 500 and 1,000. The discount schedule is all-units because the discount is applied to all of the units in an order. The order cost function $C(Q)$ is defined as

$$C(Q) = \begin{cases} 0.30Q & \text{for} & 0 \leq Q < 500, \\ 0.29Q & \text{for} & 500 \leq Q < 1,000, \\ 0.28Q & \text{for} & 1,000 \leq Q \end{cases}$$

The function $C(Q)$ is pictured in Figure 4–9. In Figure 4–10, we consider the same breakpoints, but assume an incremental quantity discount schedule. ∎

FIGURE 4–9

All-units discount order cost function

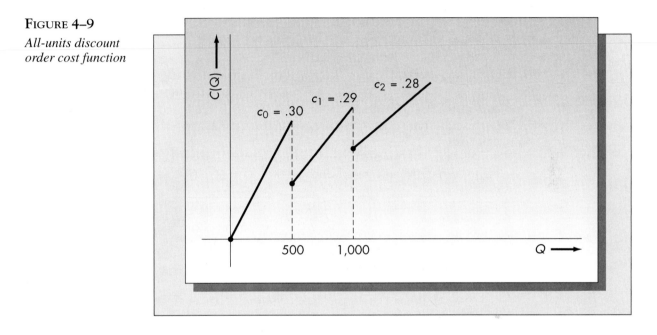

FIGURE 4–10

Incremental discount order cost function

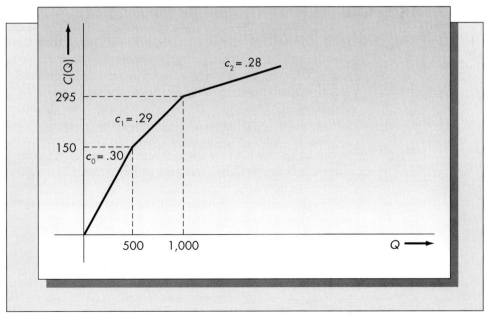

Note that the average cost per unit with an all-units schedule will be less than the average cost per unit with the corresponding incremental schedule.

The all-units schedule appears irrational in some respects. In the example above, 499 bags would cost $149.70, whereas 500 bags would cost only $145.00. Why would Weighty actually charge less for a larger order? One reason would be to provide an incentive for the purchaser to buy more. If you were considering buying 400 bags, you might choose to move up to the breakpoint to obtain the discount. Furthermore, it is possible that Weighty has stored its bags in lots of 100, so that its savings in handling costs might more than compensate for the lower total cost.

Optimal Policy for All-Units Discount Schedule

We will illustrate the solution technique using Example 4.4 above. Assume that the company considering what standing order to place with Weighty uses trash bags at a fairly constant rate of 600 per year. The accounting department estimates that the fixed cost of placing an order is $8, and holding costs are based on a 20 percent annual interest rate. From Example 4.4, $c_0 = 0.30$, $c_1 = 0.29$, and $c_2 = 0.28$ are the respective unit costs.

The first step toward finding a solution is to compute the EOQ values corresponding to each of the unit costs, which we will label $Q^{(0)}$, $Q^{(1)}$, and $Q^{(2)}$, respectively.

$$Q^{(0)} = \sqrt{\frac{2K\lambda}{Ic_0}} = \sqrt{\frac{(2)(8)(600)}{(0.2)(0.30)}} = 400,$$

$$Q^{(1)} = \sqrt{\frac{2K\lambda}{Ic_1}} = \sqrt{\frac{(2)(8)(600)}{(0.2)(0.29)}} = 406,$$

$$Q^{(2)} = \sqrt{\frac{2K\lambda}{Ic_2}} = \sqrt{\frac{(2)(8)(600)}{(0.2)(0.28)}} = 414.$$

We say that the EOQ value is realizable if it falls within the interval that corresponds to the unit cost used to compute it. Since $0 \leq 400 < 500$, Q^0 is realizable. However, neither $Q^{(1)}$ nor $Q^{(2)}$ is realizable ($Q^{(1)}$ would have to have been between 500 and 1,000, and $Q^{(2)}$ would have to have been 1,000 or more). Each EOQ value corresponds to the minimum of a different annual cost curve. In this example, if $Q^{(2)}$ were realizable, it would necessarily have to have been the optimal solution, as it corresponds to the lowest point on the lowest curve. The three average annual cost curves for this example appear in Figure 4–11. Because each curve is valid only for certain values of Q, the average annual cost function is given by the discontinuous curve shown in heavy shading. The goal of the analysis is to find the minimum of this discontinuous curve.

There are three candidates for the optimal solution: 400, 500, and 1,000. In general, the optimal solution will be either the largest realizable EOQ or one of the breakpoints that exceeds it. The optimal solution is the lot size with the lowest average annual cost.

The average annual cost functions are given by

$$G_j(Q) = \lambda c_j + \lambda K/Q + Ic_j Q/2 \qquad \text{for } j = 0, 1, \text{ and } 2.$$

FIGURE 4–11

All-units discount average annual cost function

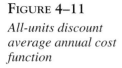

FIGURE 4–11

All-units discount average annual cost function

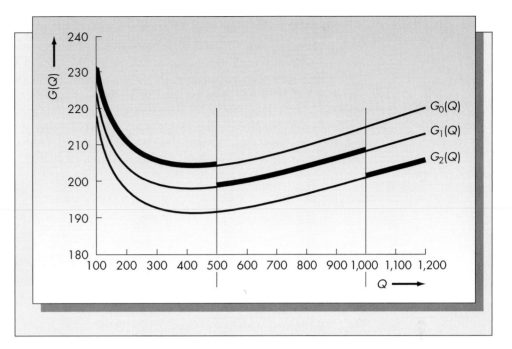

The broken curve pictured in Figure 4–11, $G(Q)$, is defined as

$$G(Q) = \begin{cases} G_0(Q) & \text{for} & 0 \le Q < 500, \\ G_1(Q) & \text{for} & 500 \le Q < 1,000, \\ G_2(Q) & \text{for} & 1,000 \le Q \end{cases}$$

Substituting Q equals 400, 500, and 1,000, and using the appropriate values of c_j, we obtain

$$G(400) = G_0(400)$$
$$= (600)(0.30) + (600)(8)/400 + (0.2)(0.30)(400)/2 = \$204.00$$

$$G(500) = G_1(500)$$
$$= (600)(0.29) + (600)(8)/500 + (0.2)(0.29)(500)/2 = \$198.10$$

$$G(1,000) = G_2(1,000)$$
$$= (600)(0.28) + (600)(8)/1,000 + (0.2)(0.28)(1,000)/2 = \$200.80.$$

Hence, we conclude that the optimal solution is to place a standing order for 500 units with Weighty at an average annual cost of $198.10.

Summary of the Solution Technique for All-Units Discounts

1. Determine the largest realizable EOQ value. The most efficient way to do this is to compute the EOQ for the lowest price first, and continue with the next higher price. Stop when the first EOQ value is realizable (that is, within the correct interval).

2. Compare the value of the average annual cost at the largest realizable EOQ and at all of the price breakpoints that are greater than the largest realizable EOQ. The optimal Q is the point at which the average annual cost is a minimum.

Incremental Quantity Discounts

Consider Example 4.4, but assume incremental quantity discounts. That is, the trash bags cost 30 cents each for quantities of 500 or fewer; for quantities between 500 and 1,000, the first 500 cost 30 cents each and the remaining amount cost 29 cents each; for quantities of 1,000 and over the first 500 cost 30 cents each, the next 500 cost 29 cents each, and the remaining amount cost 28 cents each. We need to determine a mathematical expression for the function $C(Q)$ pictured in Figure 4–10. From the figure, we see that the first price break corresponds to $C(Q) = (500)(0.30) = \$150$ and the second price break corresponds to $C(Q) = 150 + (0.29)(500) = \$295$. It follows that

$$C(Q) = \begin{cases} 0.30Q & \text{for} & 0 \leq Q < 500, \\ 150 + 0.29(Q - 500) = 5 + 0.29Q & \text{for} & 500 \leq Q < 1{,}000, \\ 295 + 0.28(Q - 1000) = 15 + 0.28Q & \text{for} & 1{,}000 \leq Q \end{cases}$$

so that

$$C(Q)/Q = \begin{cases} 0.30 & \text{for} & 0 \leq Q < 500, \\ 0.29 + 5/Q & \text{for} & 500 \leq Q < 1{,}000, \\ 0.28 + 15/Q & \text{for} & 1{,}000 \leq Q. \end{cases}$$

The average annual cost function, $G(Q)$, is

$$G(Q) = \lambda C(Q)/Q + K\lambda/Q + I[C(Q)/Q]Q/2.$$

In this example, $G(Q)$ will have three different algebraic representations ($G_0(Q)$, $G_1(Q)$, and $G_2(Q)$) depending upon into which interval Q falls. Because $C(Q)$ is continuous, $G(Q)$ also will be continuous. The function $G(Q)$ appears in Figure 4–12.

The optimal solution occurs at the minimum of one of the average annual cost curves. The solution is obtained by substituting the three expressions for $C(Q)/Q$ in the defining equation for $G(Q)$, computing the three minima of the curves, determining

FIGURE 4–12

Average annual cost function for incremental discount schedule

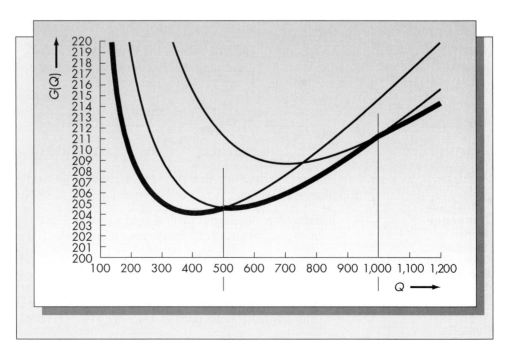

which of these minima fall into the correct interval, and, finally, comparing the average annual costs at the realizable values. We have that

$$G_0(Q) = (600)(0.30) + (8)(600)/Q + (0.20)(0.30)Q/2$$

which is minimized at

$$Q^{(0)} = \sqrt{\frac{2K\lambda}{Ic_0}} = \sqrt{\frac{(2)(8)(600)}{(0.20)(0.30)}} = 400;$$

$$\begin{aligned} G_1(Q) &= (600)(0.29 + 5/Q) + (8)(600)/Q + (0.20)(0.29 + 5/Q)(Q/2) \\ &= (0.29)(600) + (13)(600)/Q + (0.20)(0.29)Q/2 + (0.20)(5)/2 \end{aligned}$$

which is minimized at

$$Q^{(1)} = \sqrt{\frac{(2)(13)(600)}{(0.20)(0.29)}} = 519;$$

and finally

$$\begin{aligned} G_2(Q) &= (600)(0.28 + 15/Q) + (8)(600)/Q + (0.20)(0.28 + 15/Q)Q/2 \\ &= (0.28)(600) + (23)(600)/Q + (0.20)(0.28)Q/2 + (0.20)(15)/2 \end{aligned}$$

which is minimized at

$$Q^{(2)} = \sqrt{\frac{(2)(23)(600)}{(0.20)(0.28)}} = 702.$$

Both $Q^{(0)}$ and $Q^{(1)}$ are realizable. $Q^{(2)}$ is not realizable because $Q^{(2)} < 1,000$. The optimal solution is obtained by comparing $G_0(Q^{(0)})$ and $G_1(Q^{(1)})$. Substituting into the expressions above, we obtain

$$G_0(Q^{(0)}) = \$204.00,$$

$$G_1(Q^{(1)}) = \$204.58.$$

Hence, the optimal solution is to place a standing order with the Weighty Trash Bag Company for 400 units at the highest price of 30 cents per unit. The cost of using a standard order of 519 units is only slightly higher. Notice that compared to the all-units case, we obtain a smaller batch size at a higher average annual cost.

Summary of the Solution Technique for Incremental Discounts

1. Determine an algebraic expression for $C(Q)$ corresponding to each price interval. Use that to determine an algebraic expression for $C(Q)/Q$.
2. Substitute the expressions derived for $C(Q)/Q$ into the defining equation for $G(Q)$. Compute the minimum value of Q corresponding to each price interval separately.
3. Determine which minima computed in (2) are realizable (that is, fall into the correct interval). Compare the values of the average annual costs at the realizable EOQ values and pick the lowest.

Other Discount Schedules

Although all-units and incremental discount schedules are the most common, there are a variety of other discount schedules as well. One example is the carload discount schedule pictured in Figure 4–13.

FIGURE 4–13

*Order cost function
for carload discount
schedule*

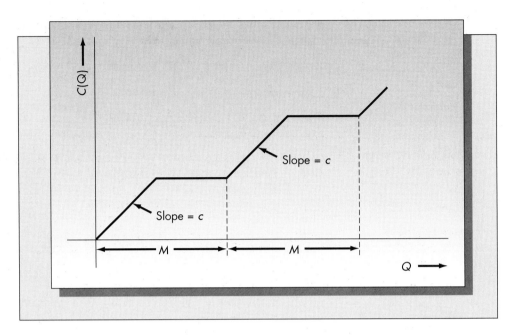

The rationale behind the carload discount schedule is the following. A carload consists of *M* units. The supplier charges a constant *c* per unit up until you have paid for the cost of a full carload, at which point there is no charge for the remaining units in that carload. Once the first carload is full, you again pay *c* per unit until the second carload is full, and so forth.

Determining optimal policies for the carload or other discount schedules could be extremely difficult. Each discount schedule has a unique solution procedure. Some can be extremely complex.

Problems for Section 7

21. Your local grocery store stocks rolls of bathroom tissue in single packages and in more economical 12-packs. You are trying to decide which to buy. The single package costs 45 cents and the 12-pack costs $5. You consume bathroom tissue at a fairly steady rate of one roll every three months. Your opportunity cost of money is computed assuming an interest rate of 25 percent and a fixed cost of $1 for the additional time it takes you to buy bathroom tissue when you go shopping. (We are assuming that you shop often enough so that you don't require a special trip when you run out.)

    a. How many single rolls should you be buying in order to minimize the annual holding and setup costs of purchasing bathroom tissue?

    b. Determine if it is more economical to purchase the bathroom tissue in 12-packs.

    c. Are there reasons other than those discussed in the problem that would motivate you not to buy the bathroom tissue in 12-packs?

22. A purchasing agent for a particular type of silicon wafer used in the production of semiconductors must decide among three sources. Source A will

sell the silicon wafers for $2.50 per wafer, independently of the number of wafers ordered. Source B will sell the wafers for $2.40 each but will not consider an order for fewer than 3,000 wafers, and Source C will sell the wafers for $2.30 each but will not accept an order for fewer than 4,000 wafers. Assume an order setup cost of $100 and an annual requirement of 20,000 wafers. Assume a 20 percent annual interest rate for holding cost calculations.

    a. Which source should be used, and what is the size of the standing order?

    b. What is the optimal value of the holding and setup costs for wafers when the optimal source is used?

    c. If the replenishment lead time for wafers is three months, determine the reorder point based on the on-hand level of inventory of wafers.

23. Assume that two years have passed, and the purchasing agent mentioned in Problem 22 must recompute the optimal number of wafers to purchase and from which source to purchase them. Source B has decided to accept any size offer, but sells the wafers for $2.55 each for orders of up to 3,000 wafers and $2.25 each for the incremental amount ordered over 3,000 wafers. Source A still has the same price schedule, and Source C went out of business. Now which source should be used?

24. In the calculation of an optimal policy for an all-units discount schedule, you first compute the EOQ values for each of the three order costs, and you obtain: $Q^{(0)} = 800$, $Q^{(1)} = 875$, and $Q^{(2)} = 925$. The all-units discount schedule has breakpoints at 750 and 900. Based on this information only, can you determine what the optimal order quantity is? Explain your answer.

25. Parasol Systems sells motherboards for personal computers. For quantities up through 25, the firm charges $350 per board; for quantities between 26 and 50, it charges $315 for each board purchased beyond 25; and it charges $285 each for the additional quantities over 50. A large communications firm expects to require these motherboards for the next 10 years at a rate of at least 140 per year. Order setup costs are $30 and holding costs are based on an 18 percent annual interest rate. What should be the size of the standing order?

---

*4.8 Resource-Constrained Multiple Product Systems

The EOQ model and its extensions apply only to single inventory items. However, these models are often used in companies stocking many different items. Although we could certainly compute optimal order quantities separately for each of the different items, there could exist constraints that would make the resulting solution infeasible. In Example 4.1, the optimal solution called for purchasing 3,870 pencils every 1.24 years. The bookstore, however, may not have allocated enough space to store that many pencils, nor enough money to purchase that many at one time.

Example 4.5

Three items are produced in a small fabrication shop. The shop management has established the requirement that the shop never have more than $30,000 invested in the inventory of these items at one time. The management uses a 25 percent annual interest charge to compute the holding cost. The relevant cost and demand parameters are given in the table below. What lot sizes should the shop be producing so that they do not exceed the budget?

	Item		
	1	*2*	*3*
Demand rate λ_j	1,850	1,150	800
Variable cost c_j	50	350	85
Setup cost K_j	100	150	50

Solution

If the budget is not exceeded when using the EOQ values of these three items, then the EOQs are optimal. Hence, the first step is to compute the EOQ values for all items to determine whether or not the constraint is active.

$$EOQ_1 = \sqrt{\frac{(2)(100)(1,850)}{(0.25)(50)}} = 172,$$

$$EOQ_2 = \sqrt{\frac{(2)(150)(1,150)}{(0.25)(350)}} = 63,$$

$$EOQ_3 = \sqrt{\frac{(2)(50)(800)}{(0.25)(85)}} = 61.$$

If the EOQ value for each item is used, the maximum investment in inventory would be

$$(172)(50) + (63)(350) + (61)(85) = \$35,835.^3$$

Because the EOQ solution violates the constraint, we need to reduce these lot sizes. But how?

The optimal solution turns out to be very easy to find in this case. We merely multiply each EOQ value by the ratio $(30,000)/(35,835) = 0.8372$. In order to guarantee that we do not exceed the $30,000 budget, we round each value *down* (adjustments can be made subsequently). Letting Q_1^*, Q_2^*, and Q_3^* be the optimal values, we obtain

$$Q_1^* = (172)(0.8372) \approx 144,$$

$$Q_2^* = (63)(0.8372) \approx 52,$$

$$Q_3^* = (61)(0.8372) \approx 51,$$

where $\approx$ should be interpreted as rounding to the next lower integer in this case.

---

[3]We are assuming for the purposes of this section that the three products are ordered simultaneously. By staggering order cycles, it is possible to meet the constraint with larger lot sizes. We will not consider that case here.

The total budget required for these lot sizes is \$29,735. The remaining \$265 can now be used to increase the lot sizes of products 1 and 3 slightly. (For example, $Q_1^* = 147$, $Q_3^* = 52$ results in a budget of \$29,970.) ∎

In general, budget- or space-constrained problems are not solved so easily. Suppose that n items have unit costs of $c_1, c_2, \ldots, c_n$, respectively, and the total budget available for them is C. Then the budget constraint can be written

$$c_1 Q_1 + c_2 Q_2 + \cdots + c_n Q_n \le C.$$

Let

$$\text{EOQ}_i = \sqrt{\frac{2 K_i \lambda_i}{h_i}} \qquad \text{for } i = 1, \ldots, n,$$

where K_i, h_i, and λ_i are the respective cost and demand parameters.

There are two possibilities: either the constraint is active or it is not. If the constraint is not active, then

$$\sum_{i=1}^{n} c_i \text{EOQ}_i \le C,$$

and the optimal solution is $Q_i = \text{EOQ}_i$. If the constraint is active, then

$$\sum_{i=1}^{n} c_i \text{EOQ}_i > C,$$

and the EOQ solution is no longer feasible. If we include the following assumption, the solution to the active case is relatively easy to find:

Assumption: $c_1/h_1 = c_2/h_2 = \cdots = c_n/h_n.$

If this assumption holds and the constraint is active, we prove in Appendix 4–A that the optimal solution is

$$Q_i^* = m\text{EOQ}_i,$$

where the multiplier m solves

$$m = C \bigg/ \left[\sum_{i=1}^{n} (c_i \text{EOQ}_i) \right].$$

Since $c_i/h_i = c_i/(I_i c_i) = 1/I_i$, the condition that the ratios be equal is equivalent to the requirement that the same interest rate be used to compute the holding cost for each item, which is reasonable in most circumstances.

Suppose that the constraint is on the available space. Let w_i be the space consumed by one unit of product i for $i = 1, 2, \ldots, n$ (this could be floor space measured, say, in square feet, or volume measured in cubic feet), and let W be the total space available. Then the space constraint is of the form

$$w_1 Q_1 + w_2 Q_2 + \cdots + w_n Q_n \le W.$$

This is mathematically of the same form as the budget constraint, so the same analysis applies. However, our condition for a simple solution now is that the ratios w_i/h_i are equal. That is, the space consumed by an item should be proportional to its holding cost. When the interest rate is fixed, this is equivalent to the requirement that the space consumed should be proportional to the value of the item. This requirement would probably be too restrictive in most cases. For example, fountain pens take up far less space than legal pads, but are more expensive.

Let us now consider the problem in which the constraint is active, but the proportionality assumption is not met. This problem is far more complex than that solved above. It requires formulating the *Lagrangian* function. The details of the formulation of this problem can be found in Appendix 4–A. As we show, the optimal lot sizes are now of the form

$$Q_i^* = \sqrt{\frac{2K_i\lambda_i}{h_i + 2\theta w_i}},$$

where θ is a constant chosen so that

$$\sum_{i=1}^{n} w_i Q_i^* = W.$$

The constant θ, known as the Lagrange multiplier, reduces the lot sizes by increasing the effective holding cost. The correct value of θ can be found by trial and error or by a search technique such as interval bisection. Note that $\theta > 0$, so that the search can be limited to positive numbers only. The value of θ can be interpreted as the decrease in the average annual cost that would result from adding an additional unit of resource. In this case, it would represent the marginal benefit of an additional square foot of space.

Example 4.6

Consider the fabrication shop of Example 4.5. In addition to the budget constraint, suppose that there are only 2,000 square feet of floor space available. Assume that the three products consume respectively 9, 12, and 18 square feet per unit.

First, we check to see if the EOQ solution is feasible. Setting $w_1 = 9$, $w_2 = 12$, and $w_3 = 18$, we find that

$$\sum \mathrm{EOQ}_i w_i = (172)(9) + (63)(12) + (61)(18) = 3,402,$$

which is obviously infeasible. Next, we check if the budget-constrained solution provides a feasible solution to the space-constrained problem. The budget-constrained solution requires $(147)(9) + (52)(12) + (52)(18) = 2,883$ square feet of space, which is also infeasible.

The next step is to compute the ratios w_i/h_i for $1 \leq i \leq 3$. These ratios turn out to be 0.72, 0.14, and 0.85 respectively. Because their values are different, the simple solution obtained by a proportional scaling of the EOQ values will not be optimal. Hence, we must determine the value of the Lagrange multiplier θ.

We can determine upper and lower bounds on the optimal value of θ by assuming equal ratios. If the ratios were equal, the multiplier, m, would be

$$m = W/\sum (\mathrm{EOQ}_i w_i) = 2,000/3,402 = 0.5879,$$

which would give the three lot sizes as 101, 37, and 36, respectively. The three values of θ that result in these lot sizes are respectively $\theta = 1.32$, $\theta = 6.86$, and $\theta = 2.33$. (These values were obtained by setting the expression for Q_1^* above equal to the lot sizes 101, 37, and 36, and solving for θ.)

The true value of θ will be between 1.32 and 6.86. If we start with $\theta = 3.5$, we obtain $Q_1^* = 70$, $Q_2^* = 45$, and $Q_3^* = 23$, and $\Sigma w_i Q_i^* = (70)(9) + (45)(12) + (23)(18) = 1,584$, which implies $\theta < 3.5$. After considerable experimentation, we finally find that the optimal value of $\theta = 1.75$, and $Q_1^* = 92$, $Q_2^* = 51$, and $Q_3^* = 31$, giving $\Sigma w_i Q_i^* = 1,998$. Notice that these lot sizes are in very different proportions from the ones obtained assuming a constant multiplier. Searching for the optimal value of the Lagrange multiplier, although tedious to do by hand, can be accomplished very efficiently using a computer. Spreadsheets are a useful means for effecting such calculations. ■

We have only touched on some of the complications that could arise when applying EOQ-type models to a realistic system with multiple products. Often, real problems are far too complex to be expressed accurately as a solvable mathematical model. For this reason, simple models such as the ones presented in this and subsequent chapters are used by practitioners. However, any solution recommended by a mathematical model must be considered in the context of the system in which it is to be used.

Problems for Section 8

26. A local outdoor vegetable stand has exactly 1,000 square feet of space to display three vegetables: tomatoes, lettuce, and zucchini. The appropriate data for these items are given in the table below.

	Item		
	Tomatoes	*Lettuce*	*Zucchini*
Annual demand (in pounds)	850	1,280	630
Cost per pound	$0.29	$0.45	$0.25

The setup cost for replenishment of the vegetables is $100 in each case, and the space consumed by each vegetable is proportional to its costs, with tomatoes requiring 0.5 square foot per pound. The annual interest rate used for computing holding costs is 25 percent. What are the optimal quantities that should be purchased of these three vegetables?

27. Suppose that the vegetables discussed in Problem 26 are purchased at different times. In what way could that have an effect on the order policy that the stand owner should use?

28. Suppose that in Problem 26 the space consumed by each vegetable is not proportional to its cost. In particular, suppose that one pound of lettuce required 0.4 square foot of space and one pound of zucchini required 1 square foot of space. Determine upper and lower bounds on the optimal values of the order quantities in this case. Test different values of the Lagrange multiplier to find the optimal values of the order quantities. (A spreadsheet is ideally suited for this kind of calculation. If you solve the problem using a spreadsheet, place the Lagrange multiplier in a cell so that its value can be changed easily.)

4.9 EOQ Models for Production Planning

Simple lot-sizing models have been successfully applied to a variety of manufacturing problems. In this section we consider an extension of the EOQ model with a finite production rate, discussed in Section 6, to the problem of producing n products on a single machine. Following the notation used in this chapter, let

λ_j = Demand rate for product j,
P_j = Production rate for product j,
h_j = Holding cost per unit per unit time for product j,
K_j = Cost of setting up the production facility to produce product j.

The goal is to determine the optimal procedure for producing n products on the machine to minimize the cost of holding and setups, and to guarantee that no stock-outs occur during the production cycle.

We require the assumption that

$$\sum_{j=1}^{n} \lambda_j / P_j \leq 1.$$

This assumption is needed to ensure that the facility has sufficient capacity to satisfy the demand for all products. Notice that this is stronger than the assumption made in Section 6 that $\lambda_j < P_j$ for each j. To see why this assumption is necessary, consider the case of two products with identical demand and production rates. In each cycle one would produce product 1 first, then product 2. Clearly $P_1 \geq 2\lambda_1$ and $P_2 \geq 2\lambda_2$, so that enough could be produced to meet the total demand for both products each cycle. This translates to $\lambda_1/P_1 \leq 0.5$ and $\lambda_2/P_2 \leq 0.5$, giving a value of the sum less than or equal to one. Similar reasoning holds for more than two products with nonidentical demand and production rates.

We also will assume that the policy used is a *rotation cycle policy*. That means that in each cycle there is exactly one setup for each product, and products are produced in the same sequence in each production cycle. The importance of this assumption will be discussed further at the end of the section.

At first, one might think that the optimal solution is to sequentially produce lot sizes for each product optimized by treating each product in isolation. From Section 6, this would result in a lot size for product j of

$$Q_j = \sqrt{\frac{2K_j\lambda_j}{h_j'}},$$

where $h_j' = h_j(1 - \lambda_j/P_j)$. The problem with this approach is that because we have only a single production facility, it is likely that some of the lot sizes Q_j will not be large enough to meet the demand between production runs for product j, thus resulting in stock-outs.

Let T be the cycle time. During time T we assume that exactly one lot of each product is produced. In order that the lot for product j be large enough to meet the demand occurring during time T, it follows that the lot size must be

$$Q_j = \lambda_j T.$$

From Section 6, the average annual cost associated with product j can be written in the form

$$G(Q_j) = K_j\lambda_j / Q_j + h_j'Q_j/2.$$

The average annual cost for all products is the sum

$$\sum_{j=1}^{n} G(Q_j) = \sum_{j=1}^{n} K_j \lambda_j / Q_j + h'_j Q_j / 2.$$

Substituting $T = Q_j / \lambda_j$, we obtain the average annual cost associated with the n products in terms of the cycle time T as

$$G(T) = \sum_{j=1}^{n} [K_j / T + h'_j \lambda_j T / 2].$$

The goal is to find T to minimize $G(T)$. The necessary condition for an optimal T is

$$\frac{dG(T)}{dT} = 0.$$

Setting the first derivative with respect to T to zero gives

$$\sum_{j=1}^{n} [-K_j / T^2 + h'_j \lambda_j / 2] = 0.$$

Solving for T, we obtain the optimal cycle time T^* as

$$T^* = \sqrt{\frac{2 \sum_{j=1}^{n} K_j}{\sum_{j=1}^{n} h'_j \lambda_j}}.$$

If setup times are a factor, we must check that there is enough time each cycle to account for both setup times and production of the n products. Let s_j be the setup time for product j. Ensuring that the total time required for setups and production each cycle does not exceed T leads to the constraint

$$\sum_{j=1}^{n} (s_j + Q_j / P_j) \leq T.$$

Using the fact that $Q_j = \lambda_j T$, this condition translates to

$$\sum_{j=1}^{n} (s_j + \lambda_j T / P_j) \leq T,$$

which gives, after rearranging terms,

$$T \geq \frac{\sum_{j=1}^{n} s_j}{1 - \sum_{j=1}^{n} (\lambda_j / P_j)} = T_{\min}.$$

Because $T_{\min}$ cannot be exceeded without compromising feasibility, the optimal solution is to choose the cycle time T equal to the *larger* of T^* and $T_{\min}$.

TABLE 4–1   **Relevant Data for Example 4.7**

Style	Annual Demand (units/year)	Production Rate (units/year)	Setup Time (hours)	Variable Cost ($/unit)
Women's pump	4,520	35,800	3.2	$40
Women's loafer	6,600	62,600	2.5	26
Women's boot	2,340	41,000	4.4	52
Women's sandal	2,600	71,000	1.8	18
Men's wingtip	8,800	46,800	5.1	38
Men's loafer	6,200	71,200	3.1	28
Men's oxford	5,200	56,000	4.4	31

Example 4.7

Bali produces several styles of men's and women's shoes at a single facility near Bergamo, Italy. The leather for both the uppers and the soles of the shoes is cut on a single machine. This Bergamo plant is responsible for seven styles and several colors in each style. (The colors are not considered different products for our purposes, because no setup is required when switching colors.) Bali would like to schedule cutting for the shoes using a rotation policy that meets all demand and minimizes setup and holding costs. Setup costs are proportional to setup times. The firm estimates that setup costs amount to an average of $110 per hour, based on the cost of worker time and the cost of forced machine idle time during setups. Holding costs are based on a 22 percent annual interest charge.

The relevant data for this problem appear in Table 4–1.

Solution

The first step is to verify that the problem is feasible. To do so we compute $\Sigma \lambda_j / P_j$. The reader should verify that this sum is equal to 0.69355. Because this is less than one, there will be a feasible solution. Next we compute the value of T^*, but to do so we need to do several intermediate calculations.

First, we compute setup costs. Setup costs are assumed to be $110 times setup times. Second, we compute modified holding costs (h'_j). This is done by multiplying the cost of each product by the annual interest rate (0.22) times the factor $1 - \lambda_j / P_j$. These calculations give

Setup Costs (K_j)	Modified Holding Costs (h'_j)
352	7.69
275	5.12
484	10.79
198	3.81
561	6.79
341	5.62
484	6.19

The sum of the setup costs is 2,695, and the sum of the products of the modified holding costs and the annual demands is 230,458.4. Substituting these figures into the formula for T^* gives the optimal cycle time as 0.1529 year. Assuming a 250-day work year, this means that the rotation cycle should repeat roughly every 38 working days. The optimal lot size for each of the shoes is found by multiplying the cycle time by the demand rate for each item. The reader should check that the following lot sizes result:

Style	Optimal Lot Sizes for Each Production Run
Women's pump	691
Women's loafer	1,009
Women's boot	358
Women's sandal	398
Men's wingtip	1,346
Men's loafer	948
Men's oxford	795

The plant would cut the soles and uppers in these lot sizes in sequence (although the sequence does not necessarily have to be this one) and would repeat the rotation cycle roughly every 38 days (0.1529 year). However, this solution can be implemented only if T^* is at least T_{min}. To determine T_{min} we must express the setup times in years. Assuming 8 working hours per day and 250 working days per year, one would divide the setup times given in hours by 2,000 (250 times 8). The reader should check that the resulting value of T_{min} is 0.04, thus making T^* feasible and, hence, optimal.

The total average annual cost of holding and setups at an optimal policy can be found by computing the value of $G(T)$ when $T = T^*$. It is \$35,244.44. It is interesting to note that if the plant manager chooses to implement this policy, the facility will be idle for a substantial portion of each rotation cycle. The total uptime each rotation cycle is found by dividing the lot sizes by the production rates for each style and summing the results. It turns out to be 0.106 year. Hence, the optimal rotation policy that minimizes total holding and setup costs results in the cutting operation remaining idle about one-third of the time. ■

The reader should be aware that the relatively simple solution to this problem was the result of two assumptions. One was that the setup costs were not sequence dependent. In the previous example, it is possible that the time required to change over from one shoe style to another could depend on the styles. For example, a changeover from a woman's style to a man's style probably takes longer than from one woman's style to another. A second assumption was that the plant used a rotation cycle policy. That is, in each cycle Bali does a single production run of each style. When demand rates and setup costs differ widely, it might be advantageous to do two or more production runs of a product in a cycle. This more general problem has not, to our knowledge, been solved. A discussion of these and related issues can be found in Maxwell (1964). Magee and Boodman (1967) provide some heuristic ways of dealing with the more general problem.

Problems for Section 9

29. A metal fabrication shop has a single punch press. There are currently three parts that the shop has agreed to produce that require the press, and it appears that they will be supplying these parts well into the future. You may assume that the press is the critical resource for these parts, so that we need not worry about the interaction of the press with the other machines in the shop. The relevant information here is

Part Number	Annual Contracted Amount (demand)	Setup Cost	Cost (per unit)	Production Rate (per year)
1	2,500	$ 80	$16	45,000
2	5,500	120	18	40,000
3	1,450	60	22	26,000

Holding costs are based on an 18 percent annual interest rate, and the products are to be produced in sequence on a rotation cycle. Setup times can be considered negligible.
a. What is the optimal time between setups for part number 1?
b. What percentage of the time is the punch press idle, assuming an optimal rotation cycle policy?
c. What are the optimal lot sizes of each part put through the press at an optimal solution?
d. What is the total annual cost of holding and setup for these items on the punch press, assuming an optimal rotation cycle?

30. Tomlinson Furniture has a single lathe for turning the wood for various furniture pieces, including bedposts, rounded table legs, and other items. Four forms are turned on the lathe and produced in lots for inventory. To simplify scheduling, one lot of each type will be produced in a cycle, which may include idle time. The four products and the relevant information concerning them appears in the table below.

Piece	Monthly Requirements	Setup Time (hours)	Unit Cost	Production Rate (units/day)
J–55R	125	1.2	$20	140
H–223	140	0.8	35	220
K–18R	45	2.2	12	100
Z–344	240	3.1	45	165

Worker time for setups is valued at $85 per hour, and holding costs are based on a 20 percent annual interest charge. Assume 20 working days per month and 12 months per year for your calculations.
a. Determine the optimal length of the rotation cycle.
b. What are the optimal lot sizes for each product?

c. What are the percentages of uptime and downtime for the lathe, assuming that it is not used for any other purpose?

d. Draw a graph showing the change in the inventory level over a typical cycle for each product.

e. Discuss why the solution you obtained might not be feasible for the firm, or why it might not be desirable even when it is feasible.

4.10 Software for Inventory Control

Prior to the proliferation of the personal computer in the 1980s, there were only a limited number of software products available for inventory control applications. In many cases, these products were sold by the vendors of mainframe computer systems such as IBM, Control Data, DEC, and others. The prototype mainframe inventory control system was IBM's IMPACT, developed by IBM (purportedly with the assistance of R. G. Brown). IMPACT used exponential smoothing to track demand (as described in Chapter 2) and established reorder points based on the distribution of forecast errors (discussed in both Chapters 2 and 5). IBM has since developed and marketed several variants of IMPACT to serve specific markets. An example is INFOREM, designed for retailing.

To this writer's knowledge, the inventory control systems currently sold by IBM are designed to run on mainframe computers only. Because the cost of personal computers has dropped so dramatically in recent years, and the power of these machines has continued to improve, virtually every small business uses them. This means computerized inventory control is accessible to everyone.

Most inventory control software products for the personal computer are modules in an integrated accounting system. Some of these accounting packages are expensive, costing orders of magnitude more than the computers on which they run. Typically, low-end programs provide only basic record-keeping functions. The kinds of models discussed in Chapter 2 for forecasting, and for inventory control discussed here and in Chapter 5, are incorporated into high-end systems only. Some products also support just-in-time inventory management methods.

A comprehensive review of inventory control software for the personal computer was reported by Cohn (1994). The products discussed there include 25 software packages, most of which are part of integrated accounting systems. These packages offer quite a range of inventory control features. On the low end, for example, Microsoft Profit provides only the most basic inventory control features, including stock level reports and item reports. Most of the other packages reviewed include features such as LIFO and FIFO costing methods, multiple locations, multiple pricing for items, and ABC inventory analysis.

Some developers target products for specific markets only. EFC Systems markets Kairos and Vanguard for agricultural applications primarily. Other firms offer multiple products for several market segments. One of the high-end products reviewed is offered by Realworld Corporation, which sells both Realworld Inventory Control and Inventory Plus. Realworld Inventory Control can handle over 1,000 warehouses, multiple pricing structures, and quantity discounts. Inventory Plus adds serialization, lot control, and bar coding. This product is designed for retailing while the former is for wholesaling.

Very few of the products discussed at this end of the spectrum provide for item forecasting, lot sizing, and reorder point calculations. One exception is Integrated

SNAPSHOT APPLICATION

Mervyn's Recognized for State-of-the-Art Inventory Control System

Mervyn's, a division of Dayton-Hudson Corporation, is a dry goods retailer based in Hayward, California. In 1994 Mervyn's was recognized for its outstanding computer-based forecasting and inventory control system (Donnelly, 1994).

Since the late 1980s *Computerworld* magazine and the Smithsonian Institute have awarded public and private organizations for effective uses of information technology. Nominees for the award are selected by a 100-member committee of executives in the information technology industry. Winners are then chosen by smaller panels in each field. Fields in which awards are made include finance, government, manufacturing, medicine, and business.

In 1994 Mervyn's was selected as the winner in the Business and Related Services category for its planned store inventory (PSI) management system. PSI is a highly sophisticated system that coordinates sales data from 286 stores located nationwide. According to Mervyn's management, PSI was responsible for reducing inventories by an estimated $60 million with no degradation of in-store service levels. At the heart of PSI is a forecasting module that allows store managers to predict demand for specific colors, sizes, and patterns at individual stores. One of the unique features of the system that are forecasting takes into account is geographic location and market demographics to distinguish sales patterns at stores located in different parts of the country.

Each store carries an average of 50,000 SKUs, (stockkeeping units, or individual part numbers), so coordinating data from the 286 stores is a massive information-processing job. The system uses the latest client–server technology, linking 486 PCs to a UNIX-based multiprocessor database server.

The most serious obstacle to implementing the PSI system was the store managers themselves. Mervyn's stores pride themselves on providing high levels of customer service. They target at 95 percent service levels (that is, being able to satisfy customer demand 95 percent of the time from in-store stock). Store managers were skeptical that they could continue to provide these levels of service after cutting inventory levels. In fact, PSI recommended reductions of in-store inventories of up to 50 percent for some items.

Mervyn's system uses a combination of mathematical modeling techniques and client–server technology to guide inventory levels for individual stores and items with ongoing adjustments based on observations of sales. Inventories are lowered by applying a just-in-time approach to stock replenishment. According to Mervyn's, competitors' systems are geared toward overall inventory levels and adjusted according to predetermined factors.

Mervyn's, one of the founding members of Santa Clara University's Retail Management Institute, has always maintained a strong logistics support function. Management has worked closely with university faculty on other projects to be sure its forecasting and inventory control methods utilize state-of-the-art theory and technology. Retailing is a highly competitive industry, and the successful players are the ones that recognize that seat-of-the-pants management techniques are no longer adequate. (See the Snapshot Application on Wal-Mart in Chapter 6. Wal-Mart is another retailer with a strong emphasis on logistics and operations.)

Accounting System sold by CODA, Inc. This system is a higher-end system designed to run on DEC, IBM AS/400, and HP computers. It has an inventory management function that supports multiple warehouses, and performs reorder point and order quantity calculations. According to Tom Wittman, a user of the system for his commercial fishing company, this system allows him to keep track of the inventory status in five locations simultaneously.

Most of the systems reviewed would be adequate for small and medium-sized businesses. However, they don't include the kinds of features that larger businesses may require, such as (1) point-of-sale scanning of items, (2) networking of several locations, (3) a central database that can be accessed by satellite locations, (4) implementation of EDI (electronic data interchange), (5) demand forecasting, (6) lot sizing, and (7) reorder point and service level calculations. Systems with these features are typically turnkey systems. That is, they are integrated hardware and software sold as a unit. Mainframe and workstation manufacturers, such as HP and IBM, and some specialty vendors offer such systems. We review one such set of products in the Snapshot Application in Chapter 5.

4.11 Historical Notes and Additional Topics

The interest in using mathematical models to control the replenishment of inventories dates back to the early part of the 20th century. Ford Harris (1915) is generally credited with the development of the simple EOQ model. R. H. Wilson (1934) is also recognized for his analysis of the model. The procedure we suggest for the all-units quantity discount model appears to be due to Churchman, Ackoff, and Arnoff (1957), and the incremental discount model appears to be due to Hadley and Whitin (1963). Kenneth Arrow provides an excellent discussion of the economic motives for holding inventories in Chapter 1 of Arrow, Karlin, and Scarf (1958).

Hadley and Whitin (1963) also seem to have been the first to consider budget- and space-constrained problems, although several researchers have studied the problem subsequently. Rosenblatt (1981) derives results similar to those of Section 8 and considers several extensions not treated in this section.

Academic interest in inventory management problems took a sudden upturn in the late 1950s and early 1960s with the publication of a number of texts in addition to those mentioned above, including Whitin (1957); Magee and Boodman (1967); Bowman and Fetter (1961); Fetter and Dalleck (1961); Hanssmann (1961); Star and Miller (1962); Wagner (1962); and Scarf, Gilford, and Shelly (1963).

By and large, the vast majority of the published research on inventory systems since the 1960s has been on stochastic models, which will be the subject of the next chapter. Extensions of the EOQ model have been considered more recently by Barbosa and Friedman (1978) and Schwarz and Schrage (1971), for example.

4.12 Summary

This chapter presented several popular models used to control inventories when the demand is known. We discussed the various *economic motives for holding inventories,* which include economies of scale, uncertainties, speculation, and logistics. In addition, we mentioned some of the *physical characteristics of inventory systems* that are important in determining the complexity and applicability of the models to real problems. These include demand, lead time, review time, back-order/lost-sales assumptions, and the changes that take place in the inventory over time.

There are three significant classes of costs in inventory management. These are *holding or carrying costs, order or setup costs,* and *penalty cost* for not meeting demand. The holding cost is usually expressed as the product of an interest rate and the cost of the item.

The grandfather of all inventory control models is the simple *EOQ model,* in which demand is assumed to be constant, no stock-outs are permitted, and only holding and order costs are present. The optimal batch size is given by the classic square root formula. The first extension of the EOQ model we considered was to the case in which items are produced internally at a finite production rate. We showed that the optimal batch size in this case could be obtained from the EOQ formula with a modified holding cost.

Two types of *quantity discounts* were considered: *all-units,* in which the discounted price is valid on all of the units in an order, and *incremental,* in which the discount is applied only to the additional units beyond the breakpoint. The optimal solution to the all-units case will often occur at a breakpoint, whereas the optimal solution to the incremental case will almost never occur at a breakpoint.

In most real systems, the inventory manager cannot ignore the interactions that exist between products. These interactions impose *constraints* on the system; these might arise because of limitations in the space available to store inventory or in the budget available to purchase items. We considered both cases, and showed that when the ratio of the item value or space consumed by the item over the holding cost is the same for all items, a solution to the constrained problem can be obtained easily. When this condition is not met, the formulation requires introducing a *Lagrange multiplier.* The correct value of the Lagrangian multiplier can be found by trial and error or by some type of search. The presence of the multiplier reduces lot sizes by effectively increasing holding costs.

The finite production rate model of Section 6 was extended to multiple products under the assumption that all the products are produced on a single machine. Assuming a *rotation cycle policy* in which one lot of each product is produced each cycle, we showed how to determine the cycle time minimizing the sum of annual setup and holding costs for all products. Rotation cycles provide a straightforward way to schedule production on one machine, but may be suboptimal when demand or production rates differ widely or setup costs are sequence dependent.

The chapter concluded with a brief overview of several commercial inventory control systems. There are dozens of products available ranging in price from under $100 to tens of thousands of dollars. Many of the products designed to run on personal computers are parts of integrated accounting systems. Most of these products do not include item forecasting, lot sizing, and reorder point calculations.

Additional Problems on Deterministic Inventory Models

31. Peet's Coffees in Menlo Park, California, sells Melitta Number 101 coffee filters at a fairly steady rate of about 60 boxes of filters monthly. The filters are ordered from a supplier in Trenton, New Jersey. Peet's manager is interested in applying some inventory theory to determine the best replenishment strategy for the filters.

    Peet's pays $2.80 per box of filters and estimates that fixed costs of employee time for placing and receiving orders amount to about $20. Peet's uses a 22 percent annual interest rate to compute holding costs.

    a. How large a standing order should Peet's have with its supplier in Trenton, and how often should these orders be placed?

    b. Suppose that it takes three weeks to receive a shipment. What inventory of filters should be on hand when an order is placed?

    c. What are the average annual holding and fixed costs associated with these filters, assuming they adopt an optimal policy?

d. The Peet's store in Menlo Park is rather small. In what way might this affect the solution you recommended in part (*a*)?

32. A local supermarket sells a popular brand of shampoo at a fairly steady rate of 380 bottles per month. The cost of each bottle to the supermarket is 45 cents, and the cost of placing an order has been estimated at $8.50. Assume that holding costs are based on a 25 percent annual interest rate. Stock-outs of the shampoo are not allowed.
    a. Determine the optimal lot size the supermarket should order and the time between placements of orders for this product.
    b. If the procurement lead time is two months, find the reorder point based on the on-hand inventory.
    c. If the item sells for 99 cents, what is the annual profit (exclusive of overhead and labor costs) from this item?

33. Diskup produces a variety of personal computer products. High density 3.5-inch disks are produced at a rate of 1,800 per day and are shipped out at a rate of 800 per day. The disks are produced in batches. Each disk costs the company 20 cents, and the holding costs are based on an 18 percent annual interest rate. Shortages are not permitted. Each production run of a disk type requires recalibration of the equipment. The company estimates that this step costs $180.
    a. Find the optimal size of each production run and the time between runs.
    b. What fraction of the time is the company producing high density 3.5-inch disks?
    c. What is the maximum dollar investment in inventory that the company has in these disks?

34. Berry Computer is considering moving some of its operations overseas in order to reduce labor costs. In the United States, its main circuit board costs Berry $75 per unit to produce, while overseas it costs only $65 to produce. Holding costs are based on a 20 percent annual interest rate, and the demand has been a fairly steady 200 units per week. Assume that setup costs are $200 both locally and overseas. Production lead times are one month locally and six months overseas.
    a. Determine the average annual costs of production, holding, and setup at each location, assuming that an optimal solution is employed in each case. Based on these results only, which location is preferable?
    b. Determine the value of the pipeline inventory in each case. (The pipeline inventory is the inventory on order.) Does comparison of the pipeline inventories alter the conclusion reached in part (*a*)?
    c. Might considerations other than cost favor local over overseas production?

35. A large producer of household products purchases a glyceride used in one of its deodorant soaps from outside of the company. It uses the glyceride at a fairly steady rate of 40 pounds per month, and the company uses a 23 percent annual interest rate to compute holding costs. The chemical can be purchased from two suppliers, A and B. A offers the following all-units discount schedule:

Order Size	Price per Pound
$0 \le Q < 500$	$1.30
$500 \le Q < 1,000$	1.20
$1,000 \le Q$	1.10

whereas B offers the following incremental discount schedule: $1.25 per pound for all orders less than or equal to 700 pounds, and $1.05 per pound for all incremental amounts over 700 pounds. Assume that the cost of order processing for each case is $150. Which supplier should be used?

36. The president of Value Filters became very enthusiastic about using EOQs to plan the sizes of his production runs, and instituted lot sizing based on EOQ values before he could properly estimate costs. For one particular filter line, which had an annual demand of 1,800 units per year and which was valued at $2.40 per unit, he assumed a holding cost based on a 30 percent annual interest rate and a setup cost of $100. Some time later, after the cost accounting department had time to perform an analysis, it found that the appropriate value of the interest rate was closer to 20 percent and the setup cost was about $40. What was the additional average annual cost of holding and setup incurred from the use of the wrong costs?

37. Consider the carload discount schedule pictured in Figure 4–13 (page 220). Suppose $M = 500$ units, $C = 10 per unit, and a full carload of 500 units costs $3,000.
   a. Develop a graph of the average cost per unit, $C(Q)/Q$, assuming this schedule.
   b. Suppose that the units are consumed at a rate of 800 per week, order setup cost is $2,500, and holding costs are based on an annual interest charge of 22 percent. Graph the function $G(Q) = \lambda C(Q)/Q + K\lambda/Q + I(C(Q)/Q)Q/2$ and find the optimal value of Q. (Assume that 1 year = 50 weeks.)
   c. Repeat part (b) for $\lambda = 1,000$ per week and $K = $1,500$.

38. Harold Gwynne is considering starting a sandwich-making business from his dormitory room in order to earn some extra income. However, he has only a limited budget of $100 to make his initial purchases. Harold divides his needs into three areas: breads, meats and cheeses, and condiments. He estimates that he will be able to use all of the products he purchases before they spoil, so perishability is not a relevant issue. The demand and cost parameters are given below.

	Breads	Meats and Cheeses	Condiments
Weekly demand	6 packages	12 packages	2 pounds
Cost per unit	$.85	$3.50	$1.25
Fixed order cost	$12	$8	$10

The choice of these fixed costs is based on the fact that these items are purchased at different locations in town. They include the cost of Harold's time in making the purchase. Assume that holding costs are based on an annual interest rate of 25 percent.
   a. Find the optimal quantities that Harold should purchase of each type of product so that he does not exceed his budget.
   b. If Harold could purchase all of the items at the same location, would that alter your solution? Why?

39. Mike's Garage, a local automotive service and repair shop, uses oil filters at a fairly steady rate of 2,400 per year. Mike estimates that the cost of his time to make and process an order is about $50. It takes one month for the supplier to

deliver the oil filters to the garage, and each one costs Mike $5. Mike uses an annual interest rate of 25 percent to compute his holding cost.

a. Determine the optimal number of oil filters that Mike should purchase, and the optimal time between placement of orders.

b. Determine the level of on-hand inventory at the time a reorder should be placed.

c. Assuming that Mike uses an optimal inventory control policy for oil filters, what is the annual cost of holding and order setup for this item?

40. An import/export firm has leased 20,000 cubic feet of storage space to store six items that it imports from the Far East. The relevant data for the six items they plan to store in the warehouse are

Item	Annual Demand (units)	Cost per Unit	Space per Unit (feet^3)
VCR	800	$200.00	12
13-inch color TV	1,600	150.00	18
Videocassettes (box of 10)	8,000	30.00	3
Audiocassettes (box of 10)	12,000	18.00	2
Compact stereo	400	250.00	24
Telephone	1,200	12.50	3

Setup costs for each product amount to $2,000 per order, and holding costs are based on a 25 percent annual interest rate. Find the optimal order quantities for these six items so that the storage space is never exceeded. (Hint: Use a cell location for the Lagrange multiplier and experiment with different values until the storage constraint is satisfied as closely as possible.)

41. A manufacturer of greeting cards must determine the size of production runs for a certain popular line of cards. The demand for these cards has been a fairly steady 2 million per year, and the manufacturer is currently producing the cards in batch sizes of 50,000. The cost of setting up for each production run is $400.

Assume that for each card the material cost is 35 cents, the labor cost is 15 cents, and the distribution cost is 5 cents. The accounting department of the firm has established an interest rate to represent the opportunity cost of alternative investment and storage costs at 20 percent of the value of each card.

a. What is the optimal value of the EOQ for this line of greeting cards?

b. Determine the additional annual cost resulting from using the wrong production lot size.

42. Suppose that in Problem 41 the firm decides to account for the fact that the production rate of the cards is not infinite. Determine the optimal size of each production run assuming that cards are produced at the rate of 75,000 per week.

43. Pies 'R' Us bakes its own pies on the premises in a large oven that holds 100 pies. They sell the pies at a fairly steady rate of 86 per month. The pies cost $2 each to

make. Prior to each baking, the oven must be cleaned out, which requires one hour's time for four workers, each of whom is paid $8 per hour. Inventory costs are based on an 18 percent annual interest rate. The pies have a shelf life of three months.

a. How many pies should be baked for each production run? What is the annual cost of setup and holding for the pies?

b. The owner of Pies 'R' Us is thinking about buying a new oven that requires one-half the cleaning time of the old oven and has a capacity twice as large as the old one. What is the optimal number of pies to be baked each time in the new oven?

c. The net cost of the new oven (after trading in the old oven) is $350. How many years would it take for the new oven to pay for itself?

44. The Kirei-Hana Japanese Steak House in San Francisco consumes 3,000 pounds of sirloin steak each month. Yama Hirai, the new restaurant manager, recently completed an MBA degree. He learned that the steak was replenished using an EOQ value of 2,000 pounds. The EOQ value was computed assuming an interest rate of 36 percent per year. Assume that the current cost of the sirloin steak to the steak house is $4 per pound.

a. What is the setup cost used in determining the EOQ value?

Mr. Hirai received an offer from a meat wholesaler in which a discount of 5 percent would be given if the steak house purchased the steak in quantities of 3,000 pounds or more.

b. Should Mr. Hirai accept the offer from the wholesaler? If so, how much can be saved?

c. Because of Mr. Hirai's language problems, he apparently misunderstood the offer. In fact, the 5 percent discount is applied only to the amounts ordered over 3,000 pounds. Should this offer be accepted, and if so, how much is now saved?

45. Green's Buttons of Rolla, Missouri, supplies all the New Jersey Fabrics stores with eight different styles of buttons for men's dress shirts. The plastic injection molding machine can produce only one button style at a time and requires substantial time to reconfigure the machine for different button styles. As Green's has contracted to supply fixed quantities of buttons for the next three years, its demand can be treated as fixed and known. The relevant data for this problem appear in the table below.

Button Type	Annual Sales	Production Rate (units/day)	Setup Time (hours)	Variable Cost
A	25,900	4,500	6	$0.003
B	42,000	5,500	4	0.002
C	14,400	3,300	8	0.008
D	46,000	3,200	4	0.002
E	12,500	1,800	3	0.010
F	75,000	3,900	6	0.005
G	30,000	2,900	1	0.004
H	18,900	1,200	3	0.007

Assume 250 working days per year. Green's accounting department has established an 18 percent annual interest rate for the cost of capital and a 3 percent annual interest rate to account for storage space. Setup costs are $20 per hour required to reconfigure the equipment for a new style. Suppose that the firm decides to use a rotation cycle policy for production of the buttons.

a. What is the optimal rotation cycle time?
b. How large should the lots be?
c. What is the average annual cost of holding and setups at the optimal solution?
d. What contractual obligations might Green's have with New Jersey Fabrics that would prevent them from implementing the policy you determined in parts (a) and (b)? More specifically, if Green's agreed to make three shipments per year for each button style, what production policy would you recommend?

APPENDIX 4–A MATHEMATICAL DERIVATIONS FOR MULTIPRODUCT CONSTRAINED EOQ SYSTEMS

Consider the standard multiproduct EOQ system with a budget constraint, which was discussed in Section 8. Mathematically, the problem is to find values of the variables $Q_1, Q_2, \ldots, Q_n$ to

$$\text{Minimize} \sum_{i=1}^{n} \left[\frac{h_i Q_i}{2} + \frac{K_i \lambda_i}{Q_i} \right]$$

subject to

$$\sum_{i=1}^{n} c_i Q_i \leq C.$$

Let EOQ_i be the respective unconstrained EOQ values. Then there are two possibilities:

$$\sum_{i=1}^{n} c_i EOQ_i \leq C, \tag{1}$$

$$\sum_{i=1}^{n} c_i EOQ_i > C. \tag{2}$$

If (1) holds, the optimal solution is the trivial solution; namely, set $Q_i = EOQ_i$. If (2) holds, then we are guaranteed that the constraint is binding at the optimal solution. That means that the constraint may be written

$$\sum_{i=1}^{n} c_i Q_i = C.$$

In this case, we introduce the Lagrange multiplier θ, and the problem is now to find $Q_1, Q_2, \ldots, Q_n$, and θ to solve the unconstrained problem:

$$\text{Minimize } G(Q_1, Q_2, \ldots, Q_n, \theta) = \sum_{i=1}^{n} \left(\frac{h_i Q_i}{2} + \frac{K_i \lambda_i}{Q_i} \right) + \theta \sum_{i=1}^{n} (c_i Q_i - C).$$

Necessary conditions for optimality are that

$$\frac{\partial G}{\partial Q_i} = 0 \qquad \text{for } i = 1, \ldots, n$$

and

$$\frac{\partial G}{\partial \theta} = 0.$$

The first n conditions give

$$\frac{h_i}{2} - \frac{K_i \lambda_i}{Q_i^2} + \theta c_i = 0 \qquad \text{for } i = 1, \ldots, n.$$

Rearranging terms, we get

$$Q_i = \sqrt{\frac{2K_i \lambda_i}{h_i + 2\theta c_i}} \qquad \text{for } i = 1, \ldots, n,$$

and we also have the final condition

$$\sum_{i=1}^{n} c_i Q_i = C.$$

Now consider the case where $c_i/h_i = c/h$ independent of i. By dividing the numerator and the denominator by h_i, we may write

$$Q_i = \sqrt{\frac{2K_i \lambda_i}{h_i}} \sqrt{\frac{1}{1 + 2\theta c/h}}$$

$$= \text{EOQ}_i \sqrt{\frac{1}{1 + 2\theta c/h}}$$

$$= \text{EOQ}_i m,$$

where

$$m = \sqrt{\frac{1}{1 + 2\theta c/h}}.$$

Substituting this expression for Q_i into the constraint gives

$$\sum_{i=1}^{n} c_i \text{EOQ}_i m = C$$

or

$$m = C \Big/ \left(\sum_{i=1}^{n} c_i \text{EOQ}_i \right).$$

APPENDIX 4–B   GLOSSARY OF NOTATION FOR CHAPTER 4

c = Proportional order cost.
EOQ = Economic order quantity (optimal lot size).
$G(Q)$ = Average annual cost associated with lot size Q.
h = Holding cost per unit time.
h' = Modified holding cost for finite production rate model.
I = Annual interest rate used to compute holding cost.
K = Setup cost or fixed order cost.

λ = Demand rate (units per unit time).

P = Production rate for finite production rate model.

Q = Lot size or size of the order.

s_i = Setup time for product i (refer to Section 9).

T = Cycle time; time between placement of successive orders.

τ = Order lead time.

θ = Lagrange multiplier for space-constrained model (refer to Section 8).

w_i = Space consumed by one unit of product i (refer to Section 8).

W = Total space available (refer to Section 8).

Bibliography

Arrow, K. A.; S. Karlin; and H. Scarf, eds. *Studies in the Mathematical Theory of Inventory Production.* Stanford, CA: Stanford University Press, 1958.

Barbosa, L. C., and M. Friedman. "Deterministic Inventory Lot Size Models—A General Root Law." *Management Science* 23 (1978), pp. 820–29.

Bowman, E. H., and R. B. Fetter. *Analysis for Production Management.* New York: McGraw-Hill/Irwin, 1961.

Churchman, C. W.; R. L. Ackoff; and E. L. Arnoff. *Introduction to Operations Research.* New York: John Wiley & Sons, 1957.

Cohn, M. "Keeping Track of the Pieces." *Accounting Technology* 10, no. 2 (1994), pp. 18–37.

Donnelly, H. "Technology Awards: Recognizing Retailing's Best." *Stores* 76, no. 10 (1994), pp. 52–56.

Fetter, R. B., and W. C. Dalleck. *Decision Models for Inventory Management.* New York: McGraw-Hill/Irwin, 1961.

Hadley, G. J., and T. M. Whitin. *Analysis of Inventory Systems.* Englewood Cliffs, NJ: Prentice Hall, 1963.

Hanssmann, F. "A Survey of Inventory Theory from the Operations Research Viewpoint." In *Progress in Operations Research* 1, ed. R. L. Ackoff. New York: John Wiley & Sons, 1961.

Harris, F. W. *Operations and Cost* (Factory Management Series). Chicago: Shaw, 1915.

Johnson, L. A., and D. C. Montgomery. *Operations Research in Production Planning, Scheduling, and Inventory Control.* New York: John Wiley & Sons, 1974.

Magee, J. F., and D. M. Boodman. *Production Planning and Inventory Control.* 2nd ed. New York: McGraw-Hill, 1967.

Maxwell, W. L. "The Scheduling of Economic Lot Sizes." *Naval Research Logistics Quarterly* 11 (1964), pp. 89–124.

Plossl, G. W., and O. W. Wight. *Production and Inventory Control.* Englewood Cliffs, NJ: Prentice Hall, 1967.

Rosenblatt, M. J. "Multi-Item Inventory System with Budgetary Constraint: A. Comparison between the Lagrangian and the Fixed Cycle Approach." *International Journal of Production Research* 19 (1981), pp. 331–39.

Scarf, H. E.; D. M. Gilford; and M. W. Shelly. *Multistage Inventory Models and Techniques.* Stanford, CA: Stanford University Press, 1963.

Schwarz, L. B., and L. Schrage. "Optimal and Systems Myopic Policies for Multiechelon Production/Inventory Assembly Systems." *Management Science* 21 (1971), pp. 1285–94.

Starr, M. K., and D. W. Miller. *Inventory Control: Theory and Practice.* Englewood Cliffs, NJ: Prentice Hall, 1962.

Wagner, H. M. *Statistical Management of Inventory Systems.* New York: John Wiley & Sons, 1962.

Whitin, T. M. *The Theory of Inventory Management.* Rev. ed. Princeton, NJ: Princeton University Press, 1957.

Wilson, R. H. "A Scientific Routine for Stock Control." *Harvard Business Review* 13 (1934), pp. 116–28.

$$Q = \sqrt{\dfrac{2K\lambda}{h}}$$

One learns by doing

CHAPTER 5

INVENTORY CONTROL SUBJECT TO UNCERTAIN DEMAND

The management of uncertainty plays an important role in the success of any firm. What are the sources of uncertainty that affect a firm? A partial list includes uncertainty in consumer preferences and trends in the market, uncertainty in the availability and cost of labor and resources, uncertainty in vendor resupply times, uncertainty in weather and its ramifications on operations logistics, uncertainty of financial variables such as stock prices and interest rates, and uncertainty of demand for products and services.

The uncertainty of demand and its effect on inventory management strategies are the subjects of this chapter. When a quantity is uncertain, it means that we cannot predict its value exactly in advance. For example, a department store cannot exactly predict the sales of a particular item on any given day. An airline cannot exactly predict the number of people that will choose to fly on any given flight. How, then, can these firms choose the number of items to keep in inventory or the number of flights to schedule on any given route?

Although exact sales of an item or numbers of seats booked on a plane cannot be predicted in advance, one's past experience can provide useful information for planning. As shown in the next section, previous observations of any random phenomenon can be used to estimate its *probability distribution*. By properly quantifying the consequences of incorrect decisions, a well-thought-out mathematical model of the system being studied will result in intelligent strategies. When uncertainty is present, the objective is almost always to minimize expected cost or maximize expected profits.

As the economic recession of the early 1990s started to take hold, some businesses that relied on direct consumer spending, such as retailing, suffered severe losses. Both Sears and Macy's Department Stores, long-standing successes in the American retail market, posted poor earnings in 1991; in fact, Macy's was on the verge of bankruptcy. However, several retailers enjoyed dramatic successes during these difficult economic times. In the fashion business, both The Gap, headquartered in San Francisco, and The Limited, headquartered in Columbus, Ohio, did very well. Wal-Mart Stores continued its ascendancy, surpassing Sears as the largest retailer in the United States. Demand uncertainty, particularly in the fashion industry, is a significant factor. Intelligent inventory management in the face of uncertainty certainly played a key role in the success of these firms.

As some level of demand uncertainty seems to characterize almost all inventory management problems in practice, one might question the value of the deterministic inventory control models discussed in Chapter 4. There are two reasons for studying deterministic models. One is that they provide a basis for understanding the fundamental trade-offs encountered in inventory management. Another is that they may be good approximations depending on the degree of uncertainty in the demand.

To better understand the second point, let D be the demand for an item over a given period of time. We express D as the sum of two parts, D_{Det} and D_{Ran}. That is,

$$D = D_{Det} + D_{Ran},$$

where

$$D_{Det} = \text{Deterministic component of demand}$$

and

$$D_{Ran} = \text{Random component of demand.}$$

There are a number of circumstances under which it would be appropriate to treat D as being deterministic even though D_{Ran} is not zero. Some of these are

1. When the variance of the random component, D_{Ran}, is small relative to the magnitude of D.
2. When the predictable variation is more important than the random variation.
3. When the problem structure is too complex to include an explicit representation of randomness in the model.

An example of (2) occurs in the aggregate planning problem. Although the forecast error of the aggregate demands over the planning horizon may not be zero, we are more concerned with planning for the anticipated changes in the demand than for the unanticipated changes. An example of (3) occurs in material requirements planning (treated in detail in Chapter 7). The intricacies of the relationships among various component levels and end items make it difficult to incorporate demand uncertainty into the analysis.

However, for many items, the random component of the demand is too significant to ignore. As long as the expected demand per unit time is relatively constant and the problem structure not too complex, explicit treatment of demand uncertainty is desirable. In this chapter we will examine several of the most important stochastic inventory models and the key issues surrounding uncertainty.[1]

Overview of Models Treated in This Chapter

Inventory control models subject to uncertainty are basically of two types: (1) **periodic review** and (2) **continuous review.** (Recall the discussion at the start of Chapter 4. Periodic review means that the inventory level is known at discrete points in time only, and continuous review means that the inventory level is known at all times.) Periodic review models may be for one planning period or for multiple planning periods. For one-period models, the objective is to properly balance the costs of overage (ordering too much) and underage (ordering too little). Single-period models are useful in several contexts: planning for initial shipment sizes for high fashion items, ordering policies for food products that perish quickly, or determining run sizes for items with short useful lifetimes, such as newspapers.

---

[1] For those unfamiliar with it, the word *stochastic* is merely a synonym for *random*.

Because of this last application, the single-period stochastic inventory model has come to be known as the **newsboy model.**[2] The newsboy model will be the first one considered in this chapter.

From this writer's experience, the vast majority of computer-based inventory control systems on the market use some variant of the continuous review models treated in the remainder of this chapter. They are, in a sense, extensions of the EOQ model to incorporate uncertainty. Their popularity in practice is attributable to several factors. First, the policies are easy to compute and easy to implement. Second, the models accurately describe most systems in which there is ongoing replenishment of inventory items under uncertainty. A detailed discussion of service levels is included as well. Because estimating penalty costs is difficult in practice, service level approaches are more frequently implemented than penalty cost approaches.

Multiperiod stochastic inventory models dominate the professional literature on inventory theory. There are enough results in this fascinating research area to compose a volume in its own right. However, our goal in this book is to concentrate on methodology that has been applied in the real world. Although they provide insight, multiperiod stochastic models are rarely implemented. In addition, the level of mathematical sophistication they require is beyond that of this book. For these two reasons, multiperiod inventory models subject to uncertainty are not considered here.

5.1 The Nature of Randomness

In order to clarify what the terms *randomness* and *uncertainty* mean in the context of inventory control, we begin with an example.

Example 5.1

On consecutive Sundays, Mac, the owner of a local newsstand, purchases a number of copies of *The Computer Journal,* a popular weekly magazine. He pays 25 cents for each copy and sells each for 75 cents. Copies he has not sold during the week can be returned to his supplier for 10 cents each. The supplier is able to salvage the paper for printing future issues. Mac has kept careful records of the demand each week for the *Journal.* (This includes the number of copies actually sold plus the number of customer requests that could not be satisfied.) The observed demands during each of the last 52 weeks were

15	19	9	12	9	22	4	7	8	11
14	11	6	11	9	18	10	0	14	12
8	9	5	4	4	17	18	14	15	8
6	7	12	15	15	19	9	10	9	16
8	11	11	18	15	17	19	14	14	17
13	12								

There is no discernible pattern to these data, so it is difficult to predict the demand for the *Journal* in any given week. However, we can represent the demand

---

[2]I hope no one is offended by this label. Some authors have used the terms News Person or News Vendor model, but I choose to retain the original nomenclature for historical accuracy.

FIGURE 5–1

*Frequency
histogram for a
52-week history
of sales of* The
Computer Journal
at Mac's

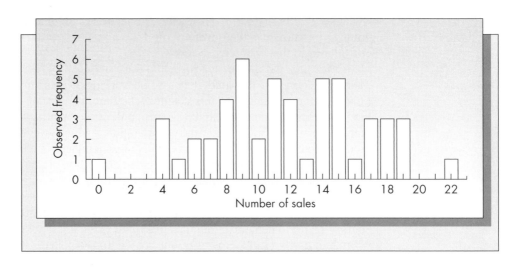

experience of this item as a frequency histogram, which gives the number of times
each weekly demand occurrence was observed during the year. The histogram for
this demand pattern appears in Figure 5–1.

One uses the frequency histogram to estimate the probability that the number of
copies of the *Journal* sold in any week is a specific value. These probability estimates
are obtained by dividing the number of times that each demand occurrence was
observed during the year by 52. For example, the probability that demand is 10 is
estimated to be $2/52 = .0385$, and the probability that the demand is 15 is
$5/52 = .0962$. The collection of all of the probabilities is known as the *empirical
probability distribution*. Cumulative probabilities also can be estimated in a
similar way. For example, the probability that there are nine or fewer copies of the
Journal sold in any week is $(1 + 0 + 0 + 0 + 3 + 1 + 2 + 2 + 4 + 6) = 19/52 = .3654$. ■

Although empirical probabilities can be used in subsequent analysis, they are
inconvenient for a number of reasons. First, they require maintaining a record of the demand history for every item. This can be costly and cumbersome. Second, the
distribution must be expressed (in this case) as 23 different probabilities. Other items
may have an even wider range of past values. Finally, it is more difficult to compute optimal inventory policies with empirical distributions.

For these reasons, we generally approximate the demand history using a
continuous distribution. The form of the distribution chosen depends upon the history of past demand and its ease of use. By far the most popular distribution for
inventory applications is the normal. One reason is the frequency with which it
seems to accurately model demand fluctuations. Another is its convenience. The normal model of demand must be used with care, however, as it admits the possibility
of negative values. When using the normal distribution to describe a nonnegative
phenomenon such as demand, the likelihood of a negative observation should be
sufficiently small (less than .01 should suffice for most applications) so as not to be
a factor.

A normal distribution is determined by two parameters: the mean μ and the variance σ^2. These can be estimated from a history of demand by the sample mean $\overline{D}$
and the sample variance s^2. Let $D_1, D_2, \ldots, D_n$ be n past observations of demand.

FIGURE 5–2

*Frequency
histogram and
normal
approximation*

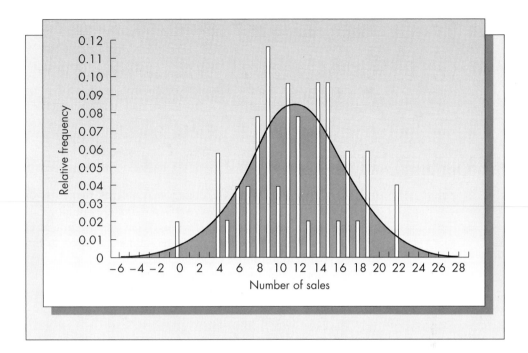

Then

$$\overline{D} = \frac{1}{n} \sum_{i=1}^{n} D_i,$$

$$s^2 = \frac{1}{n-1} \sum_{i=1}^{n} (D_i - \overline{D})^2.$$

For the data pictured in Figure 5–1 we obtain

$$\overline{D} = 11.73,$$
$$s = 4.74.$$

The normal density function, $f(x)$, is given by the formula

$$f(x) = \frac{1}{\sigma\sqrt{2\pi}} \exp\left[-\frac{1}{2}\left(\frac{x - \mu}{\sigma}\right)^2\right] \qquad \text{for } -\infty < x < +\infty.$$

We substitute $\overline{D}$ as the estimator for μ and s as the estimator for σ.

The *relative frequency histogram* is the same as the frequency histogram pictured in Figure 5–1, except that the y axis entries are divided by 52. In Figure 5–2 we show the normal density function that results from the above substitution, superimposed on the relative frequency histogram.

In practice, exponential smoothing is used to recursively update the estimates of the mean and the standard deviation of demand. The standard deviation is estimated using the MAD (mean absolute deviation). Both exponential smoothing and MAD are discussed in detail in Chapter 2. Let $\overline{D}_t$ be the estimate of the mean after observing demand D_t, and let MAD_t be the estimate of the MAD. Then

$$\overline{D}_t = \alpha D_t + (1 - \alpha)\overline{D}_{t-1},$$

$$\text{MAD}_t = \alpha|D_t - \overline{D}_{t-1}| + (1 - \alpha)\text{MAD}_{t-1},$$

where $0 < \alpha < 1$ is the smoothing constant. For normally distributed demand

$$\sigma \approx 1.25 * \text{MAD}.$$

A smoothing constant of $\alpha \approx .1$ is generally used to ensure stability in the estimates (see Chapter 2 for a more complete discussion of the issues surrounding the choice of the smoothing constant).

5.2 Optimization Criterion

In general, optimization in production problems means finding a control rule that achieves minimum cost. However, when demand is random, the cost incurred is itself random, and it is no longer obvious what the optimization criterion should be. Virtually all stochastic optimization techniques applied to inventory control assume that the goal is to minimize *expected* costs.

The motivation for using the expected value criterion is that inventory control problems are generally ongoing problems. Decisions are made repetitively. The law of large numbers from probability theory says that the arithmetic average of many observations of a random variable will converge to the expected value of that random variable. In the context of the inventory problem, if we follow a control rule that minimizes expected costs, then the arithmetic average of the actual costs incurred over many periods will also be a minimum.

In certain circumstances, the expected value may not be the best optimization criterion. When a product is acquired only once and not on an ongoing basis, it is not clear that minimizing expected costs is appropriate. In such a case, maximizing the probability of some event (such as satisfying a proportion of the demand) is generally more suitable. However, because of the ongoing nature of most production problems, the expected value criterion is used in virtually all stochastic inventory control applications.

Problems for Sections 1 and 2

1. Suppose that Mac only kept track of the number of magazines sold. Would this give an accurate representation of the demand for the *Journal?* Under what circumstances would the actual demand and the number sold be close, and under what circumstances would they differ by a substantial amount?

2. What is the difference between deterministic and random variations in the pattern of demands? Provide an example of a real problem in which predictable variation would be important and an example in which random variation would be important.

3. Oakdale Furniture Company uses a special type of woodworking glue in the assembly of its furniture. During the past 36 weeks, the following amounts of glue (in gallons) were used by the company:

25	38	26	31	21	46	29	19	35	39	24	21
17	42	46	19	50	40	43	34	31	51	36	32
18	29	22	21	24	39	46	31	33	34	30	30

    a. Compute the mean and the standard deviation of this sample.

    b. Consider the following class intervals for the number of gallons used each week:

Less than 20.
20–27.
28–33.
34–37.
38–43.
More than 43.

      Determine the proportion of data points that fall into each of these intervals. Compare these proportions to the probabilities that a normal variate with the mean and the standard deviation you computed in part (*a*) falls into each of these intervals. Based on the comparison of the observed proportions and those obtained from assuming a normal distribution, would you conclude that the normal distribution provides an adequate fit of these data? (This procedure is essentially the same as a chi-square goodness-of-fit test.)

    c. Assume that the numbers of gallons of glue used each week are independent random variables, having the normal distribution with mean and standard deviation computed in part (*a*). What is the probability that the total number of gallons used in six weeks does not exceed 200 gallons? (Hint: The mean of a sum of random variables is the sum of the means and the *variance* of a sum of *independent* random variables is the sum of the variances.)

4. In Problem 3, what other probability distributions might accurately describe Oakdale's weekly usage of glue?

5. Rather than keeping track of each demand observation, Betty Sucasas, a member of the marketing staff with a large company that produces a line of switches, has kept only grouped data. For switch C9660Q, used in small power supplies, she has observed the following numbers of units of the switch shipped over the last year.

Units Shipped	Number of Weeks
0–2,000	3
2,001–5,000	6
5,001–9,000	12
9,001–12,000	17
12,001–18,000	10
18,001–20,000	4

      Based on these observations, estimate the mean and the standard deviation of the weekly shipments. (Hint: This is known as grouped data. For the purposes of your calculation, assume that all observations occur at the midpoint of each interval.)

6. *a.* Consider the Oakdale Furniture Company described in Problem 3. Under what circumstances might the major portion of the usage of the glue be predictable?

> b. If the demand were predictable, would you want to use a probability law to describe it? Under what circumstances might the use of a probability model of demand be justified even if the demand could be predicted exactly?

5.3 The Newsboy Model

Let us return to Example 5.1, in which Mac wishes to determine the number of copies of *The Computer Journal* he should purchase each Sunday. A study of the historical data showed that the demand during any week is a random variable that is approximately normally distributed, with mean 11.73 and standard deviation 4.74. Each copy is purchased for 25 cents and sold for 75 cents, and he is paid 10 cents for each unsold copy by his supplier. One obvious solution is that he should buy enough to meet the mean demand, which is approximately 12 copies. There is something wrong with this solution. Suppose Mac purchases a copy that he does not sell. His out-of-pocket expense is 25 cents − 10 cents = 15 cents. Suppose, on the other hand, that he is unable to meet the demand of a customer. In that case, he loses 75 cents −25 cents = 50 cents profit. Hence, there is a significantly greater penalty for not having enough than there is for having too much. If he only buys enough to satisfy mean demand, he will stock-out with the same frequency that he has an oversupply. Our intuition tells us that he should buy more than the mean, but how much more? This question is answered in this section.

Notation

This problem is an example of the newsboy model, in which a single product is to be ordered at the beginning of a period and can be used only to satisfy demand during that period. Assume that all relevant costs can be determined on the basis of ending inventory. Define

c_o = Cost per unit of positive inventory remaining at the end of the period (known as the *overage cost*).

c_u = Cost per unit of unsatisfied demand. This can be thought of as a cost per unit of negative ending inventory (known as the *underage cost*).

In the development of the model, we will assume that the demand D is a continuous nonnegative random variable with density function $f(x)$ and cumulative distribution function $F(x)$. (A brief review of probability theory is given in Appendix 5–A. In particular, both $F(x)$ and $f(x)$ are defined there.)

The *decision variable Q* is the number of units to be purchased at the beginning of the period. The goal of the analysis is to determine Q to minimize the expected costs incurred at the end of the period.

Development of the Cost Function

A general outline for analyzing most stochastic inventory problems is the following:

1. Develop an expression for the cost incurred as a function of both the random variable D and the decision variable Q.
2. Determine the expected value of this expression with respect to the density function or probability function of demand.
3. Determine the value of Q that minimizes the expected cost function.

Define $G(Q, D)$ as the total overage and underage cost incurred at the end of the period when Q units are ordered at the start of the period and D is the demand. If Q units are purchased and D is the demand, $Q - D$ units are left at the end of the period as long as $Q \geq D$. If $Q < D$, then $Q - D$ is negative and the number of units remaining on hand at the end of the period is 0. Notice that

$$\max\{Q - D, 0\} = \begin{cases} Q - D & \text{if } Q \geq D, \\ 0 & \text{if } Q \leq D. \end{cases}$$

In the same way, $\max\{D - Q, 0\}$ represents the excess demand over the supply, or the unsatisfied demand remaining at the end of the period. For any realization of the random variable D, either one or the other of these terms will be zero.

Hence, it now follows that

$$G(Q, D) = c_o \max(0, Q - D) + c_u \max(0, D - Q).$$

Next, we derive the expected cost function. Define

$$G(Q) = E(G(Q, D)).$$

Using the rules outlined in Appendix 5–A for taking the expected value of a function of a random variable, we obtain

$$G(Q) = c_o \int_0^\infty \max(0, Q - x) f(x)\, dx + c_u \int_0^\infty \max(0, x - Q) f(x)\, dx$$

$$= c_o \int_0^Q (Q - x) f(x)\, dx + c_u \int_Q^\infty (x - Q) f(x)\, dx.$$

Determining the Optimal Policy

We would like to determine the value of Q that minimizes the expected cost $G(Q)$. In order to do so, it is necessary to obtain an accurate description of the function $G(Q)$. We have that

$$\frac{dG(Q)}{dQ} = c_o \int_0^Q 1 f(x)\, dx + c_u \int_Q^\infty (-1) f(x)\, dx$$

$$= c_o F(Q) - c_u (1 - F(Q)).$$

(This is a result of Leibniz's rule, which indicates how one differentiates integrals. Leibniz's rule is stated in Appendix 5–A.)

It follows that

$$\frac{d^2 G(Q)}{dQ^2} = (c_o + c_u) f(Q) \geq 0 \qquad \text{for all } Q \geq 0.$$

Because the second derivative is nonnegative, the function $G(Q)$ is said to be *convex* (bowl shaped). We can obtain additional insight into the shape of $G(Q)$ by further analysis. Note that

$$\left. \frac{dG(Q)}{dQ} \right|_{Q=0} = c_o F(0) - c_u(1 - F(0))$$

$$= -c_u < 0 \qquad \text{since } F(0) = 0.$$

FIGURE 5–3

Expected cost function for newsboy model

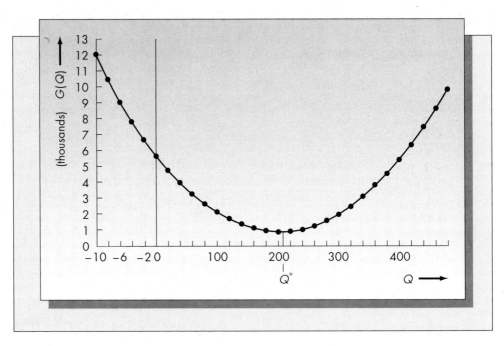

Since the slope is negative at $Q = 0$, $G(Q)$ is decreasing at $Q = 0$. The function $G(Q)$ is pictured in Figure 5–3.

It follows that the optimal solution, say Q^*, occurs where the first derivative of $G(Q)$ equals zero. That is,

$$G'(Q^*) = (c_o + c_u)F(Q^*) - c_u = 0.$$

Rearranging terms gives

$$F(Q^*) = c_u/(c_o + c_u).$$

We refer to the right-hand side of the last equation as the *critical ratio*. Because c_u and c_o are positive numbers, the critical ratio is strictly between zero and one. This implies that for a continuous demand distribution this equation is always solvable.

As $F(Q^*)$ is defined as the probability that the demand does not exceed Q^*, the critical ratio is the probability of satisfying all the demand during the period if Q^* units are purchased at the start of the period. It is important to understand that this is *not* the same as the proportion of satisfied demands. When underage and overage costs are equal, the critical ratio is exactly one-half. In that case Q^* corresponds to the *median* of the demand distribution. When the demand density is symmetric (such as the normal density), the mean and the median are the same.

Example 5.1 (continued)

Consider the example of Mac's newsstand. From past experience, we saw that the weekly demand for the *Journal* is approximately normally distributed with mean $\mu = 11.73$ and standard deviation $\sigma = 4.74$. Because Mac purchases the magazines for 25 cents and can salvage unsold copies for 10 cents, his overage cost is $c_o = 25 - 10 = 15$ cents. His underage cost is the profit on each sale, so that

FIGURE 5–4

*Determination of the
optimal order
quantity for newsboy
example*

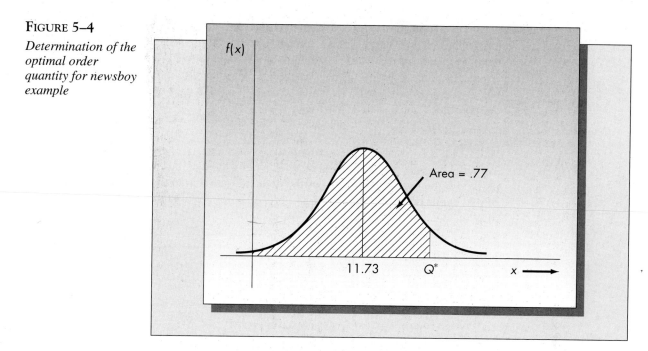

$c_u = 75 - 25 = 50$ cents. The critical ratio is $c_u/(c_o + c_u) = 0.50/0.65 = .77$.
Hence, he should purchase enough copies to satisfy all of the weekly demand with
probability .77. The optimal Q^* is the 77th percentile of the demand distribution (see
Figure 5–4).

Using either Table A–1 or Table A–4 at the back of the book, we obtain a
standardized value of $z = 0.74$. The optimal Q is

$$Q^* = \sigma z + \mu = (4.74)(0.74) + 11.73$$
$$= 15.24 \approx 15.$$

Hence, he should purchase 15 copies every week. ■

Optimal Policy for Discrete Demand

Our derivation of the newsboy formula was based on the assumption that the demand
in the period was described by a continuous probability distribution. We noted a number of reasons above for the desirability of working with continuous distributions. However, in some cases, and particularly when the mean demand is small, it may not be possible to obtain an accurate representation of the observed pattern of demand using a
continuous distribution. For example, the normal approximation of the 52-week history
of the demand for the *Journal* pictured in Figure 5–2 may not be considered sufficiently
accurate for our purposes.

The procedure for finding the optimal solution to the newsboy problem when
the demand is assumed to be discrete is a natural generalization of the continuous case. In the continuous case, the optimal solution is the value of Q that makes
the distribution function equal to the critical ratio $c_u/(c_u + c_o)$. In the discrete
case, the distribution function increases by jumps; it is unlikely that any of its values exactly equal the critical ratio. The critical ratio will generally fall between two

values of $F(Q)$. The optimal solution procedure is to locate the critical ratio between two values of $F(Q)$ and choose the Q corresponding to the *higher* value. (The fact that you always round Q up rather than simply round it off can be easily proven mathematically. This is different from assuming that units are ordered in discrete amounts but demand is continuous, in which case Q is rounded to the closest integer, as we did above.)

Example 5.2

We will solve the problem faced by Mac's newsstand using the empirical distribution derived from a one-year history of the demand, rather than the normal approximation of that demand history. The empirical probabilities are obtained from Figure 5–1 by dividing each of the heights by 52. We obtain

Q	$f(Q)$	$F(Q)$	Q	$f(Q)$	$F(Q)$
0	1/52	1/52 (.0192)	12	4/52	30/52 (.5769)
1	0	1/52 (.0192)	13	1/52	31/52 (.5962)
2	0	1/52 (.0192)	14	5/52	36/52 (.6923)
3	0	1/52 (.0192)	15	5/52	41/52 (.7885)
4	3/52	4/52 (.0769)	16	1/52	42/52 (.8077)
5	1/52	5/52 (.0962)	17	3/52	45/52 (.8654)
6	2/52	7/52 (.1346)	18	3/52	48/52 (.9231)
7	2/52	9/52 (.1731)	19	3/52	51/52 (.9808)
8	4/52	13/52 (.2500)	20	0	51/52 (.9808)
9	6/52	19/52 (.3654)	21	0	51/52 (.9808)
10	2/52	21/52 (.4038)	22	1/52	52/52 (1.0000)
11	5/52	26/52 (.5000)			

The critical ratio for this problem was .77, which corresponds to a value of $F(Q)$ between $Q = 14$ and $Q = 15$. Because we round up, the optimal solution is $Q^* = 15$. Notice that this is exactly the same order quantity obtained using the normal approximation. ■

Extension to Include Starting Inventory

In the derivation of the newsboy model, we assumed that the starting inventory at the beginning of the period was zero. Suppose now that the starting inventory is some value u and $u > 0$. The optimal policy for this case is a simple modification of the case $u = 0$. Note that this extension would not apply to newspapers, but would be appropriate for a product with a shelf life that exceeds one period.

Consider the expected cost function $G(Q)$, pictured in Figure 5–3. If $u > 0$, it just means that we are starting at some point other than 0. We still want to be at Q^* after ordering, as that is still the lowest point on the cost curve. If $u < Q^*$, this is accomplished by ordering $Q^* - u$. If $u > Q^*$, we are past where we want to be on the curve. Ordering any additional inventory merely moves us up the cost curve, which results in higher costs. In this case it is optimal to simply not order.

Hence the optimal policy when there is a starting inventory of $u > 0$ is

Order $Q^* - u$ if $u < Q^*$.
Do not order if $u \geq Q^*$.

Note that Q^* should be interpreted as the order-up-to point rather than the order quantity when $u > 0$. It is also known as a target or base stock level.

Example 5.2 (continued)

Let us suppose that in Example 5.2, Mac has received 6 copies of the *Journal* at the beginning of the week from another supplier. The optimal policy still calls for having 15 copies on hand after ordering, so now he would order the difference $15 - 6 = 9$ copies. (Set $Q^* = 15$ and $u = 6$ to get the order quantity of $Q^* - u = 9$.) ∎

Extension to Multiple Planning Periods

The underlying assumption made in the derivation of the newsboy model was that the item "perished" quickly and could not be used to satisfy demand in subsequent periods. In most industrial and retail environments, however, products are durable and inventory left at the end of a period can be stored and used to satisfy future demand.

This means that the ending inventory in any period becomes the starting inventory in the next period. In the previous section, we indicated how the optimal policy is modified when starting inventory is present. However, when the number of periods remaining exceeds one, the value of Q^* also must be modified. In particular, the interpretation of both c_o and c_u will be different. We consider only the case in which there are infinitely many periods remaining. The optimal value of the order-up-to point when a finite number of periods remain will fall between the one-period and the infinite-period solutions.

In our derivation and subsequent analysis of the EOQ formula in Chapter 4, we saw that the variable order cost c only entered into the optimization to determine the holding cost ($h = Ic$). In addition, we saw that all feasible operating policies incurred the same average annual cost of replenishment, λc. It turns out that essentially the same thing applies in the infinite horizon newsboy problem. As long as excess demand is back-ordered, all feasible policies will just order the demand over any long period of time. Similarly, as long as excess demand is back-ordered, the number of units sold will just be equal to the demand over any long period of time. Hence, both c_u and c_o will be independent of both the proportional order cost c and the selling price of the item. Interpret c_u as the loss-of-goodwill cost and c_o as the holding cost in this case. That this is the correct interpretation of the underage and overage costs is established rigorously in Appendix 5–B.

Example 5.3

Let us return to Mac's newsstand, described in Examples 5.1 and 5.2. Suppose that Mac is considering how to replenish the inventory of a very popular paperback thesaurus that is ordered monthly. Copies of the thesaurus unsold at the end of a month are still kept on the shelves for future sales. Assume that customers who request copies of the thesaurus when they are out of stock will wait until the following month. Mac buys the thesaurus for $1.15 and sells it for $2.75. Mac estimates a loss-of-goodwill cost of 50 cents each time a demand for a thesaurus must be back-ordered. Monthly demand for the book is fairly closely approximated by a normal distribution with mean 18 and standard deviation 6. Mac uses a 20 percent annual interest rate to determine his holding cost. How many copies of the thesaurus should he purchase at the beginning of each month?

Solution

The overage cost in this case is just the cost of holding, which is (1.15) $(0.20)/12 = 0.0192$. The underage cost is just the loss-of-goodwill cost, which is assumed to be 50 cents. Hence, the critical ratio is $0.5/(0.5 + 0.0192) = .9630$. From Table A–1 at the back of this book, this corresponds to a z value of 1.79. The optimal value of the order-up-to point $Q^* = \sigma z + \mu = (6)(1.79) + 18 = 28.74 \approx 29$. ∎

Example 5.3 (continued)

Assume that a local bookstore also stocks the thesaurus and that customers will purchase the thesaurus there if Mac is out of stock. In this case excess demands are lost rather than back-ordered. The order-up-to point will be different from that obtained assuming full back-ordering of demand. In Appendix 5–B we show that in the lost sales case the underage cost should be interpreted as the loss-of-goodwill cost plus the lost profit. The overage cost should still be interpreted as the holding cost only. Hence, the lost sales solution for this example gives $c_u = 0.5 + 1.6 = 2.1$. The critical ratio is $2.1/(2.1 + 0.0192) = .9909$, giving a z value of 2.36. The optimal value of Q in the lost sales case is $Q^* = \sigma z + \mu = (6)(2.36) + 18 = 32.16 \approx 32$. ∎

Although the multiperiod solution appears to be sufficiently general to cover many types of real problems, it suffers from one serious limitation: there is no fixed cost of ordering. This means that the optimal policy, which is to order up to Q^*, requires that ordering take place in every period. In most real systems, however, there are fixed costs associated with ordering, and it is not optimal to place orders each period. Unfortunately, if we include a fixed charge for placing an order, it becomes extremely difficult to determine optimal operating policies. For this reason, we approach in a different way the problem of random demand when a fixed charge for ordering is present. We will assume that inventory levels are reviewed continuously and develop a generalization of the EOQ analysis presented in the previous chapter. This analysis is presented in the next section.

Problems for Section 3

7. A newsboy keeps careful records of the number of papers he sells each day and the various costs that are relevant to his decision regarding the optimal number of newspapers to purchase. For what reason might his results be inaccurate? What would he need to do in order to accurately measure the daily demand for newspapers?

8. Billy's Bakery bakes fresh bagels each morning. The daily demand for bagels is a random variable with a distribution estimated from prior experience

given by

Number of Bagels Sold in One Day	Probability
0	.05
5	.10
10	.10
15	.20
20	.25
25	.15
30	.10
35	.05

The bagels cost Billy's 8 cents to make, and they are sold for 35 cents each. Bagels unsold at the end of the day are purchased by a nearby charity soup kitchen for 3 cents each.

a. Based on the discrete distribution above, how many bagels should Billy's bake at the start of each day? (Your answer should be a multiple of 5.)

b. If you were to approximate the discrete distribution with a normal distribution, would you expect the resulting solution to be close to the answer that you obtained in part (a)? Why or why not?

c. Determine the optimal number of bagels to bake each day using a normal approximation. (Hint: You must compute the mean μ and the variance σ^2 of the demand from the discrete distribution above.)

9. The Crestview Printing Company prints a particularly popular Christmas card once a year and distributes the cards to stationery and gift shops throughout the United States. It costs Crestview 50 cents to print each card, and the company receives 65 cents for each card sold.

Because the cards have the current year printed on them, those cards that are not sold are generally discarded. Based on past experience and forecasts of current buying patterns, the probability distribution of the number of cards to be sold nationwide for the next Christmas season is estimated to be

Quantity Sold	Probability
100,000–150,000	.10
150,001–200,000	.15
200,001–250,000	.25
250,001–300,000	.20
300,001–350,000	.15
350,001–400,000	.10
400,001–450,000	.05

Determine the number of cards that Crestview should print this year.

10. Happy Henry's car dealer sells an imported car called the EX123. Once every three months, a shipment of the cars is made to Happy Henry's. Emergency shipments can be made between these three-month intervals to resupply the cars

when inventory falls short of demand. The emergency shipments require two weeks and buyers are willing to wait this long for the cars, but will generally go elsewhere before the next three-month shipment is due.

From experience, it appears that the demand for the EX123 over a three-month interval is normally distributed with a mean of 60 and a variance of 36. The cost of holding an EX123 for one year is $500. Emergency shipments cost $250 per car over and above normal shipping costs.

a. How many cars should Happy Henry's be purchasing every three months?

b. Repeat the calculations, assuming that excess demands are back-ordered from one three-month period to the next. Assume a loss-of-goodwill cost of $100 for customers having to wait until the next three-month period and a cost of $50 per customer for bookkeeping expenses.

c. Repeat the calculations, assuming that when Happy Henry's is out of stock of EX123s, the customer will purchase the car elsewhere. In this case, assume that the cars cost Henry an average of $10,000 and sell for an average of $13,500. Ignore loss-of-goodwill costs for this calculation.

11. Irwin's sells a particular model of fan, with most of the sales being made in the summer months. Irwin's makes a one-time purchase of the fans prior to each summer season at a cost of $40 each and sells each fan for $60. Any fans unsold at the end of the summer season are marked down to $29 and sold in a special fall sale. Virtually all marked-down fans are sold. The following is the number of sales of fans during the past 10 summers: 30, 50, 30, 60, 10, 40, 30, 30, 20, 40.

a. Estimate the mean and the variance of the demand for fans each summer.

b. Assume that the demand for fans each summer follows a normal distribution, with mean and variance given by what you obtained in part (a). Determine the optimal number of fans for Irwin's to buy prior to each summer season.

c. Based on the observed 10 values of the prior demand, construct an empirical probability distrib1ution of summer demand and determine the optimal number of fans for Irwin's to buy based on the empirical distribution.

d. Based on your results for parts (b) and (c), would you say that the normal distribution provides an adequate approximation?

12. The buyer for Needless Markup, a famous "high end" department store, must decide on the quantity of a high-priced woman's handbag to procure in Italy for the following Christmas season. The unit cost of the handbag to the store is $28.50 and the handbag will sell for $150.00. Any handbags not sold by the end of the season are purchased by a discount firm for $20.00. In addition, the store accountants estimate that there is a cost of $0.40 for each dollar tied up in inventory, as this dollar invested elsewhere could have yielded a gross profit. Assume that this cost is attached to unsold bags only.

a. Suppose that the sales of the bags are equally likely to be anywhere from 50 to 250 handbags during the season. Based on this, how many bags should the buyer purchase? (Hint: This means that the correct distribution of demand is uniform. You may solve this problem assuming either a discrete or a continuous uniform distribution.)

b. A detailed analysis of past data shows that the number of bags sold is better described by a normal distribution, with mean 150 and standard deviation 20. Now what is the optimal number of bags to be purchased?

c. The expected demand was the same in parts (a) and (b), but the optimal order quantities should have been different. What accounted for this difference?

5.4 Lot Size–Reorder Point Systems

The form of the optimal solution for the simple EOQ model with a positive lead time analyzed in Chapter 4 is: When the level of on-hand inventory hits R, place an order for Q units. In that model the only independent decision variable was Q, the order quantity. The value of R was determined from Q, λ, and τ. In what follows, we also assume that the operating policy is of the (Q, R) form. However, when generalizing the EOQ analysis to allow for random demand, we treat Q and R as independent decision variables.

The multiperiod newsboy model was unrealistic for two reasons: it did not include a setup cost for placing an order and it did not allow for a positive lead time. In most real systems, however, both a setup cost and a lead time are present. For these reasons, the kinds of models discussed in this section are used much more often in practice and, in fact, form the basis for the policies used in many commercial inventory systems.

Note that Q in this section is the amount ordered, whereas Q in the last section was the order-up-to point.

We make the following assumptions:

1. The system is continuous review. That is, demands are recorded as they occur, and the level of on-hand inventory is known at all times.
2. Demand is random and stationary. That means that although we cannot predict the value of demand, the *expected* value of demand over any time interval of fixed length is constant. Assume that the expected demand rate is λ units per year.
3. There is a fixed positive lead time τ for placing an order.
4. The following costs are assumed:

   Setup cost at $\$K$ per order.

   Holding cost at $\$h$ per unit held per year.

   Proportional order cost of $\$c$ per item.

   Stock-out cost of $\$p$ per unit of unsatisfied demand. This is also called the shortage cost or the penalty cost.

Describing Demand

In the newsboy problem, the appropriate random variable is the demand during the period. One period is the amount of time required to effect a change in the on-hand inventory level. This is known as the response time of the system. In the context of our current problem, the response time is the reorder lead time τ. Hence, the random variable of interest is the demand during the lead time. We will assume that the demand during the lead time is a continuous random variable D with probability density function (or pdf) $f(x)$ and cumulative distribution function (or cdf) $F(x)$. Let $\mu = E(D)$ and $\sigma = \sqrt{\text{var}(D)}$ be the mean and the standard deviation of demand during the lead time.

Decision Variables

There are two decision variables for this problem, Q and R, where Q is the lot size or order quantity and R is the reorder level in units of inventory.[3] Unlike the EOQ model,

---

[3] When lead times are very long, it may happen that an order should be placed again before a prior order arrives. In that case, the reorder decision variable R should be interpreted as the inventory position (on-hand plus on-order) when a reorder is placed, rather than the inventory level.

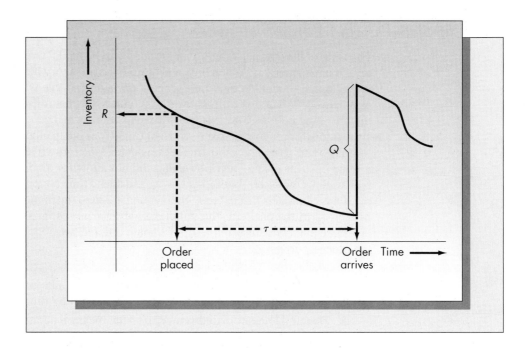

this problem treats Q and R as independent decision variables. The policy is implemented as follows: when the level of on-hand inventory reaches R, an order for Q units is placed that will arrive in τ units of time. The operation of this system is pictured in Figure 5–5.

Derivation of the Expected Cost Function

The analytical approach we will use to solve this problem is, in principle, the same as that used in the derivation of the newsboy model. Namely, we will derive an expression for the expected average annual cost in terms of the decision variables (Q, R) and search for the optimal values of (Q, R) to minimize this cost.

The Holding Cost.   We assume that the mean rate of demand is λ units per year. The expected inventory level varies linearly between s and $Q + s$. We call s the safety stock; it is defined as the expected level of on-hand inventory just before an order arrives, and is given by the formula $s = R - \lambda\tau$. The expected inventory level curve appears in Figure 5–6.

We estimate the holding cost from the average of the expected inventory curve. The average of the function pictured in Figure 5–6 is $s + Q/2 = R - \lambda\tau + Q/2$. An important point to note here is that this computation is only an approximation. When computing the average inventory level, we include both the cases when inventory is positive and the cases when it is negative. However, the holding cost should *not* be charged against the inventory level when it is negative, so that we are underestimating the true value of expected holding cost. An exact expression for the true average inventory is quite complex and has been derived only for certain specific demand distributions. In most real systems, however, the proportion of time spent out of stock is generally small, so this approximation should be reasonably accurate.

FIGURE 5–6

*Expected inventory
level for (Q, R)
inventory model*

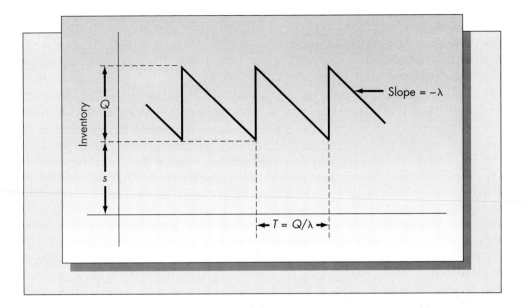

Setup Cost.   A cycle is defined as the time between the arrival of successive orders of size Q. Consistent with the notation used in Chapter 4, let T represent the expected cycle length. Because the setup cost is incurred exactly once each cycle, we need to obtain an expression for the average length of a cycle in order to accurately estimate the setup cost per unit of time.

There are a number of ways to derive an expression for the expected cycle length. From Figure 5–6, we see that the distance between successive arrivals of orders is Q/λ. Another argument is the following. The expected demand during T is clearly λT. However, because the number of units that are entering inventory each cycle is Q and there is conservation of units, the number of units demanded each cycle on average also must be Q. Setting $Q = \lambda T$ and solving for T gives the same result.

It follows, therefore, that the average setup cost incurred per unit time is $K/T = K\lambda/Q$.

Penalty Cost.   From Figure 5–5 we see that the only portion of the cycle during which the system is exposed to shortages is between the time that an order is placed and the time that it arrives (the lead time). The number of units of excess demand is simply the amount by which the demand over the lead time, D, exceeds the reorder level, R. It follows that the expected number of shortages that occur in one cycle is given by the expression

$$E(\max(D - R, 0)) = \int_R^\infty (x - R)f(x)\, dx,$$

which is defined as $n(R)$.

Note that this is essentially the same expression that we derived for the expected number of stock-outs in the newsboy model. As $n(R)$ represents the expected number of stock-outs incurred in a cycle, it follows that the expected number of stock-outs incurred per unit of time is $n(R)/T = \lambda n(R)/Q$.

Proportional Ordering Cost Component.   Over a long period of time, the number of units that enter inventory and the number that leave inventory must be the same. This means that every feasible policy will necessarily order a number of units equal to the demand over any long interval of time. That is, every feasible policy, on average, will replenish inventory at the rate of demand. It follows that the expected proportional order cost per unit of time is λc. Because this term is independent of the decision variables Q and R, it does not affect the optimization. We will henceforth ignore it.

It should be pointed out, however, that the proportional order cost will generally be part of the optimization in an indirect way. The holding cost h is usually computed by multiplying an appropriate value of the annual interest rate I by the value of the item c. For convenience we use the symbol h to represent the holding cost, but keep in mind that it could also be written in the form Ic.

The Cost Function

Define $G(Q, R)$ as the expected average annual cost of holding, setup, and shortages. Combining the expressions derived above for each of these terms gives

$$G(Q, R) = h(Q/2 + R - \lambda\tau) + K\lambda/Q + p\lambda n(R)/Q.$$

The objective is to choose Q and R to minimize $G(Q, R)$. We present the details of the optimization in Appendix 5–C. As shown, the optimal solution is to iteratively solve the two equations

$$Q = \sqrt{\frac{2\lambda[K + pn(R)]}{h}} \tag{1}$$

$$1 - F(R) = Qh/p\lambda. \tag{2}$$

The solution procedure requires iterating between Equations (1) and (2) until two successive values of Q and R are (essentially) the same. The procedure is started by using $Q_0 = $ EOQ (as defined in Chapter 4). One then finds R_0 from (2). That value of R is used to compute $n(R)$, which is substituted into Equation (1) to find Q_1, which is then substituted into (2) to find R_1, and so on. Convergence generally occurs within two or three iterations. When units are integral, the computations should be continued until successive values of both Q and R are within a single unit of their previous values. When units are continuous, a convergence requirement of less than one unit may be required depending upon the level of accuracy desired.

When the demand is normally distributed, $n(R)$ is computed by using the standardized loss function. The standardized loss function $L(z)$ is defined as

$$L(z) = \int_{z}^{\infty} (t - z)\phi(t)\, dt$$

where $\phi(t)$ is the standard normal density. If lead time demand is normal with mean μ and standard deviation σ, then it can be shown that

$$n(R) = \sigma L\left(\frac{R - \mu}{\sigma}\right) = \sigma L(z).$$

The standardized variate z is equal to $(R - \mu)/\sigma$. Calculations of the optimal policy are carried out using Table A–4 at the back of this book.

Example 5.4

Harvey's Specialty Shop is a popular spot that specializes in international gourmet foods. One of the items that Harvey sells is a popular mustard that he purchases from an English company. The mustard costs Harvey $10 a jar and requires a six-month lead time for replenishment of stock. Harvey uses a 20 percent annual interest rate to compute holding costs and estimates that if a customer requests the mustard when he is out of stock, the loss-of-goodwill cost is $25 a jar. Bookkeeping expenses for placing an order amount to about $50. During the six-month replenishment lead time, Harvey estimates that he sells an average of 100 jars, but there is substantial variation from one six-month period to the next. He estimates that the standard deviation of demand during each six-month period is 25. Assume that demand is described by a normal distribution. How should Harvey control the replenishment of the mustard?

Solution

We wish to find the optimal values of the reorder point R and the lot size Q. In order to get the calculation started we need to find the EOQ. However, this requires knowledge of the annual rate of demand, which does not seem to be specified. But notice that if the order lead time is six months and the mean lead time demand is 100, that implies that the mean yearly demand is 200, giving a value of $\lambda = 200$. It follows that the EOQ $= \sqrt{2K\lambda/h} = \sqrt{(2)(50)(200)/(0.2)(10)} = 100$.

The next step is to find R_0 from Equation (2). Substituting $Q = 100$, we obtain

$$1 - F(R_0) = Q_0 h/p\lambda = (100)(2)/(25)(200) = .04.$$

From Table A–4 we find that the z value corresponding to a right tail of .04 is $z = 1.75$. Solving $R = \sigma z + \mu$ gives $R = (25)(1.75) + 100 = 144$. Furthermore, $z = 1.75$ results in $L(z) = 0.0162$. Hence, $n(R) = \sigma L(z) = (25)(0.0162) = 0.405$. We can now find Q_1 from Equation (1):

$$Q_1 = \sqrt{\frac{(2)(200)}{2}[50 + (25)(0.405)]} = 110.$$

This value of Q is compared with the previous one, which is 100. They are not close enough to stop. Substituting $Q = 110$ into Equation (2) results in $1 - F(R_1) = (110)(2)/(25)(200) = .044$. Table A–4 now gives $z = 1.70$ and $L(z) = 0.0183$. Furthermore, $R_1 = (25)(1.70) + 100 = 143$. We now obtain $n(R_1) = (25)(0.0183) = 0.4575$, and $Q_2 = \sqrt{(200)[50 + (25)(0.4575)]} = 110.85 \approx 111$. Substituting $Q_2 = 111$ into Equation (2) gives $1 - F(R_2) = .0444$, $z = 1.70$, and $R_2 = R_1 = 143$. Because both Q_2 and R_2 are within one unit of Q_1 and R_1, we may terminate computations.

We conclude that the optimal values of Q and R are $(Q, R) = (111, 143)$. Hence, each time that Harvey's inventory of this type of mustard hits 143 jars, he should place an order for 111 jars. ∎

Example 5.4 (continued)

For the same example, determine the following:

1. Safety stock.
2. The average annual holding, setup, and penalty costs associated with the inventory control of the mustard.

3. The average time between placement of orders.
4. The proportion of order cycles in which no stock-outs occur.
5. The proportion of demands that are not met.

Solution

1. The safety stock is $s = R - \mu = 143 - 100 = 43$ jars.
2. We will compute average annual holding, setup, and penalty costs separately.

   The holding cost is $h[Q/2 + R - \mu] = 2[111/2 + 143 - 100] = \$197$ per year.
   The setup cost is $K\lambda/Q = (50)(200)/111 = \$90.09$ per year.
   The stock-out cost is $p\lambda n(R)/Q = (25)(200)(0.4575)/111 = \$20.61$ per year.

   Hence, the total average annual cost associated with the inventory control of the mustard, assuming an optimal control policy, is \$307.70 per year.
3. $T = Q/\lambda = 111/200 = 0.556$ year $= 6.7$ months.
4. Here we need to compute the probability that no stock-out occurs in the lead time. This is the same as the probability that the lead time demand does not exceed the reorder point. We have $P\{D \leq R\} = F(R) = 1 - .044 = .956$. We conclude that there will be no stock-outs in 95.6 percent of the order cycles.
5. The expected demand per cycle must be Q (see the argument in the derivation of the expected setup cost). The expected number of stock-outs per cycle is $n(R)$. Hence, the proportion of demands that stock out is $n(R)/Q = 0.4575/111 = 0.004$. Another way of stating this result is that on average 99.6 percent of the demands are satisfied as they occur. ∎

It should be noted in ending this section that Equations (1) and (2) are derived under the assumption that all excess demand is back-ordered. That is, when an item is demanded that is not immediately available, the demand is filled at a later time. However, in many competitive situations, such as retailing, a more accurate assumption is that excess demand is lost. This case is known as *lost sales*. As long as the likelihood of being out of stock is relatively small, Equations (1) and (2) will give adequate solutions to both lost sales and back-order situations. If this is not the case, a slight modification to Equation (2) is required. The lost sales version of Equation (2) is

$$1 - F(R) = Qh/(Qh + p\lambda). \tag{2'}$$

The effect of solving (1) and (2') simultaneously rather than (1) and (2) will be to increase the value of R slightly and decrease the value of Q slightly.

5.5 Service Levels in (Q, R) Systems

Although the inventory model described above is quite realistic for describing many real systems, managers often have a difficult time determining an exact value for the stock-out cost p. In many cases, the stock-out cost includes intangible components such as loss of goodwill and potential delays to other parts of the system. A common substitute for a stock-out cost is a service level. Although there are a number of different definitions of service, it generally refers to the probability that a demand or a collection of demands is met. Service levels can be applied in both periodic review and (Q, R) systems. The application of service levels to periodic-review systems will be discussed in the next section. Service levels for continuous-review systems are considered below.

Two types of service are considered, labeled Type 1 and Type 2, respectively.

Type 1 Service

In this case we specify the probability of not stocking out in the lead time. We will use the symbol α to represent this probability. As specification of α completely determines the value of R, the computation of R and Q can be decoupled. The computation of the optimal (Q, R) values subject to a Type 1 service constraint is very straightforward.

    a. Determine R to satisfy the equation $F(R) = \alpha$.

    b. Set $Q = \text{EOQ}$.

Interpret α as the proportion of cycles in which no stock-out occurs. A Type 1 service objective is appropriate when a shortage occurrence has the same consequence independent of its time or amount. One example would be where a production line is stopped whether 1 unit or 100 units are short. However, Type 1 service is not how service is interpreted in most applications. Usually when we say we would like to provide 95 percent service, we mean that we would like to be able to fill 95 percent of the demands when they occur, not fill all of the demands in 95 percent of the order cycles. Also, as different items have different cycle lengths, this measure will not be consistent among different products, making the proper choice of α difficult.

Type 2 Service

Type 2 service measures the proportion of demands that are met from stock. We will use the symbol β to represent this proportion. As we saw in part 5 of Example 5.4, $n(R)/Q$ is the average fraction of demands that stock out each cycle. Hence, specification of β results in the constraint $n(R)/Q = 1 - \beta$.

This constraint is more complex than the one arising from Type 1 service, as it involves both Q and R. It turns out that although the EOQ is not optimal in this case, it usually gives pretty good results. If we use the EOQ to estimate the lot size, then we would find R to solve $n(R) = \text{EOQ}(1 - \beta)$.

Example 5.5

Consider again Harvey's Specialty Shop, described in Example 5.4. Harvey feels uncomfortable with the assumption that the stock-out cost is $25 and decides to use a service level criterion instead. Suppose that he chooses to use a 98 percent service objective.

1. *Type 1 service.* If we assume an α of .98, then we find R to solve $F(R) = 0.98$. From Table A–1 or A–4, we obtain $z = 2.05$. Setting $R = \sigma z + \mu$ gives $R = 151$.

2. *Type 2 service.* Here $\beta = 0.98$. We are required to solve the equation

$$n(R) = \text{EOQ}(1 - \beta),$$

which is equivalent to

$$L(z) = \text{EOQ}(1 - \beta)/\sigma.$$

Substituting EOQ $= 100$ and $\beta = .98$, we obtain

$$L(z) = (100)(0.02)/25 = 0.08.$$

From Table A–4 of the unit normal partial expectations, we obtain $z = 1.02$. Setting $R = \sigma z + \mu$ gives $R = 126$. Notice that the same values of α and β give *considerably* different values of R. ∎

In order to understand more clearly the difference between these two measures of service, consider the following example. Suppose that we have tracked the demands and stock-outs over 10 consecutive order cycles with the following results:

Order Cycle	Demand	Stock-Outs
1	180	0
2	75	0
3	235	45
4	140	0
5	180	0
6	200	10
7	150	0
8	90	0
9	160	0
10	40	0

Based on a Type 1 measure of service, we find that the fraction of periods in which there is no stock-out is $8/10 = 80$ percent. That is, the probability that all of the demands are met in a single order cycle is 0.8, based on these observations. However, the Type 2 service provided here is considerably better. In the example above, the total number of demands over the 10 periods is 1,450 (the sum of the numbers in the second column), and the total number of demands that result in a stock-out is 55. Hence, the number of satisfied demands is $1,450 - 55 = 1,395$. The proportion of satisfied demands is $1,395/1,450 = 0.9621$, or roughly 96 percent.

The term *fill rate* is often used to describe Type 2 service, and is generally what most managers mean by service. (The fill rate in the example above is 96 percent.) We saw in the example that there is a significant difference between the proportion of cycles in which all demands are satisfied (Type 1 service) and the fill rate (Type 2 service). *Even though it is easier to determine the best operating policy satisfying a Type 1 service objective, this policy will not accurately approximate a Type 2 service objective and should not be used in place of it.*

Optimal (Q, R) Policies Subject to Type 2 Constraint

Using the EOQ value to estimate the lot size gives reasonably accurate results when using a fill rate constraint, but the EOQ value is only an approximation of the optimal lot size. A more accurate value of Q can be obtained as follows. Consider the pair of equations (1) and (2) we solved for the optimal values of Q and R when a stock-out cost was present. Solving for p in Equation (2) gives

$$p = Qh/[(1 - F(R))\lambda],$$

which now can be substituted for p in Equation (1), resulting in

$$Q = \sqrt{\frac{2\lambda\{K + Qhn(R)/[(1 - F(R))\lambda]\}}{h}},$$

which is a quadratic equation in Q. It can be shown that the positive root of this equation is

$$Q = \frac{n(R)}{1 - F(R)} + \sqrt{\frac{2K\lambda}{h} + \left(\frac{n(R)}{1 - F(R)}\right)^2}. \tag{3}$$

Equation (3) will be called the SOQ formula (for service level order quantity).[4] This equation is solved simultaneously with

$$n(R) = (1 - \beta)Q \tag{4}$$

to obtain optimal values of (Q, R) satisfying a Type 2 service constraint.

The reader should note that the version of Equation (4) used in the calculations is in terms of the standardized variate z and is given by

$$L(z) = (1 - \beta)Q/\sigma.$$

The solution procedure is essentially the same as that required to solve Equations (1) and (2) simultaneously. Start with $Q_0 = $ EOQ, find R_0 from (4), use R_0 in (3) to find Q_1, and so on, and stop when two successive values of Q and R are sufficiently close (within one unit is sufficient for most problems).

Example 5.5 (continued)

Returning to Example 5.5, $Q_0 = 100$ and $R_0 = 126$. Furthermore, $n(R_0) = (0.02)(100) = 2$. Using $z = 1.02$ gives $1 - F(R_0) = 0.154$. Continuing with the calculations,

$$Q_1 = \frac{2}{0.154} + \sqrt{(100)^2 + \left(\frac{2}{0.154}\right)^2}$$
$$= 114.$$

Solving Equation (4) gives $n(R_1) = (114)(0.02)$, which is equivalent to

$$L(z) = (114)(0.02)/25 = 0.0912.$$

From Table A–4, $z = 0.95$, so that

$$1 - F(R_1) = 0.171$$

and

$$R_1 = \sigma z + \mu = 124.$$

Carrying the computation one more step gives $Q_2 = 114$ and $R_2 = 124$. As both Q and R are within one unit of their previous values, we terminate computations. Hence, we conclude that the optimal values of Q and R satisfying a 98 percent fill rate constraint are $(Q, R) = (114, 124)$.

Consider the cost error resulting from the EOQ substituted for the SOQ. In order to compare these policies, we compute the average annual holding and setup costs (notice that there is no stock-out cost) for the policies $(Q, R) = (100, 126)$ and $(Q, R) = (114, 124)$.

Recall the formulas for average annual holding and setup costs.

Holding cost $= h(Q/2 + R - \mu)$.
Setup cost $= K\lambda/Q$.

---

[4]The SOQ formula also could have been derived by more conventional Lagrange multiplier techniques. We include this derivation to demonstrate the relationship between the fill rate objective and the stock-out cost model.

For (100, 126):

$$\left.\begin{array}{lll} \text{Holding cost} = 2(100/2 + 126 - 100) = \$152 \\ \text{Setup cost} \quad = (50)(200)/100 \qquad\qquad = \$100 \end{array}\right\} \text{Total} = \$252$$

For (114, 124):

$$\left.\begin{array}{lll} \text{Holding cost} = 2(114/2 + 124 - 100) = \$162 \\ \text{Setup cost} \quad = (50)(200)/114 \qquad\qquad = \$88 \end{array}\right\} \text{Total} = \$250$$

We see that the EOQ approximation gives costs close to the optimal in this case. ∎

Imputed Shortage Cost

Consider the solutions that we obtained for (Q, R) in Example 5.5 when we used a service level criterion rather than a shortage cost. For a Type 2 service of $\beta = 0.98$ we obtained the solution (114, 124). Although no shortage cost was specified, this solution clearly corresponds to *some* value of p. That is, there is some value of p such that the policy (114, 124) satisfies Equations (1) and (2). This particular value of p is known as the imputed shortage cost.

The imputed shortage cost is easy to find. One solves for p in Equation (2) to obtain $p = Qh/[(1 - F(R))\lambda]$. The imputed shortage cost is a useful way to determine whether the value chosen for the service level is appropriate.

Example 5.5 (continued)

Consider again Harvey's Specialty Shop. Using a value of $\alpha = .98$ (Type 1 service), we obtained the policy (100, 151). The imputed shortage cost is $p = (100)(2)/[(0.02)(200)] = \$50$.

Using a value of $\beta = 0.98$ (Type 2 service) we obtained the policy (114, 124). In his case the imputed cost of shortage is $p = (114)(2)/[(0.171)(200)] = \$6.67$. ∎

Scaling of Lead Time Demand

In all previous examples the demand during the lead time was given. However, in most applications demand would be forecast on a periodic basis, such as monthly. In such cases one would need to convert the demand distribution to correspond to the lead time.

Assume that demands follow a normal distribution. Because sums of independent normal random variables are also normally distributed, the form of the distribution of lead time demand is normal. Hence, all that remains is to determine the mean and the standard deviation. Let periodic demand have mean λ and standard deviation ν, and let τ be the lead time in periods. As both the means and the *variances* (not standard deviations) are additive, the mean demand during lead time is $\mu = \lambda\tau$ and the variance of demand during lead time is $\nu^2\tau$. Hence, the standard deviation of demand during lead time is $\sigma = \nu\sqrt{\tau}$ (although the square root may not always be appropriate).[5]

---

[5]Often in practice it turns out that there is more variation in the demand process than is described by a pure normal distribution. For that reason the standard deviation of demand is generally expressed in the form $\nu\tau^q$ where the correct value of q, generally between 0.5 and 1, must be determined for each item or group of items by an analysis of historical data.

Example 5.6

Weekly demand for a certain type of automotive spark plug in a local repair shop is normally distributed with mean 34 and standard deviation 12. Procurement lead time is six weeks. Determine the lead time demand distribution.

Solution

The demand over the lead time is also normally distributed with mean $(34)(6) = 204$ and standard deviation $(12)\sqrt{6} = 29.39$. These would be the values of μ and σ that one would use for all remaining calculations. ∎

Estimating Sigma when Inventory Control and Forecasting Are Linked

Thus far in this chapter we have assumed that the distribution of demand over the lead time is known. In practice, one assumes a *form* for the distribution, but its parameters must be estimated from real data. Assuming a normal distribution for lead time demand (which is the most common assumption), one needs to estimate the mean and the standard deviation. When a complete history of past data is available, the standard statistical estimates for the mean and the standard deviation (i.e., those suggested in Section 1) are fine. However, most forecasting schemes do *not* use all past data. Moving averages use only the past N data values and exponential smoothing places declining weights on past data.

In these cases, it is unclear exactly what are the right estimators for the mean and the standard deviation of demand. This issue was discussed in Section 2.12. The best estimate of the mean is simply the forecast of demand for the next period. For the variance, one should use the estimator for the variance of forecast error. The rationale for this is rarely understood. The variance of forecast error and the variance of demand are *not* the same thing. This was established rigorously in Appendix 2–A.

Why is it appropriate to use the standard deviation of forecast error to estimate σ? The reason is that it is the forecast that we are using to estimate demand. Safety stock is held to protect against errors in forecasting demand. In general, the variance of forecast error will be higher than the variance of demand. This is the result of the additional sampling error introduced by a forecasting scheme that uses only a portion of past data.[6]

R. G. Brown (1959) was apparently the first to recommend using the standard deviation of forecast error in safety stock calculations. The method he recommended, which is still in widespread use today, is to track the MAD (mean absolute deviation) of forecast error using the formula

$$\text{MAD}_t = \alpha \, \text{MAD}_{t-1} + (1 - \alpha)|F_t - D_t|$$

where F_t is the forecast of demand at time t and D_t is the actual observed demand at time t. The estimator for the standard deviation of forecast error at time t is 1.25 MAD_t. While this method is very popular in commercial inventory control systems, apparently few realize that by using this approach they are estimating not demand variance but forecast error variance, and that these are not the same quantities.

---

[6] In cases where the underlying demand process is nonstationary, that is, where there is trend or seasonality, the variance of demand due to *systematic* changes could be higher than the variance of forecast error. All of our analysis assumes stationary (constant mean and variance) demand patterns, however.

Lead Time Variability

Thus far, we have assumed that the lead time τ is a known constant. However, lead time uncertainty is common in practice. For example, the time required to transport commodities, such as oil, that are shipped by sea depends upon weather conditions. In general, it is very difficult to incorporate the variability of lead time into the calculation of optimal inventory policies. The problem is that if we assume that successive lead times are independent random variables, then it is possible for lead times to cross; that is, two successive orders would not necessarily be received in the same sequence that they were placed.

Order crossing is unlikely when a single supplier is used. If we are willing to make the simultaneous assumptions that orders do not cross and that successive lead times are independent, the variability of lead time can be easily incorporated into the analysis. Suppose that the lead time τ is a random variable with mean μ_τ and variance σ_τ^2. Furthermore, suppose that demand in any time t has mean λt and variance $\nu^2 t$. Then it can be shown that the demand during lead time has mean and variance[7]

$$\mu = \lambda \mu_\tau,$$

$$\sigma^2 = \mu_\tau \nu^2 + \lambda^2 \sigma_\tau^2.$$

Example 5.7

Harvey Gold, the owner of Harvey's Specialty Shop, orders an unusual olive from the island of Santorini, off the Greek coast. Over the years, Harvey has noticed considerable variability in the time it takes to receive orders of these olives. On average, the order lead time is four months and the standard deviation is six weeks (1.5 months). Monthly demand for the olives is normally distributed with mean 15 (jars) and standard deviation 6.

Setting $\mu_\tau = 4$, $\sigma_\tau = 1.5$, $\lambda = 15$, and $\nu = 6$, we obtain

$$\mu = \mu_\tau \lambda = (4)(15) = 60,$$

$$\sigma^2 = \mu_\tau \nu^2 + \lambda^2 \sigma_\tau^2 = (4)(36) + (225)(2.25) = 650.25.$$

One would proceed with the calculations of optimal inventory policies using $\mu = 60$ and $\sigma^2 = 650.25$ as the mean and the variance of lead time demand. ∎

Calculations in Excel

The standardized loss function, $L(z)$, can be computed in Excel. To do so, write $L(z)$ in the following way:

$$L(z) = \int_z^\infty (t - z)\phi(t)\, dt = \int_z^\infty t\phi(t)\, dt - z(1 - \Phi(z)) = \phi(z) - z(1 - \Phi(z)).$$

The first equality results from the definition of the cumulative distribution function, and the second equality is a consequence of a well-known property of the standard normal distribution (see, for example, Hadley and Whitin, 1963, p. 444). To program this formula into Excel, use the definition of the standard normal density function, which is

$$\phi(z) = \frac{1}{\sqrt{2\pi}} \exp(-0.5z^2),$$

---

[7]Hadley and Whitin (1963), p. 153.

and the built-in Excel function *normsdist( )*, which returns the value of $\Phi( )$. In this way one could compute (Q, R) policies within the spreadsheet. Alternatively, one could build a table of $L(z)$, which can then be embedded into a search routine.

Problems for Sections 4 and 5

13. An automotive warehouse stocks a variety of parts that are sold at neighborhood stores. One particular part, a popular brand of oil filter, is purchased by the warehouse for $1.50 each. It is estimated that the cost of order processing and receipt is $100 per order. The company uses an inventory carrying charge based on a 28 percent annual interest rate.

    The monthly demand for the filter follows a normal distribution with mean 280 and standard deviation 77. Order lead time is assumed to be five months.

    Assume that if a filter is demanded when the warehouse is out of stock, then the demand is back-ordered, and the cost assessed for each back-ordered demand is $12.80. Determine the following quantities:
    a. The optimal values of the order quantity and the reorder level.
    b. The average annual cost of holding, setup, and stock-out associated with this item assuming that an optimal policy is used.
    c. Evaluate the cost of uncertainty for this process. That is, compare the average annual cost you obtained in part (*b*) with the average annual cost that would be incurred if the lead time demand had zero variance.

14. Weiss's paint store uses a (Q, R) inventory system to control its stock levels. For a particularly popular white latex paint, historical data show that the distribution of monthly demand is approximately normal, with mean 28 and standard deviation 8. Replenishment lead time for this paint is about 14 weeks. Each can of paint costs the store $6. Although excess demands are back-ordered, the store owner estimates that unfilled demands cost about $10 each in bookkeeping and loss-of-goodwill costs. Fixed costs of replenishment are $15 per order, and holding costs are based on a 30 percent annual rate of interest.
    a. What are the optimal lot sizes and reorder points for this brand of paint?
    b. What is the optimal safety stock for this paint?

15. After taking a production seminar, Al Weiss, the owner of Weiss's paint store mentioned in Problem 14, decides that his stock-out cost of $10 may not be very accurate and switches to a service level model. He decides to set his lot size by the EOQ formula and determines his reorder point so that there is *no stock-out* in 90 percent of the order cycles.
    a. Find the resulting (Q, R) values.
    b. Suppose that, unfortunately, he really wanted to satisfy 90 percent of his demands (that is, achieve a 90 percent fill rate). What fill rate did he actually achieve from the policy determined in part (*a*)?

16. Suppose that in Problem 13 the stock-out cost is replaced with a Type 1 service objective of 95 percent. Find the optimal values of (Q, R) in this case.

17. Suppose that in Problem 13 a Type 2 service objective of 95 percent is substituted for the stock-out cost of $12.80. Find the resulting values of Q and R. Also, what is the imputed cost of shortage for this case?

18. Suppose that the warehouse mistakenly used a Type 1 service objective when it really meant to use a Type 2 service objective (see Problems 16 and 17). What is the additional holding cost being incurred each year for this item because of this mistake?

19. Disk Drives Limited (DDL) produces a line of internal Winchester disks for microcomputers. The drives use a 3.5-inch platter that DDL purchases from an outside supplier. Demand data and sales forecasts indicate that the weekly demand for the platters seems to be closely approximated by a normal distribution with mean 38 and variance 130. The platters require a three-week lead time for receipt. DDL has been using a 40 percent annual interest charge to compute holding costs. The platters cost $18.80 each, order cost is $75.00 per order, and the company is currently using a stock-out cost of $400.00 per platter. (Because the industry is so competitive, stock-outs are very costly.)

   a. Due to a prior contractual agreement with the supplier, DDL must purchase the platters in lots of 500. What is the reorder point that it should be using in this case?

   b. When DDL renegotiates its contract with the supplier, what lot size should it write into the agreement?

   c. How much of a penalty in terms of setup, holding, and stock-out cost is DDL paying for contracting to buy too large a lot?

   d. DDL's president is uncomfortable with the $400 stock-out cost and decides to substitute a 99 percent fill rate criterion. If DDL used a lot size equal to the EOQ, what would its reorder point be in this case? Also, find the imputed cost of shortage.

20. Bobbi's Restaurant in Boise, Idaho, is a popular place for weekend brunch. The restaurant serves real maple syrup with french toast and pancakes. Bobbi buys the maple syrup from a company in Maine that requires three weeks for delivery. The syrup costs Bobbi $4 a bottle and may be purchased in any quantity. Fixed costs of ordering amount to about $75 for bookkeeping expenses, and holding costs are based on a 20 percent annual rate. Bobbi estimates that the loss of customer goodwill for not being able to serve the syrup when requested amounts to $25. Based on past experience, the weekly demand for the syrup is normal with mean 12 and variance 16 (in bottles). For the purposes of your calculations, you may assume that there are 52 weeks in a year and that all excess demand is backordered.

   a. How large an order should Bobbi be placing with her supplier for the maple syrup, and when should those orders be placed?

   b. What level of Type 1 service is being provided by the policy you found in part (a)?

   c. What level of Type 2 service is being provided by the policy you found in part (a)?

   d. What policy should Bobbi use if the stock-out cost is replaced with a Type 1 service objective of 95 percent?

   e. What policy should Bobbi use if the stock-out cost is replaced with a Type 2 service objective of 95 percent? (You may assume an EOQ lot size.)

   f. Suppose that Bobbi's supplier requires a minimum order size of 500 bottles. Find the reorder level that Bobbi should use if she wishes to satisfy 99 percent of her customer demands for the syrup.

5.6 Additional Discussion of Periodic-Review Systems

(s, S) Policies

In our analysis of the newsboy problem, we noted that a severe limitation of the model from a practical viewpoint is that there is no setup cost included in the formulation. The (Q, R) model treated in the preceding sections included an order setup cost, but assumed that the inventory levels were reviewed continuously; that is, known at all times. How should the system be managed when there is a setup cost for ordering, but inventory levels are known only at discrete points in time?

The difficulty that arises from trying to implement a continuous-review solution in a periodic-review environment is that the inventory level is likely to overshoot the reorder point R during a period, making it impossible to place an order the instant the inventory reaches R. To overcome this problem, the operating policy is modified slightly. Define two numbers, s and S, to be used as follows: When the level of on-hand inventory is *less than or equal to s,* an order for the difference between the inventory and S is placed. If u is the starting inventory in any period, then the (s, S) policy is

If $u \leq s$, order $S - u$.

If $u > s$, do not order.

Determining optimal values of (s, S) is extremely difficult, and for that reason few real operating systems use optimal (s, S) values. Several approximations have been suggested. One such approximation is to compute a (Q, R) policy using the methods described earlier, and set $s = R$ and $S = R + Q$. This approximation will give reasonable results in many cases, and is probably the most commonly used. The reader interested in a comprehensive comparison of several approximate (s, S) policies should refer to Porteus (1985).

*Service Levels in Periodic-Review Systems

Service levels also may be used when inventory levels are reviewed periodically. Consider first a Type 1 service objective. That is, we wish to find the order-up-to point Q so that all of the demand is satisfied in a given percentage of the periods. Suppose that the value of Type 1 service is α. Then Q should solve the equation

$$F(Q) = \alpha.$$

This follows because $F(Q)$ is the probability that the demand during the period does not exceed Q. Notice that one simply substitutes α for the critical ratio in the newsboy model. To find Q to satisfy a Type 2 service objective of β, it is necessary to obtain an expression for the fraction of demands that stock out each period. Using essentially the same notation as that used for (Q, R) systems, define

$$n(Q) = \int_Q^\infty (x - Q)f(x)\,dx.$$

Note that $n(Q)$, which represents the expected number of demands that stock out at the end of the period, is the same as the term multiplying c_u in the expression for the expected cost function for the newsboy model discussed in Section 3. As the demand per period is μ, it follows that the proportion of demands that stock out each period is $n(Q)/\mu$. Hence, the value of Q that meets a fill rate objective of β solves

$$n(Q) = (1 - \beta)\mu.$$

The specification of either a Type 1 or Type 2 service objective completely determines the order quantity, independent of the cost parameters.

Example 5.8

Mac, the owner of the newsstand described in Example 5.1, wishes to use a Type 1 service level of 90 percent to control his replenishment of *The Computer Journal*. The z value corresponding to the 90th percentile of the unit normal is $z = 1.28$. Hence,

$$Q^* = \sigma z + \mu = (4.74)(1.28) + 11.73 = 17.8 \approx 18.$$

Using a Type 2 service of 90 percent, we obtain

$$n(Q) = (1 - \beta)\mu = (0.1)(11.73) = 1.173.$$

It follows that $L(z) = n(Q)/\sigma = 1.173/4.74 = 0.2475$. From Table A–4 at the back of this book, we find

$$z \approx 0.35;$$

then

$$Q^* = \sigma z + \mu = (4.74)(0.35) + 11.73 = 13.4 \approx 13.$$

As with (Q, R) models, notice the striking difference between the resulting values of Q^* for the same levels of Type 1 and Type 2 service. ∎

Problems for Section 6

21. Consider the Crestview Printing Company mentioned in Problem 9. Suppose that Crestview wishes to produce enough cards to satisfy all Christmas demand with probability 90 percent. How many cards should they print? Suppose the probability is 97 percent. What would you recommend in this case? (Your answer will depend upon the assumption you make concerning the shape of the cumulative distribution function.)

22. Consider Happy Henry's car dealer described in Problem 10.
    a. How many EX123s should Happy Henry's purchase to satisfy all of the demand over a three-month interval with probability .95?
    b. How many cars should be purchased if the goal is to satisfy 95 percent of the demands?

23. For the problem of controlling the inventory of white latex paint at Weiss's paint store, described in Problem 14, suppose that the paint is reordered on a monthly basis rather than on a continuous basis.
    a. Using the (Q, R) solution you obtained in part (a) of Problem 14, determine appropriate values of (s, S).
    b. Suppose that the demands during the months of January to June were

Month	Demand	Month	Demand
January	37	April	31
February	33	May	14
March	26	June	40

If the starting inventory in January was 26 cans of paint, determine the number of units of paint ordered in each of the months January to June following the (s, S) policy you found in part (a).

5.7 Multiproduct Systems

ABC Analysis

One issue that we have not discussed is the cost of implementing an inventory control system and the trade-offs between the cost of controlling the system and the potential benefits that accrue from that control. In multiproduct inventory systems, not all products are equally profitable. Control costs may be justified in some cases and not in others. For example, spending $200 annually to monitor an item that contributes only $100 a year to profits is clearly not economical.

For this reason, it is important to differentiate profitable from unprofitable items. To do so, we borrow a concept from economics. The economist Vilfredo Pareto, who studied the distribution of wealth in the 19th century, noted that a large portion of the wealth was owned by a small segment of the population. This *Pareto effect* also applies to inventory systems: a large portion of the total dollar volume of sales is often accounted for by a small number of inventory items. Assume that items are ranked in decreasing order of the dollar value of annual sales. The *cumulative* value of sales generally results in a curve much like the one pictured in Figure 5–7.

Typically, the top 20 percent of the items account for about 80 percent of the annual dollar volume of sales, the next 30 percent of the items for the next 15 percent of sales, and the remaining 50 percent for the last 5 percent of dollar volume. These figures are only approximate and will vary slightly from one system to another. The three item groups are labeled A, B, and C, respectively. When a finer distinction is

FIGURE 5–7

Pareto curve:
Distribution of
inventory by value

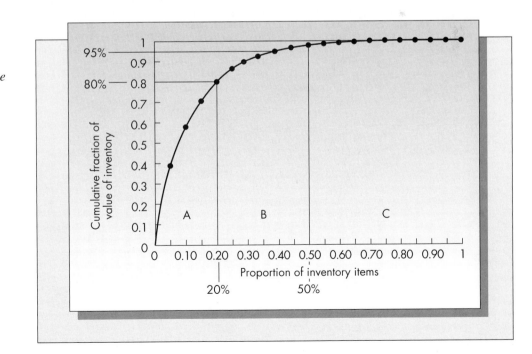

needed, four or five categories could be used. Even when using only three categories, the percentages used in defining A, B, and C items could be different from the 80 percent, 15 percent, and 5 percent recommended above.

Because A items account for the lion's share of the yearly revenue, these items should be watched most closely. Inventory levels for A items should be monitored continuously. More sophisticated forecasting procedures might be used and more care would be taken in the estimation of the various cost parameters required in calculating operating policies. For B items inventories could be reviewed periodically, items could be ordered in groups rather than individually, and somewhat less sophisticated forecasting methods could be used. The minimum degree of control would be applied to C items. For very inexpensive C items with moderate levels of demand, large lot sizes are recommended to minimize the frequency that these items are ordered. For expensive C items with very low demand, the best policy is generally not to hold any inventory. One would simply order these items as they are demanded.

Example 5.9

A sample of 20 different stock items from Harvey's Specialty Shop is selected at random. These items vary in price from $0.25 to $24.99 and in average yearly demand from 12 to 786. The results of the sampling are presented in Table 5–1. In Table 5–2 the items are ranked in decreasing order of the annual dollar volume of sales. Notice that only 4 of the 20 stock items account for over 80 percent of the annual dollar volume generated by the entire group. Also notice that there are high-priced items in both categories A and C.

TABLE 5–1   **Performance of 20 Stock Items Selected at Random**

Part Number	Price	Yearly Demand	Dollar Volume
5497J	$ 2.25	260	$ 585.00
3K62	2.85	43	122.55
88450	1.50	21	31.50
P001	0.77	388	298.76
2M993	4.45	612	2,723.40
4040	6.10	220	1,342.00
W76	3.10	110	341.00
JJ335	1.32	786	1,037.52
R077	12.80	14	179.20
70779	24.99	334	8,346.66
4J65E	7.75	24	186.00
334Y	0.68	77	52.36
8ST4	0.25	56	14.00
16113	3.89	89	346.21
45000	7.70	675	5,197.50
7878	6.22	66	410.52
6193L	0.85	148	125.80
TTR77	0.77	690	531.30
39SS5	1.23	52	63.96
93939	4.05	12	48.60

TABLE 5–2   **Twenty Stock Items Ranked in Decreasing Order of Annual Dollar Volume**

Part Number	Price	Yearly Demand	Dollar Volume	Cumulative Dollar Volume	
70779	$24.99	334	$8,346.66	$ 8,346.66	A items:
45000	7.70	675	5,197.50	13,544.16	20% of items
2M993	4.45	612	2,723.40	16,267.56	account for 80.1%
4040	6.10	220	1,342.00	17,609.56	of total value.
JJ335	1.32	786	1,037.52	18,647.08	B items:
5497J	2.25	260	585.00	19,232.08	30% of items
TTR77	0.77	690	531.30	19,763.38	account for 14.8%
7878	6.22	66	410.52	20,173.90	of total value.
16113	3.89	89	346.21	20,520.11	
W76	3.10	110	341.00	20,861.11	
P001	0.77	388	298.76	21,159.87	C items:
4J65E	7.75	24	186.00	21,345.87	50% of items
R077	12.80	14	179.20	21,525.07	account for 5.1%
6193L	0.85	148	125.80	21,650.87	of total value.
3K62	2.85	43	122.55	21,773.42	
39SS5	1.23	52	63.96	21,837.38	
334Y	0.68	77	52.36	21,889.74	
93939	4.05	12	48.60	21,938.34	
88450	1.50	21	31.50	21,969.84	
8ST4	0.25	56	14.00	21,983.84	

    This report was very illuminating to Harvey, who had assumed that R077, a packaged goat cheese from the south of France, was a profitable item because of its cost, and had been virtually ignoring TTR77, a domestic chocolate bar. ∎

Exchange Curves

Much of our analysis assumes a single item in isolation and that the relevant cost parameters K, h, and p (or just K and h in the case of service levels) are constants with "correct" values that can be determined. However, it may be more appropriate to think of one or all of the cost parameters as policy variables. The correct values are those that result in a control system with characteristics that meet the needs of the firm and the goals of management. In a typical multiproduct system, the same values of setup cost K and interest rate I are used for all items. We can treat the ratio K/I as a policy variable; if this ratio is large, lot sizes will be larger and the average investment in inventory will be greater. If this ratio is small, the number of annual replenishments will increase.

    To see exactly how a typical exchange curve is derived, consider a deterministic system consisting of n products with varying demand rates $\lambda_1, \ldots, \lambda_n$ and item values $c_1, \ldots, c_n$. If EOQ values are used to replenish stock for each item, then

$$Q_i = \sqrt{\frac{2K\lambda_i}{Ic_i}} \qquad \text{for } 1 \leq i \leq n.$$

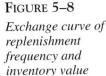

FIGURE 5–8

*Exchange curve of
replenishment
frequency and
inventory value*

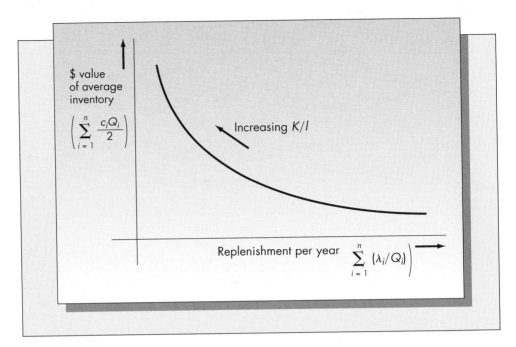

For item i, the cycle time is Q_i/λ_i, so that λ_i/Q_i is the number of replenishments in one year. The total number of replenishments for the entire system is $\Sigma\lambda_i/Q_i$. The average on-hand inventory of item i is $Q_i/2$, and the value of this inventory in dollars is $c_iQ_i/2$. Hence, the total value of the inventory is $\Sigma c_iQ_i/2$.

Each choice of the ratio K/I will result in a different value of the number of replenishments per year and the dollar value of inventory. As K/I is varied, one traces out a curve such as the one pictured in Figure 5–8. An exchange curve such as this one allows management to easily see the trade-off between the dollar investment in inventory and the frequency of stock replenishment.

Exchange curves also can be used to compare various safety stock and service level strategies. As an example, consider a system in which a fill rate constraint is used (i.e., Type 2 service) for all items. Furthermore, suppose that the lead time demand distribution for all items is normal, and each item gets equal service. The dollar value of the safety stock is $\Sigma c_i(R_i - \mu_i)$, and the annual value of back-ordered demand is $\Sigma c_i\lambda_i n(R_i)/Q_i$. A fixed value of the fill rate β will result in a set of values of the control variables $(Q_1, R_1), \ldots, (Q_n, R_n)$, which can be computed by the methods discussed earlier in the chapter. Each set of (Q, R) values yields a pair of values for the safety stock and the back-ordered demand. As the fill rate is increased, the investment in safety stock increases and the value of back-ordered demand decreases. The exchange curve one would obtain is pictured in Figure 5–9. Such an exchange curve is a useful way for management to assess the dollar impact of various service levels.

Example 5.10

Consider the 20 stock items listed in Tables 5–1 and 5–2. Suppose that Harvey, the owner of Harvey's Specialty Shop, is reconsidering his choices of the setup cost of $50 and interest charge of 20 percent. Harvey uses the EOQ formula to compute lot sizes for the 20 items for a range of values of K/I from 50 to 500. The resulting exchange curve appears in Figure 5–10.

FIGURE 5–9

Exchange curve of the investment in safety stock and β

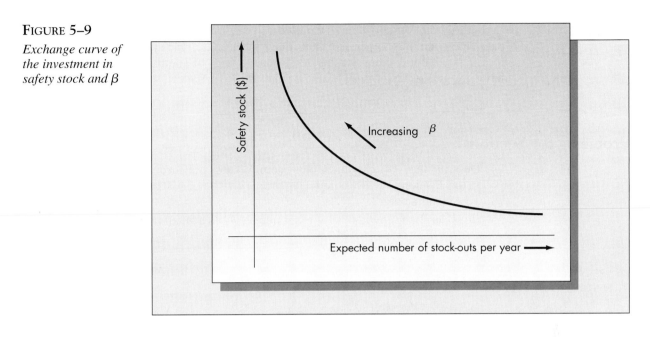

FIGURE 5–10

Exchange curve: Harvey's Specialty Shop

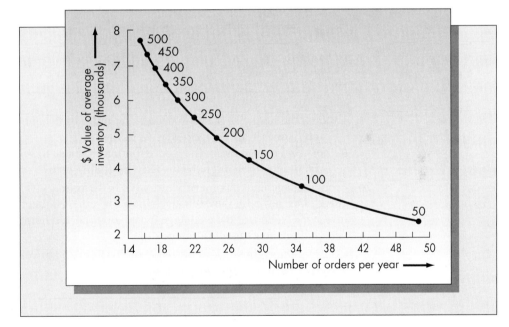

Harvey is currently operating at $K/I = 50/0.2 = 250$, which results in approximately 22 orders per year and an average inventory cost of $5,447 annually. By reducing K/I to 100, the inventory cost for these 20 items is reduced to $3,445 and the order frequency is increased to 34 orders a year. After some thought, Harvey decides that the additional time and bookkeeping expenses required to track an additional 12 orders annually is definitely worth the $2,000 savings in inventory cost. (He is fairly comfortable with the 20 percent interest rate, which means that the true value of his setup cost for ordering is closer to $20 than $50.

In this way the exchange curve can assist in determining the correct value of cost parameters that may otherwise be difficult to estimate.) He also considers moving to $K/I = 50$, but decides that the additional savings of about $1,000 are not worth having to process almost 50 orders a year. ■

Problems for Section 7

24. Describe the ABC classification system. What is the purpose of classifying items in this fashion? What would be the primary value of ABC analysis to a retailer? To a manufacturer?

25. Consider the following list of retail items sold in a small neighborhood gift shop.

Item	Annual Volume	Average Profit per Item
Greeting cards	3,870	$ 0.40
T-shirts	1,550	1.25
Men's jewelry	875	4.50
Novelty gifts	2,050	12.25
Children's clothes	575	6.85
Chocolate cookies	7,000	0.10
Earrings	1,285	3.50
Other costume jewelry	1,900	15.00

   a. Rank the item categories in decreasing order of the annual profit. Classify each in one of the categories as A, B, or C.
   b. For what reason might the store proprietor choose to sell the chocolate cookies even though they might be his least profitable item?

26. From management's point of view, what is the primary value of an exchange curve? Discuss both the exchange curve for replenishment frequency and inventory value and the exchange curve for expected number of stock-outs per year and the investment in safety stock.

27. Consider the eight stock items listed in Problem 25 above. Suppose that the average costs of these item categories are

Item	Cost
Greeting cards	$ 0.50
T-shirts	3.00
Men's jewelry	8.00
Novelty gifts	12.50
Children's clothes	8.80
Chocolate cookies	0.40
Earrings	4.80
Other costume jewelry	12.00

Compare the total number of replenishments per year and the dollar value of the inventory of these items for the following values of the ratio of K/I: 100, 200, 500, 1,000. From the four points obtained, estimate the exchange curve of replenishment frequency and inventory value.

---

*5.8 Overview of Advanced Topics

This chapter has treated only a small portion of inventory models available. Considerable research has been devoted to analyzing far more complex stochastic inventory control problems, but most of this research is beyond the scope of our coverage. This section presents a brief overview of two areas not discussed in this chapter that account for a large portion of the recent research on inventory management.

Multiechelon Systems

Firms involved in the manufacture and distribution of consumer products must take into account the interactions of the various levels in the distribution chain. Typically, items are produced at one or more factories, shipped to warehouses for intermediate storage, and subsequently shipped to retail outlets to be sold to the consumer. We refer to the levels of the system as *echelons*. For example, the factory–warehouse–retailer three-echelon system is pictured in Figure 5–11.

A system of the type pictured in Figure 5–11 is generally referred to as a *distribution* system. In such a system, the demand arises at the lowest echelon (the retailer in this case) and is transmitted up to the higher echelons. Production plans at the factory must be coordinated with orders placed at the warehouse by the retailers. Another type of multiechelon system arising in manufacturing contexts is known as an *assembly* system. In an assembly system, components are combined to form subassemblies, which are eventually combined to form end items (that is, final products). A typical assembly system is pictured in Figure 5–12.

In the assembly system, external demand originates at the end-item level only. Demand for the components arises only when "orders" are placed at higher levels of the system, which are the consequence of end-item production schedules. Materials

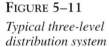

FIGURE 5–11

Typical three-level distribution system

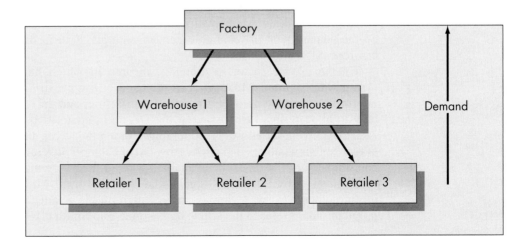

FIGURE 5–12

*Typical three-level
assembly system*

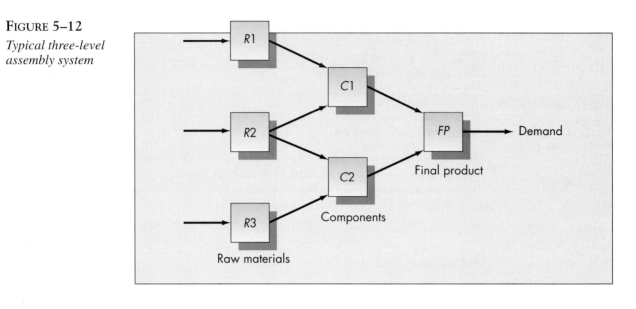

requirements planning methodology is designed to model systems of this type, but does not consider the effect of uncertainty on the final demand. Although there has been some research on this problem, precisely how the uncertainty in the final demand affects the optimal production and replenishment policies of components remains an open question.

One multiechelon system that has received a great deal of attention in the literature (Sherbrooke, 1968) was designed for an application in military logistics. Consider the replenishment problem that arises in maintaining a supply of spare parts to support jet aircraft engine repair. When an engine (or engine component) fails, it is sent to a central depot for repair. The depot maintains its own inventory of engines, so that it can ship a replacement to the base before the repair is completed. The problem is to determine the correct levels of spares at the base and at the depot given the conflicting goals of budget limitations and the desirability of a high system fill rate.

The U.S. military's investment in spare parts is enormous. As far back as 1968 it was estimated that the U.S. Air Force alone has $10 billion invested in repairable item inventories. Considering the effects of inflation and the sizes of the other armed services, the current investment in repairable items in the U.S. military probably exceeds $100 billion. Mathematical inventory models have made a significant impact on the management of repairable item inventories, both in military and nonmilitary environments.

In the private sector, retailing is an area in which multiechelon distribution systems are common. Most major retail chains utilize a distribution center (DC) as an intermediate storage point between the manufacturer and the retail store. Determining how to allocate inventory between the DC and the stores is an important strategic issue. DC inventory allows "risk pooling" among the stores and facilitates redistribution of store inventories that might grow out of balance. Several commercial computer-based inventory packages are available for retail inventory management, perhaps the best known being Inforem, designed and marketed by IBM. Readers interested in a recent comprehensive review of the research on the retailer inventory management problem should refer to the article by Nahmias and Smith (1992).

Perishable Inventory Problems

Although certain types of decay can be accounted for by adjustments to the holding cost, the mathematical models that describe perishability, decay, or obsolescence are generally quite complex.

Typical examples of fixed-life perishable inventories include food, blood, drugs, and photographic film. In order to keep track of the age of units in stock, it is necessary to know the amount of each age level of inventory on hand. This results in a complicated multidimensional system state space. Perishability adds an additional penalty for holding inventory, thus reducing the size of orders placed. The most notable application of the theory of perishable inventory models has been to the management of blood banks.

A somewhat simpler but related problem is that of exponential decay, in which it is assumed that a fixed fraction of the inventory in stock is lost each period. Exponential decay is an accurate description of the changes that take place in volatile liquids such as alcohol and gasoline, and is an exact model of decline in radioactivity of radioactive pharmaceuticals and nuclear fuels.

Obsolescence has been modeled by assuming that the length of the planning horizon is a random variable with a known probability distribution. In practice, such models are valuable only if the distribution of the useful lifetime of the item can be accurately estimated in advance.

A related problem is the management of style goods, such as fashion items in the garment industry. The style goods problem differs from the inventory models considered in this chapter in that not only is the demand itself uncertain, but the distribution of demand is uncertain as well. This is typical of problems for which there is no prior demand history. The procedures suggested to deal with the style goods problem generally involve some type of Bayesian updating scheme to combine current observations of demand with prior estimates of the demand distribution.

5.9 Electronic Data Interchange

An important development in inventory control and supply chain management that is gaining ever wider acceptance is *electronic data interchange* (EDI). EDI is a catchall term for the electronic links between warehouses and stores, vendors and purchasers, and so on, that allow inventory status information to be exchanged on an instantaneous basis. EDI systems have been common in some industries for years and are only beginning to make inroads into other industries. In retailing, for example, EDI systems link stores and central warehouses. Sales information is transmitted on a real-time basis using point-of-sale scanners. Managers are able to determine in-store inventory levels instantly. Links to central warehouses eliminate the need for manual stocktaking and replenishment decisions. In more advanced systems, EDI links vendors directly with customers to facilitate JIT purchasing.

Although EDI has made its mark in inventory management, there is still a long way to go. Industry watchers estimate that about 80,000 American companies with revenues exceeding $5 to $6 million use EDI (Mohan, 1995). This is only about 1 percent of the potential users. Nonusers fear that costs of implementing EDI systems will outweigh the benefits. Also, they are concerned that customers prefer paper-based systems.

SNAPSHOT APPLICATION

Triad's Inventory Systems Meet Markets' Needs

Triad Systems Corporation, headquartered in Livermore, California, develops and markets a range of inventory control and information management products. The company serves every tier of the automotive aftermarket as well as several other wholesale and retail markets. Triad's primary products are turnkey systems. That is, both hardware and software are marketed as an integrated system. Triad's systems (as of this writing) are UNIX-based, utilizing 486 and Pentium processors, and include various peripherals such as POS (point-of-sale) scanners, label and report printers, and bar code scanners.

A typical customer is Bill Clark of Golden Big A Auto Parts in Golden, Colorado. He claims that in less than 15 minutes a week, Triad's system allows him to review the status of 72,000 parts records. One of the features that he finds particularly valuable is the ability of the system to identify "slow movers" so that overstock situations do not occur frequently.

Source: 1994 Annual Report, promotional literature, and personal conversations with Triad employees.

The goal of Triad, founded in the mid 1970s, was to find an application that utilized the Z–80 chip. (The Z–80 chip predated the microprocessor of the first personal computers, which was the 8088 for the IBM PC.) The founders soon began developing inventory systems for the automotive aftermarket, which is still the firm's primary market niche.

One might think a specialized product in a specialized market would have only limited revenue potential. However, Triad's revenues have exceeded $150 million annually since 1992. The company sees growth coming from other parts of the automotive aftermarket (such as service dealers) and expanded sales into other markets and other locations (particularly Europe). In addition to automotive wholesalers, Triad has sold systems to hardware stores, home decorating retailers, and lumber yards.

Triad's new generation of UNIX-based inventory control systems utilize many of the forecasting and inventory control techniques discussed in this book. They include exponential smoothing for tracking the mean and the variance of demand, ABC analysis to identify fast and slow movers, and reorder point and lot sizing calculations to provide minimum cost and a desired level of service.

There are many documented successes of EDI implementation. For example, La Strada Sportswear, a maker of women's active wear that ships more than 50,000 garments weekly, was forced by client demands to enter the EDI arena in the late 1980s. The company reports that EDI facilitates communications, cuts processing times, and streamlines warehousing and distribution functions (AIN, 1995). The grocery industry is one that has been slow to embrace EDI. One exception is the 156-store Giant Food chain, which developed a store delivery system with Nabisco and Frito-Lay driven solely by EDI. The new system covers all phases of store replenishment, including order placement and payment of invoices (Garry, 1995).

What does the theory reveal about the advantages of EDI for reducing inventory-related costs? First, EDI could reduce order lead times. Safety stock is directly proportional to order lead time, so when lead times are reduced, so is safety stock. Second, and perhaps more importantly, EDI should reduce fixed ordering costs. Lot sizes increase as the square root of fixed costs, so that lower fixed costs mean smaller lot sizes. Smaller lot sizes result in more frequent ordering. Consequently, average inventories are reduced. The result is more of a just-in-time–type system.

As EDI systems gain in popularity, legal, as well as logistical, issues need to be addressed. As Murray (1995) points out, the Uniform Commercial Code requires that a signed contract be available for any purchase of goods valued at $500 or more. This means that purchase orders generated electronically via EDI would have to be printed out and signed, thus eliminating some of the advantages of the paperless system. While the EDI transaction is stored in an electronic file, the courts have yet to determine if this constitutes a sufficient record to satisfy the code.

5.10 Historical Notes and Additional Readings

Research on stochastic inventory models was sparked by the national war effort and appears to date back to the 1940s, although the first published material appeared in the early 1950s (see Arrow, Harris, and Marschak, 1951; and Dvorestsky, Kiefer, and Wolfowitz, 1952). The important text by Arrow, Karlin, and Scarf (1958) produced considerable interest in the field.

The model we discuss of a lot size–reorder point system with stock-out cost criterion seems to be attributed to Whitin (1957). The extensions to service levels are attributed to Brown (1967). Brown also treats a variety of other topics of practical importance, including the relationship between forecasting models and inventory control. An excellent discussion of exchange curves and ABC analysis, and more in-depth coverage of a number of the topics we discuss, can be found in Silver and Peterson (1985). Love (1979) provides an excellent summary of many of the issues we treat. Hadley and Whitin (1963) also deal with these topics at a more sophisticated mathematical level.

There has been considerable interest in the development of approximate (s, S) policies. Scarf (1960), and later Iglehart (1963), proved the optimality of (s, S) policies, and Veinott and Wagner (1965) considered methods for computing optimal (s, S) policies. A number of approximation techniques have been suggested. Ehrhardt (1979) considers using regression analysis to fit a grid of optimal policies, whereas Freeland and Porteus (1980) adapt dynamic programming methods to the problem. Porteus (1985) numerically compares the effectiveness of various approximation techniques. Some of the issues regarding lead time uncertainty are treated by Kaplan (1970) and Nahmias (1979), among others.

The original formulation of the standard multiechelon inventory problem came from Clark and Scarf (1960). Extensions have been developed by a variety of researchers, most notably Bessler and Veinott (1966), and, more recently, Federgruen and Zipkin (1984). Retailer-warehouse systems have been studied by Deuermeyer and Schwarz (1981) and Eppen and Schrage (1981). Schmidt and Nahmias (1985) analyzed an assembly system when demand for the final product is random.

The classic work on the type of multiechelon model that has become the basis for the inventory control systems implemented in the military was done by Sherbrooke (1968). A recent extension of Sherbrooke's analysis is considered by Graves (1985). Muckstadt and Thomas (1980) discuss the advantages of implementing multiechelon models in an industrial environment. A comprehensive review of models for managing repairables can be found in Nahmias (1981).

Interest in perishable inventory control models appears to stem from the problem of blood bank inventory control, although food management has a far greater economic impact. Most of the mathematical models for perishables assume that the inventory is issued from stock on an oldest-first basis (although a notable exception is the study by Cohen and Pekelman, 1978). Nahmias (1982) provides a comprehensive review of the research on perishable inventory control.

The style goods problem has been studied by a variety of researchers. Representative work in this area includes that by Murray and Silver (1966), Hartung (1973), and Hausman and Peterson (1972). Those interested in additional readings on stochastic inventory models should refer to the various review articles in the field. Some of the most notable include those of Scarf (1963), Veinott (1966), and Nahmias (1978). A recent collection of up-to-date reviews can be found in Graves et al. (1992).

5.11 Summary

This chapter presented an overview of several inventory control methods when the demand for the item is random. The *newsboy model* is based on the assumption that the product has a useful life of exactly one planning period. We assumed that the demand during the period is a continuous random variable with cumulative distribution function $F(x)$, and that there are specified overage and underage costs of c_o and c_u charged against the inventory remaining on hand at the end of the period or the excess demand, respectively. The optimal order quantity Q^* solves the equation

$$F(Q^*) = \frac{c_u}{c_u + c_o}.$$

Extensions to discrete demand and multiple planning periods also were considered.

From a practical standpoint, the newsboy model has a serious limitation: it does not allow for a positive order setup cost. For that reason, we considered an extension of the EOQ model known as a *lot size–reorder point* model. The key random variable for this case was the demand during the lead time. We showed how optimal values of the decision variables Q and R could be obtained by the iterative solution of two equations. The system operates in the following manner. When the level of on-hand inventory hits R, an order for Q units is placed (which will arrive after the lead time τ). This policy is the basis of many commercial inventory control systems.

Service levels provide an alternative to stock-out costs. Two service levels were considered: Type 1 and Type 2. Type 1 service is the probability of not stocking out in any order cycle, and Type 2 service is the probability of being able to meet a demand when it occurs. Type 2 service, also known as the *fill rate*, is the more natural definition of service for most applications.

Several additional topics for (Q, R) systems also were considered. The *imputed shortage cost* is the effective shortage cost resulting from specification of the service level. Assuming normality, we showed how one transforms the distribution of periodic demand to lead time demand. Finally, we considered the effects of *lead time variability*.

If a setup cost is included in the multiperiod version of the newsboy problem, the optimal form of the control policy is known as an (s, S) *policy*. This means that if the starting inventory in a period, u, is less than or equal to s, an order for $S - u$ units is placed. An effective approximation for the optimal (s, S) policy can be obtained by solving the problem as if it were continuous review to obtain the corresponding (Q, R) policy, and setting $s = R$ and $S = Q + R$. We also discussed the application of service levels to periodic-review systems.

Most real inventory systems involve the management of more than a single product. We discussed several issues that arise when managing a multiproduct inventory system. One of these is the amount of time and expense that should be allocated to the control of each item. The *ABC system* is a method of classifying items by their annual volume of sales. Another issue concerns the correct choice of the cost parameters used in computing inventory control policies. Because many of the cost parameters used in inventory analysis involve managerial judgment and are not easily measured, it would be useful if management could compare the effects of various parameter settings on the performance of the system. A convenient technique for making these comparisons is via *exchange curves*. We discussed two of the most popular exchange curves: (1) the trade-off between the investment in inventory and the frequency of stock replenishment and (2) the trade-off between the investment in safety stock and service levels.

Finally, we concluded the chapter with a discussion of *EDI* (electronic data interchange). EDI refers to the electronic linking of different parts of a logistics system, such as warehouses and stores. By reducing order costs and order lead times, EDI can result in significantly lower inventory levels with no loss of service. Some industries, such as retailing, have embraced EDI, while others, such as groceries, have been slow to adopt the new technology.

Additional Problems on Stochastic Inventory Models

28. An artist's supply shop stocks a variety of different items to satisfy the needs of both amateur and professional artists. In each case described below, what is the appropriate inventory control model that the store should use to manage the replenishment of the item described? Choose your answer from the following list and be sure to explain your answer in each case:

Simple EOQ	Newsboy with service level
Finite production rate	(Q, R) model with stock-out cost
EOQ with quantity discounts	(Q, R) model with Type 1 service level
Resource-constrained EOQ	(Q, R) model with Type 2 service level
Newsboy model	Other type of model

   a. A highly volatile paint thinner is ordered once every three months. Cans not sold during the three-month period are discarded. The demand for the paint thinner exhibits considerable variation from one three-month period to the next.
   b. A white oil-base paint sells at a fairly regular rate of 600 tubes per month and requires a six-week order lead time. The paint store buys the paint for $1.20 per tube.
   c. Burnt sienna oil paint does not sell as regularly or as heavily as the white. Sales of the burnt sienna vary considerably from one month to the next. The useful lifetime of the paint is about two years, but the store sells almost all of the paint prior to the two-year limit.
   d. Synthetic paint brushes are purchased from an East Coast supplier who charges $1.60 for each brush in orders of under 100 and $1.30 for each brush in orders of 100 or greater. The store sells the brushes at a fairly steady rate of 40 per month for $2.80 each.
   e. Camel hair brushes are purchased from the supplier of part (d), who offers a discount schedule similar to the one for the synthetic brushes. The camel hair brushes, however, exhibit considerable sales variation from month to month.

29. Annual demand for number 2 pencils at the campus store is normally distributed with mean 1,000 and standard deviation 250. The store purchases the pencils for 6 cents each and sells them for 20 cents each. There is a two-month lead time from the initiation to the receipt of an order. The store accountant estimates that the cost in employee time for performing the necessary paperwork to initiate and receive an order is $20, and recommends a 22 percent annual interest rate for determining holding cost. The cost of a stock-out is the cost of lost profit plus an additional 20 cents per pencil, which represents the cost of loss of goodwill.

   a. Find the optimal value of the reorder point R assuming that the lot size used is the EOQ.

   b. Find the simultaneous optimal values of Q and R.

   c. Compare the average annual holding, setup, and stock-out costs of the policies determined in parts (a) and (b).

   d. What is the safety stock for this item at the optimal solution?

30. Consider the problem of satisfying the demand for number 2 pencils faced by the campus store mentioned in Problem 29.

   a. Resolve the problem, substituting a Type 1 service level criterion of 95 percent for the stock-out cost.

   b. Resolve the problem, substituting a Type 2 service level criterion of 95 percent for the stock-out cost. Assume that Q is given by the EOQ.

   c. Find the simultaneous optimal values of Q and R assuming a Type 2 service level of 95 percent.

31. Answer the following true or false.

   a. The lead time is always less than the cycle time.

   b. The optimal lot size for a Type 1 service objective of X percent is always less than the optimal lot size for a Type 2 service objective of X percent for the same item.

   c. The newsboy model does not include a fixed order cost.

   d. ABC analysis ranks items according to the annual value of their demand.

   e. For a finite production rate model, the optimal lot size to produce each cycle is equal to the maximum inventory each cycle.

32. One of the products stocked at Weiss's paint store, mentioned in Problem 14, is a certain type of highly volatile paint thinner that, due to chemical changes in the product, has a shelf life of exactly one year. Al Weiss purchases the paint thinner for $20 a gallon can and sells it for $50 a can. The supplier buys back cans not sold during the year for $8 for reprocessing. The demand for this thinner generally varies from 20 to 70 cans a year. Al assumes a holding cost for unsold cans at a 30 percent annual interest rate.

   a. Assuming that all values of the demand from 20 to 70 are equally likely, what is the optimal number of cans of paint thinner for Al to buy each year?

   b. More accurate analysis of the demand shows that a normal distribution gives a better fit of the data. The distribution mean is identical to that used in part (a), and the standard deviation estimator turns out to be 7. What policy do you now obtain?

33. Semicon is a start-up company that produces semiconductors for a variety of applications. The process of burning in the circuits requires large amounts of nitric acid, which has a shelf life of only three months. Semicon estimates that it will need between 1,000 and 3,000 gallons of acid for the next three-month period and assumes that all values in this interval are equally likely. The acid costs them

$150 per gallon. The company assumes a 30 percent annual interest rate for the money it has invested in inventory, and the acid costs the company $35 a gallon to store. (Assume that all inventory costs are attached to the end of the three-month period.) Acid that is left over at the end of the three-month period costs $75 per gallon to dispose of. If the company runs out of acid during the three-month period, it can purchase emergency supplies quickly at a price of $600 per gallon.

a. How many gallons of nitric acid should Semicon purchase? Experience with the marketplace later shows that the demand is closer to a normal distribution, with mean 1,800 and standard deviation 480.

b. Suppose that now Semicon switches to a 94 percent fill rate criterion. How many gallons should now be purchased at the start of each three-month period?

34. *Newsboy simulator.* In order to solve this problem, your spreadsheet program will need to have a function that produces random numbers (@RAND in Lotus 1–2–3 and RAND ( ) in Excel). The purpose of this exercise is to construct a simulation of a periodic-review inventory system with random demand. We assume that the reader is familiar with the fundamentals of Monte Carlo simulation.

Your spreadsheet should allow for cell locations for storing values of the holding cost, the penalty cost, the proportional order cost, the order-up-to point, the initial inventory, and the mean and the standard deviation of periodic demand.

An efficient means of generating an observation from a standard normal variate is the formula

$$Z = [-2 \ln(U_1)]^{0.5} \cos(2\pi U_2),$$

where U_1 and U_2 are two independent draws from a (0, 1) uniform distribution. (See, for example, Fishman, 1973.) Note that two independent calls of @RAND are required. Since Z is approximately standard normal, the demand X is given by

$$X = \sigma Z + \mu$$

where μ and σ are the mean and the standard deviation of one period's demand.

A suggested layout of the spreadsheet is

NEWSBOY SIMULATOR

Holding cost =   Mean demand =

Order cost =   Std. dev. demand =

Penalty cost =   Initial inventory =

                        Order-up-to point =

Period	Starting Inventory	Order Quantity	Demand	Ending Inventory	Holding Cost	Penalty Cost	Order Cost
1							
2							
3							
.							
.							
.							
20							
Totals							

Each time you recalculate, the simulator will generate a different sequence of demands. A set of suggested parameters is

$$h = 2,$$
$$c = 5,$$
$$p = 20,$$
$$\mu = 100,$$
$$\sigma = 20,$$
$$I_0 = 50,$$

Order-up-to point $= 150$.

35. *Using the simulator for optimization.* You will be able to do this problem only if the program you are using does not restrict the size of the spreadsheet. Assume the parameters given in Problem 34 but use $\sigma = 10$. Extend the spreadsheet from Problem 34 to 1,000 rows or more and compute the average cost per period. If the cost changes substantially as you recalculate the spreadsheet, you need to add additional rows.

    Now, experiment with different values of the order-up-to point to find the one that minimizes the average cost per period. Compare your results to the theoretical optimal solution.

36. *Newsboy calculator.* Design a spreadsheet to calculate the optimal order-up-to point for a newsboy problem with normally distributed demands. In order to avoid table look-ups, use the approximation formula

$$Z = 5.0633[F^{0.135} - (1 - F)^{0.135}]$$

    for the inverse of the standard normal distribution function. (This formula is from Ramberg and Schmieser, 1972.) The optimal order-up-to point, y^*, is of the form $y^* = \sigma Z + \mu$.
    a. Assume that all parameters are as given in Problem 34. Graph y^* as a function of p, the penalty cost, for $p = 10, 15, \ldots, 100$. By what percentage does y^* increase if p increases from 10 to 20? from 50 to 100?
    b. Repeat part (a) with $\sigma = 35$. Comment on the effect that the variance in demand has on the sensitivity of y^* to p.

37. A large national producer of canned foods plans to purchase 100 combines that are to be customized for its needs. One of the parts used in the combine is a replaceable blade for harvesting corn. Spare blades can be purchased at the time the order is placed for $100 each, but will cost $1,000 each if purchased at a later time because a special production run will be required.

    It is estimated that the number of replacement blades required by a combine over its useful lifetime can be closely approximated by a normal distribution with mean 18 and standard deviation 5.2. The combine maker agrees to buy back unused blades for $20 each. How many spare blades should the company purchase with the combines?

38. Crazy Charlie's, a discount stereo shop, uses simple exponential smoothing with a smoothing constant of $\alpha = .2$ to track the mean and the MAD of monthly item demand. One particular item, a stereo receiver, has experienced the following sales over the last three months: 126, 138, 94.

    Three months ago the computer had stored values of mean $= 135$ and MAD $= 18.5$.

   a. Using the exponential smoothing equations given in Section 1 of this chapter, determine the current values for the mean and the MAD of monthly demand. (Assume that the stored values were computed *prior* to observing the demand of 126.)

   b. Suppose that the order lead time for this particular item is 10 weeks (2.5 months). Assuming a normal distribution for monthly demand, determine the current estimates for the mean and the standard deviation of lead time demand.

   c. This particular receiver is ordered directly from Japan, and as a result there has been considerable variation in the replenishment lead time from one order cycle to the next. Based on an analysis of past order cycles, it is estimated that the standard deviation of the lead time is 3.3 weeks. All other relevant figures are given in parts (*a*) and (*b*). Find the mean and the standard deviation of lead time demand in this case.

   d. If Crazy Charlie's uses a Type 1 service objective of 98 percent to control the replenishment of this item, what is the value of the reorder level? (Assume that the lead time demand has a normal distribution.)

39. The home appliance department of a large department store is using a lot size–reorder point system to control the replenishment of a particular model of FM table radio. The store sells an average of 10 radios each week. Weekly demand follows a normal distribution with variance 26.

     The store pays $20 for each radio, which it sells for $75. Fixed costs of replenishment amount to $28. The accounting department recommends a 20-percent interest rate for the cost of capital. Storage costs amount to 3 percent and breakage to 2 percent of the value of each item.

     If a customer demands the radio when it is out of stock, the customer will generally go elsewhere. Loss-of-goodwill costs are estimated to be about $25 per radio. Replenishment lead time is three months.

   a. If lot sizes are based on the EOQ formula, what reorder level should be used for the radios?

   b. Find the optimal values of (Q, R).

   c. Compare the average annual costs of holding, ordering, and stock-out for the policies that you found in parts (*a*) and (*b*).

   d. Re-solve the problem using Equations (1) and (2′) rather than (1) and (2). What is the effect of including lost sales explicitly?

40. Re-solve the problem faced by the department store mentioned in Problem 39, replacing the stock-out cost with a 96 percent Type 1 service level.

41. Re-solve the problem faced by the department store mentioned in Problem 39, replacing the stock-out cost with a 96 percent Type 2 service level. What is the imputed shortage cost?

42. Consider the equation giving the expected average annual cost of the policy (Q, R) in a continuous-review inventory control system from Section 4:

$$G(Q, R) = h\left(\frac{Q}{2} + r - \lambda T\right) + \frac{K\lambda}{Q} + \frac{P\lambda n(R)}{Q}.$$

     Design a spreadsheet to compute $G(Q, R)$ for a range of values of $Q \geq$ EOQ and $R \geq \mu$. Use the following approximation formula for $L(z)$ to avoid table look-ups:

$$L(z) = \exp(-0.92 - 1.19z - 0.37z^2).$$

(This formula is from Parr, 1972.) Store the problem parameters c, h, p, μ, σ, and λ in cell locations. Visually search through the tabled values of $G(Q, R)$ to discover the minimum value and estimate the optimal (Q, R) values in this manner. Compare your results to the true optimal found from manual calculation.

a. Solve Example 5.4.

b. Solve Problem 13 in this manner.

c. Solve Problem 14 in this manner.

43. The daily demand for a spare engine part is a random variable with a distribution, based on past experience, given by

Number of Demands per Day	Probability
0	.21
1	.38
2	.19
3	.14
4	.08

The part is expected to be obsolete after 400 days. Assume that demands from one day to the next are independent. The parts cost $1,500 each when acquired in advance of the 400-day period and $5,000 each when purchased on an emergency basis during the 400-day period. Holding costs for unused parts are based on a daily interest rate of 0.08 percent. Unused parts can be scrapped for 10 percent of their purchase price. How many parts should be acquired in advance of the 400-day period? (Hint: Let $D_1, D_2, \ldots, D_{400}$ represent the daily demand for the part. Assume each D_i has mean μ and variance σ^2. The central limit theorem says that the total demand for the 400-day period, ΣD_i is approximately normally distributed with mean 400μ and variance $400\sigma^2$.)

44. Cassorla's Clothes sells a large number of white dress shirts. The shirts, which bear the store label, are shipped from a manufacturer in New York City. Hy Cassorla, the proprietor, says, "I want to be sure that I never run out of dress shirts. I always try to keep at least a two months' supply in stock. When my inventory drops below that level, I order another two-month supply. I've been using that method for 20 years, and it works."

The shirts cost $6 each and sell for $15 each. The cost of processing an order and receiving new goods amounts to $80, and it takes three weeks to receive a shipment. Monthly demand is approximately normally distributed with mean 120 and standard deviation 32. Assume a 20 percent annual interest rate for computing the holding cost.

a. What value of Q and R is Hy Cassorla using to control the inventory of white dress shirts?

b. What fill rate (Type 2 service level) is being achieved with the current policy?

c. Based on a 99 percent fill rate criterion, determine the optimal values of Q and R that he should be using. (Assume four weeks in a month for your calculations.)

d. Determine the difference in the average annual holding and setup costs between the policies in (b) and (c).

   e. Estimate how much time would be required to pay for a $25,000 inventory control system, assuming that the dress shirts represent 5 percent of Hy's annual business and that similar savings could be realized on the other items as well.

APPENDIX 5–A   NOTATIONAL CONVENTIONS AND PROBABILITY REVIEW

Demand will be denoted by D, which is assumed to be a random variable. The cumulative distribution function (cdf) of demand is $F(x)$ and is defined by

$$F(x) = P\{D \le x\} \qquad \text{for } -\infty < x < +\infty.$$

When D is continuous, the probability density function (pdf) of demand, $f(x)$, is defined by

$$f(x) = \frac{dF(x)}{dx}.$$

When D is discrete, $f(x)$ is the probability function (pf) defined by

$$f(x) = P\{X = x\} = F(x) - F(x - 1).$$

Note that in the continuous case the density function is not a probability and the value of $f(x)$ is not necessarily less than 1, although it is always true that $f(x) \ge 0$ for all x.

The expected value of demand, $E(D)$, is defined as

$$E(D) = \int_{-\infty}^{+\infty} x f(x)\, dx$$

in the continuous case, and

$$E(D) = \sum_{x=-\infty}^{+\infty} x f(x)$$

in the discrete case.

We use the symbol μ to represent the expected value of demand ($E(D) = \mu$). In what follows we assume that D is continuous; similar formulas hold in the discrete case. Let $g(x)$ be any real-valued function of the real variable x. Then

$$E(g(D)) = \int_{-\infty}^{+\infty} g(x) f(x)\, dx.$$

In particular, let $g(D) = \max(0, Q - D)$. Then

$$E(g(D)) = \int_{-\infty}^{+\infty} \max(0, Q - x) f(x)\, dx.$$

Because demand is nonnegative, it must be true that $f(x) = 0$ for $x < 0$. Furthermore, when $x > Q$, $\max(0, Q - x) = 0$, so we may write

$$E(g(D)) = \int_{0}^{Q} (Q - x) f(x)\, dx.$$

In the analysis of the newsboy model, Leibniz's rule is used to determine the derivative of $G(Q)$. According to Leibniz's rule:

$$\frac{d}{dy} \int_{a_1(y)}^{a_2(y)} h(x, y)\, dx = \int_{a_1(y)}^{a_2(y)} [\partial h(x, y)/\partial y]\, dx + h(a_2(y), y)a_2'(y) - h(a_1(y), y)a_1'(y).$$

APPENDIX 5–B   ADDITIONAL RESULTS AND EXTENSIONS FOR THE NEWSBOY MODEL

1. Interpretation of the Overage and Underage Costs for the Single Period Problem

Define

S = Selling price of the item.

c = Variable cost of the item.

h = Holding cost per unit of inventory remaining in stock at the end of the period.

p = Loss-of-goodwill cost plus bookkeeping expense (charged against the number of back orders on the books at the end of the period).

We show how c_u and c_o should be interpreted in terms of these parameters. As earlier, let Q be the order quantity and D the demand during the period. Assume without loss of generality that starting inventory is zero. Then the cost incurred at the end of the period is

$$cQ + h \max(Q - D, 0) + p \max(D - Q, 0) - S \min(Q, D).$$

The expected cost is

$$G(Q) = cQ + h\int_0^Q (Q - x)f(x)\, dx + p \int_Q^\infty (x - Q)f(x)\, dx$$
$$- S \int_0^Q xf(x)\, dx - SQ \int_Q^\infty f(x)\, dx.$$

Using

$$\int_0^Q xf(x)\, dx = \int_0^\infty xf(x)\, dx - \int_Q^\infty xf(x)\, dx = \mu - \int_Q^\infty xf(x)\, dx,$$

the expected cost may be written

$$G(Q) = cQ + h\int_0^Q (Q - x)f(x)\, dx + (p + S) \int_Q^\infty (x - Q)f(x)\, dx - S\mu.$$

The optimal order quantity satisfies

$$G'(Q) = 0$$

or

$$c + hF(Q) - (p + S)(1 - F(Q)) = 0,$$

which results in

$$F(Q) = \frac{p + S - c}{p + S + h}.$$

Setting $c_u = p + S - c$ and $c_o = h + c$ gives the critical ratio in the form $c_u/(c_u + c_o)$.

2. Extension to Infinite Horizon Assuming Full Back-Ordering of Demand

Let $D_1, D_2, \ldots$ be an infinite sequence of demands. Assume that the demands are independent identically distributed random variables having common distribution function $F(x)$ and density $f(x)$. The policy is to order up to Q each period. As all excess demand is back-ordered, the order quantity in any period is exactly the previous period's demand. The number of units sold will also equal the demand. In order to see this, consider the number of units sold in successive periods:

Number of units sold in period 1 $= \min(Q, D_1)$,
Number of units sold in period 2 $= \max(D_1 - Q, 0) + \min(Q, D_2)$,
Number of units sold in period 3 $= \max(D_2 - Q, 0) + \min(Q, D_3)$,

and so on.

These relationships follow because back ordering of excess demand means that the sales are made in the subsequent period. Now, notice that

$$\min(Q, D_i) + \max(D_i - Q, 0) = D_i \qquad \text{for } i = 1, 2, \ldots,$$

which follows from considering the cases $Q < D_i$ and $Q \geq D_i$.

Hence, the expected cost over n periods is

$$cQ + (c - S)E(D_1 + D_2 + \cdots + D_{n-1}) - (S)E[\min(Q, D_n)] + nL(Q)$$
$$= cQ + (c - S)(n - 1)\mu - (S)E[\min(Q, D_n)] + nL(Q)$$

where

$$L(Q) = h \int_0^Q (Q - x)f(x)\,dx + p \int_Q^\infty (x - Q)f(x)\,dx.$$

Dividing by n and letting $n \to \infty$ gives the average cost over infinitely many periods as

$$(c - S)\mu + L(Q).$$

The optimal value of Q occurs where $L'(Q) = 0$, which results in

$$F(Q) = p/(p + h).$$

3. Extension to Infinite Horizon Assuming Lost Sales

If excess demand is lost rather than back-ordered, then the argument above is no longer valid. The number of units sold in period 1 is $\min(Q, D_1)$, which is also the number of units ordered in period 2; the number of units sold in period 2 is $\min(Q, D_2)$ (since there is no back-ordering of excess demand), which is also the number of units ordered in period 3; and so on. As shown in Section 1 of this appendix,

$$E[\min(Q, D)] = \mu - \int_Q^\infty (x - Q)f(x)\,dx.$$

Hence, it follows that the expected cost over n periods is given by

$$cQ + [(n-1)c - nS]\left[\mu - \int_Q^\infty (x - Q)f(x)\,dx\right] + nL(Q).$$

If we divide by n and let $n \to \infty$, we obtain the following expression for the average cost per period:

$$(c - S)\left[\mu - \int_Q^\infty (x - Q)f(x)\,dx\right] + L(Q).$$

Differentiating with respect to Q and setting the result equal to zero gives the following condition for the optimal Q:

$$F(Q) = \frac{p + S - c}{p + S + h - c}.$$

Setting $c_u = p + S - c$ and $c_o = h$ gives the critical ratio in the form $c_u/(c_u + c_o)$. Hence, we interpret c_u as the cost of the loss of goodwill plus the lost profit per sale, and c_o as the cost of holding only.

APPENDIX 5–C   DERIVATION OF THE OPTIMAL (Q, R) POLICY

From Section 4, the objective is to find values of the variables Q and R to minimize the function

$$G(Q, R) = h(Q/2 + R - \lambda\tau) + K\lambda/Q + p\lambda n(R)/Q. \tag{1}$$

Because this function is to be minimized with respect to the two variables (Q, R), a necessary condition for optimality is that $\partial G/\partial Q = \partial G/\partial R = 0$. The two resulting equations are

$$\frac{\partial G}{\partial Q} = \frac{h}{2} - \frac{K\lambda}{Q^2} - \frac{p\lambda n(R)}{Q^2} = 0, \tag{2}$$

$$\frac{\partial G}{\partial R} = h + p\lambda n'(R)/Q = 0. \tag{3}$$

Note that since $n(R) = \int_R^\infty (x - R)f(x)\,dx$, one can show that

$$n'(R) = -(1 - F(R)).$$

From Equation (2) we obtain

$$\frac{1}{Q^2}[K\lambda + p\lambda n(R)] = \frac{h}{2}$$

or

$$Q^2 = \frac{2K\lambda + 2p\lambda n(R)}{h},$$

which gives

$$Q = \sqrt{\frac{2\lambda[K + pn(R)]}{h}}. \tag{4}$$

From Equation (3) we obtain

$$h + p\lambda[-(1 - F(R))]/Q = 0,$$

which gives

$$1 - F(R) = Qh/p\lambda. \tag{5}$$

APPENDIX 5–D   PROBABILITY DISTRIBUTIONS FOR INVENTORY MANAGEMENT

In this chapter we have made frequent reference to the normal distribution as a model for demand uncertainty. Although the normal distribution certainly dominates applications, it is not the only choice available. In fact, it could be a poor choice in some circumstances. In this appendix we discuss other distributions for modeling demand uncertainty.

1. The Poisson Distribution as a Model of Demand Uncertainty

One situation in which the normal distribution may be inappropriate is for slow-moving items, that is, ones with small demand rates. Because the normal is an infinite distribution, when the mean is small it is possible that a substantial portion of the density curve extends into the negative half line. This could give poor results for safety stock calculations. A common choice for modeling slow movers is the Poisson distribution. The Poisson is a discrete distribution defined on the positive half line only. Let X have the Poisson distribution with parameter μ. Then

$$f(x) = \frac{e^{-\mu}\mu^x}{x!} \qquad \text{for } x = 0, 1, 2, \ldots.$$

(The derivation of the Poisson distribution and its relationship to the exponential distribution are discussed in detail in Section 3 of Chapter 12.)

An important feature of the Poisson distribution is that both the mean and the variance are equal to μ (giving $\sigma = \sqrt{\mu}$). Hence, it should be true that the observed standard deviation of periodic demand (or lead time demand) is close to the square root of the mean periodic demand (or lead time demand) for the Poisson to be an appropriate choice. Table A–3 at the back of the book is a table of the complementary cumulative Poisson distribution. This table allows one to compute optimal policies for the newsboy and the (Q, R) models assuming that demand follows the Poisson distribution.

We show how to find optimal policies for both the newsboy and the (Q, R) models when demand is Poisson. For the newsboy case, one simply applies the method outlined in Section 3 for discrete demand and obtains the probabilities from Table A–3. We illustrate with an example.

Example 5D.1

Consider Mac's newsstand, discussed in several examples in Chapter 5. A few regular customers have asked Mac to stock a particular stereo magazine. Mac has agreed even though sales have been slow. Based on past data, Mac has found that a Poisson distribution with mean 5 closely fits the weekly sales pattern for the magazine. Mac purchases the magazines for $0.50 and sells them for $1.50. He returns unsold copies to his supplier, who pays Mac $0.10 for each. Find the number of these magazines he should purchase from his supplier each week.

Solution

The overage cost is $c_o = \$0.50 - \$0.10 = \$0.40$ and the underage cost is $c_u = \$1.50 - \$0.50 = \$1.00$. It follows that the critical ratio is $c_u/(c_o + c_u) = 1/1.4 = .7143$. As Table A–3 gives the complementary cumulative distribution, we subtract table entries from 1 to obtain values of the cumulative distribution function. From the table we see that

$$F(5) = 1 - .5595 = .4405,$$
$$F(6) = 1 - .3840 = .6160,$$
$$F(7) = 1 - .2378 = .7622.$$

Because the critical ratio is between $F(6)$ and $F(7)$, we move to the larger value, thus giving an optimal order quantity of 7 magazines.

It is not difficult to calculate optimal (Q, R) policies when the demand follows a Poisson distribution, but the process requires an iterative solution similar to that for the normal distribution described in Section 5 of this chapter. Define $P(x)$ as the complementary cumulative probability of the Poisson. That is,

$$P(x) = \sum_{k=x}^{\infty} f(k).$$

Table A–3 gives values of $P(x)$. It can then be shown that for the Poisson distribution

$$n(R) = \mu P(R) - RP(R + 1),$$

which means that all the information required to compute optimal policies appears in Table A–3.[1] That is, special tables for computing $n(R)$ are not required in this case. This relationship allows one to compute the (Q, R) policy using the pair of equations (1) and (2) from Section 4 of this chapter. ■

Example 5D.2

A department store uses a (Q, R) policy to control its inventories. Sales of a pocket FM radio average 1.4 radios per week. The radio costs the store $50.00, and fixed costs of replenishment amount to $20.00. Holding costs are based on a 20 percent annual interest charge, and the store manager estimates a $12.50 cost of lost profit and loss of goodwill if the radio is demanded when out of stock. Reorder lead time is 2 weeks, and statistical studies have shown that demand over the lead time is accurately described by a Poisson distribution. Find the optimal lot sizes and reorder points for this item.

Solution

The relevant parameters for this problem are

$\lambda = (1.4)(52) = 72.8$.
$h = (50)(0.2) = 10.0$.
$K = \$20$.
$\mu = (1.4)(2) = 2.8$.
$p = \$12.50$.

---

[1]See Hadley and Whitin (1963), p. 441.

To start the solution process, we compute the EOQ. It is

$$\text{EOQ} = \sqrt{\frac{2K\lambda}{h}} = \sqrt{\frac{(2)(20)(72.8)}{10}} = 17.1.$$

The next step is to find R solving

$$P(R) = Qh/p\lambda.$$

Substituting the EOQ for Q and solving gives $P(R_0) = .1868$. From Table A–3, we see that $R = 4$ results in $P(R) = .3081$ and $R = 5$ results in $P(R) = .1523$. Assuming the conservative strategy of rounding to the *larger R*, we would choose $R = 5$ and $P(R) = .1523$. It now follows that

$$n(R_0) = (2.8)(.1523) - (5)(.0651) = 0.1009.$$

Hence Q_1 is

$$Q_1 = \sqrt{\frac{(2)(72.8)[20 + (12.5)(0.1009)]}{10}} = 17.6.$$

It follows that $P(R_1) = .1934$, giving $R_1 = R_0 = 5$. Hence the solution has converged, because successive R (and, hence, Q) values are the same. The optimal solution is $(Q, R) = (18, 5)$. ■

2. The Laplace Distribution

A continuous distribution, which has been suggested for modeling slow-moving items or ones with more variance in the tails than the normal, is the Laplace distribution. The Laplace distribution has been called the pseudo-exponential distribution, as it is mathematically an exponential distribution with a symmetric mirror image. (The exponential distribution is discussed at length in Chapter 12 in the context of reliability management.)

The mathematical form of the Laplace pdf is

$$f(x) = \frac{1}{2\theta} \exp(-|x - \mu|/\theta) \qquad \text{for } -\infty < x < +\infty.$$

Because the pdf is symmetric around μ, the mean is μ. The variance is $2\theta^2$. The Laplace distribution is also a reasonable model for slow-moving items, and is an alternative to the normal distribution for fast-moving items when there is more spread in the tails of the distribution than the normal distribution gives. As far as inventory applications are concerned, Presutti and Trepp (1970) noticed that it significantly simplified the calculation of the optimal policy for the (Q, R) model.

One can show that for any value of $R > \mu$, the complementary cumulative distribution ($P(R)$) and the loss integral ($n(R)$) are given by

$$P(R) = 0.5 \exp(-[(R - \mu)/\theta])$$
$$n(R) = 0.5\theta \exp(-[(R - \mu)/\theta])$$

so that the ratio $n(R)/P(R) = \theta$, independent of R. This fact results in a very simple solution for the (Q, R) model. Recall that the two equations defining the optimal policy were

$$Q = \sqrt{\frac{2\lambda[K + pn(R)]}{h}}$$

$$P(R) = Qh/p\lambda \qquad \text{(where } P(R) = 1 - F(R)).$$

The simplification is achieved by using the SOQ formula presented in Section 5 of this chapter. The SOQ representation is an alternative representation for Q that does not include the stock-out cost, p. Using $P(R) = 1 - F(R)$, the SOQ formula is

$$Q = \frac{n(R)}{P(R)} + \sqrt{\frac{2K\lambda}{h} + \left(\frac{n(R)}{P(R)}\right)^2}$$

$$= \theta + \sqrt{\frac{2K\lambda}{h} + \theta^2}$$

independently of R! Hence, the optimal Q and R can be found in a simple one-step calculation. When using a cost model, find the value of $P(R)$ from $P(R) = Qh/p\lambda$. Then, using the representation $P(R) = \exp[-(R - \mu)/\theta)]/2$, it follows that $R = -\theta \ln(2P(R)) + \mu$. Using a service level model, one simply uses the formulas for R given in Section 5. We illustrate with an example.

Example 5D.3

Consider Example 5D.2 above, but suppose that we wish to use the Laplace distribution to compute the optimal policy rather than the Poisson distribution. As both the mean and the variance of the lead time demand are 2.8, we set $\mu = 2.8$ and $2\theta^2 = 2.8$, giving $\theta = 1.1832$. It follows that

$$Q = 1.1832 + \sqrt{\frac{(2)(20)(72.8)}{10} + (1.1832)^2} = 18.3.$$

As with the previous example, we obtain $P(R) = .1934$. Using $R = -\theta \ln(2P(R)) + \mu$ and substituting $P(R) = .1934$ gives $R = 3.92$, which we round to 4. Notice that this solution differs slightly from the one we obtained assuming Poisson demand. However, recall that using the Poisson distribution we found that the optimal R was between 4 and 5, which we chose to round to 5 to be conservative. ∎

3. Other Lead Time Demand Distributions

Many other probability distributions have been recommended for modeling lead time demand. Some of these include the negative binomial distribution, the gamma distribution, the logarithmic distribution, the Pearson distribution, and others. (See Silver and Peterson, 1985, p. 289, for a list of articles that discuss these distributions in the context of inventory management.) The normal distribution probably accounts for the lion's share of applications, with the Poisson accounting for almost all the rest. We included the Laplace distribution because of the interesting property that optimal (Q, R) policies can be found without an iterative solution, and because it could be a good alternative to the Poisson for low-demand items.

APPENDIX 5–E GLOSSARY OF NOTATION FOR CHAPTER 5

α = Desired level of Type 1 service.
β = Desired level of Type 2 service.
c_o = Unit overage cost for newsboy model.
c_u = Unit underage cost for newsboy model.

D = Random variable corresponding to demand. It is the one-period demand for the newsboy model and the lead time demand for the (Q, R) model.

EOQ = Economic order quantity.

$F(t)$ = Cumulative distribution function of demand. Values of the standard normal CDF appear in Table A–4 at the end of the book.

$f(t)$ = Probability density function of demand.

$G(Q)$ = Expected one-period cost associated with lot size Q (newsboy model).

$G(Q, R)$ = Expected average annual cost for the (Q, R) model.

h = Holding cost per unit per unit time.

I = Annual interest rate used to compute holding cost.

K = Setup cost or fixed order cost.

λ = Expected demand rate (units per unit time).

$L(z)$ = Normalized loss function. Used to compute $n(R) = \sigma L(z)$. Tabled values of $L(z)$ appear in Table A–4 at the end of the book.

μ = Mean demand (lead time demand for (Q, R) model).

$n(R)$ = Expected number of stock-outs in the lead time for (Q, R) model.

p = Penalty cost per unit for not satisfying demand.

Q = Lot size or size of the order.

S = Safety stock; $S = R - \lambda\tau$ for (Q, R) model.

SOQ = Service order quantity.

T = Expected cycle time; mean time between placement of successive orders.

τ = Order lead time.

Type 1 service = Proportion of cycles in which all demand is satisfied.

Type 2 service = Proportion of demands satisfied.

Bibliography

AIN. "EDI Keeps Sportswear Shipments in Top Shape." *Automatic I.D. News,* January 1995, pp. 38–39.

Arrow, K. A.; T. E. Harris; and T. Marschak. "Optimal Inventory Policy." *Econometrica* 19 (1951), pp. 250–72.

Arrow, K. A.; S. Karlin; and H. E. Scarf, eds. *Studies in the Mathematical Theory of Inventory and Production.* Stanford, CA: Stanford University Press, 1958.

Bessler, S. A., and A. F. Veinott Jr. "Optimal Policy for a Dynamic Multiechelon Inventory Model." *Naval Research Logistics Quarterly* 13 (1966), pp. 355–89.

Brown, R. G. *Statistical Forecasting for Inventory Control.* New York: McGraw-Hill, 1959.

Brown, R. G. *Decision Rules for Inventory Management.* Hinsdale, IL: Dryden Press, 1967.

Clark, A., and H. E. Scarf. "Optimal Policies for a Multiechelon Inventory Problem." *Management Science* 6 (1960), pp. 475–90.

Cohen, M. A., and D. Pekelman. "LIFO Inventory Systems." *Management Science* 24 (1978), pp. 1150–62.

Deuermeyer, B. L., and L. B. Schwarz. "A Model for the Analysis of System Service Level in Warehouse-Retailer Distribution Systems: Identical Retailer Case." In *Multilevel Production/Inventory Control Systems: Theory and Practice,* ed. L. B. Schwarz, pp. 163–94. Amsterdam: North Holland, 1981.

Dvoretsky, A.; J. Kiefer; and J. Wolfowitz. "The Inventory Problem: I. Case of Known Distributions of Demand." *Econometrica* 20 (1952), pp. 187–222.

Ehrhardt, R. "The Power Approximation for Computing (s, S) Inventory Policies." *Management Science* 25 (1979), pp. 777–86.

Eppen, G., and L. Schrage. "Centralized Ordering Policies in a Multi-Warehouse System with Lead Times and Random Demand." In *Multilevel Production/Inventory Control Systems: Theory and Practice,* ed. L. B. Schwarz, pp. 51–68. Amsterdam: North Holland, 1981.

Federgruen, A., and P. Zipkin. "Computational Issues in an Infinite Horizon Multi-Echelon Inventory Model." *Operations Research* 32 (1984), pp. 818–36.

Fishman, G. S. *Concepts and Methods in Discrete Event Digital Simulation.* New York: John Wiley & Sons, 1973.

Freeland, J. R., and E. L. Porteus. "Evaluating the Effectiveness of a New Method of Computing Approximately Optimal (s, S) Inventory Policies." *Operations Research* 28 (1980), pp. 353–64.

Garry, M. "Completing the Loop." *Progressive Grocer* 74 (February 1995), pp. 75–80.

Graves, S. C. "A Multiechelon Inventory Model for a Repairable Item with One for One Replenishment." *Management Science* 31 (1985), pp. 1247–56.

Graves, S. C.; A. H. G. Rinnooy Kan; and P. Zipkin, eds. *Handbooks in Operations Research and Management Science.* Volume 4, *Logistics of Production and Inventory.* Amsterdam: Elsevier Science Publishers, 1992.

Hadley, G. J., and T. M. Whitin. *Analysis of Inventory Systems.* Englewood Cliffs, NJ: Prentice Hall, 1963.

Hartung, P. "A Simple Style Goods Inventory Model." *Management Science* 19 (1973), pp. 1452–58.

Hausman, W. H., and R. Peterson. "Multiproduct Production Scheduling for Style Goods with Limited Capacity, Forecast Revisions, and Terminal Delivery." *Management Science* 18 (1972), pp. 370–83.

Iglehart, D. L., "Optimality of (s, S) Inventory Policies in the Infinite Horizon Dynamic Inventory Problem." *Management Science* 9 (1963), pp. 259–67.

Kaplan, R. "A Dynamic Inventory Model with Stochastic Lead Times." *Management Science* 16 (1970), pp. 491–507.

Love, S. F. *Inventory Control.* New York: McGraw-Hill, 1979.

Mohan, S. "EDI's Move to Prime Time Stalled by Cost Perception." *Computerworld* 29 (February 20, 1995), p. 91.

Muckstadt, J. M., and L. J. Thomas. "Are Multiechelon Inventory Models Worth Implementing in Systems with Low Demand Rate Items?" *Management Science* 26 (1980), pp. 483–94.

Murray, G. R., and E. A. Silver. "A Bayesian Analysis of the Style Goods Inventory Problem." *Management Science* 12 (1966), pp. 785–97.

Murray, J. E. "The EDI Explosion." *Purchasing* 118 (February 16, 1995), pp. 28–30.

Nahmias, S. "Inventory Models." In *The Encyclopedia of Computer Science and Technology, Volume 9,* ed. J. Belzer, A. G. Holzman, and A. Kent, pp. 447–83. New York: Marcel Dekker, 1978.

Nahmias, S. "Simple Approximations for a Variety of Dynamic Lead Time Lost-Sales Inventory Models." *Operations Research* 27 (1979), pp. 904–24.

Nahmias, S. "Managing Reparable Item Inventory Systems: A Review." In *Multilevel Production/Inventory Control Systems: Theory and Practice,* ed. L. B. Schwarz, pp. 253–77. Amsterdam: North Holland, 1981.

Nahmias, S. "Perishable Inventory Theory: A Review." *Operations Research* 30 (1982), pp. 680–708.

Nahmias, S., and S. Smith. "Mathematical Models of Retailer Inventory Systems: A Review." In *Perspectives in Operations Management: Essays in Honor of Elwood S. Buffa,* ed. R. K. Sarin. Boston: Kluwer, 1992.

Parr, J. O. "Formula Approximations to Brown's Service Function." *Production and Inventory Management* 13 (1972), pp. 84–86.

Porteus, E. L. "Numerical Comparisons of Inventory Policies for Periodic Review Systems." *Operations Research* 33 (1985), pp. 134–52.

Presutti, V., and R. Trepp. "More Ado about EOQ." *Naval Research Logistics Quarterly* 17 (1970), pp. 243–51.

Ramberg, J. S., and B. W. Schmeiser. "An Approximate Method for Generating Symmetric Random Variables." *Communications of the ACM* 15 (1972), pp. 987–89.

Scarf, H. E. "The Optimality of (s, S) Policies in the Dynamic Inventory Problem." In *Mathematical Methods in the Social Sciences,* ed. K. J. Arrow, S. Karlin, and P. Suppes. Stanford, CA: Stanford University Press, 1960.

Scarf, H. E. "Analytical Techniques in Inventory Theory." In *Multi-Stage Inventory Models and Techniques,* ed. H. E. Scarf, D. M. Gilford, and M. W. Shelly. Stanford, CA: Stanford University Press, 1963.

Schmidt, C. P., and S. Nahmias. "Optimal Policy for a Single Stage Assembly System with Stochastic Demand." *Operations Research* 33 (1985), pp. 1130–45.

Schwarz, L. B., ed. *Multilevel Production/Inventory Control Systems: Theory and Practice.* Amsterdam: North Holland, 1981.

Sherbrooke, C. C. "METRIC: Multiechelon Technique for Recoverable Item Control." *Operations Research* 16 (1968), pp. 122–41.

Silver, E. A., and R. Peterson. *Decision Systems for Inventory Management and Production Planning.* 2nd ed. New York: John Wiley & Sons, 1985.

Veinott, A. F. "The Status of Mathematical Inventory Theory." *Management Science* 12 (1966), pp. 745–77.

Veinott, A. F., and H. M. Wagner. "Computing Optimal (s, S) Inventory Policies." *Management Science* 11 (1965), pp. 525–52.

Whitin, T. M. *The Theory of Inventory Management.* Rev. ed. Princeton, NJ: Princeton University Press, 1957.

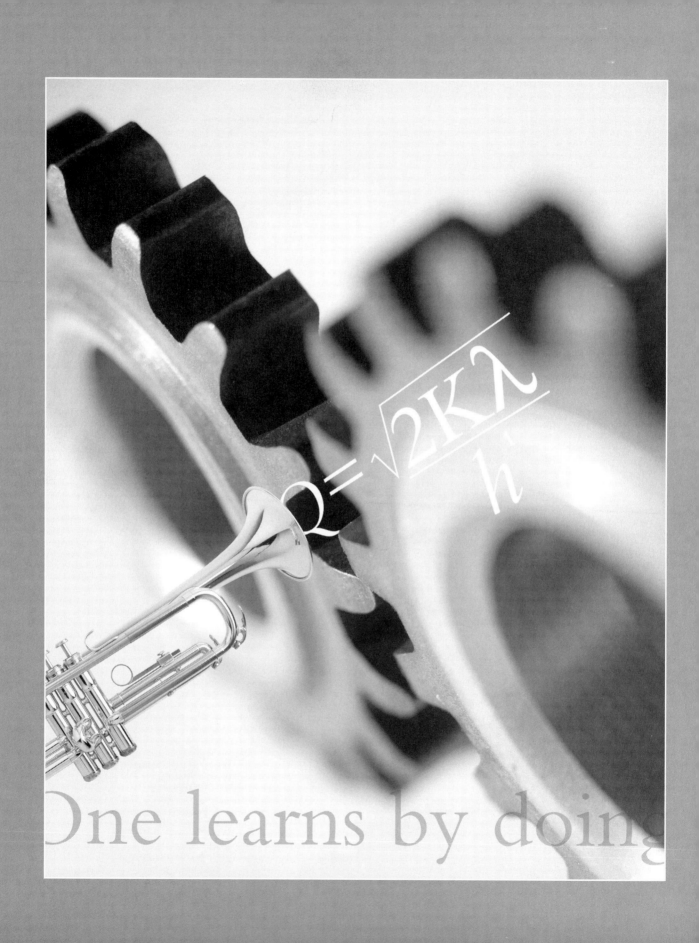

$$Q = \sqrt{\frac{2K\lambda}{h}}$$

One learns by doing

CHAPTER 6

SUPPLY CHAIN MANAGEMENT

The term *supply chain management* (SCM) seems to have emerged in the late 1980s and continues to gain interest at an increasing rate. The trade literature abounds with book titles and articles relating to some aspect of SCM. Software and consulting firms specializing in supply chain management solutions are now commonplace. These companies have grown at remarkable rates and include giants SAP and Oracle, which offer SCM solutions as part of comprehensive information retrieval systems. Although the supply chain management label is relatively new, the problems considered are not. Nearly all the material in the four preceding chapters of this book involves supply chain management. So what is new about SCM? The simple answer is that SCM looks at the problem of managing the flow of goods as an integrated system. Many definitions of SCM have been proposed, and it is instructive to examine some of them. The simplest and most straightforward appears at the Stanford Supply Chain Forum (1999) website and is probably due to Hau Lee, the head of the Forum. It is

> Supply chain management deals with the management of materials, information and financial flows in a network consisting of suppliers, manufacturers, distributors, and customers.

While short, this definition is fairly complete. It indicates that it is not only the flow of goods that is important, but the flow of information and money as well.

A definition due to Simchi-Levi et al. (1999, p. 1) that focuses on only the flow of goods is

> Supply chain management is a set of approaches utilized to efficiently integrate suppliers, manufacturers, warehouses, and stores, so that merchandise is produced and distributed at the right quantities, to the right locations, and at the right time, in order to minimize systemwide costs while satisfying service level requirements.

Since this definition focuses on the flow of goods, it implies that SCM is relevant only to manufacturing firms. Is SCM also relevant to service organizations? Where do issues of strategy play into these definitions? Clearly, an all-encompassing definition is difficult to devise. However, too much generality obscures meaning. Consider, for example, the following definition by Ross (1998):

> Supply chain management is a continuously evolving management philosophy that seeks to unify the collective productive competencies and resources of the business functions found both within the enterprise and outside the firm's allied business partners located along intersecting supply channels into a highly competitive, customer-enriching supply system focused on developing innovative solutions and synchronizing the flow of marketplace products, services, and information to create unique, individualized sources of customer value.

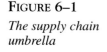

FIGURE 6–1

*The supply chain
umbrella*

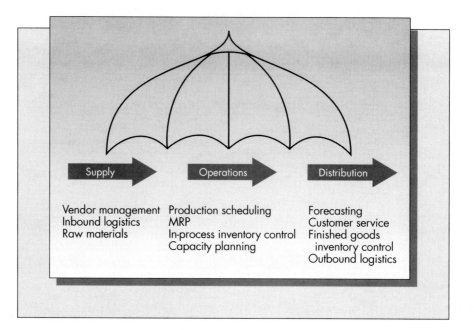

Perhaps the clearest description appeared in a *Fortune* magazine article devoted to the subject:

> Call it distribution or logistics or supply chain management. By whatever name, it is the sinuous, gritty, and cumbersome process by which companies move material, parts, and products to customers. In industry after industry, from cars and clothing to computers and chemicals, executives have plucked this once dismal discipline off the loading dock and placed it near the top of the corporate agenda. Hard-pressed to knock out competitors on quality or price, companies are trying to gain an edge through their ability to deliver the right stuff in the right amount at the right time. (Henkoff, 1994)

Most writers agree that "logistics" deals with essentially the same issues as supply chain management. The term *logistics* originated in the military and concerned problems of moving men and materiel in critical times. It was later adopted by business and became a common moniker for professional societies and academic programs. One might argue that although classical logistics treated the same set of problems as SCM, it did not consider the supply chain as an integrated system (Copacino, 1997). A summary of the "umbrella" of activities composing SCM appears in Figure 6–1.

In a sense, one might consider this entire text to deal with some aspect of SCM if one takes the broad view of the field depicted in Figure 6–1. However, most of the chapters focus on specific topics, like inventory control or job shop scheduling. SCM, on the other hand, treats the *entire* product delivery system from suppliers of raw materials to distribution channels of finished goods. While important and useful in many contexts, simple formulas such as the EOQ are unlikely to shed much light on effective management of complex supply chains.

The Supply Chain as a Strategic Weapon

Where does the supply chain fit into the overall business strategy of the firm? In Chapter 1, it was noted that an important part of the firm's competitive edge is its strategic positioning in the marketplace. Examples of strategic positioning include being the

SNAPSHOT APPLICATION

Wal-Mart Wins with Solid Supply Chain Management

Although there are many examples of companies winning (or losing) because of either good or bad supply chain strategies, perhaps none is more dramatic than the stories of Wal-Mart and Kmart. Both companies were founded in the same year, 1962. In only a few years, Kmart had become a household term, while Wal-Mart was largely unknown except for a few communities in the South. Wal-Mart stores were typically located in rural areas, so they rarely competed head-on with the larger, better-known chains like Sears and Kmart, most of whose stores were located in large cities and surrounding suburbs. In 1987 Kmart's sales were almost twice those of Wal-Mart (about $25 billion annually versus about $15 billion annually). By 1990 Wal-Mart had overtaken Kmart and in 1994 Wal-Mart's annual sales almost tripled Kmart's (about $80 billion versus about $27 billion)! Wal-Mart is now the largest discount retailer in the United States, surpassing Sears as well as Kmart. What could have possibly accounted for this dramatic turn-around?

While one can point to several factors leading to Wal-Mart's success, there is no question that the firm's emphasis on solid supply chain management was one of the most important. According to Duff and Ortega (1995) in comparing the management decisions of Kmart's former CEO, Joseph Antonini, and Wal-Mart's Sam Walton:

> When Mr. Antonini took the reins of Kmart in 1987, he had his hands full . . . Also, his predecessors neglected to implement the sophisticated computer systems that were helping Wal-Mart track and replenish its merchandise swiftly and efficiently.
>
> A self-promoter with a boisterous voice and wide smile, Mr. Antonini invested heavily in national television campaigns and glamorous representatives such as Jaclyn Smith . . . Mr. Walton avoided publicity. And instead of marketing, he became obsessed with operations. He invested tens of millions of dollars in a companywide computer system linking cash registers to headquarters, enabling him to quickly restock goods selling off the shelves. He also invested heavily in trucks and distribution centers. Besides enhancing his control, these moves sharply reduced costs.
>
> Mr. Antonini tried bolstering growth by overseeing the purchase of other types of retailers: the Sports Authority sporting goods chain, OfficeMax office supply stores, Borders bookstores, and Pace Membership Warehouse clubs . . .
>
> But the least visible difference between Wal-Mart and Kmart was beginning to matter a lot. Wal-Mart's incredibly sophisticated distribution inventory and scanner systems meant that customers almost never encountered depleted shelves or price-check delays at the cash register.
>
> The halls of Kmart, meanwhile, were filled with distribution horror stories. Joseph R. Thomas, who oversaw distribution, said that in retrospect, he should have smelled trouble when he found warehouses stuffed full of merchandise on December 15, the height of the Christmas season.

There are some important lessons here. These lessons go to the core of what's been wrong with American business thinking for decades. When problems arose, how did Kmart's management respond? By putting money into marketing and acquisitions. Wal-Mart, on the other hand, invested in better stores and better logistics systems for making sure the shelves in those stores were stocked with what customers wanted. Today, Wal-Mart has one of the most sophisticated inventory control and EDI (electronic data interchange) systems in the industry.

The story of Kmart and Wal-Mart is reminiscent of the story of the American and Japanese auto industries. In the 1950s and 1960s when American auto makers had a virtual monopoly on much of the world's auto market, where did the Big Three put their resources? Into producing better engineered cars at lower cost? No. Into slick marketing campaigns and annual cosmetic makeovers. The Japanese, on the other hand, invested in the latest automotive technology and sophisticated logistics systems such as just-in-time. The rest, as they say, is history.

low-cost provider (such as Hyundai automobiles) or the high-quality provider (such as Mercedes-Benz) or exploiting market segmentation to achieve shares of both markets (as does General Motors with different brands aimed at different market segments). The design of the supply chain also reflects a firm's strategic positioning.

In a supply chain, the primary trade-off is between cost and response time. It is less costly to transport product by boat or truck, but air freight can get the product from point A to point B much more quickly. Will deliveries be more reliable if the product is moved via the firm's internal system, or would it be better to subcontract the logistics operation to a third party? Third-party logistics (abbreviated 3PL) is becoming more common. Why? For the same reason that third-party manufacturing has become so widespread. Companies such as Silicon Valley–based Solectron Corporation are able to achieve significant economies of scale by providing manufacturing services to large numbers of firms producing similar products. In this way, it can actually be less expensive to subcontract manufacturing than to do it in-house. Cost is not the only issue. For example, the Trimble Corporation, producers of state-of-the-art global positioning systems, had to come to grips with the fact that product design, not manufacturing, was the firm's core competency. In the late 1990s, Trimble closed their manufacturing facilities and today subcontract most of their manufacturing to Solectron. (Perhaps coincidentally, Trimble's share price rose significantly after the company made that change.)

The manufacturing function has become very efficient in recent years. Firms have either established manufacturing as a core competency or outsourced it. For that reason, manufacturing generally does not provide an opportunity for significant cost reduction. Where, then, will the cost reductions come, if not from manufacturing? The answer is from the supply chain. The firms that are able to move product quickly and efficiently will gain an edge over their competitors. New technologies, such as web-based supply management systems (discussed later in this chapter), and the intelligent application of mathematical modeling, are ways that firms will trim costs from the supply chain without compromising service.

The Snapshot Application featured in this section highlights Wal-Mart, whose extraordinary success was achieved to some extent because of its sophisticated supply chain strategies.

There are two goals for this chapter. The first is to provide the reader with an appreciation of the most important issues encountered in managing complex supply chains. The second is to present a sampling of the types of mathematical models used in SCM analysis.

6.1 The Transportation Problem

The transportation problem is a mathematical model for optimally scheduling the flow of goods from production facilities to distribution centers. Assume that a fixed amount of product must be transported from a group of sources (plants) to a group of sinks (warehouses). The unit cost of transporting from each source to each sink is assumed to be known. The goal is to find the optimal flow paths and the amounts to be shipped on those paths to minimize the total cost of all shipments.

The transportation problem can be viewed as a prototype supply chain problem. Although most real-world problems involving the shipment of goods are more complex, the model provides an illustration of the issues and methods one would encounter in practice.

Example 6.1

The Pear Disk Drive Corporation produces several capacities of Winchester drives for personal computers. In 1999, Pear produced drives with capacities from 2 to 16 gigabytes, all in the 3.5-inch form factor. The most popular product is the 8 GB drive, which is sold to several computer manufacturers. Pear produces the drives in three plants located in Sunnyvale, California; Dublin, Ireland; and Bangkok, Thailand. Periodically, shipments are made from these three production facilities to four distribution warehouses located in the United States in Amarillo, Texas; Teaneck, New Jersey; Chicago, Illinois; and Sioux Falls, South Dakota. Over the next month, it has been determined that these warehouses should receive the following proportions of the company's total production of the 8 GB drives:

Warehouse	Percentage of Total Production
Amarillo	31
Teaneck	30
Chicago	18
Sioux Falls	21

The production quantities at the factories in the next month are expected to be (in thousands of units)

Plant	Anticipated Production (in 1,000s of units)
Sunnyvale	45
Dublin	120
Bangkok	95

Since the total production at the three plants is 260 units, the amounts shipped to the four warehouses will be (rounded to the nearest unit)

Warehouse	Total Shipment Quantity (1000s)
Amarillo	80
Teaneck	78
Chicago	47
Sioux Falls	55

While the shipping cost may be lower between certain plants and distribution centers, Pear has established shipping routes between every plant and every warehouse. This is in case of unforeseen problems such as a forced shutdown at a plant, unanticipated swings in regional demands, or poor weather along some routes. The unit costs for shipping 1,000 units from each plant to each warehouse is given in the table below:

Shipping Costs per 1,000 Units in $

		TO			
		Amarillo	*Teaneck*	*Chicago*	*Sioux Falls*
F	Sunnyvale	250	420	380	280
R O	Dublin	1280	990	1440	1520
M	Bangkok	1550	1420	1660	1730

The goal is to determine a pattern of shipping that minimizes the total transportation cost from plants to warehouses. The network representation of Pear's distribution problem appears in Figure 6–2. ∎

FIGURE 6–2

Pear Disk Drive transportation problem

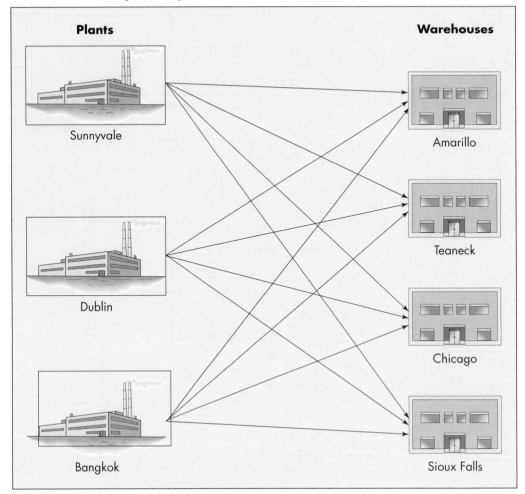

The Greedy Heuristic

The transportation problem can be formulated as a linear program, and thus can be solved using any linear programming code, such as Solver in Excel. To gain some intuition about the structure of the problem, however, we will consider a simple heuristic that usually gives good, but possibly suboptimal, solutions. To implement the heuristic, we construct a transportation tableau.

Example 6.1 (continued)

The transportation tableau is similar to the table above that gives the shipping costs from plants to warehouses. It is

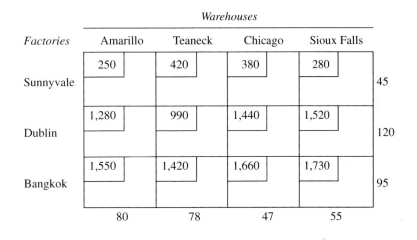

Note the structure of the tableau. Rows correspond to supply sources (factories) and columns to sinks (warehouses). The numbers we will place in the cells of the tableau will be the value of the flow of product from each source to each sink. To implement the greedy heuristic, we search the tableau for the minimum unit cost, which is $250 and corresponds to the Sunnyvale/Amarillo cell. Into this cell we place the minimum of the following two quantities: the availability at Sunnyvale (45) and the requirement at Amarillo (80). Hence, we place a 45 into this cell. At this point, we've saturated the first row (meaning no more can be shipped out of Sunnyvale), so we draw a line through the first row. The tableau now becomes

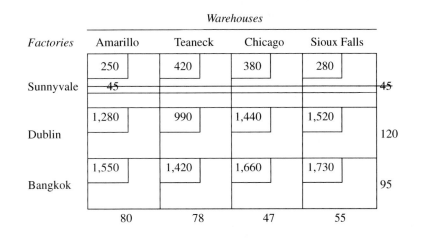

Of the uncovered cells, the least-cost cell is Dublin/Teaneck at $990 (i.e., cell (2,2)). In this cell, we can assign a maximum flow of 78 units (the minimum of 78 and 120), which we do. This saturates the second column and so we draw a line through the second column to indicate this. The next least-cost cell is Dublin/Amarillo at $1,280. The capacity of this cell is minimum (80 − 45, 120 − 78), since 45 is already allocated to the first column and 78 to the second row. Hence, we allocate min(35,42) = 35 to this cell, which saturates the first column. The tableau at this stage is

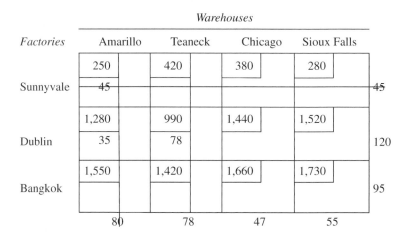

The greedy heuristic continues in this manner until all rows and columns are saturated. The final solution is.

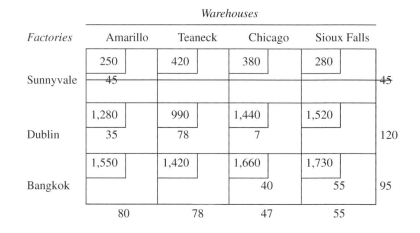

If we let x_{ij} be the amount of flow from source i to sink j, then the solution shown in the last tableau is

$$x_{11} = 45, \quad x_{21} = 35, \quad x_{22} = 78, \quad x_{23} = 7,$$
$$x_{33} = 40, \quad \text{and} \quad x_{34} = 55,$$

and all other $x_{ij} = 0$. The total cost of this solution is (45)(250) + (35)(1,280) + · · · = $304,900. As we will see in the next section, this solution is suboptimal for this problem. ∎

6.2 Solving Transportation Problems with Linear Programming

Several heuristics for solving transportation problems have been proposed, such as the one illustrated in the previous section. However, it is unlikely that anyone with a real problem would use a heuristic, since optimal solutions can be found efficiently by linear programming. In fact, because of the special structure of the transportation problem, today's specialized codes can solve problems with millions of variables. The greedy heuristic was presented in the last section to give the reader a feel for how flows are assigned in networks and how one could obtain reasonable solutions quickly.

Let m be the number of sources and n the number of sinks. (In Example 6.1, $m = 3$ and $n = 4$.) Recall the definition of the decision variables from the previous section:

x_{ij} = the flow from source i to sink j for $1 \leq i \leq m$ and $1 \leq j \leq n$,

and define c_{ij} as the cost of shipping one unit from i to j. It follows that the total cost of making all shipments is

$$\sum_{i=1}^{m} \sum_{j=1}^{n} c_{ij} x_{ij}$$

For the case of Pear Company, described in Example 6.1, the objective function is

$$250x_{11} + 420x_{12} + 380x_{13} + 280x_{14} + \cdots + 1730x_{34}.$$

Since many routes are obviously not economical, it is likely that many of the decision variables will equal zero at the optimal solution.

The constraints are designed to ensure that the total amount shipped out of each source equals the amount available at that source, and the amount shipped into any sink equals the amount required at that sink. Since there are m sources and n sinks, there are a total of $m + n$ constraints (excluding nonnegativity constraints). Let a_i be the total amount to be shipped out of source i and b_j the total amount to be shipped into sink j. Then linear programming constraints may be written:

$$\sum_{j=1}^{n} x_{ij} = a_i \qquad \text{for } 1 \leq i \leq m$$

$$\sum_{i=1}^{m} x_{ij} = b_j \qquad \text{for } 1 \leq j \leq n$$

$$x_{ij} \geq 0 \qquad \text{for } 1 \leq i \leq m \text{ and } 1 \leq j \leq n.$$

For the Pear Disk Drive Company problem, we obtain the following seven constraints:

$$x_{11} + x_{12} + x_{13} + x_{14} = 45 \qquad \text{(shipments out of Sunnyvale)}$$
$$x_{21} + x_{22} + x_{23} + x_{24} = 120 \qquad \text{(shipments out of Dublin)}$$
$$x_{31} + x_{32} + x_{33} + x_{34} = 95 \qquad \text{(shipments out of Bangkok)}$$

$$x_{11} + x_{21} + x_{31} = 80 \qquad \text{(shipments into Amarillo)}$$
$$x_{12} + x_{22} + x_{32} = 78 \qquad \text{(shipments into Teaneck)}$$
$$x_{13} + x_{23} + x_{33} = 47 \qquad \text{(shipments into Chicago)}$$
$$x_{14} + x_{24} + x_{34} = 55 \qquad \text{(shipments into Sioux Falls)}$$

FIGURE 6–3

Solution of Pear's transportation problem using Excel Solver

	A	B	C	D	E	F	G	H	I	J	K	L	M	N	O	P
1							**Solution of Example 6.1 (Pear's Transportation Problem)**									
2																
3	Variables	×11	×12	×13	×14	×21	×22	×23	×24	×31	×32	×33	×34	Operator	Value	RHS
4																
5	Values	0	0	0	45	42	78	0	0	38	0	47	10			
6																
7	Objective Coeff	250	420	380	280	1280	990	1440	1520	1550	1420	1660	1730	Min	297800	
8	st															
9	Constraint 1	1	1	1	1									=	45	45
10	Constraint 2					1	1	1	1					=	120	120
11	Constraint 3									1	1	1	1	=	95	95
12	Constraint 4	1			1				1					=	80	80
13	Constraint 5		1			1				1				=	78	78
14	Constraint 6			1			1				1			=	47	47
15	Constraint 7				1			1				1		=	55	55
16																
17																
18		Notes:	Formula for Cell O9: =SUMPRODUCT (B9:M9,$B$5:$M$5). Copied to O10 to O15.													
19			Changing cells for Solver are $B$5:$M$5.													
20																

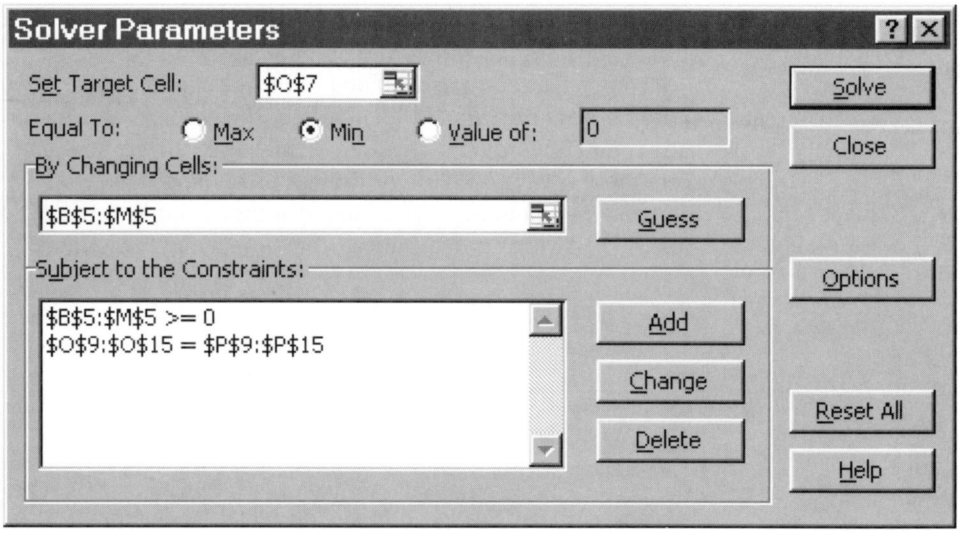

Solver Parameters dialog box:

Set Target Cell: $O$7

Equal To: ○ Max　⦿ Min　○ Value of: 0

By Changing Cells: $B$5:$M$5

Subject to the Constraints:
$B$5:$M$5 >= 0
$O$9:$O$15 = $P$9:$P$15

Buttons: Solve, Close, Guess, Options, Add, Change, Delete, Reset All, Help

and the nonnegativity constraints required in linear programming:

$$x_{ij} \geq 0 \qquad \text{for } 1 \leq i \leq 3 \text{ and } 1 \leq j \leq 4.$$

The problem was entered in Excel Solver. The spreadsheet used and the solution appear in Figure 6–3. The solution obtained is

$$x_{14} = 45, \qquad x_{21} = 42, \qquad x_{22} = 78, \qquad x_{31} = 38,$$
$$x_{33} = 47, \quad \text{and} \quad x_{34} = 10,$$

with all other values equaling zero. The total cost of this solution is $297,800. Notice that this is not the same as the solution obtained from the greedy heuristic. As noted earlier, the cost of the greedy heuristic is $304,900, which differs from the optimal cost by only 2.4 percent.

6.3 Generalizations of the Transportation Problem

The Pear Company example is the simplest type of transportation problem. Every link from source to sink is feasible, and the total amount available from the sources exactly equals the total demand at the sinks. Several of these requirements can be relaxed without making the problem significantly more complicated.

Infeasible Routes

Suppose in the example that the firm has decided to eliminate the routes from Dublin to Chicago and from Bangkok to Sioux Falls. This would be accounted for by placing very high costs on these arcs in the network. Traditionally, uppercase M has been used to signify a very high cost. In practice, of course, one would have to assign a number to these locations. As long as that number is much larger than the other costs, an optimal solution will never assign flow to these routes. For the example, suppose we assign costs of $1,000,000 to each of these routes and re-solve the problem. The reader can check that one now obtains the following solution:

$$x_{14} = 45, \qquad x_{21} = 32, \qquad x_{22} = 78, \qquad x_{31} = 48, \qquad \text{and} \qquad x_{33} = 47,$$

with all other values equaling zero. The cost of the new solution is $298,400, only slightly larger than the cost obtained when all of the routes were feasible.

Unbalanced Problems

An unbalanced transportation problem is one in which the total amount shipped from the sources is not equal to the total amount required at the sinks. This can arise if the demand exceeds the supply or vice versa. There are two ways of handling unbalanced problems. One is to add either a dummy row or a dummy column to absorb the excess supply or demand. A second method for solving unbalanced problems is to alter the appropriate set of constraints to either ≤ or ≥ form. Both methods will be illustrated.

Suppose in Example 6.1 that the demand for the drives was higher than anticipated. Suppose that the respective requirements at the four warehouses are now Amarillo, 90; Teaneck, 78; Chicago, 55; and Sioux Falls, 55. This means that the total demand is 278 and the total supply is 260. To turn this into a balanced problem, we add an additional fictitious factory to account for the 18-unit shortfall. This can be labeled as a dummy row in the transportation tableau and all entries for that row assigned an arbitrarily large unit cost. The greedy heuristic could then be applied to the new tableau, which will have four rows and four columns. We will not present the details, but the reader should check that the greedy heuristic applied to this unbalanced problem assigns all of the shortfall to the Sioux Falls warehouse. Note that when supply exceeds demand and one adds a dummy column, the costs in the dummy column do *not* have to be very large numbers, but they do all have to be the same. (In fact, one can assign zero to all costs in the dummy column.)

FIGURE 6–4

Solution of Example 6.1 with excess demand and dummy row

	X11	X12	X13	X14	X21	X22	X23	X24	X31	X32	X33	X34	X41	X42	X43	X44	Oper	Value	RHS
Solution of Example 6.1 with Excess Demand and a Dummy Row																			
Variables	X11	X12	X13	X14	X21	X22	X23	X24	X31	X32	X33	X34	X41	X42	X43	X44	Oper	Value	RHS
Values	0	0	0	45	42	78	0	0	48	0	47	0	0	0	8	10			
Obj Func	250	420	380	280	1280	990	1440	1520	1550	1420	1660	1730	1.E+06	1.E+06	1.E+06	1.E+06	Min	2E+07	
st																			
Constraint 1	1	1	1	1													=	45	60
Constraint 2					1	1	1	1									=	120	130
Constraint 3									1	1	1	1					=	95	95
Constraint 4													1	1	1	1	=	18	18
Constraint 5	1				1				1				1				=	90	90
Constraint 6		1				1				1				1			=	78	78
Constraint 7			1				1				1				1		=	55	55
Constraint 8				1				1				1				1	=	55	55

The method of adding a dummy row or column also works when finding optimal solutions via linear programming. In the example, we assigned a cost of 10^6 to each cell in the dummy row. The resulting Excel spreadsheet and Solver solution appear in Figure 6–4. The optimal solution calls for assigning the shortfall to two warehouses: 8 units to Chicago and 10 units to Sioux Falls.

Unbalanced transportation problems also can be formulated as linear programs by using inequality constraints. In the example above, where there is excess demand, one would use equality for the first three constraints, to be sure that all of the supply is shipped, and less than or equal to constraints for the last four, with the slack accounting for the shortfall. The reader should check that one obtains the same solution by doing so as was obtained by adding a dummy row. This method has the advantage of giving an accurate value for the objective function. (In the case where the supply exceeds the demand, the principle is the same, but the details differ. The first three supply constraints are converted to *greater than or equal to* form, while the last four demand constraints are still equality constraints. The slack in the first three constraints corresponds to the excess supply.)

6.4 More General Network Formulations

The transportation problem is a special type of network where all nodes are either supply nodes (also called sources) or demand nodes (also called sinks). Linear programming also can be used to solve more complex network distribution problems as well. One example is the *transshipment problem*. In this case, one or more of the nodes in the network are transshipment points rather than supply or demand points. Note that a transshipment node also can be either a supply or a demand node as well (but no node is both a supply and a demand node).

For general network flow problems, we use the following balance of flow rules:

If	Apply the Following Rule at Each Node:
1. Total supply > Total demand	Inflow − Outflow ≥ Supply or demand
2. Total supply < Total demand	Inflow − Outflow ≤ Supply or demand
3. Total supply = Total demand	Inflow − Outflow = Supply or demand

The decision variables are defined in the same way as with the simple transportation problem. That is, x_{ij} represents the total flow from node i to node j. For general network flow problems, we represent the supply as a negative number attached to that node and the demand as a positive number attached to that node. This convention along with the rules above will result in the correct balance-of-flow equations.

Example 6.2

Consider the example of Pear Disk Drives. The company has decided to place a warehouse in Sacramento to be used as a transshipment node and has expanded the Chicago facility to also allow for transshipments. Suppose that in addition to being transshipment nodes, both Chicago and Sacramento are also demand nodes. The new network is pictured in Figure 6–5. Note that several of the old routes have been eliminated in the new configuration.

We define a decision variable for each arc in the network. In this case, there are a total of 10 decision variables. The objective function is

$$\text{minimize} \quad 250x_{16} + 76x_{14} + 380x_{15} + 1440x_{25} + 1660x_{35}$$
$$+ 110x_{46} + 95x_{48} + 180x_{56} + 120x_{57} + 195x_{58}$$

FIGURE 6–5

Pear Disk Drive problem with transshipment nodes

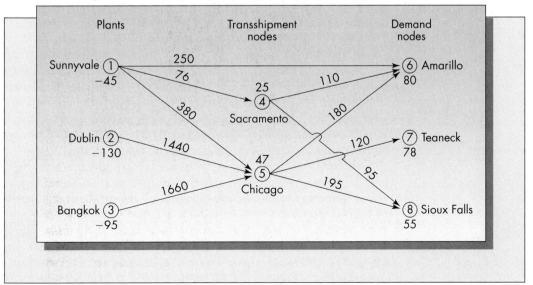

The total supply available is still 260 units, but the demand is 285 units (due to the additional 25 units demanded at Sacramento). Hence, this corresponds to case 2 above in which total demand exceeds total supply. Applying rule 2 to each node gives the following eight constraints for this problem:

Node 1: $-x_{14} - x_{15} - x_{16} \leq -45$

Node 2: $-x_{25} \leq -120$

Node 3: $-x_{35} \leq -95$

Node 4: $x_{14} - x_{46} - x_{48} \leq 25$

Node 5: $x_{16} + x_{46} + x_{56} - x_{56} - x_{57} - x_{58} \leq 47$

Node 6: $x_{16} + x_{46} + x_{56} \leq 80$

Node 7: $x_{57} \leq 78$

Node 8: $x_{48} + x_{58} \leq 55.$

For some linear programming codes, all right-hand sides would have to be nonnegative. In those cases, one would multiply the first three constraints by -1. However, this is not required in Excel, so we can enter the constraints just as they appear above.

The Excel spreadsheet and solution for Pear's transshipment problem appears in Figure 6–6. Note that the solution calls for shipping all units from the sources. Since this problem had more supply than demand, it is interesting to see where the shortfall occurs. The amount shipped into Sacramento is 45 units (all from Sunnyvale), and the total shipped out of Sacramento is 20 units. Thus, all the demand (25 units) is satisfied in Sacramento. For the other transshipment point, Chicago, the shipments into Chicago are 120 units from Dublin and 95 units from Bangkok, and the shipments out of Chicago are 48 units to Amarillo and 120 units to Teaneck. The difference is $120 + 95 - (48 + 120) = 47$ units. Hence, there is also no shortage in Chicago. Total shipments into the demand nodes at Amarillo, Teaneck, and Sioux Falls are respectively 80, 78, and 30 units. Hence, all of the shortage (25 units) is absorbed at the Sioux Falls location at the optimal solution. ∎

Real networks can be extremely complex by virtue of their sheer magnitude. As Simchi-Levi et al. (1999) note, a typical soft drink distribution system could involve anywhere between 10,000 and 120,000 accounts. Some retailers have thousands of stores and hundreds of thousands of products. For this reason, efficient data aggregation may be required to solve problems of this magnitude. Customer aggregation is generally accomplished by combining accounts that are nearby. Combining customers with like or similar zip codes is a common means of geographic aggregation. Product aggregation is discussed in detail in Chapter 3 in the context of manufacturing. Product aggregation rules for a supply chain are likely to be different from those discussed in Chapter 3. For example, one might aggregate products according to where they are picked up or where they are delivered. We refer the interested reader to Simchi-Levi et al. (1999) for a more comprehensive discussion of the practical issues surrounding implementation of supply chain networks.

Mathematical modeling has been used successfully in many supply chain applications. Sophisticated models lie at the heart of several commercial software products such as those offered by Texas-based i2 Technologies. The Snapshot Application for this section discusses a successful application of advanced inventory control models to IBM's supply chain for spare parts.

FIGURE 6–6

Excel spreadsheet for Pear transshipment problem in Example 6.2

	A	B	C	D	E	F	G	H	I	J	K	L	M	N
1					Excel Spreadsheet for Pear Transshipment Problem in Example 6.2									
2														
3	Variables	×14	×15	×16	×25	×35	×46	×48	×56	×57	×58	Operator	Value	RHS
4														
5	Values	45	0	0	120	95	0	20	80	78	10			
6														
7	Obj Funct	76	380	250	1440	1660	110	95	180	120	195	Min	361530	
8	st													
9	Node 1	−1	−1	−1								<=	−45	−45
10	Node 2				−1							<=	−120	−120
11	Node 3					−1						<=	−95	−95
12	Node 4	1					−1	−1				<=	25	25
13	Node 5		1		1	1			−1	−1	−1	<=	47	47
14	Node 6			1			1		1			<=	80	80
15	Node 7									1		<=	78	78
16	Node 8							1			1	<=	30	55

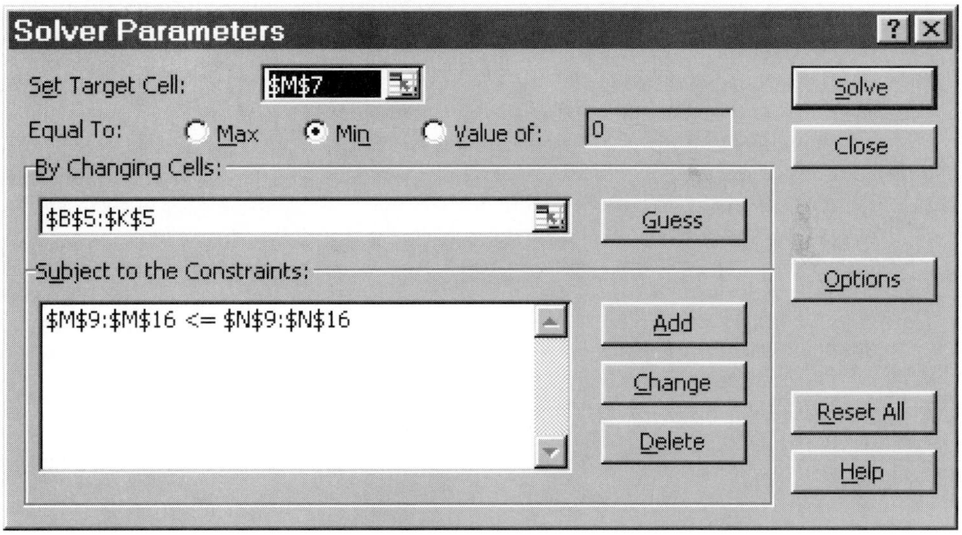

Problems for Sections 1–4

1. Consider Example 6.1 of the Pear Company assuming the following supplies and
   demands:

Plant	Production	Warehouse	Requirement
Sunnyvale	60	Amarillo	100
Dublin	145	Teaneck	84
Bangkok	125	Chicago	77
		Sioux Falls	69

SNAPSHOT APPLICATION

IBM Streamlines Its Supply Chain for Spare Parts Using Sophisticated Mathematical Models

The IBM Corporation must manage a highly complex network of warehouses and service centers to provide spare parts support to its customer base. IBM's success has been due, in part, to its ability to provide quick and efficient service to its customers. As the installed base and assortment of products grew, management of its spare parts inventory and distribution grew more complex. As of 1990, IBM had approximately 1,000 different products in service and more than 10 million of these at user locations worldwide. Approximately 200,000 distinct part numbers were required to support these products. Parts would be shipped from the manufacturer or vendor to large national and regional warehouses. From there they would go to smaller warehouses, and then to local stations and ultimately to installed machines. A system in which parts are stored at different classes of locations is referred to as a multiechelon inventory system. The echelons refer to the hierarchy of storage locations.

In the mid 1980s, IBM's National Service Division (NSD) approached a team of researchers from the Wharton School to consider ways of improving the performance of the spare parts delivery system while lowering operating costs. The team built a sophisticated mathematical model and, using effective heuris-

Source: Cohen et al., 1990.

tics, was able to provide significant improvements in the performance of the system. Their system developed recommendations for setting inventory stocking policies for all parts at all stocking locations. It was integrated into a more comprehensive system consisting of modules for accounting, central distribution, real-time system, planning and control, and field systems. The underlying mathematics used concepts of service, costs, order levels, lead times, and demand uncertainty that are discussed in this book in Chapter 5.

What improvements were observed after the implementation of the team's approach? First, the optimizer recommended inventory levels that were 20 to 25 percent below the then-current levels. Second, the inventory reduction was accompanied by a demonstrable improvement in service to the customer and an increased responsiveness of the control system. Concrete measures of improvement included a 10 percent improvement in parts availability at the lower echelons in the system and savings of approximately $20 million annually.

What is the lesson learned from this case? As product delivery systems get larger and more complex, more and more sophisticated methods will be required to manage those systems. Generic software products may not be able to provide sufficient power and customization to be effective in such environments. IBM's experience is only one example of how the management of complex supply chain structures can be improved with the aid of the modeling methods discussed in this text.

    a. What is the solution obtained using the greedy heuristic?
    b. Use Excel's Solver (or other linear programming code) to determine the optimal solution.
    c. What is the percentage error of the solution you obtained in part (*a*)?

2. Resolve Pear's transportation problem assuming that the following routes are eliminated: Sunnyvale to Teaneck, Dublin to Chicago, and Bangkok to Amarillo. What is the percentage increase in total shipping costs at the optimal solution due to the elimination of these routes?

3. Resolve Pear's transshipment problem assuming that an additional transshipment point is located at Oklahoma City. Assume that the unit costs of shipping from the three plants to Oklahoma City are Sunnyvale, $170; Dublin,

$1,200; and Bangkok, $1,600; and the respective costs of shipping from Oklahoma City to the demand nodes are Amarillo, $35; Teaneck, $245; and Sioux Falls, $145. Assume that Oklahoma City is only a transshipment point and has no demand of its own. Find the new shipping pattern with the addition of the new transshipment point and the savings, if any, of introducing this additional node.

4. Major Motors produces its Trans National model in three plants located in Flint, Michigan; Fresno, California; and Monterrey, Mexico. Dealers receive cars from regional distribution centers located in Phoenix, Arizona; Davenport, Iowa; and Columbia, South Carolina. Anticipated production at the plants over the next month (in 100s of cars) is 43 at Flint, 26 at Fresno, and 31 at Monterrey. Based on firm orders and other requests from dealers, Major Motors has decided that they need to have the following numbers of cars at the regional distribution centers at month's end: Phoenix, 26; Davenport, 28; and Columbia, 30. Suppose that the cost of shipping 100 cars from each plant to each distribution center is given in the following matrix (in $1,000s):

		TO		
		Phoenix	Davenport	Columbia
F R O M	Flint	12	8	17
	Fresno	7	14	21
	Monterrey	18	22	31

a. Convert the problem to a balanced problem by adding an appropriate row or column and find the solution suggested by the greedy heuristic.
b. Find the optimal solution using Solver and inequality constraints.
c. What is the percentage error of the greedy heuristic for this problem?
d. Suppose that the route between Monterrey and Columbia is no longer available due to a landslide on a key road in Mexico. What solution do you now obtain?

5. Consider the problem of Major Motors described in Problem 4. In order to be able to deal more effectively with unforeseen events (such as the road closing), Major Motors has established two transshipment points between the factories and the regional distribution centers at Santa Fe, New Mexico, and Jefferson City, Missouri. The cost of shipping 100 cars to the transshipment points is (in $1,000s).

		TO	
		Santa Fe	Jefferson City
F R O M	Flint	8	6
	Fresno	6	9
	Monterrey	9	14

while the cost of shipping from the transshipment points to the distribution centers is

		TO		
		Phoenix	*Davenport*	*Columbia*
F				
R	*Santa Fe*	3	8	10
O	*Jefferson City*	5	5	9
M				

Assuming that none of the direct routes between the factories and the distribution centers is available, find the optimal flow of cars through the transshipment points that minimizes the total shipping costs.

6. Toyco produces a line of Bonnie dolls and accessories at its plants in New York and Baltimore that must be shipped to distribution centers in Chicago and Los Angeles. The company uses Air Freight, Inc., to make its shipments. Suppose that they can ship directly or through Pittsburgh and Denver. The daily production rates at the plants are respectively 5,000 and 7,000 units daily, and the demands at the distribution centers are respectively 3,500 and 8,500 units daily. The costs of shipping 1,000 units are given in the table below. Find the optimal shipping routes and the associated cost.

		TO			
		Pittsburgh	*Denver*	*Chicago*	*Los Angeles*
F	*New York*	$182	$375	$285	$460
R	*Baltimore*	77	290	245	575
O	*Pittsburgh*	—	275	125	380
M	*Denver*	—	—	90	110

7. Reconsider Problem 6 if there is a drop in the demand for dolls to 3,000 at Chicago and 7,000 at Los Angeles. Find the optimal shipping pattern in this case. How much of the total decrease in demand of 2,000 units is absorbed at each factory at the optimal solution?

8. Reconsider Problem 6 assuming that the maximum amount that can be shipped from either New York or Baltimore through Pittsburgh is 2,000 units due to the size of the plane available for this route.

6.5 Distribution Resource Planning

Distribution resource planning (DRP) is a term coined by Martin (1990). In brief, it uses the MRP logic developed for manufacturing systems to manage distribution systems. The primary advantage of DRP over traditional ROP (reorder point systems, discussed in detail in Chapter 5) is that it can deal with changing demand patterns. However, while ROP explicitly incorporates demand uncertainty (i.e., forecast error), DRP does not. DRP is very much like MRP in concept and execution. Since MRP will be discussed in detail in Chapter 7, we provide only a brief overview of the concept.

Example 6.3

The campus bookstore has decided to implement a DRP approach for reordering blue books because of the wide demand variation that has been observed. Sales tend to be high just prior to exam periods and very low during winter and spring breaks. Blue books are shipped in packages of 20. Suppose that forecasted sales over the next eight weeks (measured in packages) are 22, 35, 60, 12, 0, 19, 85, 33. The supplier has shipped 40 packages, which are in transit and awaiting delivery. Furthermore, there are currently 26 packages on order scheduled to arrive in two weeks and 44 packages in inventory. At the current time, the DRP order profile for blue books is

DRP-Based Order Profile for Blue Books

On hand	26
Order leadtime	2 weeks
Safety stock	0

Weeks:	0	1	2	3	4	5	6	7	8
Sales forecasts		22	35	60	12	0	19	85	33
In transit		40							
On order			26						
Projected balance	26	44	35	−25					

The DRP profile shows that without any further action, the bookstore will stock out of 25 packages in week 3. To avoid this shortage, an order needs to be placed in advance of week 3 (namely in week 1). ■

By representing the inventory profile in this fashion, the magnitude and the timing of shortages becomes clear before these shortages occur. When using DRP, one does not plan orders for too many weeks into the future, since sales forecasts could change as new information becomes available. Furthermore, we also expect that actual sales will be different from projected sales. The clear representation of order history, arrivals, and anticipated shortages has advantages over traditional ROP reporting. According to Martin (1990):

> Obviously, the DRP inventory model is vastly superior. . . . The data come alive, tell a story, and demand action. They project and anticipate future shortages and recommend action in the form of suggested order quantities in a time-phased manner. The result is that purchasing activity is neither premature nor overdue, but just in time and in the right quantities. As conditions change (and they will), this inventory modeling approach updates itself dynamically, identifies potential future shortages, and sends appropriate action messages to its users.

Example 6.3 (continued)

Let's continue with the bookstore example. At this point, it is clear that an order needs to be placed in week 1 to avoid a shortage in week 3. This is shown in the following table.

DRP-Based Order Profile for Blue Books

On hand	26
Order leadtime	2 weeks
Safety stock	0

Weeks:	0	1	2	3	4	5	6	7	8
Sales forecasts		22	35	60	12	0	19	85	33
In transit		40							
On order			26						
Projected balance	26	44	35	12	0	0	−19		
Planned shipments order date		37							
Planned shipments receipt date				37					

The table shows a shipment of 37 packages scheduled to arrive in week 3. This averts the anticipated shortage in this week and also avoids the need for the bookstore to have to reorder again until week 4 (to avert the anticipated shortage in week 6). ■

Although Martin does not discuss the issue, one might ask how large the planned shipments should be. If we view the forecasts as a deterministic stream of demands, then the problem of determining optimal order sizes is precisely the same as the problem to be addressed in Chapter 7 of determining optimal production lot sizes in MRP. We discuss several lot sizing methods in Chapter 7 that are also applicable in the DRP context: the Wagner-Whitin method, part period balancing, and the Silver-Meal heuristic. To apply any of these methods, one would have to know the holding costs and the order setup costs.

Although not shown in this example, the DRP approach can incorporate positive safety stock. Orders would be planned so anticipated on-hand inventory never drops below the safety stock. Alternatively, one can replace forecasted demand with forecasted demand plus safety stock (say "effective demand") and determine order sizes to meet effective demand. However, DRP does not recommend safety stock levels. Computing appropriate safety stock levels requires an assessment of the uncertainty of demand and an estimate of stock-out costs or specification of service levels. This problem was discussed in detail in Chapter 5.

Although simple, DRP could be an effective tool, especially for complex multilevel distribution systems. Such systems typically include one or more layers of intermediate DCs (distribution centers). Note that there is an extensive theoretical literature on mathematical models for managing multilevel inventory systems (known as multiechelon systems in the literature). However, most of these models are quite complex mathematically and are rarely suited for implementation in large-scale systems.

Problems for Section 5

9. Consider Example 6.3 of this section. Based on the forecasted sales for the next eight weeks given in the example, determine the average sales over this period. Suppose that the fixed cost of placing an order is $40 and the holding cost per

package is \$0.25. Use these numbers to determine a value for the EOQ (refer to Chapter 4). Assuming that the EOQ is ordered so that anticipated inventories never drop below zero, determine the DRP order profile for the blue books over the next eight weeks. (Assume the on-hand and in-transit figures given in the problem.)

10. Using the data from Problem 9, determine the lot sizing based on the Silver-Meal heuristic (discussed in Chapter 7) and show the DRP order profile in this case.

11. Re-solve Problem 10 assuming that you wish to maintain a safety stock of 30 units, and show the DRP profile in this case.

12. Discuss the advantages and disadvantages of DRP compared to ROP (reorder point systems), which are discussed in detail in Chapter 5.

6.6 Determining Delivery Routes in Supply Chains

An important aspect of supply chain logistics is efficiently moving product from one place to another. The transportation and transshipment problems discussed earlier in this chapter deal with this problem at a macro or firmwide level. At a micro level, deliveries to customers also must be planned efficiently. Because of the scale of the problem, efficient delivery schedules can have a very significant impact on the bottom line. As a result, they become an important part of designing the entire supply chain.

Determining optimal delivery schedules turns out to be a very difficult problem in general, rivaling the complexity of job shop scheduling problems discussed in detail in Chapter 8. Vehicle scheduling is closely related to a classical operations research problem known as the traveling salesman problem. The problem is described in the following way. A salesman starts at his home base, labeled city 1. He must then make stops at $n - 1$ other cities, visiting each city exactly once. The problem is to determine the optimal sequence in which to visit the cities to minimize the total distance traveled. Although this problem is easy to state, it turns out to be very hard to solve. If the number of cities is small, it is possible to enumerate all the possible tours. There are $n!$ orderings of n objects.[1] For modest values of n, one can enumerate all the tours and compute their distances directly. For example, for $n = 5$ there are 120 sequences. This number grows very fast, however. For $n = 10$, the number of sequences grows to over 3 million, and for $n = 25$ it grows to more than 1.55×10^{25}. To get some idea of how large this number is, suppose that we could evaluate 1 trillion sequences per second on a supercomputer. Then, for a 25-city problem, it would take nearly 500,000 years to evaluate all the sequences!

Total enumeration is hopeless for solving all but the smallest traveling salesman problems. Problems such as this are known in mathematics as *NP hard*. The NP stands for no polynomial, meaning that the time required to solve such problems is an exponential function of the number of cities rather than a polynomial function. We will not dwell on the traveling salesman problem here, but note that methods of solution have been proposed that are vast improvements over total enumeration. However, finding optimal solutions to even moderate-sized problems is still difficult.

Finding optimal routes in vehicle scheduling is a similar, but more complex, problem. Assume that there is a central depot with one or more delivery vehicles and n customer locations, each having a known requirement. The question is how to assign

---

[1] $n!$ is equal to n times $(n - 1)$ times $(n - 2) \ldots$ times 1.

vehicles to customer locations to meet customer demand and satisfy whatever constraints there might be at minimum cost. More real vehicle scheduling problems are too large and complex to solve optimally.

Because optimality may be impossible to achieve, methods for determining "good" solutions are important. We will discuss a simple technique for finding good routes, known as the savings method and developed by Clarke and Wright (1964).

Suppose that there is a single depot from which all vehicles depart and return. Customers' locations and needs are known. Identify the depot as location 0 and the customers as locations 1, 2, ..., n. We assume that there are known costs of traveling from the depot to each customer location, given by

$$c_{0j} = \text{Cost of making one trip from the depot to customer } j.$$

To implement the method, we will also need to know the costs of trips between customers. This means that we will assume that the following constants are known as well:

$$c_{ij} = \text{Cost of making a trip from customer location } i \text{ to customer location } j.$$

For our purposes we consider only the case in which $c_{ij} = c_{ji}$ for all $1 \le i, j \le n$. This does not necessarily hold in all situations, however. For example, if there are one-way streets, the distance from i to j may be different from the distance from j to i. The method proceeds as follows: Suppose initially that there is a separate vehicle assigned to each customer location. Then the initial solution consists of n separate routes from the depot to each customer location and back. It follows that the total cost of all round trips for the initial solution is

$$2 \sum_{j=1}^{n} c_{0j}.$$

Now, suppose that we link customers i and j. That is, we go from the depot to i to j and back to the depot again. In doing so, we would save one trip between the depot and location i and one trip between the depot and location j. However, there would be an added cost of c_{ij} for the trip from i to j (or vice versa). Hence, the savings realized by linking i and j is

$$s_{ij} = c_{0i} = c_{0j} - c_{ij}.$$

The method is to compute s_{ij} for all possible pairs of customer locations i and j, and then rank the s_{ij} in decreasing order. One then considers each of the links in descending order of savings and includes link (i, j) in a route if it does not violate feasibility constraints. If including the current link violates feasibility, one goes to the next link on the list and considers including that on a single route. One continues in this manner until the list is exhausted. Whenever link (i, j) is included on a route, the cost savings is s_{ij}.

The total number of calculations of s_{ij} required is

$$\binom{n}{2} = \frac{n!}{2!(n-2)!} = \frac{n(n-1)}{2}.$$

(When c_{ij} and c_{ji} are not equal, twice as many savings terms must be computed.)

The savings method is feasible to solve by hand for only small values of n. For example, for $n = 10$ there are 45 terms, and for $n = 100$ there are nearly 5,000 terms. However, as long as the constraints are not too complex, the method can be implemented easily on a computer.

We illustrate the method with the following example.

Example 6.4

Whole Grains is a small bakery that supplies five major customers with bread each morning. If we locate the bakery at the origin of a grid (i.e., at the point (0, 0)), then the five customer locations and their daily requirements are

Customer	Location	Daily Requirements (loaves)
1	(15, 30)	85
2	(5, 30)	162
3	(10, 20)	26
4	(5, 5)	140
5	(20, 10)	110

The relative locations of the Whole Grains bakery and its five customers are shown in Figure 6–7. The bakery has several delivery trucks, each having a capacity of 300 loaves. We shall assume that the cost of traveling between any two locations is simply the straight-line or Euclidean distance between the points. The formula for the straight-line distance separating the points (x_1, y_1) and (x_2, y_2) is

$$\sqrt{(x_1 - x_2)^2 + (y_1 - y_2)^2}.$$

The goal is to find a delivery pattern that both meets customer demand and minimizes delivery costs, subject to not exceeding the capacity constraint on the size of the delivery trucks.

FIGURE 6–7

Customer locations in Example 6.4

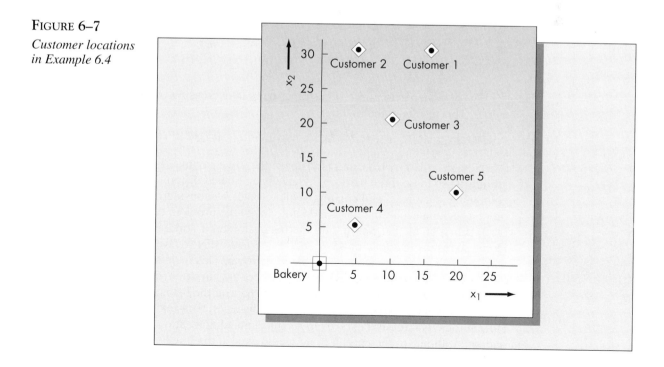

Solution

The first step is to compute the cost for each pair (i, j) where i and j vary from 0 to 5. We are assuming that this cost is the straight-line distance between the points representing customer locations. The straight-line distances are given in the matrix below.

Cost Matrix (c_{ij})

		TO					
		0	1	2	3	4	5
FROM	0		33.5	30.4	22.4	7.1	22.4
	1			10.0	11.2	26.9	20.6
	2				11.2	25.0	25.0
	3					15.8	14.1
	4						15.8

Next, we compute the savings for all pairs (i, j), $1 \leq i < j \leq 5$. There are a total of 10 savings terms to compute for this example:

$$s_{12} = c_{01} + c_{02} - c_{12} = 33.5 + 30.4 - 10 = 53.9,$$
$$s_{13} = c_{01} + c_{03} - c_{13} = 33.5 + 22.4 - 11.2 = 44.7.$$

The remaining terms are computed in the same way, with the results

$$s_{14} = 13.7, \qquad s_{25} = 27.8,$$
$$s_{15} = 35.3, \qquad s_{34} = 13.7,$$
$$s_{23} = 41.6, \qquad s_{35} = 30.7,$$
$$s_{24} = 12.5, \qquad s_{45} = 13.7.$$

The next step is to rank the customer pairs in decreasing order of their savings values. This results in the ranking

(1, 2), (1, 3), (2, 3), (1, 5), (3, 5), (2, 5), (1, 4), (3, 4), (4, 5), and (2, 4).

Note that (1, 4), (3, 4), and (4, 5) have the same savings. Ties are broken arbitrarily, so these three pairs could have been ranked differently. We now begin combining customers and creating vehicle routes by considering the pairs in ranked order, checking each time that we do not violate the problem constraints. Because, (1, 2) is first on the list, we first try linking customers 1 and 2 on the same route. Doing so results in a load of $85 + 162 = 247$ loaves. Next, we consider combining 1 and 3, which means including 3 on the same route. This results in a load of $247 + 26 = 273$, which is still feasible. Hence, we have now constructed a route consisting of customers 1, 2, and 3. The next pair on the list is (2, 3). However, 2 and 3 are already on the same route. Next on the list is (1, 5). Linking customer 5 to the current route is infeasible, however. As the demand at location 5 is 110 loaves, adding location 5 to the current route would exceed the truck's capacity. The next

FIGURE 6–8

Vehicle routing found from the savings method for Example 6.4

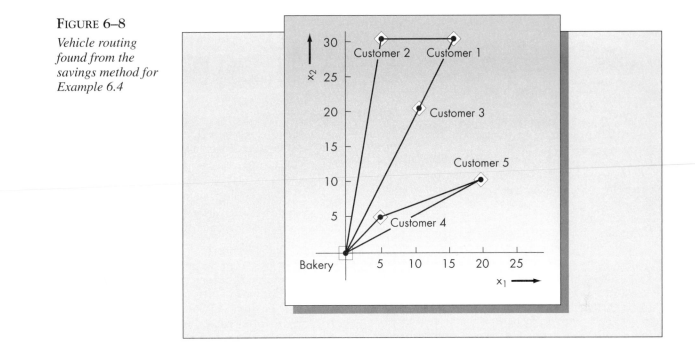

feasible pair on the list is (4, 5) which we make into a new route. The solution recommended by the savings method consists of two routes, as shown in Figure 6–8.

We should point out that the savings method is only a heuristic. It does not necessarily produce an optimal routing. The problem is that forcing the choice of a highly ranked link may preclude other links that might have slightly lower savings but might be better choices in a global sense by allowing other links to be chosen downstream. Several authors have suggested modifications of the savings method to attempt to overcome this difficulty. The interested reader should refer to Eilon et al. (1971) for a discussion on these methods. However, the authors point out that these modifications do not always result in a more cost-effective solution. ∎

Practical Issues in Vehicle Scheduling

We may classify vehicle scheduling problems as one of two types: arc-based or node-based. Arc-based problems are ones in which the goal is to cover a certain collection of arcs in a network. Typical examples are snow removal and garbage collection. The type of distribution problems we have discussed in this section are node-based problems. The objective is to visit a specified set of locations. Problems also may be a combination of both of these.

Most real vehicle scheduling problems are much more complex than that described in Example 6.4. Schrage (1981) lists 10 features that make real problems difficult to solve. These include the following seven features:

1. *Frequency requirements.* Visits to customers may have to occur at a certain frequency, and that frequency may vary from customer to customer. In our example, bread deliveries are made daily to each customer, so that frequency is not an issue. Consider, however, the problem of delivering oil or gas for

SNAPSHOT APPLICATION

Air Products Saves Big with Routing and Scheduling Optimizer

The Air Products Corporation is a producer of industrial gases located in Allentown, Pennsylvania. Products include liquid oxygen (steel production, sanitation) and liquid nitrogen (quick freezing of food, grinding operations), among others. Air Products has always been a leader in the field and pioneered innovative distribution strategies, such as building generating facilities near large customers to avoid costly and dangerous movement of gases. By the 1970s, the plants, trucks, and other equipment used by Air Products were highly automated. This was in sharp contrast to the manual system for scheduling the delivery of its products. During this period, management recognized that in order for the firm to remain competitive, it would have to improve its delivery system. Scheduling delivery of industrial gases has several features that make the problem much more complex than the types of vehicle scheduling problems discussed earlier in this section. These added complications include

- Individual deliveries can be split among multiple trucks, since only the total delivery quantity is important.

- Deliveries must be made within a certain time window only. Customers can accept deliveries only on certain days and at certain times.
- Trucks differ in capacities and operating costs. Furthermore, capacities are mandated by state law and can vary from one state to another.
- Some trucks cannot serve some customers. They may be too big or may require external power.

Bell et al. (1983) report on a complex routing and scheduling optimizer developed for the company. Schedules are found by solving an extremely large (up to 800,000 variables and 200,000 constraints) mixed-integer program to near optimality using a method known as Lagrangian relaxation. The authors report significant savings to Air Products from this system. The most tangible benefit was a reduction in delivery expenses resulting from the increased productivity of the drivers and the vehicles. Mileage savings were estimated to be about 10 percent of costs, which translated to several million dollars for a small subsystem of 16 depots. This application shows that sophisticated operations research techniques can be successfully implemented in a real environment.

residential use. The frequency of delivery depends on the usage rate, so delivery frequency will vary from customer to customer.

2. *Time windows.* This refers to the requirement that visits to customer locations be made at specific times. Dial-a-ride systems and postal and bank pickups and deliveries are typical examples.

3. *Time-dependent travel time.* When deliveries are made in urban centers, rush-hour congestion can be an important factor. This is an example of the case in which travel time (and hence the cost associated with a link in the network) depends on the time of day.

4. *Multidimensional capacity constraints.* There may be constraints on weight as well as on volume. This can be a thorny issue, especially when the same vehicles are used to transport a variety of different products.

5. *Vehicle types.* Large firms may have several vehicle types from which to choose. Vehicle types may differ according to capacity, the cost of operation, and whether the vehicle is constrained to closed trips only (where it must return

to the depot after making deliveries). When several types of vehicles are available, the number of feasible alternatives increases dramatically.

6. *Split deliveries.* If one customer has a particular requirement, it could make sense to have more than one vehicle assigned to that customer.

7. *Uncertainty.* Routing algorithms invariably assume that all the information is known in advance. In truth, however, the time required to cross certain portions of a network could be highly variable, depending on factors such as traffic conditions, weather, and vehicle breakdowns.

Problems for Section 6

13. Re-solve Example 6.4 assuming that the capacity of the vehicles is only 250 loaves of bread.

14. Re-solve Example 6.4 assuming that the distance between any two locations is the rectangular distance rather than the Euclidean distance. (See Chapter 10, Section 7, for a definition of rectangular distance.)

15. Add the following customer locations and requirements to Example 6.4 and resolve:

Customer	Location	Daily Requirement
6	(12, 12)	78
7	(23, 3)	126

16. Suppose that one wishes to schedule vehicles from a central depot to five customer locations. The cost of making trips between each pair of locations is given in the following matrix. (Assume that the depot is location 0.)

Cost Matrix (c_{ij})

			TO			
		0	*1*	*2*	*3*	*4*	*5*
F	*0*		20	75	33	10	30
R	*1*			35	5	20	15
O	*2*				18	58	42
M	*3*					40	20
	4						25

Assume that these costs correspond to distances between locations and that each vehicle is constrained to travel no more than 50 miles on each route. Find the routing suggested by the savings method.

17. All-Weather Oil and Gas Company is planning delivery routes to six natural gas customers. The customer locations and gas requirements (in gallons) are

given below.

Customer	Location	Requirements (gallons)
1	(5, 14)	550
2	(10, 25)	400
3	(3, 30)	650
4	(35, 12)	250
5	(10, 7)	300

Assume that the depot is located at the origin of the grid and that the delivery trucks have capacity of 1,200 gallons. Also assume that the cost of travel between any two locations is the straight-line (Euclidean) distance between them. Find the route schedule obtained from the savings method.

6.7 Designing Products for Supply Chain Efficiency

Product design was traditionally a function that was totally divorced from more mundane operations management issues. Designers would be concerned primarily with aesthetics and marketability. Little attention would be given to nuts-and-bolts concerns such as manufacturing and logistics at the design stage. As quality and reliability advanced to the forefront, it became clear that product reliability and product design are closely linked. The design for manufacturability (DFM) movement developed out of a need to know why products fail and how those failures could be minimized. Once it was understood that reliability could be factored in at the design stage, the link between design and manufacturing was forged. Another term for DFM is concurrent engineering.

In recent years, firms have realized that the logistics of managing the supply chain can have as much impact on the bottom line as product design and manufacturing. More than ever, we are seeing innovative designs that take supply chain considerations into account. One way of describing this concept is design for logistics (DFL). Another is three-dimensional concurrent engineering (3-DCE), a term adopted by Fine (1998). The three dimensions here are product, process, and supply chains. It carries the concept of concurrent engineering one step further. Concurrent engineering means that product-related issues (functionality, marketability) and process-related issues (how the product is produced, reliability and quality of the final product) are joint considerations in the design phase. Three-dimensional concurrent engineering means that supply chain logistics is also considered in the product design phase.

Two significant ways that logistics considerations enter into the product design phase are

1. Product design for efficient transportation and shipment.
2. Postponement of final product configuration (also called delayed differentiation).

Products that can be packed, moved, and stored easily streamline both manufacturing and logistics. Buyers prefer products that are easy to store and easy to move. Some products tend to be large and bulky and present a special challenge in this regard. An example is furniture. Swedish-based Ikea certainly did an excellent job of designing products that are modular and easily stored. Furniture is sold in simple-to-assemble kits that allow Ikea retailers to store furniture in the same warehouselike locations at which they are displayed. Simchi-Levi et al. (1999) discuss several other examples of products whose success is partially based on their ease of shipment and storage.

Postponement in Supply Chains

One of the key findings from supply chain research in recent years is the value of postponing the configuration of the final product as long as possible. The first application of this principle known to this writer was implemented by Benetton (Signorelli and Heskett, 1984). The Benetton Group is an Italian-based maker of fashion clothes that had, by 1982, become the world leader in the field of knitwear. About 60 percent of the garments sold by the firm are made of wool. Traditionally, wool is dyed before it is knitted. In 1972 Benetton undertook a unique strategy: to dye the garments *after* they were knitted. One might well question this strategy since labor and production costs for garments dyed after manufacture are about 10 percent higher than for garments knitted from dyed thread.

The advantage of reversing the order of the dying and the knitting operations is that it provides additional time before committing to the final mix of colors. This time gives the firm a chance to gain additional data on consumer preferences for colors. Benetton's knitwear included nearly 500 color and style combinations. Undyed garments are referred to as *gray stock*. Keeping inventories of gray stock rather than dyed garments had several advantages. First, if a specific color became more popular than anticipated, Benetton could meet the demand for that color. Second, the company would run less risk of having large unsold stockpiles of garments in unpopular colors. These advantages more than offset the higher costs of dying the knitted garments rather than the raw wool.

How does postponement correlate with inventory management theory? As we saw in Chapter 5, safety stock is retained to protect against demand uncertainty over the replenishment lead time. The lead time for garments of a specific color is reduced by postponing the dying operation. It follows that the uncertainty is reduced as well, thus achieving comparable service levels with less safety stock.

Postponement has become a key strategy in many diverse industries. Another example discussed in the literature is that of Hewlett-Packard (HP) (Lee, Billington, and Carter, 1994). HP is one of the world's leading producers of inkjet and laser printers, among other products. Printers are sold worldwide. While the basic mechanisms of the printers sold overseas are the same as the American versions, subassemblies, such as power supplies, must be customized for local markets. HP's original strategy was to configure printers for local requirements (i.e., localize them) at the factory. That is, printers with the correct power supplies, plugs, and manuals would be produced at the factory, sorted, and shipped overseas as final products. The result was that HP needed to carry large safety stocks of all printer configurations.

In order to reduce inventories and improve the service provided by the distribution centers (DCs) to retail customers, HP adopted a strategy similar to Benetton's. Printers sent from the factory would be generic or gray stock. Localization would be done at the DC level rather than at the factory. As with Benetton, this had the ultimate effect of reducing necessary safety stocks while improving service. Printers are shipped to overseas DCs by boat, requiring one-month transit time. With local customization of the product, the replenishment lead time for locally configured printers was dramatically reduced. Also, the demand on the factory is now the aggregate of the demands at the DCs. As we saw in Chapters 2 and 5, aggregate demands have relatively smaller variation. Lee, Billington, and Carter showed that DC localization for HP Deskjet-Plus printers would lead to an 18 percent reduction in inventories with no reduction in service levels.

According to Harold E. Edmondson, an HP vice president (as quoted in Lee, Billington, and Carter, 1994):

> The results of this model analysis confirmed the effectiveness of the strategy for localizing the printers at remote distribution centers. Such a design strategy has significant benefits in terms of increased flexibility to meet customer demands, as well as savings in both inventory and transportation costs . . . I should add that the design for localization concept is now part of our manufacturing and distribution strategy.

The notion of postponing customization of a product appears to be gaining acceptance. For example, semiconductors are produced in generic form when possible and customized through programming or other means after firm orders are received (Barrone, 1996). Firms in many industries are becoming aware of the risk-pooling and lead time reduction benefits from postponement and local customization of products.

Additional Issues in Supply Chain Design

While products can be better designed for efficient supply chain operation, there are several important issues to consider in the design of the supply chain itself. Fine (1998) refers to supply chain design as the "ultimate core competency." Three important relevant issues are

- The configuration of the supplier base.
- Outsourcing arrangements.
- Channels of distribution.

Configuration of the Supplier Base.   The number of suppliers, their locations, and their sizes are important considerations for efficient supply chain design. In recent years, the trend has been to reduce the number of suppliers and develop long- term arrangements with the existing supplier base. An example is the Xerox Corporation. Jacobson and Hillkirk (1986) discuss several reasons for the impressive turnaround of Xerox during the 1980s. One of the company's strategies was to streamline the supply chain on the procurement side by reducing the number of their suppliers from 5,000 to only about 400. Those 400 suppliers included a mix of both overseas and local suppliers. The overseas suppliers were chosen primarily on a cost basis, while the local suppliers could provide more timely deliveries when necessary.

Cooperative efforts between manufacturers and suppliers (as well as manufacturers and retailers on the other end of the chain) have been gaining popularity. Traditionally this relationship was an adversarial one. Today, with the advent of arrangements like vendor-managed inventories and MRP II, suppliers and manufacturers work closely and often share what was once proprietary data to improve performance on both ends.

Outsourcing Arrangements.   Outsourcing of manufacturing has become a popular trend in recent years. Successful contract manufacturers such as the Solectron Corporation experienced rapid growth in the past decade. Many firms are outsourcing

SNAPSHOT APPLICATION

Dell Computer Designs the Ultimate Supply Chain

Dell Computer was one of the biggest success stories of the 1990s. How successful? The stock price increased a whopping 27,000 percent during the decade. There is no question that a sleek, efficient supply chain design was a contributing factor to Dell's phenomenal success.

Michael Dell, the firm's founder, began by assembling and reselling IBM PCs from his dorm room at the University of Texas in the early 1980s. Later, in the mid-1980s, he formed PC's Limited, which offered one of the first of the mail order "clones." The computers were clones of the IBM XT, which was selling for several thousand dollars at the time. PC's Limited sold a basic box consisting of a motherboard with an Intel 8088 chip, a power supply, a floppy drive, and a controller for $795. The buyer had to add a graphics card, a monitor, and a hard drive to make the system functional, but the final product cost less than half as much as an equivalent IBM XT and ran faster. (This was, in fact, the first computer purchased by this writer.) From this modest beginning grew the multibillion-dollar Dell Computer Corporation which dominates the PC market today.

To understand Dell's success, one must understand the nature of the personal computer marketplace. The central processor is really the computer, and its power determines the power of the PC. In the PC marketplace, Intel has been the leader in designing every new generation of processor chip. As of this writing, Intel is selling the sixth generation of processors. The progression of microprocessors from the first IBM PC until the present is the 8088 (and the similar 8086), 286, 386, 486, Pentium, and Pentium II. A few companies, such as AMD and Texas Instruments, have had limited success in this market, but the relentless new-product introduction by Intel has resulted in competitors' processors becoming obsolete before they can garner significant market share. Each new generation of microprocessors renders old technology obsolete. For that reason, computers and computer components have a relatively short life span. As new chips are developed at an ever-increasing pace, the obsolescence problem becomes even more critical.

How does one avoid getting stuck with obsolete inventory? Dell's solution was simple and elegant. Don't keep *any* inventory! All PCs purchased from Dell are made to order. Dell does not build to stock. They do store components, however. Although Dell can't guarantee all components are used, their market strategy is designed to move as much of the component inventory as possible. Dell focuses only on the latest technology, and both systems and components are priced to sell quickly. Furthermore, because of their high volume, they can demand quantity discounts from suppliers. Dell targets high-end users who demand the latest technology and are willing to pay for it. Profit margins are higher in this market segment, thus accounting for Dell's exceptional performance in an industry where margins are very slim.

As the so-called clockspeed of the industry (a term coined by Fine, 1998) increases, Dell's advantage over manufacturers with more traditional supply chain designs also increases. For example, Compaq also has been successful over the past decade by selling through Sears, Circuit City, and other major retailers. However, Compaq runs the risk of having obsolete product in the reseller's distribution channels. Why don't traditional PC manufacturers copy the Dell approach? As Fine points out, switching from one system to another isn't so easy. The changeover period could be fatal if consumers experience out-of-stock situations at their local retailers. (However, as of this writing, many other computer manufacturers, including Compaq and Gateway, offer PCs directly to the consumer via the web.)

---

supply chain functions as well. Third-party logistics (3PL) is becoming big business. For example, Saturn Corporation now outsources a major portion of its logistics operation to Ryder Trucks.

Dreifus (1992) describes a case of a smaller firm that decided to outsource purchasing. The Singer Furniture Company contracted with the Florida-based IMX Corporation to handle their procurements. In the agreement, IMX handles all negotiations from consolidated shipments, takes care of the paperwork, conducts quality-control inspection, and searches for new suppliers in exchange for a fixed commission. In a trial run, IMX suggested a switch from a Taiwanese supplier of bedposts to one based in the Caribbean, which resulted in a 50 percent reduction in price to Singer. From Singer's point of view, the primary advantages of entering into this arrangement were

Buying economies of scale. IMX buys for several clients (including other furniture makers), which allows them to negotiate lower unit costs on larger orders. There are also economies of scale in shipping costs from overseas. IMX claims that it saves at least 20 percent for clients in this way.

Reduced management overhead. Singer was able to eliminate its international purchasing unit and several related management functions.

Lower warehousing costs. IMX bears much of the risk on several lines of product by absorbing local storage costs.

Above-board accounting. The firms agreed on open procedures and disclosures to be sure that the quality programs are functioning and charges are in line with costs.

Channels of Distribution.   Failure to establish solid channels of distribution can spell the end for many firms. The design of the distribution channel includes the number and the configuration of distribution centers; arrangements with third-party wholesalers; licensing and other agreements with retailers, including vendor-managed inventory programs; and establishment of the means to move product quickly through the channel. Direct selling to consumers has been an enormously successful strategy for some retailers, especially for products that become obsolete quickly. By eliminating middlemen, manufacturers reduce the risk of product dying in the supply channel. This has been a successful strategy for computer makers Dell and Gateway and for the bookseller Amazon.com. Dell Computer's phenomenal success in the 1990s can be attributed partially to their outstanding supply chain design, highlighted in the Snapshot Application in this section.

Problems for Section 7

18. Describe the concept of postponement in supply chains. If you were planning a trip to a distant place, what decisions would you want to postpone as long as possible?

19. Many automobiles can be ordered in one of two engine sizes (examples are the Lincoln LS, the Lexus Coupe, and the Jaguar S Type) but are virtually identical in every other way. How might these automakers use the concept of postponement in their production planning?

20. Discuss why having too many suppliers can be troublesome. Can having too few suppliers also be a problem?

21. Many large companies that have their own manufacturing facilities and logistics organizations outsource a portion of their production (such as IBM) or their supply chain operations (such as Saturn). Why do you suppose this is?

6.8 The Role of Information in the Supply Chain

How often have we heard it said that we are living in the "information age"? The availability of information is increasing at an exponential rate. New sources of information in the form of academic and trade journals, magazines, newsletters, and so on are introduced every day. The explosion of information availability on the web has been truly phenomenal. Web searches are now the first place many go for information on almost anything.

Knowledge is power. In supply chains, information is power. It provides the decision maker the power to get ahead of the competition, the power to run a business smoothly and efficiently, and the power to succeed in an ever more complex environment. Information plays a key role in the management of the supply chain. As we saw in the earlier chapters of this book, many aspects of operations planning start with the forecast of sales and build a plan for manufacturing or inventory replenishment from that forecast. Forecasts, of course, are based on information.

An excellent example of the role of information in supply chains is the Harvard Business School cases Barilla SpA (A and B) (1994) written by Jan Hammond. In many operations management courses, these cases provide an introduction to the role of information in supply chains. Barilla is an Italian company that specializes in the production of pasta. In the late 1980s, Barilla's head of logistics tried to introduce a new approach to dealing with distributors, which he called just-in-time distribution (JITD). Briefly, the idea was to obtain sales data directly from the distributors (Barilla's customers) and use these data to allow Barilla to determine when and how large the deliveries should be. At that time, Barilla was operating in the traditional fashion. Distributors would independently place weekly orders based on standard reorder point methods. This led to wide swings in the demand on Barilla's factories due to the "bullwhip effect," which is discussed in detail below. The JITD idea met with staunch resistance, both within and outside Barilla. The Barilla sales and marketing organizations, in particular, were most threatened by the proposed changes. Distributors were concerned that their prerogatives would be compromised.

Without going into the details (which can be found in Barilla SpA (B)), management eventually did prevail, and the JITD system was implemented with several of Barilla's largest distributors. The results were striking. Delivery reliability improved, and the variance of orders placed on the factory were substantially reduced. The program proved a win-win for Barilla and its customers.

Barilla's success with its JITD program is one example of what today is known as vendor-managed inventory (VMI). VMI programs have been the mainstay of many successful retailers in the United States. For example, Procter & Gamble has assumed responsibility for keeping track of inventories for several of its major clients, such as Wal-Mart. Years ago, grocery store managements were responsible for keeping track of their inventories. Today, it is commonplace to see, in our local grocery stores, folks checking shelves who are not store employees but employees of the manufacturers. With VMI programs, it becomes the manufacturer's responsibility to keep the shelves stocked. While stock-outs certainly hurt the retailer, they hurt the manufacturer even more. When a product stocks out, customers typically will substitute another, so the store makes the sale anyway. It is the manufacturer that really suffers the penalty of lost sales, so that the manufacturer has a strong incentive to keep shelves stocked.

The Bullwhip Effect

Barilla's experience prior to the implementation of their VMI program is one example of the bullwhip effect. It has become a topic of considerable interest among both practitioners and academics alike. The history of bullwhip appears to be the following.

Executives at Procter & Gamble (P&G) were studying replenishment patterns for one of their best-selling products: Pampers disposable diapers. They were surprised to see that the orders placed by distributors had much more variation than sales at retail stores. Furthermore, orders of materials to suppliers had even greater variability. Demand for diapers is pretty steady, so one would assume that variance would be low in the entire supply chain. However, this was clearly not the case. P&G coined the term "bullwhip" effect for this phenomenon. It also has been referred to as the "whiplash" or "whipsaw" effect.

This phenomenon was observed by other firms as well. HP experienced the bullwhip effect in patterns of sales of its printers. Orders placed by a reseller exhibited wider swings than retail sales, and orders placed by the printer division to the company's integrated circuit division had even wider swings. Figure 6–9 shows how variance increases as one moves up the supply chain.

Where does the bullwhip effect come from? One cause stems from a phenomenon discussed in Chapter 7 in MRP systems. Consider a basic two-level MRP system in which the pattern of final demand is fixed and constant. Since items are produced in batches of size Q, the demand pattern one level down is much spikier. The tendency for lower levels in the supply chain to batch orders in this way is one of the root causes of the bullwhip effect.

High levels of inventories, caused in part by the bullwhip effect, are common in the grocery industry as well. To address this problem, the industry has adopted the efficient consumer response (ECR) initiative (see, for example, Crawford, 1994). The total food delivery supply chain, from the point at which products leave the manufacturer to the point when they are stocked on the shelves at the retailer, has an average of over 100 days of supply. The stated goal of the ECR initiative is to save $30 billion annually by streamlining food delivery logistics.

Another example of the bullwhip effect is the popular "beer game" due to Sterman (1989). Participants play the roles of retailers, wholesalers, and manufacturers of beer. Communication among participants is prohibited: each player must make ordering decisions based on what is demanded from the downstream player only. What one observes are wild swings in orders placed downstream even when the original demands are fairly stable. This is a consequence of the bullwhip effect.

FIGURE 6–9

Increasing variability of orders up the supply chain

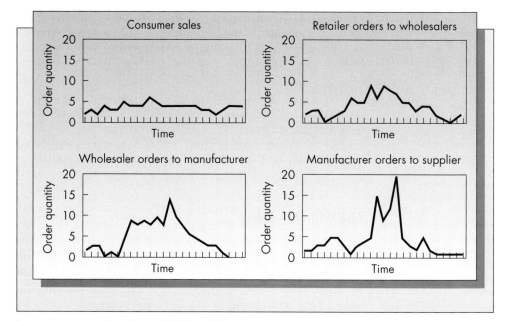

Source: H. L. Lee, P. Padmanabhan, and S. Whang, 1997.

The telescoping variation in the demand patterns in a supply chain results in a planning nightmare for many industries. What can be done to alleviate these effects? First, we need to understand the causes of this phenomenon. According to Lee, Padmanabhan and Whang (1997), there are four primary causes of the bullwhip effect:

- Demand forecast updating.
- Order batching.
- Price fluctuations.
- Shortage gaming.

We consider each of these effects in turn. *Demand forecasts* at each stage of the supply chain are the result of demands observed one level downstream (as in the beer game). Only at the final stage of the chain (the retailer) are consumer demands observed directly. When each individual in a serial supply chain determines his or her demand forecasts individually, the bullwhip effect results. The retailer builds in safety stocks to protect against uncertainty in consumer demand. These safety stocks cause the retailers' orders to have greater variance than the consumers'. The distributor observes these swings in the orders of the retailer and builds in even larger safety stocks, and so on.

Order batching is the phenomenon we saw in MRP systems that results in smooth demand patterns being translated to spiky demand patterns at lower levels of the product structure. The natural tendency to save fixed costs by ordering more frequently (which is the idea behind the EOQ formula) gives rise to order batching. Given the cost structure at each level, this is certainly a reasonable response.

When *prices fluctuate,* there is a speculative motive for holding inventories. (This was first discussed in the inventory management context by Arrow, 1958, and relates to the motivations for holding cash postulated by the economist John Maynard Keynes.) In the food industry, the majority of transactions from the manufacturer to distributors are made under a "forward buy" arrangement. This refers to the practice of buying in advance of need because of an attractive price offered by manufacturers. Such practices also contribute to the bullwhip effect. Large orders are placed when promotions are offered.

Shortage gaming occurs when product is in short supply and manufacturers place customers on allocation. When customers find out that they may not get all the units they request, they simply inflate orders to make up for the anticipated shortfall. For example, if a computer manufacturer expects to receive only half a request for a CPU in short supply, they can simply double the size of their order. If anticipated demands don't materialize, orders can be canceled. The result is that the manufacturer gets an inflated picture of the real demand for the product. This could have dire consequences for a manufacturer placing large amounts of capital into capacity expansion based on "phantom" demands.

Clearly, the bullwhip effect is not the result of poor planning or irrational behavior on the part of players in the supply chain. Each individual acts to optimize his or her position. What, then, can be done to alleviate the situation? There are several potential remedies, which we discuss below. However, these remedies must take into account that people behave selfishly. The motivation for bullwhip-inducing behavior must be eliminated.

Lee, Padmanabhan, and Whang (1997) recommend four initiatives. They are

1. Information sharing.
2. Channel alignment.
3. Price stabilization.
4. Discouragement of shortage gaming.

1. *Information sharing* means that all parties involved share information on POS (point-of-sale) data and base forecasts on these data only. Information sharing can be accomplished by several techniques, one of which is EDI (electronic data interchange), discussed in detail below. The trend toward information sharing is beginning to take hold. Computer manufacturers, for example, are now requiring sell-through data from resellers (these are data on the withdrawal of stocks from the central warehouse). At some point, we expect to see manufacturers linked directly to sources of POS data.

2. *Channel alignment* is the coordination of efforts in the form of pricing, transportation, inventory planning, and ownership between upstream and downstream sites in the supply chain. One of the tendencies that defeats channel alignment is order batching. As we noted above, fixed costs motivate order batching. Reducing fixed costs will result in smaller order sizes. One of the important components of fixed costs is the paperwork required to process an order. With new technologies, such as EDI, these costs could be reduced substantially. Another factor motivating large batches is transportation economies of scale. It is cheaper on a per-unit basis to order a full truckload than a partial one. If manufacturers allow customers to order an assortment of items in a single truckload, the motivation to order large batches of single items is reduced. Another trend encouraging small batch ordering is the outsourcing of logistics to third parties. Logistics companies can consolidate loads from multiple suppliers. Outsourcing of logistics (like outsourcing of manufacturing) is expanding rapidly.

3. Pricing promotions motivate customers to buy in large batches and store items for future use. As noted earlier, this is very common in the grocery industry. By *stabilizing pricing,* sales patterns will have less variation. In retailing, the effect of stable pricing is evident by comparing sales patterns at retailers such as Macy's that run frequent promotions and warehouse stores such as Price/Costco that offer everyday low pricing. The warehouse stores experience steadier sales than do the department stores. In fact, for many department stores, promotional sales account for most of their business. Major grocery manufacturers such as P&G, Kraft, and Pillsbury are moving toward a value-pricing strategy and away from promotional pricing for the same reason.

4. One way to minimize excessive orders as a result of *shortage gaming* is to allocate based on past sales records rather than on orders. This will reduce the tendency of customers to exaggerate orders. Several companies (including General Motors) are moving in this direction.

In summary, the bullwhip effect is the consequence of individual agents in the supply chain acting in their own best interests. To mitigate bullwhip effects, several changes must be made. Incentives must be put in place to reduce demand forecast errors, reduce excessive order sizes in allocation situations, and encourage information sharing and system alignment. As these initiatives become policy, everyone, especially the consumer, will benefit from reduced costs and improved supply chain efficiency.

Electronic Commerce

Electronic commerce is a catch-all term for a wide range of methods for effecting business transactions without the traditional paper-based systems. It includes EDI (electronic data interchange), e-mail, electronic funds transfers, electronic publishing, image processing, electronic bulletin boards, shared databases, and all manner of web-based business systems (Handfield and Nichols, 1999). The point-of-sale barcode system, now common in supermarkets and retailers, is another type of electronic commerce. Many well-known American companies currently use multiple forms of electronic commerce. For example, Wilder and McGee (1997) report that the General Electric Corporation, one of the world's largest diversified manufacturers of a wide variety of products, uses web-based transactions systems and EDI as a regular part of its business practices in most divisions. As well as being a user, GE is also a developer of electronically based supply chain systems. Its Information Services division is working with Netscape to develop a system that will allow EDI users to do business on the Internet. It is not only the corporate behemoths like GE that use the Internet to do business. Many smaller businesses are moving their supply chain systems to some form of electronic commerce. As an example, Palevich (1999) reports that Do It Best Corporation, a small distributor of hardware products, building materials, and lumber, uses EDI, Internet-based systems, electronic forecasting, and warehouse management systems to gain competitiveness.

The growth of Internet-based retailing has been legend. Today anyone with a personal computer and a modem has access to an enormous quantity of information as well as the opportunity to purchase almost anything, often at substantial discounts. The rash of IPOs (initial public offerings) and the meteoric rise of the equity prices of the so-called dot-com companies in recent years is without precedent. Equity prices rose even as most Internet-based retailers continued to post losses. Consider the example of Amazon.com, one of the earliest and best-known Internet retailers. The company originally started as an Internet-based bookseller. Amazon's model was to offer substantial discounts (typically 40 percent on hardcover books plus shipping) and the convenience of shopping on the Web. By selling out of a single location, they could reap the benefits of stock centralization (discussed in detail in the next section) and save costs on traditional "bricks and mortar" stores and personnel. However, even with these advantages, it is still not clear when e-retailing will be profitable. As of this writing in early 2000, Amazon (and many other e-retailers) continued to lose money as share prices soared.

The model for web-based retailing is essentially the same as that for catalog shopping. Catalog retailers have been around for many years, with Lands' End and L.L. Bean being perhaps the most successful. However, catalogs need to be developed, printed, and mailed regularly, which is quite expensive, especially considering that most catalogs are thrown away. For this reason, the web could have significant advantages over the traditional mail-order catalog businesses. On the other hand, just because one develops a website doesn't guarantee that it will be visited. Some of the primary portals (AOL and Yahoo, for example) require big money to allow direct access to an e-commerce site.

Although web-based retailing has received the most publicity in recent years, it is only a small part of the activities that compose the vast arena of electronic commerce. Business-to-business web-based systems account for a much greater share of the electronic commerce marketplace than do web-based retailers. For example, companies like Cisco Systems have seen the web-based portion of their business grow from 5 percent in 1996 to more than 80 percent in 2000. The Web has the potential of changing the entire face of supply chain management in coming years.

Electronic Data Interchange

Electronic data interchange (EDI) is one of the enabling technologies for streamlining supply chain operations. EDI refers to the electronic transmission of standard business documents in a predetermined format from one company's business computer to its trading partners' computer (Cannon 1993). What separates EDI from other forms of electronic commerce is that EDI relies on standardized formats rather than proprietary formats. At the present time, the two standards are ANSI and EDIFACT.

EDI is one way that a firm can transmit information, such as point-of-sale demand information, purchase orders, and inventory status information, to users within the firm and to customers and trading partners. EDI could serve as one means of ameliorating the bullwhip effect discussed above, since it provides a means for players in the supply chain to share information. From the point of view of managing the supply chain, the main advantage of EDI is speed. By transmitting information quickly, lead times are reduced. As noted in Chapter 5, it is the uncertainty of demand over the replenishment lead time that results in the need to carry safety stock.

EDI systems have been implemented in a wide variety of settings, but generally by larger firms. The reason is that EDI can be expensive. It requires dedicated software and more advanced hardware than many small firms possess. One area in which EDI has been implemented is VMI (vendor-managed inventory) systems. Gerber Foods, for example, uses VMI with several of its customers and has implemented a consistent EDI system with these retailers to be sure that it can communicate efficiently with them (Cooke, 1999).

Web-Based Transactions Systems

The web provides an attractive alternative to firms that are considering investing in EDI. The web (also known as the world wide web) is a vast collection of connected computers and servers, which can be accessed from anywhere in the world via a gateway or portal. One enters the web with a browser, and its use is facilitated via a search engine. Netscape/AOL and Microsoft produce the most popular browsers, and Yahoo appears to be the leading search engine. The use of the web for both B2C (business to consumer) and B2B (business to business) is growing quickly. The web will have significant implications for supply chain management in the coming years.

During 1999 and the early part of 2000, there was enormous interest in so-called dot-com companies. The vast majority of these are retail operations that allow consumers to order products directly via their computers. One of the most notable dot-com companies is Amazon.com, which was originally designed to be a discount bookseller. Today, Amazon (which has yet to be profitable as of this writing) has expanded into many other markets, and continues to grow. Interest in "e-tailers" (B2C companies) waned somewhat in the latter half of 2000 after a precipitous drop in the Nasdaq, fueled primarily by a drop in this sector.

While B2C companies have gotten the most visibility, it is B2B that holds the most potential for the future. B2B refers to web-based transactions systems that allow firms to share electronic information such as order quantities, forecasts, and future needs. Many software firms offer web-based systems. Some, like Oracle and SAP, bundle the software as part of their comprehensive information storage and retrieval systems. Several other firms offer products specifically designed only for B2B supply chains. Three notable ones are Dallas-based i2 Technologies, and Silicon Valley–based Agile Software and Ariba. Based on information obtained from their websites, i2 offers products for many different industry sectors, including automotive, retailing, energy and chemicals, computers and electronics, metals, pharmaceuticals, semiconductors, soft goods, and transportation and logistics. Agile Software has designed its products to take advantage of the trend toward the outsourcing of manufacturing. As Agile enlists more original equipment manufacturers (OEMs), their suppliers and enablers are encouraged also to use Agile's suite to remain compatible. Ariba indicates that they supply buyers, market makers, and suppliers.

As of this writing in 2000, there are more than 100 different firms offering web-based supply chain solutions. Many tailor their products to specific industry segments, such as Printnation.com (printing), Rightfreight.com (airfreight carriers), and Supply Solution, Inc. (automotive aftermarket). As with any new industry, most of these firms will not survive, and a few market leaders will emerge.

What advantages do web-based supply chain systems offer? For one, virtually everyone has access to the web. The speed of web access is increasing at a steady rate with the advent of faster computers and faster modes of transmitting data. Second, the web does not require the complex protocols that characterize EDI systems. Third, web-based systems can be set up and implemented at a fraction of the cost of EDI.

One might think that web-based systems dominate EDI systems based on these advantages, but this has not been the case up to now. Many firms are still concerned about security. The complex EDI protocols provide much tighter security than do web-based systems. Some companies are exploring hybrid systems in which EDI documents are exchanged over the Internet. This provides larger firms that have heavy investment in legacy EDI systems the means to integrate smaller customers into their existing systems (Sliwa, 1999). However, one has to believe that with all the advantages that web-based systems offer, they will begin to replace EDI systems in the coming years.

Problems for Section 8

22. What is the bullwhip effect? What is the origin of the term?

23. Do you think that "eliminating the middleman" would always get rid of the bullwhip effect? Under what circumstances might it not work?

24. Discuss the four initiatives recommended for ameliorating the bullwhip effect in supply chains.

25. What are the advantages and disadvantages of web-based transactions systems compared to EDI?

26. Discuss the advantages and disadvantages of e-retailing as compared to traditional retailing.

6.9 Multilevel Distribution Systems

In many large commercial and military inventory systems, it is common for stock to be stored at multiple locations. For example, inventory may flow from manufacturer to regional warehouses, from regional warehouses to local warehouses, and from local warehouses to stores or other point-of-use locations. Determining the allocation of inventory among these multiple locations is an important strategic consideration, and can have a significant impact on the bottom line. A typical multilevel distribution system is pictured in Figure 6–10.

In the parlance of traditional inventory theory, such systems are referred to as multiechelon inventory systems. The interest in this area was originally sparked by a seminal paper by Clark and Scarf (1960) that was part of a major initiative at the Rand Corporation to study logistics and inventory management problems arising in the military. This paper laid the theoretical framework for what later became a very large body of research in multiechelon inventory theory. (There is a brief overview of multiechelon inventory models in Chapter 5.)

Including intermediate storage locations in a large-scale distribution system has advantages and disadvantages. The advantages include

1. Risk pooling.
2. Distribution centers that can be designed to meet local needs.
3. Economies of scale in storage and movement of goods.
4. Faster response time.

FIGURE 6–10

Typical multilevel distribution system

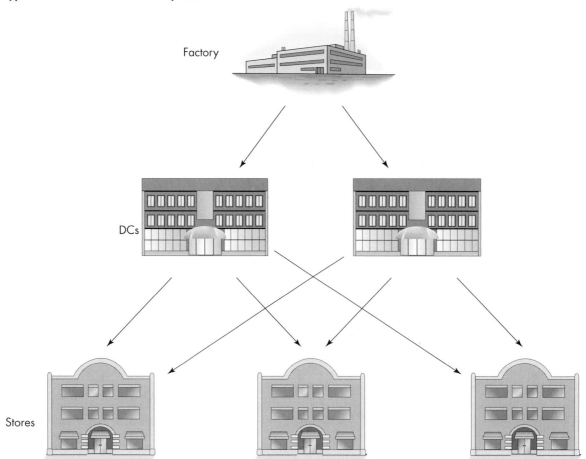

Factory

DCs

Stores

Risk pooling derives from the property that in most cases aggregated demands have lower relative variation than individual demands. (This property also was noted in the context of forecasting in Chapter 2.) Suppose a distribution center serves 50 retail outlets. By holding the majority of the stock at the distribution center rather than at the stores, the same systemwide service level could be achieved with less total inventory. We show below with a simple example how this comes about.

In retailing, the mix of products sold depends on the location of the store. For example, one sells more short-sleeve shirts in Arizona than in Maine. Hence, *distribution centers* located in the Southwest would stock a different assortment of products than distribution centers located in the Northeast. Distribution centers could be designed to take these differences into account.

Distribution centers allow for *economies of scale* in the following way. If all product were shipped directly from the factory to the local outlet, shipment sizes would be relatively small. Large bulk shipments from factories located overseas could be made to distribution centers at a lower unit cost. In that way, the smaller shipments would be made over shorter distances from distribution centers to stores.

Finally, because distribution centers can be located closer to the customer than can factories, demands can be met more quickly. Retailers try to locate distribution centers in the United States within one day's drive away from any store. (This is why Ohio is a popular place for retail distribution centers, since this is the geographic center of the country.) In this way, store inventories can be replenished overnight if necessary.

Multilevel distribution systems could have several disadvantages as well. These include that they

1. May require more inventory than simpler distribution systems.
2. May increase total order lead times from the plant to the customer.
3. Could ultimately result in greater costs for storage and movement of goods.
4. Could contribute to the bullwhip effect.

Depending on the number of distinct locations used, it is likely that a multilevel system requires more overall total inventory than a single-level system, since safety stock is likely built in at each level. This means that more money is tied up in the pipeline.

When distribution centers experience stock-outs, the response time of the system from manufacturer to point of sale could actually be worse in a multilevel system. This is a consequence of the fact that in the multilevel case, the total lead time from plant to store is the sum of the lead times at each level. (However, in a well-managed system, stock-outs at the distribution center should be very rare occurrences.)

One must not ignore the cost of the distribution center itself. Building a modern, large-scale facility could cost upward of $500 million. In addition, there are recurring costs such as rent and labor. As a result, building and maintaining a multilocation distribution system is expensive.

The *bullwhip effect,* discussed in detail in the previous section of this chapter, is the propensity of the variance of orders to increase as one moves up the supply chain. Adding additional levels to a distribution system could cause a bullwhip effect.

Perhaps the most significant advantage of centralized storage is the opportunity for risk pooling, as noted above. To understand how risk pooling works, consider the following simple scenario. Assume n independent retail locations stock similar items. For example, these could be Macy's department stores located in different cities. Let us further assume that the stock level for a particular item is determined from the newsboy model. Referring to the discussion in Chapter 5, Section 5, there are known values of the unit overage cost, c_o, and the unit underage cost, c_u. In Chapter 5 it was proven that the optimal stocking level is the $c_u/(c_u + c_o)$ fractile of the demand distribution. (If we assume that stocking levels are determined based on service levels instead of costs, then the critical ratio *is* the service level and c_o and c_u need not be known.)

To simplify the analysis, let us suppose that the demand for this item follows the same normal distribution at each store with mean μ and standard deviation σ, and the demands are independent from store to store. Let z^* be the value of the standard normal variate that corresponds to a left-tail probability equal to the critical ratio. Then, as is shown in Section 5.3, the optimal policy is to order up to $Q = m + \sigma z^*$ at each location. The safety stock held at each location is σz^*, so the total safety stock in the system is $n\sigma z^*$. Refer to this case as the decentralized system.

Alternatively, suppose that all inventory for this item is held at a single distribution center and shipped overnight on an as-needed basis to the stores. Let's consider the amount of safety stock needed in this case to provide the same level of service as with the decentralized system. Since store demands are independent normal random variables with mean μ and variance σ^2, the aggregated demand from all n stores is also normal, but with mean $n\mu$ and variance $n\sigma^2$. The standard deviation of the aggregated demand therefore has standard deviation $\sigma\sqrt{n}$. This means that to achieve the same level of service, the warehouse needs to stock up to the level Q_w given by $Q_w = n\mu + z^*\sigma\sqrt{n}$. The total safety stock is now $z^*\sigma\sqrt{n}$. This corresponds to a centralized system.

Forming the ratio of the safety stock in the decentralized system over the safety stock in the centralized system gives $z^*\sigma n / z^*\sigma\sqrt{n} = \sqrt{n}$. Hence, the decentralized system will have to hold $\sqrt{n}$ times more safety stock than the centralized system to achieve the same level of service. Even for small values of n, this difference is significant, and for large retailers such as Wal-Mart that have thousands of stores, this difference is enormous.

Of course, this example is a simplification of reality. For most products, it is impractical for every sale to come from a distribution center, and it is impractical for a single distribution center to serve the entire country. Furthermore, the assumption that demands at different stores for the same item are independent is also unlikely to be true. Demands for some items, such as desirable fashion items, are likely to be positively correlated, while others, such as seasonal items like bathing suits, might be negatively correlated. However, even with these caveats, the advantages of centralization are substantial and account for the fact that multilevel distribution systems are widespread in many industries, especially retailing. These results are based on the work of Eppen (1979) (who allowed for correlated demands). This model was extended by Eppen and Schrage (1981) and Erkip, Hausman, and Nahmias (1990) to more general settings.

Several authors have examined the issue of how item characteristics affect the optimal breakdown of DC versus in-store inventory. Muckstadt and Thomas (1980) showed that high-cost, low-demand items derived the greatest benefit from centralized stocking, while Nahmias and Smith (1994) showed how other factors, such as the probability of a lost sale and the frequency of shipments from the DC to the stores, also affect the optimal breakdown between store and DC inventory. A comprehensive review of inventory control models for retailing can be found in Nahmias and Smith (1993), and an excellent collection of articles on general multiechelon inventory models in Schwarz (1981).

Problems for Section 9

27. Describe how risk pooling works in a multiechelon inventory system. Under what circumstances does one derive the greatest benefit from risk pooling? the least benefit?

28. Why might a retailer want to consider developing a three-level multilevel system? (The three levels might be labeled National Distribution Center, Regional Distribution Center, and Store.)

29. What are the characteristics of items from which one derives the greatest benefit from centralized storage? the least benefit?

6.10 Designing the Supply Chain in a Global Environment

The economic barriers between countries are disappearing at an increasing rate. Today, few industries produce in and serve only the home market. The phrase "we live in a global economy" has become a cliché, but it is certainly truer than ever. This is evidenced by the following facts, reported in Dornier et al. (1998):

- About one-fifth of the output of U.S. firms is produced abroad.
- U.S. firms currently hold about $500 billion in assets of foreign firms. This amount is growing 7 percent annually.
- Twenty-five percent of U.S. imports and exports are between foreign affiliates and parent U.S.-based firms.
- By the early 1990s, U.S. multinational firms accounted for 89 percent of the sales of all companies in the Conference Board database.
- Manufacturing assets held by U.S. firms in foreign countries is increasing (and vice versa).

One example of a dramatic shift in the marketplace occurred in the automobile industry. Consider the experience of this writer. When I was a child in the 1950s, there was no question but that our family car would be an American nameplate. In fact, almost *everybody* in the United States purchased American cars in those years. Womack et al. (1990) report that as of 1955, the share of the U.S. domestic market held by American-owned companies was nearly 100 percent. Foreign automakers began to make inroads into the American market during the late fifties and early sixties. These were primarily from European manufacturers (Volkswagen, Mercedes, MG, Volvo, Fiat, and Renault) that offered what American firms did not. By 1970, the American firms had given up about 10 percent of the domestic market to foreign competition. During the 1970s, the market share of U.S.-based firms eroded more quickly. While the American firms clung to old designs and old ways of doing business, foreign manufacturers took advantage of changing tastes. By 1989, the U.S. automakers had lost about one-third of the domestic market. The big winner was the Japanese, who saw their share of the world auto market soar from almost zero in the mid-fifties to about 30 percent by 1990 (Womack et al., 1990).

In addition to the fact that U.S. consumers are buying more foreign cars, it's also true that U.S. firms are producing more cars in other countries, and foreign competitors are producing more cars in the United States. A growing fraction of U.S. assembly plants have relocated to countries with low labor costs. GM, for example, has invested heavily in parts plants in Mexico. Thousands of GM employees are earning $1–$2 hourly for work in the Matamoros industrial park just south of the Rio Grande. At the same time, foreign automakers are building major facilities in the United States. Japanese-owned plants based in the United States produce the two most popular models sold here: the Honda Accord, assembled in Marysville Ohio, and the Toyota Camry, assembled in Georgetown, Kentucky. The automobile industry is one of many industries in which production has moved offshore and foreign competition has made significant inroads into the domestic market.

Dornier et al. (1998) identify four driving forces of the globalization process:

- Global market forces.
- Technology forces.
- Global cost forces.
- Political and macroeconomic forces.

1. Global Market Forces.   As markets mature and new entrants enter the marketplace, profit margins decline. At this point, firms turn toward newer, less-developed markets to remain profitable. Hence, one important component of global supply chain operations is establishing new markets overseas. There are several issues that must be considered when doing so. First, can the product be sold "as is" or does it need to be customized to suit the marketplace? Some products need little or no local customization. Coca-Cola is a prime example. Other products must be customized to suit regional requirements. For example, electronic products must conform to the country's wattage and electronic interface. Finally, there are products where local customization is not required but is good business. For example, both GM and Ford sell many cars in Europe. These models are designed to suit European tastes and are often significantly different from those sold at home. Because gasoline prices are far higher in Europe than in the United States and many cities still have very narrow streets, Europeans prefer smaller cars with smaller engines.

2. Technological Forces.   There can be no doubt that technological advances have contributed to globalization. The rapid growth of information technology ameliorates several disadvantages of locating operations on foreign shores. As discussed in Chapter 1, getting to the market early is an important competitive advantage. Locating production and research and development activities in different locations could be problematic. Close communication between R&D personnel and manufacturing folks is a key to successful product development. Today, however, physical distances are less of an issue. Electronic communication is now possible anywhere in the world via e-mail, fax, telephone, and web-based communications systems.

Advances in manufacturing and logistics also have contributed to globalization. Producing different products at different locations for different market segments is made easier by advances in manufacturing technology. Plants can be modularized and built relatively quickly in other countries. Product variety can be increased.

Diffusion of new technologies also contributes to globalization. Firms may want to locate manufacturing facilities close to sources of critical technologies. For example, both IBM and Xerox decided to locate manufacturing facilities for video displays in Japan, the leader in video display technology. Many European and Japanese-based firms have operations in Silicon Valley to be near the host of new technologies being developed there.

Sharing of new technology is one factor motivating strategic alliances among firms from different countries. For example, the Read-Rite Corporation of Milpitas, California, entered into a strategic alliance with Sumitomo Steel of Japan. Read-Rite is one of the largest independent producers of the Winchester drive heads, and Sumitomo was interested in acquiring some of this technology and breaking into this market. In exchange, Read-Rite received entrée into the Japanese market and a substantial amount of cash (Yansouni, 1999).

Most American investments abroad are manufacturing facilities, such as the General Motors plants in Mexico, designed to take advantage of low labor rates. However, it is becoming more common for American firms to also locate R&D facilities abroad. This is motivated primarily by the availability of low-cost, highly trained technical personnel in some countries, such as Taiwan.

3. Cost Drivers.   Another important driver of a firm's decision to globalize is the rising costs of domestic manufacturing. As profits are squeezed, firms seek not only to expand markets, but also to reduce unit costs. This is the primary reason developed countries locate facilities in less developed countries. However, in recent years, the process seems to have moved ahead much more rapidly. This is due to many factors. One is increased global competition from new international competitors. Another is the growth of new hi-tech industries that build offshore facilities as part of their initial business model. A third factor is the improvement of conditions overseas. An example is that the average educational level has improved steadily in recent years in countries like Malaysia and Singapore. Also, opportunities for both new markets and inexpensive labor in countries such as China and the former Soviet Union have become available in recent years.

However, Dornier et al. (1998) claim that this strategy could be misguided. It is based, they say, on inappropriate accounting practices that inflate the importance of direct labor rates. As labor rates change, this strategy leads to "island hopping" to seek the lowest rates. However, island hopping requires substantial capital outlays for new facilities, and there are start-up costs in the new location.

Preferred tax treatment is also an incentive to locate abroad. Most American pharmaceutical companies have their primary production facilities located in Puerto Rico because of the favorable tax status afforded by the U.S. government to American firms doing business in Puerto Rico. Ireland is another country that has provided substantial tax incentives for companies to locate there.

4. Political and Macroeconomic Forces. Major trade agreements have helped to further bring down barriers to international agreements. The two most significant are NAFTA in North America and the EEC in Europe. Political opponents to NAFTA, who made frequent reference to the "giant sucking sound," referring to the loss of domestic jobs, were proven wrong. Measuring the direct effects of trade agreements is difficult because of confounding factors, but unemployment rates certainly did not skyrocket as a result of NAFTA. The development of the European Economic Community has made it far easier for manufacturers to do business in Europe because of standardization and reduction of intracountry tariffs.

Government policies play a major role in the attractiveness of a region or of a specific country. China, in particular, provides substantial incentives for outsiders to develop strategic partnerships with domestic Chinese organizations, including state-owned enterprises, private enterprises, and collective enterprises. Other countries that offer substantial incentives for foreign investment include Ireland, Malaysia, Mexico, and Thailand. Political unrest, however, has contributed to the slow development and outside investment in Africa. Political instability can be a substantial risk of investing abroad.

One of the impediments to doing business abroad is exchange rate risk. If you build a plant in China, you pay for locally purchased materials and labor in yuan. If one sells in Germany, receipts are in deutsche marks. As the relative value of currencies shifts in the world marketplace, both costs and sales receipts also shift relative to the home currency. These changes can be dramatic. For example, up until the 1970s, the Canadian and American dollars were almost at par. Today, the exchange rate is approximately $1.5 Canadian = $1 U.S. A capital investment in Canadian-based factories lost a third of its value in U.S. dollars. Of course, this change occurred over a long period of time. But exchange rate fluctuations and currency devaluations can, and do, occur much faster. They provide an added risk for firms doing business on foreign shores.

Supply Chain Management in a Global Environment

The discussion so far has considered the motivations and problems that arise from international operations in general. Many of these issues also apply to managing the global supply chain. For example, exchange rates can play an important role in the

SNAPSHOT APPLICATION

Digital Equipment Corporation Uses Mathematical Modeling to Plan Its Global Supply Chain

The Digital Equipment Corporation (DEC) was formerly one of the world's larger vertically integrated computer companies. DEC grew rapidly in the 1970s and became one of the world's leading suppliers of minicomputers. During the 1990s DEC continued to be a major player in the minicomputer area but saw its market share erode as sophisticated workstations and PCs displaced many minicomputers. In the latter part of the 1990s, the Compac Corporation acquired DEC. Prior to the acquisition, DEC faced the problem of managing a complex worldwide supply chain. More than half of DEC's revenues came from 81 countries outside the United States, requiring DEC, like many other global corporations, to efficiently manage a complex worldwide global supply chain. To assist DEC with this complex planning process, the firm assembled a team to consider DEC's global supply chain. Some of the issues that needed to be addressed in DEC's planning included the location of customers and

Source: Arntzen et al., 1995.

suppliers, the location and availability of inexpensive skilled labor, the length of the material pipeline, tax considerations, export regulations, and local content requirements. The research team developed a mathematical description of DEC's supply chain in the form of a mixed-integer linear program. (Linear programming is discussed in this text in Supplement 1. The transportation problem, discussed in this chapter, is an example of the types of models developed for this application.)

The mathematical model, called GSCM, was originally developed to plan for production and distribution of new products. The authors report that by spring of 1994, the model had been implemented for about 20 new products. This was done via an 18-month plan to restructure the manufacturing and distribution process to reduce costs, reduce inventories, and improve customer service. As a result of this process, DEC reduced the number of plants from 33 to 12 while maintaining increased output and revenue streams. By 1995, as a result of the implementation of this approach, DEC realized a decrease of annual manufacturing costs of $167 million and annual logistics costs of over $200 million.

design of the supply chain. While a risk, relative currency valuations and other financial factors such as local interest rates also can be viewed as an opportunity. Firms can "game" the system by maintaining manufacturing facilities in several locations and producing where costs are least.

Geographic distances also present special problems for supply chain management. While we have learned to move information at light speed, the time required to move physical goods from one place to another has not changed appreciably in the past several decades. Locating facilities offshore means longer supply lead times. Procedural requirements, bureaucratic delays, and weather conditions are all factors that increase the uncertainty of lead times.

Lean production systems such as JIT become much harder to implement when facilities and suppliers are located far apart. One of the reasons for Toyota's successful implementation of JIT in its plants in Toyoda City is that suppliers are located very close to the plant. Imagine the difficulties of trying to implement a JIT arrangement with a supplier 6,000 miles away.

Infrastructure inadequacies also present special problems for managing global supply chains. The list of problems includes lack of skilled workers, poor roads, lack of suitable airports, and undesirable living conditions for overseas management.

Inadequate communications systems such as substandard telephone systems and lack of network availability also contribute to difficulties. Such factors not only provide disincentives for locating manufacturing facilities in developing countries, they also hamper setting up channels of distribution.

As noted earlier in this chapter, product design can be affected by global supply chain considerations. Delayed differentiation is a means of improving the performance of a system in which local requirements differ. Hewlett Packard has been very successful selling printers overseas in this way. Also, the bullwhip effect is exaggerated by the long lead times experienced in global supply chains. Operating on a worldwide scale presents many challenges for managing supply chains.

Problems for Section 10

30. What do you see as the new emerging markets in the world in the next 20 years? For what reason has Africa been slow to develop as a new market and as a desirable location for manufacturing?

31. As noted earlier in this chapter, the share of U.S. sales accounted for by multinational firms is increasing. What events might reverse this trend?

32. Manufacturing has continued to move offshore at a rapid rate. Why do some writers feel that this is an inappropriate strategy? What factors might reverse this trend?

33. What difficulties for supply chain management are created by the growth of globalization?

6.11 Summary

Why the surge of interest in supply chain management in recent years? As James Alampi, president of Van Waters and Rogers, the largest U.S. chemicals distributor, says:

> Frankly, the cost of making a product is almost irrelevant. You have far more opportunity to get cost out of the supply chain than you do out of manufacturing. There's so much duplication and inefficiency. (Henkoff, 1994)

Mr. Alampi's observation is one that many others are beginning to realize. While manufacturing is still an important problem, management is taking note that better design of the supply chain is where substantial cost savings will be realized. Because of this, supply chain software is also becoming a big business. For example, Texas-based i2 Technologies produces and installs dedicated supply chain software. As of this writing, the firm is growing at a 50+ percent annual rate and counts among its clients such firms as GE, IBM, Dell, TI, and many others. i2 is not the only software firm to take advantage of the enormous interest in supply chain management. Some of the biggest names in the information system arena, including Oracle Corporation and German-based SAP, now offer supply chain modules as part of their total system solution.

Mathematical modeling can play an important role in efficient supply chain management. The transportation and transshipment problems, discussed in this chapter, are examples of the kind of mathematical optimization models that can assist large firms with determining efficient schedules for moving product from the factory to the market. There are also mathematically based techniques for efficient scheduling of delivery vehicles. The savings method, discussed in this chapter, is a simple example of one such technique.

Delivering the product to the consumer was traditionally a secondary consideration for manufacturing firms. Today, however, supply chain considerations are taken into account even at the product design level. Products that are easily stacked can be shipped and stored more easily and less expensively. The strategy of postponement has turned out to be a fundamental design principle that has proven to be highly cost effective. By designing products whose final configuration can be postponed, firms can delay product differentiation. This allows them to gain valuable time in determining the realization of demand and helps to avoid the problem of producing a product mix that doesn't match sales.

An unusual phenomenon that was observed in the late 1980s is the bullwhip effect. The variance of orders seems to increase dramatically as one moves up the supply chain. While most agree that the bullwhip effect is the result of different agents acting to optimize their own positions, solutions for this problem are not well understood. Providing incentives for the various players to share information and reduce order batching may be one way to combat the bullwhip effect, but it remains to be seen if these methods really work in practice.

An important trend is the movement toward supply chain solutions that take advantage of the enormous strides that have been made in the information industry. These include such technologies as EDI and web-based transactions systems. Point-of-sale entry systems based on bar coding are now common in most large retail operations.

Bibliography

Arntzen, B. C.; G. G. Brown; T. P. Harrison; and L. L. Trafton. "Global Supply Chain Management at Digital Equipment Corporation." *Interfaces* 25, no. 1 (January–February 1995), pp. 69–93.

Arrow, K. J. "Historical Background." Chapter 1 in *Studies in the Mathematical Theory of Inventory and Production,* ed. K. J. Arrow, S. Karlin, and H. Scarf. Stanford, CA: Stanford University Press, 1958.

Barrone, F. Private communication, 1996.

Bell, W. J. "Improving the Distribution of Industrial Gases with an Online Computerized Routing and Scheduling Optimizer." *Interfaces* **13**, no. 6 (1983), pp. 4–23.

Cannon, E. *EDI Guide. A Step by Step Approach.* New York: Van Nostrand Reinhold, 1993.

Clark, A. J., and H. E. Scarf. "Optimal Policies for a Multiechelon Inventory Problem." *Management Science* 6 (1960), pp. 475–90.

Clarke, G., and G. W. Wright. "Scheduling of Vehicles from a Central Depot to a Number of Delivery Points." *Operations Research* 12 (1964), pp. 568–81.

Cohen, M.; P. V. Kamesam; P. Kleindorfer; H. Lee; and A. Tekerian. "Optimizer: IBM's Multi-Echelon Inventory System for Managing Service Logistics." *Interfaces* 20, no. 1 (January–February 1990), pp. 65–82.

Cooke, J. A. "Software Takes Babel out of Vendor Managed Inventory." *Logistics Management & Distribution Report* 38, no. 2, February 1999, p. 87.

Copacino, W. C. *Supply Chain Management: The Basics and Beyond.* Boca Raton, FL: St. Lucie Press, 1997.

Crawford, F. A. "ECR: A Mandate for Food Manufacturers?" *Food Processing,* February 1994.

Dornier, P.-P.; R. Ernst; M. Fender; and P. Kouvelis. *Global Operations and Logistics, Text and Cases.* New York: John Wiley & Sons, 1998.

Dreifus, S. B., ed. *Business International's Global Desk Reference.* New York: McGraw-Hill, 1992.

Duff, C., and R. Ortega. "How Wal-Mart Outdid a Once-Touted Kmart in Discount Store Race." *The Wall Street Journal,* March 24, 1995.

Eilon, S.; C. D. T. Watson-Gandy; and N. Christofides. *Distribution Management: Mathematical Modeling and Practical Analysis.* London: Griffin, 1971.

Eppen, G. D. "Effects of Centralization on Expected Costs in a Multi-Location Newsboy Problem." *Management Science* 25 (1979), pp. 498–501.

Eppen, G. D., and L. Schrage. "Centralized Ordering Policies in a Multi-Warehouse System with Leadtimes and Random Demand." In *Multi-Level Production/Inventory Systems: Theory and Practice,* ed. L. B. Schwarz. New York: North Holland, 1981.

Erkip, N.; W. H. Hausman; and S. Nahmias. "Optimal Centralized Ordering Policies in Multi-Echelon Inventory Systems with Correlated Demands." *Management Science* 36 (1990), pp. 381–92.

Fine, C. H. *Clockspeed.* Reading, MA: Perseus Books, 1998.

Hammond, Jan. Barilla SpA (A and B). Copyright © 1994 by the President and Fellows of Harvard Business School.

Handfield, R. B., and E. L. Nichols Jr. *Introduction to Supply Chain Management.* Upper Saddle River, NJ: Prentice Hall, 1999.

Henkoff, R. "Delivering the Goods." *Fortune,* November 25, 1994, pp. 64–78.

Jacobson, G., and J. Hillkirk. *Xerox, American Samurai.* New York: Macmillan, 1986.

Lee, H. L.; C. Billington; and B. Carter. "Hewlett-Packard Gains Control of Inventory and Service through Design for Localization." *Interfaces* 23, no. 4 (July–August 1994), pp. 1–11.

Lee, H. L.; P. Padmanabhan; and S. Whang. "The Bullwhip Effect in Supply Chains." *Sloan Management Review,* Spring 1997, pp. 93–102.

Martin, A. J. *DRP: Distribution Resource Planning.* 2nd ed. Essex Junction, VT: Oliver Wight Limited Publications, 1990.

Muckstadt, J. A., and L. J. Thomas. "Are Multi-Echelon Inventory Methods Worth Implementing in Systems with Low Demand Rate Items?" *Management Science* 26 (1980), pp. 483–94.

Nahmias, S., and S. A. Smith. "Mathematical Models of Retailer Inventory Systems: A Review." In *Perspectives in Operations Management,* ed. R. K. Sarin. Boston: Kluwer, 1993.

Nahmias, S., and S. A. Smith. "Optimizing Inventory Levels in a Two-Echelon Retailer System with Partial Lost Sales." *Management Science* 40 (1994), pp. 582–96.

Palevich, R. F. "Supply Chain Management." *Hospital Materiel Management Quarterly* 20, no. 3 (February 1999), pp. 54–63.

Ragsdale, C. T. *Spreadsheet Modeling and Decision Analysis.* 2nd ed. Cincinnati: South-Western, 1998.

Ross, David Frederick. *Competing through Supply Chain Management: Creating Market-Winning Strategies through Supply Chain Partnerships.* New York: Chapman & Hall, 1998.

Schrage, L. "Formulation and Structure of More Complex/Realistic Routing and Scheduling Problems" *Networks* **11** (1981), pp. 229–32.

Schwarz, L. B., ed. *Multi-Level Production/Inventory Systems: Theory and Practice.* New York: North Holland, 1981.

Signorelli, S., and H. Heskett. "Benneton." Harvard Business School Case 9-685-014, 1984.

Simchi-Levi, D.; P. Kaminski; and E. Simchi-Levi. *Designing and Managing the Supply Chain: Concepts, Strategies, and Case Studies.* New York: McGraw-Hill/Irwin, 1999.

Sliwa, C. "Users Cling to EDI for Critical Transactions." *Computerworld* 33, no. 11 (March 15, 1999), p. 48.

Sterman, R. "Modeling Managerial Behavior: Misperception of Feedback in a Dynamic Decision Making Experiment." *Management Science* **35**, no. 3 (1989), pp. 321–39.

Supply Chain Forum. http://www.stanford.edu/group/scforum/, accessed August 1999.

Wilder, C., and M. K. McGee. "GE: The Net Pays Off." *Informationweek,* February 1997, pp. 14–16.

Womack, J. P.; D. T. Jones; and D. Roos. *The Machine That Changed the World.* New York: Harper Perennial, 1990.

Yansouni, C. Private communication, 1999.

CHAPTER 7

PUSH AND PULL PRODUCTION CONTROL SYSTEMS: MRP AND JIT

In this chapter we will examine two philosophies for finding rules to move product within the production facility. As we will see, the methods developed in Chapters 4 and 5 are not always appropriate for production planning purposes, and a different perspective is required. Demand can no longer be thought of as independent of our operating policies and exogenous to the system.

The two approaches we consider are *materials requirements planning (MRP)* and *just-in-time (JIT)*. These are often referred to respectively as "push" and "pull" control systems. To appreciate exactly what distinguishes push and pull systems will require an understanding of exactly how these methods work, which will be covered in detail in this chapter. The simplest definition that this writer has seen (due to Karmarkar, 1989) is that "a pull system initiates production as a reaction to present demand, while a push system initiates production in anticipation of future demand." Thus, MRP incorporates forecasts of future demand while JIT does not.

To better understand the difference between MRP and JIT, consider the following simple example. Garden spades are produced by a plant in Muncie, Indiana. Each spade consists of two parts: the metal digger and the wooden handle. The parts are connected by two screws. The plant produces spades at an average rate of 100 per week. The metal digger is produced in batches of 400 on the first two days of each month, and the handles are ordered from an outside supplier. The assembly of spades takes place during the first week of each month.

Consider now the demand pattern for the screws. Exactly 800 screws are needed during the first week of each month. Assuming four weeks per month, the weekly demand pattern for the screws is 800, 0, 0, 0, 800, 0, 0, 0, 800, 0, 0, 0, and so on. Using a weekly demand rate of 200 and appropriate holding and set-up costs, suppose that the EOQ formula gives an order quantity of 1,400. A little reflection shows that ordering the screws in lots of 1,400 doesn't make much sense. If we schedule a delivery of 1,400 screws at the beginning of a month, 800 are used immediately and 600 stored for later use. At the beginning of the next month another order for 1,400 has to be made, since the 600 screws stored are insufficient to meet the next month's requirement. It makes more sense to either order 800 screws at the beginning of each month or some multiple of 800 every several months.

The EOQ solution was clearly inappropriate here. Why? Recall that in deriving the EOQ formula, we assumed that demand was known and constant. The demand pattern in this example is known, but it is certainly not constant. In fact, it is very spiky. If we were to apply the methods of Chapter 5, we would assume that the demand was random. It is easy to show that over a one-year period the weekly demand has mean 200 and standard deviation 350. These values could be used to generate (Q, R) values assuming some form of a distribution for weekly demand. But this solution would not make any sense either. The demand pattern for the screws is not random; it is predictable, since it is a consequence of the production plan for the spades, which is known. The demand is *variable,* but it is not *random.*

We still have not solved the problem of how many screws to buy and when they should be delivered. One approach might be to just order once at the beginning of the year to meet the demand for an entire year. This would entail a one-time delivery of 10,400 screws at the start of each year (assuming 200 per week). What would be the advantage of this approach? Screws are very inexpensive items. By purchasing enough for an entire year's production, we would incur the fixed delivery costs only once.

There is a completely different way to approach this problem. One could simply decide to schedule deliveries of screws at the beginning of every month. This approach might be more expensive than the once-a-year delivery strategy, since fixed costs would be 12 times higher. However, it could have other advantages that more than compensate for the higher fixed costs. Monthly deliveries eliminate the need to store screws in the plant. If usage rates vary, delivery sizes could be adjusted to match need. Also, if a problem arose with the screws caused by either a defect in production or a design change in the spades, the company would not be stuck with a large inventory of useless items.

These two policies illustrate the basic difference between MRP and JIT (although, as we will see, there is much more to these production control philosophies than this). In an MRP system, we determine lot sizes based on forecasts of future demands and possibly on cost considerations. In a JIT system, we try to reduce lot sizes to their minimum to eliminate waste and unnecessary buildups of inventory.

MRP may be considered to be a top-down planning system in that all production quantity decisions are derived from demand forecasts. Lot-sizing decisions are found for every level of the production system. Items are produced based on this plan and *pushed* to the next level. In JIT, requests for goods originate at a higher level of the system and are *pulled* through the various levels of production. This is the basic idea behind push and pull production control systems.

MRP Basics

In general, a production plan is a complete specification of the amounts of each end item or final product and subassembly produced, the exact timing of the production lot sizes, and the final schedule of completion. The production plan may be broken down into several component parts: (1) the **master production schedule (MPS),** (2) the **materials requirements planning (MRP) system,** and (3) the detailed **job shop schedule.** Each of these parts can represent a large and complex subsystem of the entire plan.

At the heart of the production plan are the forecasts of demand for the end items produced over the planning horizon. An end item is the output of the productive system; that is, products shipped out the door. Components are items in intermediate stages of production, and raw materials are resources that enter the system. A schematic

FIGURE 7–1

Schematic of the productive system

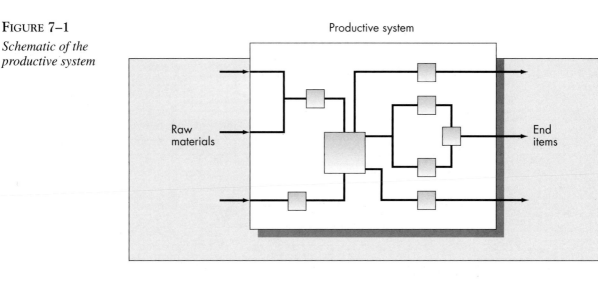

of the productive system appears in Figure 7–1. It is important to bear in mind that raw materials, components, and end items are defined in a relative and not an absolute sense. Hence, we may wish to isolate a portion of a company's operation as a productive system. End items associated with one portion of the company may be raw materials for another portion. A single productive system may be the entire manufacturing operation of the firm or only a small part of it.

The master production schedule (MPS) is a specification of the exact amounts and timing of production of each of the end items in a productive system. The MPS refers to *unaggregated* items. As such, the inputs for determining the MPS are forecasts for future demand by item rather than by aggregate items, as discussed in Chapter 3. The MPS is then broken down into a detailed schedule of production for each of the components that comprise an end item. The materials requirements planning (MRP) system is the means by which this is accomplished. Finally, the results of the MRP are translated into specific shop floor schedules (using methods such as those discussed in Chapter 8) and requirements for raw materials.

The data sources for determining the MPS include the following:

1. Firm customer orders.
2. Forecasts of future demand by item.
3. Safety stock requirements.
4. Seasonal plans.
5. Internal orders from other parts of the organization.

An important part of the success of MRP is the integrity and timeliness of the data. The information system that supports the MRP receives inputs from the production, marketing, and finance departments of the firm. A smooth flow of information among these three functional areas is a key ingredient to a successful production planning system.

We may consider the control of the production system to be composed of three major phases. Phase 1 is the gathering and coordinating of the information required to develop the master production schedule. Phase 2 is the determination of planned order releases using MRP, and phase 3 is the development of detailed shop floor schedules

FIGURE 7–2

*The three major
control phases of
the productive
system*

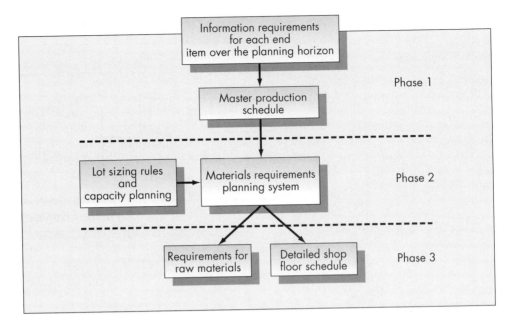

and resource requirements from the MRP planned order releases. Figure 7–2 is a schematic of these three control phases of the productive system.

This chapter is concerned with the way that the MPS is used as input to the MRP system. We will show in detail exactly how the MRP calculus works; that is, how product structures are converted to parent–child relationships between levels of the production system, how production lead times are used to obtain time-phased requirements, and how lot-sizing methods result in specific schedules. In the lot-sizing section we will consider both optimal and heuristic lot-sizing techniques for uncapacitated systems and a straightforward heuristic technique for capacitated lot sizing.

JIT Basics

The just-in-time approach has its roots in the kanban system of material flow pioneered by Toyota. We will discuss the mechanics of kanban later in this chapter. The notion of JIT has grown significantly from its roots as a material flow technology. It has important strategic implications for firms, not just in manufacturing, but also in managing the supplier base and in distribution management.

The fundamental ideas behind JIT are

1. *Work-in-process (WIP) inventory is reduced to a bare minimum.* The amount of WIP inventory allowed is a measure of how tightly the JIT system is tuned. The less WIP designed in the system, the better balanced the various steps in the process need to be.

2. *JIT is a pull system.* Production at each stage is initiated only when requested. The flow of information in a JIT system proceeds sequentially from level to level.

3. *JIT extends beyond the plant boundaries.* Special relationships with suppliers must be in place to assure that deliveries are made on an as-needed basis. Suppliers and manufacturers must be located in close proximity if the JIT design is to include the suppliers.

4. *The benefits of JIT extend beyond savings of inventory-related costs.* Plants can be run efficiently without the clutter of inventory of raw material and partially finished goods clogging the system. Quality problems can be identified before they build up to unmanageable proportions. Rework and inspection of finished goods are minimized.

5. *The JIT approach requires a serious commitment from top management and workers alike.* Workers need to maintain an awareness of their systems and products, and need to be empowered to stop the flow of production if they see something wrong. Management must allow these workers to have that flexibility.

Lately, the term "lean production" has been used to describe JIT. The term appears to have been coined by Womack et al. (1990) in their landmark study of the automobile industry, *The Machine That Changed the World.* In comparing the worst of American mass production and the best Japanese lean production, the authors show just how effective a properly implemented JIT philosophy can be. They described their experience at General Motors' Framingham, Massachusetts, plant in 1986:

> Next we looked at the line itself. Next to each work station were piles—in some cases weeks' worth—of inventory. Littered about were discarded boxes and other temporary wrapping material. On the line itself the work was unevenly distributed with some workers running madly to keep up and others finding time to smoke and even read a newspaper. . . . At the end of the line we found what is perhaps the best evidence of old-fashioned mass production: an enormous work area full of finished cars riddled with defects. All these cars needed further repair before shipment, a task that can prove enormously time-consuming and often fails to fix fully the problems now buried under layers of parts and upholstery.

Now contrast this with their experience at Toyota's Takaoka plant in Toyoda City:

> The differences between Takaoka and Framingham are striking to anyone who understands the logic of lean production. For a start hardly anyone was in the aisles. The armies of indirect workers so visible at GM were missing, and practically every worker in sight was actually adding value to the car . . . The final assembly line revealed further differences. Less than an hour's worth of inventory was next to each worker at Takaoka. The parts went on more smoothly and the work tasks were better balanced, so that every worker worked at about the same pace. . . . At the end of the line, the difference between lean and mass production was even more striking. At Takaoka we observed almost no rework area at all. Almost every car was driven directly from the line to the boat or the trucks taking cars to the buyer.

These differences are exciting and dramatic. We should note that GM has since closed Framingham and that the plants run by the "big three" (namely, GM, Ford, and the Chrysler division of Daimler-Chrysler) are far more efficient and better managed than Framingham. Still, we have a ways to go to duplicate the phenomenal success that the Japanese have had with lean production systems.

This chapter begins with a discussion of the basic explosion calculus of MRP and how lot-sizing strategies other than lot-for-lot are incorporated into a basic single-level MRP solution.

7.1 The Explosion Calculus

Explosion calculus is a term that refers to the set of rules by which gross requirements at one level of the product structure are translated into a production schedule at that level and requirements at lower levels. At the heart of any MRP system is the product

FIGURE 7–3

Typical product structure diagram

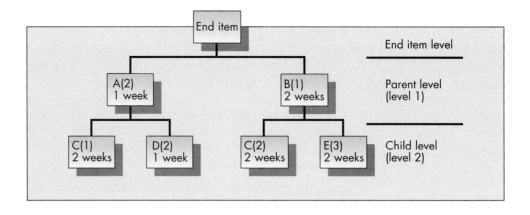

structure. The product structure refers to the relationship between the components at adjacent levels of the system. The product structure diagram details the parent–child relationship of components and end items at each level, the number of periods required for production of each component, and the number of components required at the child level to produce one unit at the parent level.

A typical product structure appears in Figure 7–3. In order to produce one unit of the end item, two units of A and one unit of B are required. Assembly of A requires one week, and assembly of B requires two weeks. A and B are "children" of the end item. In order to produce A, one unit of C and two units of D are required. In order to produce B, two units of C and three units of E are required. The respective production lead times also appear on the product structure diagram. Product structure diagrams can be quite complex, with as many as 15 or more levels in some industries.

The explosion calculus (also known as the bill-of-materials explosion) follows a set of rules that translate the planned order releases for end items and components into production schedules for lower-level components. The method involves properly phasing requirements in time and accounting for the number of components required at the child level to produce a single parent item. The method is best illustrated by example.

Example 7.1

The Harmon Music Company produces a variety of wind instruments at its plant in Joliet, Illinois. Because the company is relatively small, it would like to minimize the amount of money tied up in inventory. For that reason production levels are set to match predicted demand as closely as possible. In order to achieve this goal, the company has adopted an MRP system to determine production quantities.

One of the instruments produced is the model 85C trumpet. The trumpet retails for $800 and has been a reasonably, if not spectacularly, profitable item for the company. Based on orders from music stores around the country, the production manager receives predictions of future demand for about four months into the future.

Figure 7–4 shows the trumpet and its various subassemblies. Figure 7–5 gives the product structure diagram for the construction of the trumpet. The bell section and the lead pipe and valve sections are welded together in final assembly. Before the welding, three slide assemblies and three valves are produced and fitted to the valve casing assembly. The forming and shaping of the bell section requires two weeks, and the forming and shaping of the lead pipe and valve sections require four weeks. The valves require three weeks to produce, and the slide assemblies two weeks.

FIGURE 7–4

*Trumpet and
subassemblies*

FIGURE 7–4

*Trumpet and
subassemblies*

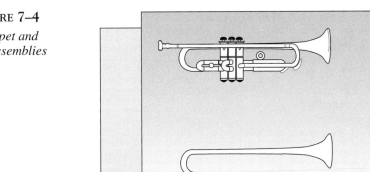

FIGURE 7–5

*Product structure
diagram for
Harmon trumpet*

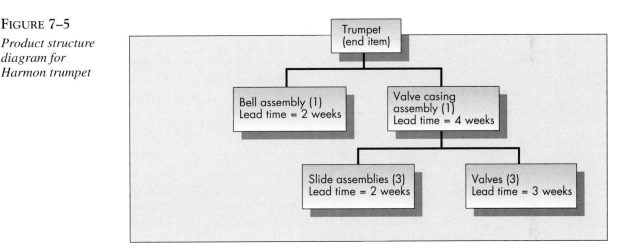

The trumpet assembly problem is a three-level MRP system. Level 0 corresponds to the final product or end item, which is the completed trumpet. Level 1, the child level relative to the trumpet, corresponds to the bell and valve casing assemblies. Level 2 corresponds to the slide and valve assemblies. The information in the product structure diagram is often represented as an indented bill of materials (BOM), which is a more convenient representation for preparation of computer input. The indented BOM for the trumpet is[1]

  1 Trumpet
    1 Bell assembly
    1 Valve assembly
      3 Slide assemblies
      3 Valves

---

[1]The astute reader will notice that the valves and the slides are not identical. Hence, each valve and each slide should be treated as a separate item. However, if we agree that slides and valves correspond to matching groups of three, our approach is valid. This allows us to demonstrate the multiplier effect when multiple components are needed for a single end item.

It takes seven weeks to produce a trumpet. Hence, the company must begin production now on trumpets to be shipped in seven weeks. For that reason we will consider only forecasts for demands that are at least seven weeks into the future. If we label the current week as week 1, then Harmon requires forecasts for the sales of trumpets for weeks 8 to 17. Suppose that the predicted demands for those weeks are

Week	8	9	10	11	12	13	14	15	16	17
Demand	77	42	38	21	26	112	45	14	76	38

These forecasts represent the numbers of trumpets that the firm would like to have ready to ship in the specified weeks. Harmon periodically receives returns from its various suppliers. These are instruments that are defective for some reason or are damaged in shipping. Once the necessary repairs are completed, the trumpets are returned to the pool of those ready for shipping. Based on the current and anticipated returns, the company expects the following schedule of receipts to the inventory:

Week	8	9	10	11
Scheduled receipts	12		6	9

In addition to the scheduled receipts, the company expects to have 23 trumpets in inventory at the end of week 7. The MPS for the trumpets is now obtained by netting out the inventory on hand at the end of week 7 and the scheduled receipts, in order to obtain the net predicted demand:

Week	8	9	10	11	12	13	14	15	16	17
Net predicted demand	42	42	32	12	26	112	45	14	76	38

Having determined the MPS for the end product, we must translate it into a production schedule for the components at the next level of the product structure. These are the bell assembly and the valve casing assembly. Consider first the bell assembly. The first step is to translate the MPS for trumpets into a set of gross requirements by week for the bell assembly. Because there is exactly one bell assembly used for each trumpet, this is the same as the MPS. The next step is to subtract any on-hand inventory or scheduled receipts to obtain the net requirements (here there are none). The net requirements are then translated back in time by the production lead time, which is two weeks for the bell assembly, to obtain the time-phased requirements. Finally, the lot-sizing algorithm is applied to the time-phased requirements to obtain the planned order release by period. Assuming a lot-for-lot production rule, we obtain the following MRP calculations for the bell assembly:

Week	6	7	8	9	10	11	12	13	14	15	16	17
Gross requirements			42	42	32	12	26	112	45	14	76	38
Net requirements			42	42	32	12	26	112	45	14	76	38
Time-phased net requirements	42	42	32	12	26	112	45	14	76	38		
Planned order release (lot for lot)	42	42	32	12	26	112	45	14	76	38		

Lot for lot means that the production quantity each week is just the time-phased net requirement. A lot-for-lot production rule means that no inventory is carried from one period to another. As we will see later, lot for lot is rarely an optimal production rule. Optimal and heuristic production scheduling rules will be examined in a later portion of this chapter.

The calculation is essentially the same for the valve casing assembly, except that the production lead time is four weeks rather than two weeks. The calculations for the valve casing assembly are

Week	4	5	6	7	8	9	10	11	12	13	14	15	16	17
Gross requirements					42	42	32	12	26	112	45	14	76	38
Net requirements					42	42	32	12	26	112	45	14	76	38
Time-phased net requirements	42	42	32	12	26	112	45	14	76	38				
Planned order release (lot for lot)	42	42	32	12	26	112	45	14	76	38				

Now consider the MRP calculations for the valves. Let us assume that the company expects an on-hand inventory of 186 valves at the end of week 3 and a receipt from an outside supplier of 96 valves at the start of week 5. There are three valves required for each trumpet. (Note that the valves are not identical, and hence are not interchangeable. We could display three separate sets of MRP calculations, but this is unnecessary because each trumpet has exactly one valve of each type.) One obtains gross requirements for the valves by multiplying the production schedule for the valve casing assembly by 3. Net requirements are obtained by subtracting on-hand inventory and scheduled receipts. The MRP calculations for the valves are

Week	2	3	4	5	6	7	8	9	10	11	12	13
Gross requirements			126	126	96	36	78	336	135	42	228	114
Scheduled receipts				96								
On-hand inventory		186	60	30								
Net requirements			0	0	66	36	78	336	135	42	228	114
Time-phased net requirements		66	36	78	336	135	42	228	114			
Planned order release (lot for lot)		66	36	78	336	135	42	228	114			

Net requirements are obtained by subtracting on-hand inventory and scheduled receipts from gross requirements. Because the on-hand inventory of 186 in period 3 exceeds the gross requirement in period 4, the net requirements for period 4 are 0. The remaining 60 units (186 − 126) are carried into period 5. In period 5 the scheduled receipt of 96 is added to the starting inventory of 60 to give 156 units. The gross requirements for period 5 are 126, so the net requirements for period 5 are 0, and the additional 30 units are carried over to period 6. Hence, the resulting net requirements for period 6 are 96 − 30 = 66.

The net requirements are phased back three periods in order to obtain the time-phased net requirements and the production schedule. Note that the valves are produced internally. The scheduled receipt of 96 corresponds to defectives that were sent out for rework. A similar calculation is required for the slide assemblies. ■

Example 7.1 represents the essential elements of the explosion calculus. Note that we have assumed for the sake of the example that the production scheduling rule is lot for lot. That is, in each period the production quantity is equal to the net requirements for that period. However, such a policy may be suboptimal and even infeasible. For example, the schedule requires the delivery of 336 valves in week 9. However, suppose that the plant can produce only 200 valves in one week. If that is the case, a lot-for-lot scheduling rule is infeasible.

Problems for Section 1

1. The inventory control models discussed in Chapters 4 and 5 are often labeled *independent* demand models and MRP is often labeled a *dependent* demand system. What do the terms *independent* and *dependent* mean in this context?
2. What information is contained in a product structure diagram?
3. For the example of Harmon Music presented in this section, determine the planned order release for the slide assemblies. Assume lot-for-lot scheduling.
4. The Noname Computer Company builds a computer designated model ICU2. It imports the motherboard of the computer from Taiwan, but the company inserts the sockets for the chips and boards in its plant in Lubbock, Texas. Each computer requires a total of 90 64K DRAM (dynamic random access memory) chips. Noname sells the computers with three add-in boards and two disk drives. The company purchases both the DRAM chips and the disk drives from an outside supplier. The product structure diagram for the ICU2 computer is given in Figure 7–6.

   Suppose that the forecasted demands for the computer for weeks 6 to 11 are 220, 165, 180, 120, 75, 300. The starting inventory of assembled computers in week 6 will be 75, and the production manager anticipates returns of 30 in week 8 and 10 in week 10.
   a. Determine the MPS for the computers.
   b. Determine the planned order release for the motherboards assuming a lot-for-lot scheduling rule.
   c. Determine the schedule of outside orders for the disk drives.
5. For the previous problem, suppose that Noname has 23,000 DRAM chips in inventory. It anticipates receiving a lot of 3,000 chips in week 3 from another firm

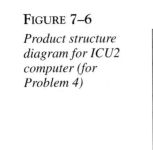

FIGURE 7–6

Product structure diagram for ICU2 computer (for Problem 4)

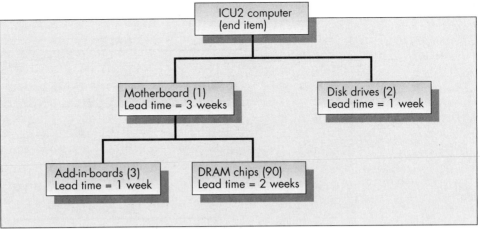

that has gone out of business. At the current time, Noname purchases the chips from two vendors, A and B. A sells the chips for less, but will not fill an order exceeding 10,000 chips per week.

a. If Noname has established a policy of inventorying as few chips as possible, what order should it be placing with vendors A and B over the next six weeks?

b. Noname has found that not all the DRAM chips purchased function properly. From past experience it estimates an 8 percent failure rate for the chips purchased from vendor A and a 4 percent failure rate for the chips purchased from vendor B. What modification in the order schedule would you recommend to compensate for this problem?

6. Consider the product structure diagram given in Figure 7–3. Assume that the MPS for the end item for weeks 10 through 17 is

Week	10	11	12	13	14	15	16	17
Net requirements	100	100	40	40	100	200	200	200

Assume that lot-for-lot scheduling is used throughout. Also assume that there is no entering inventory in period 10 and no scheduled receipts.

a. Determine the planned order release for component A.

b. Determine the planned order release for component B.

c. Determine the planned order release for component C. (Hint: Note that C is required for both A and B.)

7. What alternatives are there to lot-for-lot scheduling at each level? Discuss the potential advantages and disadvantages of other lot-sizing techniques.

8. One of the inputs to the MRP system is the forecast of demand for the end item over the planning horizon. From the point of view of production, what advantages are there to a forecasting system that smooths the demand (that is, provides forecasts that are relatively constant) versus one that achieves greater accuracy but gives "spiky" forecasts that change significantly from one period to the next?

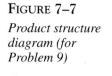

FIGURE 7–7

*Product structure
diagram (for
Problem 9)*

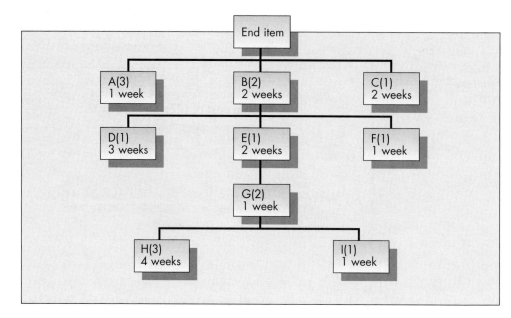

9. An end item has the product structure diagram given in Figure 7–7.
  a. Write the product structure diagram as an indented bill-of-materials list.
  b. Suppose that the MPS for the end item is

Week	30	31	32	33	34	35
MPS	165	180	300	220	200	240

   If production is scheduled on a lot-for-lot basis, find the planned order release for component F.
  c. Using the data in part (*b*), find the planned order release for component I.
  d. Using the data in part (*b*), find the planned order release for component H.

7.2 Alternative Lot-Sizing Schemes

In Example 7.1 we assumed that the production scheduling rule was lot for lot. That is, the number of units scheduled for production each period was the same as the net requirements for that period. In fact, this policy is assumed for convenience and ease of use only. It is, in general, not optimal. The problem of finding the best (or near best) production plan can be characterized as follows: we have a known set of time-varying demands and costs of setup and holding. What production quantities will minimize the total holding and setup costs over the planning horizon? Note that neither the methods of Chapter 4 (which assumes known but constant demands) nor those of Chapter 5 (which assumes random demands) are appropriate.

In this section we will discuss several popular heuristic (i.e., approximate) lot-sizing methods that easily can be incorporated into the MRP calculus.

EOQ Lot Sizing

To apply the EOQ formula, we need three inputs: the average demand rate, λ; the holding cost rate, h; and the setup cost, K. Consider the valve casing assembly in Example 7.1. Suppose that the setup operation for the machinery used in this assembly operation takes two workers about three hours. The workers average $22 per hour. That translates to a setup cost of $(22)(2)(3) = \$132$.

The company uses a holding cost based on a 22 percent annual interest rate. Each valve casing assembly costs the company $141.82 in materials and value added for labor. Hence, the holding cost amounts to $(141.82)(0.22)/52 = \$0.60$ per valve casing assembly per week.

The planned order release resulting from a lot-for-lot policy requires production in each week. Consider the total holding and setup cost incurred from weeks 6 through 15 when using this policy. If we adopt the convention that the holding cost is charged against the inventory each week, then the total holding cost over the 10-week horizon is zero. As there is one setup incurred each week, the total setup cost incurred over the planning horizon is $(132)(10) = \$1,320$.

This cost can be reduced significantly by producing larger amounts less often. As a "first cut" we can use the EOQ formula to determine an alternative production policy. The total of the time-phased net requirements over weeks 8 through 17 is 439, for an average of 43.9 per week. Using $\lambda = 43.9$, $h = 0.60$, and $K = 132$, the EOQ formula gives

$$Q = \sqrt{\frac{2K\lambda}{h}} = \sqrt{\frac{(2)(132)(43.9)}{0.6}} = 139.$$

If we schedule the production in lot sizes of 139 while guaranteeing that all net requirements are filled, the resulting MRP calculations for the valve casing assembly are

Week	4	5	6	7	8	9	10	11	12	13	14	15	16	17
Net requirements					42	42	32	12	26	112	45	14	76	38
Time-phased net requirements	42	42	32	12	26	112	45	14	76	38				
Planned order release (EOQ)	139	0	0	0	139	0	139	0	0	139				
Planned deliveries					139	0	0	0	139	0	139	0	0	139
Ending inventory					97	55	23	11	124	12	106	92	16	117

One finds the ending inventory each period from the formula

$$\frac{\text{Ending}}{\text{inventory}} = \frac{\text{Beginning}}{\text{inventory}} + \frac{\text{Planned}}{\text{deliveries}} - \frac{\text{Net}}{\text{requirements}}.$$

Consider the cost of using EOQ lot sizing rather than lot for lot. During periods 8 through 17 there are a total of four setups, resulting in a total setup cost of $(132)(4) = \$528$. The most direct way to compute the holding cost is to simply accumulate the ending inventories for the 10 periods and multiply by h. The cumulative ending inventory is $97 + 55 + 23 + \cdots + 117 = 653$. Hence, the total holding cost incurred over the 10 periods is $(0.6)(653) = \$391.80$. The total holding and setup cost when lot sizes are computed from the EOQ formula is $\$528 + \$391.80 = \$919.80$. This is a considerable improvement over the cost of $1,320 obtained when using lot-for-lot production scheduling. (However, this savings does not consider the cost impact that lot sizing at this level may have upon lower levels in the product tree. It is possible, though unlikely, that in a global sense lot for lot could be more cost effective than EOQ. This point will be explored in more depth later in the chapter.) Note that the use of the EOQ to set production quantities results in an entirely different pattern of gross requirements for the valve and slide assemblies one level down. In particular, the gross requirements for the valves are now

Week	4	5	6	7	8	9	10	11	12	13
Gross requirements	417	0	0	0	417	0	417	0	0	417

In the remainder of this section, we discuss three other popular lot-sizing schemes when demand is known and time is varying. It should be pointed out that the problem of determining lot sizes subject to time-varying demand occurs in contexts other than MRP. We have included it here to illustrate how these methods can be linked to the MRP explosion calculus.

The Silver-Meal Heuristic

The Silver-Meal heuristic (named for Harlan Meal and Edward Silver) is a forward method that requires determining the average cost per period as a function of the number of periods the current order is to span, and stopping the computation when this function first increases.

Define $C(T)$ as the average holding and setup cost per period if the current order spans the next T periods. As above, let $(r_1, \ldots, r_n)$ be the requirements over the n-period horizon. Consider period 1. If we produce just enough in period 1 to meet the demand in period 1, then we just incur the order cost K. Hence,

$$C(1) = K.$$

If we order enough in period 1 to satisfy the demand in both periods 1 and 2, then we must hold r_2 for one period. Hence,

$$C(2) = (K + hr_2)/2.$$

Similarly,

$$C(3) = (K + hr_2 + 2hr_3)/3$$

and, in general,

$$C(j) = (K + hr_2 + 2hr_3 + \cdots + (j - 1)hr_j)/j.$$

Once $C(j) > C(j - 1)$, we stop and set $y_1 = r_1 + r_2 + \cdots + r_{j-1}$, and begin the process again starting at period j.

Example 7.2

A machine shop uses the Silver-Meal heuristic to schedule production lot sizes for computer casings. Over the next five weeks the demands for the casings are $\mathbf{r} = (18, 30, 42, 5, 20)$. The holding cost is $2 per case per week, and the production setup cost is $80. Find the recommended lot sizing.

Solution

Starting in period 1:

$C(1) = 80,$
$C(2) = (80 + (2)(30))/2 = 70,$
$C(3) = (80 + (2)(30) + (2)(2)(42))/3 = 102.67.$ Stop because $C(3) > C(2)$.

Set $y_1 = r_1 + r_2 = 18 + 30 = 48$.
  Starting in period 3:

$C(1) = 80,$
$C(2) = (80 + (2)(5))/2 = 45,$
$C(3) = (80 + (2)(5) + (2)(2)(20))/3 = 56.67.$ Stop.

Set $y_3 = r_3 + r_4 = 42 + 5 = 47$.
  Because period 5 is the final period in the horizon, we do not need to start the process again. We set $y_5 = r_5 = 20$. Hence, the Silver-Meal heuristic results in the policy $\mathbf{y} = (48, 0, 47, 0, 20)$. (Hint: You can streamline calculations by noting that $C(j + 1) = (j/(j + 1))(C(j) + hr_{j+1})$.) ∎

To show that the Silver-Meal heuristic will not always result in an optimal solution, consider the following counterexample.

Example 7.3

Let $\mathbf{r} = (10, 40, 30)$, $K = 50$, and $h = 1$. The Silver-Meal heuristic gives the solution $\mathbf{y} = (50, 0, 30)$, but the optimal solution is $(10, 70, 0)$. ∎

In closing this section, we note that Silver and Peterson (1985, p. 238) recommend conditions under which the Silver-Meal heuristic should be used instead of the EOQ. The condition is based on the variance of periodic demand: the higher the variance, the better the improvement the heuristic gives. However, our feeling is that given today's computing technology and the ease with which the heuristic solution can be found, the additional computational costs of using Silver-Meal (or one of the two methods described below) instead of EOQ are minimal and not an important consideration.

Least Unit Cost

The least unit cost (LUC) heuristic is similar to the Silver-Meal method except that instead of dividing the cost over j periods by the number of periods, j, we divide it by the total number of units demanded through period j, $r_1 + r_2 + \cdots + r_j$. We choose the order horizon that minimizes the cost per unit of demand rather than the cost per period.
  Define $C(T)$ as the average holding and setup cost per unit for a T-period order horizon. Then,

$C(1) = K/r_1,$
$C(2) = (K + hr_2)/(r_1 + r_2),$
$$\vdots$$
$C(j) = (K + hr_2 + 2hr_3 + \cdots + (j - 1)hr_j)/(r_1 + r_2 + \cdots + r_j).$

As with the Silver-Meal heuristic, this computation is stopped when $C(j) > C(j-1)$, and the production level is set equal to $r_1 + r_2 + \cdots + r_{j-1}$. The process is then repeated, starting at period j and continuing until the end of the planning horizon is reached.

Example 7.4

Assume the same requirements schedule and costs as given in Example 7.2.
Starting in period 1:

$C(1) = 80/18 = 4.44,$
$C(2) = (80 + (2)(30))/(18 + 30) = 2.92,$
$C(3) = (80 + (2)(30) + (2)(2)(42))/(18 + 30 + 42) = 3.42.$

Because $C(3) > C(2)$, we stop and set $y_1 = r_1 + r_2 = 48$.
Starting in period 3:

$C(1) = 80/42 = 1.90,$
$C(2) = (80 + (2)(5))/(42 + 5) = 1.92.$

Because $C(2) > C(1)$, stop and set $y_3 = r_3 = 42$.
Starting in period 4:

$C(1) = 80/5 = 16,$
$C(2) = (80 + (2)(20))/(5 + 20) = 4.8.$

As we have reached the end of the horizon, we set $y_4 = r_4 + r_5 = 5 + 20 = 25$. The solution obtained by the LUC heuristic is $\mathbf{y} = (48, 0, 42, 25, 0)$. It is interesting to note that the policy obtained by this method is different from that for the Silver-Meal heuristic. It turns out that the S-M method gives the optimal policy, with cost \$310, whereas the LUC gives a suboptimal policy, with cost \$340. ■

Part Period Balancing

Another approximate method for solving this problem is part period balancing. Although the Silver-Meal technique seems to give better results in a greater number of cases, part period balancing seems to be more popular in practice.

The method is to set the order horizon equal to the number of periods that most closely matches the total holding cost with the setup cost over that period. The order horizon that exactly equates holding and setup costs will rarely be an integer number of periods (hence the origin of the name of the method).

Example 7.5

Again consider Example 7.2. Starting in period 1, we find

Order Horizon	Total Holding Cost
1	0
2	60
3	228

Because 228 exceeds the setup cost of 80, we stop. As 80 is closer to 60 than to 228, the first order horizon is two periods. That is, $y_1 = r_1 + r_2 = 18 + 30 = 48$. We start the process again in period 3.

Order Horizon	Total Holding Cost
1	0
2	10
3	90

We have exceeded the setup costs of 80, so we stop. Because 90 is closer to 80 than is 10, the order horizon is three periods. Hence $y_3 = r_3 + r_4 + r_5 = 67$. The complete part period balancing solution is $\mathbf{y} = (48, 0, 67, 0, 0)$, which is different from both the Silver-Meal and LUC solutions. This solution is optimal, as it also has a total cost of $310. ■

All three of the methods discussed in this section are heuristic methods. That is, they are reasonable methods based on the structure of the problem but don't necessarily give the optimal solution. In Appendix 7–A, we discuss the Wagner-Whitin algorithm that guarantees an optimal solution to the problem of production planning with time-varying demands. While tedious to solve by hand, the Wagner-Whitin algorithm can be implemented easily on a computer and solved quickly and efficiently.

Problems for Section 2

10. Perform the MRP calculations for the valves in the example of this section, using the gross requirements schedule that results from EOQ lot sizing for the valve casing assemblies. Use $K = \$150$ and $h = 0.4$.

11. *a.* Determine the planned order release for the motherboards in Problem 4 assuming that one uses the EOQ formula to schedule production. Use $K = \$180$ and $h = 0.40$.
    b. Using the results from part (*a*), determine the gross requirements schedule for the DRAM chips, which are ordered from an outside supplier. The order cost is $25.00, and the holding cost is $0.01 per chip per week. What order schedule with the vendor results if the EOQ formula is used to determine the lot size?
    c. Repeat the calculation of part (*b*) for the add-in boards. Use the same value of the setup cost and a holding cost of 28 cents per board per week.

12. *a.* Discuss why the EOQ formula may give poor results for determining planned order releases.
    b. If the forecasted demand for the end item is the same each period, will the EOQ formula result in optimal lot sizing at each level of the product structure? Explain.

13. The problem of lot sizing for the valve casing assembly described for Harmon Music Company in Section 2 was solved using the EOQ formula. Determine the

lot sizing for the 10 periods using the following methods instead:
a. Silver-Meal.
b. Least unit cost.
c. Part period balancing.
d. Which lot-sizing method resulted in the lowest cost for the 10 periods?

14. A single inventory item is ordered from an outside supplier. The anticipated demand for this item over the next 12 months is 6, 12, 4, 8, 15, 25, 20, 5, 10, 20, 5, 12. Current inventory of this item is 4, and ending inventory should be 8. Assume a holding cost of $1 per period and a setup cost of $40. Determine the order policy for this item based on
   a. Silver-Meal.
   b. Least unit cost.
   c. Part period balancing.
   d. Which lot-sizing method resulted in the lowest cost for the 12 periods?

15. For the counterexample (Example 6.3), which shows that the Silver-Meal heuristic may give a suboptimal solution, do either the least unit cost or the part period balancing heuristics give the optimal solution?

16. Discuss the advantages and disadvantages of the following lot-sizing methods in the context of an MRP scheduling system: lot for lot, EOQ, Silver-Meal, least unit cost, and part period balancing.

17. The time-phased net requirements for the base assembly in a table lamp over the next six weeks are

Week	1	2	3	4	5	6
Requirements	335	200	140	440	300	200

The setup cost for the construction of the base assembly is $200, and the holding cost is $0.30 per assembly per week.
a. What lot sizing do you obtain from the EOQ formula?
b. Determine the lot sizes using the Silver-Meal heuristic.
c. Determine the lot sizes using the least unit cost heuristic.
d. Determine the lot sizes using part period balancing.
e. Compare the holding and setup costs obtained over the six periods using the policies found in parts (a) through (d) with the cost of a lot-for-lot policy.

Problems 18–22 are based on the material appearing in Appendix 7–A.

18. Anticipated demands for a four-period planning horizon are 23, 86, 40, and 12. The setup cost is $300 and the holding cost is $h = 3 per unit per period.
   a. Enumerate all the exact requirements policies, compute the holding and setup costs for each, and find the optimal production plan.
   b. Solve the problem by backward dynamic programming.

19. Anticipated demands for a five-period planning horizon are 14, 3, 0, 26, 15. Current starting inventory is four units, and the inventory manager would like to have eight units on hand at the end of the planning horizon. Assume that $h = 1$ and $K = 30$. Find the optimal production schedule. (Hint: Modify the first and the last period's demands to account for starting and ending inventories.)

20. A small manufacturing firm that produces a line of office furniture requires casters at a fairly constant rate of 75 per week. The MRP system assumes a six-week planning horizon. Assume that it costs $266 to set up for production of the casters and the holding cost amounts to $1 per caster per week.
    a. Compute the EOQ and determine the number of periods of demand to which this corresponds by forming the ratio (EOQ)/(demand per period). Let T be this ratio rounded to the nearest integer. Determine the policy that produces casters once every T periods.
    b. Using backward dynamic programming with $N = 6$ and $\mathbf{r} = (75, 75, \ldots, 75)$, find the optimal solution. (Refer to Appendix 7–A) Does your answer agree with what you obtained in part (a)?

21. a. Based on the results of Problem 20, suggest an approximate lot-sizing technique. Under what circumstances would you expect this method to give good results?
    b. Use this method to solve Example 7A.2 (see Appendix 7–A). By what percentage does the resulting solution differ from the optimal?

22. Solve Problem 17 using the Wagner-Whitin algorithm. (Refer to Appendix 7–A.)

7.3 Incorporating Lot-Sizing Algorithms into the Explosion Calculus

Example 7.6

Let us return to Example 7.1 concerning the Harmon Music Company and consider the impact of lot sizing on the explosion calculus. Consider first the valve casing assembly. The time-phased net requirements for the valve casing assembly are

Week	4	5	6	7	8	9	10	11	12	13
Time-phased net requirements	42	42	32	12	26	112	45	14	76	38

The setup cost for the valve casing assembly is $132, and the holding cost is $h = \$0.60$ per assembly per week. We will determine the lot sizing by the Silver-Meal heuristic.

Starting in week 4:

$$C(1) = 132,$$

$$C(2) = \frac{132 + 0.6(42)}{2} = 78.6,$$

$$C(3) = \frac{132 + 0.6[42 + (2)(32)]}{3} = 65.2,$$

$$C(4) = \frac{132 + 0.6[42 + (2)(32) + (3)(12)]}{4} = 54.3,$$

$$C(5) = \frac{132 + 0.6[42 + (2)(32) + (3)(12) + (4)(26)]}{5} = 55.92.$$

Since $C(5) > C(4)$, we terminate computations and set $y_4 = 42 + 42 + 32 + 12 = 128$.

Starting in week 8:

$C(1) = 132,$

$$C(2) = \frac{132 + 0.6(112)}{2} = 99.6,$$

$$C(3) = \frac{132 + 0.6[112 + (2)(45)]}{3} = 84.4,$$

$$C(4) = \frac{132 + 0.6[112 + (2)(45) + (3)(14)]}{4} = 69.6,$$

$$C(5) = \frac{132 + 0.6[112 + (2)(45) + (3)(14) + (4)(76)]}{5} = 92.16.$$

Hence, $y_8 = 26 + 112 + 45 + 14 = 197$.

Production occurs next in week 12. It is easy to show that $y_{12} = 76 + 38 = 114$.

A summary of the MRP calculations using the Silver-Meal heuristic to determine lot sizes for the valve casing assembly appears below.

Week	4	5	6	7	8	9	10	11	12	13	14	15	16	17
Net requirements					42	42	32	12	26	112	45	14	76	38
Time-phased net requirements	42	42	32	12	26	112	45	14	76	38				
Planned order release (S-M)	128	0	0	0	197	0	0	0	114	0				
Planned deliveries					128	0	0	0	197	0	0	0	114	0
Ending inventory					86	44	12	0	171	59	14	0	38	0

It is interesting to compare the holding and setup cost of the policy using the Silver-Meal heuristic with the previous solutions using lot for lot and the EOQ formula. There are exactly three setups in the solution above, resulting in a total setup cost of $(132)(3) = \$396$. The sum of the ending inventories each week is $86 + 44 + \cdots + 38 + 0 = 424$. The total holding cost is $(0.6)(424) = \$254.40$. Hence, the total cost of the Silver-Meal solution for this assembly amounts to $\$650.40$. Compare this to the costs of $\$1,320$ using lot for lot and $\$919.80$ using the EOQ solution.

It is also interesting to note what would have been the result if we had employed the Wagner-Whitin algorithm to find the true optimal solution. The optimal solution for this problem turns out to be $y_4 = 154$, $y_9 = 171$, and $y_{12} = 114$, with a total cost of $\$610.20$, which is only a slight improvement over the Silver-Meal heuristic.

We will now consider how the planned order release for the valve casing assembly affects the scheduling for lower-level components. In particular, if we consider the MRP calculations for the valves and assume that the lot sizing for the valves is determined by the Silver-Meal heuristic as well, we obtain

Week	1	2	3	4	5	6	7	8	9	10	11	12	13
Gross requirements				384	0	0	0	591	0	0	0	342	0
Scheduled receipts				96									
On-hand inventory			186	0	96	96	96	0					
Net requirements				198	0	0	0	495	0	0	0	342	0
Time-phased net requirements	198	0	0	0	495	0	0	0	342	0			
Planned order release (S-M)	198	0	0	0	495	0	0	0	342	0			

Note in this calculation summary that the scheduled receipts in period 5 must be held until period 8 before they can be used to offset demand. This results from the zero gross requirements in periods 5 through 8.

The calculation of the planned order release is based on a setup cost of $80 and a holding cost of $0.07 per valve per week. It is interesting to note that the Silver-Meal heuristic resulted in a lot-for-lot production rule in this case. This results from the lumpy demand pattern caused by the lot sizing applied at a higher level of the product structure. Both Silver-Meal and Wagner-Whitin give the same results for this example. ■

Problems for Section 3

23. If we were to solve Example 7.6 using the Wagner-Whitin algorithm (described in Appendix 7–A), we would obtain (154, 0, 0, 0, 0, 171, 0, 0, 114, 0) as the planned order release for the valve casing assembly. What are the resulting planned order releases for the valves?

24. Consider the example of Noname Computer Company presented in Problem 4. Suppose that the setup cost for the production of the motherboards is $180 and the holding cost is $h = \$0.40$ per motherboard per week. Using part period balancing, determine the planned order release for the motherboards and the resulting gross requirements schedule for the DRAM chips. (Hint: Use the net demand for computers after accounting for starting inventory and returns.)

25. For the example presented in Problem 6, assume that the setup cost for both components A and B is $100 and that the holding costs are respectively $0.15 and $0.25 per component per week. Using the Silver-Meal algorithm, determine the planned order releases for both components A and B and the resulting gross requirements schedules for components C, D, and E.

7.4 Lot Sizing with Capacity Constraints

We consider a variant of the problem treated above. Assume that in addition to known requirements $(r_1, \ldots, r_n)$ in each period, there are also production capacities $(c_1, \ldots, c_n)$. Hence, we now wish to find the optimal production quantities $(y_1, \ldots, y_n)$ subject to the constraints $y_i \leq c_i$, for $1 \leq i \leq n$.

The introduction of capacity constraints clearly makes the problem far more realistic. As lot-sizing algorithms can be incorporated into an MRP planning system, production capacities will be an important part of any realizable solution. However, they also make the problem more complex. The rather neat result that optimal policies always order exact requirements is no longer valid. Determining true optimal policies is difficult and time-consuming, and is probably not practical for most real problems.

Even finding a feasible solution may not be obvious. Consider our simple four-period example with vector $\mathbf{r} = (52, 87, 23, 56)$, but now suppose that the production capacity in each period is $\mathbf{c} = (60, 60, 60, 60)$. First we must determine if the problem is feasible; that is, whether at least one solution exists. On the surface the problem looks solvable, as the total requirement over the four periods is 218 and the total capacity is 240. But this problem is infeasible; the most that can be produced in the first two periods is 120, but the requirements for those periods sum to 139.

We have the following feasibility condition:

$$\sum_{i=1}^{j} c_i \geq \sum_{i=1}^{j} r_i \qquad \text{for } j = 1, \ldots, n.$$

Even when the feasibility condition is satisfied, it is not obvious how to find a feasible solution. Consider the following example.

Example 7.7

$$\mathbf{r} = (20, 40, 100, 35, 80, 75, 25),$$

$$\mathbf{c} = (60, 60, 60, 60, 60, 60, 60).$$

Checking for feasibility, we have that:

$r_1 = 20,$	$c_1 = 60;$
$r_1 + r_2 = 60,$	$c_1 + c_2 = 120;$
$r_1 + r_2 + r_3 = 160,$	$c_1 + c_2 + c_3 = 180;$
$r_1 + r_2 + r_3 + r_4 = 195,$	$c_1 + c_2 + c_3 + c_4 = 240;$
$r_1 + r_2 + r_3 + r_4 + r_5 = 275,$	$c_1 + c_2 + c_3 + c_4 + c_5 = 300;$
$r_1 + r_2 + r_3 + r_4 + r_5 + r_6 = 350,$	$c_1 + c_2 + c_3 + c_4 + c_5 + c_6 = 360;$
$r_1 + r_2 + r_3 + r_4 + r_5 + r_6 + r_7 = 375,$	$c_1 + c_2 + c_3 + c_4 + c_5 + c_6 + c_7 = 420.$

The feasibility test is satisfied, so we know at least that a feasible solution exists. However, it is far from obvious how we should go about finding one. Scheduling on a lot-for-lot basis is not going to work because of the capacity constraints in periods 3, 5, and 6.

We will present an approximate lot-shifting technique to obtain an initial feasible solution. The method is to back-shift demand from periods in which demand exceeds capacity to prior periods in which there is excess capacity. This process is repeated for each period in which demand exceeds capacity until we construct a new requirements schedule in which lot-for-lot is feasible. In the example, the first period in which demand exceeds capacity is period 3. We replace r_3 with c_3. The difference of 40 units must now be redistributed back to periods 1 and 2. We consider the first prior period, which is period 2. There are 20 units of excess capacity in period 2, which we absorb. We still have 20 units of demand from period 3 that is not yet accounted for; this is added to the requirement for period 1. Summarizing the results up until this point, we have

$$\begin{array}{ccc} 40 & 60 & 60 \\ \end{array}$$
$$\mathbf{r}' = (\cancel{20}, \cancel{40}, \cancel{100}, 35, 80, 75, 25),$$
$$\mathbf{c} = (60, 60, \quad 60, 60, 60, 60, 60).$$

The next period in which demand exceeds capacity is period 5. The excess demand of 20 units can be back-shifted to period 4. Finally, the 15 units of excess demand in period 6 can be back-shifted to periods 4 (5 units) and 1 (10 units). The feasibility condition guarantees that this process leads to a feasible solution.

This leads to

$$50 \qquad 60$$
$$4\!\!\!/0 \;\; 60 \quad 60 \;\; 5\!\!\!/5 \;\; 60 \;\; 60$$
$$\mathbf{r}' = (2\!\!\!/0, 4\!\!\!/0, 1\!\!\!/0\!\!\!/0, 3\!\!\!/5, 8\!\!\!/0, 7\!\!\!/5, 25),$$
$$\mathbf{c} = (60, 60, \;\; 60, 60, 60, 60, 60).$$

Hence, the modified requirements schedule obtained is

$$\mathbf{r}' = (50, 60, 60, 60, 60, 60, 25).$$

Setting $\mathbf{y} = \mathbf{r}'$ gives a feasible solution to the original problem. ∎

The Improvement Step.   We have that lot for lot for the modified requirements schedule $\mathbf{r}'$ is feasible for the original problem. Next we would like to see if we can discover an improvement; that is, another feasible policy that has lower cost. There are a variety of reasonable improvement rules that one can use. We will employ the following one.

For each lot that is scheduled, starting from the last and working backward to the beginning, determine whether it is cheaper to produce the units composing that lot by shifting production to prior periods of excess capacity. By eliminating a lot, one reduces setup cost in that period to zero, but shifting production to prior periods increases the holding cost. The shift is made only if the additional holding cost is less than the setup cost. We illustrate the process with an example.

Example 7.8

Assume that $K = \$450$ and $h = \$2$.

$$\mathbf{r} = (100, 79, 230, 105, 3, 10, 99, 126, 40),$$
$$\mathbf{c} = (120, 200, 200, 400, 300, 50, 120, 50, 30).$$

Computing the cumulative sum of requirements and capacities for each period makes it easy to see the problem as feasible. However, because requirements exceed capacities in some periods, lot for lot is not feasible. We back-shift excess demand to prior periods in order to obtain the modified requirements schedule $\mathbf{r}' = (100, 109, 200, 105, 28, 50, 120, 50, 30)$.

Lot for lot for the modified requirements schedule $\mathbf{r}'$ is feasible for the original problem. The initial feasible solution requires nine setups at a total setup cost of $9 \times 450 = \$4,050$. The holding cost of the initial policy is $2(0 + 30 + 0 + 0 + 25 + 65 + 86 + 10) = \$432$.

In order to do the improvement step, it is convenient to arrange the data in a table.

	1	2	3	4	5	6	7	8	9
$\mathbf{r}'$	100	109	200	105	28	50	120	50	30
c	120	200	200	400	300	50	120	50	30
y	100	109	200	105	28	50	120	50	30
Excess capacity	20	91	0	295	272	0	0	0	0

Starting from the last period, consider the final lot of 30 units. There is enough excess capacity in prior periods to consider shifting this lot. The latest period that this lot can be scheduled for is period 5. The extra holding cost incurred by making the shift is $2 \times 30 \times 4 = \$240$. As this is cheaper than the setup cost of $450, we make the shift and increase y_5 from 28 to 58 and reduce the excess capacity in period 5 from 272 to 242.

Now consider the lot of 50 units scheduled in period 8. This lot can also be shifted to period 5 with a resulting additional holding cost of $2 \times 50 \times 3 = \$300$. Again this is cheaper than the setup cost, so we make the shift. At this point we have $y_5 = 108$, and the excess capacity in period 5 is reduced to 192.

The calculations are summarized on our table in the following way:

	1	2	3	4	5	6	7	8	9
r′	100	109	200	105	28	50	120	50	30
c	120	200	200	400	300	50	120	50	30
					108				
					~~58~~			0	0
y	100	109	200	105	~~28~~	50	120	~~50~~	~~30~~
					192				
					~~242~~				
Excess capacity	20	91	0	295	~~272~~	0	0	0	0

Next consider the lot of 120 units scheduled in period 7. At this point, we still have 192 units of excess capacity in period 5. The additional holding cost of shifting the lot of 120 from period 7 to period 5 is $2 \times 120 \times 2 = \$480$. This exceeds the setup cost of $450, so we do not make the shift.

It is clearly advantageous to shift the lot of 50 units in period 6 to period 5, thus reducing the excess capacity in period 5 to 142 and increasing the lot size in period 5 from 108 to 158. Doing so results in the following:

	1	2	3	4	5	6	7	8	9
r′	100	109	200	105	28	50	120	50	30
c	120	200	200	400	300	50	120	50	30
					158				
					~~108~~				
					~~58~~			0	0
y	100	109	200	105	~~28~~	0	120	~~50~~	~~30~~
					142				
					~~192~~				
					~~242~~				
Excess capacity	20	91	0	295	~~272~~	0	0	0	0

At this point it may seem that we are done. However, there is enough capacity in period 4 to shift the entire lot of 158 units from period 5 to period 4. The additional holding cost of this shift is $2 \times 158 = \$316$. Because this is cheaper than the setup cost, we make the shift.

Summarizing these calculations on the table gives

	1	2	3	4	5	6	7	8	9
r′	100	109	200	105	28	50	120	50	30
c	120	200	200	400	300	50	120	50	30
					0				
					~~158~~				
					~~108~~				
				263	~~58~~	0		0	0
y	100	109	200	~~105~~	~~28~~	~~50~~	120	~~50~~	~~30~~
					300				
					~~142~~				
					~~192~~				
				137	~~242~~				
Excess capacity	20	91	0	~~295~~	~~272~~	0	0	0	0

At this point, no additional lot shifting is possible. The solution we have obtained is

$$\mathbf{y} = (100, 109, 200, 263, 0, 0, 120, 0, 0)$$

for the original requirements schedule

$$\mathbf{r} = (100, 79, 230, 105, 3, 10, 99, 126, 40).$$

We will compute the cost of this solution and compare it with that of our initial feasible solution. There are five setups at a total setup cost of $5 \times 450 = \$2,250$. The holding cost is $2(0 + 30 + 0 + 158 + 155 + 145 + 166 + 40 + 0) = 2 \times 694 = \$1,388$. The total cost of this policy is \$3,638, compared with \$4,482 for our initial feasible policy. For this example the improvement step resulted in a cost reduction of close to 20 percent. ■

Problems for Section 4

26. Consider the example presented in Section 2 of this chapter of scheduling the production of the valve casing assembly.
    a. Suppose that the production capacity in any week is 100 valve casings. Using the algorithm presented in this section, determine the planned order release for the valve casings.
    b. What gross requirements schedule for the valves does the lot sizing you obtained in part (a) give?
    c. Suppose that the production capacity for the valves is 200 valves per week. Is the gross requirements schedule from part (b) feasible? If not, suggest a modification in the planned order release computed in part (a) that would result in a feasible gross requirements schedule for the valves.
27. Solve Problem 14 assuming a maximum order size of 20 per month.

28. *a.* Solve Problem 17 assuming the following production capacities:

Week	1	2	3	4	5	6
Capacity	600	600	600	400	200	200

   b. On a percentage basis, how much larger is the total holding and setup cost in the capacitated case than in the solutions obtained from parts (*b*), (*c*), and (*d*) of Problem 17?

29. The method of rescheduling the production of a lot to one or more prior periods if the increase in the holding cost is less than the cost of a setup also can be used when no capacity constraints exist. This method is an alternative heuristic lot scheduling technique for the uncapacitated problem. For Problem 14, start with a lot-for-lot policy and consider shifting lots backward as we have done in this section, starting with the final period and ending with the first period. Compare the total cost of the policy that you obtain with the policies derived in Problem 14.

7.5 Shortcomings of MRP

MRP is a closed production system with two major inputs: (1) the master production schedule for the end item and (2) the relationships between the various components, modules, and subassemblies composing the production process for that end item. The method is logical and seemingly sensible for scheduling production lot sizes. However, many of the assumptions made are unrealistic. In this section, we will discuss some of these assumptions, the problems that arise as a result of them, and the means for dealing with these problems.

Uncertainty

Underlying MRP is the assumption that all required information is known with certainty. However, uncertainties do exist. The two key sources of uncertainty are the forecasts for future sales of the end item and the estimation of the production lead times from one level to another. Forecast uncertainty usually means that the realization of demand is likely to be different from the forecast of that demand. In the production planning context, it also could mean that updated forecasts of future demands are different from earlier forecasts of those demands. Forecasts must be revised when new orders are accepted, prior orders are canceled, or new information about the marketplace becomes available. That has two implications in the MRP system. One is that *all* of the lot-sizing decisions that were determined in the last run of the system could be incorrect, and, even more problematic, former decisions that are currently being implemented in the production process may be incorrect.

The analysis of stochastic inventory models in Chapter 5 showed that an optimal policy included safety stock to protect against the uncertainty of demand. That is, we would order to a level exceeding expected demand. The same logic can be applied to MRP systems. The manner in which uncertainty transmits itself through a complex multilevel production system is not well understood. For that reason, it is not recommended to include independent safety stock at all levels of the system. Rather, by using the methods in Chapter 5, suitable safety levels can be built into the forecasts for the end item. These will be transmitted automatically down through the system to the lower levels through the explosion calculus.

Example 7.9

Consider Example 7.1 on the Harmon Music Company. Suppose that the firm wishes to incorporate uncertainty into the demand forecasts for weeks 8 through 17. Based on historical records of trumpet sales maintained by the firm, an analyst finds that the ratio of the standard deviation of the forecast error to the mean demand each week is near 0.3.[2] Furthermore, weekly demand is closely approximated by a normal distribution. Harmon has decided that it would like to produce enough trumpets to meet all the demand each week with probability .90. (In the terminology used in Chapter 5, this means that they are using a Type 1 service level of 90 percent for the trumpets.)

The safety stock is of the form σz where z is the appropriate cut-off point from the normal table. Here $z = 1.28$. Incorporating safety stock into the demand forecasts, we obtain

Week	8	9	10	11	12	13	14	15	16	17
Predicted demand (μ)	77	42	38	21	26	112	45	14	76	38
Standard deviation (σ)	23.1	12.6	11.4	6.3	7.8	33.6	13.5	4.2	22.8	11.4
Mean demand plus safety stock ($\mu + \sigma z$)	107	58	53	29	36	155	62	19	105	53

Of course, this is not the only way to compute safety stock. Alternatives are to employ a Type 2 service criterion or to use a stock-out cost model instead of a service level model. The next step is to net out the scheduled receipts and anticipated on-hand inventory to arrive at a revised MPS for the trumpets. The explosion calculus would now proceed as before, except that the safety stock that is included in the revised MPS would automatically be transmitted to the lower-level assemblies.

Safety lead times are used to compensate for the uncertainty of production lead times in MRP systems. Simply put, this means that the estimates for the time required to complete a production batch at one level and transport it to the next level would be multiplied by some safety factor. If a safety factor of 1.5 were used at Harmon Music Company, the lead times for the components would be revised as follows: bell assembly, 3 weeks; valve casing assembly, 6 weeks; slide assemblies, 3 weeks; valves, 4.5 weeks. Conceptually, safety lead times make sense if the essential uncertainty is in the production times from one level to the next, and safety stocks make sense if the essential uncertainty is in the forecast of the demand for the end item. In practice, both sources of uncertainty are generally present and some mixture of both safety stocks and safety lead times is used. ■

Capacity Planning

Another important issue that is not treated explicitly by MRP is the capacity of the production facility. The type of capacitated lot-sizing method we discussed earlier will deal with production capacities at one level of the system but will not solve the overall capacity problem. The problem is that even if lot sizes at some level do not exceed the production capacities, there is no guarantee that when these lot sizes are translated to gross requirements at a lower level, these requirements also can be satisfied with the existing capacity. That is, a feasible production schedule at one level may result in an infeasible requirements schedule at a lower level.

---

[2]In symbols, this is written σ/μ and is known as the coefficient of variation.

Capacity requirements planning (CRP) is the process by which the capacity requirements placed on a work center or group of work centers are computed by using the output of the MRP planned order releases. If the planned order releases result in an infeasible requirements schedule, there are several possible corrective actions. One is to schedule overtime at the bottleneck locations. Another is to revise the MPS so that the planned order releases at lower levels can be achieved with the current system capacity. This is clearly a cumbersome way to solve the problem, requiring an iterative trial-and-error process between the CRP and the MRP.

As an example of CRP, consider the manufacture of the trumpet discussed in Example 7.1 and throughout the rest of this chapter. Suppose that the valves are produced in three work centers: 100, 200, and 300. At work center 100, the molten brass is poured into the form used to shape the valve. At work center 200, the holes are drilled in the appropriate positions in the valves (there are three hole configurations, depending upon whether the valve is number 1, 2, or 3). Finally, at work center 300, the valve is polished and the surface appropriately graded to ensure that the valve does not stick in operation. A summary of the appropriate information for the work centers is given below.

	Worker Time Required to Produce One Unit (hours/unit)	*Machine Throughput (units/day)*
Work center 100	0.1	120
Work center 200	0.25	100
Work center 300	0.15	160

According to this information, there would be a total of six minutes (0.1 hour) of worker time required to produce a single valve at work center 100, and the existing equipment can support a maximum throughput of 120 valves per day. Consider the planned order releases obtained for the valves resulting from the Silver-Meal lot scheduling rule given in Section 3:

Week	2	3	4	5	6	7	8	9	10	11
Planned order release (S-M)	198	0	0	0	495	0	0	0	342	0

This planned order release translates to the following capacity requirements at the three work centers:

Week	2	3	4	5	6	7	8	9	10	11
Labor time requirements (hours):										
Work center 100	19.8	0	0	0	49.5	0	0	0	34.2	0
Work center 200	49.5	0	0	0	123.75	0	0	0	85.5	0
Work center 300	29.7	0	0	0	72.25	0	0	0	51.3	0
Machine time requirements (days):										
Work center 100	1.65	0	0	0	4.125	0	0	0	2.85	0
Work center 200	1.98	0	0	0	4.95	0	0	0	3.42	0
Work center 300	1.24	0	0	0	3.09	0	0	0	2.14	0

The capacity requirements show whether the planned order release obtained from the MRP is feasible. For example, suppose that the requirement of 123.75 labor hours in week 6 at work center 200 exceeds the capacity of this work center. This means that the current lot sizing is infeasible and some corrective action is required. One possibility is to split the lot scheduled for week 6 by producing some part of it in a prior week. Another is to adjust the lot sizing for the valve casing assembly at the next higher level of the product structure to accommodate the capacity constraints at the current level. In either case, substantial changes in the initial production plan may be required.

This example suggests an interesting speculation. Would it not perhaps make more sense to determine where the bottlenecks occur *before* attempting to explode the MPS through the various levels of the system? In this way, a feasible production plan could be found that would meet capacity constraints. Additional refinements could then be considered.

Rolling Horizons and System Nervousness

Thus far, our view of MRP is that it is a static system. Given known requirements for the end items over a specified planning horizon, one determines both the timing and the sizes of the production lot sizes for all the lower-level components. In practice, however, the production planning environment is dynamic. The MRP system may have to be rerun each period and the production decisions reevaluated. Often it is the case that only the lot-sizing decisions for the current planning period need to be implemented. We use the term *rolling horizons* to refer to the situation in which only the first-period decision of an N-period problem is implemented. The full N-period problem is rerun each period to determine a new first-period decision.

When using rolling horizons, the planning horizon should be long enough to guarantee that the first-period decision does not change. Unfortunately, certain demand patterns are such that even for long planning horizons, the first-period decision does not remain constant. Consider the following simple example (from Carlson, Beckman, and Kropp, 1982).

Example 7.10

Suppose that the demand follows the cyclic pattern 190, 210, 190, 210, 190. . . . For a five-period planning horizon, the requirements schedule for periods 1 to 5 is

$$\mathbf{r} = (190, 210, 190, 210, 190).$$

Furthermore, suppose that $h = 1$ and $K = 400$. The optimal solution for this problem obtained from the Wagner-Whitin algorithm is

$$\mathbf{y} = (190, 400, 0, 400, 0).$$

However, suppose that the planning horizon is chosen to be six periods instead of five periods. The requirements schedule for a six-period planning horizon is

$$\mathbf{r} = (190, 210, 190, 210, 190, 210).$$

The optimal solution in this case is

$$\mathbf{y} = (400, 0, 400, 0, 400, 0).$$

That is, the first-period production quantity has changed from 190 to 400. If we go to a seven-period planning horizon, then y_1 will be 190. With an eight-period planning horizon, y_1 again becomes 400. One might think that this cycling of the value of y_1 would continue indefinitely. It turns out that this is *not* the case, however. For planning horizons of $n \geq 21$ periods, the value of y_1 remains fixed at 190.[3] However,

---

[3] I am grateful to Lawrence Robinson of Cornell University for pointing out that this anomaly does not continue for all values of n as was claimed in past editions and by Carlson, Beckman, and Kropp (1982).

even when there is eventual convergence of y_1, as in this example, the cycling for the first 20 periods could be troublesome when using rolling planning horizons. ∎

Another common problem that results when using MRP is "nervousness." The term was coined by Steele (1973), who used it to refer to the changes that can occur in a schedule when the horizon is moved forward one period. Some of the causes of nervousness include unanticipated changes in the MPS because of updated forecasts, late deliveries of raw materials, failure of key equipment, absenteeism of key personnel, and unpredictable yields.

There has been some analytical work on the nervousness problem. Carlson, Jucker, and Kropp (1979 and 1983) use the term *nervousness* specifically to mean that a revised schedule requires a setup in a period in which the prior schedule did not. They have proposed an interesting technique to reduce this particular type of nervousness: Let $(\hat{y}_1, \hat{y}_2, \ldots, \hat{y}_N)$ be the existing production schedule and $(y_1, y_2, \ldots, y_N)$ be a revised schedule based on new demand information. Suppose that besides the usual costs of holding and setup, there is an additional cost of v if the new schedule $\mathbf{y}$ calls for a setup in a period that the old schedule $\hat{\mathbf{y}}$ did not. This means that there is an additional setup cost associated with the new schedule in those periods in which no setup was called for in the old schedule. Their method is to increase the setup cost from K to $K + v$ if $\hat{y}_k = 0$ prior to determining the new schedule $\mathbf{y}$. Resolving the problem with the modified setup costs using any of the lot-sizing algorithms previously discussed now will result in fewer setup revisions. The advantages of the various lot-sizing methods in this context are discussed in the two references above. The revision cost v reflects the relative importance of the cost of nervousness.

Additional Considerations

Although MRP would seem to be the most logical way to schedule production in batch type operations, as we saw above the basic method has some very serious shortcomings. Other difficulties are considered below.

Lead Times Dependent on Lot Sizes. The MRP calculus assumes that the production lead time from one level to the next is a fixed constant independent of the size of the lot. In many contexts this assumption is clearly unreasonable. One would expect the lead time to increase if the lot size increases. Including a dependence between the production lead time and the size of the production run into the explosion calculus seems to be extremely difficult.

MRP II: Manufacturing Resource Planning. As we have noted, MRP is a closed production planning system that converts an MPS into planned order releases. Manufacturing resource planning (MRP II) is a philosophy that attempts to incorporate the other relevant activities of the firm into the production planning process. In particular, the financial, accounting, and marketing functions of the firm are tied to the operations function. As an example of the difference between the perspectives offered by MRP and MRP II, consider the role of the master production schedule. In MRP, the MPS is treated as input information. In MRP II, the MPS would be considered a part of the system and, as such, would be considered a decision variable as well. Hence, the production control manager would work with the marketing manager to determine when the production schedule should be altered to incorporate revisions in the forecast and new order commitments. Ultimately, all divisions of the company would work together to find a production schedule consistent with the overall business plan and long-term financial strategy of the firm.

Another important aspect of MRP II is the incorporation of CRP (capacity resource planning). Capacity considerations are not explicitly accounted for in MRP. MRP II is a closed-loop cycle in which lot sizing and the associated shop floor schedules are compared to capacities and recalculated to meet capacity restrictions. However, capacity issues continue to be an important issue in both MRP and MRP II operating systems.

Obviously, such a global approach to the production scheduling problem is quite ambitious. Whether such a philosophy can be converted to a workable system in a particular operating environment remains to be seen.

Imperfect Production Processes.   An implicit assumption made in MRP is that there are no defective items produced. Because requirements for components and subassemblies are computed to exactly satisfy end-item demand forecasts, losses due to defects can seriously upset the balance of the production plan. In some industries, such as semiconductors, yield rates can be as low as 10 to 20 percent for new products. As long as yields are stable, incorporating yield losses into the MRP calculus is not difficult. One computes net demands and lot sizes in the same way, and in the final step divides the planned order release figures by the average yield. For example, if a particular process has a yield rate of 78 percent, one would multiply the planned order releases by $1/0.78 = 1.28$. The problem is much more complex if yields are random and variances are too large to be ignored. Using mean yields would result in substantial stockouts. One would have to develop a kind of newsboy model that balanced the cost of producing too many and too few and determine an appropriate safety factor. Because of the dependencies of successive levels, this would be a difficult problem to model mathematically. Monte Carlo computer simulation would be a good alternative. To this writer's knowledge, no one has attempted to develop a mathematical model of random yields in the context of MRP systems. (Random yields, however, are discussed by several researchers. For example, Nahmias and Moinzadeh, 1996, consider a mathematical model of random yields in a single-level lot-sizing problem.)

Data Integrity.   An MRP system can function effectively only if the numbers representing inventory levels are accurate. It is easy for incorrect data to make their way into the scheduling system. This can occur if a shipment is not recorded or is recorded incorrectly at some level, items entering inventory for rework are not included, scrap rates are higher than anticipated, and so on. In order to ensure the integrity of the data used to determine the size and the timing of lot sizes, physical stock-taking may be required at regular intervals. An alternative to complex physical stock-taking is a technique known as *cycle counting*. Cycle counting simply means directly verifying the on-hand levels of the various inventories comprising the MRP system. For example, are the 45 units of part A557J indicated on the current record the actual count of this part number?

Efficient cycle counting can be achieved in a variety of ways. Stockrooms may have containers that only hold a fixed number or weight of items. Coded shelving systems could be used to more easily identify items with part numbers. Certain areas could be made accessible only to specific personnel. Cycle counting systems can be based on number or on weight. Furthermore, an error in the inventory level must be considered in relative terms. Based on the importance of the item, different percentage errors may be considered acceptable. Different error tolerances should be applied to weigh-counted items versus hand-counted items. If MRP is to have a positive impact

SNAPSHOT APPLICATION

Raymond Corporation Builds World-Class Manufacturing with MRP II

The Raymond Corporation is a major materials handling equipment manufacturer headquartered in Greene, New York. In order to achieve an estimated payback of 10 times in reduced inventories and organizational efficiencies, the organization underwent a process of implementation of MRP II that spanned over two years. The success of their effort was based on several basic principles. The first was that top management had to be committed to the process of change. For that reason, the first step was the education of the top management utilizing a "canned" educational package available from an outside consulting firm. The CEO facilitated on-site training for the vice presidents, who in turn were responsible for training the middle management implementation team. With top management on board, middle management could not simply ignore the issue and wait for it to go away. Training continued throughout the implementation phase of the project and ultimately included employees at every level of the company.

Once training was far enough along to begin thinking about implementation, a strategy for implementation was laid out. The first step was to take a hard look at the data that would be used as inputs to the system. As a result of this effort, stockroom reporting accuracy went from 66 percent to over 95 percent in about 16 months. Every inventory status report was measured for accuracy and corrected when necessary. Setting up standards and systems that give accurate information can be the lion's share of the benefit from an effort such as this. An early payoff was that reporting accuracy led to reduced need to closely monitor

purchasing. Inventory status reports were used for ABC classification, and more time and energy were devoted to managing the "A" items.

In order to determine the effectiveness of new systems, the implementation team would meet weekly to review the progress of internal measures. An attempt was made to determine the cause of problems or lack of progress without pointing fingers and assigning blame. Performance measurements become an important part of the success of the implementation effort. Accurate and up-to-date performance measurements are the cornerstone of any systems change effort, but they must be put in place without threatening workers.

Finally, the author recommends that the *final* step should be the purchase of new software. If the underlying data and measurements systems in place are not sound, the software, no matter how sophisticated, won't help. Many firms believe that purchasing an expensive MRP software system is all that's necessary to achieve reduced inventory, lower materials costs, improved on-time delivery, and so on. However, without employees that are on-board, accurate data, and performance measurements, new software can be more of a hindrance than a help.

Aside from reaching class A status, what benefits did Raymond see from this process? Sheldon (1994) claimed elimination of overtime, elimination of shortages, improved on-time deliveries, reductions in setup times and costs, reductions in rework and scrap rates, and lower inventory and material handling costs. While it is hard to believe that the improvements were as dramatic as this, a carefully planned implementation of MRP II clearly paid off for the Raymond Corporation.

Source: This description is based on Sheldon (1994).

on the overall production scheduling problem, the inventory records must be an accurate reflection of the actual state of the system.

Order Pegging. In some complex systems, a single component may be used in more than one item at the next higher level of the system. For example, a company producing many models of toy cars may use the same-sized axle in each of the cars. Gross requirements for axles would be the sum of the gross requirements generated by the MPS for each model of car. Hence, when one component is used in several items, the gross requirements schedule for this component comes from several sources. If a shortage of this component occurs, it is useful for the firm to be able to identify the particular items higher in the tree structure that would be affected. In order to do this, the gross requirements schedule is broken down by the items that generate it and each requirement is "pegged" with the part number of the source of the requirement. Pegging adds considerable complexity to the information storage requirements of the system and should only be considered when the additional information is important in the decision-making process.

Problems for Section 5

30. MRP systems have been used with varying degrees of success. Describe under what circumstances MRP might be successful and under what circumstances it would not be successful.

31. Discuss the advantages and disadvantages of including safety stock in MRP lot-sizing calculations. Do you think that a production control manager would be reluctant to build to safety stock if he is behind schedule?

32. For what reason is the capacitated lot-sizing method discussed in Section 5 not adequate for solving the overall capacity problem?

33. Planned order releases (POR) for three components, A, B, and C, are given below. Suppose that the yields for these components are respectively 84 percent, 92 percent, and 70 percent. Assuming lot-for-lot scheduling, how should these planned order releases be adjusted to account for the fact that the yields are less than 100 percent?

Week	6	7	8	9	10	11	12	13	14	15	16	17
POR(A)				200	200	80	80	200	400	400	400	
POR(B)			100	100	40	40	100	200	200	200		
POR(C)	200	400	280	100	280	600	800	800	400			

34. Define the terms *rolling horizons* and *system nervousness* in the context of MRP systems.

7.6 JIT Fundamentals

Just-in-time, lean production, and zero inventories are all names for essentially the same thing: a system of moving material through a plant that requires a minimum of inventory. Some have speculated that the roots of the system go back to the situation in postwar Japan. Devasted by the war, the Japanese firms were cash poor and did not have the luxury of investing cash in excess inventories. Thus, lean production was born from necessity. However, as Japanese cars started gaining in popularity in the United States, it quickly became clear that they were far superior to American- and European-made cars in terms of quality, value, efficiency, and reliability. We know today that JIT and the quality initiatives of the 1950s played important roles in this success.

Two developments were key to the success of this new approach to mass production: the kanban system and SMED (which stands for single minute exchange of dies).

Kanban is a Japanese word for *card* or *ticket.* It is a manual information system developed and used by Toyota for implementing just-in-time.

The Mechanics of Kanban

There are a variety of different types of kanban tickets, but two are the most prevalent. These are withdrawal kanbans and production ordering kanbans. A withdrawal kanban is a request for parts to a work center from the prior level of the system. A production ordering kanban is a signal for a work center to produce additional lots. The manner in which these two kanban tickets are used to control the flow of production is depicted in Figure 7–8.

The process is as follows: Parts are produced at work center 1, stored in an intermediate location (known as the store), and subsequently transported to work center 2. Parts are transported in small batches represented by the circles in the figure.

FIGURE 7–8

Kanban system for two production centers

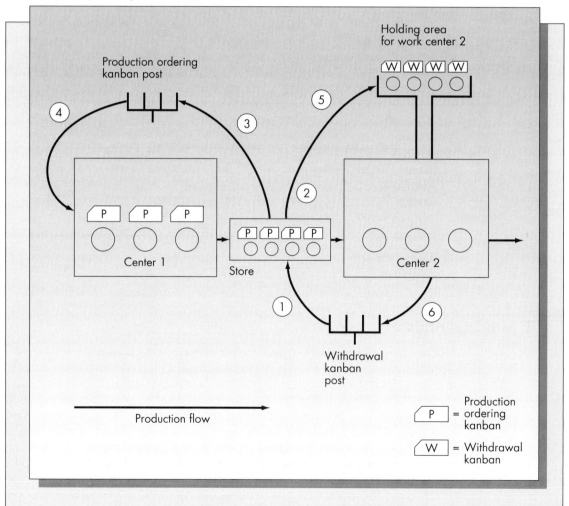

Production flows from left to right in the diagram. The detailed steps in the process are as follows (the numbers below appear in the appropriate locations in Figure 7–8):

1. When the number of tickets on the withdrawal kanban reaches a predetermined level, a worker takes these tickets to the store location.

2. If there are enough canisters available at the store, the worker compares the part number on the production ordering kanbans at the store with the part number on the withdrawal kanbans.

3. If the part numbers match, the worker removes the production ordering kanbans from the containers, places them on the production ordering kanban post, and places the withdrawal kanbans in the containers.

4. When a specified number of production ordering kanbans have accumulated, work center 1 proceeds with production.

5. The worker transports parts picked up at the store to work center 2 and places them in a holding area until they are required for production.

6. When the parts enter production at work center 2, the worker removes the withdrawal kanbans and places them on the withdrawal kanban post. (Note that production ordering kanbans for work center 2 are then attached to the parts produced at that work center. These kanban tickets are not shown in Figure 7–8.)

One computes the number of kanban tickets in the system in advance. Toyota uses the following formula (Monden, 1981b):

$$y = \frac{\overline{D}L + w}{a},$$

where

y = Number of kanbans.
$\overline{D}$ = Expected demand per unit of time.
L = Lead time (processing time + waiting time between processes + conveyance time).
w = Policy variable specifying the level of buffer stock, generally around 10 percent of $\overline{D}L$.
a = Container capacity (usually no more than 10 percent of daily demand).

This formula implies that the maximum level of inventory is given by $ay = \overline{D}L + w$. The ideal value of w is zero. However, it is difficult to balance a system so perfectly that buffer stock is eliminated entirely.

As mentioned earlier, the kanban system is a manual information system for implementing just-in-time. JIT systems also can be implemented in other ways, which may be more efficient than the kanban method. More will be said about this later in the section.

Single Minute Exchange of Dies

One of the key components of the success of Toyota's production system was the concept of single minute exchange of dies (SMED), championed by Shigeo Shingo. Shingo is generally credited with developing and implementing SMED at Toyota in 1970, which has become an important part of the overall Toyota production system. The basic theory developed in Chapter 4 tells us that small lot sizes will be optimal only if fixed costs are small (as K decreases in the EOQ formula, so does the value of Q). The most significant component of the cost of setting up for a new operation in a plant is the time required to change over the machinery for that operation, since the production line must be frozen during the changeover operation. This requires changing some set of tools and/or dies required in the process, hence the origin of the term SMED. (A die is a tool used for shaping or impressing an object or material.)

Die-changing operations are required in automotive plants when switching over the production line from one car model to another. These operations typically required about four hours. The idea behind SMED is that a significant portion of the die-changing operation can be done off-line while the previous dies are still in place and the line continues to operate. According to Shingo (1981), this is accomplished by dividing the die-changing operation into two components: IED (inside exchange of die) and OID (outside exchange of die). The OID operation is to be performed while the line is running in advance of the actual exchange. The goal is to structure the die change so that there are as many steps as possible in the OID portion of the operation.

While this idea sounds simple, it has led to dramatic improvements in throughput rates in many circumstances, both within and outside the automotive industry. Shingo (1981) describes some of the successes:

- At Toyota, an operation that required eight hours for exchange of dies and tools for a bolt maker was reduced over a year's time to only 58 seconds.
- At Mitsubishi Heavy Industry, a tool-changing operation for a boring machine was reduced from 24 hours to only 2 minutes and 40 seconds.
- At H. Weidmann Company of Switzerland, changing the dies for a plastic molding machine was reduced from 2.5 hours to 6 minutes and 35 seconds.
- At Federal-Mogul Company of the United States, the time required to exchange the tools for a milling machine was reduced from 2 hours to 2 minutes.

Of course, SMED cannot be applied in all manufacturing contexts. Even in contexts where it can be applied, the benefits of the die-changing time reduction can be realized only if the process is integrated into a carefully designed and implemented overall manufacturing control system.

Advantages and Disadvantages of the Just-in-Time Philosophy

Champions of just-in-time would have one believe that all other production planning systems are now obsolete. The zeal with which they promote the method is reminiscent of the enthusiasm that heralded MRP in the early 1970s. At that time, some claimed that classical inventory analysis was no longer valid. However, each new production method should be viewed as an addition to, rather than a replacement for, what is already known. Just-in-time can be a useful tool in the right circumstances, but is far from a panacea for the problems facing industry today.

As we saw in the previous section, just-in-time and EOQ are not mutually exclusive. Setup time and, consequently, setup cost reduction result in smaller lot sizes. Smaller lot sizes require increased efficiency and reliability of the production process, but considerably less investment in raw materials, work-in-process inventories, and finished-goods inventories.

Kanban is a manual information system used to support just-in-time inventory control. Just-in-time and kanban are not necessarily wedded. In future years, we expect to see just-in-time systems based on more current and sophisticated information transfer

FIGURE 7–9

Kanban information system versus centralized information system

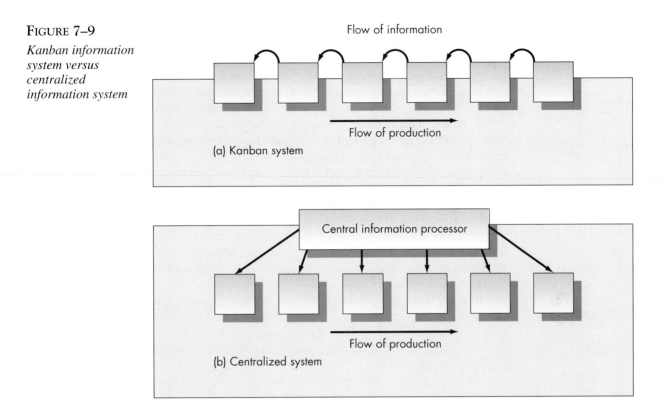

Flow of information

Flow of production

(a) Kanban system

Central information processor

Flow of production

(b) Centralized system

technology. A shortcoming of kanban is the time required to transmit new information through the system. Figure 7–9 considers a schematic of a serial production process with six levels. With the kanban system, the direction of the flow of information is opposite to the direction of the flow of the production. Consider the consequences of a sudden change in the requirements at level 6. This change is transmitted first to level 5, then to level 4, and so on. There could be a substantial time lag from the instant that the change occurs at level 6 until the information reaches level 1.

A centralized information processing system will help to alleviate this problem. If there are sudden changes in the requirements at one end of the system resulting from unplanned changes in demand or breakdowns of key equipment, these changes will be instantly transmitted to the entire system.

MRP has an important advantage over kanban in this regard. One of the strengths of MRP is its ability to react to forecasted changes in the pattern of demand. The MRP system recomputes production quantities based on these changes and makes this information available to all levels simultaneously. MRP allows planning to take place at all levels in a way that just-in-time, and especially kanban, will not. As Meal (1984) points out:

> Just-in-time production works well when the overall production rate is constant, but it is unsatisfactory for communicating basic changes in production rate to earlier stages in the process. . . . On the other hand using the HPP [hierarchical production planning] approach, plant managers do not rely on their short term signals to establish their early stage production rates.

Just-in-time is most efficient when the pattern of demand is stable and predictable. Changes in the demand may result from predictable causes or random fluctuations or both. MRP makes use of forecasts of anticipated changes in demand and transmits this information to all parts of the productive system. However, neither MRP nor just-in-time is designed to protect against random fluctuations of demand. Both methods could be unstable in the face of high demand variance.

Another potential shortcoming of just-in-time is the idle time that may result when unscheduled breakdowns occur. Part of the Japanese philosophy is that workers should have a familiarity with more than one portion of the productive process. If a breakdown occurs, then workers' attention can be immediately focused on the problem. However, if workers are familiar with only their own operation, there will be significant worker idle time when a breakdown occurs. This is consistent with the trade-off curve presented in Figure 8–4 of Chapter 8 on shop floor control and sequence scheduling. Buffer inventories between successive operations provide a means of smoothing production processes. However, buffer inventories also have disadvantages. They can mask underlying problems. A popular analogy is to compare a production process with a river and the level of inventory with the water level in the river. When the water level is high, the water will cover the rocks. Likewise, when inventory levels are high, problems are masked. However, when the water level (inventory) is low, the rocks (problems) are evident (see Figure 7–10).

Because items are moved through the system in small batches, 100 percent inspection is feasible. Seen in this light, just-in-time can be incorporated easily into an overall quality control strategy. Total quality management (TQM), discussed in Chapter 11, and JIT can work together not only to reduce inventory costs, but also to bring about significant improvements in product quality.

As users of just-in-time point out, it is not simply an inventory control system. For just-in-time to work properly, it must be coordinated with the purchasing system and with purchasing strategies. One complaint about just-in-time is that it merely pushes system uncertainty and higher inventories onto the supplier. There is no doubt that greater flexibility on the part of the suppliers is necessary; they must be able to react quickly and provide sufficiently reliable parts to relieve the manufacturer of the necessity to inspect all incoming lots. Furthermore, multiple sourcing becomes difficult under such a system. That is, the firm may be forced to deal with a single supplier in order to develop the close relationship that the system requires. Single sourcing presents risks for both suppliers and manufacturers. The manufacturer faces the risk that the supplier will be unable to supply parts when they are needed, and the supplier faces the risk that the manufacturer will suffer reverses and demand will drop.

FIGURE 7–10

River/inventory analogy illustrating the advantages of just-in-time

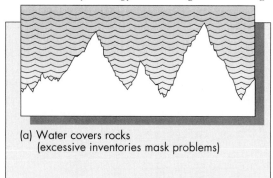

(a) Water covers rocks
(excessive inventories mask problems)

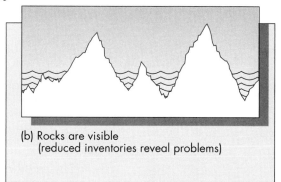

(b) Rocks are visible
(reduced inventories reveal problems)

TABLE 7–1   **Summary of Advantages and Disadvantages of Just-in-Time and Kanban**

Feature	Advantages	Disadvantages
Small work-in-process inventories	1. Decreases inventory costs. 2. Improves production efficiency. 3. Points out quality problems quickly.	1. May result in increased worker idle time. 2. May decrease the production rate.
Kanban information flow system	1. Provides for efficient lot tracking. 2. Inexpensive means of implementing just-in-time. 3. Allows for predetermined level of WIP inventory by presenting number of kanban tickets.	1. Slow to react to changes in demand. 2. Ignores known information about future demand patterns.
Coordinated inventory and purchasing	1. Inventory reduction. 2. Improved coordination of different systems. 3. Improved relationships with vendors.	1. Decreased opportunity for multiple sourcing. 2. Suppliers must react more quickly. 3. Improved reliability required of suppliers.

Table 7–1 briefly summarizes the primary advantages and disadvantages discussed in this section.

We close by noting that JIT and MRP are certainly not the only ways to approach manufacturing control. In their book, Hopp and Spearman (1996) explore a production planning system based on fixing WIP inventory. If reduction of work-in-process is a goal of JIT, then why not design a system from scratch that forces a desired level of WIP? They have designed just such a system, which they call CONWIP (for *CONstant Work-In-Process*). In a CONWIP system, each time an item is completed (exits the production line), initiation of production of a new item is begun at the start of the line. In this way the WIP stays fixed. One can adjust the size of the WIP based on variability and cost considerations. CONWIP is similar in principle to JIT in that both are pull systems. However, with CONWIP, the information flow is not from a stage in the process to the previous stage, but from the end of the process to the beginning. The method sounds intriguing, but still needs to stand the test of implementation in the real world.

Implementation of JIT in the United States

A dramatic example of successful implementation of JIT in the United States is the Harley-Davidson Motorcycle Company. Until the early 1980s Harley-Davidson was owned by American Foundry Company (AMF). Harley, well known as a manufacturer of large-displacement motorcycles, faced severe competition from the Japanese. Honda, traditionally a manufacturer of small-displacement motorcycles, was beginning to make inroads into Harley's market. It appeared that motorcycles would become another consumer product produced only by foreign companies.

The first step in Harley's recovery was the purchase of the company from AMF by a group of employees. Shortly after the buyout, top management traveled to Japan to see how its competitors' factories were run. According to Willis (1986),

> The real "secret," the executives discovered, lay not in robotics or high-tech systems but in the intelligent, efficient organization of the company's employees and production systems.

As a result of these visits, Harley-Davidson was completely reorganized. Traditional management structure was replaced by a system in which each employee was given ownership of his or her area of the line. A large proportion of staff jobs were cut, resulting in a much shallower organization chart similar to that in Japanese companies. In this way, problems could not become buried in bureaucracy. The firm also instituted a quality circles program with active participation of the line personnel (see the discussion of quality circles in Chapter 11).

The firm invested in state-of-the-art CAD/CAM (computer-aided design/computer-aided manufacturing) and robotics equipment. The physical layout of the plant was restructured into work cells using a group technology layout (see Chapter 10). Motorcycles were no longer manufactured on a traditional assembly line.

A key change was the manner in which material was moved through the plant. Harley adopted the term MAN (materials as needed) to describe its system, but this was clearly just another name for JIT. To make JIT work, it instituted a program to reduce the number of setups each day. Prior to the restructuring, it produced 13 to 14 different models each day. Under the new system, improved forecasting allowed it to implement two-hour repeating cycles.

In order to make JIT work, Harley needed a commitment from its vendors as well as its own employees. To encourage its vendors to "buy in" to the program, Harley offered training sessions on the principles of JIT. These programs became so popular that Harley started to offer them to firms that were neither vendors nor clients. What is the result of this effort? Today Harley-Davidson's motorcycles are 99 percent defect-free, as compared to only 50 percent in 1982. In addition to achieving dramatic improvements in product quality, Harley developed new employee benefit programs and expanded its product line. All this was accomplished with a simultaneous reduction in costs.

Harley-Davidson is not the only example of successful implementation of JIT in the United States. Jacobson and Hillkirk (1986) describe the turnaround of Xerox, which was beleaguered by competition from Kodak and several Japanese firms (especially Canon) in the early 1980s. Xerox embarked on a self-examination program based on competitive benchmarking and undertook a complete restructuring of its business. Implementation of JIT systems played a major role in its turnaround. For example, it reduced the number of vendors supplying the company from 5,000 to only 400, which included a mix of sources overseas as low-cost suppliers and sources close by to provide timely deliveries when needed. Its systems allowed it to reduce copier parts inventories by $240 million.

The American auto industry has made significant moves toward embracing JIT, but there is still a long way to go. There are many examples of plants in the United States that have adopted the JIT philosophy. Two General Motors ventures are the NUMMI plant in northern California, a joint venture with Toyota, and the Saturn plant in Tennessee. Ironically, Ford, which pioneered the assembly line for mass production of automobiles, has been the most successful American firm at adopting Japanese manufacturing philosophies. Ford's assembly plant near Atlanta, Georgia, at which the Ford Taurus and the Mercury Sable are manufactured, is impressive on most productivity and quality measures, even when compared with the Japanese. When GM attempted to isolate the factors responsible for Ford's significantly better productivity in this plant as compared to GM's Fairfax, Kansas, plant, which makes the Pontiac Grand Prix, it found that 41 percent of the difference could be attributed to the manufacturability of the car's designs and 48 percent to factory practice (Womack et al., 1990, p. 96).

Although U.S. firms have certainly made significant progress toward adopting JIT principles, the conversion to "lean production" has not been as quick as one would have hoped. This is partly due to an environment that is less conducive to JIT. For example, in the U.S. auto industry, assembly plants and suppliers are often separated by large distances. Japanese firms can have daily and even hourly deliveries from suppliers, because those suppliers are located close to assembly plants. Also, JIT works best when product demand is relatively stable. Sales of automobiles in the United States have been far more cyclic than in Japan (Womack et al., 1990, p. 247), making a pure JIT system more difficult to implement here. Hence, the U.S. auto industry faces a more difficult environment for successful implementation of JIT. Even with these difficulties, both U.S. and European manufacturers will have to make the transition to lean production methods if they hope to remain competitive with the Japanese.

Problems for Section 6

35. Discuss the concepts of push versus pull and how they relate to just-in-time.

36. What is the difference between a just-in-time system and a kanban system? Can just-in-time be implemented without kanban?

37. A regional manufacturer of table lamps plans on using a manual kanban information system. On average the firm produces 1,200 lamps monthly. Production lead time is 18 days and they plan to have 15 percent buffer stock. Assume 20 working days per month.
    a. If each container holds 15 lamps, what will be the total number of kanban tickets required? (Use the formula given in this section.)
    b. What is the maximum WIP inventory the company can expect to have using this system?
    c. Suppose that each lamp costs the firm $30 to produce. If carrying costs are based on a 20 percent annual interest rate, what annual carrying cost for WIP inventory is the company incurring? (You may wish to review the discussion of holding costs in Chapter 4.)

38. What is SMED? In what way can die-changing operations be reduced by orders of magnitude? Why is it an advantage to do so?

39. Explain how the dual-card kanban system operates.

40. Discuss the advantages and disadvantages of each of the following features of just-in-time:
    a. Small lot sizes.
    b. Integrated purchasing and inventory control.
    c. Kanban information system.

7.7 A Comparison of MRP and JIT

MRP and JIT are fundamentally different systems for manufacturing control. As we noted earlier, MRP is a push system and JIT is a pull system. JIT is a reactive system. If a problem develops and the line is shut down, JIT reacts immediately, since requests for new material are discontinued. In this way, one might say that JIT reacts to uncertainties and MRP does not. However, JIT is clearly not going to work well when it is known that demands will vary significantly over time. MRP builds this information into the planning structure while JIT does not.

For most manufacturing environments, implementing a pure JIT system is simply not feasible. Suppliers may not be located in close enough proximity to allow inputs to be delivered according to a rigid schedule. Demands for products may be highly variable, making it impractical to ignore this information in the planning process. It may be difficult to move products in small batches. Implementing SMED may not be possible in some environments. When setup costs are very high, it makes economic sense to produce in large lots and store items rather than change over production processes frequently. With that said, however, major reductions in WIP inventories can be achieved in the vast majority of traditional manufacturing plants. Plants that run lean run better.

Toyota's enormous success in reducing inventory-related costs while producing high-quality products with high throughput rates has formed the basis for much of the support for JIT-based systems. However, it is unclear if JIT is primarily responsible for Toyota's success. Toyota's way of doing business differs from that of American auto makers in many dimensions. Is its success a direct result of JIT methods, or can it be attributed to other factors that might be more difficult to emulate? In order to determine under what circumstances JIT would be most advantageous, Krajewski et al. (1987) developed a large-scale simulator to compare JIT, MRP, and ROP (reorder point) manufacturing environments. Their comparison included 36 distinct factors aggregated into eight major categories, with each factor varied from one to five levels. The eight categories considered and a brief summary of the factors in these categories included

1. *Customer influence.* Demand forecast error.
2. *Vendor influence.* Vendor reliability (order size received compared to order size requested, and time received compared to time requested).
3. *Buffer mechanisms.* Capacity buffer and safety stock and safety lead times.
4. *Product structure.* Considered pyramid (few end items) and inverted pyramid (many end items) structures.
5. *Facility design.* Routing patterns, shop emphasis, and length of routings.
6. *Process.* Scrap rates, equipment failures, worker flexibility, and capacity imbalances.
7. *Inventory.* Reporting accuracy, lot-sizing rules, and setup times.
8. *Other factors.* Number of items, number of work stations, and several other factors.

The results obtained were very interesting. JIT worked well only in favorable manufacturing environments. This means little or no demand variability, reliable vendors, and small setup times for production. JIT showed poor performance when one or more of these factors became unfavorable. In fact, in a favorable environment both ROP and MRP gave good results as well. This suggests that the primary benefit comes from creating a favorable manufacturing environment, as opposed to simply implementing a JIT system. Perhaps greater benefit would result from carefully evaluating and (where possible) rectifying the key problems affecting manufacturing than by blindly implementing a new production-planning system. Alternatively, perhaps it is the *process* of implementing a JIT system from which the greatest benefit is obtained.

New approaches to manufacturing control are quickly tagged with three-letter acronyms and proselytized with the zealousness of a religion. JIT is no exception. At its best, JIT is a set of methods that should be implemented on a continuous improvement basis for reducing inventories at every level of the supply chain. At its worst, JIT is a romantic idea that when applied blindly can be very damaging to worker morale, relationships with suppliers, and ultimately the bottom line. Inventory savings can often be an illusion, as illustrated by Chhikara and Weiss (1995). They show in three case studies that if inventory reductions are not tied to accounting systems and cash flows, inventory reductions do not translate to reduced inventory costs. JIT for its own sake does not make sense. It must be carefully integrated into the entire supply and manufacturing chain in order to realize its benefits.

Zipkin (1991) offers a very thoughtful piece on the "JIT revolution." His main point is that JIT can have real benefits, but if the pragmatism is not distinguished from the romance, the consequence could be disaster. He cites an example of a manager of a computer assembly plant who was ordered to reduce his WIP inventory to zero in six months. This was the result, apparently, of the experiences of the CFO, who had attended a JIT seminar that inspired him to promote massive inventory reductions.

Finally, as noted by Karmarker (1989), the issue is not to make a choice between MRP and JIT, but to make the best use of both techniques. JIT reacts very slowly to sudden changes in demand, while MRP incorporates demand forecasts into the plan. Does that mean that Toyota, famous for their JIT system, ignores demand forecasts in their manufacturing planning and control? Very unlikely. Understanding what different methodologies offer and their limitations leads to a manufacturing planning and control system that is well designed and efficient. Improvements should be incremental, not revolutionary.

7.8 MRP Software: A Big Business

Software for manufacturing planning and control is a big business and continues to expand rapidly. Materials requirements planning software began to surface in the 1970s. These systems were designed for mainframe computers, often bundling hardware and software into a single package. However, it soon became clear that stand-alone manufacturing planning programs were falling short of the mark because they failed to make the necessary linkages with other parts of the firm's operations. The second generation of MRP software (MRP II or manufacturing resource planning) attempted to incorporate the *resources* that link to the manufacturing plan, including projected sales, personnel, plant capacity, and distribution capacity.

How big is this market? LaPlante (1994) reported that projected revenues for such systems (both MRP and MRP II) for 1994 was $2.8 billion in the United States, which represented a 40 percent growth over 1993. In fact, the sales rate continued to climb in 1995 with no signs of slowing down. Blood (1994) reported that more than 70,000 such systems have been implemented in the United States.

With this rapid growth and tens of thousands of users in the United States alone, one would assume that customers are happy with their experiences with this software. However, this seems to be far from the truth. Blood (1994) claims that two-thirds of MRP users are dissatisfied with the product. According to George Stalk, Jr.: "Factory automation efforts that reduce costs and incrementally improve efficiency capture only the extremely low hanging fruit. Most companies are correcting the inefficiencies caused by inattention. They utterly are failing to get the really big returns" (quoted in LaPlante, 1994).

Even picking the "low hanging" fruit is not guaranteed. New software can make things worse rather than better. Implementation problems come from two directions. Users expect that a day after a new system is put into place, inventories will drop, on-time deliveries will improve, and everything will run smoothly. Contrast this attitude with the careful preparation of the Raymond Corporation discussed earlier in the Snapshot Application. Raymond undertook a multiyear training effort prior to implementing new software.

The other problem arises from the limitations in the systems themselves. Most of the MRP and MRP II systems on the market still utilize mainframe computers and cumbersome source codes that provide no links with accounting or marketing databases. While many of these have been very successful in the past (such as the MANMAN system from Ask Systems of Mountain View, California), the real growth in the market is coming from another source. Several companies offer more modern configurations based on client–server systems. In a client–server arrangement, satellite machines (usually PCs) are linked via a server to access common databases. The PC interface facilitates data entry and retrieval via familiar environments such as the spreadsheet. It also allows various parts of the organization to access information about the status of critical orders.

Firms offering open client–server systems include Avalon, Baan, qad.inc, Oracle, SAP, Symix, and Triton. While all of these firms have had some measure of success, the one that seems to be making the fastest inroads into the manufacturing market is German-based SAP. The firm, which markets hardware, software, and related consulting services on a worldwide basis, projected revenues of $9.3 billion in 1995. SAP systems link the factory, warehouse, sales offices, accounting, and corporate headquarters in a comprehensive information system. The firm claims some 4,500 customers throughout the world (Lieber, 1995).

With SAP, the goal is to require that data be entered into the system only once. By linking all the functional areas of the firm, these data become available where and when they are needed. SAP (and its competitors) are, of course, more than just manufacturing planning systems. They are integrated management information systems with multiple planning functions. These systems do not come cheaply. A large-scale system including hardware and ongoing consulting for implementation could cost in the tens of millions of dollars. Rarely can one purchase an SAP product for less than $100,000. For that reason, there will continue to be a market for lower-priced alternatives.

Below are several quick studies of firms that have had positive experiences with implementing the new generation of client–server manufacturing software. (The first four are described in LaPlante, 1994, and the last in Lieber, 1995.)

1. Northrup Grumann of Long Beach, California, scrapped a mainframe-based system that was costing the firm $20 million annually in maintenance and licensing fees and replaced it with a client–server system from Oracle. Even though the Oracle system cost $14 million, overall costs were reduced immediately and delivery lead time was reduced from 18 months to only 7 months.

2. Coors Ceramics of Golden, Colorado, suffered from the maladies that many manufacturing firms have. Numbers on computer printouts were not trusted, the status of orders could not be tracked, and lead times for delivery were too long. Coors replaced an old system running on a DEC PDP 11 with a client–server system from qad that linked manufacturing, field service, and finance. The result was that product cycles, which ranged from 10 to 14 weeks, were reduced to 6 to 10 weeks, and on-time deliveries improved to over 95 percent.

3. Phoenix Designs is a made-to-order manufacturer of office systems and furniture. Phoenix implemented a Symix information system on an IBM RS/6000 server linked to a custom CAD (computer-aided design) program located directly at dealers' outlets. The system permitted dealers to put together a Phoenix system on a computer screen using a trackball and mouse. It utilized a rule-based application of artificial

intelligence principles. The software is used by 236 dealers and linked to the firm's manufacturing information system. Prior to the installation of this system, it required four to six weeks from the time an initial order was placed until the product was delivered to the customer. With the new system, a made-to-order system can be delivered in two weeks.

4. Red Devil, a $50 million producer of sealants, caulks, and hand tools, found that employees had developed a deep-seated mistrust of its manufacturing planning system, and for good reason: the numbers were invariably wrong. Red Devil went to Avalon software on an NCR Unix server and achieved immediate savings in MIS costs as well as a healthy $2 million savings in inventory costs. Customer service went from a 95 percent fill rate in five days to a 98 percent fill rate in two days.

5. Food giant Borden found that its implementation of SAP software and hardware solved several problems. In 1994 the firm undertook an implementation of SAP for order processing. With the new system in place, employees from any location in the country could track the status of an order while a customer waited. In the past, it often took an entire day to track down this information.

As information systems for manufacturing planning become more sophisticated and user friendly, firms will be able to realize the kinds of efficiencies that have eluded them in the past. With the information available, these systems might someday incorporate the kinds of mathematical models discussed in this book to achieve even greater efficiencies.

7.9 Historical Notes

The specific term *MRP* is relatively new, although the concept of materials planning based on predicted demand and product structures appears to have been around for quite a while. In fact, there is a reference to this approach by Alfred Sloan (1964), who refers to a purely technical calculation by analysts at General Motors in 1921 of the quantity of materials needed for production of a given number of cars. The term *BOM* (bill of materials) *explosion* was commonly used to describe what is now called MRP.

The books by Orlicky (1975) and New (1974) did a great deal to legitimize MRP as a valid and identifiable technique, although the term seems to have been around since the mid-1960s. In addition, the well-known practitioners George Plossl and Oliver Wight also must be given credit for popularizing the method (see, for example, Plossl and Wight, 1971). In Anderson, Schroeder, Tupy, and White (1982) the authors state that the first computerized MRP systems were implemented close to 1970. The number of installed systems has increased since that time at an exponential rate.

Interestingly, much of the work on optimal and suboptimal lot-sizing methods predates the formal recognition of MRP. The seminal paper in this area was by Wagner and Whitin (1958), who first recognized the optimality of an exact requirements policy for periodic-review inventory control systems with time-varying demand, and developed the dynamic programming algorithm described in this chapter. The Silver-Meal heuristic when opportunities for replenishment are at the start of periods appeared in Silver and Meal (1973). DeMatteis (1968) is generally credited with the part period balancing approach. However, the paper by Gorham (1968) refers to both part period balancing (called the least total cost approach) and least unit cost methods, suggesting that these methods were well known at the time. It is likely that both methods were developed by practitioners before 1968 but were not reported in the literature.

The lot-shifting algorithm for the capacitated problem outlined in Section 4 is very similar to one developed by Karni (1981). Similar ideas were explored by Dixon and

Silver (1981) as well. Lot sizing is not always used by practitioners in operational MRP systems because of the effect that small errors at a high level of the product structure are telescoped into large errors at a lower level. Furthermore, lot-sizing algorithms require estimation of the holding and setup costs and more calculations than the simple lot-for-lot policy.

The historical references to JIT are contained within the chapter. It is unclear who coined the term JIT, but the concept is clearly derived from Toyota's kanban system. SMED, single minute exchange of dies, has played an important role in the success of the Japanese lean production methods. Shigeo Shingo is generally credited with its development.

7.10 Summary

Materials requirements planning (MRP) is a set of procedures for converting forecast demand for a manufactured product into a requirements schedule for the components, subassemblies, and raw materials comprising that product. A closely related concept is that of the *master production schedule* (MPS), which is a specification of the projected needs of the end product by time period. The *explosion calculus* represents the set of rules and procedures for converting the MPS into the requirements at lower levels. The information required to do the explosion calculus is contained in the *product structure diagram* and the *indented bill-of-materials list.* The two key pieces of information contained in the product structure diagram are the production lead times needed to produce the specific component and the multiplier giving the number of units of the component required to produce one item at the next higher level of the product structure.

Many MRP systems are based on a *lot-for-lot* production schedule. That is, the number of units of a component produced in a period is the same as the requirements for that component in that period. However, if setup and holding costs can be estimated accurately, it is possible to find other lot-sizing rules that are more economical. The optimal lot-sizing procedure is the Wagner-Whitin algorithm. However, the method is rarely used in practice, largely because of the relative complexity of the calculations required (although such calculations take very little time on a computer). We explored three heuristic methods that require fewer calculations than the Wagner-Whitin algorithm, although none of these methods will necessarily give an optimal solution. They are the *Silver-Meal heuristic,* the *least unit cost heuristic,* and *part period balancing.*

We also treated the dynamic lot-sizing problem when capacity constraints exist. One of the limitations of MRP is that capacities are ignored. This is especially important if lot sizing is incorporated into the system. Finding optimal solutions to a capacity-constrained inventory system subject to time-varying demand is an extremely difficult problem. (For those of you familiar with the term, the problem is said to be NP complete, which is a reference to the level of difficulty.) A straightforward heuristic method for obtaining a solution to the capacitated lot-sizing problem was presented. However, incorporating such a method into the MRP system will, in and of itself, *not* solve the complete capacitated MRP problem, because even if a particular lot-sizing schedule is feasible at one level, there is no guarantee that it will result in a feasible requirements schedule at a lower level.

Truly optimal lot-sizing solutions for an MRP system would require formulating the problem as an integer program in order to determine the optimal decisions for all levels simultaneously. For real assembly systems, which can be as many as 10 to 15 levels deep, this would result in an enormous mathematical programming problem.

In view of the host of other issues concerned with implementing MRP, the marginal benefit one might achieve from multilevel optimization would probably not justify the effort involved.

System nervousness is one problem that arises when implementing an MRP system. The term refers to the unanticipated changes in a schedule that result when the planning horizon is rolled forward by one period. Another difficulty is that in many circumstances, production lead times depend on the lot sizes: MRP assumes that production lead times are fixed. Still another problem is that the yields at various levels of the process may not be perfect. If the yield rates can be accurately estimated in advance, these rates can be factored into the calculations in a straightforward manner. However, in many industries the yield rates may be difficult to estimate in advance.

Manufacturing resource planning (MRP II) attempts to deal with some of the problems of implementing MRP by integrating the financial, accounting, and marketing functions into the production-planning function.

The MRP software industry continues to grow at a rapid rate. Worldwide sales of manufacturing planning systems are in the tens of billions of dollars. Traditional MRP systems were mainframe-based and operated with no link to other databases within the firm. The newest generation of manufacturing software links all parts of the firm's activities and is based on client–server systems rather than mainframe computers. Although there has been a lot of talk about JIT replacing MRP, the evidence is that the MRP software industry continues to expand. MRP-type manufacturing planning systems play an important role in factory management methods.

In this chapter we also discussed just-in-time. Just-in-time is an inventory control system whose goal is to reduce work-in-process to a bare minimum. The concepts are based on the production control systems used by Toyota in Japan in the 1970s. JIT is a pull system, whereas MRP is a push system. Parts are transferred from one level to the next only when requested. In addition to reduced inventories, this approach allows workers to quickly locate quality control problems. Because parts are moved through the system in small batches, defects can be identified quickly. By the nature of the system, stopping production at one location automatically stops production on the entire line so that the source of the problem can be identified and corrected before defective inventories build up.

An important aspect of JIT is reduction of production lot sizes. To make small lot sizes economical, it is necessary to reduce fixed costs of changeover. This is the goal of SMED, or single minute exchange of dies. By dividing the die- or tool-changing operation into portions that can be done off-line and those that must be done on-line, enormous reductions in setup times can be achieved.

Kanban and JIT are closely related. Kanban is a manual information system for implementing JIT that relies on cards or tickets to signal the need for more product. While not very "high tech," the method has worked well in practice.

JIT has several disadvantages vis-à-vis MRP as well as several advantages. JIT will react more quickly if a problem develops. However, JIT systems are very slow to react to changes in the pattern of demand. MRP, on the other hand, builds forecasts directly into the explosion calculus.

Additional Problems for Chapter 7

41. CalcIt produces a line of inexpensive pocket calculators. One model, IT53, is a solar-powered scientific model with an LCD display. The calculator is pictured in Figure 7–11.

FIGURE 7–11

*CalcIt Model IT53
Scientific Calculator
(Problem 41)*

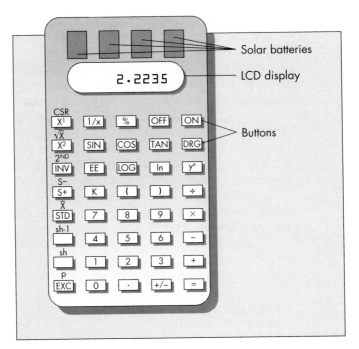

Each calculator requires 4 solar cells, 40 buttons, 1 LED display, and 1 main processor. All parts are ordered from outside suppliers, but final assembly is done by CalcIt. The processors must be in stock three weeks before the anticipated completion date of a batch of calculators to allow enough time to set the processor in the casing, connect the appropriate wiring, and allow the setting paste to dry. The buttons must be in stock two weeks in advance and are set by hand into the calculators. The LCD displays and the solar cells are ordered from the same supplier and need to be in stock one week in advance.

Based on firm orders that CalcIt has obtained, the master production schedule for IT53 for a 10-week period starting at week 8 is given by

Week	8	9	10	11	12	13	14	15	16	17
MPS	1,200	1,200	800	1,000	1,000	300	2,200	1,400	1,800	600

Determine the gross requirements schedule for the solar cells, the buttons, the LCD display, and the main processor chips.

42. Consider the example of the CalcIt Company for Problem 41. Suppose that the buttons used in the calculators cost $0.02 each and the company estimates a fixed cost of $12 for placing and receiving orders of the buttons from an outside supplier. Assume that holding costs are based on a 24 percent annual interest rate and that there are 48 weeks to a year. Using the gross requirements schedule for the buttons determined in Problem 41, what order policy does the Silver-Meal heuristic recommend for the buttons? (Hint: Express h as a cost per 10,000 units and divide each demand by 10,000.)

43. Solve Problem 42 using part period balancing and least unit cost. Compare the costs of the resulting solutions to the cost of the solution obtained by using the Silver-Meal heuristic.

44. *Work-in-process* (WIP) *inventory* is a term that refers to the inventory of components and subassemblies in a manufacturing process. Assuming lot-for-lot scheduling at all levels, theoretically the WIP inventory should be zero. Do you think that this is likely to be the case in a real manufacturing environment? Discuss the possible reasons for large WIP inventories occurring even when an MRP system is used.

45. Vivian Lowe is planning a surprise party for her husband's 50th birthday. She has decided to serve shish kabob. The recipe that she is using calls for two pineapple chunks for each shrimp. She plans to size the kabobs so that each has three shrimp. She estimates that a single pineapples will yield about 50 chunks, but from past experience, about 1 out of every 10 pineapples is bad and has to be thrown out. She has invited 200 people and expects that about half will show up. Each person generally eats about 2 kabobs.
   a. How many pineapples should she buy?
   b. Suppose that the number of guests is a random variable having the normal distribution with mean 100 and variance 1,680. If she wants to make enough kabobs to feed all the guests with probability 95 percent, how many pineapples should she buy?

46. In this chapter, we assumed that the "time bucket" was one week. This implies that forecasts are reevaluated and the MRP system rerun on a weekly basis.
   a. Discuss the potential advantages of using a shorter time bucket, such as a day.
   b. Discuss the potential advantages of using a longer time bucket, such as two weeks or a month.

47. Develop a spreadsheet that reproduces the calculations for Example 7.1. As in the example, the spreadsheet should include the net predicted demand for trumpets. The columns should correspond to weeks and should be labeled 1 to 18. Below the net predicted demand for trumpets should be the calculations for the valve casing assembly, and below that the calculations for the valves. For each component, include rows for the following information: (1) gross requirements, (2) scheduled receipts, (3) on-hand inventory, (4) time-phased net requirements, and (5) lot-for-lot planned order release. Your spreadsheet should automatically update all calculations if the net predicted demand for trumpets changes.

48. Two end products, EP1 and EP2, are produced in the Raleigh, North Carolina, plant of a large manufacturer of furniture products located in the Southeast. The product structure diagrams for these products appear in Figure 7–12.
   Suppose that the master production schedules for these two products are

Week	18	19	20	21	22	23	24
EP1	120	112	76	22	56	90	210
EP2	62	68	90	77	26	30	54

   Assuming lot-for-lot production, determine the planned order releases for components F, G, and H.

FIGURE 7–12

*Product structure
diagrams (for
Problem 48)*

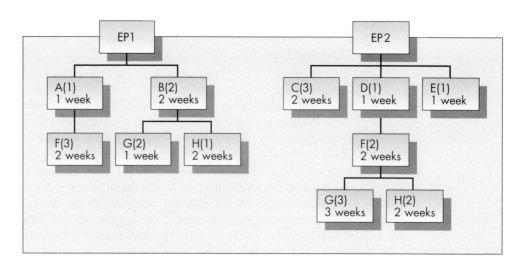

49. A component used in a manufacturing facility is ordered from an outside
    supplier. Because the component is used in a variety of end products, the demand
    is high. Estimated demand (in thousands) over the next 10 weeks is

Week	1	2	3	4	5	6	7	8	9	10
Demand	22	34	32	12	8	44	54	16	76	30

The components cost 65 cents each and the interest rate used to compute the
holding cost is 0.5 percent per week. The fixed order cost is estimated to be
$200. (Hint: Express h as the holding cost per thousand units.)

a. What ordering policy is recommended by the Silver-Meal heuristic?
b. What ordering policy is recommended by the part period balancing heuristic?
c. What ordering policy is recommended by the least unit cost heuristic?
d. Which method resulted in the lowest-cost policy for this problem?

50. A popular heuristic lot-sizing method is known as period order quantity (POQ).
    The method requires determining the average number of periods spanned by the
    EOQ and choosing the lot size to equal this fixed period supply. Let λ be the total
    demand over an N-period planning horizon ($\lambda = (\Sigma\, r_i/n)$) and assume that EOQ
    is computed as described in Section 2. Then $P = EOQ/\lambda$ rounded to the nearest
    integer. For the example in Section 2, $P = 139/43.9 = 3.17$, which is rounded
    to 3. The POQ would call for setting the lot size equal to three periods of
    demand. For the example in Section 2, the resulting planned order release would
    be 116, 0, 0, 150, 0, 0, 135, 0, 0, 38.

a. Compare the cost of the policy obtained by this method for the example in
   Section 2 to that obtained using the EOQ.
b. What are the advantages of this approach over EOQ?
c. Do you think that this method will be more cost effective in general than the
   heuristic methods discussed in Section 2?
d. Solve Problem 17 using this method, and compare the total holding and setup
   cost with that obtained by the other methods.

e. Solve Problem 49 using this method, and compare the total holding and setup cost with that obtained by the other methods.

51. The campus store at a large Midwestern university sells writing tablets to students, faculty, and staff. They sell more tablets near exam time. During a typical 10-week quarter, the pattern of sales is 2,280, 1,120, 360, 3,340, 1,230, 860, 675, 1,350, 4,600, 1,210. The pads cost the store $1.20 each, and holding costs are based on a 30 percent annual interest rate. The cost of employee time, paperwork, and handling amounts to $30 per order. Assume 50 weeks per year.

a. What is the optimal order policy during the 10-week quarter based on the Silver-Meal heuristic? Using this policy, what are the total holding and ordering costs incurred over the 10-week period?

b. The bookstore manager has decided that it would be more economical if the demand were the same each week. In order to even out the demand he limits weekly sales (to the annoyance of his clientele). Hence, assume that the total demand for the 10-week quarter is still the same, but the sales are constant from week to week. Determine the optimal order policy in this case and compare the total holding and ordering cost over the 10 weeks to the answer you obtained in part (*a*). (You may assume continuous time for the purposes of your calculations, so that the optimal lot size is the EOQ.)

c. Based on the results of parts (*a*) and (*b*), do you think that it is more economical in general to face a smooth or a spiky demand pattern?

52. Develop a spreadsheet template for finding Silver-Meal solutions for general lot-sizing problems. Store the holding and setup cost parameters in separate cell locations so that they can be inputted and changed easily. Allow for 30 periods of demand to be inputted in column 2 of the spreadsheet. List the period numbers 1, 2, . . . , 30 in column 1. Work out the logic that gives $C(j)$ in column 3.

One would use such a spreadsheet in the following way: Input requirements period by period until an increase is observed in column 3. This identifies the first forecast horizon. Now replace entries in column 2 with zeros and input requirements starting at the current period. Continue until the next forecast horizon is identified. Repeat this process until the end of the planning horizon. Use this method to find the Silver-Meal solution for the following production-planning problems:

a. Solve Problem 14 in this manner.

b. Weekly demands for 2-inch rivets at a division of an aircraft company are predicted to be (in gross)

Week	1	2	3	4	5	6	7	8	9	10	11	12
Demand	240	280	370	880	950	120	135	450	875	500	400	200

Week	13	14	15	16	17	18	19	20	21	22	23	24
Demand	600	650	1,250	250	800	700	750	200	100	900	400	700

Setup costs for ordering the rivets are estimated to be $200, and holding costs amount to 10 cents per gross per week. Find the lot sizing given by Silver-Meal.

53. Along the lines described in Problem 52, construct a spreadsheet for finding the least unit cost lot-sizing rule.
   a. Solve both parts (a) and (b) of Problem 52.
   b. Which method, least unit cost or Silver-Meal, gave the more cost-effective solution?

APPENDIX 7–A   OPTIMAL LOT SIZING FOR TIME-VARYING DEMAND

The techniques considered in Section 3 are easy to use and give lot sizes with costs that are generally near the true optimal. This appendix considers how one would go about computing true optimal lot sizes. Optimal in this context means the policy that minimizes the total holding and setup cost over the planning horizon. This appendix shows how optimal policies can be determined by casting the problem as a shortest-path problem. It also shows how dynamic programming can be used to find the shortest path.

Assume that:

1. Forecasted demands over the next n periods are known and given by the vector $\mathbf{r} = (r_1, \ldots, r_n)$.

2. Costs are charged against holding at $\$h$ per unit per period and $\$K$ per setup. We will assume that the holding cost is charged against ending inventory each period.

In order to get some idea of the potential difficulty of this problem, consider the following simple example.

Example 7A.1

The forecast demand for an electronic assembly produced at Hi-Tech, a local semiconductor fabrication shop, over the next four weeks is 52, 87, 23, 56. There is only one setup each week for production of these assemblies, and there is no back-ordering of excess demand. Assume that the shop has the capacity to produce any number of the assemblies in a week.

Consider the total number of feasible production policies for Hi-Tech over the four-week period. Let $y_1, \ldots, y_4$ be the order quantities in each of the four weeks. Clearly $y_1 \geq 52$ in order to guarantee that we do not stock out in period 1. If we assume that ending inventory in period 4 is zero (which will be easy to show is optimal), then $y_1 \leq 218$, the sum of all the demands. Hence y_1 can take any one of 167 possible values. Consider y_2. The number of feasible values of y_2 depends upon the value of y_1. As no stock-out is permitted to occur in period 2, $y_1 + y_2 \geq 52 + 87 = 139$. If

$$y_1 \leq 139, \qquad \text{then } 139 - y_1 \leq y_2 \leq 218 - y_1,$$

and if

$$y_1 > 139, \qquad \text{then } 0 \leq y_2 \leq 218 - y_1.$$

With a little effort, one can show that this results in a total of 10,200 different values of just the pair (y_1, y_2). It is thus clear that for even moderately sized problems the number of feasible solutions is enormous. ∎

Searching all of the feasible policies is unreasonable. However, an important discovery by Wagner and Whitin reduces considerably the number of policies one must consider as candidates for optimality.

The Wagner-Whitin algorithm is based on the following observation:

Result. An optimal policy has the property that each value of **y** is exactly the sum of a set of future demands. (We will call this an exact requirements policy.) That is,

$$y_1 = r_1, \quad \text{or } y_1 = r_1 + r_2, \ldots, \quad \text{or } y_1 = r_1 + r_2 + \cdots + r_n.$$
$$y_2 = 0, \quad \text{or } y_2 = r_2, \quad \text{or } y_2 = r_2 + r_3, \ldots, \quad \text{or}$$
$$y_2 = r_2 + r_3 + \cdots + r_n$$
$$\vdots$$
$$y_n = 0 \quad \text{or } y_n = r_n.$$

An exact requirements policy is completely specified by designating the periods in which ordering is to take place. The number of exact requirements policies is much smaller than the total number of feasible policies.

Example 7A.1 (continued)

Consider the four-period scheduling problem described above. Because y_1 must satisfy exact requirements, we see that it can assume values of 52, 139, 162, or 218 only; that is, only four distinct values. Ignoring the value of y_1, y_2 can assume values 0, 87, 110, 166. It is easy to see that every exact requirements policy is completely determined by specifying in what periods ordering should take place. That is, each such policy is the form $(i_1, \ldots, i_n)$, where the values of i_j are either 0 or 1. If $i_j = 1$, then production takes place in period j. Note that $i_1 = 1$ because we must produce in period 1 to avoid stocking out, whereas $i_2, \ldots, i_n$ will each be either 0 or 1. For example, the policy $(1, 0, 1, 0)$ means that production occurs in periods 1 and 3 only. It follows that $\mathbf{y} = (139, 0, 79, 0)$. For this example, there are exactly $2^3 = 8$ distinct exact requirements policies. ∎

A convenient way to look at the problem is as a one-way network with the number of nodes equal to exactly one more than the number of periods. Every path through the network corresponds to a specific exact requirements policy. The network for the four-period problem discussed above appears in Figure 7A–1.

For any pair (i, j) with $i < j$, if the arc (i, j) is on the path, it means that ordering takes place in period i and the order size is equal to the sum of the requirements in periods i, $i + 1, \ldots, j - 1$. Period j is the next period of ordering. Note that all paths end at period $n + 1$. The policy of ordering in periods 1 and 3 only would correspond to the path 1–3–5. The path 1–2–4–5 means that ordering is to take place in periods 1, 2, and 4.

The next step is to assign a value to each arc in the network. The value or "length" of the arc (i, j), called c_{ij}, is defined as the setup and holding cost of ordering in period

FIGURE 7A–1

Network representation for lot scheduline (Example 7A.1)

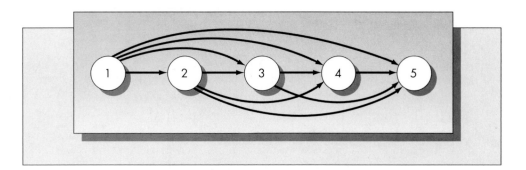

i to meet requirements through period $j - 1$. For example, c_{15} = the cost of ordering in period 1 to satisfy the demands in periods 1 through 4.

Finally, we would like to determine the minimum-cost production schedule, or shortest path through the network. As we will see in the next section, *dynamic programming* is one method of solving this problem. However, for a small problem, the optimal policy can be found by simply enumerating the paths through the network and choosing one with minimum cost.

Example 7A.2

We will solve Example 7A.1 using path enumeration. Recall that $\mathbf{r} = (52, 87, 23, 56)$. In addition, assume that there is a cost of holding of $h = \$1$ per unit per period and a cost of $K = \$75$ per setup.

The first step is to compute c_{ij} for $1 \le i \le 4$ and $i + 1 \le j \le 5$.

$c_{12} = 75$ (setup cost only).
$c_{13} = 75 + 87 = 162$.
$c_{14} = 75 + (23 \times 2) + 87 = 208$.
$c_{15} = 75 + (56 \times 3) + (23 \times 2) + 87 = 376$.
$c_{23} = 75$.
$c_{24} = 75 + 23 = 98$.
$c_{25} = 75 + 23 + (56 \times 2) = 210$.
$c_{34} = 75$.
$c_{35} = 75 + 56 = 131$.
$c_{45} = 75$.

Summarizing these costs in matrix form gives the following:

j / i	1	2	3	4	5
1		75	162	208	376
2			75	98	210
3				75	131
4					75

As there are only eight exact requirements policies, we can solve this problem by enumerating the policies and comparing the costs.

Path	Cost
1–2–3–4–5	$300
1–2–4–5	248
1–2–5	285
1–2–3–5	281
1–3–4–5	312
1–3–5	293
1–4–5	283
1–5	376

It follows that the optimal path is 1–2–4–5 at a cost of \$248. This corresponds to ordering in periods 1, 2, and 4 only. The optimal ordering policy is $y_1 = 52$, $y_2 = 110$, $y_3 = 0$, $y_4 = 56$. ∎

*Solution by Dynamic Programming

The total number of exact requirements policies for a problem of n periods is 2^{n-1}. As n gets large, total enumeration is not efficient. Dynamic programming is a recursive solution technique that can significantly reduce the number of computations required, although it too can be quite cumbersome.

Dynamic programming is based on the *principle of optimality*. One version of this principle is that if a problem consists of exactly n stages and there are $r < n$ stages remaining, the optimal policy for the remaining stages is independent of policy adopted in the previous stages. Because dynamic programming is not used anywhere else in this text, we will not present a detailed discussion of it. The interested reader should refer to Hillier and Lieberman (1990) for a brief overview or Nemhauser (1966) for a more in-depth treatment at a mathematical level consistent with ours.

Define f_k as the minimum cost starting at node k, assuming that an order is placed in period k. The principle of optimality for this problem results in the following system of equations:

$$f_k = \min_{j>k} (c_{kj} + f_j) \qquad \text{for } k = 1, \ldots, n.$$

The initial condition is $f_{n+1} = 0$.

Example 7A.3

We will solve Example 7A.1 by dynamic programming in order to illustrate the technique. One starts with the initial condition and works backward from period $n + 1$ to period 1.[4] In each period one determines the value of j that achieves the minimum.

$$f_5 = 0.$$
$$f_4 = \min_{j>4} (c_{4j} + f_j)$$
$$= 75 \qquad \text{at } j = 5 \text{ (the only possible value of } j\text{)}.$$
$$f_3 = \min_{j>3} (c_{3j} + f_j) = \min \begin{Bmatrix} c_{34} + f_4 \\ c_{35} + f_5 \end{Bmatrix} = \min \begin{Bmatrix} 75 + 75 \\ 131 + 0 \end{Bmatrix} = \min \begin{Bmatrix} 150 \\ 131 \end{Bmatrix}$$
$$= 131 \qquad \text{at } j = 5.$$
$$f_2 = \min_{j>2} (c_{2j} + f_j) = \min \begin{Bmatrix} c_{23} + f_3 \\ c_{24} + f_4 \\ c_{25} + f_5 \end{Bmatrix} = \min \begin{Bmatrix} 75 + 131 \\ 98 + 75 \\ 210 + 0 \end{Bmatrix} = \min \begin{Bmatrix} 206 \\ 173 \\ 210 \end{Bmatrix}$$
$$= 173 \qquad \text{at } j = 4.$$

Finally,

$$f_1 = \min_{j>1} (c_{1j} + f_j) = \min \begin{Bmatrix} c_{12} + f_2 \\ c_{13} + f_3 \\ c_{14} + f_4 \\ c_{15} + f_5 \end{Bmatrix} = \min \begin{Bmatrix} 75 + 173 \\ 162 + 131 \\ 208 + 75 \\ 376 + 0 \end{Bmatrix} = \min \begin{Bmatrix} 248 \\ 293 \\ 283 \\ 376 \end{Bmatrix}$$
$$= 248 \qquad \text{at } j = 2.$$

---

[4]Actually, the original Wagner-Whitin (1958) algorithm is based on a forward dynamic programming formulation. Although the forward formulation has some advantages for planning horizon analysis, we feel that backward recursion is more natural and more intuitive.

To determine the optimal order policy, we retrace the solution back from the beginning. In period 1 the optimal value of j is $j = 2$. This means that the production level in period 1 is equal to the demand in period 1, so that $y_1 = r_1 = 52$. The next order period is period 2. The optimal value of j in period 2 is $j = 4$, which implies that the production quantity in period 2 is equal to the sum of the demands in periods 2 and 3, or $y_2 = r_2 + r_3 = 110$. The next period of ordering is period 4. The optimal value of j in period 4 is $j = 5$. This gives $y_4 = r_4 = 56$. Hence, the optimal order policy is $\mathbf{y} = (52, 110, 0, 56)$. ∎

APPENDIX 7–B   GLOSSARY OF NOTATION FOR CHAPTER 7

$C(T)$ = Average holding and setup cost per period (for Silver-Meal heuristic) or per unit (LUC heuristic) if the current order spans T periods.

c_i = Production capacity in period i.

c_{ij} = Cost associated with arc (i, j) in network representation of lot scheduling problem used for Wagner-Whitin algorithm.

f_j = Minimum cost from period i to the end of the horizon (refer to the dynamic programming algorithm for Wagner-Whitin).

h = Holding cost per unit per time period.

K = Setup cost for initiating an order.

r_i = Requirement for period i.

y_i = Production lot size in period i.

Bibliography

Afentakis, P.; B. Gavish; and U. Karmarkar. "Computationally Efficient Optimal Solutions to the Lot-Sizing Problem in Multistage Assembly System." *Management Science* 30 (1984), pp. 222–39.

Anderson, J. C.; R. G. Schroeder; S. E. Tupy; and E. M. White. "Material Requirements Planning Systems: The State of the Art." *Production and Inventory Management* 23 (1982), pp. 51–66.

Blood, B. E. "Read My Lips—No More Late Deliveries." *Hospital Materiel Management Quarterly,* 15, no. 4 (1994), pp. 53–55.

Carlson, R. C.; S. L. Beckman; and D. H. Kropp. "The Effectiveness of Extending the Horizon in Rolling Production Scheduling." *Decision Sciences* 13 (1982), pp. 129–46.

Carlson, R. C.; J. V. Jucker; and D. H. Kropp. "Less Nervous MRP Systems: A Dynamic Economic Lot-Sizing Approach." *Management Science* 25 (1979), pp. 754–61.

Carlson, R. C.; J. V. Jucker; and D. H. Kropp. "Heuristic Lot Sizing Approaches for Dealing with MRP System Nervousness." *Decision Sciences* 14 (1983), pp. 156–69.

Chhikara, J., and E. N. Weiss. "JIT Savings—Myth or Reality?" *Business Horizons* 38 (May–June 1995), pp. 73–78.

DeMatteis, J. J. "An Economic Lot Sizing Technique: The Part-Period Algorithm." *IBM Systems Journal* 7 (1968), pp. 30–38.

Dixon, P. S., and E. A. Silver. "A Heuristic Solution Procedure for the Multi-Item, Single-Level, Limited Capacity Lot Sizing Problem." *Journal of Operations Management* 2 (1981), pp. 23–39.

Gorham, T. "Dynamic Order Quantities." *Production and Inventory Management* 9 (1968), pp. 75–81.

Hillier, F. S., and G. J. Lieberman. *Introduction to Operations Research.* 5th ed. New York: McGraw-Hill, 1990.

Hopp, W., and M. Spearman. *Factory Physics.* Burr Ridge, IL: Richard D. Irwin, 1996.

Jacobson, G., and J. Hillkirk. *Xerox, American Samurai.* New York: MacMillan, 1986.

Karmarkar, U. "Getting Control of Just-In-Time." *Harvard Business Review* 67 (September–October 1989), pp. 122–31.

Karni, R. "Maximum Part Period Gain (MPG)—A Lot Sizing Procedure for Unconstrained and Constrained Requirements Planning Systems." *Production and Inventory Management* 22 (1981), pp. 91–98.

Krajewski, L. J.; B. E. King; L. P. Ritzman; and D. S. Wong. "Kanban, MRP, and Shaping the Manufacturing Environment." *Management Science* 33 (1987), pp. 39–57.

LaPlante, A. "New Software = Faster Factories." *Forbes ASAP Supplement,* October 10, 1994, pp. 36–41.

Lieber, R. B. "Here Comes SAP." *Fortune* 132, no. 7 (October 10, 1995), pp. 122–24.

Love, Stephen. *Inventory Control.* New York: McGraw-Hill, 1979.

McClaren, B. J. "A Study of Multiple Level Lot Sizing Techniques for Material Requirements Planning Systems." Unpublished Ph.D dissertation, Purdue University, 1976.

McLeavey, D. W., and S. L. Narasimhan. *Production Planning and Inventory Control.* Boston: Allyn & Bacon, 1985.

Meal, H. "Putting Production Decisions Where They Belong." *Harvard Business Review* 62 (1984), pp. 102–11.

Monden, Y. "What Makes the Toyota Production System Really Tick?" *Industrial Engineering* 13, no. 1 (1981a), pp. 36–46.

Monden, Y. "Adaptable Kanban System Helps Toyota Maintain Just-in-Time Production." *Industrial Engineering* 13, no. 5 (1981b), pp. 28–46.

Nahmias, S., and K. Moinzadeh. "Lot Sizing with Randomly Graded Yields." *Operations Research* 46, no. 6 (1997), pp. 974–86.

Nemhauser, G. L. *Introduction to Dynamic Programming.* New York: John Wiley & Sons, 1966.

New, C. *Requirements Planning.* Essex, England: Gower Press, 1974.

Orlicky, J. *Materials Requirements Planning.* New York: McGraw-Hill, 1975.

Plossl, G., and O. Wight. *Materials Requirements Planning by Computer.* Washington, DC: American Production and Inventory Control Society, 1971.

Sheldon, D. "MRP II Implementation: A Case Study." *Hospital Materiel Management Quarterly* 15, no. 4 (1994), pp. 48–52.

Shingo, S. *Study of "Toyota" Production System from Industrial Engineering Viewpoint.* Tokyo: Japan Management Association, 1981.

Silver, E. A., and H. C. Meal. "A Heuristic for Selecting Lot Size Quantities for the Case of a Deterministic Time-Varying Demand Rate and Discrete Opportunities for Replenishment." *Production and Inventory Management* 14 (1973), pp. 64–74.

Silver, E. A., and R. Peterson. *Decision Systems for Inventory Management and Production Planning.* 2nd ed. New York: John Wiley & Sons, 1985.

Sloan, A. *My Years with General Motors.* Garden City, NY: Doubleday, 1964.

Steele, D. C. "The Nervous MRP System: How to Do Battle." *Production and Inventory Management* 16 (1973), pp. 83–89.

Steinberg, E., and H. A. Napier. "Optimal Multi-Level Lot Sizing for Requirements Planning Systems." *Management Science* 26 (1980), pp. 1258–71.

Vollman, T. E.; W. L. Berry; and D. C. Whybark. *Manufacturing and Control Systems.* 3rd ed. New York: McGraw-Hill/Irwin, 1992.

Wagner, H. M., and T. M. Whitin. "Dynamic Version of the Economic Lot Size Model." *Management Science* 5 (1958), pp. 89–96.

Willis, R. "Harley Davidson Comes Roaring Back." *Management Review* 75 (March 1986), pp. 20–27.

Womack, J. P.; D.T. Jones; and D. Roos. *The Machine That Changed the World: The Story of Lean Production.* New York: Harper Perennial, 1990.

Zipkin, P. "Does Manufacturing Need a JIT Revolution?" *Harvard Business Review* 69 (January–February 1991), pp. 40–50.

$$Q = \sqrt{\frac{2K\lambda}{h}}$$

One learns by doing

CHAPTER 8

OPERATIONS SCHEDULING

Scheduling is an important aspect of operations control in both manufacturing and service industries. With increased emphasis on time to market and time to volume as well as improved customer satisfaction, efficient scheduling will gain increasing emphasis in the operations function in the coming years.

In some sense, much of what has been discussed so far in this text can be considered a subset of production scheduling. Aggregate planning, treated in Chapter 3, is aimed at macroscheduling of workforce levels and overall production levels for the firm. Detailed inventory control, discussed in Chapters 4 and 5, concerns methods of scheduling production at the individual item level; Chapter 6 treated vehicle scheduling; and MRP, discussed in Chapter 7, provides production schedules for end items and subassemblies in the product structure.

There are many different types of scheduling problems faced by the firm. A partial list includes

1. *Job shop scheduling.* Job shop scheduling, known more commonly in practice as shop floor control, is the set of activities in the shop that transform inputs (a set of requirements) to outputs (products to meet those requirements). Much of this chapter will be concerned with sequencing issues on the shop floor, and more will be said about this problem in the next section.

2. *Personnel scheduling.* Scheduling personnel is an important problem for both manufacturing and service industries. Although shift scheduling on the factory floor may be considered one of the functions of shop floor control, personnel scheduling is a much larger problem. Scheduling health professionals in hospitals and other health facilities is one example. Determining whether to meet peak demand with overtime shifts, night shifts, or subcontracting is another example of a personnel scheduling problem.

3. *Facilities scheduling.* This problem is particularly important when facilities become a bottleneck resource. Scheduling operating rooms at hospitals is one example. As the need for health care increases, some hospitals and HMOs (health maintenance organizations) find that facilities are strained. A similar problem occurs in colleges and universities in which enrollments have increased without commensurate increases in the size of the physical plant.

4. *Vehicle scheduling.* Manufacturing firms must distribute their products in a cost-efficient and timely manner. Some service operations, such as dial-a-ride systems, involve pick-ups and deliveries of goods and/or people. Vehicle routing is a problem that arises in many contexts. Problems such as scheduling snow removal equipment, postal and bank deliveries, and

shipments to customers with varying requirements at different locations are some examples. Vehicle scheduling is discussed in Section 6 of Chapter 6.

5. *Vendor scheduling.* For firms with just-in-time (JIT) systems, scheduling deliveries from vendors is an important logistics issue. Purchasing must be coordinated with the entire product delivery system to ensure that JIT production systems function efficiently. Vollman et al. (1992, p. 191) discuss the application of vendor scheduling to the JIT system at Steelcase. (JIT is discussed in Chapters 1 and 7 of this book.)

6. *Project scheduling.* A project may be broken down into a set of interrelated tasks. Although some tasks can be done concurrently, many tasks cannot be started until others are completed. Complex projects may involve thousands of individual tasks that must be coordinated for the project to be completed on time and within budget. Project scheduling is an important component of the planning function, which we treat in detail in the next chapter.

7. *Dynamic versus static scheduling.* Most scheduling theory that we review in this chapter views the scheduling problem as a static one. Numerous jobs arrive simultaneously to be processed on a set of machines. In practice, many scheduling problems are dynamic in the sense that jobs arrive continuously over time. One example is the problem faced by an air traffic controller who must schedule runways for arriving planes. The problem is a dynamic one in that planes arrive randomly and runways are freed up and committed randomly over time. Dynamic scheduling problems, treated in Section 9 of this chapter, are analyzed using the tools of queuing theory (discussed in detail in Supplement 2, which follows this chapter).

Scheduling is a complex but extremely important operations function. The purpose of this chapter is to give the reader the flavor of the kinds of results one can obtain using analytical models, and to show how these models can be used to solve certain classes of scheduling problems. Our focus is primarily on job shop scheduling, but we consider several other scheduling problems as well.

8.1 Production Scheduling and the Hierarchy of Production Decisions

Crucial to controlling production operations is the detailed scheduling of various aspects of the production function. We may view the production function in a company as a hierarchical process. First, the firm must forecast demand for aggregate sales over some predetermined planning horizon. These forecasts provide the input for determining the aggregate production and workforce levels for the planning horizon. Aggregate planning techniques are discussed in Chapter 3. The aggregate production plan then must be translated into the master production schedule (MPS). The MPS results in specific production goals by product and time period.

Materials requirements planning (MRP), treated in detail in Chapter 7, is one method for meeting specific production goals of finished-goods inventory generated by the MPS. The MRP system "explodes" the production levels one obtains from the MPS analysis back in time to obtain production targets at each level of assembly by time period. The result of the MRP analysis is specific planned order releases for final products, subassemblies, and components.

Finally, the planned order releases must be translated into a set of tasks and the due dates associated with those tasks. This level of detailed planning results in the shop floor schedule. Because the MRP or other lot scheduling system usually recommends revisions in the planned order releases, shop floor schedules change frequently. The hierarchy of production planning decisions is shown schematically in Figure 8–1.

Shop floor control means scheduling personnel and equipment in a work center to meet the due dates for a collection of jobs. Often, jobs must be processed through the machines in the work center in a unique order or sequence. Figure 8–2 shows the layout of a typical job shop.

FIGURE 8–1

Hierarchy of production decisions

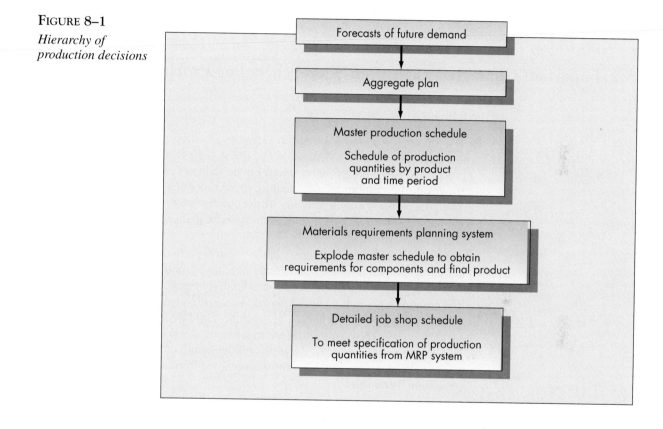

FIGURE 8–2

Typical job shop layout

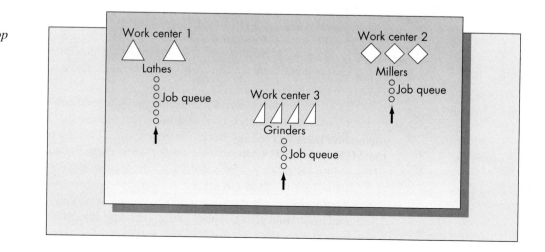

Both jobs and machines are treated as indivisible. Jobs must wait, or queue up, for processing when machines are busy. This is referred to as discrete processing. Production scheduling in continuous-process industries, such as sugar or oil refining, has a very different character.

Although there are many problems associated with operations scheduling, our concern in this chapter will be job sequencing. Given a collection of jobs remaining to be processed on a collection of machines, the problem is how to sequence these jobs to optimize some specified criterion. Properly choosing the sequencing rule can effect dramatic improvements in the throughput rate of the job shop.

8.2 Important Characteristics of Job Shop Scheduling Problems

Significant issues for determining optimal or approximately optimal scheduling rules are the following:

1. *The job arrival pattern.* We often view the job shop problem as a static one in which we take a "snapshot" of the system at a point in time and proceed to solve the problem based on the value of the current state. However, the number of jobs waiting to be processed is constantly changing. Hence, although many of the solution algorithms we consider view the problem as being static, most practical shop scheduling problems are dynamic in nature.

2. *Number and variety of machines in the shop.* A particular job shop may have unique features that could make implementing a solution obtained from a scheduling algorithm difficult. For example, it is generally assumed that all machines of a given type are identical. This is not always the case, however. The throughput rate of a particular machine could depend upon a variety of factors, such as the condition of the machine or the skill of the operator. Depending on the layout of the shop and the nature of the jobs, constraints might exist that would make solutions obtained from an "all purpose" procedure infeasible.

3. *Number of workers in the shop.* Both the number of workers in the shop and the number and variety of machines in the shop determine the shop's capacity. Capacity planning is an important aspect of production planning. Many control systems, such as traditional MRP discussed in Chapter 7, do not explicitly incorporate capacity considerations. Furthermore, capacity is dynamic. A breakdown of a single critical machine or the loss of a critical employee could result in a bottleneck and a reduction in the shop's capacity.

4. *Particular flow patterns.* The solutions obtained from the scheduling algorithms to be presented in this chapter require that jobs be completed in a fixed order. However, each sequence of jobs through machines results in a pattern of flow of materials through the system. Because materials-handling issues often are treated separately from scheduling issues, infeasible flow patterns may result.

5. *Evaluation of alternative rules.* The choice of objective will determine the suitability and effectiveness of a sequencing rule. It is common for more than

one objective to be important, so that it may be impossible to determine a unique optimal rule. For example, one may wish to minimize the time required to complete all jobs, but also may wish to limit the maximum lateness of any single job.

Objectives of Job Shop Management

One of the difficulties of scheduling is that many, often conflicting, objectives are present. The goals of different parts of the firm are not always the same. Some of the most common objectives are

1. Meet due dates.
2. Minimize work-in-process (WIP) inventory.
3. Minimize the average flow time through the system.
4. Provide for high machine/worker time utilization. (Minimize machine/worker idle time.)
5. Provide for accurate job status information.
6. Reduce setup times.
7. Minimize production and worker costs.

It is obviously impossible to optimize all seven objectives simultaneously. In particular, (1) and (3) are aimed primarily at providing a high level of customer service, and (2), (4), (6), and (7) are aimed primarily at providing a high level of plant efficiency. Determining the trade-off between cost and quality is one of the most important strategic issues facing a firm today.

Some of these objectives conflict. If the primary objective is to reduce work-in-process inventory (as, for example, with just-in-time inventory control systems, discussed in Chapter 7), it is likely that worker idle time will increase. As the system tightens up by reducing the inventory within and between manufacturing operations, differences in the throughput rate from one part of the system to another may force the faster operations to wait. Although not recommended by those espousing the just-in-time philosophy, buffer inventories between operations can significantly reduce idle time.

As an example, consider the simple system composed of two operations in series, pictured in Figure 8–3. If work-in-process inventory is zero, then the throughput of the system at any point in time is governed by the smaller of the throughputs of the two operations. If operation 1 is temporarily halted by a machine failure, then operation 2 also must remain idle. However, if there is a buffer inventory placed between the operations, then 2 can continue to operate while 1 is undergoing repair or recalibration.

Finding the proper mix between WIP inventory and worker idle time is equivalent to choosing a point on the trade-off curve of these conflicting objectives. (Trade-off, or

FIGURE 8–3

A process composed of two operations in series

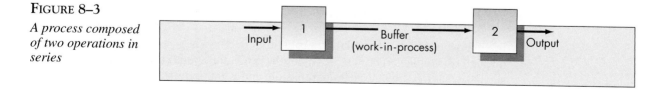

FIGURE 8–4

Conflicting objectives in job shop management

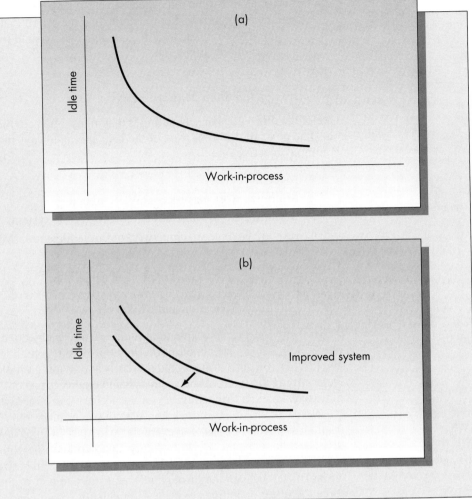

exchange, curves were discussed in Chapter 5 in the context of multi-item inventory control.) Such a curve is pictured in Figure 8–4*a*. A movement from one point to another along such a curve does not necessarily imply that the system has improved, but rather that different weights are being applied to the two objectives. A true improvement in the overall system would mean that the entire trade-off curve undergoes a downward shift, such as that pictured in Figure 8–4*b*.

8.3 Job Shop Scheduling Terminology

In general, a job shop scheduling problem is one in which *n* jobs must be processed through *m* machines. The complexity of the problem depends upon a variety of factors, such as what job sequences are permissible and what optimization criteria are chosen.

In this section we define some of the terms that will be used throughout this chapter.

1. *Flow shop.* In a flow shop each of the n jobs must be processed through the m machines in the same order, and each job is processed exactly once on each machine. This is what we typically think of as an assembly line.

2. *Job shop.* A general job shop differs from a flow shop in that not all jobs are assumed to require exactly m operations, and some jobs may require multiple operations on a single machine. Furthermore, in a job shop each job may have a different required sequencing of operations. General job shop problems are extremely complex. All-purpose solution algorithms for solving general job shop problems do not exist.

3. *Parallel processing versus sequential processing.* Most of the problems that we will consider involve sequential processing. This means that the m machines are distinguishable, and different operations are performed by different machines. In parallel processing we assume that the machines are identical, and any job can be processed on any machine. An example of parallel processing occurs in a phone switching center, in which calls are processed through the next available server. Parallel processing is discussed in the context of stochastic scheduling in Section 8.

4. *Flow time.* The flow time of job i is the time that elapses from the initiation of the first job on the first machine to the completion of job i. Equivalently, it is the amount of time that job i spends in the system. The *mean flow time,* which is a common measure of system performance, is the arithmetic average of the flow times for all n jobs.

5. *Makespan.* The makespan is the flow time of the job that is completed last. It is also the time required to complete all n jobs. Minimizing the makespan is a common objective in multiple-machine sequencing problems.

6. *Tardiness and lateness.* Tardiness is the positive difference between the completion time (flow time) and the due date of a job. A tardy job is one that is completed after its due date. Lateness refers to the difference between the job completion time and its due date, and differs from tardiness in that lateness can be either positive or negative. Minimizing the average tardiness and the maximum tardiness is also a common scheduling objective.

8.4 A Comparison of Specific Sequencing Rules

In the comparison and evaluation of sequencing rules, we consider the job shop at a fixed point in time. This section will focus on a single machine only. Assume that there is a collection of jobs that must be processed on the machine and that each job has associated with it a processing time and a due date. We compare the performance of four sequencing rules commonly used in practice. The purpose of this section is to illustrate how these sequencing rules affect various measures of system performance.

We compare the following four sequencing rules:

1. *FCFS (first-come, first-served).* Jobs are processed in the sequence in which they entered the shop.

2. *SPT (shortest processing time).* Jobs are sequenced in increasing order of their processing times. The job with the shortest processing time is first, the job with the next shortest processing time is second, and so on.

3. *EDD (earliest due date).* Jobs are sequenced in increasing order of their due dates. The job with the earliest due date is first, the job with the next earliest due date is second, and so on.

4. *CR (critical ratio).* Critical ratio scheduling requires forming the ratio of the processing time of the job, divided by the remaining time until the due date, and scheduling the job with the largest ratio next.

We compare the performance of these four rules for a specific case based on mean flow time, average tardiness, and number of tardy jobs. The purpose of the next example is to help the reader develop an intuition for the mechanics of scheduling before presenting formal results.

Example 8.1

A machining center in a job shop for a local fabrication company has five unprocessed jobs remaining at a particular point in time. The jobs are labeled 1, 2, 3, 4, and 5 in the order that they entered the shop. The respective processing times and due dates are given in the table below.

Job Number	Processing Time	Due Date
1	11	61
2	29	45
3	31	31
4	1	33
5	2	32

First-Come, First-Served

Because the jobs are assumed to have entered the shop in the sequence that they are numbered, FCFS scheduling means that the jobs are scheduled in the order 1, 2, 3, 4, 5. This results in

Sequence	Completion Time	Due Date	Tardiness
1	11	61	0
2	40	45	0
3	71	31	40
4	72	33	39
5	74	32	42
Totals	268		121

Mean flow time $= 268/5 = 53.6$.

Average tardiness $= 121/5 = 24.2$.

Number of tardy jobs $= 3$.

The tardiness of a job is equal to zero if the job is completed prior to its due date and is equal to the number of days late if the job is completed after its due date.

Shortest Processing Time

Here jobs are sequenced in order of increasing processing time.

Job	Processing Time	Completion Time	Due Date	Tardiness
4	1	1	33	0
5	2	3	32	0
1	11	14	61	0
2	29	43	45	0
3	31	74	31	43
Totals		135		43

$$\text{Mean flow time} = 135/5 = 27.0.$$
$$\text{Average tardiness} = 43/5 = 8.6.$$
$$\text{Number of tardy jobs} = 1.$$

Earliest Due Date

Here jobs are completed in the order of their due dates.

Job	Processing Time	Completion Time	Due Date	Tardiness
3	31	31	31	0
5	2	33	32	1
4	1	34	33	1
2	29	63	45	18
1	11	74	61	13
Totals		235		33

$$\text{Mean flow time} = 235/5 = 47.0.$$
$$\text{Average tardiness} = 33/5 = 6.6.$$
$$\text{Number of tardy jobs} = 4.$$

Critical Ratio Scheduling

After each job has been processed, we compute

$$\frac{\text{Due date} - \text{Current time}}{\text{Processing time}}$$

known as the critical ratio and schedule the next job in order to minimize the value of the critical ratio. The idea behind critical ratio scheduling is to provide a balance between SPT, which only considers processing time, and EDD, which only considers due dates. The ratio will grow smaller as the current time approaches the due date, and more priority will be given to those jobs with longer processing times. One disadvantage of the method is that the critical ratios need to be recalculated each time a job is scheduled.

It is possible that the numerator will be negative for some or all of the remaining jobs. When that occurs it means that the job is late, and we will assume that late jobs are automatically scheduled next. If there is more than one late job, then the late jobs are scheduled in SPT sequence.

First we compute the critical ratios starting at time $t = 0$.

Current time: $t = 0$

Job	Processing Time	Due Date	Critical Ratio
1	11	61	61/11 (5.545)
2	29	45	45/29 (1.552)
3	31	31	31/31 (1.000)
4	1	33	33/1 (33.00)
5	2	32	32/2 (16.00)

The minimum value corresponds to job 3, so job 3 is performed first. As job 3 requires 31 units of time to process, we must update all of the critical ratios in order to determine the next job to process. We move the clock to time $t = 31$ and recompute the critical ratios.

Current time: $t = 31$

Job	Processing Time	Due Date − Current Time	Critical Ratio
1	11	30	30/11 (2.727)
2	29	14	14/29 (0.483)
4	1	2	2/1 (2.000)
5	2	1	1/2 (0.500)

The minimum is 0.483, which corresponds to job 2. Hence, job 2 is scheduled next. Since job 2 has a processing time of 29, we update the clock to time $t = 31 + 29 = 60$.

Current time: $t = 60$

Job	Processing Time	Due Date − Current Time	Critical Ratio
1	11	1	1/11 (.0909)
4	1	−27	−27/1 < 0
5	2	−28	−28/2 < 0

Jobs 4 and 5 are now late so they are given priority and scheduled next. Since they are scheduled in SPT order, they are done in the sequence job 4, then job 5. Finally, job 1 is scheduled last.

Summary of the Results for Critical Ratio Scheduling

Job	Processing Time	Completion Time	Tardiness
3	31	31	0
2	29	60	15
4	1	61	28
5	2	63	31
1	11	74	13
Totals		289	87

Mean flow time = 289/5 = 57.8.

Average tardiness = 87/5 = 17.4.

Number of tardy jobs = 4.

We summarize the results of this section for all four scheduling rules:

Summary of the Results of Four Scheduling Rules

Rule	Mean Flow Time	Average Tardiness	Number of Tardy Jobs
FCFS	53.6	24.2	3
SPT	27.0	8.6	1
EDD	47.0	6.6	4
CR	57.8	17.4	4

8.5 Objectives in Job Shop Management: An Example

Example 8.2

An air traffic controller is faced with the problem of scheduling the landing of five aircraft. Based on the position and runway requirements of each plane, he estimates the following landing times:

Plane:	1	2	3	4	5
Time (in minutes):	26	11	19	16	23

Only one plane may land at a time. The problem is essentially the same as that of scheduling five jobs on a single machine, with the planes corresponding to the jobs, the landing times to the processing times, and the runway to the machine.

1. Two reasonable objectives given the information above would be to minimize the total time required to land all planes (i.e., the makespan) or the average time required to land the planes (the mean flow time). The makespan for any sequence is clearly 95 minutes, the sum of the landing times. However, as we saw in the previous example, the mean flow time is not sequence independent. As with the previous example, the shortest-processing-time rule minimizes the mean flow time. We will show in the next section that SPT is the optimal sequencing rule for minimizing mean flow time for a single machine in general.

2. An alternative objective might be to land as many people as quickly as possible. In this case we also would need to know the number of passengers on each plane. Suppose that these numbers are as follows:

Plane	1	2	3	4	5
Landing time	26	11	19	16	23
Number of passengers	180	12	45	75	252

The appropriate objective in this case might be to minimize the *weighted* makespan or the weighted sum of the completion times, where the weights would correspond to the number of passengers in each plane. Notice that the objective function would now be in units of passenger-minutes.

3. An issue that we have not yet addressed is the time that each plane is scheduled to arrive. Assume the following data:

Plane	1	2	3	4	5
Landing time	26	11	19	16	23
Scheduled arrival time	5:30	5:45	5:15	6:00	5:40

Sequencing rules that ignore due dates could give very poor results in terms of meeting the arrival times. Some possible objectives related to due dates include minimizing the average tardiness and minimizing the maximum tardiness.

4. Thus far we have ignored special conditions that favor some planes over others. Suppose that plane number 4 has a critically low fuel level. This would probably result in plane 4 taking precedence. Priority constraints could arise in other ways as well: planes that are scheduled for continuing flights or planes carrying precious or perishable cargo also might be given priority. ■

The purpose of this section was to demonstrate the difficulties of choosing an objective function for job sequencing problems. The optimal sequence is highly sensitive to the choice of the objective, and the appropriate objective is not always obvious.

Problems for Sections 1–5

1. Discuss each of the objectives listed below and the relationship each has with job shop performance.
   a. Reduce WIP inventory.
   b. Provide a high level of customer service.

    c. Reduce worker idle time.

    d. Improve factory efficiency.

2. In Problem 1, why are (*a*) and (*c*) conflicting objectives, and why are (*b*) and (*d*) conflicting objectives?

3. Define the following terms:

    a. Flow shop.

    b. Job shop.

    c. Sequential versus parallel processing.

    d. Makespan.

    e. Tardiness.

4. Four trucks, 1, 2, 3, and 4, are waiting on a loading dock at XYZ Company that has only a single service bay. The trucks are labeled in the order that they arrived at the dock. Assume the current time is 1:00 P.M. The times required to unload each truck and the times that the goods they contain are due in the plant are given in the table below.

Truck	Unloading Time	Time Material Is Due
1	20 minutes	1:25 P.M.
2	14 minutes	1:45 P.M.
3	35 minutes	1:50 P.M.
4	10 minutes	1:30 P.M.

    Determine the schedules that result for each of the rules FCFS, SPT, EDD, and CR. In each case compute the mean flow time, average tardiness, and number of tardy jobs.

5. Five jobs must be scheduled for batch processing on a mainframe computer system. The processing times and the promised times for each of the jobs are listed below.

Job	1	2	3	4	5
Processing time	40 min.	2.5 hr.	20 min.	4 hr.	1.5 hr.
Promised time	11:00 A.M.	2:00 P.M.	2:00 P.M.	1:00 P.M.	4:00 P.M.

Assume that the current time is 10:00 A.M.

    a. If the jobs are scheduled according to SPT, find the tardiness of each job and the mean tardiness of all jobs.

    b. Repeat the calculation in part (*a*) for EDD scheduling.

8.6 An Introduction to Sequencing Theory for a Single Machine

Assume that *n* jobs are to be processed through one machine. For each job *i*, define the following quantities:

$$t_i = \text{Processing time for job } i,$$
$$d_i = \text{Due date for job } i,$$

W_i = Waiting time for job i,
F_i = Flow time for job i,
L_i = Lateness of job i,
T_i = Tardiness of job i,
E_i = Earliness of job i.

The processing time and the due date are constants that are attached to the description of each job. The waiting time for a job is the amount of time that the job must wait before its processing can begin. For the cases that we consider, it is also the sum of the processing times for all of the preceding jobs. The flow time is simply the waiting time plus the job processing time ($F_i = W_i + t_i$). The flow time of job i and the completion time of job i are the same. We will define the lateness of job i as $L_i = F_i - d_i$, and assume that lateness can be either a positive or a negative quantity. Tardiness is the positive part of lateness ($T_i = \max[L_i, 0]$), and earliness is the negative part of lateness ($E_i = \max[-L_i, 0]$).

Other related quantities are maximum tardiness T_{max}, given by the formula

$$T_{max} = \max\{T_1, T_2, \ldots, T_n\},$$

and the mean flow time F', given by the formula

$$F' = \frac{1}{n}\sum_{i=1}^{n}F_i.$$

As we are considering only a single machine, every schedule can be represented by a permutation (that is, ordering) of the integers $1, 2, \ldots, n$. There are exactly $n!$ different permutation schedules ($n! = n(n-1)\cdots(2)(1)$).

Shortest-Processing-Time Scheduling

We have the following result:

Theorem 8.1

The scheduling rule that minimizes the mean flow time F' is SPT.

Theorem 8.1 is easy to prove. Let $[1], [2], \ldots, [n]$ be any permutation of the integers $1, 2, 3, \ldots, n$. The flow time of the job that is scheduled in position k is given by

$$F_{[k]} = \sum_{i=1}^{k} t_{[i]}.$$

It follows that the mean flow time is given by

$$F' = \frac{1}{n}\sum_{k=1}^{n}F_{[k]} = \frac{1}{n}\sum_{k=1}^{n}\sum_{i=1}^{k}t_{[i]}.$$

The double summation term may be written in a different form. Expanding the double summation, we obtain

$$k = 1: t_{[1]}$$

$$k = 2: t_{[1]} + t_{[2]}$$

$$\vdots$$

$$k = n: t_{[1]} + t_{[2]} + \cdots + t_{[n]}.$$

By summing down the column rather than across the row, we may rewrite F' in the form

$$nt_{[1]} + (n - 1)t_{[2]} + \cdots + t_{[n]},$$

which is clearly minimized by setting

$$t_{[1]} \leq t_{[2]} \leq \cdots \leq t_{[n]},$$

which is exactly the SPT sequencing rule.

We have the following corollary to Theorem 8.1.

Corollary 8.1

The following measures are equivalent:

1. **Mean flow time.**
2. **Mean waiting time.**
3. **Mean lateness.**

Taken together, Corollary 8.1 and Theorem 8.1 establish that SPT minimizes mean flow time, mean waiting time, and mean lateness for single-machine sequencing.

Earliest-Due-Date Scheduling

If the objective is to minimize the maximum lateness, then the jobs should be sequenced according to their due dates. That is, $d_{[1]} \leq d_{[2]} \leq \cdots \leq d_{[n]}$. We will not present a proof of this result. The idea behind the proof is to choose some schedule that does not sequence the jobs in order of their due dates; that implies that there is some value of k such that $d_{[k]} > d_{[k+1]}$. One shows that by interchanging the positions of jobs k and $k + 1$, the maximum lateness is reduced.

Minimizing the Number of Tardy Jobs

There are many instances in which the penalty for a late (tardy) job remains the same no matter how late it is. For example, any delay in the completion of all tasks required for preparation of a space launch would cause the launch to be aborted, independent of the length of the delay.

We will describe an algorithm from Moore (1968) that minimizes the number of tardy jobs for the single machine problem.

Step 1. Sequence the jobs according to the earliest due date to obtain the initial solution. That is $d_{[1]} \leq d_{[2]} \leq \cdots \leq d_{[n]}$.

Step 2. Find the first tardy job in the current sequence, say job $[i]$. If none exists go to step 4.

Step 3. Consider jobs $[1], [2], \ldots, [i]$. Reject the job with the largest processing time. Return to step 2.

Step 4. Form an optimal sequence by taking the current sequence and appending to it the rejected jobs. The jobs appended to the current sequence may be scheduled in any order because they constitute the tardy jobs.

Example 8.3

A machine shop processes custom orders from a variety of clients. One of the machines, a grinder, has six jobs remaining to be processed. The processing times and promised due dates (both in hours) for the six jobs are given below.

Job	1	2	3	4	5	6
Due date	15	6	9	23	20	30
Processing time	10	3	4	8	10	6

The first step is to sequence the jobs according to the EDD rule.

Job	2	3	1	5	4	6
Due date	6	9	15	20	23	30
Processing time	3	4	10	10	8	6
Completion time	3	7	17	27	35	41

We see that the first tardy job is job 1, and there are a total of four tardy jobs. We now consider jobs 2, 3, and 1 and reject the job with the longest processing time. This is clearly job 1. At this point, the new current sequence is

Job	2	3	5	4	6
Due date	6	9	20	23	30
Processing time	3	4	10	8	6
Completion time	3	7	17	25	31

The first tardy job in the current sequence is now job 4. We consider the sequence 2, 3, 5, 4, and reject the job with the longest processing time, which is job 5. The current sequence is now

Job	2	3	4	6
Due date	6	9	23	30
Processing time	3	4	8	6
Completion time	3	7	15	21

Clearly there are no tardy jobs at this stage. The optimal sequence is 2, 3, 4, 6, 5, 1 or 2, 3, 4, 6, 1, 5. In either case the number of tardy jobs is exactly 2. ∎

Precedence Constraints: Lawler's Algorithm

Lawler's algorithm (Lawler, 1973) is a powerful technique for solving a variety of constrained scheduling problems. The objective function is assumed to be of the form

$$\min \max_{1 \le i \le n} g_i(F_i)$$

where g_i is any nondecreasing function of the flow time F_i. Furthermore, the algorithm handles *any* precedence constraints. Precedence constraints occur when certain jobs must be completed before other jobs can begin; they are quite common in scheduling problems. Some examples of functions g_i that one might consider are $g_i(F_i) = F_i - d_i = L_i$, which corresponds to minimizing maximum lateness, or $g_i(F_i) = \max(F_i - d_i, 0)$, which corresponds to minimizing maximum tardiness.

The Algorithm. Lawler's algorithm first schedules the job to be completed last, then the job to be completed next to last, and so on. At each stage one determines the set of jobs not required to precede any other. Call this set V. Among the set V, choose the job k that satisfies

$$g_k(\tau) = \min_{i \in V}(g_i(\tau)),$$

where $\tau = \Sigma_{i=1}^n t_i$ and corresponds to the processing time of the current sequence.

Job k is now scheduled last. Consider the remaining jobs and again determine the set of jobs that are not required to precede any other remaining job. After scheduling job k, this set may have changed. The value of τ is reduced by t_k and the job scheduled next to last is now determined. The process is continued until all jobs are scheduled. Note that as jobs are scheduled, some of the precedence constraints may be relaxed, so the set V is likely to change at each iteration.

Example 8.4

Tony D'Amato runs a local body shop that does automotive painting and repairs. On a particular Monday morning he has six cars waiting for repair. Three (1, 2, and 3) are from a car rental company and he has agreed to finish these cars in the order of the dates that they were promised. Cars 4, 5, and 6 are from a retail dealer who has requested that car 4 be completed first because a customer is waiting for it. The resulting precedence constraints can be represented as two disconnected networks, as pictured in Figure 8–5.

FIGURE 8–5

Precedence constraints for Example 8.4

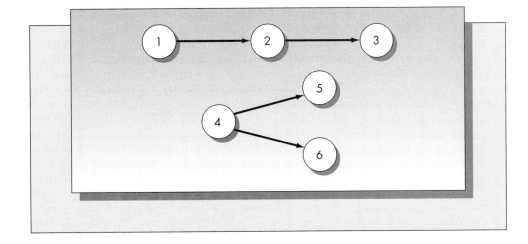

The times required to repair each of the cars (in days) and the associated promised completion dates are

Job	1	2	3	4	5	6
Processing time	2	3	4	3	2	1
Due date	3	6	9	7	11	7

Determine how the repair of the cars should be scheduled through the shop in order to minimize the maximum tardiness.

Solution

1. First we find the job scheduled last (sixth). Among the candidates for the last position are those jobs that are not predecessors of other jobs. These are 3, 5, and 6. The total processing time of all jobs is $2 + 3 + 4 + 3 + 2 + 1 = 15$. (This is the current value of τ.) As the objective is to minimize the maximum tardiness, we compare the tardiness of these three jobs and pick the one with the smallest value. We obtain $\min\{15 - 9, 15 - 11, 15 - 7\} = \min\{6, 4, 8\} = 4$, corresponding to job 5. Hence job 5 is scheduled last (position 6).

2. Next we find the job scheduled fifth. The candidates are jobs 3 and 6 only. At this point the value of τ is $15 - 2 = 13$. Hence, we find $\min\{13 - 9, 13 - 7\} = \min\{4, 6\} = 4$, which corresponds to job 3. Hence, job 3 is scheduled in the fifth position.

3. Find the job scheduled fourth. Because job 3 is no longer on the list, job 2 now becomes a candidate. The current value of $\tau = 13 - 4 = 9$. Hence, we compare $\min\{9 - 6, 9 - 7\} = \min\{3, 2\} = 2$, which corresponds to job 6. Schedule job 6 in the fourth position.

4. Find the job scheduled third. Job 6 has been scheduled, so job 4 now becomes a candidate along with job 2, and $\tau = 9 - 1 = 8$. Hence, we look for $\min\{8 - 6, 8 - 7\} = \min\{2, 1\} = 1$, which occurs at job 4.

5. At this point we would find the job scheduled second. However, we are left with only jobs 1 and 2, which, because of the precedence constraints, must be scheduled in the order 1–2.

Summarizing the results above, the optimal sequence to repair the cars is 1–2–4–6–3–5.

In order to determine the value of the objective function, the maximum tardiness, we compute the flow time for each job and compare it to the due date. We have

Job	Processing Time	Flow Time	Due Date	Tardiness
1	2	2	3	0
2	3	5	6	0
4	3	8	7	1
6	1	9	7	2
3	4	13	9	4
5	2	15	11	4

Hence, the maximum tardiness is four days. The reader should convince him- or herself that any other sequence results in a maximum tardiness of at least four days. ■

Problems for Section 6

6. Consider the information given in Problem 4. Determine the sequence that the trucks should be unloaded in order to minimize
   a. Mean flow time.
   b. Maximum lateness.
   c. Number of tardy jobs.

7. On May 1, a lazy MBA student suddenly realizes that he has done nothing on seven different homework assignments and projects that are due in various courses. He estimates the time required to complete each project (in days) and also notes their due dates:

Project	1	2	3	4	5	6	7
Time (days)	4	8	10	4	3	7	14
Due date	4/20	5/17	5/28	5/28	5/12	5/7	5/15

Because projects 1, 3, and 5 are from the same class, he decides to do those in the sequence that they are due. Furthermore, project 7 requires results from projects 2 and 3, so 7 must be done after 2 and 3 are completed. Determine the sequence in which he should do the projects in order to minimize the maximum lateness.

8. Eight jobs are to be processed through a single machine. The processing times and due dates are given below.

Job	1	2	3	4	5	6	7	8
Processing time	2	3	2	1	4	3	2	2
Due date	5	4	13	6	12	10	15	19

Furthermore, assume that the following precedence relationships must be satisfied:

$$2 \rightarrow 6 \rightarrow 3.$$
$$1 \rightarrow 4 \rightarrow 7 \rightarrow 8.$$

Determine the sequence in which these jobs should be done in order to minimize the maximum lateness subject to the precedence restrictions.

9. Jane Reed bakes breads and cakes in her home for parties and other affairs on a contract basis. Jane has only one oven for baking. One particular Monday morning she finds that she has agreed to complete five jobs for that day. Her husband John will make the deliveries, which require about 15 minutes each.

Suppose that she begins baking at 8:00 A.M.

Job	Time Required	Promised Time
1	1.2 hr.	11:30 A.M.
2	40 min.	10:00 A.M.
3	2.2 hr.	11:00 A.M.
4	30 min.	1:00 P.M.
5	3.1 hr.	12:00 NOON
6	25 min.	2:00 P.M.

Determine the sequence in which she should perform the jobs in order to minimize

a. Mean flow time.
b. Number of tardy jobs.
c. Maximum lateness.

10. Seven jobs are to be processed through a single machine. The processing times and due dates are given below.

Job	1	2	3	4	5	6	7
Processing time	3	6	8	4	2	1	7
Due date	4	8	12	15	11	25	21

Determine the sequence of the jobs in order to minimize

a. Mean flow time.
b. Number of tardy jobs.
c. Maximum lateness.
d. What is the makespan for any sequence?

8.7 Sequencing Algorithms for Multiple Machines

We now extend the analysis of the previous section to the case in which several jobs must be processed on more than one machine. Assume that n jobs are to be processed through m machines. The number of possible schedules is staggering, even for moderate values of both n and m. For each machine, there are $n!$ different orderings of the jobs. If the jobs may be processed on the machines in any order, it follows that there are a total of $(n!)^m$ possible schedules. For example, for $n = 5$, $m = 5$, there are $2,4883 \times 10^{10}$, or about 25 billion, possible schedules. Even with the availability of inexpensive computing today, enumerating all feasible schedules for even moderate-sized problems is impossible or, at best, impractical.

In this section we will present some known results for scheduling jobs on more than one machine. A convenient way to represent a schedule is via a Gantt chart. As an example, suppose that two jobs, I and J, are to be scheduled on two machines, 1 and 2.

The processing times are

	Machine 1	Machine 2
Job I	4	1
Job J	1	4

Assume that both jobs must be processed first on machine 1 and then on machine 2. The possible schedules appear in four Gantt charts in Figure 8–6. The first two schedules are known as permutation schedules. That means that the jobs are processed in the same sequence on both machines. Clearly, for this example, the permutation schedules provide better system performance in terms of both total and average flow time.

Recall that the total flow time (or makespan) is the total elapsed time from the initiation of the first job on the first machine until the completion of the last job on the last machine. For the schedules above, the makespans (total flow time) are respectively 9, 6, 10, and 10.

The mean flow time is also used as a measure of system performance. For the first schedule in the example above, the mean flow time is $(5 + 9)/2 = 7$. For the second schedule, it is $(5 + 6)/2 = 5.5$, and so on.

A third possible objective is minimization of the mean idle time in the system. The mean idle time is the arithmetic average of the idle times for each machine. In schedule 1, we see that machine 1 is idle for 4 units of time (between times 5 and 9) and machine 2 is idle for 4 units of time as well (between times 0 and 4). Hence the mean

FIGURE 8–6

All possible schedules for two jobs on two machines

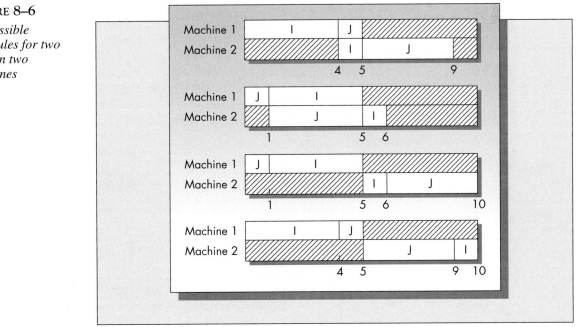

idle time for schedule 1 is 4. In schedule 2, both machines 1 and 2 are idle for 1 unit of time, giving a mean idle time of 1. The mean idle times for schedules 3 and 4 are 5 units of time.

Scheduling n Jobs on Two Machines

Assume that n jobs must be processed through two machines and that each job must be processed in the order machine 1 then machine 2. Furthermore, assume that the optimization criterion is to minimize the makespan. The problem of scheduling on two machines turns out to have a relatively simple solution.

Theorem 8.2

The optimal solution for scheduling n jobs on two machines is always a permutation schedule.

This theorem means that one can restrict attention to schedules in which the sequence of jobs is the same on both machines. This result can be demonstrated as follows. Consider a schedule for n jobs on two machines in which the sequencing of the jobs on the two machines is different. That is, the schedule looks as follows:

Machine 1	$\cdots$	I	$\cdots$	J				
Machine 2					$\cdots$	J	$\cdots$	I

By reversing the position of these jobs on either machine, the flow time decreases. By scheduling the jobs in the order I–J on machine 2 the pair (I, J) on machine 2 may begin after I is completed on machine 1, rather than having to wait until J is completed on machine 1.

Because the total number of permutation schedules is exactly $n!$, determining optimal schedules for two machines is roughly of the same level of difficulty as determining optimal schedules for one machine.

A very efficient algorithm for solving the two-machine problem was discovered by Johnson (1954). Following Johnson's notation, denote the machines by A and B. It is assumed that the jobs must be processed first on machine A and then on machine B. Suppose that the jobs are labeled i, for $1 \leq i \leq n$, and define

A_i = Processing time of job i on machine A.
B_i = Processing time of job i on machine B.

Johnson's result is that the following rule is optimal for determining an order in which to process the jobs on the two machines.

Rule: Job i precedes job $i + 1$ if $\min(A_i, B_{i-1}) < \min(A_{i+1}, B_i)$.

An easy way to implement this rule is as follows:

1. List the values of A_i and B_i in two columns.
2. Find the smallest remaining element in the two columns. If it appears in Column A, then schedule that job next. If it appears in column B, then schedule that job last.
3. Cross off the jobs as they are scheduled. Stop when all jobs have been scheduled.

FIGURE 8–7

Gantt chart for the optimal schedule for Example 8.5

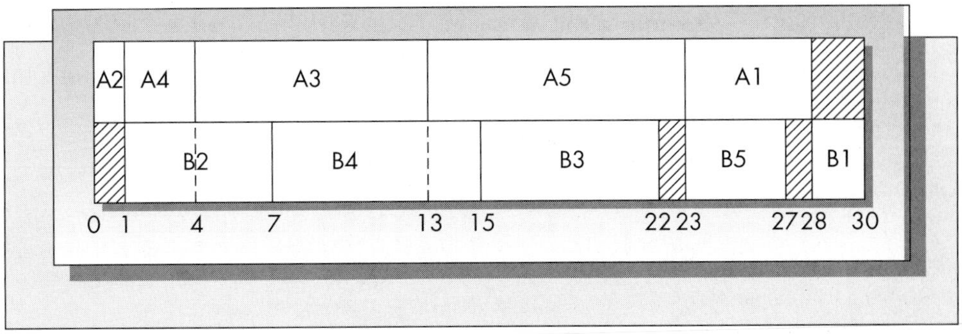

Example 8.5

Five jobs are to be scheduled on two machines. The processing times are

Job	Machine A	Machine B
1	5	2
2	1	6
3	9	7
4	3	8
5	10	4

The first step is to identify the minimum job time. It is 1, for job 2 on machine A. Because it appears in column A, job 2 is scheduled first and row 2 is crossed out. The next smallest element is 2, for job 1 on machine B. This appears in the B column, so job 1 is scheduled last. The next smallest element is 3, corresponding to job 4 in column A, so that job 4 is scheduled next. Continuing in this fashion, we obtain the optimal sequence

$$2\text{–}4\text{–}3\text{–}5\text{–}1.$$

The Gantt chart for the optimal schedule is pictured in Figure 8–7. Note that there is no idle time between jobs on machine A. This is a feature of all optimal schedules. ■

Extension to Three Machines

The problem of scheduling jobs on three machines is considerably more complex. If we restrict attention to total flow time only, it is still true that a permutation schedule is optimal (this is not necessarily the case for average flow time). Label the machines A, B, and C. The three-machine problem can be reduced to (essentially) a two-machine problem if the following condition is satisfied:

$$\min A_i \geq \max B_i \qquad \text{or} \qquad \min C_i \geq \max B_i$$

It is only necessary that *either one* of these conditions be satisfied. If that is the case, then the problem is reduced to a two-machine problem in the following way.

Define $A_i' = A_i + B_i$, and define $B_i' = B_i + C_i$. Now solve the problem using the rules described above for two machines, treating A_i' and B_i' as the processing times. The resulting permutation schedule will be optimal for the three-machine problem.

Example 8.6

Consider the following job times for a three-machine problem. Assume that the jobs are processed in the sequence A–B–C.

	Machine		
Job	A	B	C
1	4	5	8
2	9	6	10
3	8	2	6
4	6	3	7
5	5	4	11

Checking the conditions, we find

$\min A_i = 4,$
$\max B_i = 6,$
$\min C_i = 6,$

so that the required condition is satisfied. We now form the two columns A′ and B′.

	Machine	
Job	A′	B′
1	9	13
2	15	16
3	10	8
4	9	10
5	9	15

The problem is now solved using the two-machine algorithm. The optimal solution is

$$1–4–5–2–3.$$

Note that because of ties in column A, the optimal solution is not unique. ∎

If the conditions for reducing a three-machine problem to a two-machine problem are not satisfied, this method will usually give reasonable, but possibly suboptimal, results. As long as the objective is to minimize the makespan or total flow time, a permutation schedule is optimal for scheduling on three machines. (It is not necessarily true, however, that a permutation schedule is optimal for three machines when using an average flow time criterion.)

Note that we assume that the machines are different and the processing proceeds sequentially: all jobs are assumed to be processed first on machine 1, then on machine 2.

For example, machine 1 might be a drill press and machine 2 a lathe. A related problem that we discuss in the context of stochastic scheduling is that of parallel processing on identical machines. In this case the machines are assumed to perform the same function, and any job may be assigned to any machine. For example, a collection of 10 jobs might require processing on either one of two drill presses. The results for parallel processing suggest that SPT is an effective rule for minimizing mean flow time, but LPT (longest processing time first) is often more effective for minimizing total flow time or makespan. We will discuss parallel processing in the context of random job times in a later section.

The Two-Job Flow Shop Problem

Assume that two jobs are to be processed through m machines. Each job must be processed by the machines in a particular order, but the sequences for the two jobs need not be the same. We present a graphical procedure for solving this problem developed by Akers (1956).

1. Draw a cartesian coordinate system with the processing times corresponding to the first job on the horizontal axis and the processing times corresponding to the second job on the vertical axis. On each axis, mark off the operation times in the order in which the operations must be performed for that job.
2. Block out areas corresponding to each machine at the intersection of the intervals marked for that machine on the two axes.
3. Determine a path from the origin to the end of the final block that does not intersect any of the blocks and that minimizes the vertical movement. Movement is allowed only in three directions: horizontal, vertical, and 45-degree diagonal. The path with minimum vertical distance will indicate the optimal solution. Note that this will be the same as the path with minimum horizontal distance.

This procedure is best illustrated by example.

Example 8.7

A regional manufacturing firm produces a variety of household products. One is a wooden desk lamp. Prior to packing, the lamps must be sanded, lacquered, and polished. Each operation requires a different machine. There are currently shipments of two models awaiting processing. The times required for the three operations for each of the two shipments are

Job 1		Job 2	
Operation	*Time*	*Operation*	*Time*
Sanding (A)	3	A	2
Lacquering (B)	4	B	5
Polishing (C)	5	C	3

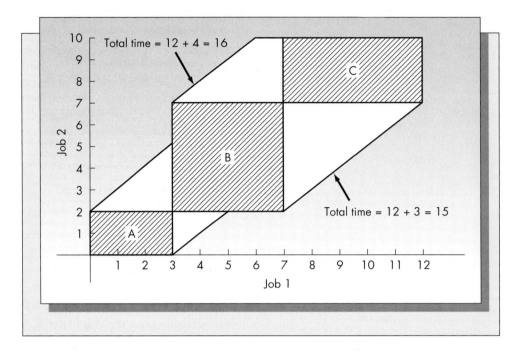

The first step is to block out the job times on each of the axes. Refer to Figure 8–8 for this step. Every feasible schedule is represented by a line connecting the origin to the tip of block C, with the condition that the line not go through a block. Only three types of movement are allowed: horizontal, vertical, and 45-degree diagonal. Horizontal movement implies that only job 1 is being processed, vertical movement implies that only job 2 is being processed, and diagonal movement implies that both jobs are being processed. Minimizing the flow time is the same as maximizing the time that both jobs are being processed. This is equivalent to finding the path from the origin to the end of block C that maximizes the diagonal movement and therefore minimizes either the horizontal or the vertical movement.

Two feasible schedules for this problem are represented in Figure 8–8. The total time required by any feasible schedule can be obtained in two ways: it is either the total time represented on the horizontal axis (12 in this case) plus the total vertical movement (4 and 3 respectively), or the total time on the vertical axis (10 in this case) plus the total horizontal movement (6 and 5 respectively). Schedule 1 has total time 16, and schedule 2 has total time 15. Schedule 2 turns out to be optimal for this problem.

The Gantt chart for the optimal schedule appears in Figure 8–9. ∎

We should note that this method does *not* require the two jobs to be processed in the same sequence on the machines. We present another example to illustrate the case in which the sequence of the jobs is different.

Example 8.8

Reggie Sigal and Bob Robinson are roommates who enjoy spending Sunday mornings reading the Sunday newspaper. Reggie likes to read the main section first,

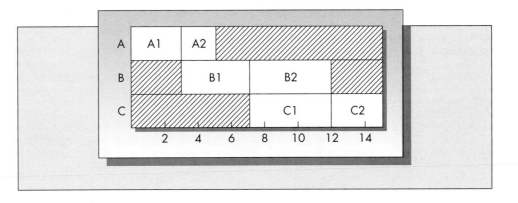

followed by the sports section, then the comics, and finally the classifieds. Bob also starts with the main section, but then goes directly to the classifieds, followed by the sports section and finally the comics. The times required (in tenths of an hour) for each to read the various sections are

Reggie		Bob	
Required Sequence	*Time*	*Required Sequence*	*Time*
Main section (A)	6	Main section (A)	4
Sports (B)	1	Classifieds (D)	3
Comics (C)	5	Sports (B)	2
Classifieds (D)	4	Comics (C)	5

The goal is to determine the order of the sections read for each to minimize the total time required to complete reading the paper. In this problem we identify Reggie as job 1 and Bob as job 2. The sections of the paper correspond to the machines.

Solution

In order to obtain the optimal solution, we first block out the processing times for each of the jobs. Assume that job 1 (Reggie) is blocked out on the *x* axis and job 2 (Bob) on the *y* axis. The processing times are sequenced on each axis in the order stated above. The graphical representation for this problem is given in Figure 8–10. In the figure, two different feasible schedules are represented as paths from the origin to the point (16, 14). The top path represents a schedule that calls for Bob to begin reading the main section first (job 2 is processed first on machine A), and the lower path calls for Reggie to begin first. The lower path turns out to be optimal for this problem with a processing time of 20.

The optimal processing time for this problem is 20. As the time units are in tenths of an hour, this is exactly 2 hours. One converts the lower path in Figure 8–10 to a Gantt chart in the following way: From time 0 to 6, Reggie reads A and Bob is idle. Between times 6 and 12, the 45-degree line indicates that both Bob and Reggie are reading. Reggie reads B for 1 time unit and C for 5 time units, and Bob reads A for 4 time units and D for 3 time units. Reggie is now idle for one time unit because the

FIGURE 8–10

Graphical solution of Example 8.8

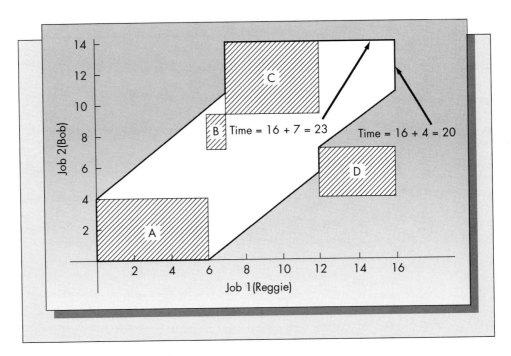

FIGURE 8–11

Gantt chart for optimal solution to Example 8.8

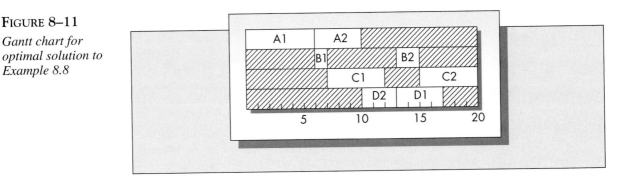

path is vertical at this point, and begins reading D at time 13. When Reggie completes D, he is done. Starting at time 15, Bob reads C and completes his reading at time 20. Figure 8–11 shows the Gantt chart indicating the optimal solution. ∎

Problems for Section 7

11. Consider Example 8.6, illustrating the use of Johnson's algorithm for three machines. List all optimal solutions for this example.

12. Suppose that 12 jobs must be processed through six machines. If the jobs may be processed in any order, how many different possible schedules are there? If you were to run a computer program that could evaluate 100 schedules every second, how much time would the program require to evaluate all feasible schedules?

13. Two law students, John and Marsha, are planning an all-nighter to prepare for their law boards the following day. Between them they have one set of materials

in the following five subjects: contracts, torts, civil law, corporate law, and patents. Based on their previous experience, they estimate that they will need the following amount of time (in hours) with each set of materials:

	Contracts	Torts	Civil	Corporate	Patents
John	1.2	2.2	0.7	0.5	1.5
Marsha	1.8	0.8	3.1	1.1	2.3

They agree that Marsha will get the opportunity to see each set of notes before John. Assume that they start their studying at 8 P.M. Determine the exact times that each will begin and end studying each subject in order to minimize the total time required for both to complete studying all five subjects.

14. The following four jobs must be processed through a three-machine flow shop.

	Machine		
Job	A	B	C
1	4	2	6
2	2	3	7
3	6	5	6
4	3	4	8

Find the optimal sequencing of the jobs in order to minimize the makespan. What is the makespan at the optimal solution? Draw a Gantt chart illustrating your solution.

15. Mary and Marcia Brown are two sisters currently attending university together. Each requires advising in five subjects: history, English, mathematics, science, and religion. They estimate that the time (in minutes) that each will require for advising is

	Mary	Marcia
Math	40	20
History	15	30
English	25	10
Science	15	35
Religion	20	25

They think that the five advisers will be available all day. Mary would like to visit the advisers in the order above, and Marcia would prefer to see them in the order math, religion, English, science, and history. At what times should each plan to see the advisers in order to minimize the total time for both to complete their advising?

16. Two jobs must be processed through four machines in the same order. The processing times in the required sequence are

Job 1		Job 2	
Machine	*Time*	*Machine*	*Time*
A	5	A	2
B	4	B	4
C	6	C	3
D	3	D	5

Determine how the two jobs should be scheduled in order to minimize the total makespan, and draw the Gantt chart indicating the optimal schedule.

17. Peter Minn is planning to go to the Department of Motor Vehicles to have his driver's license renewed. His friend, Patricia, who is accompanying him, is applying for a new license. In both cases there are five steps that are required: (A) having a picture taken, (B) signing a signature verification form, (C) passing a written test, (D) passing an eye test, and (E) passing a driving test.

For renewals the steps are performed in the order A, B, C, D, and E with average times required respectively of 0.2, 0.1, 0.3, 0.2, and 0.6 hour. In the case of new applications, the steps are performed in the sequence D, B, C, E, and A, with average times required of 0.3, 0.2, 0.7, 1.1, and 0.2 hour respectively. Peter and Pat go on a day when the department is essentially empty. How should they plan their schedule in order to minimize the time required for both to complete all five steps?

8.8 Stochastic Scheduling: Static Analysis

Single Machine

An issue we have not yet addressed is uncertainty of the processing times. In practice it is possible and even likely that the exact completion time of one or more jobs may not be predictable. It is of interest to know whether or not there are some results concerning the optimal sequencing rules when processing times are uncertain. We assume that processing times are independent of one another.

In the case of processing on a single machine, most of the results are quite similar to those discussed earlier for the deterministic case. Suppose that n jobs are to be processed through a single machine. Assume that the job times, $t_1, t_2, \ldots, t_n$, are random variables with known distribution functions. The goal is to minimize the *expected* average weighted flow time; that is,

$$\text{Minimize } E\left(\frac{1}{n}\sum_{i=1}^{n} u_i F_i\right),$$

where u_i are the weights and F_i is the (random) flow time of job i.

Rothkopf (1966) has shown that the optimal solution is to sequence the jobs so that job i precedes job $i + 1$ if

$$E(t_i)/u_i < E(t_{i+1})/u_{i+1}.$$

Notice that if we set all the weights $u_i = 1$, then this rule is simply to order the jobs according to the minimum expected processing time; that is, it is essentially the same as SPT.

In the case of due date scheduling with random processing times, results are also similar to the deterministic case. Banerjee (1965) shows that if the objective is to minimize the maximum over all jobs of the probability that a job is late, then the optimal schedule is to order the jobs according to earliest due date (or according to earliest expected due date when due dates are themselves random).

Multiple Machines

Somewhat more interesting results exist for scheduling jobs with random job times on multiple machines. In fact, there are results available for this case that are *not* true for the deterministic problem.

An assumption that is usually made for the multiple-machine problem is that the distribution of job times is exponential. This assumption is needed because the exponential distribution is the only one having the *memoryless property.* (This property is discussed in detail in Chapter 12 on reliability and maintenance.) The requirement that job times be exponentially distributed is severe in the context of scheduling. In most job shops, if processing times cannot be predicted accurately in advance, it is unlikely that they would be exponentially distributed. Why? The memoryless property tells us that the probability that a job is completed in the next instant of time is independent of the length of time already elapsed in processing the job. Certain applications, such as telephone systems and shared computer applications, may be accurately modeled in this manner, but by and large the exponential law would not accurately describe processing times for most manufacturing job shops.

With these caveats in mind, we present several results for scheduling jobs on multiple machines with random job times. Consider the following problem: n jobs are to be processed through two identical parallel machines. Each job needs to be processed only once on either machine. The objective is to minimize the expected time that elapses from time zero until the last job has completed processing. This is known as the expected makespan. We assume that the n jobs have processing times $t_1, t_2, \ldots, t_n$, which are exponential random variables with rates $\mu_1, \mu_2, \ldots, \mu_n$. This means that the expected time required to complete job i, $E(t_i)$, is $1/\mu_i$.

Parallel processing is different from flow shop processing. In flow shop processing, the jobs are processed first on machine 1 and then on machine 2. In parallel processing, the jobs need to be processed on only one machine, and any job can be processed on either machine. Assume that at time $t = 0$ machine 1 is occupied with a prior job, job 0, and the remaining processing time of job 0 is t_0, which could be a random or deterministic variable. The remaining jobs are processed as follows: let [1], [2], . . . , [n] be a permutation of the n jobs. Job [1] is scheduled on the vacant machine. Job [2] follows either job 0 on machine 1 or job [1] on machine 2, depending on which is completed first. Each successive job is then scheduled on the next available machine.

Let $T_0 \leq T_1 \leq \cdots \leq T_n$ be the completion times of the successive jobs. The makespan is the time of completion of the last job, which is T_n. The expected value of the makespan is minimized by using the longest-expected-processing-time-first rule (LEPT).[1] Note that this is exactly the opposite of the SPT rule for a single machine.

---

[1] However, if we consider flow time, then SEPT (shortest expected processing time first) minimizes the expected flow time on two machines.

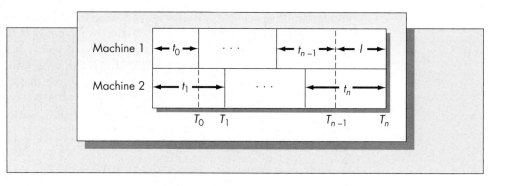

The optimality of scheduling the jobs in decreasing order of their expected size rather than in increasing order (as the SPT rule does) is more likely a result of parallel processing than of randomness of the job times.

We intuitively show the optimality of LEPT for this problem as follows: Consider the schematic diagram in Figure 8–12, which gives a particular realization of the processing times for an arbitrary sequencing of the jobs. In Figure 8–12 the random variable I corresponds to the idle time of the machine that does not process the job completed last. Intuitively, we would like to make I as small as possible in order to minimize the expected makespan. This can be shown more rigorously as follows.

From the picture, it is clear that

$$T_n + T_{n-1} = \sum_{i=0}^{n} t_i$$

and

$$T_n = T_{n-1} - I.$$

Solving for T_{n-1} in the second equation and substituting into the first gives

$$T_n + T_n - I = \sum_{i=0}^{n} t_i$$

or

$$2T_n = \sum_{i=0}^{n} t_i + I.$$

As Σt_i is fixed independent of the processing sequence, it follows that minimizing $E(T_n)$ is equivalent to minimizing $E(I)$. Because I is minimized when the processing time of the last job is minimized, we schedule the jobs in order of decreasing expected processing time. Note that this result does *not* necessarily carry over to the case of parallel processing on two machines with deterministic processing times. However, in the deterministic case, scheduling the longest job first will generally give good results when minimizing total makespan. SPT is superior for minimizing mean flow time.

A class of problems we will not discuss, but for which there are several interesting results, are problems in which jobs are to be processed through m nonidentical processors and the processing time does not depend on the job. See Righter (1988) for the characterization of the optimal scheduling rule for this problem.

The Two-Machine Flow Shop Case

An interesting question is whether there is a stochastic analog to Johnson's algorithm for scheduling n jobs on two machines in a flow shop setting; that is, where each job must be processed through machine 1 first, then through machine 2. Johnson's algorithm tells us that job i precedes job $i + 1$ if

$$\min(A_i, B_{i+1}) < \min(A_{i+1}, B_i)$$

in order to minimize the makespan.

Now suppose that $A_1, A_2, \ldots, A_n$ and $B_1, B_2, \ldots, B_n$ are exponential random variables with respective rates $a_1, a_2, \ldots, a_n$ and $b_1, b_2, \ldots, b_n$. We now wish to minimize the expected value of the makespan. As the minimum of two exponential random variables has a rate equal to the sum of the rates, it follows that

$$E[\min(A_i, B_{i+1})] = \frac{1}{a_i + b_{i+1}}.$$

$$E[\min(A_{i+1}, B_i)] = \frac{1}{a_{i+1} + b_i}.$$

It follows that Johnson's condition translates in the stochastic case to the condition that

$$a_i - b_i \geq a_{i+1} - b_{i+1},$$

so that the jobs should be scheduled in the order of decreasing values of the difference in the rates.

Example 8.9

Consider Example 8.5, used to illustrate Johnson's algorithm, but let us assume that the job times are random variables having the exponential distribution with mean times given in the example. Hence, we have the following:

Job	Expected Times		Rates		Differences
	A	B	A	B	
1	5	2	0.20	0.500	-0.30
2	1	6	1.00	0.170	0.83
3	9	7	0.11	0.140	-0.03
4	3	8	0.33	0.125	0.21
5	10	4	0.10	0.250	-0.15

Ordering the jobs according to decreasing values of the differences in the final column results in the sequence

$$2\text{--}4\text{--}3\text{--}5\text{--}1.$$

which is exactly the same sequence we found in the deterministic case using Johnson's algorithm. ∎

This section considered several solution procedures when job times are random variables. Even when job times are known with certainty, randomness resulting from other sources may still be present. For example, when considering scheduling as a dynamic problem, one must determine the pattern of arrivals to the system. It is common for jobs to arrive according to some random process and queue up for service. Queuing theory and simulation are useful tools for dealing with randomness of this type. Conway, Maxwell, and Miller (1967) discuss the application of both simulation and queuing to operations scheduling problems.

Problems for Section 8

18. Consider Example 8.2 in Section 5 on determining the optimal sequence to land the planes. Suppose that the landing times are random variables with standard deviation equal to one-third of the mean in each case.
    a. In what sequence should the planes be landed in order to minimize the expected average weighted flow time, if the weights to be used are the reciprocals of the number of passengers on each plane?
    b. For the sequence you found in part (a), what is the probability that all planes are landed within 100 minutes? Assume that the landing times are independent normally distributed random variables. Will your answer change if the planes are landed in a different sequence?

19. A computer center has two identical computers for batch processing. The computers are used as parallel processors. Job times are estimated by the user, but experience has shown that an exponential distribution gives an accurate description of the actual job times. Suppose that at a point in time there are eight jobs remaining to be processed with the following expected job times (expressed in minutes):

Job	1	2	3	4	5	6	7	8
Expected time	4	8	1	50	1	30	20	6

    a. In what sequence should the jobs be processed in order to minimize the expected completion time of all eight jobs (i.e., the makespan)?
    b. Assume that computer A is occupied with a job that has exactly two minutes of processing time remaining and computer B is idle. If job times are deterministic, show the start and end times of each job on each computer using the sequence derived in part (a).

20. Six ships are docked in a harbor awaiting unloading. The times required to unload the ships are random variables with respective means of 0.6, 1.2, 2.5, 3.5, 0.4, and 1.8 hours. The ships are given a priority weighting based on tonnage. The respective tonnages are 12, 18, 9, 14, 4, and 10. In what sequence should the ships be unloaded in order to minimize the expected weighted time?

21. Solve Problem 13 assuming that the times required by John and Marsha are exponentially distributed random variables with expected times given in Problem 13.

22. Five sorority sisters plan to attend a social function. Each requires hair styling and fitting for a gown. Assume that the times required are independent exponentially distributed random variables with the mean times for the fittings of 0.6, 1.2, 1.5, 0.8, and 1.1 hours respectively, and mean times for the stylings of 0.8, 1.6, 1.0, 0.7, and 1.3 hours respectively. Assume that the fittings are done before the stylings and that there is only a single hair stylist and a single seamstress available. In what sequence should they be scheduled in order to minimize the total expected time required for fittings and stylings?

8.9 Stochastic Scheduling: Dynamic Analysis

The scheduling algorithms discussed thus far in this chapter are based on the assumption that all jobs arrive for processing simultaneously. In practice, however, scheduling jobs on machines is a dynamic problem. We use the term *dynamic* here to mean that jobs are arriving randomly over time, and decisions must be made on an ongoing basis as to how to schedule those jobs.

Queuing theory provides a means of modeling some dynamic scheduling problems. Supplement 2 at the end of this chapter provides a review of basic queuing theory. In this section familiarity with the results presented in Supplement 2 is assumed.

Consider the following problem. Jobs arrive completely at random to a single machine. This means that the arrival process is a *Poisson* process. Assume that the mean arrival rate is λ. We initially will assume that processing times are exponentially distributed with mean $1/\mu$. This means that the average processing rate is μ and processing times are independent identically distributed exponential random variables. Finally, we assume that jobs are processed on a first-come, first-served (FCFS) basis. In queuing terminology, we are assuming an M/M/1/FCFS queue. Other processing sequences also will be considered.

Basic queuing theory answers several questions about the performance characteristics of this scheduling problem. First, the probability distribution of the number of jobs in the system (the number waiting to be processed plus the number being processed) in the steady state is known to be geometric with parameter $\rho = \lambda/\mu$. That is, if L is the number of jobs in the system in steady state, then

$$P\{L = i\} = \rho^i(1 - \rho) \qquad \text{for } i = 0, 1, 2, 3, \ldots$$

The expected number of jobs in the system is $\rho/(1 - \rho)$. This implies that a solution exists only for $\rho < 1$. Intuitively this makes sense: the rate at which jobs arrive in the system must be less than the rate at which they are processed to guarantee that the queue does not grow without bound.[2]

Minimizing mean flow time is a common objective not only in static scheduling, but also in dynamic scheduling. The flow time of a job begins the instant the job joins the queue of unprocessed jobs and continues until its processing is completed. For the

---

[2]What is not obvious, however, is what happens at the boundary when $\lambda = \mu$ or $\rho = 1$. It turns out that when the processing and the arrival rates are equal (and interarrival times and job processing times are random), the queue still grows without bound. The reason for this is certainly not obvious, but it appears to be a consequence of the randomness of the arrivals and processing times.

dynamic scheduling problem, the flow time of a job is a random variable; it depends on the realization of the processing times of preceding jobs as well as its own processing time. The queuing term for the flow time of a job is the waiting time in the system and is denoted by the symbol W. Supplement 2 shows that the distribution of the flow time for the M/M/1/FCFS queue is exponential with parameter $\mu - \lambda$. That is,

$$P\{W > t\} = e^{-(\mu-\lambda)t} \qquad \text{for all } t > 0.$$

Also derived are the distribution for the waiting time in the queue and the expected number of jobs in the queue waiting to be processed.[3] We can see the application of these formulas to the dynamic scheduling problem in the following example.

Example 8.10

A student computer laboratory has a single laser printer. Jobs queue up on the printer from the network server in the lab and are completed on a first-come, first-served basis. The average printing job requires 4 minutes, but the times vary considerably. Experience has shown that the distribution of times closely follows an exponential distribution. At peak times, about 12 students per hour require use of the printer, but the arrival of jobs to the printer can be assumed to occur completely at random.

Assuming a peak traffic period, determine the following:

   a. The average number of jobs in the printer queue.
   b. The average flow time of a job.
   c. The probability that a job will wait more than 30 minutes before it begins processing.
   d. The probability that there are more than 6 jobs in the system.

Solution

First we must determine the service and arrival rates. As each printing job requires an average of 4 minutes, it follows that the service *rate* is $\mu = 1/4$ per minute or 15 per hour. The arrival rate is given as $\lambda = 12$ per hour. The traffic intensity $\rho = 12/15 = .8$.

   a. The average number of jobs in the queue is L_q, which is given by (refer to Supplement 2)

$$L_q = \rho^2/(1 - \rho) = .64/.2 = 3.2.$$

   b. The average flow time of a job is the same as the waiting time in the system. From Supplement 2.

$$W = \rho/(\lambda(1 - \rho)) = .8/((12)(.2)) = 0.3333 \text{ hour (20 minutes)}.$$

   c. Here we are interested in $P\{W_q > 0.5\}$. The distribution of W_q is essentially exponential with parameter $\mu - \lambda$, but with positive mass at zero.

$$P\{W_q > t\} = \rho e^{-(\mu-\lambda)t} = .8e^{-(3)(0.5)} = .1785.$$

---

[3]For convenience we use the same symbol for a random variable and its mean, so that W will be used for both the waiting time in the system and the expected waiting time in the system. The same will be true for L. The meaning will be clear from the context.

d. Here we wish to determine $P\{L > 6\}$. From the solution of Problem S2.1 in Supplement 2, we showed that

$$P\{L > k\} = \rho^{k+1} = .8^7 = .2097. \blacksquare$$

Selection Disciplines Independent of Job Processing Times

Although our sense of fair play says that the service discipline should be FCFS, there are many occasions when jobs are processed in other sequences. For example, consider a manufacturing process in which parts are stacked as they are completed. The next stage in the process may simply take the parts from the top of the stack, resulting in a LCFS (last-come, first-served) discipline. Similarly, one could envision a situation in which completed parts are thrown into a bin and taken out at random, resulting in service occurring in a random order.

Queuing theory tells us that as long as the selection discipline *does not depend on the processing times,* the mean flow times (and hence mean numbers in the system and mean queue lengths) are the same.[4] However, the **variance** of the flow times does depend on the selection discipline. The flow time variance is greatest with LCFS and least with FCFS among the three disciplines FCFS, LCFS, and random.[5] The second moments of the flow time are given by

$$E_{\text{LCFS}}(W^2) = \frac{1}{1 - \rho} E_{\text{FCFS}}(W^2),$$

$$E_{\text{RANDOM}}(W^2) = \frac{1}{1 - \rho/2} E_{\text{FCFS}}(W^2).$$

(The second equation has been proved for exponential processing times only, but it has been conjectured that it holds in general.) Recall that the variance of a random variable is the second moment minus the mean squared, so we can obtain the variance of W directly from these formulas.

Example 8.10 (continued)

Determine the variance of the flow times for the laser printer during peak hours assuming (1) FCFS, (2) LCFS, and (3) random selection disciplines.

Solution

Because under FCFS the flow time is exponentially distributed with parameter $\mu - \lambda$, it follows that the mean flow time is $1/(\mu - \lambda)$ and the variance of the flow time is $1/(\mu - \lambda)^2$.

$$E_{\text{FCFS}}(W) = 1/(\mu - \lambda) = 1/(15 - 12) = 1/3 \text{ hour.}$$
$$Var_{\text{FCFS}}(W) = 1/(3)^2 = 1/9.$$

---

[4]Many of the results quoted in this section are discussed more fully in Chapter 8 of Conway, Maxwell, and Miller (1967). In fact, even though this book is over 30 years old, it still provides the most comprehensive treatment of the application of queuing theory to scheduling problems.

[5]This is strictly true only if we eliminate the possibility of preemption. Preemption means that a newly arriving job is allowed to interrupt the service of a job already in progress. We will not treat preemptive disciplines here.

Because $Var(W) = E(W^2) - (E(W))^2$, it follows that

$$E_{\text{FCFS}}(W^2) = Var_{\text{FCFS}}(W) + (E_{\text{FCFS}}(W))^2 = 1/9 + (1/3)^2 = 2/9.$$

From the results above, we obtain

$$E_{\text{LCFS}}(W^2) = (1/(1 - .8))(2/9) = 10/9, \qquad \text{giving}$$

$$Var_{\text{LCFS}}(W) = 10/9 - (1/3)^2 = 1.0.$$

Similarly,

$$E_{\text{RANDOM}}(W^2) = (1/(1 - .4))(2/9) = 0.3704, \qquad \text{giving}$$

$$Var_{\text{RANDOM}}(W) = 0.3704 - (1/3)^2 = 0.2593.$$

Hence, we see that if the jobs are processed on the printer on an LCFS basis, the variance of the flow time is 1.0, as compared to 2/9 for FCFS. Processing jobs in a random order also increases the variance of the flow time over FCFS, but by a much smaller degree. Intuitively, the variance is larger for LCFS than for FCFS, because when jobs are processed in the opposite order of their arrival, it is likely that a job that has been in the queue for a while will continue to be "bumped" by newly arriving jobs. The result will be a very long flow time. A similar phenomenon occurs in the random case, but the effect is not as severe. ∎

Selection Disciplines Dependent on Job Processing Times

One of the goals of research into dynamic queuing models is to discover optimal selection disciplines. In the previous section, we stated that the average measures of performance (W, W_q, L, and L_q) are independent of the selection discipline as long as the selection discipline does not depend on the job processing times. However, consider the case in which job processing times are realized at the instant the job joins the queue. This assumption is reasonable for most industrial scheduling applications. For machine processing problems, the work content is likely to be a multiple of the number of parts that have to be processed, so that the processing time is known at the instant a job joins the queue. An example in which this does not hold would be a bank. It is generally not possible to tell how long a customer will require for service until he or she enters service and reveals the nature of the transaction.

When job processing times are realized when a job joins the queue, it is possible to use a selection discipline that is dependent on job times. One such discipline, discussed at length earlier in this chapter, is the SPT rule: The next job processed is the one with the shortest processing time. It turns out that the SPT rule is effective for the dynamic problem as well as the static problem.

SPT scheduling can significantly reduce the size of the queue in the dynamic scheduling problem. In Figure 8–13 we show just how dramatic this effect can be. This figure assumes an M/M/1 queue. Define the relative flow time as the ratio

$$\frac{E_{\text{SPT}}(\text{Flow Time})}{E_{\text{FCFS}}(\text{Flow Time})}.$$

We see that as the traffic intensity increases, the advantage of SPT at reducing flow time (and hence the number in the system and the queue length) improves. The queue is reduced by "cleaning out" the shorter jobs. For values of ρ near one, this ratio could be as low as 0.2. Because of Little's formula (see Supplement 2), the expected number in the system and the flow time are proportional, so this curve also represents the ratio of the expected numbers in the system for the respective selection disciplines.

TABLE 8–1   **Variance of the Flow Time under FCFS and SPT**

ρ	FCFS	SPT
.1	1.2345	1.179
.2	1.5625	1.482
.3	2.0408	1.896
.4	2.6666	2.563
.5	4.0000	3.601
.6	6.2500	5.713
.7	11.1111	12.297
.8	25.	32.316
.9	100.	222.2
.95	400.	1596.5
.98	2500.	22096.
.99	10000.	161874.

Source: Conway, Maxwell, and Miller (1967), p. 189.

FIGURE 8–13

Relative flow times:
SPT versus FCFS

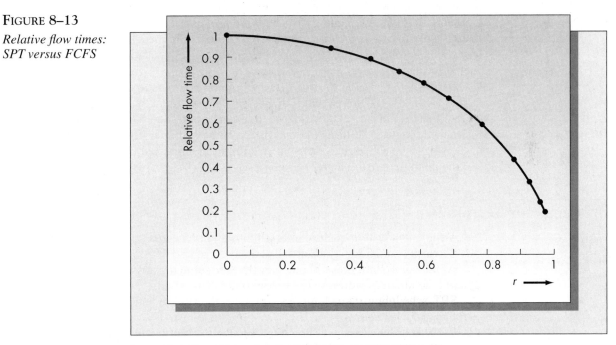

Source: Adapted from tabled results in Conway, Maxwell, and Miller (1967), p. 184.

An interesting question is what happens to the variance of the performance measures under each selection discipline. Again assuming exponential service times, Table 8–1 gives the variances of the flow times for SPT and FCFS as a function of the traffic intensity, ρ.

What we see from this table is that for low values of the traffic intensity (less than about .7), SPT has slightly lower variance than FCFS. However, as the traffic intensity approaches 1, the variance under SPT increases dramatically. For $\rho = .99$, the variance of the flow time under SPT is more than 16 times that for FCFS. Hence, the reduction in the mean flow time achieved by SPT comes at the cost of possibly increasing the variance.

A large portion of the research in dynamic scheduling considers priority disciplines, which means that incoming jobs are classified into groups and priority is given to certain groups over others. Priority scheduling is common in hospital emergency rooms. It also occurs in scheduling batch jobs on a mainframe computer, as certain users may be given priority over others. Priorities may be preemptive or nonpreemptive. Preemptive priority means that a newly arriving job with a higher priority than the job currently being processed is permitted to interrupt the service of the job in process. The interrupted job may continue where it left off at a later time, or it may have to start processing again from scratch. We will not pursue this complex area, but note that SPT is quite robust; it is optimal for a large class of priority scheduling problems.

The $c\mu$ Rule

Consider the following scheduling problem. Jobs arrive randomly to a single machine with exponential processing times. We allow the jobs in the queue to have different service rates μ_i. That is, at any moment, suppose that there are n jobs waiting to be processed. We index the jobs $1, 2, 3, \ldots, n$ and assume that the time required to complete job i has the exponential distribution with mean $1/\mu_i$. In addition, suppose that there is a return of c_i if job i is completed by some fixed time t. The issue is, what is the best choice for the next job to be processed if the objective is to maximize the total expected earnings?

Derman et al. (1978) showed that the optimal policy is to choose the job with the largest value of $c_i\mu_i$. Notice that if the weights are set equal to 1, the $c\mu$ rule is exactly the same as SPT in expectation. Hence, this can be considered to be a type of weighted SPT scheduling rule. It turns out that the $c\mu$ rule is optimal for several other versions of the stochastic scheduling problem. We refer the interested reader to Pinedo (1983) and the references listed there.

Problems for Section 9

23. A computer system queues up batch jobs and processes them on an FCFS basis. Between 2 and 5 P.M., jobs arrive at an average rate of 30 per hour and require an average of 1.2 minutes of computer time. Assume the arrival process is Poisson and the processing times are exponentially distributed.
    a. What is the expected number of jobs in the system and in the queue in the steady state?
    b. What are the expected flow time and the time in the queue in the steady state?
    c. What is the probability that the system is empty?
    d. What is the probability that the queue is empty?
    e. What is the probability that the flow time of a job exceeds 10 minutes?

24. Consider the computer system of the previous problem.
    a. Compute the variance of the flow times assuming FCFS, LCFS, and random selection disciplines.
    b. Using a normal approximation of the flow time distribution for the LCFS and random cases, estimate the probability that the flow time in the system exceeds 10 minutes in each case.

25. A critical resource in a manufacturing operation experiences a very high traffic intensity during the shop's busiest periods. During these periods the arrival rate is approximately 57 jobs per hour. Job processing times are approximately exponentially distributed with mean 1 minute.

    a. Compute the expected flow time in the system assuming an FCFS processing discipline, and the expected flow time under SPT using Figure 8–13.

    b. Compute the probability that a job waits more than 30 minutes for processing under FCFS.

    c. Using a normal approximation, estimate the probability that a job waits more than 30 minutes for processing under LCFS.

8.10 Assembly Line Balancing

The problem of balancing an assembly line is a classic industrial engineering problem. Even though much of the work in the area goes back to the mid-50s and early 60s, the basic structure of the problem is relevant to the design of production systems today, even in automated plants. The problem is characterized by a set of n distinct tasks that must be completed on each item. The time required to complete task i is a known constant t_i. The goal is to organize the tasks into groups, with each group of tasks being performed at a single workstation. In most cases, the amount of time allotted to each workstation is determined in advance, based on the desired rate of production of the assembly line. This is known as the cycle time and is denoted by C. A schematic of a typical assembly line is given in Figure 8–14. In the figure, circles represent tasks to be done at the corresponding stations.

Assembly line balancing is traditionally thought of as a facilities design and layout problem, and one might argue that it would be more appropriately part of Chapter 10. Assigning tasks to workstations has traditionally been a one-shot decision made at the time the plant is constructed and the line is set up. However, the nature of the modern factory is changing. New plants are being designed with flexibility in mind, allowing new lines to be brought up and old ones restructured on a continuous basis. In such an environment, line balancing is more like a dynamic scheduling problem than a one-shot facilities layout problem.

There are a variety of factors that contribute to the difficulty of the problem. First, there are precedence constraints: some tasks may have to be completed in a particular sequence. Another problem is that some tasks cannot be performed at the same workstation. For example, it might not be possible to work on the front end and the back end of a large object such as an automobile at the same workstation. This is known as a

FIGURE 8–14

Schematic of a typical assembly line

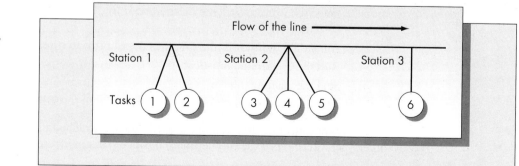

zoning restriction. Still other complications might arise. For example, certain tasks may have to be completed at the same workstation, and other tasks may require more than one worker.

Finding the optimal balance of an assembly line is a difficult combinatorial problem even when the problems described above are not present. Several relatively simple heuristics have been suggested for determining an approximate balance. Many of these methods require few calculations and make it possible to solve large problems by hand.

Let $t_1, t_2, \ldots, t_n$ be the time required to complete the respective tasks. The total work content associated with the production of an item, say T, is given by

$$T = \sum_{i=1}^{n} t_i.$$

For a cycle time of C, the minimum number of workstations possible is $[T/C]$, where the brackets indicate that the value of T/C is to be rounded to the next larger integer. Because of the discrete and indivisible nature of the tasks and the precedence constraints, it is often true that more stations are required than this ideal minimum value. If there is leeway in the choice of the cycle time, it is advisable to experiment with different values of C to see if a more efficient balance can be obtained.

We will present one heuristic method from Helgeson and Birnie (1961) known as the *ranked positional weight technique.* The method places a weight on each task based on the total time required by all of the succeeding tasks. Tasks are assigned sequentially to stations based on these weights. We illustrate the method by example.

Example 8.11

The final assembly of Noname personal computers, a generic mail-order PC clone, requires a total of 12 tasks. The assembly is done at the Lubbock, Texas, plant using various components imported from the Far East. The tasks required for the assembly operations are

1. Drill holes in the metal casing and mount the brackets to hold disk drives.
2. Attach the motherboard to the casing.
3. Mount the power supply and attach it to the motherboard.
4. Place the main processor and memory chips on the motherboard.
5. Plug in the graphics card.
6. Mount the floppy disk drives. Attach the floppy drive controller and the power supply to the drives.
7. Mount the hard disk drive. Attach the hard disk controller and the power supply to the hard drive.
8. Set switch settings on the motherboard for the specific configuration of the system.
9. Attach the monitor to the graphics board prior to running system diagnostics.
10. Run the system diagnostics.
11. Seal the casing.
12. Attach the company logo and pack the system for shipping.

The holes must be drilled and the motherboard attached to the casing prior to any other operations. Once the motherboard has been mounted, the power supply,

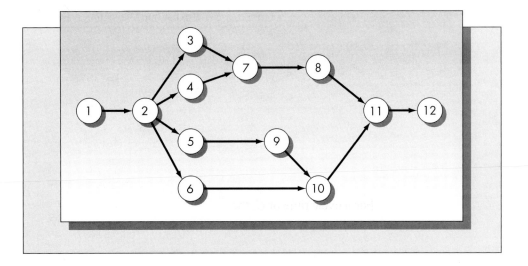

memory, processor chips, graphics card, and disk controllers can be installed. The floppy drives are placed in the unit prior to the hard drive and require that the power supply and controller be in place first. Based on the memory configuration and the choice of graphics adapter, the switch settings on the motherboard are determined and set. The monitor must be attached to the graphics board so that the results of the diagnostic tests can be read. Finally, after all other tasks are completed, the diagnostics are run and the system is packed for shipping. The job times and precedence relationships for this problem are summarized in the table below. The network representation of this particular problem is given in Figure 8–15.

Task	Immediate Predecessors	Time
1	—	12
2	1	6
3	2	6
4	2	2
5	2	2
6	2	12
7	3, 4	7
8	7	5
9	5	1
10	9, 6	4
11	8, 10	6
12	11	7

Suppose that the company is willing to hire enough workers to produce one assembled machine every 15 minutes. The sum of the task times is 70, which means that the minimum number of workstations is the ratio $70/15 = 4.67$ rounded to the next larger integer, which is 5. This does not mean that a five-station balance necessarily exists.

TABLE 8–2   **Positional Weights for Example 8.11**

Task	Positional Weight
1	70
2	58
3	31
4	27
5	20
6	29
7	25
8	18
9	18
10	17
11	13
12	7

The solution procedure requires determining the positional weight of each task. The positional weight of task i is defined as the time required to perform task i plus the times required to perform all tasks having task i as a predecessor. As task 1 must precede all other tasks, its positional weight is simply the sum of the task times, which is 70. Task 2 has positional weight 58. From Figure 8–15 we see that task 3 must precede tasks 7, 8, 11, and 12, so that the positional weight of task 3 is $t_3 + t_7 + t_8 + t_{11} + t_{12} = 31$. The other positional weights are computed similarly. The positional weights are listed in Table 8–2.

The next step is to rank the tasks in the order of decreasing positional weight. The ranking for this case is 1, 2, 3, 6, 4, 7, 5, 8, 9, 10, 11, 12. Finally, the tasks are assigned sequentially to stations in the order of the ranking, and assignments are made only as long as the precedence constraints are not violated.

Let us now consider the balance obtained using this technique assuming a cycle time of 15 minutes. Task 1 is assigned to station 1. That leaves a slack of three minutes at this station. However, because task 2 must be assigned next, in order not to violate the precedence constraints, and the sum $t_1 + t_2$ exceeds 15, we close station 1. Tasks 2, 3, and 4 are then assigned to station 2, resulting in an idle time of only one minute at this station. Continuing in this manner, we obtain the following balance for this problem:

Station	1	2	3	4	5	6
Tasks	1	2, 3, 4	5, 6, 9	7, 8	10, 11	12
Idle time	3	1	0	3	5	8

Notice that although the minimum possible number of stations for this problem is five, the ranked positional weight technique results in a six-station balance. As the method is only a heuristic, it is possible that there is a solution with five stations. In this case, however, the optimal balance requires six stations when $C = 15$ minutes.

The head of the firm assembling Noname computers is interested in determining the minimum cycle time that would result in a five-station balance. If we increase the cycle time from $C = 15$ to $C = 16$, then the balance obtained is

Station	1	2	3	4	5
Tasks	1	2, 3, 4, 5	6, 9	7, 8, 10	11, 12
Idle time	4	0	3	0	3

This is clearly a much more efficient balance: The total idle time has been cut from 20 minutes per unit to only 10 minutes per unit. The number of stations decreases by about 16 percent while the cycle time increases by only about 7 percent. Assuming that a production day is seven hours, a value of $C = 15$ minutes would result in a daily production level of 28 units per assembly operation and a value of $C = 16$ minutes would result in a daily production level of 26.25 units per assembly operation. Management would have to determine whether the decline in the production rate of 1.75 units per day per operation is justified by the savings realized with five rather than six stations.

An alternative choice is to stay with the six stations, but see if a six-station balance can be obtained with a cycle time less than 15 minutes. It turns out that for values of the cycle time of both 14 minutes and 13 minutes, the ranked positional weight method will give six-station balances. The $C = 13$ solution is

Station	1	2	3	4	5	6
Tasks	1	2, 3	6	4, 5, 7, 9	8, 10	11, 12
Idle time	1	1	1	1	4	0

Thirteen minutes appear to be the minimum cycle time with six stations. The total idle time of eight minutes resulting from the balance above is two minutes less than that achieved with five stations when $C = 16$. The production rate with six stations and $C = 13$ would be 32.3 units per day per operation. Increasing the number of stations from five to six results in a substantial improvement in the throughput rate. ■

In this section we presented the ranked positional weight heuristic for solving the assembly line balancing problem. Other heuristic methods exist as well. One is COMSOAL, a computer-based heuristic developed by Arcus (1966). The method is efficient for large problems involving many tasks and workstations. Kilbridge and Wester (1961) suggest a method similar to the ranked positional weight technique.

There are optimal procedures for solving the line balancing problem, but the calculations are complex and time-consuming, requiring either dynamic programming (Held et al., 1963) or integer programming (Thangavelu and Shetty, 1971). More recent interest in the line balancing problem has focused on issues relating to uncertainty in the performance times for the individual tasks. (See Hillier and Boling, 1986, and the references contained there.)

Virtually all assembly line balancing procedures assume that the objective is to minimize the total idle time at all workstations. However, as we saw in this section, an optimal balance for a fixed cycle time may not be optimal in a global sense. Carlson and Rosenblatt (1985) suggest that most assembly line balancing procedures are based on an incorrect objective. The authors claim that maximizing profit (rather than minimizing idle time) would give a different solution to most assembly line balancing problems and present several models in which both numbers of stations and cycle time are decision variables.

Problems for Section 10

26. Consider the example of Noname computers presented in this section.
    a. What is the minimum cycle time that is possible? What is the minimum number of stations that would theoretically be required to achieve this cycle time?
    b. Based on the ranked positional weight technique, how many stations are actually required for the cycle time indicated in part (*a*)?
    c. Suppose that the owner of the company that sells Noname computers finds that he is receiving orders for approximately 100 computers per day. How many separate assembly lines are required assuming (*i*) the best five-station balance, (*ii*) the best six-station balance (both determined in the text), and (*iii*) the balance you obtained in part (*b*)? Discuss the trade-offs involved with each choice.

27. A production facility assembles inexpensive telephones on a production line. The assembly requires 15 tasks with precedence relationships and activity times as shown in Figure 8–16. The activity times appear next to the node numbers in the network.
    a. Determine positional weights for each of the activities.
    b. For a cycle time of 30 units, what is the minimum number of stations that could be achieved? Find the $C = 30$ balance obtained using the ranked positional weight technique.
    c. Is there a solution with the same number of stations you found in part (*b*) but with a lower cycle time? In particular, what appears to be the minimum

FIGURE 8–16
(For Problem 27)

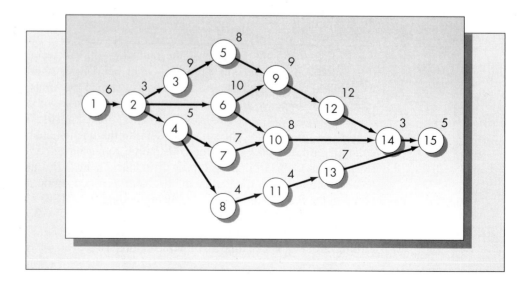

SNAPSHOT APPLICATION

Manufacturing Divisions Realize Savings with Scheduling Software

Motorola Goes for MOOPI

Motorola Corporation has adopted MOOPI from Berclain Group of Canada to schedule the remanufacture of 1,000 engine controllers each month in the Automotive and Industrial Electronics Group in its Seguin, Texas, plant. Utilization of used engine controllers has improved from 35 percent to over 70 percent and the time required for scheduling has reduced from 100 to only 8 hours per month as a consequence of implementing the software. The problem of scheduling cores remanufacture is a very complex one, requiring five or six employees when it was done manually. According to Eileen Svoboda, a process improvement manager:

> Remanufacturing in itself is a unique and complex process because the raw material—the core—arrives randomly through time, and the material required to remanufacture it is unknown until after the core is inspected. Right now in our business we have 7,000 Motorola core model numbers that can be remanufactured into approximately 800 different customer models. To keep all this straight in the old manual system we relied on paper trails, as well as the memory and expertise of a few key individuals in Texas, Michigan, and Illinois.

The software allows the group to schedule one month's worth of production in detail, as well as some additional planning functions for a 26-week time horizon. Motorola considered four different scheduling systems. MOOPI was chosen because it could both handle complex material assignment needs and easily run "what if" simulations. Although the simulation feature played an important part in the decision, MOOPI is basically an optimization package, and, according to Berclain, Motorola's primary goal was optimization of materials utilization.

The system's database incorporates manufacturing data, including sequence of operations or routings, bills of materials, and setup sequencing information. It distinguishes raw materials, subassembly, and finished goods inventory levels. It can exchange data with other inventory control systems, and it details production orders for all work centers.

H.P. Implements FastMAN for Materials Scheduling

The Hewlett-Packard Corporation of Palo Alto, California, has experienced enormous growth in the highly competitive PC business. It has been reported to be the second-fastest–growing personal computer company in the world. Part of this success is attributable to careful management of the materials in the PC product plants.

H.P.'s scheduling system is PC-based and is used in conjunction with their MRP II system. One post-MRP scheduler used by H.P. is FastMAN, produced by the company with the same name based in Chicago, Illinois. Today, H.P. has eight installations of FastMAN and three with APS, a workstation-based system recently made available on PCs. According to Dr. Lee Seaton, a senior industry analyst with H.P.:

> The materials content of a personal computer constitutes over 90 percent of its cost. Unfortunately, we frequently had too much of one thing and too little of something else. And we knew it was only going to get worse, since product life cycles continue to accelerate, recently down from 15 months to under one year. We had to reduce product write-offs and have a smoother cut-off for a given product set. Too often, we'd launch a successful product, but either not have enough to sell or at the end of the cycle have excessive material write-offs.

While material control is normally the domain of MRP II, H.P. was not having the desired level of success with MRP II. Dr. Seaton claims that 30 percent of the MRP runs were considered useless. Updates of inventory levels or bills of materials were not entered into the system in a timely fashion, making the results out of date. The new scheduling tools allowed H.P. employees to verify the validity of MRP runs with desktop systems. Also, the software provided the capability for determining the consequences of business decisions in a spreadsheet or graphical form. Engineering changes could more easily be incorporated into the materials plan, and excess material could be utilized more effectively by constructing "matched sets." The key to H.P.'s success was "getting information into the hands of the decision makers," says Seaton.

Source: These two applications are discussed in Parker (1995).

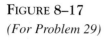

FIGURE 8–17

(For Problem 29)

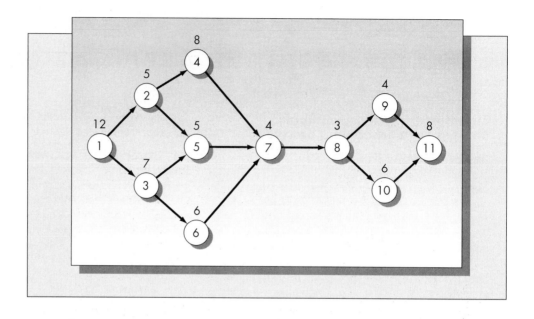

cycle time that gives a balance with the same number of stations you found in part *(b)*?

28. For the data given in Problem 27, determine by experimentation the minimum cycle time for a three-station balance.

29. Consider the assembly line balancing problem represented by the network in Figure 8–17. The performance times are shown above the nodes.
   a. Determine a balance for $C = 15$.
   b. Determine a balance for $C = 20$.
   c. What appears to be the minimum cycle time that can be achieved using the number of stations you obtained in part *(a)*?
   d. What appears to be the minimum cycle time that can be achieved using the number of stations you obtained in part *(b)*?

8.11 Simulation: A Valuable Scheduling Tool

Real-world scheduling problems are often too complex to be amenable to mathematical analysis. Computer-based simulation is a valuable tool for comparing various scheduling strategies and scenarios. Some of the earliest applications of computer-based operations research applications involved simulations.

A simulation is a model or a re-creation of a real situation that allows the user to examine different scenarios in a laboratory environment. For example, wind tunnels simulate conditions under which planes fly, allowing an airplane designer to test different wing configurations without crashing planes. Flight simulators are valuable teaching tools. Students can make "fatal" errors in a simulator and live to talk about it.

These are physical simulators. Production planning applications utilize computer-based simulators. A computer simulation is a computer program that accurately reflects a real-world situation. As with a mathematical model, variables are defined to represent real quantities and expressions developed to describe the relations among these variables. Although both deterministic and stochastic problems are amenable to simulation, the method has been applied more in problems with some element of randomness. Simulators are particularly valuable for modeling uncertainty. Computer scientists have expended substantial effort in developing so-called random number generators. These are mathematical recursions that create strings of numbers that appear to be completely random (i.e., drawn from a uniform distribution). Random numbers can be transformed into observations on random variables having almost any desired distribution using relationships from probability theory. Simulations that include some element of uncertainty are referred to as Monte Carlo simulations. (See, for example, Fishman, 1973, for an excellent summary of the concepts and methods used to build simulation programs and analyze the output of these programs.)

Simulation as a modeling tool is gaining in popularity. Spreadsheets include random number generators, allowing one to construct simple simulators in this way. More sophisticated spreadsheet models can be created with third-party add-in products (such as @Risk for Lotus 1–2–3 and Excel). While most simulation programs can easily be written directly in a source language (such as Basic or C), many dedicated simulation packages are available. Examples of early products are SLAM and GPSS. Since the 1960s Pritsker Associates of West Lafayette, Indiana, has been marketing network-based simulators specifically designed for production planning applications. One of the most popular general purpose simulators available for the PC is SIMAN. However, the market is expanding rapidly. In a survey of simulation packages, Swain (1993) compared 55 different simulation software packages, all designed for PCs and workstations.

Most of the scheduling software discussed in the next section incorporate a simulation component. Real-world simulation–based scheduling applications abound. For example, Hutchison et al. (1993) used computer-based simulation to analyze dispatch rules for gears at the Jeffrey Mining Equipment Division of Dresser Industries. They found that EDD scheduling coupled with flexible capacity planning resulted in a potential improvement of nearly 40 percent under high-demand conditions. In another application, SGS Thomson applied simulation for finite capacity planning and scheduling of its semiconductor test and finish areas (Nichols, 1994). In still another study, simulation was used to compare three scheduling policies for a unit load automated storage and retrieval system with the goal of maximizing total throughput (Randhawa and Shroff, 1995). This is only a small sampling of the vast literature on simulation-based applications in scheduling.

8.12 Post-MRP Production Scheduling Software

While the worldwide market for integrated MRP systems is enormous (nearing $10 billion), there is a growing market for scheduling software specifically designed to pick up where MRP systems fall short. According to Parker (1995):

> Finite capacity scheduling systems appeared on the scene because MRP II systems assume infinite capacity. Planning systems which can work alongside MRP II or can even replace key MRP II modules, optimize for both material and capacity constraints. But the latest planning and scheduling tools offer two additional things. The first is that they are said to be dynamic, in that they are meant to cope with continuous change. . . . The second thing the systems offer is decision support.

Parker (1994) estimated that the domestic market for these tools was about $100 million and growing rapidly. While full-blown MRP systems are generally mainframe or client–server based, these post-MRP schedulers (which are also used as stand-alone systems) are generally designed to run on PCs or workstations. Most of the systems offer some combination of optimization for determining schedules and Monte Carlo simulation to allow the user to explore "what if" scenarios.

A sampling of providers of these products include i2 Technologies, Dallas, Texas; The Berclain Group, Quebec City; Red Pepper Software, San Mateo, California; Bridgeware, Hayward, California; FastMAN, Chicago, Illinois; Orissa International, Torrance, California; and Enterprise Planning Systems, Wheaton, Illinois. Several of the providers of MRP software discussed in Chapter 7 also are incorporating some of the features of these schedulers, including simulation and decision support, directly into the MRP systems. New on the market is the Preactor series of shop floor scheduling software by Systems Modeling Corporation (Vasilash, 1995).

An important distinction between planners and schedulers is the length of planning horizon. According to Sanjeev Gupta, president of Orissa International, "Planning is done over a six- to nine-month period, while finite capacity scheduling looks at a two- to three-month period, although it depends on what the product cycle time is. Dispatching, which is dependent on real-time information, is done on a two-to three-day period."

Although the value of spin-off technology from military spending is coming under closer scrutiny, the technology for finite capacity schedulers may benefit from work done at NASA for scheduling space systems. The Advanced Scheduling Environment, now sold commercially by Avyx, Inc., of Englewood, Colorado, can generically describe a wide range of scheduling applications. The system embodies a set of data structures that accommodate an extremely general description of a factory's resources, activities, and constraints. The software makes use of tree-like data structures to solve scheduling problems, rather than the more traditional relational databases that characterize the vast majority of scheduling and MRP systems.

8.13 Historical Notes

Interest in studying the effects of different sequencing strategies in the job shop is relatively recent. One of the earliest monographs in the open literature that considered sequencing issues is by Salveson (1952). The first significant published work to present analytical results for optimal sequencing strategies is Johnson (1954). Bellman's (1956) review article discusses sequence scheduling (among other topics) and presents an interesting proof of the optimality of Johnson's algorithm based on dynamic programming arguments. Bellman appears to be the first to have recognized the optimality of a permutation schedule for scheduling n jobs on three machines.

Johnson's work spawned considerable interest in scheduling. The excellent monograph by Conway, Maxwell, and Miller (1967), for example, lists over 200 publications up until 1967. It is likely that considerably more than 200 papers have been published since that time. Much of the recent research in sequence scheduling has focused on stochastic scheduling. Weiss (1982) provides an interesting synopsis of work in this area. That jobs should be scheduled in order of decreasing expected processing time when scheduling is on two parallel processors and the objective is to minimize the expected makespan appears to have been discovered at about the same time by Bruno and Downey (1977) and Pinedo and Weiss (1979). The elegant method of proof we present is from Pinedo and Weiss (1979).

Much of the work on dynamic scheduling under uncertainty appears in the Conway, Maxwell, and Miller (1967) monograph. The $c\mu$ rule, apparently discovered by Derman et al. (1978) in an entirely different context, led to several extensions. See, for example, Righter and Shanthikumar (1989).

8.14 Summary

Scheduling problems arise in many contexts in managing operations. One of the areas explored in this chapter was optimal sequencing rules for scheduling jobs on machines, a problem that arises in *shop floor control*. We also considered dynamic scheduling problems, which are more typical of the kinds of scheduling problems that one encounters in management of service systems.

Much of the emphasis of the chapter was on determining efficient *sequencing rules*. The form of the optimal sequencing rule depends on several factors, including the pattern of arrivals of jobs, the configuration of the job shop, constraints, and the optimization objectives. Typical objectives in job shop management include meeting due dates, minimizing work-in-process, minimizing the average flow time, and minimizing machine idle time.

We discussed four sequencing rules: FCFS (first-come, first-served), SPT (shortest processing time first), EDD (earliest due date), and CR (critical ratio). For sequencing jobs on a single machine, we showed that *SPT* optimized several objectives, including mean flow time. However, SPT sequencing could be problematic. Long jobs would be constantly pushed to the rear of the job queue and might never be processed. For that reason, pure SPT scheduling is rarely used in practice. *Critical ratio scheduling* attempts to balance the importance placed on processing time and the remaining time until the due date. However, there is little evidence to suggest that critical ratio scheduling performs well relative to the common optimization criteria such as mean flow time. As one would expect, *earliest due date* scheduling performs best when the goal is to minimize the maximum tardiness.

We considered a variety of algorithms for *single-machine sequencing,* including Moore's (1968) algorithm for minimizing the number of tardy jobs and Lawler's (1973) algorithm for minimizing any nondecreasing function of the flow subject to precedence constraints. An excellent summary of algorithms for sequence scheduling can be found in French (1982). Several techniques for *multiple-machine scheduling* also were presented in this chapter. Johnson's (1954) classic work on scheduling *n jobs through two machines* is discussed, as well as Akers's (1956) graphical method for *two jobs through m machines.*

Stochastic scheduling problems were treated in two contexts: static and dynamic. Static problems are stochastic counterparts to the problems treated earlier in the chapter, except that job completion times are random variables. Most of the simple results for static stochastic scheduling problems require the assumption that job times have the exponential distribution. Because of the memoryless property of the exponential distribution, this assumption is probably not satisfied in most manufacturing job shops. The dynamic case occurs when jobs arrive randomly over time. *Queuing theory,* discussed in detail in Supplement 2, is the means for analyzing such problems.

The *assembly line balancing problem* is one in which a collection of tasks must be performed on each item. Furthermore, the tasks must be performed in a specified sequence. The problem is to allot the tasks to workstations on an assembly line. The quality of the balance is measured by the idle time remaining at each station. Determining the best mix of cycle time (amount of time allotted to each station) and number of stations is an extremely difficult analytical problem. We discuss one very simple heuristic solution method from Helgeson and Birnie (1961), known as the ranked positional weight technique, which provides an efficient balance quickly.

While most of this chapter concerns analytical methods for developing schedules, it is important to keep in mind that many real scheduling problems are too complex to be modeled mathematically. In such cases, *simulation* can be a very valuable tool. A computer-based simulation is a model of a real process expressed as a computer program. By running simulations under different scenarios, consequences of alternative strategies can be evaluated easily. Simulations are particularly effective when significant randomness or variation exists.

Job shop *scheduling software* is a growing business. Programs designed to run on PCs and workstations are available from several vendors. Most of these programs provide convenient interfaces to existing MRP systems. While the market for the post-MRP schedulers is much smaller than the market for full-blown integrated MRP II systems, it is growing and probably exceeds $100 million in the United States alone. These programs are designed to run in conjunction with an MRP system, and generally incorporate some combination of optimization and simulation. Many companies have been successful at implementing these programs on the factory floor.

There are a number of excellent texts on scheduling. For further reading in this area, we would suggest the books by Baker (1974), French (1982), and Conway, Maxwell, and Miller (1967). Pinedo (1995) provides the most up-to-date and comprehensive coverage.

Additional Problems on Scheduling

30. Mike's Auto Body Shop has five cars waiting to be repaired. The shop is quite small, so only one car can be repaired at a time. The number of days required to repair each car and the promised date for each are given in the table below.

Cars	Repair Time (days)	Promised Date
1	3	5
2	2	6
3	1	9
4	4	11
5	5	8

Mike has agreed to provide a rental car to each customer whose car is not repaired on time. Compare the performance of the four sequencing rules FCFS, SPT, EDD, and CR relative to minimizing average tardiness.

31. For each of the problems below, indicate precisely what or who would correspond to jobs and who or what would correspond to machines. In each case discuss what objectives might be appropriate and special priorities that might exist.
    a. Treating patients in a hospital emergency room.
    b. Unloading cargo from ships at port.
    c. Serving users on a time-shared computer system.
    d. Transferring long-distance phone calls from one city to another.

32. Six patients are waiting in a hospital emergency room to receive care for various complaints, ranging from a sprained ankle to a gunshot wound. The patients are numbered in the sequence that they arrived and are given a priority weighting

based on the severity of the problem. There is only one doctor available.

Patient	1	2	3	4	5	6
Time required	20 min.	2 hr.	30 min.	10 min.	40 min.	1 hr.
Priority	1	10	3	5	2	2

    a. Suppose that the patients are scheduled on an FCFS basis. Compute the mean flow time and the *weighted* mean flow time, where the weight is the priority number associated with each patient.

    b. Perform the calculations in part (*a*) for SPT sequencing.

    c. Determine a sequence different from those in (*a*) and (*b*) that achieves a lower value of the weighted mean flow time.

33. Consider Mike's Auto Body Shop mentioned in Problem 30. What sequencing of the jobs minimizes the

    a. Mean flow time?

    b. Maximum lateness?

    c. Number of tardy jobs?

34. Consider the situation of the emergency room mentioned in Problem 32. Determine the sequence in which patients should be treated in order to minimize the *weighted* value of the mean flow time.

35. Barbara and Jenny's Ice Cream Company produces four different flavors of ice cream: vanilla, chocolate, strawberry, and peanut fudge. Each batch of ice cream is produced in the same large vat, which must be cleaned prior to switching flavors. One day is required for the cleaning process.

    At the current time they have the following outstanding orders of ice cream:

Flavor	Order Size	Due Date
Vanilla	385 gallons	3
Chocolate	440 gallons	8
Strawberry	200 gallons	6
Peanut fudge	180 gallons	12

    It takes one day to produce the ice cream and a maximum of 100 gallons can be produced at one time. Cleaning is required only when flavors are switched. The production for one flavor will be completed prior to beginning production for another flavor. Cleaning is always started at the beginning of a day.

    Treating each ice cream flavor as a different job, find the following:

    a. The sequence in which the ice cream flavors should be produced in order to minimize the mean flow time for all of the flavors.

    b. The optimal sequence to produce the flavors in order to minimize the number of flavors that are late.

36. Consider Barbara and Jenny's Ice Cream Company, mentioned in Problem 35. Suppose that if vanilla or strawberry is produced after chocolate or peanut fudge, an extra day of cleaning is required. For that reason they decide that the vanilla and strawberry will be produced before the chocolate and peanut fudge.

    a. Find the optimal sequencing of the flavors to minimize the maximum lateness using Lawler's algorithm.

b. Enumerate all of the feasible sequences and determine the sequence that minimizes the maximum lateness by evaluating and comparing the objective function value for each case.

37. Irving Bonner, an independent computer programming consultant, has contracted to complete eight computer programming jobs. Some jobs must be completed in a certain sequence because they involve program modules that will be linked.

Job	Time Required	Due Date
1	4 days	June 8
2	10 days	June 15
3	2 days	June 10
4	1 day	June 12
5	8 days	July 1
6	3 days	July 6
7	2 days	June 25
8	6 days	June 29

Precedence restrictions:

$$1 \rightarrow 2 \rightarrow 5 \rightarrow 6.$$
$$4 \rightarrow 7 \rightarrow 8.$$

Assume that the current date is Monday, June 1, and that Bonner does not work on weekends. Using Lawler's algorithm, find the sequence in which he should be performing the jobs in order to minimize maximum lateness subject to the precedence constraints.

38. William Beebe owns a small shoe store. He has 10 pairs of shoes that require resoling and polishing. He has a machine that can resole one pair of shoes at a time, and the time required for the operation varies with the type and condition of the shoe and the type of sole that is used. Shoes are polished on a machine dedicated to this purpose as well, and polishing is always done after resoling. His assistant generally does the polishing while Mr. Beebe does the resoling. The resoling and polishing times (in minutes) are

Shoes	Resoling Time	Polishing Time
1	14	3
2	28	1
3	12	2
4	6	5
5	10	10
6	14	6
7	4	12
8	25	8
9	15	5
10	10	5

In what order should the shoes be repaired in order to minimize the total makespan for these 10 jobs?

39. A leatherworks factory has two punch presses for punching holes in the various leather goods produced in the factory prior to sewing. Suppose that 12 different jobs must be processed on one or the other punch press (i.e., parallel processing). The processing times (in minutes) for these 12 jobs are given below.

Job	1	2	3	4	5	6	7	8	9	10	11	12
Time	26	12	8	42	35	30	29	21	25	15	4	75

Assume that both presses are initially idle. Compare the performance of SPT and LPT (longest processing time) rules for this example. (In parallel processing the next job is simply scheduled on the next available machine.)

40. An independent accountant is planning to prepare tax returns for six of his clients. Prior to his actually preparing each return, his secretary checks the client's file to be sure all the necessary documentation is there and obtains all of the tax forms needed for the preparation of the return. Based on past experience with the clients, she estimates that the following times (in hours) are required for her preparation and for the accountant to complete the necessary paperwork prior to filing each return:

Client	Secretary Time	Accountant Time
1	1.2	2.5
2	1.6	4.5
3	2.0	2.0
4	1.5	6.0
5	3.1	5.0
6	0.5	1.5

In what order should the work be completed in order to minimize the total time required for all six clients?

41. Five Hong Kong tailors—Simon, Pat, Choon, Paul, and Wu—must complete alterations on a suit for the duke and a dress for the duchess as quickly as possible. On the dress, Choon must first spend 45 minutes cutting the fabric, then Pat will spend 75 minutes sewing the bodice, Simon will need 30 minutes stitching the sleeves, Paul 2 hours lacing the hem, and finally Wu will need 80 minutes for finishing touches. As far as the suit is concerned, Pat begins with shortening the sleeves, which requires 100 minutes. He is followed by Paul, who sews in the lining in 1.75 hours; Wu, who spends 90 minutes sewing on the buttons and narrowing the lapels; and finally Choon, who presses and cleans the suit in 30 minutes.

Determine precisely when each tailor should be performing each task in order to minimize the total time required to complete the dress and the suit. Assume that the tailors start working at 9 A.M. and take no breaks. Draw the Gantt chart indicating your solution.

42. The assembly of a transistorized clock radio requires a total of 11 tasks. The task times and predecessor relationships are given in the following table:

Task	Time (seconds)	Immediate Predecessors
1	4	—
2	38	—
3	45	—
4	12	1, 2
5	10	2
6	8	4
7	12	5
8	10	6
9	2	7
10	10	8, 9
11	34	3, 10

a. Develop a network for this assembly operation.

b. What is the minimum cycle time that could be considered for this operation? What is the minimum number of stations that could be used with this cycle time?

c. Using the ranked positional weight technique, determine the resulting balance using a cycle time of 45 seconds.

d. Determine by experimentation the minimum cycle time that results in a four-station balance.

e. What is the daily production rate for this product if the company adopts the balance you determined in part (c)? (Assume a six-hour day for your calculations.) What would have to be done if the company wanted a higher rate of production?

43. Suppose in Problem 42 that additional constraints arise from the fact that certain tasks cannot be performed at the same station. In particular suppose that the tasks are zoned in the following manner:

Zone 1	Tasks 2, 3, 1, 4, 6
Zone 2	Tasks 5, 8, 7, 9
Zone 3	Tasks 10, 11

Assuming that only tasks in the same zone category can be performed at the same station, determine the resulting line balance for Problem 42 based on a 45-second cycle time.

44. The Southeastern Sports Company produces golf clubs on an assembly line in its plant in Marietta, Georgia. The final assembly of the woods requires the eight

operations below.

Task	Time Required (min.)	Immediate Predecessors
1. Polish shaft	12	—
2. Grind the shaft end	14	—
3. Polish club head	6	—
4. Imprint number	4	3
5. Connect wood to shaft	6	1, 2, 4
6. Place and secure connecting pin	3	5
7. Place glue on other end of shaft	3	1
8. Set in grips and balance	12	6, 7

    a. Draw a network to represent the assembly operation.

    b. What is the minimum cycle time that can be considered? Determine the balance that results from the ranked positional weight technique for this cycle time.

    c. By experimentation, determine the minimum cycle time that can be achieved with a three-station balance.

45. Develop a template that computes several measures of performance for first-come, first-served job sequencing. Allow for up to 20 jobs so that column 1 holds the numbers 1, 2, . . . , 20. Column 2 should be the processing times to be inputted by the user, and column 3 the due dates also inputted by the user. Column 4 should be the tardiness and column 5 the flow time. Develop the logic to compute the mean flow time, the average tardiness, and the number of tardy jobs. (When computing the average of a column, be sure that your spreadsheet does not treat blanks as zeros.)

    a. Use your template to find the mean flow time, average tardiness, and number of tardy jobs for Problem 7 (Section 6), assuming FCFS sequencing.

    b. Find the mean flow time, average tardiness, and number of tardy jobs for Problem 8, assuming FCFS sequencing.

    c. Find the mean flow time, average tardiness, and number of tardy jobs assuming FCFS sequencing for the following 20-job problem:

Job	Processing Time	Due Date	Job	Processing Time	Due Date
1	10	34	11	17	140
2	24	38	12	8	120
3	16	60	13	23	110
4	8	52	14	25	160
5	14	25	15	40	180
6	19	95	16	19	140
7	26	92	17	6	130
8	24	61	18	23	190
9	4	42	19	25	220
10	12	170	20	14	110

46. *a.* Solve Problem 45(*a*) assuming SPT sequencing. To do this you may use the spreadsheet developed in Problem 45 and simply sort the data in the first three columns, using column 2 (the processing time) as a sort key.
    b. Solve Problem 45(*b*) assuming SPT sequencing.
    c. Solve Problem 45(*c*) assuming SPT sequencing.

47. *a.* Solve Problem 45(*a*) assuming EDD sequencing. In this case one sorts on the due date column.
    b. Solve Problem 45(*b*) assuming EDD sequencing.
    c. Solve Problem 45(*c*) assuming EDD sequencing.

Bibliography

Akers, S. B. "A Graphical Approach to Production Scheduling Problems." *Operations Research* 4 (1956), pp. 244–45.

Arcus, A. L. "COMSOAL: A Computer Method for Sequencing Operations for Assembly Lines." *International Journal of Production Research* 4 (1966), pp. 259–77.

Baker, K. R. *Introduction to Sequencing and Scheduling.* New York: John Wiley & Sons, 1974.

Banerjee, B. P. "Single Facility Sequencing with Random Execution Times." *Operations Research* 13 (1965), pp. 358–64.

Bellman, R. E. "Mathematical Aspects of Scheduling Theory." *SIAM Journal of Applied Mathematics* 4 (1956), pp. 168–205.

Bruno, J., and P. Downey. "Sequencing Tasks with Exponential Service Times on Two Machines." Technical Report, Department of Electrical Engineering and Computer Science, University of California at Santa Barbara, 1977.

Carlson, R., and M. Rosenblatt. "Designing a Production Line to Maximize Profit." *IIE Transactions* 17 (1985), pp. 117–22.

Conway, R. W.; W. L. Maxwell; and L. W. Miller. *Theory of Scheduling.* Reading, MA: Addison Wesley, 1967.

Derman, C.; G. L. Lieberman; and S. M. Ross. "A Renewal Decision Problem." *Management Science* 24 (1978), pp. 554–61.

Eilon, S.; C. D. T. Watson-Gandy; and N. Christofides. *Distribution Management: Mathematical Modeling and Practical Analysis.* London: Griffin, 1971.

Fisher, M.; A. J. Greenfield; R. Jaikumar; and J. T. Uster III. "A Computerized Vehicle Routing Application." *Interfaces* 12 (1982), pp. 42–52.

Fishman, G. S. *Concepts and Methods in Discrete Event Digital Simulation.* New York: John Wiley & Sons, 1973.

French, S. *Sequencing and Scheduling: An Introduction to the Mathematics of the Job Shop.* Chichester, England: Ellis Horwood Limited, 1982.

Graves, S. C. "A Review of Production Scheduling." *Operations Research* 29 (1981), pp. 646–75.

Held, M.; R. M. Karp; and R. Shareshian. "Assembly Line Balancing—Dynamic Programming with Precedence Constraints." *Operations Research* 11 (1963), pp. 442–59.

Helgeson, W. P., and D. P. Birnie. "Assembly Line Balancing Using the Ranked Positional Weight Technique." *Journal of Industrial Engineering* 12 (1961), pp. 394–98.

Hillier, F. S., and R. W. Boling. "On the Optimal Allocation of Work in Symmetrically Unbalanced Production Line Systems with Variable Operation Times." *Management Science* 25 (1986), pp. 721–28.

Hutchison, J.; G. Leong; and P. T. Ward. "Improving Delivery Performance in Gear Manufacturing at Jeffrey Division of Dresser Industries." *Interfaces* 23, no. 2 (March–April 1993), pp. 69–83.

Johnson, S. M. "Optimal Two and Three Stage Production Schedules with Setup Times Included." *Naval Research Logistics Quarterly* 1 (1954), pp. 61–68.

Kilbridge, M. D., and L. Wester. "A Heuristic Method of Line Balancing." *Journal of Industrial Engineering* 12 (1961), pp. 292–98.

Lawler, E. L. "Optimal Sequencing of a Single Machine Subject to Precedence Constraints." *Management Science* 19 (1973), pp. 544–46.

Melnyk, S. A.; P. L. Carter; D. M. Dilts; and D. M. Lyth. *Shop Floor Control*. Homewood IL: Dow Jones–Irwin, 1985.

Moore, J. M. "An *n*-job, One Machine Sequencing Algorithm for Minimizing the Number of Late Jobs." *Management Science* 15 (1968), pp. 102–109.

Nichols, J. C. "Planning for Real World Production." *Production* 106, no. 8 (August 1994), pp. 18–20.

Parker, K. "What New Tools Will Best Tame Time." *Manufacturing Systems* 12, no. 1 (January 1994), pp. 16–22.

Parker, K. "Dynamism and Decision Support." *Manufacturing Systems* 13, no. 4 (April 1995), pp. 12–24.

Pinedo, M. "Stochastic Scheduling with Release Dates and Due Dates." *Operations Research* 31 (1983), pp. 554–72.

Pinedo, M. *Scheduling, Theory, Algorithms and Systems*. Englewood Cliffs, NJ: Prentice Hall, 1995.

Pinedo, M., and G. Weiss. "Scheduling Stochastic Tasks on Two Parallel Processors." *Naval Research Logistics Quarterly* 26 (1979), pp. 527–36.

Randhawa, S. U., and R. Shroff. "Simulation-Based Design Evaluation of Unit Load Automated Storage/Retrieval Systems." *Computers and Industrial Engineering* 28, no. 1 (January 1995), pp. 71–79.

Righter, R. "Job Scheduling to Minimize Weighted Flowtime on Uniform Processors." *Systems and Control Letters* 10 (1988), pp. 211–16.

Righter, R., and J. G. Shanthikumar. "Scheduling Multiclass Single-Server Queuing Systems to Stochastically Maximize the Number of Successful Departures." *Probability in the Engineering and Informational Sciences* 3 (1989), pp. 323–33.

Rothkopf, M. S. "Scheduling with Random Service Times." *Management Science* 12 (1966), pp. 707–13.

Salveson, M. E. "Production Planning and Scheduling." *Econometrica* 20 (1952), pp. 554–90.

Schrage, L. "Formulation and Structure of More Complex/Realistic Routing and Scheduling Problems." *Networks* 11 (1981), pp. 229–32.

Swain, J. J. "Flexible Tools for Modeling." *OR/MS Today* 20, no. 6 (1993), pp. 62–78.

Thangavelu, S. R., and C. M. Shetty. "Assembly Line Balancing by Zero One Integer Programming." *AIIE Transactions* 3 (1971), pp. 61–68.

Vasilash, G. "Scheduling for the Shop Floor." *Production* 107, no. 6 (June 1995), pp. 46–47.

Vollman, T. E.; W. L. Berry; and D. C. Whybark. *Manufacturing Planning and Control Systems*. 3rd ed. Homewood, IL: Richard D. Irwin, 1992.

Weiss, G. "Multiserver Stochastic Scheduling." In *Deterministic and Stochastic Scheduling*. Proceedings of the NATO Advanced Study and Research Institute on Theoretical Approaches to Scheduling Problems. Norwell, MA: D. Reidel, 1982.

QUEUING THEORY

1 Introduction

Queuing theory is the study of waiting line processes. A *queue* is the British version of the American waiting line. Virtually all the results in queuing theory assume that both the arrival and the service processes are random. It is the interaction between these two processes that makes queuing an interesting and challenging area. In Figure S2–1 we show a typical queuing system. Customers arrive at one or more service facilities. If other customers are already waiting for service, depending on the service discipline, newly arriving customers would wait their turn for the next available server and would exit the system when service is completed.

Queuing problems are common in operations management. In the context of manufacturing, the job shop environment can be thought of as a complex interrelated network of queues. As jobs are completed at a work center, they queue up for processing at the next work center. When scheduling in a job shop environment is viewed as a dynamic rather than a static problem, flow rates and other measures of performance should be found by using the results from queuing theory.

Queuing problems occur most frequently in service systems. Each of us experiences queues in service systems almost every day. In fact, according to Michael Fortino, of Priority Management in Pittsburgh, most people spend about five years of their lives waiting in lines and six months sitting at traffic lights (*Chicago Tribune*, June 21, 1988).

We queue up to wait our turn at banks, supermarkets, hair stylists, highway tollbooths, and restaurants. Scheduling the landings of planes on a particular runway is a queuing problem: think of the planes in the air as customers waiting to be served and the runway as the server. Telephone traffic systems are examples of complex queuing networks. Telephone calls are routed through switching systems, where they queue up until they are either switched to the next switching station or routed to their final destination. In fact, it was A. K. Erlang, a Danish telephone engineer, who was responsible for many of the early theoretical developments in the area of queuing.

There are relatively simple formulas available for most important measures of performance for a certain class of queues. For this class, both the service and the arrival patterns are completely random (what "completely random" means in this context will be discussed later). However, when the system exhibits behavior that is not completely random, useful results are harder to find. For that reason, Monte Carlo simulation is a common tool for analyzing complex queuing problems.

FIGURE S2–1

Typical single-server queuing system

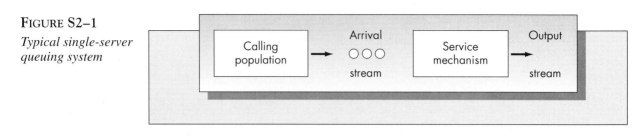

---

2 Structural Aspects of Queuing Models

1. *Arrival process.* This is the process describing arrivals of customers to the system. The arrival process is characterized by the distribution of interarrival times. The simplest case is when arrivals occur one at a time, completely at random. This case gives rise to an exponential distribution for interarrival times. Other possibilities include general interarrival distributions or more than one arrival at a time (known in queuing as batch arrivals).

2. *Service process.* The service process is characterized by the distribution of the time required to serve a customer. Again, the easiest case to analyze is when the distribution of service times is exponential. Other service time distributions give rise to more complex queuing models.

3. *Service discipline.* This is the rule by which customers in the queue are served. Most queuing problems occurring in service systems are FCFS (first-come, first-served). This is the rule we usually think of as "fair." However, other service disciplines are also common. When we buy milk, we may check the dates of the cartons and buy the one with the latest expiration date. Thinking of the milk as the queue, this means that the service discipline is LCFS (last-come, first-served). Hospital emergency rooms will give priority to patients with a life-threatening condition, such as trauma from an automobile accident, over patients with less severe problems. This would be referred to as a priority service discipline.

4. *Capacity of the queue.* In some cases, the size of the queue might be limited. For example, restaurants and movie theaters can accommodate only a limited number of customers. From a mathematical point of view, the simplest assumption is that the queue size is unlimited. Even where there is a finite capacity, it is reasonable to ignore the capacity constraint if the queue is unlikely to fill.

5. *Number of servers.* Queues may be either single-server or multiserver. A bank is the most common example of a multiserver queue. Customers form a single line and are served by the next available server. By contrast, the checkout area of a typical supermarket is *not* a multiserver queue. Because a shopper must commit to a specific checkout line, this is a parallel system of (possibly dependent) single-server queues. Another example of a multiserver queue is the landing area of the airport; planes may take off or land on one of several runways.

6. *Network structure.* A network of queues results when the output of one queue forms the input of another queue. Most manufacturing processes are generally some form of a queuing network. Highway systems and telephone switching systems are other examples. Network queuing structures are often too complex to analyze mathematically.

3 Notation

Summary.

$N(t)$ = Number of customers in the system at time t.
$P_n(t) = P\{N(t) = n \mid N(0) = 0\}$.
P_n = Steady-state probability of n customers in the system.
λ_n = Arrival rate when there are n customers in the system.
μ_n = Service rate when there are n customers in the system.
c = Number of servers.
ρ = Utilization rate.
L = Expected number of customers in the system in steady state.
L_q = Expected number of customers in the queue in steady state.
W = Expected waiting time of a customer in steady state.
W_q = Expected waiting time in the queue in steady state.
K = Maximum number in the system for a finite capacity system.

Discussion. For the cases we consider, $N(t)$ is the state of the system at time t. This means that, given knowledge of the current value of $N(t)$, the likelihood of future values of $N(t)$ is independent of past values (this is known as the Markovian property). For the systems that we will examine, defining $N(t)$ as the total number of customers in the system at time t will guarantee that $N(t)$ has the Markovian property and qualifies as the definition of the state of the system. The number of customers in the system includes both the number of customers waiting for service (that is, in the queue) *and* the number being served. We let n be a particular fixed value of $N(t)$.

$P_n(t)$ is the *transient* probability that there are n customers in the system at time t, given that the system is empty at time 0. For our purposes, we are interested only in the steady state or stationary probability of n customers in the system, P_n, given by

$$P_n = \lim_{t \to \infty} P_n(t).$$

Let c be the number of servers. The case $c > 1$ corresponds to multiple servers in parallel. Most of our applications will be for the case $c = 1$.

The arrival rate, λ_n, is the expected or average rate of customers arriving to the system expressed as the number of arrivals per unit time. Notice that allowing the arrival rate to depend on n means that we are able to consider cases in which customers might choose not to join the queue based on its size. Usually, we assume $\lambda_n = \lambda$ independent of n (that is, the arrival rate is not affected by the number of customers in the system).

Similarly, μ_n, the mean service rate when there are n customers in the system, can depend on n as well. In this case, including the dependence on n allows us to consider situations in which the server might speed up or slow down depending on the number of customers waiting to be served.

When λ and μ are independent of n, we define the utilization rate of the system, ρ, by the formula

$$\rho = \lambda/c\mu.$$

The utilization rate is interpreted as the proportion of time that each server is busy or the expected number of customers in service. The utilization rate must be less than 1 to ensure that the size of the queue does not grow without limit.

4 Little's Formula

In this section we show some useful relationships between the steady-state expected values L, L_q, and W, and W_q. Because W_q is the expected time in the queue only, whereas W is the expected time in the queue plus the expected time in service, it follows that W and W_q differ by the expected time in service. That is,

$$W = W_q + 1/\mu.$$

(If the mean service rate is μ, it follows that the mean service time is $1/\mu$.)

Little's formula is named for John D. C. Little of MIT, who proved that it holds under very general circumstances. It is a simple but very useful relationship between the L's and the W's. Assume that the arrival and the service rates are constant; that is, $\lambda_n = \lambda$ and $\mu_n = \mu$ for all values of n. The basic result is

$$L = \lambda W.$$

We will not present a formal proof of this result, but provide only the following intuitive explanation. Consider a customer who joins the queue in steady state. At the instant the customer is about to complete service, he[1] looks over his shoulder at the customers who have arrived behind him. There will be, on average, L customers in the system. The expected amount of time that has elapsed since he joined the queue is, by definition, W. Because customers arrive at a constant rate λ, it follows that during a time W, on average, there will have been λW arrivals, giving $L = \lambda W$. For example, if customers arrive at the rate of 2 per minute and each spends an average of 5 minutes in the system, this formula tells us that there will be 10 customers in the system on average.

Another version of Little's formula is

$$L_q = \lambda W_q.$$

The argument here is essentially the same, except that the customer looks over his shoulder as he enters service, rather than when completing service.

5 The Exponential and Poisson Distributions in Queuing

In Chapter 12 on reliability modeling, we discuss the memoryless property of the exponential distribution and its relationship to the Poisson process and the Poisson distribution. Both the exponential and the Poisson distributions play a key role in queuing theory, just as they do in reliability theory. When we talk about purely random arrivals in queuing, we mean that the arrival process is a Poisson process. A purely random service process means that service times have the exponential distribution. We recommend that the reader review Sections 1 and 3 of Chapter 12, which discuss the exponential distribution and the Poisson process, before proceeding with this supplement.

If arrivals follow a Poisson process, it means that interarrival times have the exponential distribution. Because of the memoryless property of the exponential distribution, we refer to the Poisson process as a purely random arrival process. If T is a

---

[1]We use the masculine pronoun for convenience. Please read this and similar references as he or she.

ASIDE

random variable representing the time between successive arrivals, then

$$P\{T > t\} = \exp(-\lambda t).$$

Furthermore, the number of arrivals in any time t, say $A(t)$, has the Poisson distribution with parameter λt.[2] That is,

$$P\{A(t) = n\} = \exp(-\lambda t)(\lambda t)^n/n! \qquad \text{for } n = 0, 1, 2, \ldots$$

The exponential distribution has a property related to the memoryless property that is particularly useful in queuing analysis. It has to do with what are known as forward and backward recurrence times. Let $N(t)$ be a Poisson process with rate λ, and T_1, $T_2, \ldots$ be successive interarrival times. Consider some deterministic time t that falls between two successive interarrival times, say T_{i-1} and T_i. The forward recurrence time is the random variable $T_i - t$, or the time that elapses from t until the next arrival. The exponential distribution is the only distribution that has the property that the distribution of the forward recurrence time also has the exponential distribution with rate λ *independent of t*. In queuing, this means that if a server is busy when a customer arrives, the amount of time that elapses until the completion of service is still exponential with rate μ. For example, suppose that cabs arrive at a cab stand according to a Poisson process. You arrive at the stand at some random time and wait for the next cab. Your waiting time is exponential, with exactly the same distribution as the time between two successive arrivals of cabs. This leads to an apparent paradox (see the aside above).

These properties of the Poisson process and the exponential distribution allow for relatively straightforward analysis of queuing systems. When either the arrival process or the service process is not purely random (i.e., exponential), the mathematics becomes

---

[2]In Chapter 12 we use $N(t)$ for the Poisson process. However, $N(t)$ is the state of the queuing system, so we use $A(t)$ for the Poisson process here.

much more complex and the results available become more sparse. We discuss the implication of these issues to the application of queuing in real life later in this supplement.

An important question is, how realistic are the assumptions of Poisson arrivals and exponential service? In practice, when the customer population is large and arrivals occur one at a time, Poisson arrivals are a reasonable assumption. However, assuming exponentially distributed service times is more questionable. We would expect under most circumstances that the more time someone has spent in service, the more likely it is that the person will complete service soon. We will discuss general service distributions at the end of the supplement.

6 Birth and Death Analysis for the M/M/1 Queue

A shorthand notation for queuing problems is of the form

Label 1/Label 2/Number,

where Label 1 is an abbreviation for the arrival process, Label 2 is an abbreviation for the service process, and Number indicates the number of servers.[3] The letter M is used to denote pure random arrivals or pure random service. This means that interarrival times are exponential or service times are exponential. The M stands for Markovian, a reference to the memoryless property of the exponential distribution. The simplest queuing problem is the one labeled M/M/1. Another symbol we consider is G, for general distribution. Hence G/G/s would correspond to a queuing problem in which the interarrival distribution was general, the service distribution was general, and there were s servers. There are other labels for other distributions, but we do not consider them here. The remainder of this section assumes an M/M/1 queue.

The process $A(t)$, the number of arrivals up until time t, is a pure birth process. It increases by one at each arrival. The process $N(t)$ is known as a birth and death process because it both increases and decreases. It increases by one at each arrival and decreases by one at each completion of a service. A realization of $N(t)$ is shown in Figure S2–2. Notice that the state of the system either increases by one or decreases by one. The intensity or rate at which the system state increases is λ and the intensity at which the system state decreases is μ.[4] This means that we can represent the rate at which the system changes state by the diagram in Figure S2–3.

Let us suppose that the system has evolved to a steady-state condition. That means that the state of the system is independent of the starting state. Because we are in steady state, we consider only the stationary probabilities P_n. The following derivation is based on the Balance Principle:

> **Balance Principle:** In the steady state, the rate of entry into a state must equal the rate of entry out of a state if a steady state probability distribution exists.

Consider the application of the Balance Principle to state 0. We enter state 0 only from state 1. Given that we are in state 1, we move from state 1 to state 0 at a rate μ (see Figure S2–3). The probability of being in state 1 is P_1. It follows that the rate we move

---

[3]More complex notations exist that include capacity restrictions and specification of the queuing discipline. See, for example, Gross and Harris (1985), p. 9.

[4]At this point we consider only the case in which the arrival and the service rates are fixed and independent of the state of the system. The extension to the more general case will be considered below.

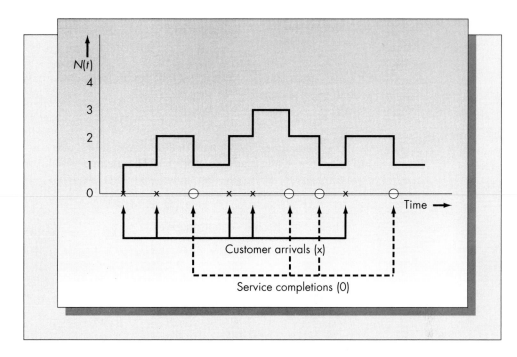

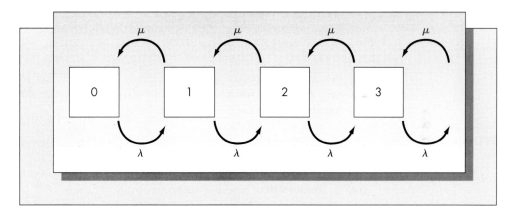

into state 0 is μP_1. Consider the rate at which we move out of state 0. When we are in state 0, we can only move to state 1, which we do (when a customer arrives) at rate λ. As the probability of being in state 0 is P_0, it follows that the overall rate at which we move out of state 0 is λP_0. From this we obtain our first balance equation:

$$\mu P_1 = \lambda P_0.$$

Consider state 1. From Figure S2–3 we see that we can enter state 1 in two ways: from state 0 or from state 2. Given that we are in state 0, we enter state 1 at rate λ, and given that we are in state 2, we enter state 1 at rate μ. If follows that the rate at which we enter state 1 is $\lambda P_0 + \mu P_2$. We can leave state 1 by going either to state 0 if an arrival occurs or state 2 if a service completion occurs. Hence, the rate at which we leave state 1 is $\lambda P_1 + \mu P_1 = (\lambda + \mu)P_1$. It follows that the second balance equation is

$$\mu P_2 + \lambda P_0 = (\lambda + \mu)P_1.$$

The form of the remaining balance equations is essentially the same as that of the second balance equation. In general,

$$\mu P_{i+1} + \lambda P_{i-1} = (\lambda + \mu)P_i \qquad \text{for } 1 \leq i \leq \infty.$$

These equations, along with one other condition, allow us to obtain an explicit solution for the steady-state probabilities. The method of solution is to first express each P_i in terms of P_0. From the first balance equation we have

$$P_1 = (\lambda/\mu)P_0.$$

The second balance equation gives

$$\mu P_2 = (\lambda + \mu)P_1 - \lambda P_0 = (\lambda + \mu)(\lambda/\mu)P_0 - \lambda P_0$$
$$= (\lambda^2/\mu)P_0 + \lambda P_0 - \lambda P_0 = (\lambda^2/\mu)P_0.$$

Dividing both sides by μ gives

$$P_2 = (\lambda/\mu)^2 P_0.$$

Similarly, we will find in general that

$$P_i = (\lambda/\mu)^i P_0.$$

The solution is obtained by using the condition that

$$\sum_{i=0}^{\infty} P_i = 1,$$

since $P_0, P_1, P_2, \ldots$ forms a probability distribution on the states of the system.
Substituting for each P_i, we have

$$\sum_{i=0}^{\infty} (\lambda/\mu)^i P_0 = 1.$$

Let $\rho = \lambda/\mu$ be the utilization rate. For a solution to exist, it must be true that $\rho < 1$. In that case

$$\sum_{i=0}^{\infty} \rho^i = 1/(1 - \rho),$$

known as the geometric series, from which we obtain

$$P_0 = (1 - \rho)$$

and

$$P_i = \rho^i(1 - \rho) \qquad \text{for } i = 1, 2, 3, \ldots$$

(This formula is also valid when $i = 0$.)

This distribution, known as the geometric distribution, is pictured in Figure S2–4. There are several aspects to this result that are both interesting and surprising. First, the geometric distribution is the discrete analog of the exponential distribution. (The exponential is the distribution of both interarrival times and service times.) Second, the probability of state i is a decreasing function of i, as pictured in Figure S2–4, as long

FIGURE S2–4

*Geometric
distribution of
number in system
for M/M/1 queue
($\rho = .8$)*

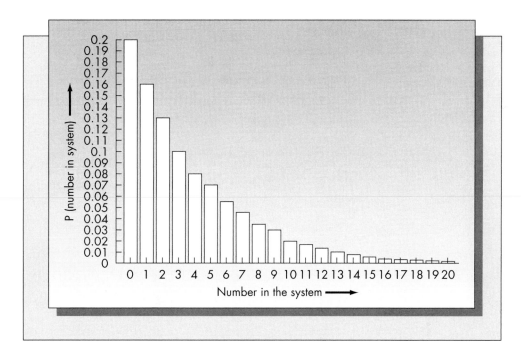

as $\rho < 1$. As ρ gets close to one, the variance increases and the distribution "spreads out" (large values become more likely), and as ρ gets close to zero the probabilities associated with larger values drop off to zero more quickly. This means that the most likely state is *always* state 0 (as long as $\rho < 1$)! This is an extremely surprising result. As ρ approaches one, the queues get longer and longer. One would have thought that the probability of n in the system for some large value of n would be higher than the probability of zero in the system when ρ is near one. This turns out not to be the case. What is true is that for ρ close to one, the probability that the system is in state zero is close to the probability that the system is in state 1 or state 2, for example, whereas for ρ close to zero the probability that the system is in state zero is much larger than the probability that it is in state 1 or 2. This phenomenon holds *only* for exponential services and exponential arrivals, however.

7 Calculation of the Expected System Measures for the M/M/1 Queue

In this section we calculate L, L_q, W, and W_q. The expected value of a random variable is the sum of its outcomes weighted by the probabilities of those outcomes. It follows that the average number of customers in the system in steady state, L, is

$$L = \sum_{i=0}^{\infty} i P_i = \sum_{i=0}^{\infty} i(1 - \rho)\rho^i = (1 - \rho)p \sum_{i=0}^{\infty} i\rho^{i-1}.$$

To complete the calculation, we use the fact that

$$\sum_{i=0}^{\infty} i\rho^{i-1} = \frac{d}{d\rho}\left(\sum_{i=0}^{\infty} \rho^i\right) = \frac{d}{d\rho}(1/(1 - \rho)) = 1/(1 - \rho)^2.$$

It follows that

$$L = \frac{(1 - \rho)\rho}{(1 - \rho)^2} = \rho/(1 - \rho).$$

For the case of L_q, we note that the number in the queue is exactly one less than the number in the system as long as there is at least one in the system. It follows that

$$L_q = \sum_{i=1}^{\infty}(i - 1)P_i = \sum_{i=1}^{\infty}iP_i - \sum_{i=1}^{\infty}P_i$$
$$= L - (1 - P_0) = L - \rho = \rho^2/(1 - \rho).$$

Given knowledge of L and L_q, we can obtain W and W_q directly from Little's formula. From Little's formula $W = L/\lambda$, giving

$$W = \rho/(\lambda(1 - \rho)).$$

Similarly, $W_q = L_q/\lambda$, which gives

$$W_q = \rho^2/(\lambda(1 - \rho)).$$

8 The Waiting Time Distribution

In this section, we derive the distribution of the waiting time W for a random customer joining the queue in the steady state. For convenience, we use the symbol W in this section to refer to the random variable, rather than its expectation. The appropriate interpretation of W will be clear from the context.

To derive the distribution of W we condition on the number of customers in the system at steady state, n, and uncondition by multiplying by the probability P_n. Suppose that a customer joining the queue at a random point in time finds n customers already in the system. Then that customer must wait for n service completions before entering service himself. As W is the total time in the system, this means that in this case W will be the sum of $n + 1$ service completions. Let $S_1, S_2, \ldots$ be the times of the successive services. By assumption, these random variables are mutually independent and exponentially distributed with common mean μ. The time for $n + 1$ service completions is $S_1 + S_2 + \cdots + S_{n+1}$, which we know has the Erlang distribution with parameters μ and $n + 1$ (see Section 3 in Chapter 12). That is,

$$P\{W > t \,|\, n \text{ in the system}\} = \sum_{k=0}^{n} \frac{e^{-\mu t}(\mu t)^k}{k!}.$$

We know from the previous section that the unconditional probability of n in the system in the steady state, P_n, has the geometric distribution. Substituting $\rho = \lambda/\mu$, we may write P_n in the form

$$P_n = \left(\frac{\mu - \lambda}{\mu}\right)\left(\frac{\lambda}{\mu}\right)^n.$$

Unconditioning on P_n gives

$$P\{W > t\} = \sum_{n=0}^{\infty} \sum_{k=0}^{n} \frac{e^{-\mu t}(\mu t)^k}{k!} \left(\frac{\mu - \lambda}{\mu} \right) \left(\frac{\lambda}{\mu} \right)^n$$

$$= \sum_{k=0}^{\infty} \sum_{n=k}^{\infty} \frac{e^{-\mu t}(\mu t)^k}{k!} \left(\frac{\mu - \lambda}{\mu} \right) \left(\frac{\lambda}{\mu} \right)^n$$

$$= \frac{\mu - \lambda}{\mu} e^{-\mu t} \sum_{k=0}^{\infty} \frac{(\mu t)^k}{k!} \sum_{n=k}^{\infty} \left(\frac{\lambda}{\mu} \right)^n.$$

Using the fact that

$$\sum_{n=k}^{\infty} \left(\frac{\lambda}{\mu} \right)^n = \left(\frac{\lambda}{\mu} \right)^k \frac{1}{1 - \lambda/\mu},$$

and substituting this into the above gives, after simplifying,

$$P\{W > t\} = e^{-\mu t} \sum_{k=0}^{\infty} \frac{(\lambda t)^k}{k!} = e^{-\mu t} e^{+\lambda t} = e^{-(\mu - \lambda)t}.$$

The summation term equals $e^{+\lambda t}$ because it is the Taylor series expansion for e (and because Poisson probabilities sum to one). What we have shown is the surprising result that W has the negative exponential distribution with parameter $\mu - \lambda$. This implies that W has the memoryless property. That is, suppose that a customer has already been waiting for s units of time. The probability that he will have to wait at least an additional t units of time is the same as the probability that a newly joining customer waits at least t units of time. This result is not intuitive and is rather depressing for the poor customer, who has already spent a substantial amount of time waiting for service!

We will not present the derivation (it is similar to the one above), but state that the distribution of W_q is essentially exponential with the complementary cumulative distribution function

$$P\{W_q > t\} = \rho e^{-(\mu - \lambda)t} \qquad \text{for all } t \geq 0.$$

Note that the probability that the waiting time in the queue is zero is positive. It is equal to the probability that the system is empty, P_0. That is,

$$P\{W_q = 0\} = P_0 = 1 - \rho.$$

Example S2.1

Customers arrive one at a time, completely at random, at an ATM at the rate of 6 per hour. Customers take an average of four minutes to complete their transactions, and historical data have shown that the service times closely follow the negative exponential distribution. Customers queue up on a first-come, first-served basis. Assume that there is only one ATM.

    a. Find the following expected measures of performance for this system: the expected number of customers in the system, the expected number of customers waiting for service, the expected time in the system, and the expected time in the queue.

b. What is the probability that there are more than five people in the system at a random point in time?

c. What is the probability that the waiting time in the queue exceeds 10 minutes?

d. Given these results, do you think that management should consider adding another ATM?

Solution

The statement that customers arrive one at a time completely at random implies that the input process is a Poisson process. The arrival rate is $\lambda = 6$ per hour. The mean service time is 4 minutes $= 1/15$ hour, so that the service rate is $\mu = 15$ per hour. The utilization rate is $\rho = \lambda/\mu = 6/15 = 2/5 = 0.4$.

a. $L = \rho/(1-\rho) = (2/5)/(3/5) = 2/3 \ (= 0.6667)$.
$L_q = \rho L = (2/5)(2/3) = 4/15 \ (= 0.2667)$.
$W = L/\lambda = (2/3)/6 = 2/18 = 1/9$ hour (6.6667 minutes).
$W_q = L_q/\lambda = (4/15)/6 = 4/90 = 2/45$ hour (2.6667 minutes).

b. Here we are interested in $P\{L > 5\}$.
In general,

$$P\{L > k\} = \sum_{n=k+1}^{\infty} P_n = \sum_{n=k+1}^{\infty} (1-\rho)\rho^n = (1-\rho) \sum_{n=k+1}^{\infty} \rho^n$$
$$= (1-\rho)\rho^{k+1}(1/(1-\rho)) = \rho^{k+1}.$$

Hence, $P\{L > 5\} = \rho^6 = (0.4)^6 = 0.0041$.

c. Here we are interested in $P\{W_1 > 1/6\}$.

$$P\{W_q > t\} = \rho e^{-(\mu-\lambda)t} = 0.4e^{-(15-6)(1/6)} = 0.4e^{-1.5} = 0.0892.$$

d. The answer is not obvious. Looking at the expected measures of performance, it would appear that the service provided is reasonable. The expected number of customers in the system is fewer than 1 and the average waiting time in the queue is less than 3 minutes. However, from part (c) we see that the proportion of customers having to wait more than 10 minutes for service is almost 10 percent. This means that there are probably plenty of irate customers, even though on average the system looks good. This illustrates a pitfall of only considering expected values when evaluating queuing service systems. ■

9 Solution of the General Case

In this section we consider a solution of the general Markovian queue; that is, when the times between successive arrivals are exponentially distributed and the service times are also exponentially distributed. However, we will allow *both* the arrival and the service rates to depend upon the state. Several versions of the M/M/1 model are special cases of this one.

FIGURE S2–5

State changes for the M/M/1 queue with state-dependent service and arrival rates

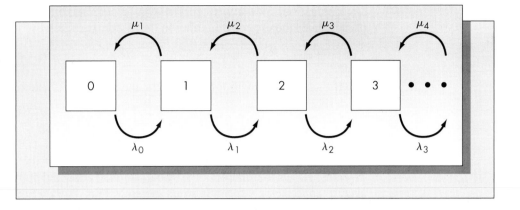

FIGURE S2–5

State changes for the M/M/1 queue with state-dependent service and arrival rates

The transition diagram is the same as that pictured in Figure S2–3, except that both μ and σ are state dependent (see Figure S2–5). The balance equation principle applied to this system yields

$$\mu_1 P_1 = \lambda_0 P_0,$$

$$\lambda_0 P_0 + \mu_2 P_2 = (\lambda_1 + \mu_1) P_1,$$

$$\lambda_1 P_1 + \mu_3 P_3 = (\lambda_2 + \mu_2) P_2,$$

and so on.

Expressing each of the state probabilities in terms of P_0 as we did earlier results in the following:

$$P_1 = \frac{\lambda_0}{\mu_1} P_0,$$

$$P_2 = \frac{\lambda_1 \lambda_0}{\mu_2 \mu_1} P_0,$$

$$P_3 = \frac{\lambda_2 \lambda_1 \lambda_0}{\mu_3 \mu_2 \mu_1} P_0,$$

and so on.

Define

$$A_n = \frac{\lambda_{n-1} \lambda_{n-2} \cdots \lambda_0}{\mu_n \mu_{n-1} \cdots \mu_1}$$

so that

$$P_n = A_n P_0 \qquad \text{for } n = 1, 2, 3, \ldots$$

Again using the fact that $P_0, P_1, \ldots$ is a probability distribution, we have that

$$\sum_{n=0}^{\infty} P_n = 1.$$

This translates to the defining condition for P_0 as

$$P_0 = \frac{1}{1 + \sum_{n=1}^{\infty} A_n}.$$

The various measures of service can be obtained by applying their definitions. In particular, L, the expected number in the system, is given by

$$L = \sum_{n=0}^{\infty} n P_n.$$

If there are assumed to be c servers, then L_q, the expected number in the queue, is given by

$$L_q = \sum_{n=c}^{\infty} (n - c) P_n.$$

Little's formula still applies, and can be used to find the expected waiting times given the expected number in the system and the expected number in the queue. However, to apply Little's formula when the arrival rate is state dependent, we must determine the *overall* average expected arrival rate, or the effective arrival rate, which we call λ_{eff}. Because the arrival rate is λ_n when the system is in state n, it follows that the effective arrival rate is

$$\lambda_{eff} = \sum_{n=0}^{\infty} \lambda_n P_n.$$

10 Multiple Servers in Parallel: The M/M/c Queue

In this section, we consider how the equations in the previous section can be used to solve a variety of configurations of the queue with exponential interarrival times and exponential service times.

First, consider the M/M/c queue; that is, the case in which there are c servers in parallel. This case is pictured in Figure S2–6. When customers arrive, they queue up in

FIGURE S2–6

c *servers in parallel*

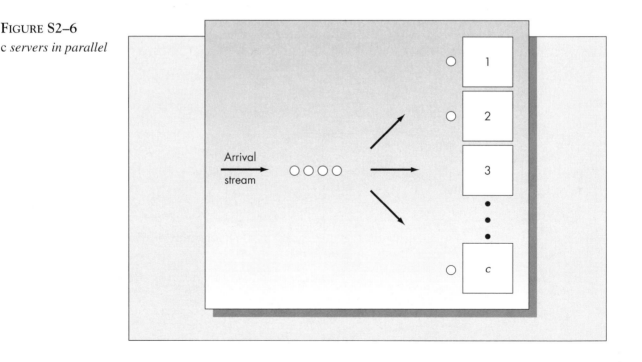

a single line. The next customer in the line is served by the next available server. It turns out that this is just a special case of the state-dependent model derived earlier, but to show that, we need to establish the following result.

Suppose that m servers are busy at a random point in time and also suppose that the service time of each of the servers is exponential with rate μ. The question is, what is the distribution of the time until the next service completion? Let $T_1, T_2, \ldots, T_m$ be the service times of the customers who are currently in service. By assumption, these are independent exponential random variables. Furthermore, if t is a random point in time, the remaining time in service from t to the end of the service completion for each of the customers is also exponential with the same distribution. (This is a consequence of the properties of the exponential distribution discussed earlier.)

It follows that the time until the next service completion, say T, is distributed as the *minimum* of $T_1, T_2, \ldots, T_m$. The result we need to analyze this case is

> **Result:** Let $T_1, T_2, \ldots, T_m$ be independent exponential random variables with common exponential distribution with rate μ, and define $T = \min(T_1, T_2, \ldots, T_n)$. Then T is also exponentially distributed, with rate $m\mu$.

(This result is proven in Chapter 12, in the context of series systems of components subject to exponential failure.)

Why is this result important? It means that the distribution between customer departures is still exponential and that the methods used in the previous section can be applied. If there are c servers, then it follows that the transition rate diagram is as pictured in Figure S2–7.

Comparing Figures S2–5 and S2–7, we see that

$$\mu_1 = \mu$$
$$\mu_2 = 2\mu$$
$$\vdots$$
$$\mu_c = c\mu$$
$$\mu_{c+1} = c\mu$$
$$\mu_{c+2} = c\mu$$

and $\lambda_i = \lambda$ for all $i = 0, 1, 2, \ldots$

FIGURE S2–7

Transition rate diagram when there are c servers in parallel

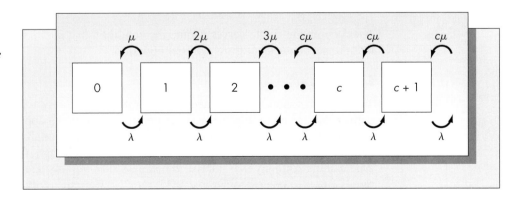

Substituting, it follows that

$$P_1 = \frac{\lambda}{\mu} P_0,$$

$$P_2 = \frac{1}{2}\left(\frac{\lambda}{\mu}\right)^2 P_0,$$

$$P_3 = \frac{1}{(3)(2)}\left(\frac{\lambda}{\mu}\right)^3 P_0,$$

$$\vdots$$

$$P_n = \frac{1}{n!}\left(\frac{\lambda}{\mu}\right)^n P_0, \qquad \text{for } 0 \leq n \leq c.$$

When $n > c$ we obtain

$$P_n = \frac{1}{c!c^{n-c}}\left(\frac{\lambda}{\mu}\right)^n P_0, \qquad \text{for } n > c.$$

Substituting these state probabilities gives the following for P_0:

$$P_0 = \left\{\sum_{n=0}^{c-1} \frac{1}{n!}\frac{\lambda^n}{\mu} + \sum_{n=c}^{\infty} \frac{1}{c!c^{n-c}}\left(\frac{\lambda}{\mu}\right)^n\right\}^{-1}.$$

This can be simplified by noting that the second term is a geometric series. Letting $\rho = \lambda/c\mu$, one can show after a bit of algebraic manipulation that

$$P_0 = \left\{\sum_{n=0}^{c-1} \frac{(c\rho)^n}{n!} + \frac{(c\rho)^c}{c!}(1-\rho)^{-1}\right\}^{-1}.$$

When computing the standard performance measures, it turns out that L_q has the simplest form. Again we will not present the details, but only the results. The derivations are similar to those presented in earlier sections.

$$L_q = \frac{c^c \rho^{c+1}}{c!(1-\rho)^2} P_0.$$

$$L = L_q + c\rho.$$

$$W_q = L_q/\lambda.$$

$$W = W_q + 1/\mu.$$

As with the single-server queue, the condition that $\rho < 1$ is required to guarantee that the queue does not grow without bound.

Example S2.2

Tony's Barbershop is run, owned, and operated by Anthony Jones, who has been cutting hair for more than 20 years. Anthony does not take appointments, so the arrival pattern of customers is essentially random. Traditionally, the arrival rate had been about one customer every 50 minutes. Two months ago, the local paper ran an article about Anthony that improved business substantially. Currently the arrival rate is closer to one customer every 35 minutes. Haircuts require an average of

25 minutes, but the times vary considerably depending on customer needs. A trim might require as little as 5 minutes, but a shampoo and styling could take as long as an hour or more. For this reason, the exponential distribution seems to provide a reasonably good fit of the service time distribution.

Anthony's customers have always been patient, but ever since business picked up, some have complained that the wait is too long. Anthony is considering taking his cousin Marvin into the business to improve customer service. Assume that Marvin cuts hair at the same rate as Anthony.

a. How much has the quality of service declined since more customers have started using the shop?

b. How much improvement in the performance of the system are customers likely to see with an additional barber in the shop?

Solution

a. First we will determine the various performance measures for the system prior to the appearance of the newspaper article. The average time between arrivals was one every 50 minutes, which gives an arrival rate of

$$\lambda = 60/50 = 1.2 \text{ arrivals per hour.}$$

Each haircut requires an average of 25 minutes, which translates to a service rate of

$$\mu = 60/25 = 2.4 \text{ haircuts per hour.}$$

It follows that $\rho = \lambda/\mu = 1.2/2.4 = 0.5$. (That is, Tony was busy half the time.)

The values of the performance measures are

$$L = \rho/(1 - \rho) = 0.5/0.5 = 1.$$
$$L_q = \rho L = 0.5.$$
$$W = L/\lambda = 1/1.2 = 0.8333 \text{ hour.}$$
$$W_q = L_q/\lambda = 0.5/1.2 = 0.4167 \text{ hour.}$$

This means that originally customers waited $(0.4167)(60) = 25$ minutes for a haircut on average.

After the article appeared, the arrival rate increased to one customer every 35 minutes. This means that λ became $60/35 = 1.7143$ and $\rho = 0.7143$. The performance measures are now

$$L = 0.7143/(1 - 0.7143) = 2.5.$$
$$L_q = \rho L = (0.7143)(2.5) = 1.7857.$$
$$W = L/\lambda = 2.5/1.7143 = 1.458 \text{ hours.}$$
$$W_q = L_q/\lambda = 1.78/1.7143 = 1.0383 \text{ hours.}$$

The customers clearly have a valid gripe. A customer has to wait an average of more than an hour before getting a haircut. In fact, because the distribution of W_q is exponential, many would have to wait quite a bit longer than this.

b. Adding an additional barber improves the system performance dramatically. With two barbers, we have

$$\rho = \lambda/(c\mu) = 1.7143/(2)(2.4) = 0.3571. \qquad (c\rho = 0.7143).$$

$$P_0 = \left\{ 1 + 0.7143 + \frac{(0.7143)^2}{2!} \frac{1}{1 - 0.3571} \right\}^{-1}$$

$$= (2.111)^{-1} = 0.4737.$$

It follows that

$$L_q = \frac{(2)^2(0.3571)^3}{2!(1 - 0.3571)^2} (0.4737) = 0.0522.$$

$$L = L_q + c\rho = 0.0522 + 0.7143 = 0.7665.$$

$$W_q = L_q/\lambda = 0.0522/1.7143 = 0.0304 \text{ hour (1.82 minutes).}$$

$$W = W_q + 1/\mu = 0.0304 + 0.4167 = 0.4471 \text{ hour (about 27 minutes).}$$

With only a single barber, customers could expect to wait more than an hour for a haircut. With the addition of another barber, time is reduced to less than 2 minutes on average! ∎

11 The M/M/1 Queue with a Finite Capacity

Another special version of the general M/M/1 queue with state-dependent service and arrival rates is the case in which there is a finite waiting area. If arrivals occur when the waiting area is full, they are turned away. Problems of this type are common in service systems such as restaurants, movie theaters, concert halls, and so on. They can also occur in manufacturing systems in which buffers between work centers have a finite capacity. This is the case, for example, with JIT systems. (See the discussion of JIT in Chapter 7.)

Suppose that the maximum number of customers permitted in the system is K. The transition rate diagram for this case is exactly the same as that pictured in Figure S2–3 except that transitions do not occur beyond state K. Because the transition rate diagram is the same up to state K, the balance equations will yield the same relationship between P_n and P_0 for $n = 1, 2, \ldots, K$. That is,

$$P_n = \rho^n P_0 \qquad \text{for } n = 1, 2, 3, \ldots, K.$$

P_0 is found from

$$\sum_{n=0}^{K} P_n = 1,$$

which gives

$$P_0 = \left(\sum_{n=0}^{K} \rho^n \right)^{-1}.$$

An explicit expression for the finite geometric series is obtained in the following way:

$$\sum_{n=0}^{K}\rho^n = \sum_{n=0}^{\infty}\rho^n - \sum_{n=k+1}^{\infty}\rho^n = \frac{1}{(1-\rho)} - \frac{\rho^K}{(1-\rho)}$$

$$= \frac{1-\rho^K}{1-\rho}.$$

It follows that

$$P_0 = \frac{1-\rho}{1-\rho^{K+1}},$$

from which we obtain

$$P_n = \frac{(1-\rho)\rho^n}{1-\rho^{K+1}} \qquad \text{for } n = 0, 1, 2, \ldots, K.$$

For the case of the finite waiting room, it is not necessary that $\rho < 1$. In fact, P_n has this value for *all* values of $\rho \neq 1$. When $\rho = 1$ it turns out that all states are equally likely, so that

$$P_n = 1/(K+1) \qquad \text{for } 0 \le n \le K \qquad \text{(when } \rho = 1 \text{ only)}.$$

Little's formula still applies, but we must use a modified value for the arrival rate because not all arriving customers are permitted to enter the system. When there are K or more in the system, the arrival rate is zero, so that the overall arrival rate is less than λ. The effective arrival rate, λ_{eff}, is computed as follows:

$$\lambda_{eff} = \lambda P\{\text{Number in the system} < K\} + 0P\{\text{Number in the system} = K\}$$

$$= \lambda(1 - P\{\text{number in the system} = K\})$$

$$= \lambda(1 - P_K).$$

The measures of performance are obtained from L, the expected number in the system in steady state. L is found from

$$L = \sum_{n=0}^{K} nP_n$$

$$= \sum_{n=0}^{K} \frac{(1-\rho)}{1-\rho^{K+1}} n\rho^n.$$

The calculation proceeds by noting that

$$\sum_{n=0}^{K} n\rho^{n-1} = \frac{d}{d\rho} \sum_{n=0}^{K} \rho^n.$$

Using the expression above for the finite geometric sum, we eventually obtain

$$L = \frac{\rho}{1-\rho} - \frac{(K+1)\rho^{K+1}}{1-\rho^{K+1}}.$$

The remaining measures of performance are found from

$$L_q = L - (1 - P_0).$$
$$W = L/\lambda_{eff}.$$
$$W_q = L_q/\lambda_{eff}.$$

Similar formulas can be derived for the case of a finite capacity queue and multiple parallel servers. (See, for example, Hillier and Lieberman, 1990.)

Example S2.3

A popular attraction at the New Jersey Shore is a street artist who will paint a caricature in about five minutes. However, because the times required for each drawing vary considerably, they are accurately described by an exponential distribution. People are willing to wait their turn, but when there are more than 10 waiting for a picture, customers are turned away and asked to return at a later time. At peak times one can expect as many as 20 customers per hour. Assume that customers arrive completely at random at the peak arrival rate.

    a. What proportion of the time is the queue at maximum capacity?
    b. How many customers are being turned away, on average? Determine the measures of performance for this queuing system.
    c. If the waiting area were doubled in size, how would that affect your answers to parts (*a*) and (*b*)?

Solution

The arrival rate is $\lambda = 20$ per hour and the service rate is $\mu = 12$ per hour, so that $\rho = 20/12 = 1.667$. The maximum number in the system is $K = 11$ (10 in the queue plus the customer being served).

    a. The probability that the system is full is P_K, which is given by

$$P_K = \frac{(1 - \rho)\rho^K}{1 - \rho^{K+1}} = \frac{(1 - 1.667)(1.667)^{11}}{1 - 1.667^{12}} = \frac{-184.25}{-459.5} = 0.40.$$

    b. As the arrival rate is 20 per hour and 40 percent of the time the system is full, there are, during peak periods, $(20)(0.40) = 8$ customers per hour being turned away. This gives $\lambda_{eff} = 12$ per hour.

$$L = \frac{\rho}{1 - \rho} - \frac{(K + 1)\rho^{K+1}}{1 - \rho^{K+1}} = \frac{1.667}{1 - 1.667} - \frac{(12)(1.667)^{12}}{1 - 1.667^{12}}$$
$$= -2.5 - (-12.03)$$
$$= 9.53.$$

We need to determine P_0 to compute L_q.

$$P_0 = 1 - \frac{1 - \rho}{1 - \rho^{K+1}} = \frac{1 - 1.667}{1 - 1.667^{12}} = 0.00145,$$

which gives $L_q = L - (1 - P_0) = 9.53 - (1 - 0.00145) = 8.53.$

Notice that the small value of P_0 means that the system is rarely empty. In particular, the artist is idle only 0.145 percent of the time!

We showed above the $\lambda_{eff} = 12$, so that

$$W = L/\lambda_{eff} = 9.53/12 = 0.7942 \text{ hour (about 48 minutes).}$$
$$W_q = L_q/\lambda_{eff} = 8.53/12 = 0.7108 \text{ hour (about 43 minutes).}$$

c. If the waiting area were doubled in size, then $K = 21$. In that case P_K is given by

$$P_{21} = \frac{(1 - 1.667)(1.667)^{21}}{1 - 1.667^{22}} = \frac{-30{,}533.28}{-76{,}309.3} = 0.40.$$

Interestingly, doubling the capacity of the queue makes no difference relative to the probability that the system is full. The reason is that because the arrival rate exceeds the service rate, the system reaches capacity quickly in either case. In both cases the effective arrival rate λ_{eff} is approximately equal to the service rate (although it will always be true that $\lambda_{eff} < \mu$). Even for much larger values of K, P_K is 0.4 in this example. ■

12 Results for Nonexponential Service Distributions

One assumption required for all the models discussed thus far is that the distribution of the service time is exponential. In many cases, this assumption is unwarranted. One would expect that service times would rarely be exponential because the exponential distribution has the memoryless property: The amount of time remaining in service would have to be independent of the time already spent. One would think that a modal distribution, such as the normal or Erlang, would be a more accurate model of service times in most circumstances.

For that reason, the M/G/1 model is of great interest. Here the G stands for general, meaning *any* service time distribution. Because we lose the memoryless property when services are not exponential, the rate diagram method no longer applies. The mathematics becomes quite a bit more complex (the interested reader should refer to Gross and Harris, 1985, for example, for a tractable presentation of this model).

The measures of performance are derived from the Pollaczek–Khintchine (P–K) formula, which gives the expected number in the queue in terms of the mean and the variance of the service time distribution. Let $E(T)$ be the expected service time and $\text{Var}(T)$ be the variance of the service time. Then the P–K formula for the expected number in the queue is

$$L_q = \frac{\lambda^2 \{\text{Var}(T) + E(T)^2\}}{2(1 - \lambda E(T))}.$$

Using our previous notation, $E(T) = 1/\mu$ and $\lambda E(T) = \lambda/\mu = \rho$, and letting $\text{Var}(T) = \sigma^2$, this formula becomes

$$L_q = \frac{\lambda^2 \sigma^2 + \rho^2}{2(1 - \rho)}.$$

Note that L_q (and hence L, W, and W_q) is increasing in σ^2. This means that congestion increases as the variance of the service time increases. Hence, the measures of performance for a system can be improved by decreasing the variance of the service time, even when the mean service time does not change. (When service times are exponentially distributed, $\sigma^2 = 1/\mu^2$ and the formula for L_q reduces to that derived for the M/M/1 queue.)

We illustrate the application of the P–K formula.

Example S2.4

A large discount warehouse is assessing the number of check stands it needs. During a period of the day when the arrival rate of customers is about one every 12 minutes, there is only one check stand open. It takes an average of 8 minutes to check out one customer. The checkout time follows a normal distribution with standard deviation 1.3 minutes. The arrival process may be assumed to be a Poisson process. Find the performance measures for this system. How far off would your calculations be if you assumed that the service distribution was exponential?

Solution

The arrival rate is $\lambda = 5$ per hour, and the service rate is $\mu = 60/8 = 7.5$ per hour, giving $\rho = 5/7.5 = 2/3$. Because both λ and μ are expressed as number of customers per hour, we must also express the variance of the service time distribution in consistent units. The standard deviation of the service time is 1.3 minutes, or $1.3/60 = 0.02167$ hour. It follows that the variance of the service time is $0.02167^2 = 4.6944 \times 10^{-4}$ hour^2. (Alternatively, we could have expressed λ and μ as number of customers per minute.) It follows from the P–K formula that

$$L_q = \frac{\lambda^2\sigma^2 + \rho^2}{2(1 - \rho)} = \frac{(25)(0.0004694) + 4/9}{(2)(1/3)} = 0.6843.$$

The remaining performance measures are

$$L = L_q + \rho = 0.6843 + 2/3 = 1.351.$$
$$W_q = L_q/\lambda = 0.13686 \text{ hour (8.21 minutes).}$$
$$W = W_q + 1/\mu = 0.13686 + 1/7.5 = 0.2702 \text{ hour (16.21 minutes).}$$

Hence, each customer should expect to wait in line about 8 minutes and expect to be in the system about twice that time. On average there is less than one customer in the queue.

Now, suppose that we assumed that the service distribution was exponential. We would have obtained the following performance measures:

$$L = \rho/(1 - \rho) = (2/3)/(1/3) = 2.$$
$$L_q = L - \rho = 2 - 2/3 = 4/3.$$
$$W = L/\lambda = 2/5 = 0.4 \text{ hour (24 minutes).}$$
$$W_q = L_q/\lambda = 1.3333/5 = 0.2667 \text{ hour (16 minutes).}$$

We see that assuming that the service distribution is exponential results in substantial errors in the system performance measures. In particular, W_q is too high by 100 percent! The M/M/1 model is very sensitive to the assumption that the service time is exponential. If the results of a queuing analysis are to be accurate, it is vitally important that any assumptions regarding the form of the service or the arrival process be verified by direct observation of the system. ■

One interesting special case of the M/G/1 queue is when the service distribution is deterministic (labeled M/D/1). The Pollaczek–Khintchine formula still applies with the variance term, $\sigma^2 = 0$. It is interesting to note that when the number of servers exceeds one, algebraic expressions for the performance measures are not known. Explicit results are available for only a few other cases, in fact. The interested reader should refer to a more complete coverage of queuing, such as that given by Gross and Harris (1985) or Kleinrock (1975). Explicit results for *many* versions of the M/M queuing model are available, however. Examples include systems with priority service, jockeying, and impatient customers, just to mention a few.

13 The M/G/∞ Queue

Another version of the queuing problem with general service distribution for which there are explicit results is the case in which there are an infinite number of servers. Customers arrive at the system completely at random according to a Poisson process with rate λ. The service time distribution is arbitrary with service rate μ. At the instant of arrival, the customer enters service. An infinite number of servers means that there is always a server available, no matter how many customers are in the system. Although this might seem unrealistic, many real problems can be modeled in this way. Because there is no queue of customers waiting for service, there is no waiting time for service. Hence, both measures of performance, L_q and W_q, are zero. However, the number of customers in the system, L, is not zero. Note that the number of customers in the system is equal to the number of busy servers. The result of interest is the following:

> **Result:** For the M/G/∞ queue with arrival rate λ and service rate μ, the distribution of the number of customers in the system (or the number of busy servers) in steady state is Poisson with rate λ/μ. That is,

$$P\{N = k\} = \frac{e^{-(\lambda/\mu)}(\lambda/\mu)^k}{k!} \qquad \text{for } k = 0, 1, 2, \ldots$$

This is a powerful result. It follows that both the mean and the variance of the number of customers in the system in steady state is λ/μ.

Example S2.5

A common inventory control policy for high-value items is the one-for-one policy, also known as the $(S - 1, S)$ policy. This means that the target stock is S and at each occurrence of a demand, a reorder for one unit is placed. Suppose that demands are generated by a stationary Poisson process with rate λ, and that the lead time required for replenishment is a random variable with arbitrary distribution with mean $1/\mu$. This problem is exactly an M/G/∞ queue. The number of customers in the system is equivalent to the number of outstanding orders, which by the result above has the Poisson distribution with mean λ/μ. If the lead time is fixed at τ, the expected number of units on order is just $\lambda\tau$. It follows that the expected number of units in stock is $S - \lambda\tau$. This result can be used to determine an expression for the expected holding, stock-out, and replenishment costs, which can then be optimized with choice of S. (See Hadley and Whitin, 1963, p. 212, for example.) ∎

Example S2.6

Personnel planning is an important function for many firms. Consider a company's department with a desired head count of 100 positions. Suppose that employees leave their positions at a rate of 3.4 per month, and that it requires an average of 4 months for the firm to fill open positions. Analysis of past data shows that the number of employees leaving the firm per month has the Poisson distribution, and that the time required to fill positions follows a Weibull distribution. What is the probability that there are more than 15 positions unfilled at any point in time? How many jobs within the department are filled on average? How many positions should the firm have in order for the head count of working employees to be 100 on average?

Solution

To determine the distribution of filled positions, we model the problem as an M/G/∞ queue. Each time that an employee leaves, his or her position enters the queue of unfilled positions. Assuming that the search for a replacement starts immediately, the correct model is an infinite number of servers. According to the theory, the expected number of unfilled positions is independent of the distribution of the time required to replace each employee. The expected number of unfilled positions is λ/μ. For this application, λ corresponds to the rate at which employees leave their positions, which is 3.4 per month, and μ the rate that jobs are filled, which is 1/4 per month. Hence the mean number of unfilled jobs is $\lambda/\mu = 3.4/(1/4) = (3.4)(4) = 13.6$. Hence there are $100 - 13.6 = 86.4$ positions filled on average.

The probability that there are more than 15 unfilled positions is the probability that a Poisson random variable with mean 13.6 exceeds 15. Interpolating from Table A–3, this probability is approximately 0.29.

It also follows that if the department were allotted 114 positions rather than 100, there would be, on average, 100 positions filled at any point in time (although the actual number is a random variable).[5] ∎

14 Optimization of Queuing Systems

Classical queuing analysis is descriptive rather than prescriptive. In practice, this means that given the various input and service distributions, one determines the measures of performance. These measures of performance do not directly translate to optimal decisions concerning the design of the system. This section shows how one would go about developing models for determining the optimal configuration of a queuing system.[6]

Let us consider some typical design problems arising in queuing service systems and how one would go about using the results of queuing theory to determine optimal system configurations.

---

[5]I am grateful to John Peterson of Smith-Kline-Beecham for bringing this application to my attention.
[6]The results of this section are based on Chapter 17 of Hillier and Lieberman (1990).

Typical Service System Design Problems

1. The State Highway Board must determine the number of toll booths to have available on a new interstate toll road. The more toll booths open at any point in time, the less wait commuters will experience. However, additional toll booths require an additional one-time cost to build and additional ongoing costs of salary for the toll taker.

2. A plant is being built by a major manufacturer of Winchester hard disk drives to produce read and write heads. The company management is considering several options for the manufacturing equipment. A new machine for winding the electronic coils has double the throughput of the conventional equipment, but at more than triple the cost. Is the investment justified?

3. A word processing service is considering how large a client base to develop. The company wishes to have a large enough number of clients to make it busy, but not so many that it cannot provide reasonable turnaround time.

Modeling Framework

1. Consider the example of the state highway board. The more time commuters spend on the highway, the less time they spend working and contributing to society. If we view the goal as societal optimization, then there is clearly a direct economic benefit to reducing commute time. Suppose that an economic analysis of the highway problem resulted in an estimate of the cost incurred when a commuter waits w units of time as the function $h(w)$. A typical case is pictured in Figure S2–8.

Let W be the waiting time of a customer chosen at random. Then W is a random variable. For the M/M/1 queue, we showed that W has the exponential distribution with parameter $\mu/(\mu - \lambda)$. Given the distribution of W, it follows that the expected

FIGURE S2–8

A typical waiting cost function

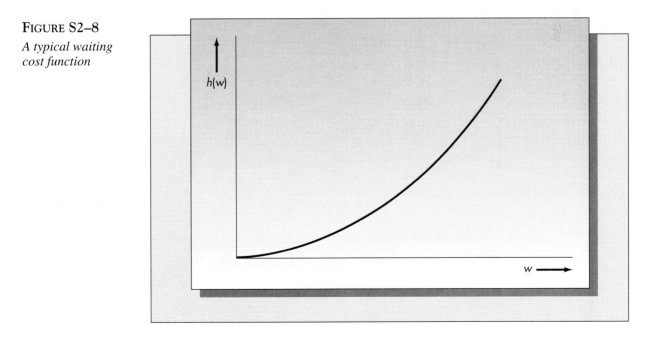

waiting cost of a customer chosen at random is

$$E(h(W)) = \int_0^\infty h(w)f(w)\,dw.$$

Because the arrival rate of customers is λ units per unit time, the overall waiting cost per unit time is $\lambda E(h(W))$. In the case of the toll booths, one would determine the distribution of W for each number of servers being considered, say W_c. If the cost per unit time for maintaining each server is S, then the objective would be to find the optimal value of c to minimize

$$cS + \lambda Eh(W_c).$$

2. Consider the example of the firm producing read and write heads for hard disk drives. The firm can purchase a larger μ (service rate), but only for an increased cost. To determine the best decision in this case, the firm would have to be able to quantify the costs associated with various levels of service. Suppose that the annual cost of the manufacturing operation when the throughput rate of the process is μ is given by the function $f(\mu)$. Because the cost decreases as μ increases, this would be a monotonically decreasing function of μ. Furthermore, suppose that the one-time cost of purchasing equipment with service rate μ is $C(\mu)$. We would expect that $C(\mu)$ would be a monotonically increasing function of μ. Let I be the annual interest rate of alternative investments. Then the total annual cost is

$$IC(\mu) + f(\mu).$$

This function will be convex in μ (see Figure S2–9), so that an optimal minimizing μ will exist and can be found easily. When there are only several possible values of μ, the objective function can be evaluated at these values and the choice yielding the lowest cost can be made.

FIGURE S2–9

Optimizing the service rate μ

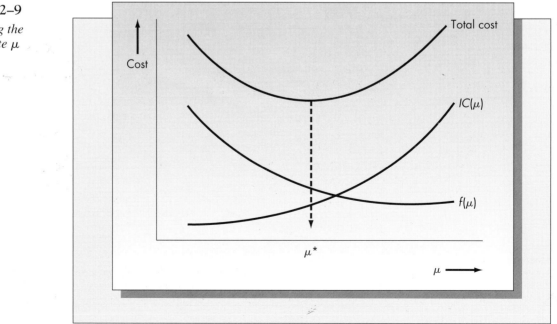

3. Consider the example of the word processing service. In this case the decision variable is the arrival rate, λ. The larger the client base, the more jobs the firm will receive and the larger the value of λ. There are several possible formulations of this problem. One would be to determine the value of the expected number in the system that allows the firm to meet its obligations. In that case, we would assume that c and μ are given and that the objective is to determine λ so that L is equal to a target value. Another approach would be to find λ so that the probability that the number of customers in the system does not exceed a target level is at least some specified probability (such as 0.95).

To illustrate this case, consider the following specific example. Mary Worth runs her own word processing service. She requires an average of 1.2 hours to complete a job, but the size of jobs varies considerably and can be described by an exponential distribution. Furthermore, the pattern of arrival of jobs appears to be completely random. Mary works eight-hour days and does not want to have more than two days of work piled up at any point in time. Suppose that she wants the likelihood of this occurring to be no more than 5 percent. This means that the number of customers in the queue should not exceed $16/1.2 = 13.33$. Hence, she wishes to have an arrival rate of jobs so that

$$P\{L > 13\} \le 0.05.$$

We showed in the solution to Example S2.1 that

$$P\{L > k\} = \rho^{k+1.}$$

Hence, we wish to find λ to solve

$$(\lambda/\mu)^{13} = 0.05.$$

Using $\mu = 1/1.2 = 0.8333$ per hour, and taking natural logarithms of both sides, we obtain

$$13 \ln(\lambda/0.8333) = \ln(0.05)$$
$$\ln(\lambda/0.8333) = -0.23044$$
$$\lambda/0.8333 = \exp(-0.23044) = 0.7942,$$

giving $\lambda = 0.662$.

The solution, then, is that she should plan to have enough clients to generate about 0.662 job each hour, or about 5.3 jobs per day.

Assigning service territories to repairmen, or deciding which region a blood bank is to cover, are other examples of problems in which the objective is to find the optimal value of the arrival rate, λ.

15 Simulation of Queuing Systems

Many queuing problems encountered in the real world are not amenable to the types of mathematical analysis discussed in this supplement. Some of the reasons include:

1. The problem involves a queuing network. Queuing networks are common in manufacturing systems where the output of one process is the input to one or more other processes. Some systems are very complex with feedback loops and other unusual features. Such systems are generally too complicated to be analyzed mathematically.

2. The interest is in transient rather than steady-state behavior. The mathematical results discussed here are for steady state only. Steady state means that the system evolves for a sufficiently long period so that initial conditions do not affect the probability distribution of system states. In many real problems, short-term behavior is an important part of proper systems management.

3. Nonexponential probability distributions of arrivals or services and/or uncommon features or constraints can make analysis intractable. Little is known about G/G/c queuing systems, especially with a finite waiting room, priority service, and so on. Even for single-server queues in the steady state, many queuing problems are too complex to analyze mathematically.

What alternatives are available when a queuing problem is not one of those for which results are known? The answer is simulation. A simulation (in our context) is a computer program that recreates the essential steps of a problem scenario. In some sense, the computer "experiences" the problem. In this way, one can see how the system responds to different parameter settings available to the system designer. Simulations evolve over time. Systems can be simulated for years in the space of a few seconds on a computer.

Problems amenable to simulation almost always involve some element of randomness. Computer-based simulators that incorporate randomness are called Monte Carlo simulators. At the heart of a Monte Carlo simulation are random numbers.

Random numbers in this context are draws from a uniform (0, 1) distribution. That is, they are numbers between zero and one with the property that every number drawn has an equal likelihood of being selected. Random number generators are algorithms that produce what appear to be independent realizations of uniform variates. The algorithms used do not cycle for a very large number of steps, thus producing number sequences that appear random. However, because the recursive algorithms used are deterministic, the resulting string of numbers is referred to as "pseudorandom" numbers. (See Fishman, 1973, for example.)

From uniform variates one obtains observations of random variables having virtually any distribution using results from probability theory. For example, the central limit theorem says that sums of independent random variables are approximately normally distributed. (This is a *very* loose statement of the central limit theorem.) Therefore, the sum of a reasonable number of independent draws from a uniform (0, 1) distribution will be approximately normal. (Convergence occurs very quickly, so the number does not have to be very large.) For example, if we let $U_1, U_2 \ldots$ be successive draws from a uniform (0, 1) distribution then

$$Z = \sum_{i=1}^{12} U_i - 6$$

is approximately standard normal. (This is meant for illustrative purposes only. In practice there are more efficient ways to generate normal variates.)

Simulation can be used to analyze many types of complex problems with uncertainty, but probably has been used most often for queuing problems. Simulators can be written in source code (such as Fortran or C), or constructed using special-purpose simulation packages, such as GPSS, Simscript, or SLAM. Many packages are specifically designed for certain types of applications such as queuing or network problems. Swain (1995) provides a comprehensive review of more than 50 packages designed for discrete event simulation on personal computers.

Most recently, two approaches for developing simulations have become popular, opening up the use of simulation to a much wider audience. Spreadsheet programs have gained a great deal of popularity in recent years. Because many spreadsheets have a random number generator built in, they can be used to construct simulators. Excel, in particular, can generate observations from many distributions. To facilitate building simulators, one might consider the add-in @RISK. @RISK includes a much wider array of distributions and convenient report generation. (Examples using @RISK can be found in Winston, 1996.)

For complex queuing networks, graphical-based simulation packages are far easier to use than spreadsheets. These programs allow the user to construct a model of the system using graphical icons. Icons represent service facilities and waiting areas and arrows represent the direction of flow. Random arrivals and random services can be incorporated easily into the model. One Windows-based product is ProModel, developed and sold through the ProModel Corporation. With a package such as this, experienced users can build simulations of complex systems very quickly. The program employs live animation to show the simulation in action, and summary statistics are collected after the simulation has run its course.

One final comment. Although we have discussed simulation in the context of queuing only, it is a powerful tool for solving many kinds of complex problems. Simulations have been used for inventory management problems, scheduling, traffic planning, resource allocation, financial modeling, and so on. However, judging from the software products available, queuing applications would appear to be the most common.

Bibliography

Feller, W. *An Introduction to Probability Theory and Its Applications*. Vol. 2. New York: John Wiley & Sons, 1966.

Fishman, G. *Concepts and Methods in Discrete Event Digital Simulation*. New York: John Wiley & Sons, 1973.

Gross, D., and C. M. Harris. *Fundamentals of Queuing Theory*. 2nd ed. New York: John Wiley & Sons, 1985.

Hadley, G., and T. M. Whitin. *Analysis of Inventory Systems*. Englewood Cliffs, NJ: Prentice Hall, 1963.

Hillier, F. S., and G. J. Lieberman. *Introduction to Operations Research*. 5th ed. New York: McGraw-Hill, 1990.

Kleinrock, L. *Queuing Systems*. Vol. 1, *Theory*. New York: Wiley Interscience, 1975.

Kleinrock, L. *Queuing Systems*. Vol. 2, *Computer Applications*. New York: Wiley Interscience, 1976.

Swain, J. J. "Simulation Survey: Tools for Process Understanding and Improvement." *OR/MS Today* 22, no. 4 (August 1995), pp. 64–79.

Winston, W. L. *Simulation Modeling Using @RISK*. Belmont, CA: Duxbury Press, 1996.

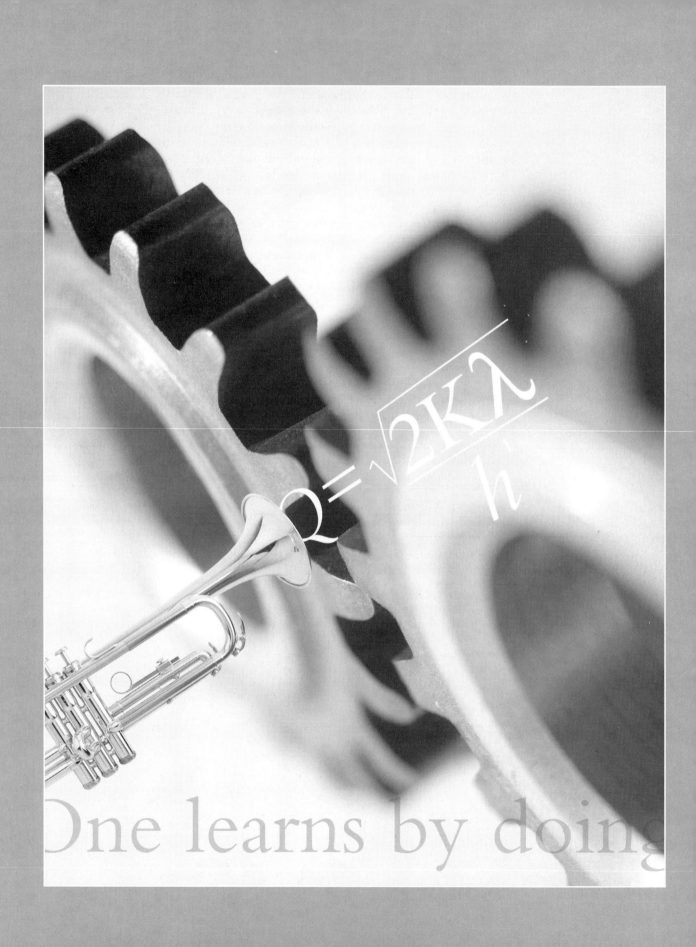

$$Q = \sqrt{\dfrac{2K\lambda}{h}}$$

One learns by doing

CHAPTER 9

PROJECT SCHEDULING

Project management is an important planning function that takes place at almost all levels of an organization. What are the consequences of poor project management? One is cost overruns. How often have we heard members of Congress express dismay at the cost overruns in military projects? In some cases these overruns could not be avoided: Unforeseen obstacles arose or technological problems could not be solved as easily as anticipated. In many cases, however, these problems were a consequence of poor project scheduling and management.

Large complex projects involving governments in partnership with business are perhaps the most vulnerable to delays and overruns. A case in point is the Trans-Alaska Pipeline System, designed to transport large quantities of oil from Prudhoe Bay on Alaska's north slope to Port Valdez on the Gulf of Alaska. Goodman (1988) describes it as one of the most complex and massive design and construction projects of recent times. Political roadblocks, environmental concerns, and contract disputes plagued the project from the start. The list of important players changed several times, resulting in some firms (Bechtel, in particular) not having enough time to do an adequate job of project planning. The Alyeska Pipeline Service Company, which was responsible for much of the actual building of the pipeline, incurred excessive cost overruns, partly due to poor project management. In retrospect, it is clear that many of the problems with the project were a consequence of the fact that there was never a single project team to oversee the entire integrated project cycle.

Project scheduling and project management methods have been an important part of doing business for many firms. For example, the Lockheed-Martin Missiles and Space Company of Mountain View, California, uses project scheduling methods not only for monitoring and control of projects, but also for the process of preparing bids and developing proposals. In fact, Lockheed was part of the team that developed PERT, a technique considered in detail in this chapter.

This chapter reviews analytical techniques for project management. Effective people management can be just as important a factor in getting projects done on time and within budget. The firm must create a structure and environment conducive to properly motivating employees. Poor organizational design and incentive structures can be just as serious a problem as poor project planning.

The project management methods considered in this chapter have been used to plan long-term projects such as launching new products, organizing research projects, and building new production facilities. The methods also have been used for smaller projects such as building of residential homes. Two techniques that are treated in detail are usually referred to as either CPM (for *critical path method*) or PERT (for *project evaluation and review technique*). Both methods were developed almost simultaneously for solving very different project management problems in the late 1950s. Although the two labels are used interchangeably today, we will retain the terminology consistent with the original intent of the methods. That is, CPM deals with purely deterministic problems, whereas PERT allows randomness in the activity times.

The primary elements of critical path analysis are:

1. *Project definition.* This is a clear statement of the project, the goals of the project, and the resources and personnel that the project requires.

2. *Activity definitions.* The project must be broken down into a set of indivisible tasks or activities. The project planner must specify the work content of each activity and estimate the activity times. Often the most difficult part of project planning is finding the best way to break down the project into a collection of distinct activities.

3. *Activity relationships.* An important part of the project planning methodology is a specification of the interrelationships among the activities. Known as *precedence constraints,* these describe the logical sequence to complete the activities comprising the project.

4. *Project scheduling.* A project schedule is a specification of the starting and ending times of *all* activities comprising the project. Using the techniques of this chapter, we will show how specification of the activity times and precedence constraints yields an efficient schedule.

5. *Project monitoring.* Once the activities have been suitably defined and a schedule determined, the proper controls must be put in place to ensure that project milestones are met. The project manager must be prepared to revise existing schedules if unforeseen problems arise.

Because of the level of detail and precision required, critical path analysis is most effective when the project can be easily expressed as a well-defined group of activities. Construction projects fall into this category. Precedence constraints are straightforward, and activity times are not difficult to estimate. For this reason, CPM has found wide acceptance in the construction industry. However, there are case studies reported in the literature of successful applications in a variety of environments, including military, government, and nonprofit.

9.1 Representing a Project as a Network

As we saw in Chapter 8 on shop floor scheduling and control, a *Gantt chart* is a convenient graphical means of picturing a schedule. A Gantt chart is a horizontal bar graph on which each activity corresponds to a separate bar. Consider the following simple example.

Example 9.1

Suppose that a project consists of five activities, A, B, C, D, and E. Figure 9–1 shows a schedule for completing this project, in the form of a Gantt chart. According to the figure, task A starts at 12:00 and finishes at 1:30. Tasks B and C start at 1:30, B finishes at 2:30, C finishes at 3:30, and so on. The project is finally completed at 6:00.

FIGURE 9–1

Gantt chart for five activities (refer to Example 9.1)

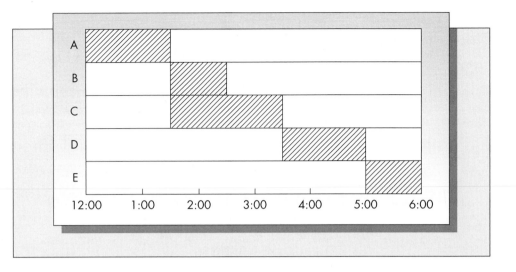

Notice that Figure 9–1 gives no information about the relationships among the activities. For example, the figure implies that E cannot start until D is completed. However, suppose that E is permitted to start any time after A, B, and C finish. Then E can start at 3:30, and the project can be completed at 5:00 rather than at 6:00. ■

Although the Gantt chart is a useful means of representing the schedule, it is not a very useful planning tool because it does not show the precedence constraints. For this reason, one graphically represents the project as a network. Networks, unlike Gantt charts, explicitly show the precedence constraints.

A network is a collection of nodes and directed arcs. In the traditional network representation of a project, a node corresponds to an event and an arc to an activity or task. Events may be (1) the start of a project, (2) the completion of a project, or (3) the completion of some collection of activities. This method of representation is known as activity-on-arrow and is historically the most common means of representing a project as a network. An alternative method of representation is activity-on-node. Although the latter method has some advantages, it is rarely used in practice. The activity-on-node method will be illustrated later in this chapter.

Example 9.2

Suppose a project consists of the five activities A, B, C, D, and E that satisfy the following precedence relationships:

1. Neither A nor B has any immediate predecessors.
2. A is an immediate predecessor of C.
3. B is an immediate predecessor of D.
4. C and D are immediate predecessors of E.

The network for this project appears in Figure 9–2. Node 1 is always the initiation of the project. All activities having no immediate predecessors emanate from node 1. As A is an immediate predecessor of C, we must have a node

FIGURE 9–2

*Project network for
Example 9.2*

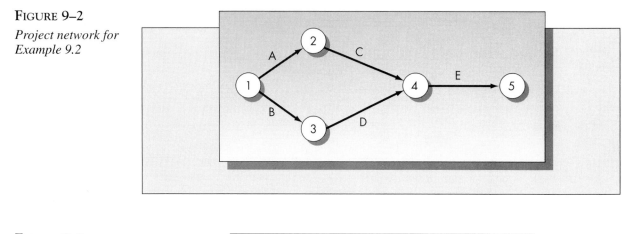

FIGURE 9–3

*Incorrect network
representation for
Example 9.3*

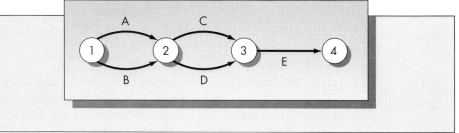

representing the completion of A. Similarly, as B is an immediate predecessor of D, there must be a node representing the completion of B. Notice that although both A and B must also be completed before E, they are not *immediate* predecessors of E. ■

Representing a project by a network is not always this straightforward.

Example 9.3

Consider Example 9.2, except now replace (3) with (3′):

   3′. A and B are immediate predecessors of D.

Try to find a network representation of this project. One might think that the correct representation is the one in Figure 9–3, as this network implies that D must wait for both A and B. However, this representation is incorrect, because it also shows that C must wait for both A and B as well.

The set of precedent relations (1), (2), (3′), and (4) require that we introduce a pseudo activity between nodes 2 and 3. The correct representation of the system is given in Figure 9–4. The pseudo activity, labeled P, is a directed arc from node 2 to node 3 and is represented by a broken line. Note that the direction of the arrow is important. With the pseudo activity, node 3 corresponds to the completion of both activities A and B, whereas node 2 still corresponds to the completion of only activity A. ■

An important aspect of critical path analysis is defining an appropriate set of tasks. A problem that may arise is that one defines the tasks too broadly. As an example, suppose in Example 9.3 that E could start when only half of C was completed. This means

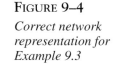

FIGURE 9–4

Correct network representation for Example 9.3

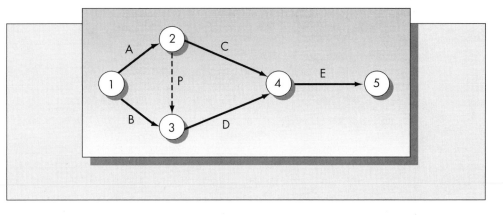

that C would have to be further broken down into two other activities, one representing the first half of C and one representing the second half of C. Conversely, it is possible to define activities too narrowly. This can result in an overly complicated network representation, with portions of the network resembling Figure 9–3.

9.2 Critical Path Analysis

Once having represented the project as a network, we can begin to explore the methods available for answering a variety of questions. For example,

1. What is the minimum time required to complete the project?
2. What are the starting and ending times for each of the activities?
3. Which activities can be delayed without delaying the project?

A path through the network is a sequence of activities from node 1 to the final node. One labels a path by the sequence of nodes visited or by the sequence of arcs (activities) traversed. In the network pictured in Figure 9–4 there are exactly three distinct paths: 1–2–4–5 (A–C–E), 1–2–3–4–5 (A–P–D–E), and 1–3–4–5 (B–D–E). Each path is a sequence of activities satisfying the precedence constraints. Because the project is completed when all activities are completed, it follows that *all* paths from the initial node to the final node must be traversed. Hence, the minimum time to complete the project must be the same as the length of the *longest* path in the network.

Consider Example 9.3 above and suppose that activity times are the same as those indicated in Figure 9–1. The activity times and the precedence constraints are

Activity	Time (hours)	Immediate Predecessors
A	1.5	—
B	1.0	—
C	2.0	A
D	1.5	A and B
E	1.0	C and D

The lengths of the three paths are

Path	Time Required to Complete (hours)
A–C–E	4.5
A–P–D–E	4.0
B–D–E	3.5

To complete the project, all three paths must be completed. The longest path for this example is obviously A–C–E. This is known as the *critical path*. The length of the critical path is the *minimum* completion time of the project. The activities that lie along the critical path are known as the critical activities. A delay in a critical activity results in a delay in the project. However, activities that do not lie along the critical path have some slack. A delay in a noncritical activity does not necessarily delay the project.

Enumerating all paths is, in general, not an efficient way to find the critical path. Later, we consider a general procedure, not requiring path enumeration, for identifying the critical path and the start and finish times for all activities comprising the project. First we consider the following case study.

Example 9.4

Simon North and Irving Bonner, computer consultants, are considering embarking on a joint project that will involve development of a relatively small commercial software package for personal computers. The program involves scientific calculations for a specialized portion of the engineering market. North and Bonner have broken down the project into nine tasks.

The first task is to undertake a market survey in order to determine exactly what the potential clientele will require and what features of the software are likely to be the most attractive. Once this stage is completed, the actual development of the programs can begin. The programming requirements fall into two broad categories: graphics and source code. Because the system will be interactive and icon driven, the first task is to identify and design the icons. After the programmers have completed the icon designs, they can proceed with the second part of the graphics development, design of the input/output screens. These include the various menus and report generators required in the system.

The second part of the project is coding the modules that do the scientific calculations. The first step is to develop a detailed flowchart of the system. After they complete the flowchart, the programmers can begin work on the modules. There are a total of four modules. The work on modules 1 and 2 can begin immediately after completion of the flowchart. Module 3 requires parts of module 1, so the work on module 3 cannot begin until module 1 is finished. The programming of module 4 cannot start until both modules 1 and 2 are completed. Once the graphics portion of the project and the modules are completed, the two separate phases of the system must be merged and the entire system tested and debugged.

North has managed to obtain some funding for the project, but his source requires that the program be completed and ready to market in 25 weeks. In order to determine whether this is feasible, the two programmers have divided the project into nine indivisible tasks and have estimated the time required for each task. The list of these tasks, the times required, and the precedence relationships are given below.

Task	Time Required (in weeks)	Immediate Predecessors
A. Perform market survey	3	—
B. Design graphic icons	4	A
C. Develop flowchart	2	A
D. Design input/output screens	6	B, C
E. Module 1 coding	5	C
F. Module 2 coding	3	C
G. Module 3 coding	7	E
H. Module 4 coding	5	E, F
I. Merge modules and graphics and test program	8	D, G, H

The total of the task times is 43 weeks. Based on this, the programmers conclude that it would be impossible to complete their project in the required 25 weeks. Fortunately, they see the error in their thinking before breaking off relations with their funding source. Since some of the activities can be done concurrently, the project should take fewer than 43 weeks. The network representation for this project is given in Figure 9–5. In order to satisfy the precedence constraints we require two pseudo activities, P_1 and P_2. We will determine the critical path *without* enumerating all paths through the network. The critical path calculations yield both the critical path *and* the allowable slack for each activity as well.

For illustrative purposes, we represent the project network using the activity-on-node format in Figure 9–6. One advantage of this format is that pseudo activities are not required, although for certain networks dummy starting and ending nodes may be needed. Given a list of activities and immediate predecessors, it should be easy to find the network representation in either format. ∎

FIGURE 9–5

Network for Example 9.4

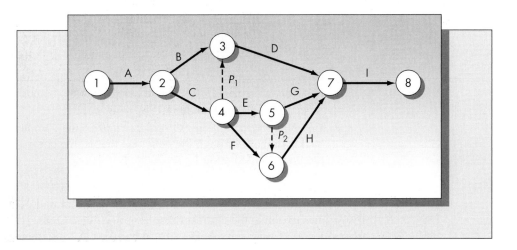

FIGURE 9–6

*Network for
Example 9.4:
Activity-on-node
representation*

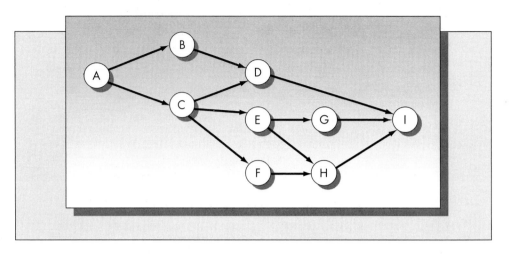

Practitioners prefer the activity-on-arrow method, believing that it is a more intuitive way to represent the project. We will use it exclusively for the remainder of the chapter.

Finding the Critical Path

We compute four quantities for each activity:

ES_i = Earliest starting time for activity i,
EF_i = Earliest finishing time for activity i,
LF_i = Latest finishing time for activity i (without delaying the project),
LS_i = Latest starting time for activity i (without delaying the project).

Suppose that t_i represents the time required to complete activity i. Then it is easy to see that

$$EF_i = ES_i + t_i$$

and

$$LS_i = LF_i - t_i.$$

The steps in the process are

1. *Compute the earliest times for each activity.* One computes the earliest times by a forward pass through the network; that is, the computations proceed from node 1 to the final node.
2. *Compute the latest times for each activity.* One computes the latest times by a backward pass through the network; that is, the computations proceed from the final node to node 1.

We will illustrate the computations using Example 9.4. The first step is to set the earliest starting times for all activities emanating from node 1 to zero and add the activity times to obtain the earliest finishing times. In this case, the only activity emanating from node 1 is activity A. Because B has A as an immediate predecessor, it follows that $ES_B = EF_A$. Similarly for C, $ES_C = EF_A$. Compute the earliest finishing times for B and C by adding the activity times to the earliest starting times: $EF_B = ES_B + t_B$, and $EF_C = ES_C + t_C$. Summarizing the calculations up to this point, we have

Activity	Time	Immediate Predecessor	ES	EF
A	3	—	0	3
B	4	A	3	7
C	2	A	3	5

Calculating the earliest starting time for D is more difficult. Activity D has two immediate predecessors, B and C. That means that D cannot start until *both* B and C have been completed. It follows that the earliest starting time for D is the *later* of the earliest finishing times for B and C. That is,

$$ES_D = \max(EF_B, EF_C) = \max(7, 5) = 7.$$

General rule: The earliest starting time of an activity is the *maximum* of the earliest finishing times of its immediate predecessors.

Continuing in this manner, we obtain the following earliest times for all remaining activities:

Activity	Time	Immediate Predecessor	ES	EF
A	3	—	0	3
B	4	A	3	7
C	2	A	3	5
D	6	B, C	7	13
E	5	C	5	10
F	3	C	5	8
G	7	E	10	17
H	5	E, F	10	15
I	8	D, G, H	17	25

At this point we have actually determined the length of the critical path. It is the maximum of the earliest finish times, or 25 weeks in this case. However, we must find the latest times before we can identify the critical activities.

The computation of the latest times proceeds by working backward through the network and applying essentially the dual of the procedure for the earliest times. The first step is to set the latest finishing time of all the activities that enter the final node to the maximum value of the earliest finishing times. In this case there is only a single ending activity, so we set

$$LF_I = 25.$$

The latest starting time is obtained by subtracting the activity time, so that

$$LS_I = 25 - 8 = 17.$$

Next we must determine the latest finishing time for all of the activities that enter node 7, which are D, G, and H. Because these activities end when I begins, we have

$$LF_D = LF_G = LF_H = LS_I = 17.$$

One finds the latest starting times for D, G, and H by simply subtracting the activity times. Summarizing the calculations up to this point, we have

Activity	Time	Immediate Predecessor	ES	EF	LS	LF
A	3	—	0	3		
B	4	A	3	7		
C	2	A	3	5		
D	6	B, C	7	13	11	17
E	5	C	5	10		
F	3	C	5	8		
G	7	E	10	17	10	17
H	5	E, F	10	15	12	17
I	8	D, G, H	17	25	17	25

Because F ends when H begins, $LF_F = LS_H = 12$, and $LS_F = 12 - 3 = 9$. Now consider activity E. From Figure 9–5, E has both G and H as immediate successors. This means that E must end prior to the time that both G and H start. Hence the latest finishing time for E is the *earlier* of the latest start times for G and H. That is, $LF_E = \min(LS_G, LS_H) = \min(10, 12) = 10$.

General Rule: The latest finishing time of an activity is the *minimum* of the latest start times of its immediate successors.

According to the network diagram of Figure 9–5, C has both E and F as immediate successors. Hence $LF_C = \min(LS_E, LS_F) = \min(5, 9) = 5$, and $LS_C = 5 - 2 = 3$. Because B has only D as an immediate successor, $LF_B = LS_D = 11$, and $LS_B = 11 - 4 = 7$. Finally, A has both B and C as immediate successors, so that $LF_A = \min(LS_B, LS_C) = \min(7, 3) = 3$, and $LS_A = 3 - 3 = 0$.

The complete summary of the calculations for this example is

Activity	Time	Immediate Predecessors	ES	EF	LS	LF	Slack
A	3	—	0	3	0	3	0
B	4	A	3	7	7	11	4
C	2	A	3	5	3	5	0
D	6	B, C	7	13	11	17	4
E	5	C	5	10	5	10	0
F	3	C	5	8	9	12	4
G	7	E	10	17	10	17	0
H	5	E, F	10	15	12	17	2
I	8	D, G, H	17	25	17	25	0

FIGURE 9–7

Gantt chart for software development project of Example 9.4

FIGURE 9–7

Gantt chart for software development project of Example 9.4

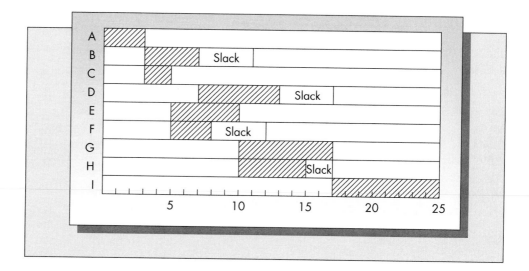

We have added a column labeled "slack." This is the difference between the columns LS and ES (it is also the difference between the columns LF and EF). The slack is the amount of time that an activity can be delayed without delaying the project. The activities with zero slack—A, C, E, G, and I—are the critical activities and constitute the critical path.

Figure 9–7 is a Gantt chart that shows the starting and the ending times for each activity. The noncritical activities are shown starting at the earliest times, although they can be scheduled at any time within the period marked by the slack.

Example 9.4 (continued)

Suppose that the two programmers begin the project on June 1. Consider the following questions:

   a. On what date should the merging of the graphics and the program modules begin in order to guarantee that they complete the project on time?

   b. What is the latest time that the screen development can be completed without delaying the project?

   c. Suppose that North discovers a bug in a coding of module 2, so that the time required to complete module 2 is longer than anticipated. Will this necessarily delay the project?

   d. Suppose that a similar problem is discovered in module 1. Will the project necessarily be delayed in this case?

   e. If Bonner is responsible for the coding of the modules and North is responsible for the screen development, and all time estimates are accurate, will the programmers be able to complete the project within 25 weeks?

Solution

   a. The merging of the graphics and the program modules is activity I. This is a critical activity, so it must be started no later than its earliest start time, which is week 17. The calendar date would be September 14.

    b. The screen development is activity D. This is not a critical activity and has latest finishing time of 17. The calendar date is again September 14.

    c. Module 2 is activity F. Because F is not critical, a delay of up to 4 weeks is permitted.

    d. Module 1 is activity E. This is a critical activity, so a delay will necessarily delay the project.

    e. The answer is no. The problem is that, assuming that Bonner can work only on a single module at one time, the current schedule requires concurrent programming of modules 1 and 2 and modules 3 and 4. The programmers would have to take in another partner or subcontract some of the module development in order to complete the project within 25 weeks. ∎

Problems for Sections 1 and 2

1. Answer the following questions for the case example presented in this section.
   a. What event is represented by node 6 in Figure 9–5?
   b. What group of activities will have to be completed by week 16 in order to guarantee that the project will not be delayed?
   c. Suppose that module 4 is coded by a subcontractor who delivers the module after 7 weeks. How much will it delay the project if the subcontractor does not start until week 11?

2. For the case example presented in this section, suppose that the programmers choose not to obtain additional help to complete the project; that is, Bonner will code the modules without outside assistance.
   a. In what way will this alter the project network and the precedence constraints?
   b. Find the critical path of the project network you obtained in part (*a*). How much longer is it than the one determined in this chapter?

3. A project consisting of eight activities satisfies the following precedence constraints:

Activity	Time (weeks)	Immediate Predecessors
A	3	—
B	5	A
C	1	A
D	4	B, C
E	3	B
F	3	E, D
G	2	F
H	4	E, D

    a. Construct a network for this project. (You should need only one pseudo activity.)

FIGURE 9–8

Project network for Problem 4

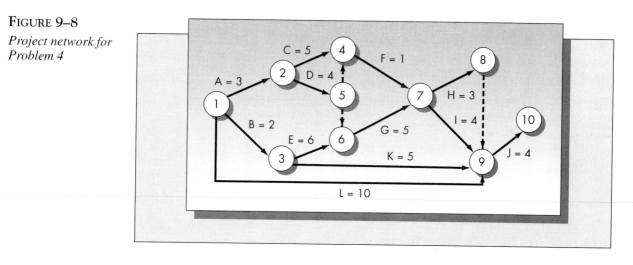

FIGURE 9–9

Network for Problem 5

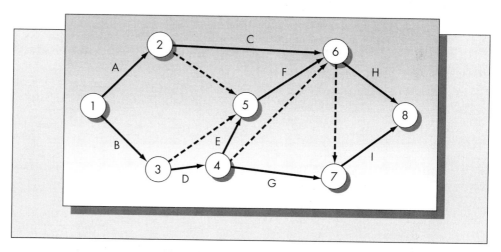

  b. Compute the earliest and the latest starting and finishing times for each activity and identify the critical path.

  c. Draw a Gantt chart of the schedule for this project based on earliest starting times.

4. For the project network pictured in Figure 9–8,

  a. List the immediate predecessors of each activity.

  b. Try to determine the critical path by enumerating all paths from node 1 to node 10.

  c. Compute the earliest starting and finishing times for all activities and identify the critical activities.

5. Consider the network pictured in Figure 9–9.

  a. Determine the immediate predecessors of each activity from the network representation. (Hint: Be certain that you consider only *immediate* predecessors.)

  b. Redraw the network based on the results of part (*a*) with only two pseudo activities.

6. A project consists of seven activities. The precedence relationships are given in the table below.

Activity	Time (days)	Immediate Predecessors
A	32	—
B	21	—
C	30	—
D	45	A
E	26	A, B
F	28	C
G	20	E, F

  a. Construct a network for this project.
  b. Compute the earliest and the latest starting and finishing times for each activity and identify the critical path.
  c. Draw a Gantt chart of the schedule for this project based on earliest starting times.
  d. What group of activities will have to be completed by day 60 in order to guarantee that the project will not be delayed?

9.3 Time Costing Methods

Besides assisting with the scheduling of large projects, CPM is a useful tool for project costing and comparing alternative schedules based on cost. This section will consider the costs of expediting activities and how one incorporates expediting costs into the project management framework.

In the previous section we assumed that the time required to complete each activity is known and fixed. In this section we will assume that activity times can be reduced at additional cost. Assume that the time specified for completing an activity is the *normal time*. The minimum possible time required is defined as the *expedited time*. Furthermore, assume that the costs of completing the activity in each of these times are known. Then the CPM assumption is that the costs of completing an activity at times between the normal and the expedited times lie along a straight line, as pictured in Figure 9–10. Assuming that the expediting cost function is linear should be reasonable in most circumstances.

There are two types of costs in most projects: direct and indirect costs. Direct costs include costs of labor, material, equipment, and so on. Indirect costs include costs of overhead such as rents, interest, utilities, and any other costs that increase with the length of the project.

Indirect costs and direct costs are respectively increasing and decreasing functions of the project completion time. When these functions are convex, the total cost function, which is their sum, also will be convex. This means that there will be a value of the project time between the normal and the expedited times that is optimal in the sense of minimizing the total cost. Convex cost functions are pictured in Figure 9–11. Figure 9–10 shows the direct cost as a function of time for a given activity.

FIGURE 9–10

The CPM cost–time linear model

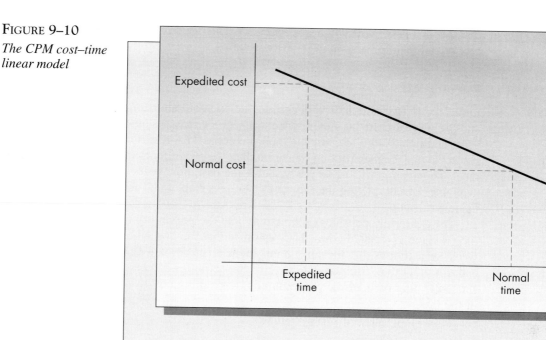

FIGURE 9–11

Optimal project completion time

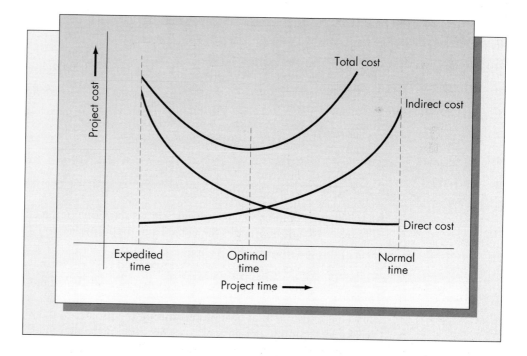

The general approach is to successively reduce the project time by one week (or in whatever unit of time activities are measured) until no further reductions are possible or until an optimal solution is identified. At each reduction, one computes the resulting additional direct cost. One continues this process until the minimum total cost solution is identified. A difficulty that arises is that as particular activities are expedited, it is possible that new paths will become critical. The procedure can be illustrated with the case example introduced in the previous section.

Example 9.5

Let us return to Example 9.4 concerning the two computer consultants, Irving Bonner and Simon North. In the previous section, we saw that it is possible for them (with some additional help) to complete their project in 25 weeks. Once they complete the project and place the program on the market, the consultants anticipate that they will receive an average of $1,500 per week for the first three years that the product is available. By completing the project earlier, they hope to be able to realize this income earlier.

They carefully consider each activity and the possibility of reducing the activity time and the associated costs. They estimate that the normal costs are $500 per week for activities that the consultants themselves do, and either more or less for activities that they contract out. Most of the activities can be expedited by subcontracting parts of the programming. Expediting costs vary based on the nature of the task.

They obtain the following estimates:

Activity	Normal Time (weeks)	Expedited Time (weeks)	Normal Cost	Expedited Cost	Cost per Week
A	3	1	$ 1,000	$ 3,000	$ 1,000
B	4	3	4,000	6,000	2,000
C	2	2	2,000	2,000	—
D	6	4	3,000	6,000	1,500
E	5	4	2,500	3,800	1,300
F	3	2	1,500	3,000	1,500
G	7	4	4,500	8,100	1,200
H	5	4	3,000	3,600	600
I	8	5	8,000	12,800	1,600

The final column, cost per week, shows the slope of the cost curve pictured in Figure 9–10. One computes the slope from the formula

$$\text{Cost per week} = \frac{\text{Expedited cost} - \text{Normal cost}}{\text{Normal time} - \text{Expedited time}}.$$

One finds the total cost of performing the project in the normal time of 25 weeks by summing the normal costs for each of the activities. This sum is $29,500. If all activities are expedited, the total cost of the project increases to $48,300. Is this additional cost worth the $1,500 per week in additional revenue they expect to realize?

If we replace the normal times with the expedited times, then using the methods of the last section (or just inspecting the network), it is easy to show that there will be two critical paths based on expedited times: A–C–E–G–I and A–C–E–H–I. The project completion time is reduced to 16 weeks. The additional income that the programmers realize by reducing the project completion time from 25 weeks to 16 weeks is (9)(1,500) = $13,500, but the additional cost of the reduction is $48,300 − $29,500 = $18,800. Hence, it is not economical to reduce all of the activities to their expedited times. (Expediting all activities means that both critical and noncritical activities are expedited. However, there is clearly no economy in expediting noncritical activities. Hence, it is likely that there is a solution with a project completion time of 16 weeks that costs less than $48,300. The preliminary analysis is used to get a ballpark estimate of the cost of reducing the project to its minimum time.)

It is likely that there is a project time between 16 and 25 weeks that is optimal. To determine the optimal project time, we will find the increase in the direct cost resulting from successive reductions of one week. If the cost increase is less than $1,500, then the reduction is clearly economical, and additional reductions should be considered. If the cost of the reduction exceeds $1,500, then further reductions are not economical and the process should be terminated.

The key point to note is that in order to reduce the time required to complete the project by one week, it is necessary to reduce the time of an activity *along the current critical path,* or activities along the current critical paths if more than one path is critical. Reducing the time of a noncritical activity will not reduce the project time.

Initially, we have the following:

Project Time	Critical Path(s)	Critical Activities	Current Time	Expedited Time	Cost to Reduce by One Week
25	A–C–E–G–I	A	3	1	$1,000
		C	2	2	—
		E	5	4	1,300
		G	7	4	1,200
		I	8	5	1,600

The least expensive activity to reduce is A. We can reduce activity A to 1 week without introducing any new critical paths. Because the cost of each weekly reduction is less than $1,500, it is economical to reduce A to its minimum time, which is one week. At that point we have the following:

Project Time	Critical Path(s)	Critical Activities	Current Time	Expedited Time	Cost to Reduce by One Week
23	A–C–E–G–I	A	1	1	—
		C	2	2	—
		E	5	4	$1,300
		G	7	4	1,200
		I	8	5	1,600

The next cheapest activity to reduce is G. The critical path will remain the same until G is reduced to five weeks. (Consider reducing G by one week at a time to be certain that no additional paths become critical.) When G is reduced to five weeks, both paths A–C–E–G–I and A–C–E–H–I are critical. Reducing G from seven weeks to five weeks results in the following:

Project Time	Critical Path(s)	Critical Activities	Current Time	Expedited Time	Cost to Reduce by One Week
21	A–C–E–G–I				
	A–C–E–H–I	A	1	1	—
		C	2	2	—
		E	5	4	$1,300
		G	5	4	1,200
		H	5	4	600
		I	8	5	1,600

In order to further reduce the project time to 20 weeks, it is necessary to be certain that we make the reduction along *both* critical paths. At first it would seem that we should reduce H because its marginal cost is least. However, this is not the case. If we reduce H to four weeks then only the critical path A–C–E–H–I is reduced and not the path A–C–E–G–I. If we reduce both H and G, the increase in the direct cost is $1,800, which is not economical. Does this necessarily mean that it is not worth reducing the project time any further? The answer is no.

Note that the activities A, C, E, and I lie simultaneously along *both* critical paths. Hence, a reduction in the activity time of any one of these four activities will result in a reduction of the project time. Among these activities, E can be reduced from five to four weeks for under $1,500. Making this reduction, we obtain

Project Time	Critical Path(s)	Critical Activities	Current Time	Expedited Time	Cost to Reduce by One Week
20	A–C–E–G–I				
	A–C–E–H–I	A	1	1	—
		C	2	2	—
		E	4	4	—
		G	5	4	$1,200
		H	5	4	600
		I	8	5	1,600

At this point the cost of reducing the project by an additional week exceeds $1,500, so we have reached the optimal solution. The reduction from 25 weeks to 20 weeks costs a total of $(1,000)(2) + (1,200)(2) + 1,300 = \$5,700$ in additional direct costs, and results in a return of $(1,500)(5) = \$7,500$ in additional revenue. If all cost and time estimates are correct, the programmers have realized a savings of $1,800 by taking costing into account. ∎

Problems for Section 3

7. Consider the case problem presented in this section in which the two programmers are to develop a commercial software package.
   a. What is the minimum project time and the total direct cost of completing the project in that time?
   b. Suppose that all of the activity times are reduced to their minimum values. Explain why the total direct cost obtained in Example 9.5 is different from the cost you obtained in part (a).

8. Consider the project described in Problem 3 with the normal and the expedited costs and times given in the table below.

Activity	Immediate Predecessor	Normal Time	Expedited Time	Normal Cost	Expedited Cost
A	—	3	2	$200	$250
B	A	5	3	600	850
C	A	1	1	100	100
D	B, C	4	2	650	900
E	B	3	2	450	500
F	E, D	3	2	500	620
G	F	2	1	500	600
H	E, D	4	2	600	900

   a. Consider successive reductions in the project time of one week and find the direct cost of the project after each reduction.
   b. Suppose that indirect costs are $150 per week. Find the optimal project completion time and the optimal total project cost.

9. Discuss the assumption that the cost–time curve is linear. What shape might be more realistic in practice?

10. Consider Problem 6. Suppose that the normal and the expedited costs and times are as given in the table below.

Activity	Normal Time	Expedited Time	Normal Cost	Expedited Cost
A	32	26	$ 200	$ 500
B	21	20	300	375
C	30	30	200	200
D	45	40	500	800
E	26	20	700	1,360
F	28	24	1,000	1,160
G	20	18	400	550

   If indirect costs amount to $100 per day, determine the optimal time to complete the project and the optimal project completion cost.

9.4 Solving Critical Path Problems with Linear Programming

Finding critical paths in project networks and finding minimum cost schedules can be accomplished by using either dedicated project scheduling software or linear programming. If one expects to solve project scheduling problems on a continuing basis, it is worth the investment in a dedicated software product. However, linear programming is a useful tool for solving a moderately sized problem on an occasional basis. Many linear programming packages, such as Excel's Solver, are widely available and easy to learn and use. This section shows how to formulate and solve critical path problems as linear programs. (In what follows, this chapter assumes that the reader is familiar with formulating problems as linear programs and interpreting computer output. Supplement 1, which follows Chapter 3, provides a discussion of linear programming and how linear programs can be solved with Excel.)

Based on the choice of objective functions, the linear programming solution will give either earliest or latest start times for each node in the project network. We first formulate the earliest start time problem. Assume that the network representation of the project consists of nodes 1 through m, with node 1 representing the starting time of the project and node m representing the ending time of the project.

Let

x_i = the earliest start time for node i.

t_{ij} = the time required to complete activity (i, j).[1]

Then the minimum project completion time is the solution to the following linear program:

$$\min \sum_{i=1}^{m} x_i$$

subject to $x_j - x_i \geq t_{ij}$ for all pairs of nodes corresponding to activity (i, j).

$$x_i \geq 0 \qquad \text{for } 1 \leq i \leq m.$$

The constraints guarantee that there is sufficient time separating each node to account for the activity times represented by the arc between the nodes. As the objective function minimizes the x_i values, the linear programming solution will give the earliest start times. The latest start times can be found by replacing the objective function with

$$\min \left\{ mx_m - \sum_{i=1}^{m-1} x_i \right\}.$$

As x_m is the project completion time, we still wish to find the smallest allowable value of x_m, so its sign must remain positive. Because each $x_i \leq x_m$, the multiplier m for the term x_m guarantees that the objective function is positive to ensure that we obtain a bounded solution. We reverse the sign of the remaining node variables so that the minimization will seek their largest values, but will still seek the minimum value for x_m. This means that the variable values will be the latest start times for the activities emanating from each of the nodes. Once the earliest and the latest start times for all the nodes have been determined, it is easy to translate this into earliest and latest start times for the activities using the network representation of the project. The resulting value of x_m, the minimum project completion time, is the same for both formulations.

---

[1] We used letters to represent activities in the earlier sections, but for most computer-based systems activities are represented in the form (i, j), where i is the origination node and j is the destination node.

Example 9.4 (continued)

We will solve Example 9.4 using linear programming. From the network representation of the project given in Figure 9–5, we see that there are a total of eight nodes. The formulation of the problem that gives the earliest start times is

$$\min \sum_{i=1}^{8} x_i$$

subject to the following:

$x_2 - x_1 \geq 3$	(A)
$x_3 - x_2 \geq 4$	(B)
$x_4 - x_2 \geq 2$	(C)
$x_3 - x_4 \geq 0$	(P_1)
$x_5 - x_4 \geq 5$	(E)
$x_6 - x_4 \geq 3$	(F)
$x_6 - x_5 \geq 0$	(P_2)
$x_7 - x_3 \geq 6$	(D)
$x_7 - x_6 \geq 5$	(H)
$x_7 - x_5 \geq 7$	(G)
$x_8 - x_7 \geq 8$	(I)
$x_i \geq 0 \quad$ for $1 \leq i \leq 8$.	

The activity giving rise to each constraint is shown in parentheses. Constraints for the pseudo activities must be included as well. Also, the constraint corresponding to P_1 is $x_3 - x_4 \geq 0$, and not vice versa, because P_1 corresponds to the directed arc from node 4 to node 3. The relevant portion of the Excel output is given below.

Target Cell (Min)

Cell	Name		Original Value	Final Value
$M$7	Min	Value	0	77

Adjustable Cells

Cell	Name	Original Value	Final Value
$B$5	x1	0	0
$C$5	x2	0	3
$D$5	x3	0	7
$E$5	x4	0	5
$F$5	x5	0	10
$G$5	x6	0	10
$H$5	x7	0	17
$I$5	x8	0	25

The minimum project completion time is the value of x_8, which is 25 weeks. The earliest times correspond to nodes rather than activities, so that these times must be converted to activity times. This is easy to do by referring to the network representation in Figure 9–5. For example, because both B and C emanate from node 2, they would have earliest start times of 3 (the value of x_2). You should satisfy yourself that the earliest start times for the other activities obtained in this way agree with the solution we obtained in Section 2.

Changing the objective function to $8x_8 - \Sigma x_i$ and rerunning Excel gives the following output:

```
Target Cell (Min)
     Cell           Name        Original Value     Final Value
     $M$7      Min  Value                  148             142


Adjustable Cells
     Cell           Name        Original Value     Final Value
     $B$5      x1                            0               0
     $C$5      x2                            3               3
     $D$5      x3                            7              11
     $E$5      x4                            5               5
     $F$5      x5                           10              10
     $G$5      x6                           10              12
     $H$5      x7                           17              17
     $I$5      x8                           25              25
```

This now gives the latest start times for all of the activities. The noncritical activities are those with the latest start times that differ from the earliest start times; the magnitude of the difference is the slack time. Again, you should assure yourself that these results agree with those obtained in Section 2. ∎

Linear Programming Formulation of the Cost–Time Problem

Linear programming also can be used to find the optimal completion time when expediting costs are included. The formulation can result in a large linear program even for moderately sized problems. We again will refer to activities by the pair (i, j), where i is the origination node and j is the destination node. Define M_{ij} as the expedited time for activity (i, j) and N_{ij} as the normal time for activity (i, j). Suppose that the cost–time function (Figure 9–10) has the representation $a_{ij} - b_{ij}t_{ij}$, where a_{ij} is the y intercept, b_{ij} the slope, and t_{ij} the activity time. Let C be the indirect cost per day. Then the linear programming formulation of the problem of finding the optimal project completion time is

$$\min \sum_{\text{all } (i,j)} [a_{ij} - b_{ij}t_{ij}] + Cx_m$$

subject to

$$x_j - x_i \geq t_{ij} \qquad \text{for all activities } (i, j),$$
$$t_{ij} \leq N_{ij},$$
$$t_{ij} \geq M_{ij},$$
$$x_i \geq 0 \qquad \text{for } 1 \leq i \leq m,$$
$$t_{ij} \geq 0 \qquad \text{for all activities } (i, j).$$

Note that the a_{ij} terms in the objective function are constants and can be eliminated without altering the solution. These constants can be added later to find the optimal value of the objective function.

The activity times, t_{ij}, are now problem variables rather than given constants. This linear programming problem is considerably larger than the one in the previous section. There is now a variable for each node in the network *and* a variable for each activity. However, the added burden is still less work than finding the minimum cost solution manually.

Example 9.5 (continued)

We will solve Example 9.5 using linear programming. The relevant data for developing the linear programming formulation are

Activity	Node Representation	a_{ij}	b_{ij}	M_{ij}	N_{ij}
A	(1,2)	4,000	1,000	1	3
B	(2,3)	12,000	2,000	3	4
C	(2,4)	—	—	2	2
D	(3,7)	12,000	1,500	4	6
E	(4,5)	9,000	1,300	4	5
F	(4,6)	6,000	1,500	2	3
G	(5,7)	12,900	1,200	4	7
H	(6,7)	6,000	600	4	5
I	(7,8)	20,800	1,600	5	8

The resulting linear program is

$$\text{Min}\{-1{,}000t_{12} - 2{,}000t_{23} - 1{,}500t_{37} - 1{,}300t_{45} - 1{,}500t_{46}$$
$$-1{,}200t_{57} - 600t_{67} - 1{,}600t_{78} + 1{,}500x_8\}$$

subject to

$$x_2 - x_1 - t_{12} \geq 0,$$
$$x_3 - x_2 - t_{23} \geq 0,$$
$$x_4 - x_2 - t_{24} \geq 0,$$
$$x_3 - x_4 \qquad \geq 0,$$
$$x_5 - x_4 - t_{45} \geq 0,$$
$$x_6 - x_4 - t_{46} \geq 0,$$
$$x_6 - x_5 \qquad \geq 0,$$
$$x_7 - x_3 - t_{37} \geq 0,$$
$$x_7 - x_6 - t_{67} \geq 0,$$
$$x_7 - x_5 - t_{57} \geq 0,$$
$$x_8 - x_7 - t_{78} \geq 0,$$
$$1 \leq t_{12} \leq 3,$$
$$3 \leq t_{23} \leq 4,$$
$$4 \leq t_{45} \leq 5,$$
$$2 \leq t_{46} \leq 3,$$
$$4 \leq t_{57} \leq 7,$$
$$4 \leq t_{67} \leq 5,$$
$$5 \leq t_{78} \leq 8,$$
$$4 \leq t_{37} \leq 6,$$
$$t_{24} = 2,$$

$$x_i \geq 0 \qquad \text{for } 1 \leq i \leq 8,$$
$$t_{ij} \geq 0 \qquad \text{for all activities } (i, j).$$

The upper and the lower bound constraints on the t_{ij} variables would have to be entered as two separate constraints into the computer, giving a total of 28 constraints. The Excel output for this problem is

Target Cell (Min)			
Cell	Name	Original Value	Final Value
$T$7	Min Value	0	-19500

Adjustable Cells			
Cell	Name	Original Value	Final Value
$B$5	x1	0	0
$C$5	x2	0	1
$D$5	x3	0	6
$E$5	x4	0	3
$F$5	x5	0	7
$G$5	x6	0	7
$H$5	x7	0	12
$I$5	x8	0	20
$J$5	t12	0	1
$K$5	t23	0	4
$L$5	t24	0	2
$M$5	t37	0	6
$N$5	t45	0	4
$O$5	t46	0	3
$P$5	t57	0	5
$Q$5	t67	0	5
$R$5	t78	0	8

The optimal length of the project is given by the value of $x_8 = 20$, and the optimal activity times are the values of t_{ij}. Notice that these times agree with the activity times we found to be optimal in Section 3. The total cost of this solution is found by adding $\Sigma\, a_{ij}$, which is \$82,700, and the cost for activity C, which was not included in the objective function. (We could have included the cost of activity C in the objective function by adding the term $1000t_{24}$, as t_{24} is fixed at 2.) The resulting value for the total cost of the project at 20 days is \$82,700 + \$2,000 − \$19,500 = \$65,200.

In general, linear programming is not an efficient way to solve large network problems. Algorithms that exploit the network structure of the problem are far more efficient. Many commercial software products based on such algorithms are available for project scheduling. Even PC-based software products are capable of solving large projects. Linear programming is a useful tool for solving moderately sized problems. However, for those with a large project scheduling problem, or with an ongoing need for project scheduling, we recommend a dedicated project scheduling program. See Section 9 for a comprehensive discussion of PC-based project management software. ■

Problems for Section 4

11. Solve Problem 3 by linear programming.
12. Solve Problem 4 by linear programming.
13. Formulate Problem 5 as a linear program.

14. Solve Problem 6 by linear programming.
15. Solve Problem 8 by linear programming.
16. Solve Problem 10 by linear programming.

9.5 PERT: Project Evaluation and Review Technique

PERT is a generalization of CPM to allow uncertainty in the activity times. When activity times are difficult to predict, PERT can provide estimates of the effect of this uncertainty on the project completion time. However, for reasons given in detail below, the results of the analysis are only approximate. Let T_i be the time required to complete activity i. In this section we will assume that T_i is a random variable. Furthermore, we assume that the collection of random variables $T_1, T_2, \ldots, T_n$ is mutually independent. The first issue addressed is the appropriate form of the distribution of these random variables. Define the following quantities:

a = Minimum activity time,
b = Maximum activity time,
m = Most likely activity time.

As an example, suppose that $a = 5$ days, $b = 20$ days, and $m = 17$ days. Then the probability density function of the activity time is as pictured in Figure 9–12. The density function should be zero for values less than 5 and more than 20, and should be a maximum at $t = 17$ days. The point where the density is a maximum is known as the *mode* in probability theory. The *beta distribution* is a type of probability distribution defined on a finite interval that may have its modal value anywhere on the interval. For this reason, the beta distribution is usually used to describe the distribution of individual activity times. The assumption of beta-distributed activity times is used to justify simple approximation formulas for the mean and the variance, but is rarely used to make probabilistic statements concerning individual activities.

FIGURE 9–12

Probability density of activity time

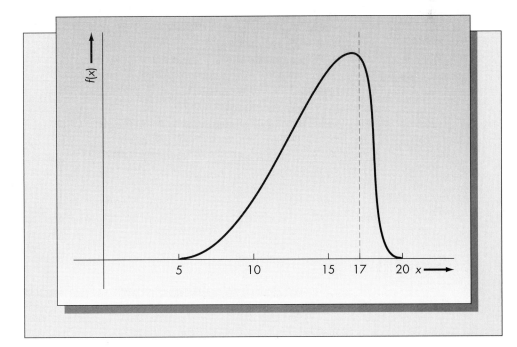

The beta distribution assumption is used to justify the approximations of the mean μ and the standard deviation σ of each activity time. The traditional PERT method is to estimate μ and σ from a, b, and m using the following formulas:

$$\mu = \frac{a + 4m + b}{6}, \qquad \sigma = \frac{b - a}{6}.$$

The formula for the standard deviation seems to be based on the following property of the normal distribution: limits at distance 3σ to either side of the mean for a normal variate include all of the population with probability exceeding .99. In view of this property, it is assumed that there are six standard deviations from a to b. The formula for the mean is obtained by assuming the variance approximation as well as the beta distribution for the activity time. Given the variance, a, b, and m, the mean is determined by solving a cubic equation. Calculating μ for various values of the other parameters and developing the best fit by linear regression results in the one-four-one weighting scheme. (See Archibald and Villoria, 1967, p. 449.)

Squaring both sides in the formula for σ gives

$$\sigma^2 = \frac{(b - a)^2}{36}$$

where σ^2 is the variance.

The uniform distribution is a special case of the beta distribution. If the job time has a uniform distribution from a to b, the density function would be rectangular, as pictured in Figure 9–13. The variance of a uniform variate is $(b - a)^2/12$. Because we would expect the variance to be less for a peaked distribution such as the one pictured in Figure 9–12, the approximation for σ^2 recommended above seems reasonable. Note that the one-four-one weighting scheme gives the correct value of the mean, $(a + b)/2$, if one substitutes $m = (a + b)/2$ as the mode.

FIGURE 9–13

Uniform density of activity times

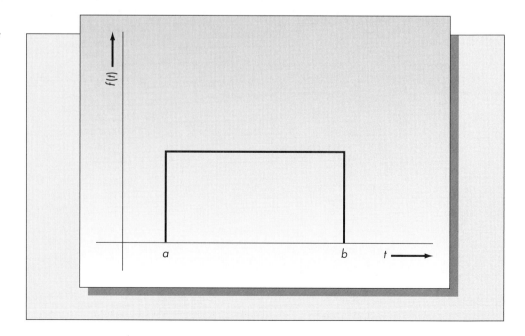

In PERT, one assumes that the distribution of the total project time is normal. The *central limit theorem* is used to justify this assumption. Roughly, the central limit theorem says that the distribution of the sum of independent random variables is approximately normal as the number of terms in the sum grows large. In most cases, convergence occurs quickly. Hence, because the total project time is the sum of the times of the activities along the critical path, it should be approximately normally distributed as long as activity times are independent.

Suppose that the critical activity times are $T_1, T_2, \ldots, T_k$. Then the total project time T is

$$T = T_1 + T_2 + \cdots + T_k.$$

It follows that the mean project time, $E(T)$, and the variance of the project time, $\text{Var}(T)$, are given by

$$E(T) = \mu_1 + \mu_2 + \cdots + \mu_k,$$
$$\text{Var}(T) = \sigma_1^2 + \sigma_2^2 + \cdots + \sigma_k^2.$$

These formulas are based on the following facts from probability theory: the expected value of the sum of *any* set of random variables is the sum of the expected values, and the variance of a sum of *independent* random variables is the sum of the variances. Independence of activity times is required in order to easily obtain the variance of the project time and to justify the application of the central limit theorem. One could modify the variance formula to incorporate correlations among the activities, but the assumption of normality of the project completion time might no longer be accurate. For these reasons, explicit treatment of the dependencies among activity times is rare in practice.

Summarizing the PERT method.

1. For each activity obtain estimates of a, b, and m. These estimates should be supplied by the project manager or by someone familiar with similar projects.
2. Using these estimates, compute the mean and the variance of each of the activity times from the formulas above.
3. Based on the mean activity times, use the methods of the previous section to determine the critical path.
4. Once the critical activities are identified, add the means and the variances of the critical activities to find the mean and the variance of the total project time.
5. The total project time is assumed to be normally distributed with the mean and the variance determined in step 4.

Using the assumption that the project time is normally distributed, we can address a variety of issues. An example will illustrate the method.

Example 9.6

Consider the case study of Example 9.4 of the two computer consultants developing a software project. Before embarking on the project, they decide that it is important to consider the uncertainty of the times required for certain tasks. As with any software project, unanticipated bugs can surface and cause significant delays. Based on their past experience, the programmers decide that the values of a, b, and m are as shown below:

Activity	Min (a)	Most Likely (m)	Max (b)	$\mu = \dfrac{a + 4m + b}{6}$	$\sigma^2 = \dfrac{(b - a)^2}{36}$
A	2	3	4	3	0.11
B	2	4	10	4.67	1.78
C	2	2	2	2	0
D	4	6	12	6.67	1.78
E	2	5	8	5	1.00
F	2	3	8	3.67	1.00
G	3	7	10	6.83	1.36
H	3	5	9	5.33	1.00
I	5	8	18	9.17	4.69

Using these values, we have computed the means and the variances of each of the activity times. Activity C, the design of the flowchart, requires precisely two days. Hence, the variance of this activity time is zero.

The traditional PERT method is to compute the critical path based on the mean activity times. In this example, the introduction of uncertainty does not alter the critical path. It is still A–C–E–G–I. (However, it is possible that in other cases the critical path based on the mean times will *not* be the same as the critical path based on the most likely times.) The expected project completion time, $E(T)$, is simply the sum of the mean activity times along the critical path. In this example,

$$E(T) = 3 + 2 + 5 + 6.83 + 9.17 = 26 \text{ weeks.}$$

Similarly, the variance of the project completion time, $\text{Var}(T)$, is the sum of the variances of the activities along the critical path. In this case,

$$\text{Var}(T) = 0.11 + 0 + 1.0 + 1.36 + 4.69 = 7.16.$$

The assumption made is that the total project completion time T is a normal random variable with mean $\mu = 26$ and standard deviation $\sigma = \sqrt{7.16} = 2.68$. We can now answer a variety of specific questions concerning this project. ∎

Example 9.7

Answer the following questions about the project scheduling problem described in Example 9.6.

1. What is the probability that the project can be completed in under 22 weeks?
2. What is the probability that the project requires more than 28 weeks?
3. Find the number of weeks required to complete the project with probability .90.

FIGURE 9–14

Answer to Example 9.7, Part 1

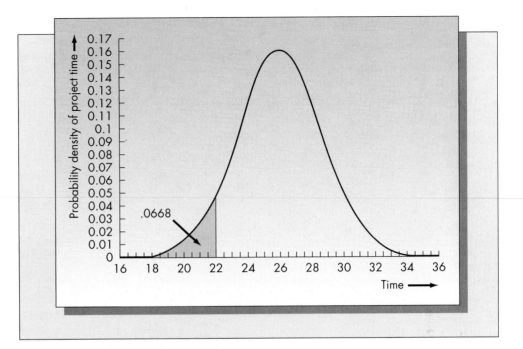

Solution

1. We wish to compute $P\{T < 22\}$.

$$P\{T < 22\} = P\left\{\frac{T - \mu}{\sigma} < \frac{22 - \mu}{\sigma}\right\} = P\left\{Z < \frac{22 - 26}{2.68}\right\}$$

$$= P\{Z < -1.5\} = .0668$$

Z is the standard normal variate. The probability is from Table A–1 of the standard normal distribution in Appendix A. The solution is pictured in Figure 9–14.

2. $P\{T > 28\} = P\left\{\dfrac{T - \mu}{\sigma} > \dfrac{28 - \mu}{\sigma}\right\} = P\left\{Z > \dfrac{28 - 26}{2.68}\right\}$

$$= P\{Z > .75\} = .2266.$$

The solution to this problem is pictured in Figure 9–15.

3. Here we wish to find the value of t such that $P\{T \le t\} = .90$:

$$.90 = P\{T \le t\} = P\left\{Z < \frac{t - \mu}{\sigma}\right\}.$$

It follows that

$$\frac{t - \mu}{\sigma} = z_{.90},$$

or $t = \mu + \sigma z_{.90} = 26 + (2.68)(1.28) = 29.43$ weeks.

The notation $z_{.90}$ stands for the 90th percentile of the standard normal distribution. That is, $P\{Z \le z_{.90}\} = .90$. Its value, 1.28, is found from Table A–1 in Appendix A at the back of the book. The solution is pictured in Figure 9–16. ■

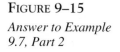

FIGURE 9–15

*Answer to Example
9.7, Part 2*

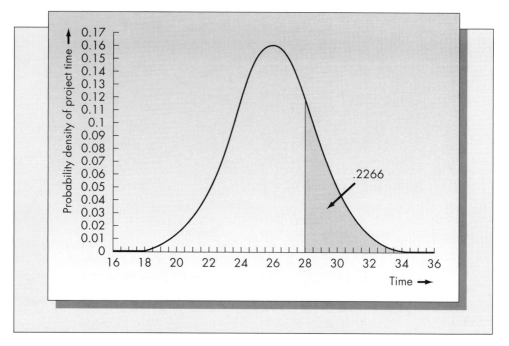

FIGURE 9–16

*Answer to Example
9.7, Part 3*

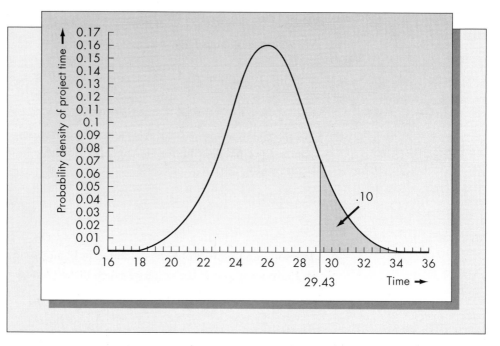

Path Independence

A serious limitation of the PERT method is the assumption that the path with the
longest expected completion time is necessarily the critical path. In most networks
there is a positive probability that a path other than the one with the longest expected
completion time, in fact, will be critical. When this is the case, the calculations presented in the previous section could be very misleading.

Consider the previous example. Suppose that the project has been completed and the following realizations of the activity times observed:

Activity	Actual Time Required to Complete	Activity	Actual Time Required to Complete
A	3.4	F	5.0
B	4.0	G	6.2
C	2.0	H	7.2
D	7.0	I	13.0
E	3.5		

With these activity times, the critical path is no longer A–C–E–G–I, but is now A–C–F–H–I. Hence, the PERT assumption that a fixed path is critical is not accurate.

Unfortunately, determining the exact distribution of the critical path is difficult. In general, this distribution is not known. The difficulty is that, because different paths include the same activities, the times required to complete different paths are *dependent* random variables.

Depending on the particular configuration of the project network, assuming independence of two or more paths may be more accurate than assuming a single critical path. Consider the following example.

Example 9.8

Consider the project pictured in Figure 9–17. In the figure we have included the means and the variances of the activity times. Assume that these times are expressed in weeks. There are exactly two paths from node 1 to node 9: A–C–E–G–I and B–D–F–H–I. The expected times of these paths are 41 and 40 weeks, respectively. Using PERT, we would assume that the critical path is A–C–E–G–I. However, there is almost an equal likelihood that path B–D–F–H–I will turn out to be critical after the activity times are realized.

In this example, the two paths have only activity I in common. Hence, the paths are almost statistically independent. Performing the calculations assuming two independent paths could give very different results from assuming a unique critical path.

FIGURE 9–17

Network for PERT Example 9.8

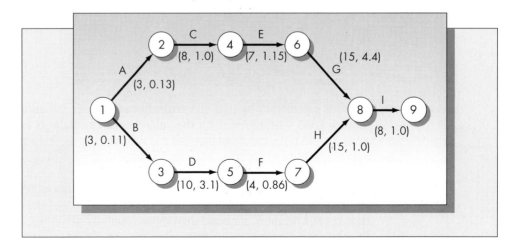

Suppose that we wish to determine the probability that the project is completed within 43 weeks. Let T_1 be the time required to complete path A–C–E–G–I and T_2 the time required to complete path B–D–F–H–I. Using the methods of the previous section, we conclude that T_1 and T_2 are approximately normally distributed, with

$$E(T_1) = 41,$$

$$\text{Var}(T_1) = 0.13 + 1 + 1.15 + 4.4 + 1 = 7.68,$$

and

$$E(T_2) = 40,$$

$$\text{Var}(T_2) = 0.11 + 3.1 + 0.86 + 3.0 + 1 = 8.07.$$

Let T be the total project time. Clearly, $T = \max(T_1, T_2)$. It follows that

$$P\{T < 43\} = P\{\max(T_1, T_2) < 43\} = P\{T_1 < 43, T_2 < 43\}$$

(the last equality follows from the fact that if the maximum of two quantities is less than a constant, then both quantities must be less than that constant)

$$= P\{T_1 < 43\}\, P\{T_2 < 43\}$$

(which follows from the assumption of path independence)

$$= P\left\{Z < \frac{43 - 41}{\sqrt{7.68}}\right\} P\left\{Z < \frac{43 - 40}{\sqrt{8.07}}\right\}$$

$$= P\{Z < 0.72\}\, P\{Z < 1.05\}$$

$$= (.7642)(.8413) = .6429.$$

The methods of the previous section would have given the estimate of completing the project within 43 weeks as .7642. For this network, .6429 is far more accurate.

Certain calculations are more complex if we assume path independence. For example, suppose that we wanted to know the number of weeks required to complete the project with probability .90. Then we wish to find t to satisfy

$$P\{T_1 < t\}\, P\{T_2 < t\} = .90$$

or

$$P\left\{Z < \frac{t - 41}{2.77}\right\} P\left\{Z < \frac{t - 40}{2.84}\right\} = .90.$$

One calculates t by trial and error. Because the probabilities are likely to be close, a good starting guess for the value of each probability is $\sqrt{.90} \approx .95$, which gives $t = 45.6$ for the first term and $t = 44.7$ for the second term. The correct value is approximately $t = 45.2$ weeks, which results in a value of .904 for the product of the two probabilities.

Because the two paths for this project have only one activity in common, the assumption of path independence is quite reasonable and the answers obtained in this manner far more accurate than those found by assuming a unique critical path. In most networks, however, paths may have many activities in common, and the assumption of path independence may be inaccurate. Consider the network for

Example 9.4 pictured in Figure 9–5. In this example, there are a total of five paths:

A–B–D–I,
A–C–P_1–D–I,
A–C–E–G–I,
A–C–E–P_2–H–I,
A–C–F–H–I.

The expected lengths of the five paths are respectively 23.51, 20.84, 26, 23, and 23.17. Unfortunately, path A–C–F–H–I contains three activities in common with the longest expected path, A–C–E–G–I. In such a case, it is not clear which choice will give more accurate results: including this path and assuming path independence or excluding it from consideration.

We will compute the probability that the project can be completed in under 22 weeks assuming path independence.

Path	Expected Completion Time	Variance of Completion Time
A–B–D–I	23.5	8.36
A–C–P_1–D–I	20.8	6.58
A–C–E–G–I	26.0	7.16
A–C–E–P_2–H–I	23.0	6.80
A–C–F–H–I	23.2	6.80

If T is the project completion time and $T_1, \ldots, T_5$ are the times required to complete each of the five paths listed above, then

$$T = \max(T_1, \ldots, T_5).$$

It follows that

$$P\{T < 22\} = P\{T_1 < 22, T_2 < 22, T_3 < 22, T_4 < 22, T_5 < 22\}$$

$$\approx P\{T_1 < 22\}\, P\{T_2 < 22\}\, P\{T_3 < 22\}\, P\{T_4 < 22\}\, P\{T_5 < 22\}.$$

Again, assuming a normal distribution for each of the path completion times, we have

$$P\{T_1 < 22\} = P\left\{Z < \frac{22 - 23.5}{\sqrt{8.36}}\right\} = P\{Z < -0.52\} = .3015,$$

$$P\{T_2 < 22\} = P\left\{Z < \frac{22 - 20.8}{\sqrt{6.58}}\right\} = P\{Z < -0.47\} = .6808,$$

$P\{T_3 < 22\} = .0668$ (from the previous section).

$$P\{T_4 < 22\} = P\left\{Z < \frac{22 - 23}{\sqrt{6.8}}\right\} = P\{Z < -0.38\} = .3520,$$

$$P\{T_5 < 22\} = P\left\{Z < \frac{22 - 23.2}{\sqrt{6.8}}\right\} = P\{Z < -0.46\} = .3228.$$

Hence, it follows that

$$P\{T < 22\} \approx (.3015)(.6808)(.0668)(.3520)(.3228) = .0016.$$

The true value of the probability will fall somewhere between .0016 and the probability computed by traditional PERT methods, .0668. Assuming path independence can have a very significant effect on the probabilities. For this example, it is safe to say that the likelihood that the consultants complete the project in less than 22 weeks is far less than 6.68 percent and is probably well under 1 percent. ■

Problems for Section 5

17. Referring to Example 9.4 and Figure 9–5, what is the probability that node 6 (i.e., the completion of activities A, C, E, and F) is reached before 12 weeks have elapsed?

18. With reference to Example 9.4 and Figure 9–5, what is the conditional probability that the project is completed by the end of week 25, given that activities A through H (node 7) are completed at the end of week 15? Assume that the time required to complete activity I is normally distributed for your calculation.

19. Consider the following PERT time estimates:

Activity	Immediate Predecessors	a	m	b
A	—	2	5	9
B	A	1	6	8
C	A	3	5	12
D	B	2	4	12
E	B, C	4	6	8
F	B	6	7	8
G	D, E	1	2	6
H	F, G	4	6	16

   a. Draw a network for this project and determine the critical path based on the *most likely times* by inspection.
   b. Assuming that the critical path is the one you identified in part (a), what is the probability that the project will be completed before 28 weeks? Before 32 weeks?
   c. Assuming that the critical path is the one you identified in part (a), how many weeks are required to complete the project with probability .95?

20. Consider the project network pictured in Figure 9–18. Assume that the times attached to each node are the means and the variances of the project completion times respectively.
   a. Identify all paths from nodes 1 to 6.
   b. Which path is critical based on the expected completion times?
   c. Determine the probability that the project is completed within 20 weeks assuming that the path identified in (b) is critical.
   d. Using independence of paths A–C–E and B–F–G only, recompute the answer to part (c).

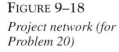

FIGURE 9–18

Project network (for Problem 20)

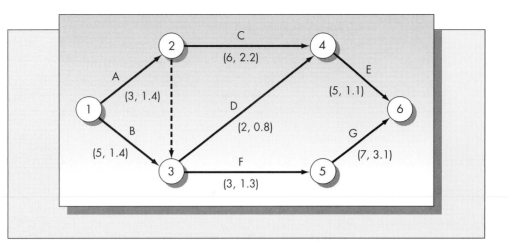

e. Recompute the answer to part (*c*) assuming independence of the paths identified in part (*a*).

f. Which answer, (*c*), (*d*), or (*e*), is probably the most accurate?

21. Consider the project described in Problem 3. Suppose that the activity times are random variables with a constant *coefficient of variation* of 0.2. (The coefficient of variation is the ratio σ/μ.) Assume that the times given in Problem 3 are the mean activity times.

a. Compute the standard deviation of each of the activity times.

b. Find the mean and the variance of the path with the longest expected completion time.

c. Using the results of part (*b*), estimate the probability that the project is completed within 20 weeks.

d. Using the results of part (*b*), estimate the number of weeks required to complete the project with probability .85.

9.6 Resource Considerations

Resource Constraints for Single-Project Scheduling

An implicit assumption made throughout this chapter is that sufficient resources are available and only the *technological constraints* (precedence relationships) are important for setting schedules. In most environments, however, resource constraints cannot be ignored. Examples of limited resources that would affect project schedules are manpower, raw materials, and equipment. Because traditional CPM ignores resource considerations, one manager described it as a "feasible procedure for producing a nonfeasible schedule."

Determining optimal schedules for complex project networks subject to resource limitations is an extremely difficult combinatorial problem. A single project may require a variety of different resources, or many different projects may compete for one or more resources. Heuristic (approximate) methods are generally used to modify schedules obtained by more conventional means.

Allocation of scarce resources among competing activities has been an area of considerable interest in recent years. This section considers the allocation process using the example of the programming project discussed earlier in the chapter.

Example 9.9

Consider Example 9.4 about the two programmers developing a software package. An analysis based only on precedence considerations showed that the project could be completed in 25 weeks without expediting any activities. This analysis, however, is based on the assumption of unlimited resources. In fact, as we noted earlier, it is not possible for the programmers to complete the project within 25 weeks without obtaining additional help. What is the minimum time required for them to complete the project if they do not obtain additional assistance?

Assume that North is responsible for activities B (design of the graphic icons) and D (design of the input/output screens), and Bonner has the expertise required for the development of modules 1 and 2 (activities E and F). Furthermore, assume that either of the programmers can perform activities G and H (modules 3 and 4), but both must work on the final merging and testing. Consider the Gantt chart pictured in Figure 9–7. It shows an infeasible schedule because activities D, E, and H must be done simultaneously between weeks 7 and 8, and activities D, G, and H must be done simultaneously between weeks 10 and 13.

If only Bonner and North are to do the programming, they must reorder the activities so as not to schedule more than two activities simultaneously. Furthermore, they must be certain that these are not two activities that can only be done by one of them. First, they try to find a feasible schedule without delaying the project by rescheduling the noncritical activities within the allowable slack. From Figure 9–7 it is clear that no matter how one rearranges the noncritical activities within the available slack, there is no way to avoid scheduling three activities simultaneously at some point.

This means that they cannot complete the project within 25 weeks without additional help. We determine a feasible schedule by considering the sequence of activities performed by each programmer. At week 3 North begins work on B and Bonner on C. At week 5 Bonner begins work on E, and at week 7 North begins work on D. Because F is module 2 coding, which must be done by Bonner, F must now follow E, so that F starts at week 10. Both G and H can begin at week 13 when both programmers are free. Finally, I starts at week 20 and ends at week 28. Hence, the project time has increased from 25 weeks to 28 weeks as a result of resource considerations.

Figure 9–19 shows the Gantt chart for the modified schedule. This schedule is feasible because there are never more than two activities scheduled at any point in time. North begins work at week 3 and works on the activities B, D, G, and I, while Bonner, who also begins working at week 3, performs activities C, E, F, H, and I. Note that activities G and H are interchangeable. ■

Including resource considerations has a number of interesting consequences. For one, the critical path is no longer the same. Activities B, D, and F no longer have any slack, so they are now critical. Furthermore, there is considerably less slack time overall.

In general, the inclusion of resource constraints has the following effects:

1. The total amount of scheduled slack is reduced.
2. The critical path may be altered. Furthermore, the zero-slack activities may not necessarily lie along one or more critical paths.
3. Earliest- and latest-start schedules may not be unique. They depend on the particular rules that are used to resolve resource limitations.

FIGURE 9–19

Gantt chart for modified schedule for Example 9.9

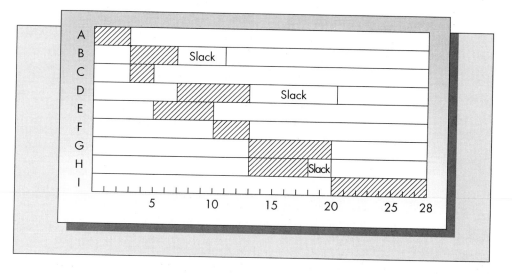

FIGURE 9–19

Gantt chart for modified schedule for Example 9.9

Several heuristic methods for solving this problem exist. Most involve ranking the activities according to some criterion and resolving resource conflicts according to the sequence of the ranking. Some of the ranking rules that have been suggested include the following:

1. *Minimum job slack.* Priority is given to activities with the smallest slack.
2. *Latest finishing times.* When resource conflicts exist, this rule assigns priority to the activity with the minimum latest finishing time.
3. *Greatest resource demand.* This rule assigns priority on the basis of the total resource requirements of all types, giving highest priority to the activities having greatest resource demand. The rationale behind this method is to give priority to potential bottleneck activities.
4. *Greatest resource utilization.* This rule gives priority to that combination of activities that results in the maximum resource utilization (minimum idle time) in any scheduling interval.

These rules (among others) were compared by Davis and Patterson (1975). Their results indicated that the first two methods on the list tended to be the best performers.

Optimal solution methods for project scheduling under resource constraints exist as well. The formulation may be an integer program requiring some type of branch and bound procedure, or may be solved by a technique known as implicit enumeration. For large networks these methods require substantial computer time, but can provide efficient solutions for moderately sized networks with few resource limitations. Patterson (1984) compares and contrasts three optimal solution methods.

Resource Constraints for Multiproject Scheduling

Dealing with resource constraints in single-product scheduling problems can be difficult, but the difficulties are magnified significantly when a common pool of resources is shared by a number of otherwise independent projects. Figure 9–20 shows this type of problem. Two projects require resources A and B. Delaying activities in order to resolve resource conflicts can have far-reaching consequences for all projects requiring the same resources. Commercial computer systems exist that have the capability of dealing with tens of projects and resource types, and possibly thousands of activities.

FIGURE 9–20

*Two projects sharing
two resources*

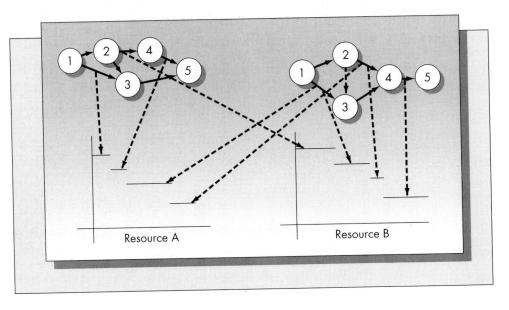

Resource Loading Profiles

Project planning is a useful means for generating schedules of interrelated activities. One significant kind of planning result involves the loading profiles of the required resources. A loading profile is a representation over time of the resources needed. As long as the requirements associated with each activity are known, one can easily obtain the resulting loading profiles for all required resources.

Example 9.10

We illustrate the procedure with the case study first introduced in Example 9.4. Let us suppose that the two programmers are developing the program on a single multiuser computer system. The system can segment the RAM (random access memory) between the two users and can segment the permanent storage on the hard disk as well. Both RAM and permanent storage are measured in megabytes (MB). The requirements for RAM memory and permanent memory for each of the activities comprising the project are given below.

Activity	RAM Required (in MB)	Permanent Storage Required (in MB)
A	0	0
B	1	60
C	0.5	5
D	2	30
E	1	10
F	3	5
G	1.5	15
H	2	10
I	4	80

FIGURE 9–21

FIGURE 9–21

Load profiles for RAM and permanent memory (refer to Example 9.10)

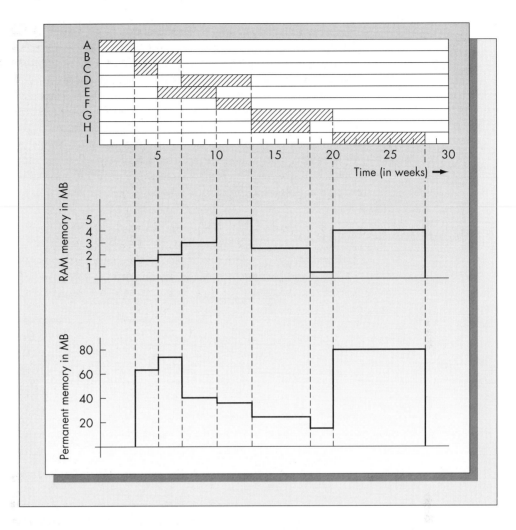

Assume that the activities comprising the project follow the starting and ending times pictured in the Gantt chart of Figure 9–19. Figure 9–21 shows the resulting load profiles of RAM and permanent memory. According to these profiles, the system will require at least 5 MB of RAM and 80 MB of permanent storage. Notice that we are assuming that these requirements are *not* cumulative. Once a portion of the project is completed, the results can be stored on tape or disk and retrieved at a later time. ∎

Resources are either consumable or nonconsumable. In Example 9.10 the resources were nonconsumable. Manpower is another example of a nonconsumable resource. Typical consumable resources are cash or fuels. An issue that arises with consumable resources is the cumulative amount of the resource consumed at each point in time. We will explore load profiles for consumable resources in the problems at the end of this section.

A desirable feature of load profiles is that they be as smooth as possible. Large variations in resource requirements make planning difficult and may result in exceeding resource availability at some point in time. The idea behind *resource leveling* is to reschedule noncritical activities within the available slack in order to smooth out the pattern of resource usage. Often it is possible to do this rescheduling by inspection. For larger networks a systematic method is desirable. Burgess and Killebrew (1962) describe a technique for leveling resource profiles that is based on reducing the value of the sum of squares of the resource profile curve by rescheduling activities within the available slack. We will not review the procedure here.

In summary, resource loading profiles provide an important means of determining the requirements imposed by any particular schedule. Given a project schedule, one can usually construct the loading profiles without the aid of a computer. Rather than a strict forecast of requirements, the profiles are often more useful as rough planning guides when significant variations in activity time are anticipated. According to Moder, Phillips, and Davis (1983), they are probably more widely used than any other resource analysis technique.

Problems for Section 6

22. For the case problem discussed in this chapter, suppose that the requirements for RAM and permanent storage are given by

Activity	RAM Required (in MB)	Permanent Storage Required (in MB)
A	0	0
B	1.5	30
C	2.5	20
D	0.5	10
E	2.0	40
F	1.5	15
G	2.0	25
H	1.5	20
I	4.0	50

a. Determine the load profiles for RAM and permanent storage assuming that the activities are scheduled according to the Gantt chart of Figure 9–7.

b. Determine the load profiles for RAM and permanent storage assuming that the activities are scheduled according to the Gantt chart of Figure 9–19.

23. Consider the project described in Problem 3. Three machines, M1, M2, and M3, are required to complete the project. The requirements for each activity are as follows:

Activity	Machines Required
A	M1, M2
B	M1, M3
C	M2
D	M1, M2
E	M2
F	M1, M3
G	M2, M3
H	M1

Determine the minimum time required to complete the project if there is only one machine of each type available. How many weeks are added to the project time when resource requirements are considered?

24. Consider the project described in Problem 6. The tasks require both welders and pipe fitters. The requirements are

Activity	Number of Welders Required	Number of Pipe Fitters Required
A	6	10
B	3	15
C	8	8
D	0	20
E	10	6
F	10	9
G	4	14

   a. Determine the load profiles for welders and pipe fitters assuming an early-start schedule.
   b. Determine the load profiles assuming a late-start schedule.

25. Consider the project network pictured in Figure 9–22. Activity times, measured in days, are shown in parentheses next to each activity label.
   a. Determine the earliest and the latest starting and finishing times for all activities. Draw a Gantt chart based on earliest starting times, but indicate activity slack where appropriate. How many days are required to complete the project?
   b. A single critical piece of equipment is required in order to complete the following activities: A, B, C, D, G, and H. Determine a feasible schedule for the project assuming that none of these activities can be done simultaneously.

FIGURE 9–22

Project network (for Problem 25)

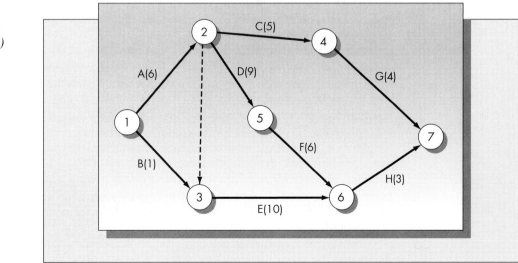

c. Two resources, R1 and R2, are used for each activity. Assume that these are both consumable resources with daily requirements as follows:

Activity	Daily Requirement of R1	Daily Requirement of R2
A	4	0
B	8	6
C	10	9
D	18	4
E	12	3
F	5	12
G	3	2
H	0	6

Determine resource loading profiles based on the schedule found in part (*b*).

d. Based on the results of part (*c*), determine the *cumulative* amounts of resources R1 and R2 consumed if the schedule found in part (*b*) is used.

9.7 Organizational Issues in Project Management

This chapter has been concerned with reviewing techniques for assisting with the project management function. Successful project management also depends on effective people management. How the organization is structured can be an important factor in whether or not a project succeeds.

The classic structure of an organization is a line organization. That means that there is a clear pyramid structure: Vice presidents report to the president, directors report to vice presidents, middle managers report to directors, and so forth. In a line organization, usually one person at the bottom is assigned to coordinate several employees who may be in other departments at the same level of the organization. The individual responsible will be given the title of project leader or project officer. The line organization is probably the weakest organizational structure for interdepartmental project management.

At the other end of the organizational spectrum is the divisional project organization. In this setting, employees are completely freed up to the project organization for the life of the project. They report only to the project leader. This is the strongest organizational structure from the point of view of promoting successful project completion. However, the divisional project organization requires frequent shifting of employees among projects, which can be disruptive for the employees and expensive for the company.

More recently, firms have been experimenting with the matrix organization. In the matrix organization, the firm is organized both horizontally and vertically. The vertical structure is the same as in the traditional line organization. The horizontal direction corresponds to individual projects that may span several functional departments. Each employee reports vertically to his or her functional superior and horizontally to his or her project leader.

The matrix organization is a compromise between the pure line organization and the project organization. However, when project teams span functional departmental boundaries, problems arise. Typically, employees' first loyalty will be to their direct functional superior. Dual subordination causes conflicts when demands are made on an employee from two directions at once. Hence, in order for the matrix concept to work, the project leader must be empowered to set priorities and provide incentives for outstanding performance from project members. In addition, there must be a shared sense of responsibility for the project among project team members. Managing split loyalties among team members is no easy task, however. Texas Instruments Corporation, for example, long advocated the matrix organization, but found that ambiguous lines of authority were causing problems.

The structure of the project team varies with the application. For example, Vollmann et al. (1992) recommend that a project team assigned the task of implementing a manufacturing planning and control system consist of five to eight employees. It would ideally be comprised of one representative from each of marketing, engineering, production planning, line manufacturing, and management information systems departments. A team member from finance also might be desirable. They recommend that the project team be freed from other responsibilities and physically isolated from their functional departments during the project's course (meaning that they advocate a project organization). The best team is comprised of experienced employees from within the company who are familiar with the business. The project team leader should be someone who will ultimately be a user of the new system, such as a professional in production planning or manufacturing.

9.8 Historical Notes

It is generally recognized that the two network-based scheduling systems discussed in this chapter, CPM and PERT, were developed almost simultaneously in the United States in the late 1950s. CPM was a result of a joint effort by the Du Pont Company and Remington Rand (Walker and Sayer, 1959). The technique resulted from a study aimed at reducing the time required to perform plant overhaul, maintenance, and construction. The CPM methodology outlined in this report included the cost–time trade-off of indirect and direct costs discussed in Section 3, as well as the methods for developing the project network and identifying the critical path. That linear programming could be used to solve project scheduling problems appears to have been first discovered by Charnes and Cooper (1962).

PERT was developed as a result of the Polaris Weapons System program undertaken by the Navy in 1958 (Department of the Navy, 1958). PERT was the result of a joint effort by Lockheed Aircraft Corporation, the Navy Special Projects Office, and the consulting firm of Booz, Allen, and Hamilton. Although the PERT system shared many features with the CPM system, the PERT project focused on activity time uncertainty rather than on project costs. An interesting issue related to the PERT approach is the justification for the approximation formulas for the mean and variance. Both Grubbs (1962) and Sasieni (1986) raised this issue. It appears that the formula for the variance is assumed (probably based on properties of the normal distribution) and the formula for the mean was obtained as a consequence of the assumption of the beta distribution and the formula assumed for the variance (see Archibald and Villoria, 1967, p. 449).

Moder, Phillips, and Davis (1983) discuss a little-known reference that precedes the development of PERT and CPM by almost 30 years and may have been the true genesis of project planning methodology. In 1931 a Polish scientist named Karol Adamiecki developed and published a planning device known as the Harmonygraph

(Adamiecki, 1931). The Harmonygraph is basically a vertical Gantt chart modified to include immediate predecessors. The technique requires the use of sliding tabs for each activity. (See Moder, Phillips, and Davis, 1983, pp. 10–12, for a more comprehensive discussion of the Harmonygraph method.)

Research into project networks continues. One area of continuing interest is resource-constrained networks (Patterson, 1984). Another is networks with random activity times. As we indicated in Section 5, the PERT methodology gives only an approximation of the distribution of the project completion time. Recent interest has focused on developing efficient simulations (Sullivan, Hayya, and Schaul, 1982) or determining the exact distribution of the project completion time assuming other than a beta distribution for the time required for each activity (Kulkarni and Adlakha, 1986).

9.9 Project Management Software for the PC

Project management software has been around almost as long as computers. When mainframe computers first became commercially viable in the early 1950s, most vendors developed and marketed some type of project management software. Today, there are many project management software packages on the market, from low-cost entry programs for personal computers to products designed for client–server systems and mainframes. As of 1993, Dataquest estimated that the PC-based project management software market alone exceeded $232 million (Wood, 1995). This number has undoubtedly grown substantially since then and does not include systems designed for more sophisticated platforms.

By the time this appears in print, the products reviewed here are likely to have been updated and new products are likely to have been announced. Most of the programs reviewed here are the industry leaders and have undergone several redesigns and upgrades from DOS to Windows and/or Unix.

Heck (1994) provides a comprehensive review of the four most popular PC-based project management software products. The software were compared along several dimensions of performance and value. All are designed for personal computers running Windows. (Note that the opinions expressed below are the reviewer's and not necessarily this writer's.)

1. Project Scheduler 6 for Windows, Scitor Corporation, Foster City, California. Rated the best of this group, this product is easy to use, providing smooth data entry and custom formulas. It allows the user to plan multiple projects easily, incorporates resource and cost tracking features competitive with the other software reviewed, and includes a convenient well-implemented report writer.

2. CA-Superproject, Computer Associates International, Inc., Islandia, New York. This program's strengths are its customizability and its ability to provide different forms of assistance to different users. The program allows for multiple projects and for cost and resource management, but the interfaces and ease of use are not up to the standard of the other three project management tools.

3. Microsoft Project, Microsoft Corporation, Redmond, Washington. This product commands a majority of the PC-based project management software market (as do many of Microsoft's products). It is one of the best available for planning single projects but falls short in features and ease of use in analyzing

multiple projects and resources. The program also lacks a comprehensive report writer and customizable formulas. (Note that a different review appearing in the Winter 1995 issue of *Massachusetts CPA Review* recommended Project very highly. The reviewer indicated that it is particularly well designed for scheduling assignments in an accounting firm.)

4. Time Line, Symantec Corp., Cupertino, California. Time Line is for corporate users. It is based on the SQL database, allowing it to share information with other compatible systems. The program has several unique features, including 24-hour shift scheduling and interchangeable resources. However, its database foundation slows the program significantly.

The following project management programs were discussed in Barney (1995), Heck (1995a and 1995b), and Vadlamudi (1995).

5. Milestones, Etc., Kidasa Software Inc., Austin, Texas. This program differentiates itself from the pack in that it uses familiar drawing tools to map out projects rather than the usual jargon. The interface is easy to use and navigate. It is available in both 16- and 32-bit versions.

6. SureTrak Project Manager for Windows, Primavera Systems, Salt Lake City, Utah. While other software does a better job with multiproject scheduling, this package shines when it comes to scheduling single projects. SureTrak emphasizes tracking the progress of a project once it is underway. Other programs, notably industry leader Microsoft Project, fall short in project tracking and monitoring functions. SureTrak can interface with the more sophisticated Primavera Project Planner.

7. In Control, Attain Corporation, Somerville, Massachusetts. In Control, which runs on both Macintosh and Windows for Workgroups, is designed to allow multiple users to provide inputs and track common projects. Files stored on the network can be shared and updated by a workgroup. If a user remotely updates project files, the changes are automatically logged when the user reconnects to the network.

8. K1 Shell, UES Inc., Dublin, Ohio. This is a high-end product designed to run on servers from Sun Microsystems, Hewlett-Packard, and IBM. K1 Shell was used to plan the building of the largest nuclear fusion reactor in the world. A history of all of the decision-making processes used in the project are stored in an Oracle database. The product can assign work and provide queuing of tasks, track work in progress, and produce status reports. It also supports business process rules.

As noted above, this review is not meant to be comprehensive. In fact, the market for project management software is so large that one software company, Welcom Software Technology of Houston, Texas, has based its product line exclusively on this application. The firm's basic product is Open Plan, but other programs provide features that complement the basic product (such as *Opera,* which incorporates uncertainty in activity times, and *Cobra,* for project costing). Even in this narrow market niche and with so much competition, the firm still has revenues in excess of $10 million worldwide. (This information was obtained from the promotional material provided by the company on the Internet.) Welcom's software was the choice of Florida Power and Light, whose successful project management function is highlighted in the Snapshot Application for this chapter.

SNAPSHOT APPLICATIONS

Project Management Helps United Stay on Schedule

United Airlines has been a long-time user of project management methods and software to help keep its business on track. United owns 549 aircraft and makes 2,000 flights every day. For an operation of this magnitude to work properly takes both a commitment from its employees and quality project management tools. Richard Gleason, MIS project office manager at United, chose Project Workbench, a PC-based software package from Advanced Business Technology Corporation in New York City, for the task.

United uses this package to integrate more than 80 mainframe applications in diverse areas such as architecture development and technical support under one operational due date. In addition, the software assists United with managing its worldwide fleet maintenance program.

The software allows United to link dependencies among an unlimited number of distinct projects and calculate schedules for these projects. The software is more expensive than most of the general-purpose software aimed at the mass market (see the discussion of project management software in Section 9). However, United feels that the additional cost is more than offset

by the flexibility in scheduling multiple projects with dependent activity links (Ouelette, 1994).

Thomas Brothers Plans Staffing with Project Management Software

Thomas Brothers Maps Inc., based in Irvine, California, is one of the country's premier makers of road maps. The firm supplies street guides and maps that cover many West Coast cities and counties. They have 230 workers and annual revenues exceeding $20 million. Bob Foster, the president of the firm, began using client–server project management software to plan a schedule for hiring and training cartographers. Foster chose PlanView, a Windows-based package, from the company with the same name based in Austin, Texas. The software handles planning for the 250 projects the company undertook in 1995. The project management package allows the president to track exactly how each worker is deployed.

One of the features of PlanView that attracted Foster and other personnel at Thomas Brothers was the ability of the software to track resources as well as schedules. PlanView looks at the labor pool in terms of resource overload rather than as a succession of tasks

In addition to a host of general-purpose project management packages, programs designed for specific applications are growing in popularity. For example, Griffith (1994) reviews nine packages that help corporate legal departments schedule their workloads. The demand for both general-purpose and special-purpose project management software continues to expand.

9.10 Summary

When large projects consist of many interrelated activities that must be completed in a specific sequence, project management techniques provide useful tools for preparing and administering schedules for these projects. Project planning takes place at many levels of an organization, for projects lasting from a few days to months or even years.

This chapter focused on the critical path method (CPM) and its extensions. *Networks* are a convenient means of representing a project. There are two ways of using networks to represent projects: activity-on-arrow and activity-on-node. Using the activity-on-arrow method, the nodes of the network correspond to completion of some and potential bottlenecks. The firm can track its progress on a large number of multiple projects by maintaining a running schedule of the workload for each cartographer. According to Foster, "We've never been able to see so far ahead so clearly."

Another feature of PlanView is that the package uses a standard SQL database and can interface easily with other database products (Oracle 7 in this case). The project management information supplied by the system can be used to keep track of other business functions as well, such as tracking the cost of sales (Wood, 1995).

Florida Power and Light Takes Project Management Seriously

Florida Power and Light (FPL), a major utility company, is responsible for managing two nuclear power–generating facilities at Turkey Point and St. Lucie. To handle the management of these facilities as well as several other functions within the firm, FPL established an independent project management department. At its peak, the department had 17 project control personnel to support more than 600 engineers and analysts. Separate dedicated groups were established for each plant.

To deal effectively with contractors, the group established project control and execution reporting requirements for all major contracted work. This process required FPL to identify their major contractors and begin negotiations with those contractors to implement the system. It turned out that four contractors were responsible for 80 percent of the workload. Within a year, each of these contractors had implemented a satisfactory project control system.

To make the system more user friendly, FPL abandoned its traditional mainframe software and replaced it with Welcom's PC-based software. While the PC-based system was not as powerful as the mainframe package for handling resource modeling issues, the feeling was that local ownership of the process afforded by PCs would compensate for the new software's limitations.

Self-assessment of FPL's project management function showed that customers were very satisfied with the project management function and scheduling support afforded by the project management group. In fact, FPL was awarded the Deming Prize several years ago for its commitment to continuous improvement and quality of service. The success of its project management initiatives played an important role in this achievement (Cooprider, 1994).

subset of activities. When nodes rather than arrows are used to represent activities, pseudo activities are not required. However, activity-on-arrow is far more common in practice and, in general, more intuitively appealing.

The *critical path* is the *longest* path or chain through the network. The length of the critical path is the minimum project completion time, and the activities that lie along the critical path are known as the critical activities. Delay in a critical activity delays the project. Noncritical activities have *slack,* which means that they can be delayed without necessarily delaying the project. Although for small networks the critical path can be identified readily by inspection, for larger networks an algorithmic procedure is required. This chapter presented a method, involving both forward and backward passes through the network, that specified the earliest and the latest starting and ending times for all activities.

One of the goals of the early development work on CPM was to consider the effect of *project costing.* We assume that costs are either direct or indirect. Direct costs include labor, material, and equipment; these costs increase if the project time decreases. Indirect costs include costs of rents, interest, and utilities; these costs increase if the project time increases. The goal of the analysis is to determine the optimal time to perform the project that minimizes the sum of indirect and direct costs.

Linear programming is one means of solving project scheduling problems. The linear programming formulations considered here solve both the CPM problem and the CPM problem with cost–time trade-offs. Although not treated in this chapter, linear programming also can be used to solve some cost–time problems when the direct costs are nonlinear functions of the activity duration times.

PERT is an extension of critical path analysis to incorporate uncertainty in the activity times. For each activity, three time estimates are required: (1) the minimum activity time (called *a*), (2) the maximum activity time (called *b*), and (3) the most likely activity time (called *m*). Based on these estimates, one approximates the mean and the standard deviation of the activity time. The project time is assumed to follow a normal distribution with mean equal to the sum of the means along the path with the longest expected completion time and the variance equal to the sum of the variances along this same path. Depending on the configuration of the network, assuming path independence of two or more paths could give more accurate results. Although the PERT approach is only approximate, it does provide a measure of the effect of uncertainty of activity times on the total project completion time.

Resource considerations also were considered. Traditional CPM and PERT methods ignore the fact that schedules may be infeasible because of insufficient resources. Typical examples of scarce resources that might give rise to an infeasible project schedule are manpower, raw materials, and equipment. When schedules are infeasible, noncritical activities should be rescheduled within the available slack if possible. If not, critical activities may have to be delayed and the project completion date moved ahead. Resource loading profiles are a useful tool for determining the requirements placed on resources by any schedule. These profiles can be used as rough planning guides as major projects evolve.

We also discuss *organizational design for effective project management*. From a project management perspective, the traditional line organization is the weakest organizational structure. On the other end of the spectrum is the project organization. Here, employees are freed up from their usual responsibilities for the life of the project. Some firms have experimented with matrix organizations, which is a hybrid of the two designs. Most companies, however, have retained the traditional line structure.

The chapter concluded with a brief overview of the software available for project management. The explosion of the personal computer has been accompanied by an explosion of software. Project management is no exception. There is an entire range of software products available. Most of the programs available are designed to run on a PC for a single user. Although many of the PC-based software can handle multiple projects and resources, very large systems require more powerful tools. There are programs available for mainframe computers and client–server systems that allow for multiple users, projects, and resources. These packages can interface with large databases and other parts of the firm's financial and production systems.

Additional Problems on Project Scheduling

26. Two brothers have purchased a small lot, in the center of town, where they intend to build a gas station. The station will have two pumps, a service area for water and tire maintenance, and a main building with rest rooms, office, and cash register area. Before they begin excavating the site, the local authorities must approve the location for a gasoline station and be certain that the placement of the storage tanks will not interfere with water, gas, and electric lines that are already in place.

Once the site has been approved, the excavation can begin. After excavation, the three primary parts of the construction can begin: laying in the gasoline tanks, building the water and tire service area (including installation of the air compressor), and constructing the main building. The surfacing can begin after all building is completed. After surfacing, the site must be cleaned and the station's signs erected. However, before the station can open for business, the air compressor must be inspected, tested, and approved.

The activities and the time required for each of them are the following:

Activity	Time Required (in weeks)
A: Obtain site approval	4
B: Begin site excavation	2
C: Place and secure gasoline tanks	3
D: Install gasoline pumps	1
E: Connect and test gasoline pumps	1
F: Construct service area	2
G: Install and connect water and air compressor	3
H: Test compressor	1
I: Construct main building including the restrooms, the office, and the cash register area	5
J: Install plumbing and electrical connections in the main building	3
K: Cover tanks and surface the area	4
L: Clean site	2
M: Erect station signs	1

    a. Based on the description of the project, determine the activity precedence relationships and develop a network for the project.

    b. Determine the critical path and the earliest and the latest starting and finishing times for each activity.

    c. Draw the Gantt chart for this project based on earliest times.

    d. Suppose that the air compressor fails to function correctly and must be replaced. It takes two weeks to obtain another compressor and test it. Will the project necessarily be delayed as a result?

    e. List the activities that must be completed by the end of the 15th week in order to guarantee that the project is not delayed.

    f. Solve this problem using linear programming.

27. A scene is being shot for a film. A total of 11 distinct activities have been identified for the filming. First, the script must be verified for continuity, the set erected and decorated, and the makeup applied to the actors. After the set is completed, the lighting is set in place. After the makeup is applied, the actors get into costume. When these five activities are completed, the first rehearsal of the scene commences, which is followed by the second scene rehearsal with the cameraman. While the rehearsals are going on, verifications of the audio and the video equipment are made. After both rehearsals and verifications are completed, the scene is shot. Afterward, it is viewed by the director to determine if it needs to be reshot.

The list of the activities and the activity times is

Activity	Time Required (in days)
A: Check script for story continuity	2.0
B: Decorate set; place necessary props	4.5
C: Check lighting of scene	1.0
D: Apply makeup to actors	0.5
E: Costumes for actors	1.5
F: First rehearsal (actors only)	2.5
G: Video verification	2.0
H: Sound verification	2.0
I: Second rehearsal (with camera and lights)	2.0
J: Shoot scene	3.5
K: Director's OK of scene	1.5

a. Develop a network for the filming of the scene.
b. Compute the earliest and the latest finishing and starting times for each of the activities, and identify the critical path.
c. Draw the Gantt chart for this project. Assume that activities with slack are scheduled so that there is equal slack before and after the activities.
d. Suppose that the video verification (G) shows that the equipment is faulty. Four additional days are required to obtain and test new equipment. How much of a delay in the total time required to film the project will result?
e. One of the costumes is damaged as it is being fitted (activity E). How much extra time is available for repair without delaying the project?
f. What kind of delays can you envision as a result of the uncertainty in the time of activity K?
g. Solve this problem by linear programming.

28. A guidance and detection system is being built as part of a large defense project. The detection portion consists of radar and sonar subsystems. Separate equipment is required for each of the subsystems. In each case, the equipment must be calibrated prior to production. After production each subsystem is tested independently. The radar and the sonar are combined to form the detection system, which also must be tested prior to integration with the guidance system. The final test of the entire system requires complex equipment. The activities and the activity times are

Activity	Time Required (in days)
A: Calibrate machine 1 (for radar)	2.0
B: Calibrate machine 2 (for sonar)	3.5
C: Calibrate machine 3 (for guidance)	1.5
D: Assemble and prepare final test gear	7.0
E: Make radar subsystem	4.5
F: Make sonar subsystem	5.0
G: Make guidance subsystem	4.5
H: Test radar subsystem	2.0
I: Test sonar subsystem	3.0
J: Test guidance subsystem	2.0
K: Assemble detection subsystem (radar and sonar)	1.5
L: Test detection subsystem	2.5
M: Final assembly of three systems	2.5
N: Testing of final assembly	3.5

a. Construct a network for this project.

b. Determine the earliest and the latest starting and finishing times for all the activities, and identify the critical path.

c. Draw a Gantt chart for this project based on the earliest times.

d. How much time is available for assembling and calibrating the final test gear without delaying the project?

e. What are the activities that must be completed by the end of 10 days to guarantee that the project is not delayed?

f. What are the activities that must be started by the end of 10 days to guarantee that the project is not delayed?

g. Solve this problem using linear programming.

29. Consider the filming of the scene described in Problem 27 above. Based on past experience, the director is not very confident about the time estimates for some of the activities. His estimates for the minimum, most likely, and maximum times for these activities are

Activity	*a*	*m*	*b*
F	2	3	12
I	1	2	8
J	3	4	10
K	1	2	7

a. Including these PERT time estimates, how long is the filming expected to take?

b. What is the probability that the number of days required to complete the filming of the scene is at least 30 percent larger than the answer you found in part (*a*)?

c. For how many days should the director plan in order to be 95 percent confident that the filming of the scene is completed?

30. Consider the following project time and cost data:

Activity	Immediate Predecessors	Normal Time	Expedited Time	Normal Cost	Expedited Cost
A	—	6	6	$ 200	$ 200
B	A	10	4	600	1,000
C	A	12	9	625	1,000
D	B	6	5	700	800
E	B	9	7	200	500
F	C, D	9	5	400	840
G	E	14	10	1,000	1,440
H	E, F	10	8	1,100	1,460

a. Develop a network for this project.

b. Compute the earliest and the latest starting and finishing times for each of the activities. Find the slack time for each activity and identify the critical path.

c. Suppose that the indirect costs of the project amount to $200 per day. Find the optimal number of days to perform the project by expediting one day at a time. What is the total project cost at the optimal solution? What savings have been realized by expediting the project?

d. Solve this problem using linear programming.

APPENDIX 9–A   GLOSSARY OF NOTATION FOR CHAPTER 9

a = Minimum activity time for PERT.

b = Maximum activity time for PERT.

EF_i = Earliest finishing time for activity i.

ES_i = Earliest starting time for activity i.

LF_i = Latest finishing time for activity i.

LS_i = Latest starting time for activity i.

m = Most likely activity time for PERT.

M_{ij} = Expedited time for activity (ij).

μ = Expected activity time estimate for PERT.

N_{ij} = Normal time for activity (ij).

σ = Estimate of the standard deviation of the activity time for PERT.

T = Project completion time for PERT. T is a random variable.

t_{ij} = Time required for activity (ij), t_{ij} is a constant in the standard formulation and a variable in the cost–time formulation.

x_i = Earliest start time for node i (linear programming formulation).

Bibliography

Adamiecki, Karol. "Harmonygraph." *Polish Journal of Organizational Review,* 1931. (In Polish.)

Archibald, R. D., and R. L. Villoria. *Network-Based Management Systems (PERT/CPM).* New York: John Wiley & Sons. 1967.

Barney, D. "Workflow Tool Manages International Nuclear Project." *Infoworld* 17, no. 2 (January 9, 1995), p. 47.

Burgess, A. R., and J. B. Killebrew. "Variational Activity Level on a Cyclic Arrow Diagram." *Journal of Industrial Engineering* 13 (1962), pp. 76–83.

Charnes, A., and W. W. Cooper. "A Network Interpretation and a Directed Subdual Algorithm for Critical Path Scheduling." *Journal of Industrial Engineering* 13 (1962), pp. 213–19.

Cooprider, D. H. "Overview of Implementing a Project Control System in the Nuclear Utility Industry." *Cost Engineering* 36, no. 3 (March 1994), pp. 21–24.

Davis, E. W., and J. H. Patterson. "A Comparison of Heuristic and Optimum Solutions in Resource-Constrained Project Scheduling." *Management Science* 21 (1975), pp. 944–55.

Department of the Navy, Special Projects Office, Bureau of Ordnance. "PERT: Program Evaluation Research Task." Phase I Summary Report. Washington, D.C., July 1958.

Goodman, L. J. *Project Planning and Management.* New York: Van Nostrand Reinhold, 1988.

Griffith, C. "Corporate Legal No Longer Being Ignored." *Information Today* 11, no. 11 (December 1994), p. 50.

Grubbs, F. E. "Attempts to Validate Certain PERT Statistics, or 'Picking on Pert.'" *Operations Research* 10 (1962), pp. 912–15.

Heck, M. "New Tools for an Old Need." *Infoworld* 16, no. 40 (October 3, 1994), pp. 68–83.

Heck, M. "Primavera's SureTrak Answers Midsize Scheduling Needs." *Infoworld* 17, no. 15 (April 10, 1995a), pp. 71–74.

Heck, M. "Milestones, Etc." *Infoworld* 17, no. 21 (May 22, 1995b), p. 94.

Kulkarni, V. G., and V. G. Adlakha. "Markov and Markov-Regenerative PERT Networks." *Naval Research Logistics Quarterly* 34 (1986), pp. 769–81.

Lockyer, K. G. *Critical Path Analysis, Problems and Solutions.* London: Isaac Pitman and Sons, Ltd., 1966.

Moder, J. J.; C. R. Phillips; and E. W. Davis. *Project Management with CPM, PERT, and Precedence Diagramming.* 3rd ed. New York: Van Nostrand Reinhold, 1983.

Ouelette, T. "Project Management Helps Airline Stick to Schedule." *Computerworld* 28, no. 45 (November 7, 1994), p. 79.

Patterson, J. H. "A Comparison of Exact Approaches for Solving the Multiple Constrained Resource Project Scheduling Problem." *Management Science* 30 (1984), pp. 854–67.

Sasieni, M. "A Note on PERT Times." *Management Science* 32 (1986), pp. 1652–53.

Sullivan, R. S.; J. C. Hayya; and R. Schaul. "Efficiency of the Antithetic Variate Method for Simulating Stochastic Networks." *Management Science* 28 (1982), pp. 563–72.

Vadlamudi, P. "Attain Project Planner Update Puts Workgroups in Control." *Infoworld* 17, no. 9 (February 1995), p. 35.

Vollmann, T. E.; W. L. Berry; and D. C. Whybark. *Manufacturing Planning and Control Systems.* 3rd ed. Homewood, IL: Richard D. Irwin, 1992.

Walker, M. R., and J. S. Sayer. "Project Planning and Scheduling." Report 6959. Wilmington, DE: E. I. du Pont de Nemours & Co., Inc., March 1959.

Wiest, J. D., and F. K. Levy. A *Management Guide to PERT/CPM,* 2nd ed. Englewood Cliffs, NJ: Prentice Hall, 1977.

Wood, L. "Perfect Harmony." *Informationweek,* no. 526 (May 8, 1995), pp. 42–54.

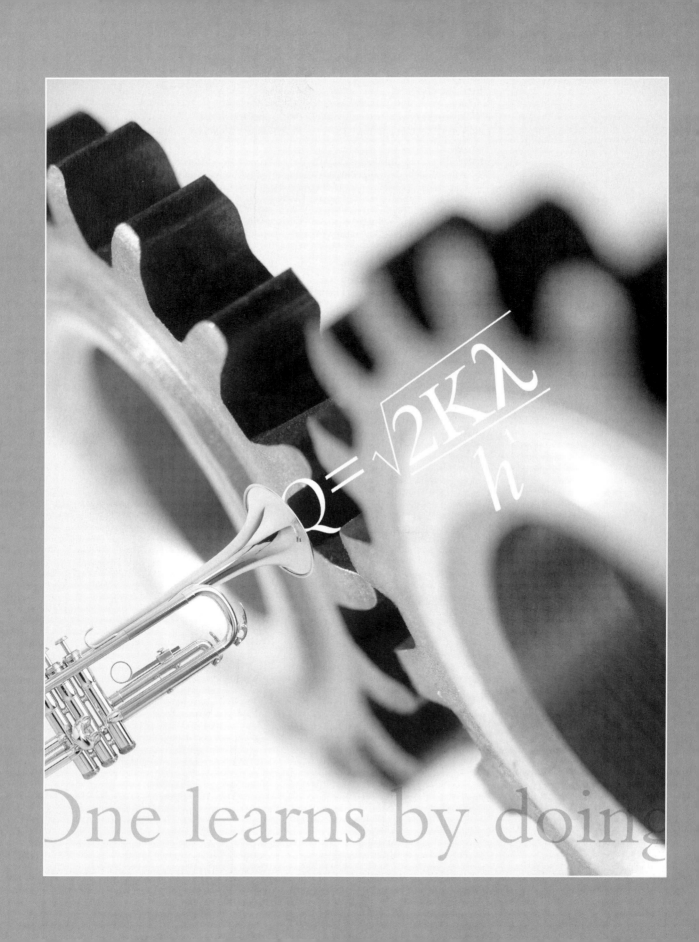

CHAPTER 10

FACILITIES LAYOUT AND LOCATION

Where to locate facilities and the efficient design of those facilities are important strategic issues for business as well as the military, nonprofit institutions, and government. During the 1980s, Northern California's Silicon Valley experienced tremendous growth in microelectronics and related industries. One can get some idea of how significant this growth was from the American Electronics Association Member Directory: more than 50 percent of the listings are in the Bay Area. Most of these firms were "start-ups," adapting a new technology to a specialized segment of the marketplace. With the surge in demand for microcomputers came support industries producing hard disk drives, floppy disks, semiconductor manufacturing equipment, local area networks, and a host of other related products. In many cases, large investments in capital equipment were required before the first unit could be sold. A typical example is the Read-Rite Corporation of Milpitas, California, the largest independent producer of thin film heads for reading from and writing to Winchester hard drives in the world. When the firm was founded in 1983, an initial capital investment of $40 million was required. The venture capital investors bore the risk that the firm would survive. Read-Rite is typical of hundreds of hi-tech start-ups.

What equipment should be purchased, how facilities should be organized, and where the facilities should be located are fundamental strategy issues facing any manufacturing organization. Service industries also are faced with the problem of finding effective layouts. Achieving efficient patient flows in hospitals is one example. Another example is the obvious care used in laying out the many theme parks across the United States. Anyone who has visited the San Diego Zoo or Disney World in Orlando, Florida, can appreciate the importance of effective layout and efficient management of people in service industries.

Tompkins and White (1984) estimated that 8 percent of the U.S. gross national product has been spent on new facilities annually since 1955. This does not include the cost of modification of existing facilities. If the authors' estimates are correct, we are spending in excess of $500 billion annually on construction and modification of facilities. The authors go on to claim that from 20 percent to 50 percent of the total operating expenses in manufacturing are attributed to materials handling costs. Effective facilities planning could reduce these costs by 10 to 30 percent annually, the authors claim.

An important part of the enormous success of Japanese companies in achieving manufacturing dominance in several key industries is efficient production. Efficient production includes efficient design of the product, employee involvement, lean inventory material management systems, and intelligent layout and organization of facilities.

Chapter 1 touched on some of the qualitative issues that management must consider when deciding on the location of new facilities. This chapter treats in greater

557

depth the issues associated with locating new facilities and the methods for determining the best layout of operations in those facilities. In addition to exploring the qualitative factors that should be taken into account, we will also consider how a manager can employ computers and quantitative techniques to assist with these complex decisions.

How a plant or workplace should be laid out is, in a sense, a special version of the location problem. Determining a suitable layout means finding the locations of departments within some specified boundary. When designing new facilities, the planner also must decide on the size and shape of the facility as well as the configuration of the departments within it.

Quantitative techniques are most useful when the goal is to minimize or maximize a single dimensional objective such as cost or profit. The objective function used in location problems generally involves either Euclidean or rectilinear distance (these terms will be defined later). However, minimizing total distance traveled may not make sense in all cases. As an extreme example, consider the problem of locating a school. A location that requires 100 students to travel 10 miles each is clearly not equally desirable to one that requires 99 students to travel 1 mile and one student to travel 901 miles.

For layout problems, the most common objective used in mathematical models is to minimize the cost of materials handling. Furthermore, such models invariably assume that the number of materials handling trips from every work center to every other work center is known with certainty. In most real environments such assumptions are naive at best. Another deficiency of mathematical models is that they ignore such factors as plant safety, flexibility of the layout for future design changes, noise, and aesthetics.

This does not imply that mathematical and computer models are not useful for solving layout and location problems. What it does mean is that quantitative solutions must not be taken on blind faith. They must be carefully considered in the context of the problem. Used properly, the results of mathematical and computer models can reduce significantly the number of alternatives that the analyst must consider.

10.1 The Facilities Layout Problem

Determining the best layout for a facility is a classical industrial engineering problem. Early industrial engineers often were known as efficiency experts and were interested in determining layouts to optimize some measure of production efficiency. For the most part, this view continues today, especially in plant layout. However, layout problems occur in many environments outside the plant. Some of these include

1. Hospitals.
2. Warehouses.
3. Schools.
4. Offices.
5. Workstations.
6. Banks.
7. Shopping centers.
8. Airports.
9. Industrial plants.

Each of these layout problems has unique characteristics. Our focus in this chapter will be on techniques for finding layouts of industrial plants, although many of the methods can be applied to other facilities as well. The objectives in a plant layout study might include one or more of the following:

1. Minimize the investment required in new equipment.
2. Minimize the time required for production.
3. Utilize existing space most efficiently.
4. Provide for the convenience, safety, and comfort of the employees.
5. Maintain a flexible arrangement.
6. Minimize the materials handling cost.
7. Facilitate the manufacturing process.
8. Facilitate the organizational structure.

Vollmann and Buffa (1966) suggest nine steps as a guide for the analysis of layout problems:

1. Determine the compatibility of the materials handling layout models with the problem under study. Find all factors that can be modeled as materials flow.
2. Determine the basic subunits for analysis. Determine the appropriate definition of a department or subunit.
3. If a mathematical or computer model is to be used, determine the compatibility of the nature of costs in the problem and in the model. That is, if the model assumes that materials handling costs are linear and incremental (as most do), determine whether or not these assumptions are realistic.
4. How sensitive is the solution to the flow data assumptions? What is the impact of random changes in these data?
5. Recognize model idiosyncrasies and attempt to find improvements.
6. Examine the long-run issues associated with the problem and the long-run implications of the proposed solution.
7. Consider the layout problem as a systems problem.
8. Weigh the importance of qualitative factors.
9. Select the appropriate tools for analysis.

10.2 Patterns of Flow

As we noted earlier, the objective used most frequently for quantitative analysis of layout problems is to minimize the materials handling cost. When minimizing the materials handling cost is the primary objective, a flow analysis of the facility is necessary. Flow patterns can be classified as horizontal or vertical. A horizontal flow pattern is appropriate when all operations are located on the same floor, and a vertical pattern is appropriate when operations are in multistory structures. Francis and White (1974) give six horizontal flow patterns and six vertical flow patterns. The six horizontal patterns appear in Figure 10–1.

The simplest pattern is (*a*), which is straight-line flow. The main disadvantage of this pattern is that separate docks and personnel are required for receiving and shipping goods. The L shape is used to replace straight-line flow when the configuration of

FIGURE 10–1

Six horizontal flow patterns

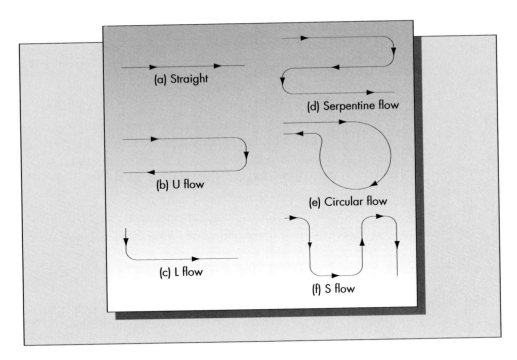

the building or the line requires it. The U shape has the advantage over the straight-line configuration of allowing shipping and receiving to be at the same location. The circular pattern is similar to the U shape. The remaining two patterns are used when the space required for production operations is too great to allow use of the other patterns.

Two charts that supply useful information regarding flows are (1) the activity relationship chart and (2) the from-to chart.

Activity Relationship Chart

An activity relationship chart (also called a rel chart for short) is a graphical means of representing the desirability of locating pairs of operations near each other. The following letter codes have been suggested for determining a "closeness" rating.

A   Absolutely necessary. Because two operations may use the same equipment or facilities, they must be located near each other.

E   Especially important. The facilities may require the same personnel or records, for example.

I   Important. The activities may be located in sequence in the normal work flow.

O   Ordinary importance. It would be convenient to have the facilities near each other, but it is not essential.

U   Unimportant. It does not matter whether the facilities are located near each other or not.

X   Undesirable. Locating a welding department near one that uses flammable liquids would be an example of this category.

The closeness ratings are represented in an activity relationship chart that specifies the appropriate rating for each pair of departments. Consider the following example.

Example 10.1

Meat Me, Inc., is a franchised chain of fast-food hamburger restaurants. A new restaurant is being located in a growing suburban community near Reston, Virginia. Each restaurant has the following departments:

1. Cooking burgers.
2. Cooking fries.
3. Packing and storing burgers.
4. Drink dispensers.
5. Counter servers.
6. Drive-up server.

The burgers are cooked on a large grill, and the fries are deep fried in hot oil. For safety reasons the company requires that these cooking areas not be located near each other. All hamburgers are individually wrapped after cooking and stored near the counter. The service counter can accommodate six servers, and the site has an area reserved for a drive-up window.

An activity relationship chart for this facility appears in Figure 10–2. In the chart, each pair of activities is given one of the letter designations A, E, I, O, U, or X. Once a final layout is determined, the proximity of the various departments can be compared to the closeness ratings in the chart. Note that Figure 10–2 gives only the closeness rating for each pair of departments. In the original conception of the chart, a number giving the reason for each closeness rating is also included in every cell. These numbers do not appear in our example. ■

From-To Chart

A from-to chart is similar to the mileage chart that appears at the bottom of many road maps and gives the mileage between selected pairs of cities. From-to charts are used to

FIGURE 10–2

Activity relationship chart for Meat Me fast-food restaurant

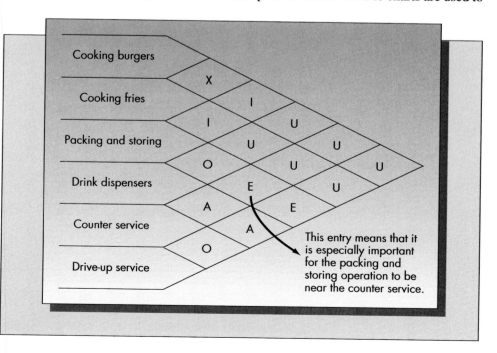

analyze the flow of materials between departments. The two most common forms are charts that show the distances between departments and charts that show the number of materials handling trips per day between departments. A from-to chart differs from an activity relationship chart in that the from-to chart is based on a specific layout. It is a convenient means of summarizing the flow data corresponding to a given layout.

Example 10.2

A machine shop has six work centers. The work centers contain one or more of the following types of machines:

1. Saws.
2. Milling machines.
3. Punch presses.
4. Drills.
5. Lathes.
6. Sanders.

The from-to chart in Figure 10–3 shows the distance in feet between centers of the six departments. Note that the chart in the example is symmetric; that is, the travel distance between A and B is the same as the travel distance between B and A. This is not necessarily always the case, however. There may be one-way lanes, which allow material flow in one direction only, or an automated materials handling system that moves pallets in one direction only.

Figure 10–4 shows a from-to chart that gives the number of materials handling trips per day. These figures could be based on a specific product mix produced in the shop or simply be representative of an average day.

Suppose that the firm's accounting department has estimated that the average cost of transporting material one foot in the machine shop is 20 cents. Using this fact, one can develop a third from-to chart that gives the average daily cost of materials handling

FIGURE 10–3

From-to chart showing distances between six department centers (measured in feet)

From \ To	Saws	Milling	Punch press	Drills	Lathes	Sanders
Saws		18	40	30	65	24
Milling	18		38	75	16	30
Punch press	40	38		22	38	12
Drills	30	75	22		50	46
Lathes	65	16	38	50		60
Sanders	24	30	12	46	60	

FIGURE 10–4

*From-to chart
showing number of
materials handling
trips per day*

From \ To	Saws	Milling	Punch press	Drills	Lathes	Sanders
Saws		43	26	14	40	
Milling			75	60		23
Punch press					45	16
Drills		22			28	
Lathes		45		30		60
Sanders		12				

FIGURE 10–5

*From-to chart
showing materials
handling cost per
day (in $)*

From \ To	Saws	Milling	Punch press	Drills	Lathes	Sanders
Saws		154.8	208	84	520	
Milling			570	900		138
Punch press					342	38.4
Drills		330			280	
Lathes		144		300		720
Sanders		72				

from every department to every other department. For example, the distance separating saws and milling in the current layout is 18 feet (from Figure 10–3) and there are an average of 43 materials handling trips between these two departments (from Figure 10–4). This translates to a total of (43)(18) = 774 feet traveled in a day or a total cost of (774)(0.2) = $154.80 per day for materials handling between saws and milling. The materials handling costs for the other pairs of departments appear in Figure 10–5. ∎

From-to charts are not a means of determining layouts, but simply a convenient way to express important flow characteristics of an existing layout. They can be useful in comparing the materials handling costs of a small number of alternatives. Because criteria other than the materials handling cost are relevant, the from-to chart should be supplemented with additional information, such as that contained in an activity relationship chart.

10.3 Types of Layouts

Different philosophies of layout design are appropriate in different manufacturing environments. Chapter 1 discussed the problem of matching the product life cycle with the process life cycle represented by the product-process matrix (see Figure 1–5 in particular). The upper left-hand corner of the product-process matrix corresponds to low-volume production and little product standardization. Such a product structure is usually characterized by a job-shop-type environment. In a job shop there are a wide variety of jobs with different flow patterns associated with each job. A commercial printer is a typical example of a jumbled flow shop such as this. As volume increases, the number of products declines and flow patterns become more standardized. For discrete parts manufacture, an auto assembly plant is a good example of this case. A different approach for designing production facilities would be appropriate in such a setting.

Fixed Position Layouts

Some products are too big to be moved, so the product remains fixed and the layout is based on the product size and shape. Examples of products requiring *fixed position* layouts are large airplanes, ships, and rockets. For such projects, once the basic frame is built, the various required functions would be located in fixed positions around the product. A project layout is similar in concept to the fixed position layout. This would be appropriate for large construction jobs such as commercial buildings or bridges. The required equipment is moved to the site and removed when the project is completed. A typical fixed position layout is shown in Figure 10–6.

FIGURE 10–6

Fixed position layout

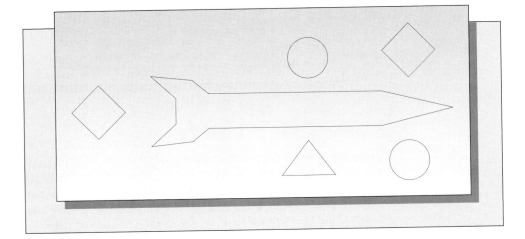

FIGURE 10–7
Product layout

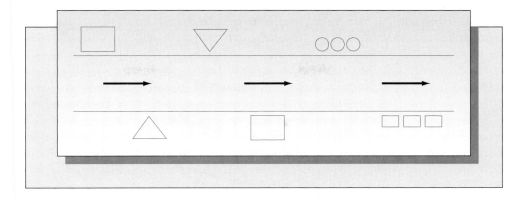

Product Layouts

In a *product layout* (or product flow layout) machines are organized to conform to the sequence of operations required to produce the product. The product layout is typical of high-volume standardized production (the lower right-hand corner in Figure 1–5 in Chapter 1). An assembly line (or transfer line) is a product layout, because assembly facilities are organized according to the sequence of steps required to produce the item. Product layouts are desirable for flow-type mass production and provide the fastest cycle times in this environment. Transfer lines are expensive and inflexible, however, and become cumbersome when changes in the product flow are required. Furthermore, a transfer line can experience significant idle time. If one part of the line stops, the entire line may have to remain idle until the problem is corrected. Figure 10–7 shows a typical product layout.

Process Layouts

Process layouts are the most common for small- to medium-volume manufacturers. A *process layout* groups similar machines having similar functions. A typical process layout would group lathes in one area, drills in one area, and so on. Process layouts are most effective when there is a wide variation in the product mix. Each product has a different routing sequence associated with it. In such an environment it would be difficult to organize the machines to conform with the production flow because flow patterns are highly variable. Process layouts have the advantage of minimizing machine idle time. Parts from multiple products or jobs queue up at each work center to facilitate high utilization of critical resources. Also, when design changes are common, parts routings will change frequently. In such an environment, a process layout affords minimal disruption. Figure 10–8 shows a typical process layout. The arrows correspond to part routings.

Layouts Based on Group Technology

With increased emphasis on automated factories and flexible manufacturing systems, *group technology layouts* have received considerable attention in recent years. To implement a group technology layout, parts must be identified and grouped based on similarities in manufacturing function or design. Parts are organized into part families. Presumably, each family requires similar processing, which suggests a layout based on

FIGURE 10–8

Process layout

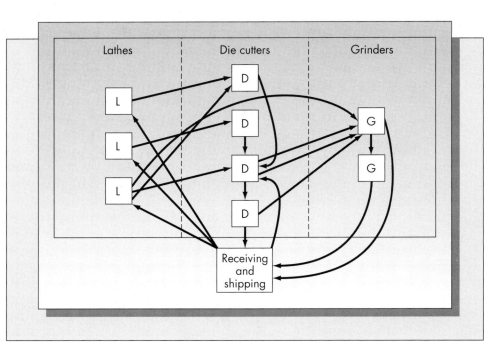

FIGURE 10–8

Process layout

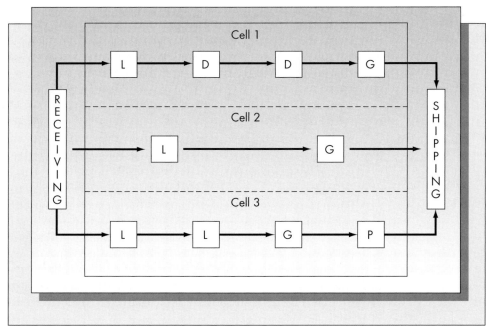

FIGURE 10–9

Group technology layout

the needs of each family. In most cases, machines are grouped into machine cells where each cell corresponds to a particular part family or a small group of part families (see Figure 10–9).

The group technology concept seems best suited for large firms that produce a wide variety of parts in moderate to high volumes. A typical firm that would consider this approach might have as many as 10,000 different part numbers, which might be grouped into 50 or so part families. Some of the advantages of using the group technology concept are

1. *Reduced work-in-process inventories.* Each manufacturing cell operates as an independent unit, allowing much tighter controls on the flow of production. Large work-in-process inventories are not needed to maintain low cycle times. A side benefit of this is reduced queues of parts and the confusion that results.

2. *Reduced setup times.* Because manufacturing cells are organized according to part types, there should not be significant variation in the machine settings required when switching from one part to another. This allows the cells to operate much more efficiently.

3. *Reduced materials handling costs.* For a firm producing 10,000 parts, a process layout would require a dizzying variety of part routings. If volumes are large, process centers would necessarily have to be separated by large distances, thus requiring substantial materials handling costs. A layout based on group technology would overcome this problem.

4. *Better scheduling.* By isolating part groupings, it is much easier to keep track of the production flow within each cell. Reduction in cycle times and work-in-process queues results in more reliable due-date schedules.

There are several disadvantages of the group technology approach. One is that it can be difficult to determine suitable part families. Parts may be grouped by size and shape or by manufacturing process requirements. The first approach is easier but not as effective for developing layouts. How parts are grouped is a function, to a large degree, of the coding system used to classify these parts. (Groover and Zimmers, 1984, discuss several parts coding systems and how they relate to the group technology concept.) Grouping part families according to the manufacturing flow requires a careful production flow analysis (Burbidge, 1975). This method is probably not feasible for a firm with a large number of parts, however.

Group technology layouts may require duplication of some machines. In order for a manufacturing cell to be self-contained, the cell must have all the machines necessary for the product being produced. Duplicating machines could be expensive and could result in greater overall idle time.

Under what circumstances would a group technology layout be preferred to a pure process or product layout? A simulation study by Sassani (1990) provides some answers. He constructed a simulation of five manufacturing cells. Initially, when the products for each cell were well defined and the cells isolated, the system ran smoothly. However, as the product mix, product design, and demand patterns started to change, the simulation showed that the efficiency of the layout deteriorated.

The vast majority of existing layouts are either process or product type. Firms producing a wide variety of parts may choose several layouts for different product lines, or may choose some hybrid approach. Product variation and annual volume are the primary determining factors for making the appropriate choice. The group technology approach is relatively new. It is slowly gaining acceptance as firms wrestle with the difficult problems of determining appropriate part groupings and cell designs. As automated factories become more widespread, however, the group technology concept could play a greater role in plant layout and process design.

Problems for Sections 1–3

1. For each of the eight objectives of a layout study listed in Section 1, give examples of situations in which that objective would be important and examples in which it would be unimportant.

2. A manufacturing facility consists of five departments: A, B, C, D, and E. At the current time, the facility is laid out as in Figure 10–10. Develop a from-to distance chart for this layout. Estimate the distance separating departments based on the flow pattern of the materials handling system in the figure. Assume that all departments are located at their centers as marked in Figure 10–10.

3. Four products are produced in the facility described in the previous problem. The routing for these products and their forecasted production rates are

Product	Routing	Forecasted Production (in units/week)
1	A–B–C–D–E	200
2	A–D–E	900
3	A–B–C–E	400
4	A–C–D–E	650

Suppose that all four products are produced in batches of size 50.

a. Convert this information into a from-to chart similar to Figure 10–4 but giving the number of materials handling trips per week between departments.

FIGURE 10–10

Layout of manufacturing facility (for Problem 2)

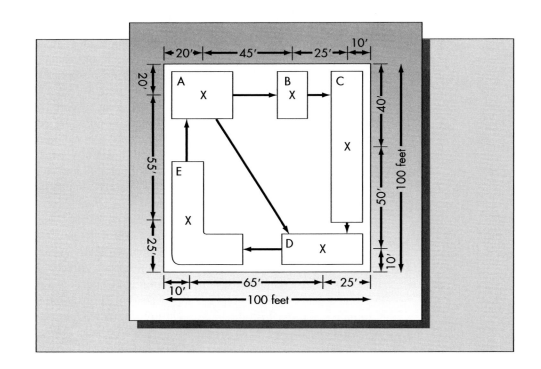

b. Suppose that the cost of transporting goods one foot is estimated to be $2. Using the from-to chart you determined in part (*a*), obtain a from-to chart giving materials handling cost per week between departments.

c. Develop an activity relationship chart for this facility based on the materials handling cost from-to chart you obtained in part (*b*). Use only the rankings A, E, I, O, U, with A being assigned to the highest cost and U the least.

d. Based on the results of parts (*b*) and (*c*), would you recommend a different layout for this facility?

4. Consider the example of the machine shop with the from-to charts given in Figures 10–3, 10–4, and 10–5. Based on the results of Figure 10–5, in what ways might the current layout be improved?

5. Suggest a layout for the Meat Me fast-food restaurant, with the relationship chart shown in Figure 10–2. Assume that the facility is 50 feet square and half the restaurant will be for customer seating.

6. Describe the differences among product, process, and group technology layouts. Describe the circumstances in which each type of layout is appropriate.

10.4 A Prototype Layout Problem and the Assignment Model

Several analytical techniques have been developed for assisting with layout problems. However, most real problems are too complex to be solved by these methods without a computer. (We will discuss the role of the computer in facilities layout later in this chapter.) In some cases, however, layout problems can be formulated as *assignment* problems, as described in this section. In particular, the assignment model is appropriate when only a discrete set of alternative locations is being considered and there are no interactions among departments. Although moderately sized assignment problems can be solved by hand, *extremely* large problems can be solved only with the aid of computers.

Example 10.3

Due to an increase in the volume of sales, Sam Melder, the owner of Melder, Inc., a small manufacturing firm located in Muncie, Indiana, has decided to expand production capacity. A new wing has been added to the plant that will house four machines: (1) a punch press, (2) a grinder, (3) a lathe, and (4) a welding machine. There are only four possible locations for these machines, say A, B, C, and D. However, the welding machine, which is the largest machine, will not fit in location B.

The plant foreman has made estimates of the impact in terms of materials handling costs of locating each of the machines in each of the possible locations. These costs, expressed in terms of dollars per hour, are represented in the table below.

		Location			
		A	*B*	*C*	*D*
	1	94	13	62	71
Machines	2	62	19	84	96
	3	75	88	18	80
	4	11	*M*	81	21

The entry M stands for a very large cost. It is used to indicate that machine 4 is not permitted in location B. As the objective is to assign the four machines to locations in order to minimize the total materials handling cost, an optimal solution will never assign machine 4 to location B.

Minimum cost assignments can rarely be found by inspection. For example, one reasonable approach might be to assign the machines to the lowest cost locations in sequence. The lowest cost in the matrix is 11, so machine 4 would be located in location A. Eliminating the last row and the first column, we see that the next remaining lowest cost is 13, so machine 1 would be assigned to location B. Now also eliminating the first row and the second column, the smallest remaining cost is 18, so machine 3 would be assigned to location C. Finally, machine 2 must be assigned to location D. The total cost of this solution is $11 + 13 + 18 + 96 = \$138$. As we will see, this solution is suboptimal. (The optimal cost is \$114. Can you find the optimal solution?) ∎

A simple approach such as this will rarely result in an optimal solution. A solution algorithm will be presented from which one can obtain an optimal solution to the assignment problem by hand for moderately sized problems.

The Assignment Algorithm

The solution algorithm presented in this section is based on the following observation:

> **Result:** If a constant is added to, or subtracted from, all entries in a row or column of an assignment cost matrix, the optimal assignment is unchanged.

In applying this result, the objective is to continue subtracting constants from rows and columns in the cost matrix until a zero cost assignment can be made. The zero cost assignment for the modified matrix is optimal for the original problem. This leads to the following algorithm for solving assignment problems:

Solution Procedure for Assignment Problems.

1. Locate the smallest number in row 1 and subtract it from all the entries in row 1. Repeat for all rows in the cost matrix.
2. Locate the smallest number in column 1 and subtract it from all the entries in column 1. Repeat for all columns in the cost matrix.
3. At this point each row and each column will have at least one zero. If it is possible to make a zero cost assignment, then do it. That will be the optimal solution. If not, go to step 4.
4. Determine the maximum number of zero cost assignments. This will equal the smallest number of lines required to cover all zeros. The lines are found by inspection and are not necessarily unique. The important point is that the number of lines drawn not exceed the maximum number of zero cost assignments.
5. Choose the smallest uncovered number and do the following:
   a. Subtract it from all other uncovered numbers.
   b. Add it to the numbers where the lines cross.
   c. Return to step 3.

The process is continued until one can make a zero cost assignment. Notice that step 5 is merely an application of the result in the following way: If the smallest uncovered number is subtracted from every element in the matrix and then added to every covered element, it will be subtracted once and added twice where lines cross.

Example 10.3 (continued)

Let us return to Example 10.3 above. The original cost matrix is

		Location			
		A	*B*	*C*	*D*
	1	94	13	62	71
Machine	2	62	19	84	96
	3	75	88	18	80
	4	11	*M*	81	21

Step 1. Subtracting the smallest number from each row gives

81	0	49	58
43	0	65	77
57	70	0	62
0	*M*	70	10

Because *M* is very large relative to the other costs, subtracting 11 from *M* still leaves a very large number, which we again denote as *M* for convenience. At this point at least one zero appears in each row, and in each column except the last.

Step 2. Subtracting 10 from every number in the final column gives

81	0	49	48
43	0	65	67
57	70	0	52
0	*M*	70	0

At this point we have at least one zero in every row and every column. However, that does not necessarily mean that a zero cost assignment is possible. In fact, it is possible to make at most three zero cost assignments at this stage, as shown below.

Step 4.

81	[0]	49	48
43	0	65	67
57	70	[0]	52
[0]	*M*	70	0

The three assignments shown above are 1–B, 3–C, and 4–A. There are other ways of assigning three locations at zero cost as well. It does not matter which we choose at this stage, only that we know three are possible. The next step is to find three lines that cover all of the zeros. These are shown below.

81	0	49	48
43	0	65	67
~~57~~	~~70~~	~~0~~	~~52~~
~~0~~	~~M~~	~~70~~	~~0~~

Again, the choice of lines is not unique. However, it is important that no more than three lines be used. Finding three lines to cover all zeros is done by trial and error.

Step 5. The smallest uncovered number, 43, is subtracted from all other uncovered numbers and added to the numbers where the lines cross. The resulting matrix is

		Location			
		A	B	C	D
Machine	1	38	0	6	5
	2	0	0	22	24
	3	57	113	0	52
	4	0	M	70	0

It is now possible to make a zero cost assignment, which is shown above. The optimal assignment is machine 1, the punch press, to location B; machine 2, the grinder, to location A; machine 3, the lathe, to location C; machine 4, the welder, to location D. The total materials handling cost per hour of the optimal solution is obtained by referring to the original assignment cost matrix. It is 13 + 62 + 18 + 21 = \$114 per hour. ■

The assignment algorithm also can be used when the number of sites is larger than the number of machines. For example, suppose that there were six potential sites for locating the four machines and the original cost matrix was

		Location					
		A	B	C	D	E	F
Machine	1	94	13	62	71	82	25
	2	62	19	84	96	24	29
	3	75	88	18	80	16	78
	4	11	M	81	21	45	14

The procedure is to add two dummy machines, 5 and 6, with zero costs. The problem is then solved using the assignment algorithm as if there were six machines and six locations. The locations to which the dummy machines are assigned are the ones that are not used.

Problems for Section 4

7. Solve the following assignment problem:

	A	B	C	D
1	21	24	26	23
2	29	27	30	29
3	24	25	34	27
4	28	26	28	25

8. Each of four machines is to be assigned to one of five possible locations. The objective is to assign the machines to locations that will minimize the materials handling cost. The machine–location costs are given below. Find the optimal assignment.

	A	B	C	D	E
1	26	20	22	21	25
2	35	31	33	40	26
3	15	18	23	16	25
4	31	34	33	30	M

9. University of the Atlantic is moving its business school into a new building, which has been designed to house six academic departments. The average time required for a student to get to and from classes in the building depends upon the location of the department in which he or she is taking the class. Based on the distribution of class loads, the dean has estimated the following mean student trip times in minutes, given the departmental locations.

		Location					
		A	B	C	D	E	F
	1	13	18	12	20	13	13
	2	18	17	12	19	17	16
Department	3	16	14	12	17	15	19
	4	18	14	12	13	15	12
	5	19	20	16	19	20	19
	6	22	23	17	24	28	25

Find the optimal assignment of departments to locations to minimize mean student trip time in and out of the building.

---

*10.5 More Advanced Mathematical Programming Formulations

The assignment model described in the previous section can be useful for determining optimal layouts for a limited number of real problems. The primary limitation of the simple assignment model is that, in most cases, the number of materials handling trips and the associated materials handling cost are assumed to be independent of the location of the other facilities. In Example 10.3, the cost of assigning the punch press to location A was assumed to be $94 per hour. However, in most cases this cost would depend upon the location of the other machines as well.

A formulation of the problem that takes this feature into account is considerably more complex. In order to avoid confusion, we will assume that the problem is to assign machines to locations. The problem could be to assign other types of subfacilities to locations, of course, but we will retain this terminology for convenience. Define the following quantities:

n = Number of machines;

c_{ij} = Cost per time period of assigning machine i to location j. This cost could be a one-time relocation cost that is converted to an annual equivalent;

d_{jr} = Cost of making a single materials handling trip from location j to location r;

f_{ik} = Mean number of trips per time period from machine i to machine k;

S_i = The set of locations to which machine i could feasibly be assigned;

$$a_{ijkr} = \begin{cases} f_{ik}d_{jr} & \text{if } i \neq k \text{ or } j \neq r, \\ c_{ij} & \text{if } i = k \text{ and } j = r; \end{cases}$$

$$x_{ij} = \begin{cases} 1 & \text{if machine } i \text{ is assigned to location } j, \\ 0 & \text{otherwise.} \end{cases}$$

Interpret a_{ijkr} as the materials handling cost per unit time when machine i is assigned to location j and machine k is assigned to location r. This cost is incurred only if both x_{ij} and x_{kr} are equal to one. Hence, it follows that the total cost of assigning machines to locations is given by

$$\frac{1}{2} \sum_{i=1}^{n} \sum_{j=1}^{n} \sum_{k=1}^{n} \sum_{r=1}^{n} a_{ijkr} x_{ij} x_{kr}. \tag{1}$$

As all indices are summed from 1 to n, each assignment will be counted twice; hence the need to multiply by $\frac{1}{2}$. Constraints are included to ensure that each machine gets assigned to exactly one location and each location is assigned exactly one machine. These are

$$\sum_{i=1}^{n} x_{ij} = 1, \qquad j = 1, \ldots, n; \tag{2}$$

$$\sum_{j=1}^{n} x_{ij} = 1, \qquad i = 1, \ldots, n; \tag{3}$$

$$x_{ij} = 0 \text{ or } 1, \qquad i = 1, \ldots, n \text{ and } j = 1, \ldots, n; \tag{4}$$

$$x_{ij} = 0, \qquad i = 1, \ldots, n \text{ and } j \notin S(i). \tag{5}$$

The mathematical programming formulation is to minimize the total materials handling cost of the assignment, (1), subject to the constraints (2), (3), (4), and (5). This formulation is known as the quadratic assignment problem. In general, such problems are extremely difficult to solve. One should consider using such a method only for moderately sized problems.

Problem for Section 5

10. Consider the following problem with two locations and three machines. Suppose that the costs of transporting a unit load from location j to location r are given in the following table:

		To location	
		A	B
From	A		6
location	B	9	

The average numbers of trips required from machine i to machine k per hour are

		To machine		
		1	2	3
	1	0	3	1
From	2	0	0	3
machine	3	0	4	0

Relocation costs are ignored. Write out the complete quadratic assignment formulation for this location problem.

10.6 Computerized Layout Techniques

The quadratic assignment formulation given above has several shortcomings. First, one must specify the costs of all materials handling trips and the expected number of trips from every department to every other department. When many departments are involved, this information could be difficult to obtain. Furthermore, an efficient solution technique for solving large quadratic assignment problems has yet to be discovered.

For these reasons there has been considerable interest in computer-aided methods. These methods are heuristics; they do not guarantee an optimal solution but generally yield efficient solutions. They can be used for solving problems that are too large to be solved analytically. The methods that we discuss in this section are CRAFT, COFAD, ALDEP, CORELAP, and PLANET.

For each of these methods the objective is to minimize the cost of materials handling. However, the solutions generated by these computer programs must be considered in the context of the problem. Issues such as plant safety, noise, and aesthetics are ignored. The layout obtained from a computer program may have to be modified in order to take these factors into account.

Computer programs for determining layouts generally fall into two classes: (1) improvement routines and (2) construction routines. An improvement routine takes an existing layout and considers the effect of interchanging the location of two or more facilities. A construction routine constructs the layout from scratch from flow data and the information in the activity relationship chart. CRAFT and COFAD are both improvement routines, and PLANET, CORELAP, and ALDEP are construction routines. The improvement routines have the disadvantage of requiring specification of an initial layout. However, improvement routines generally result in more usable layouts. Construction routines often give layouts with oddly shaped departments.

CRAFT

CRAFT (*computerized relative allocation of facilities technique*) was one of the first computer-aided layout routines developed. As noted above, CRAFT is an improvement routine and requires the user to specify an initial layout. The objective used in CRAFT is to minimize the total transportation cost of a layout, where transportation cost is defined as the product of the cost to move a unit load from department i to department j and the distance between departments i and j. To be more specific, define

n = Number of departments,
v_{ij} = Number of loads moved from department i to department j in a given time,
u_{ij} = Cost to move a unit load a unit distance from department i to department j,
d_{ij} = Distance separating departments i and j.

It follows that $y_{ij} = u_{ij}v_{ij}$ is the cost to move total product flow during the specified time interval a unit distance from i to j, and that $y_{ij}d_{ij}$ is the cost of the product flow from i to j. The total cost of the product flow between all pairs of departments is

$$\frac{1}{2} \sum_{i=1}^{n} \sum_{j=1}^{n} y_{ij}d_{ij}.$$

Note that the input information v_{ij} and u_{ij} may be represented as from-to charts, as shown in Figures 10–3 and 10–4.

Two implicit assumptions made in CRAFT are

1. Move costs are independent of the equipment utilization.
2. Move costs are a linear function of the length of the move.

When these assumptions cannot be justified, CRAFT may be used to minimize the product of flows and distances only by assigning the unit cost u_{ij} a value of 1. The entries d_{ij} are computed by the program from a specification of an initial layout. Based on an initial layout, CRAFT considers exchanging the position of adjacent departments and computes the materials handling cost of the resulting exchange. The program chooses the pairwise interchange that results in the greatest cost reduction.

Departments are assumed to be either rectangularly shaped or composed of rectangular pieces. Furthermore, departments are assumed to be located at their *centroids*. The centroid is another term for the coordinates of the center of gravity or balance point. A discussion of centroids and how they are computed for objects in the plane appears in Appendix 10–A. The accuracy of assuming that a department is located at its centroid depends upon the shape of the department. The assumption is most accurate when the shape of the department is square or rectangular, but is less accurate for oddly shaped departments.

To better understand how CRAFT determines layouts, consider the following example.

Example 10.4

A local manufacturing firm has recently completed construction of a new wing of an existing building to house four departments: A, B, C, and D. The wing is 100 feet by 50 feet. The plant manager has chosen an initial layout of the four departments. This layout appears in Figure 10–11. We have marked the centroid locations of the departments with a dot. From the figure we see that department A requires 1,800 square feet, B requires 1,200 square feet, C requires 800 square feet, and D requires 1,200 square feet.

One of the inputs required by CRAFT is the flow data, that is, the number of materials handling trips per unit time from every department to every other department. These data are given in the from-to chart appearing in Figure 10–12a. The distance between departments is assumed to be the rectilinear distance between centroid locations. From Figure 10–11, we see that the centroid locations of the initial layout are

$$(x_A, y_A) = (30, 35), \qquad (x_C, y_C) = (20, 10),$$
$$(x_B, y_B) = (80, 35), \qquad (x_D, y_D) = (70, 10).$$

The rectilinear distance between A and B, for example, is given by the formula

$$|x_A - x_B| + |y_A - y_B| = |30 - 80| + |35 - 35| = 50.$$

One computes distances between other pairs of departments in a similar way. The calculations are summarized in the from-to chart in Figure 10–12b. Notice that we have assumed that the distance from department i to department j is the same as the distance from department j to department i. As we noted earlier, this may not necessarily be true if material is transported in one direction only.

FIGURE 10–11

Initial layout for CRAFT example

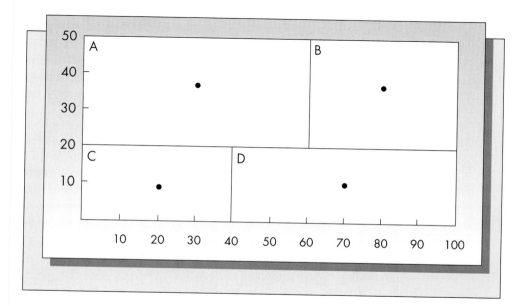

FIGURE 10–12

From-to charts for initial layout

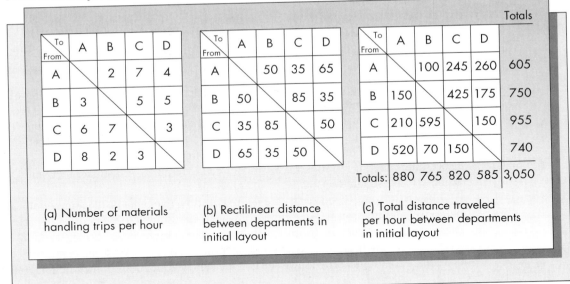

(a) Number of materials handling trips per hour

(b) Rectilinear distance between departments in initial layout

(c) Total distance traveled per hour between departments in initial layout

Normally, a third from-to chart also would be required. This would give the cost of transporting a unit load a unit distance from department *i* to department *j*. As noted above, CRAFT may be used to minimize the product of flows and distances by assigning a value of 1 to these transport costs. We will assume this to be the case in our example. Hence, one obtains the hourly cost of transporting materials to and from each of the departments by simply multiplying the entries in the from-to chart of Figure 10–12*a* by the entries in the from-to chart of Figure 10–12*b*. These calculations are summarized in Figure 10–12*c*. The total distance traveled per hour for the initial layout is 3,050 feet.

CRAFT now considers the effect of pairwise interchanges of two departments that either have adjacent borders or have the same area. (CRAFT can also consider the effect of exchanging the locations of three departments. We will not consider this option in our example.) The result of the exchange is determined by exchanging the location of the centroids. This is only an approximation, however, since exchanging the locations of two departments does *not* necessarily mean that the location of their centroids will be exchanged.

If we were to exchange the locations of A and B, for example, we would assume that $(x_A, y_A) = (80, 35)$ and $(x_B, y_B) = (30, 35)$, which would result in the new from-to distance chart appearing in Figure 10–13*a*. Multiplying the original flow data in Figure 10–12*a* by this distance chart gives a new cost chart, which appears in Figure 10–13*b*. Interchanging the centroids of A and B results in the predicted cost reduction from 3,050 to 2,950, or about 3 percent. (The actual cost of interchanging the locations of A and B would be slightly different because the centroids are not exactly exchanged.)

CRAFT considers all pairwise interchanges of adjacent departments or departments with identical areas and picks the one that results in the largest decrease in the predicted cost. We will not present the details of the calculations but merely summarize the results. Exchanging the centroids of A and C results in a total

FIGURE 10–13

New distance and cost from-to charts after exchanging centroids for A and B

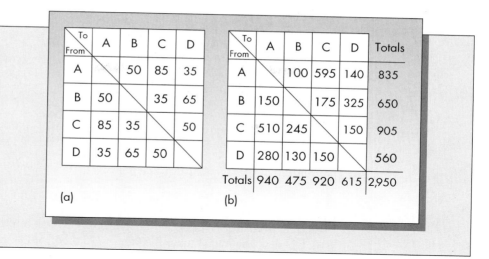

(a)

To \ From	A	B	C	D
A		50	85	35
B	50		35	65
C	85	35		50
D	35	65	50	

(b)

To \ From	A	B	C	D	Totals
A		100	595	140	835
B	150		175	325	650
C	510	245		150	905
D	280	130	150		560
Totals	940	475	920	615	2,950

FIGURE 10–14

New layout with A and C interchanged

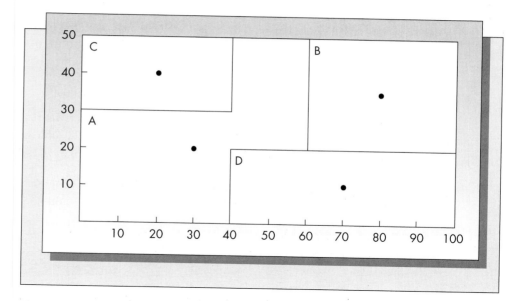

predicted cost of 2,715, and exchanging the centroids of A and D results in a total predicted cost of 3,185. Two other exchanges must be considered as well: B and D, and C and D. Notice that exchanging B and C is not considered because they do not have equal areas and do not share a common border. Exchanging the centroids of B and D results in a total predicted cost of 2,735, and exchanging the centroids of C and D in a total predicted cost of 2,830.

The maximum predicted cost reduction is achieved by exchanging A and C. The new layout with A and C exchanged appears in Figure 10–14. Because C has the smaller area, it is placed in the upper left-hand corner of the space formerly occupied by A, so that the remaining space allows A to be contiguous. Notice that A is no longer rectangular. The actual cost of the new layout is not necessarily equal to the predicted value of 2,715. The centroid of A is computed using the method outlined in Appendix 10–A. It is determined by first finding the moments M_x and M_y

given by

$$M_x = (40^2 - 0)(30 - 0)/2 + (60^2 - 40^2)(50 - 20)/2 = 54{,}000,$$

$$M_y = (30^2 - 0)(40 - 0)/2 + (50^2 - 20^2)(60 - 40)/2 = 39{,}000,$$

and dividing by the area of A to obtain

$$x_A = 54{,}000/1{,}800 = 30,$$

$$y_A = 39{,}000/1{,}800 = 21.66667.$$

The centroid of C is located at the center of symmetry, which is

$$(x_C, y_C) = (20, 40).$$

The centroids for B and D are unchanged. The cost of the new layout is actually 2,810, which is somewhat more than the 2,715 predicted at the last step, but is still less than the original cost. The process is continued until predicted cost reductions are no longer possible. At this point only three exchanges need to be considered: A and B, A and D, and B and D. Obviously, we need not consider exchanging A and C, and C and D may not be exchanged because they do not share a common border. The predicted cost resulting from exchanging the centroids of A and B is 2,763.33; of A and D, 3,641.33; and of B and D, 2,982. Clearly, the greatest predicted reduction is now achieved by exchanging the locations of A and B.

The new layout is pictured in Figure 10–15. The centroids for the respective departments are now

$$(x_A, y_A) = (70, 35), \qquad (x_C, y_C) = (20, 40),$$

$$(x_B, y_B) = (20, 15), \qquad (x_D, y_D) = (70, 10).$$

The actual cost of this layout is 2,530, which is considerably *less* than that predicted in the previous step. The process continues until no further reductions in

FIGURE 10–15

Second iteration, obtained from exchanging the locations of A and B

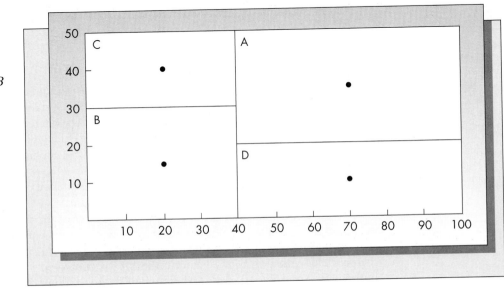

the predicted costs can be achieved. Because A and B were exchanged at the previous step, that exchange need not be considered again. Exchanging A and C results in a predicted cost of 3,175; exchanging A and D, a predicted cost of 2,753; and exchanging B and D, a predicted cost of 3,325. (We do not consider exchanging C and D at this stage because they are not adjacent and do not have equal areas.) Since none of the predicted costs is less than the current cost, we terminate calculations. The layout recommended by CRAFT, pictured in Figure 10–15, required two iterations and resulted in a reduction of total distance traveled from 3,050 feet to 2,530 feet, or about 17 percent. ■

COFAD

As noted above, CRAFT is an improvement routine. It requires the user to specify an initial layout and proceeds to improve the layout by considering pairwise interchanges of adjacent departments. An improvement routine similar to CRAFT is COFAD (for *computerized facilities design*). This chapter will not present the details of COFAD, but briefly reviews the improvements it offers over CRAFT.

COFAD is a modification of CRAFT that incorporates the choice of a materials handling system as part of the decision process. For a given layout, COFAD calculates the total move cost of the layout for a variety of materials handling alternatives and chooses the one with the minimum cost. Improvements are made by considering the equipment utilization of alternatives and exchanging assignments of poorly utilized equipment with equipment with better utilization. Once the best choice of equipment for a given layout is determined, the program considers improving the layout in much the same way as CRAFT. The program terminates when no additional improvements are possible. Once the program claims to have reached a steady state, the problem may be re-solved using from-to charts obtained from perturbing the original data from 10 to 50 percent. The purpose of re-solving the problem with the new data is to provide further assurance that the optimal solution has been reached, and to test the sensitivity of the solution to the flow data.

ALDEP

Both CRAFT and COFAD are improvement routines. That is, they start out with an initial layout and consider improvements by interchanging locations of departments. The other class of computerized layout techniques includes programs that develop the layouts essentially from scratch; they do not require an initial layout. Experience has shown that construction programs often tend to result in oddly shaped layouts, and for that reason they have not received as much attention in the literature as CRAFT.

ALDEP, which stands for *automated layout design program,* is a construction routine rather than an improvement routine. That means that layouts are determined from scratch and the program does not require the user to specify an initial layout. ALDEP makes use of the closeness ratings that appear in the activity relationship chart (rel chart for short). Figure 10–2 is an example of a rel chart. ALDEP first selects a department at random and places it in the upper left-hand corner of the layout. The next step is to scan the rel chart and place a department with a high closeness rating (A or E) in the layout adjacent to the first department. Successive departments are placed in the existing area in a top-down fashion, following a "sweep" pattern. The process is continued until all departments have been placed. At that point, a score for the layout is computed. The score is based on a numerical scale attached to the closeness ratings. ALDEP uses the following closeness rating values:

$$A = 4^3 = 64,$$
$$E = 4^2 = 16,$$
$$I = 4^1 = 4,$$
$$O = 4^0 = 1,$$
$$U = 0,$$
$$X = -4^5 = -1,024.$$

The entire process is repeated several times, and the layout with the largest score chosen. Because ALDEP tries to achieve a high closeness rating score, it often recommends layouts that have departments with very unusual shapes. The use of the sweep pattern helps to avoid this problem, but means that the resulting layout always appears as a set of adjacent strips. This type of layout may not be appropriate or desirable for some applications. The user specifies the sweep width to be used; the choice of the sweep width can have a significant effect upon the configuration of the final layout. For example, suppose that department A is 14 squares, B is 8 squares, the sweep width is 2 squares, and the facility width is 6 squares. The layout of these two departments given by ALDEP is shown in Figure 10–16*b*, assuming that the departments are placed in the order A, B.

CORELAP

Like ALDEP, CORELAP (for *computerized relationship layout planning*) is a construction routine that places departments within the layout using closeness rating

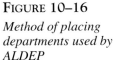

FIGURE 10–16

Method of placing departments used by ALDEP

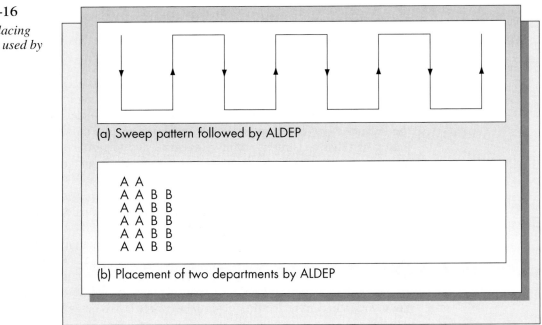

(a) Sweep pattern followed by ALDEP

```
A  A
A  A  B  B
A  A  B  B
A  A  B  B
A  A  B  B
A  A  B  B
```

(b) Placement of two departments by ALDEP

codes. The major difference between CORELAP and ALDEP is that CORELAP does not randomly select the first department to be placed. Rather, a total closeness rating (TCR) is computed for each department. The TCR is based on the numerical values $A = 6, E = 5, I = 4, O = 3, U = 2, X = 1$. Each department is compared to every other department to obtain a TCR.

For example, consider the Meat Me fast-food restaurant described in Example 10.1, with the rel chart shown in Figure 10–2. The closeness rating codes for the cooking-burgers department are X, I, U, U, and U, giving a TCR for this department of 11. The remaining values of the TCR are

Cooking fries:	X, I, U, U, U—11,
Packing and storing:	I, I, O, E, E—21,
Drink dispensers:	U, U, O, A, A—19,
Counter service:	U, U, E, A, A—21,
Drive-up service:	O, A, E, U, U—18.

CORELAP now selects the department with the highest TCR and places that department in the center of the existing facility. When there is a tie (as in this case between the packing and storing department and the counter service department), the department with the largest area is placed first. Once the initial department is selected, the rel chart is scanned to see what departments have the highest closeness rating compared with the department just placed. Suppose that counter service is placed first. Then, scanning Figure 10–2, we see that only the drink dispensers department has a closeness rating code of A relative to the counter service department. Hence, drink dispensers would be placed in the layout next. Each time that a department is chosen to be placed in the layout, alternative placements are considered and placing ratings based on user-specified numerical values for the rel codes are computed. The new department is placed to maximize the placing rating. (Detailed examples of this procedure can be found in Francis and White, 1974, and Tompkins and White, 1984.)

PLANET

PLANET, which stands for *plant layout analysis and evaluation technique,* has essentially the same required inputs as COFAD, and unlike both CORELAP and ALDEP does not use information contained in the rel chart to generate a layout. However, unlike CRAFT, PLANET is not an improvement routine. PLANET converts input data into a flow-between chart that gives the cost of sending a unit of flow between each pair of departments. In addition to the flow-between chart entries, the program also requires the user to input a priority rating for each department. The priority rating is a number from 1 to 9, with 1 representing the highest priority. The program selects departments to enter the layout in a sequential fashion, based first on the priority rating and second on the entries in the flow-between chart. Because PLANET does not restrict the final layout to conform to the shape of the building or allow fixing the location of certain departments, it suffers from the problems of other construction routines. Layouts obtained from PLANET, like those from ALDEP and CORELAP, may result in departments having unrealistic shapes. For that reason, such methods are better for providing the planner with some alternative ideas and initial layouts than for giving a final solution.

Computerized Methods versus Human Planners

An interesting debate concerning the effectiveness of computerized layout techniques versus expert human judgment appeared in the management science literature. The debate was sparked by a paper by Scriabin and Vergin (1975). In their study, the authors compared the layouts produced by three computerized layout routines, including CRAFT, and the layouts obtained without the use of these specific programs by 74 human subjects who were trained in manual layout techniques. The authors showed that the best solution obtained by groups of 20 of the 74 human subjects was better than the best solution obtained by the three computer routines. The largest difference in total cost occurred when the number of departments was the largest (20). Percentage differences ranged from 0 to 6.7. The authors concluded that humans will generally outperform computer layout routines because "in problems of a larger size the ability of man to recognize and visualize complex patterns gives him an edge over the essentially mechanical procedures followed by the computer programs."

Scriabin and Vergin's conclusions came under fire by a number of researchers. Buffa (1976) noted that the issue of flow dominance was not treated properly. Flow dominance refers to the tendency of materials to flow in the same sequence in the layout. Complete dominance occurs when a factory manufactures only a single product in an identical manner each time. In such a case, the final layout can easily be obtained by visual inspection. The other extreme occurs when products flow randomly through the factory. In that case all layouts are equivalent. (Block, 1977, gives a formal definition of flow dominance.) Buffa claimed that for problems in which flow dominance is over 200 percent, human subjects could be expected to obtain good solutions easily. The average flow dominance in the problems used by Scriabin and Vergin was apparently over 200 percent.

Another problem with their conclusions, as pointed out by Coleman (1977), was that in a typical industrial setting one does not have 20 professionals solving a layout problem. When comparing the layouts produced by the computer programs with those produced by individual human subjects, Coleman showed that the computer-generated layouts were superior. Finally, Block (1977) performed a separate experiment to test Buffa's contention that when flow dominance was under 200 percent, the computer will outperform the human. Block compared layouts produced by CRAFT with those produced by eight humans (four engineers and four laypeople) for a set of layout problems generated by Nugent, Vollmann, and Ruml (1968), in which the average flow dominance was 115 percent. He found that COFAD obtained better results in every case. The largest differences were observed when the number of departments was larger than 10.

What can we learn from these studies? It would appear that Scriabin and Vergin's conclusions are not justified. Later results showed that the computer-aided methods are, in fact, extremely useful and can result in significantly better layouts than can be obtained by simple visual or graphical methods when the number of departments is large and when material flow patterns are highly variable.

Dynamic Plant Layouts

Thus far, this chapter has considered only static layout problems. That is, we assumed that the costs appearing in the from-to charts are fixed. In some circumstances, this assumption may not be accurate, however. Nicol and Hollier (1983) note that: "Radical layout changes occur frequently and that management should therefore take this into account in their forward planning." A dynamic plant layout is based on the recognition that demands on the system, and hence costs of any given configuration, may change over time. When information on how the environment will change is available, one can incorporate this information into a model by allowing the costs in the from-to charts to change each planning period. This is precisely the scenario considered by Rosenblatt (1986). He showed how dynamic from-to charts would form the basis of a multiperiod planning model and developed a dynamic programming scheme to solve the resulting system of equations. With an example he shows how dynamic layouts can result in cost savings over fixed layouts in a changing environment. Models of this type will gain an even greater importance as firms continue to move toward structures based on agile manufacturing and increased flexibility.

Other Computer Methods

A variety of other analytical methods have been developed in addition to those discussed above. A method called bias sampling, suggested by Nugent, Vollmann, and Ruml (1968), is a straightforward modification of CRAFT that involves a randomization procedure for selecting departments to exchange. The method is considerably slower computationally but often results in better layouts. Considering the enormous strides that have been made in computing technology, the computational issues are less serious today than they were in 1968. SPACECRAFT (see Johnson, 1982) is an extension of CRAFT designed to produce layouts for multistory structures. The two primary modifications required were that (1) constraints had to be incorporated into the computational procedures that allowed certain departments to be located on specific floors and (2) the nonlinearity of the time required to go between floors required more complex cost calculations.

Not all layout problems are easily amenable to the methods reviewed here. There are applications in which assuming rectangular-shaped departments is not appropriate. Examples include the layouts of office buildings, an airplane's dashboard, a city or a neighborhood, or an integrated circuit. Drezner (1980) introduced a method called DISCON (for *dispersion and concentration*) that considers facilities as disks. The disks are first dispersed in the plane mathematically, using a system of differential equations simulating a system of springs connecting the disks. The dispersion phase provides an initial solution, which is later improved upon in the concentration phase. Drezner's approach works better than CRAFT when departments are not rectangular and when the number of facilities is large. Techniques based on graph theoretic methods (Foulds, 1983) and statistical cluster analysis (Scriabin and Vergin, 1985) also have been considered.

The "classic" software reviewed in this section illustrate the basic concepts used in most software products for facilities layout and design. However, with the proliferation of the personal computer in recent years, new lines of commercially available software products are now available for Windows-based computers, Macintosh, and UNIX workstations. Special products have been designed for distinct market segments. Office design, process piping, and industrial facilities planning are inherently very different layout problems and require specially tailored software products.

The problem of designing office facilities requires blocking out space for well-defined activities. These problems are the concern of interior designers and architects. Drafting tools with extensive libraries of symbols are the most popular for these applications. The most popular is AutoCAD, which allows for both 2-D and 3-D layouts. Process piping and layout design are typically used in conjunction with chemical plants and are almost always displayed as three-dimensional layouts. Industrial facilities layout and design usually involve some variant of the software tools discussed in this chapter, or the use of a specially tailored graphical-based simulation package, such as Pro-Model. Sly (1995) lists about 20 different software packages for facilities layout and design and describes many of their features.

Intelligent design and layout of industrial facilities is a problem that is continuing to grow in importance. For example, the cost of a new fab (fabrication facility) in the semiconductor industry typically exceeds $1 billion. Similar investments are required for new automotive assembly plants. Firms cannot afford to make mistakes when planning the organization of these enormously expensive facilities.

Problems for Section 6

11. Briefly describe each of the following computerized layout techniques. In each case, indicate whether the method is a construction or improvement method.
    a. CRAFT.
    b. COFAD.
    c. ALDEP.
    d. CORELAP.
    e. PLANET.

12. *a.* Discuss the advantages and disadvantages of using computer programs to develop layouts.
    b. Do the results of the studies discussed in Section 6 suggest that human planners or computer programs produce superior layouts? For what reasons are these studies not conclusive?

13. For the example presented in this section illustrating CRAFT, verify the values of the predicted costs (2,715, 3,185, 2,735, and 2,830) obtained in the first iteration.

14. Consider the initial layout for the CRAFT example appearing in Figure 10–11. Draw a figure showing the layout obtained from exchanging the locations of A and D, and find the centroids of A and D in the new layout.

15. Determine the centroids for the six departments in the layout pictured in Figure 10–17 using the methods outlined in Appendix 10–A.

16. Consider the initial layout and from-to chart appearing in Figure 10–18. Assuming the goal is to minimize the total distance traveled, determine the final layout recommended by CRAFT using the pairwise exchange method described in this section. Compare the predicted and the actual figures for total distance traveled in the final layout.

17. Consider the initial layout pictured in Figure 10–19 and the two from-to charts giving the flow and cost data. Use the CRAFT approach to obtain a final layout. Compare the total cost of each layout predicted by CRAFT to the actual cost of each layout.

FIGURE 10–17

Layout (for Problem 15)

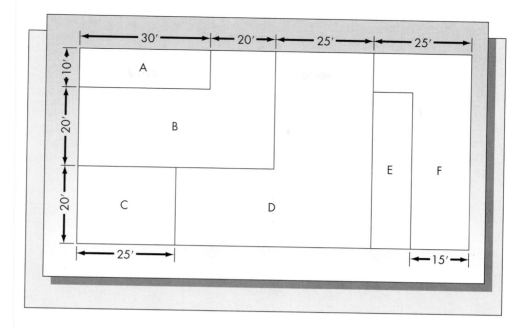

FIGURE 10–17

Layout (for Problem 15)

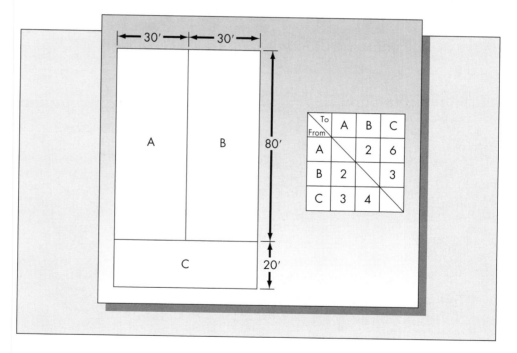

FIGURE 10–18

Layout and from-to chart (for Problem 16)

18. A facility consisting of eight departments has the activity relationship chart pictured in Figure 10–20. Compute the TCR for each department that would be used by CORELAP.

19. Six departments, A, B, C, D, E, and F, consume respectively 12, 6, 9, 18, 22, and 6 squares. Based on a sweep width of 4 and a facility size of 5 by 16, use the techniques employed by ALDEP to find a layout. Assume that the departments are placed in alphabetical order.

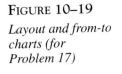

FIGURE 10–19

Layout and from-to charts (for Problem 17)

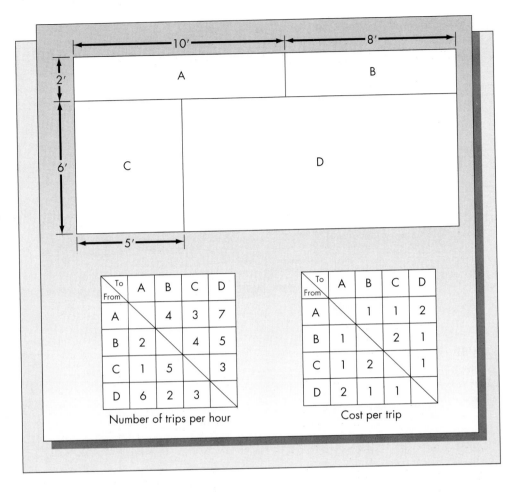

FIGURE 10–20

Rel chart (for Problem 18)

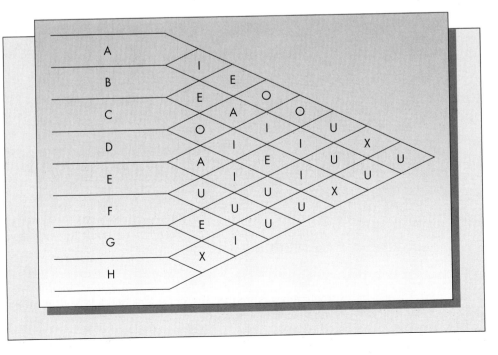

10.7 Flexible Manufacturing Systems

New technologies offer choices for the construction and design of facilities. Often these choices are conflicting. On one hand, the importance of flexibility cannot be overemphasized. In hi-tech industries in particular, manufacturing must adapt to frequent technological advances. In the personal computer industry, for example, the changing size configurations and central processor characteristics require constant redesign of manufacturing facilities.

One development that has gotten considerable attention in recent years is the *flexible manufacturing system* (FMS). An FMS is a collection of numerically controlled machines connected by a computer-controlled materials flow system. The machines are typically used for metal cutting, forming, and assembly operations, and provide the greatest benefit when a large variety of part types are required. A full-blown FMS can be extremely expensive, requiring a capital expenditure of upward of $10 million. Because of the high cost and the long payback period, firms are opting for scaled-down versions, called flexible machining cells. However, many firms feel that the capital expenditure is justified. For example, the Citroen plant in Meudon near Paris uses an FMS to produce component prototypes in batch sizes of one to about 50. The system is designed to handle a wide variety of part types (Hartley, 1984).

What advantages do such systems have over a conventional dedicated machine layout? According to Hartley (1984), they provide the opportunity to drastically slash the hidden costs of manufacturing. These include work-in-process inventory costs and overhead costs associated with indirect labor. They allow firms to quickly change tooling and product design with minimal additional investment in new equipment and personnel. However, they do require a substantial capital investment, which can be recouped only if their potential can be realized and they are used in the right environment.

The question is, under what circumstances should a firm consider employing an FMS as part of its overall layout for manufacturing? As shown in Figure 10–21, a flexible manufacturing system is appropriate when the production volume and the

FIGURE 10–21

The position of FMS in the manufacturing hierarchy

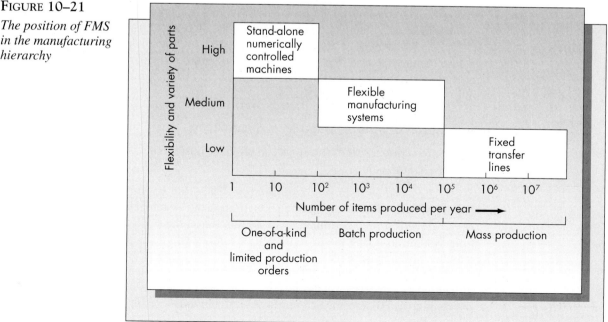

FIGURE 10–22

*A typical flexible
manufacturing
system*

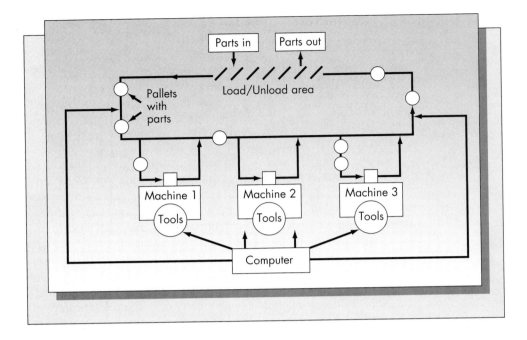

variety of parts produced are moderate. For systems with low volume and high customization, stand-alone numerically controlled machines are appropriate. These can be programmed for each individual application. For high volume production of standardized parts, fixed transfer lines are more appropriate.

The layout and structure of a typical FMS is pictured in Figure 10–22. Often the machines are controlled by a central computer, which also can be programmed for individual applications. Parts are typically loaded and unloaded at single location along the materials handling path. The materials handling system consists of pallets, which are usually metal disks two to three feet in diameter. These pallets carry work-in-process inventory and queue up at the machines for processing. Each machine may contain from 5 to 100 different tools, which are stored on a carousel. Tools can be exchanged in a matter of seconds. Note that the routing of parts must also be programmed into the system, as not all products require the same sequencing of machine operations. Although Figure 10–22 shows a centralized computer, some systems use individually programmed machines.

Advantages of Flexible Manufacturing Systems

When used correctly, these systems can provide substantial advantages over more rigid designs. These include

1. *Reduced work-in-process inventory.* The design of the system limits the number of pallets available for moving parts through the system. Hence, the WIP inventory never exceeds a predetermined level. In this sense, the FMS is similar to the just-in-time system, in which the level of WIP inventory is a decision variable whose value may be chosen in advance.

2. *Increased machine utilization.* Numerically controlled machines often have a utilization rate of 50 percent or less. However, an efficient FMS may have a utilization rate as high as 80 percent. The improved utilization is a result of both the reduction of

changeover time of machine settings and tooling, and the ability to better balance the system workload.

3. *Reduced manufacturing lead time.* Without an FMS, parts might have to be processed through several different work centers. As a result, there could be substantial transportation time between work centers and substantial queuing time at work centers. Because an FMS reduces transportation, setup, and changeover time, it results in significant reduction in lead time for production.

4. *Ability to handle a variety of different part configurations.* As noted above, the FMS is more flexible than a fixed transfer line but not as flexible as a stand-alone numerically controlled machine. Depending on the tooling available for the machines, parts may be launched into the system with little or no setup time required. Also, the FMS can process part configurations simultaneously.

5. *Reduced labor costs.* The number of workers required to manage an FMS can be as much as a factor of 10 fewer than the number required in a traditional job shop. Even when numerically controlled machines are used on a stand-alone basis, at least one worker is required per machine, and workers are required to transport the parts between machines. The automated materials handling capability of the FMS leads to significant reductions in labor requirements.

Disadvantages of Flexible Manufacturing Systems

Although there are many potential advantages of FMS, there is one factor that has delayed American acceptance of these systems. That factor is cost. Most FMS systems cost in the tens of millions of dollars. Traditional NPV (net present value) calculations often show that the investment is not justified. A well-known case is that of the Yamazaki Machinery Company in Japan. The company installed an $18 million FMS system. As a result, the number of machines was reduced from 68 to 18, the number of employees from 215 to 12, floor space from 103,000 square feet to 30,000 square feet, and the average processing time of parts from 35 days to 1.5 days (Kaplan, 1986). These figures are impressive. However, when translated to a return on investment, the story is not so rosy. The company reported a total savings of $6.9 million after two years. Including a savings of $1.5 million per year for the next 20 years, the projected total return is under 10 percent per year. In most American companies, the "hurdle rate" is generally 15 percent or higher, thus making this particular FMS operation a poor investment by NPV considerations alone. (The hurdle rate is the minimum acceptable rate of return on new projects.)

In addition to the direct costs of the equipment and the space, several indirect costs also are incurred. In order to manage the flow of materials, a sophisticated software system is required. Effective software can be extremely expensive, may require customization, often has bugs, and generally requires worker training. The cost of equipment such as feeders, conveyors, and transfer devices may not be part of the initial purchase cost. Other indirect costs include site preparation, spare parts to support the machinery, and disruptions that might result during the installation period. Furthermore, any company that purchases an FMS must anticipate a decline in productivity that accompanies the introduction of new technology.

It is no wonder that so many American companies have trouble justifying the investment in FMS. However, Kaplan (1986) argues that traditional cost accounting methods may ignore some important considerations. One is that evaluation of alternative investments based on discounted cash flow assumes a status quo. That is, NPV analysis assumes that factors such as market share, price, labor costs, and the company's competitive position in the marketplace will remain constant. However, the values of some of these variables will change, and are more likely to degenerate if the company retains outmoded production methods.

Furthermore, most firms prefer to invest in a variety of small projects rather than make a major capital outlay for a single project. Such a philosophy is safer in the short run, but could be suboptimal in the long run. An example of an industry that failed to invest in new technology is the railroad industry. Because of outmoded equipment and facilities, many firms in this industry have been unable to stay competitive and profitable.

Cost is not the only problem. The FMS may experience downtime for a variety of reasons. Planned downtime could be the result of scheduled maintenance and scheduled tool changeovers. Unplanned downtime could result from mechanical failures of the machines or electrical failures. If a single machine goes down, the system can continue to function, but if either the materials handling system or the central computer fails, the entire FMS is crippled.

Decision Making and Modeling of the FMS

Mathematical modeling can help with the decisions required to design and manage an FMS. Flexible manufacturing systems, like all job shops, must be managed on a continual basis. Some decisions, such as whether to purchase a system at all, have very far-reaching consequences. Other decisions, such as what action should be taken when a machine breaks down or a tool is damaged, may only affect the flow of materials for the next few hours. This suggests that a natural way to categorize these decisions is by the time horizon associated with their consequences. Table 10–1 summarizes the various levels of FMS decision making broken down in this manner.

Queuing-Based Models. Queuing theory is the basis for many models that can assist with long-term decisions. A queue is simply a waiting line. A single-server queuing system is characterized by an input stream (known as customers), a service discipline

---

TABLE 10–1   **Levels of Decision Making in Flexible Manufacturing Systems**

Time Horizon	Major Issues	Modeling Methods
Long term (months–years)	Design of the system Parts mix changes System modification and/or expansion	Queuing networks Simulation Optimization
Medium term (days–weeks)	Production batching Maximizing of machine utilization Planning for fluctuations in demand and/or availability of resources	Balancing methods Network simulation
Short term (minutes–hours)	Work order scheduling and dispatching Tool management Reaction to system failures	Tool operation and allocation program Work order dispatching algorithm Simulation

FIGURE 10–23

A single-server queue

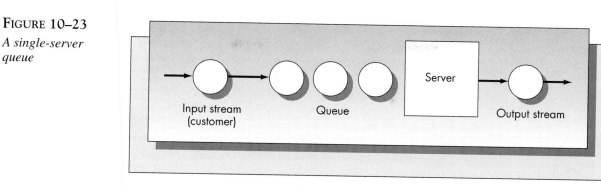

FIGURE 10–24

A network of queues

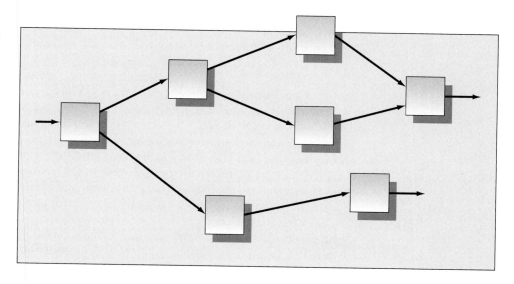

specifying the order or sequence that customers are served, and a service mechanism. Most queuing models assume that both the arrival stream of customers and the service times are random variables. It is the random nature of arrivals and services and their interactions that makes queuing problems interesting.[1] Figure 10–23 shows a simple schematic of a single-server queuing system.

In an FMS, each machine corresponds to a separate queue. The jobs correspond to the customers, and the queue is the stack of jobs waiting for processing at a machine. Because an FMS is a collection of machines connected by a materials handling system, the entering stream of customers for one machine is the result of the accumulation of departing streams from other machines. Hence, an FMS is a special kind of queuing network. A typical queuing network appears in Figure 10–24.

Queuing models are most useful in aiding with system design. They can be used to compute a variety of measures relating to machine utilization and system performance. An important issue that arises during the initial design phase is system capacity. The system capacity is the maximum production rate that the FMS can sustain.

---

[1]A summary of basic queuing theory appears in Supplement 2, which follows Chapter 8.

It depends on both the configuration of the FMS and the assumptions one makes about the nature of the arrival stream of jobs. Schweitzer's (1977) analysis shows that the output rate of a flexible manufacturing system is simply the capacity of the slowest machine in the system.

To be more precise, suppose that machine i can process jobs at a rate μ_i. If the probabilities that an entering job visits machines in a given order are known, one can find e_i, the expected number of visits of a newly arriving job to machine i, using queuing theory methods. It follows that $1/e_i$ is the rate that jobs arrive to machine i, so that μ_i/e_i is the output rate of machine i. The "slowest" machine, where slow is used in a relative sense, is that machine whose output rate is least. Hence, the output rate of the system is $\min_{1 \le i \le n}(\mu_i/e_i)$.

Such results can be very valuable when considering the design of the system and its potential utility to the firm. However, in some circumstances queuing models are only of limited value. In particular, in order to obtain explicit results for complex queuing networks, one often must assume that the arrival and the service processes are purely random. That is, both the time between the arrival of successive jobs and the time required for a machine to complete its task are exponential random variables.[2] It is unlikely that both interarrival times and job performance times in flexible manufacturing systems are accurately described by exponential distributions. There are circumstances under which exponential queuing formulas do give reasonable approximations to more complex cases, but simulation should be used to validate analytical formulas. Approximate analytical results are available for nonexponential cases as well (see Suri and Hildebrant, 1984, for example). Also, the exponential assumption is not required for some simple network configurations. A review of the analytical results for network queuing models of flexible manufacturing systems can be found in Buzacott and Yao (1986).

Even with these limitations, queuing models provide a powerful means for analyzing the performance characteristics and capacity limitations of flexible manufacturing systems. Eventually, network queuing models will accurately reflect the nature of FMSs and provide an effective means for assisting with system design and management.

Mathematical Programming Models. Stecke (1983) considers the use of mathematical programming for solving several decision problems arising in FMS management. These problems include the following:

1. *Part type selection problem.* From a given set of part types that have associated known requirements, determine a subset for immediate and simultaneous processing.
2. *Machine grouping problem.* Partition the machines into groups so that each machine in a group can perform the same set of operations.
3. *Production ratio problem.* Determine the relative ratios to produce the part types selected in step 1.
4. *Resource allocation problem.* Allocate the limited number of pallets and fixtures among the selected part types.

---

[2]The properties of the exponential distribution are discussed in detail in Chapter 2 and in Supplement 2 on queuing.

5. *Loading problem.* Allocate the operations and required tools of the selected part types among the machine groups subject to technological capacity constraints of the FMS.

We will not present the details of the mixed integer programming formulations. Our purpose is simply to note that mathematical programming is another means of helping with long-range and medium-range decisions affecting the FMS. Stecke (1983) shows how one would apply these models using actual data from an existing FMS.

Simulation Methods. Computer simulation has long been a popular and useful tool for analyzing complex manufacturing systems that are not easily analyzed by more conventional means. A computer simulation is a program designed to model the important features of a real system. Monte Carlo simulation means that there are elements of randomness in the problem. In FMS randomness arises from the variation in the input stream, the variation in the parts mix, and the possibility of failures of machines, tools, computers, or the materials handling system. Because simulation models are costly to develop and may require substantial computer time, accurate analytical models are preferred. However, often the assumptions required by queuing or mathematical programming models are too restrictive. Simulation may be the only alternative in such cases.

Simulation programs may be written in a source language such as Fortran or BASIC. However, special-purpose simulation languages exist and are more popular for manufacturing applications. Typical examples of such languages are GPSS and SLAM. SLAM is particularly well suited for simulating the kinds of networks that arise in manufacturing applications. Simulations are useful for answering "what-if" kinds of questions. Pritsker (1986) describes the use of simulation specifically for the design and evaluation of an FMS used to control machining operations on a certain type of metal casting. The objectives of the simulation study were to evaluate system balance and productivity, determine the need for additional equipment, determine which resources, if any, to eliminate, and find the number of dedicated versus flexible machines required. The simulation revealed that some of the existing equipment could be eliminated while still achieving the desired production goal.

A simulation package designed specifically for factory modeling is XCELL (see Conway et al., 1987). The factory floor is represented as a uniform grid of cells and each "element" of the factory occupies one cell. An element may be a (1) work center, (2) receiving area, (3) shipping area, (4) buffer, or (5) maintenance facility. XCELL uses symbolic graphic representations of elements. Although XCELL has yet to achieve the acceptance of GPSS and SLAM, its ease of use and flexibility for modeling factory flow make it an attractive alternative.

A new development in simulation of FMS has been to apply techniques developed for performance evaluation of computer and telecommunications networks. Bruno and Biglia (1985) explore the application of stochastic Petri nets to some management problems in FMS. Petri nets are network representations of systems with special features such as tokens indicating when certain parts of the system are blocked. In an FMS, blocking can occur if a machine breaks down or if the number of parts waiting to be processed at a machine equals the maximum queue size. The application of Petri net theory to FMS design and management problems is an open area of research. It remains to be seen whether this theory will yield any useful results for

FMS management or design that cannot be achieved by more conventional simulation methods.

The Future of FMS

There is substantial potential for the application of FMS both in the United States and abroad. It is estimated that about half of the annual expenditures on manufacturing is in the metal-working industry, and two-thirds of metal working is metal cutting (Stecke, 1983). Furthermore, about three-fourths of the dollar volume of metal-worked parts is manufactured in batches of fewer than 50 parts (Cook, 1975). Evidently, there is a huge market for FMS systems in the United States. Similar potential exists in other industrialized countries as well. In fact, the rate of growth of installed FMS systems is higher in Japan than in any other country in the world (Jaikumar, 1984).

The rate of growth of FMS in the United States has been healthy if not spectacular. In 1981 there were about 18 systems operational in the United States (Jaikumar, 1984), and by late 1986 this number had grown to about 50. Krouse (1986) estimated that there should have been 284 systems in place in the United States by 1990. As we noted earlier, the primary concern among companies considering purchasing an FMS is cost. The average cost of such systems was estimated to be $12.5 million in 1986 and is probably higher today. Also, installation generally requires from 18 to 24 months.

Both hardware and changeover costs discourage many potential users of FMS. Furthermore, many factories in this country have yet to invest in numerically controlled machine tools, let alone a full-blown FMS. As a response to this, vendors are offering scaled-down versions of the FMS known as flexible manufacturing cells. Flexible manufacturing cells are small FMS systems offering fewer machines, a less extensive materials handling system, and fewer tools per machine. Vendors anticipate healthy sales of these cells, which should result in an expanded customer base and a larger market for scaled-up systems.

Cost is not the only factor retarding acceptance. The performance of many existing FMS systems has been disappointing. Zygmont (1986) reported that some installed systems did not meet the expectations of the companies that purchased them. In one case, three years were required to debug a system purchased by Electro-Optical and Data Systems of Hughes Aircraft and in the end the system was still less flexible than originally expected. One of the problems encountered was that the level of precision required was higher than the system was capable of delivering. Similarly, Deere and Co. was disappointed with its $20 million system, and was especially disappointed with the lack of flexibility afforded by the software for handling complex and varying part sequences.

Jaikumar (1986) makes a case that the problem with FMS systems in the United States is not the systems themselves but the way they are used. In a comparison of FMSs in the United States and Japan, he found some striking differences. These differences are listed in Table 10–2.

Most telling is the fact that there are almost 10 times more part types produced on each Japanese system than on each American system. That, coupled with the enormous difference in annual volume per part, indicates that American firms are using FMSs improperly. That is, they are used as if they were simply another set of machines for high-volume standardized production rather than for producing the varied part mix for which they were designed. Another important difference is that many Japanese systems are run both day and night and often unattended. Jaikumar (1986)

TABLE 10–2   **A Comparison of FMS in the United States and Japan**

	United States	Japan
System development times (years)	25 to 3	1 to 1.25
Types of parts produced per system	10	93
Annual volume per part	1.727	258
Number of new parts introduced per year	1	22
Number of systems with unattended operations	0	18
Average metal-cutting time per day	83	202

claims that this is a consequence of the improved design and reliability of the Japanese systems.

As firms install additional FMSs, vendors will gain a better understanding of their power and their limitations. Advances in both hardware and software should make future systems more capable of dealing with the problems discussed above. As both the systems and the understanding of the circumstances under which they can be most effective improve, we should see many more companies installing FMSs in the next decade.

Problems for Section 7

20. In each case below, state whether the factor listed is an advantage or a disadvantage of FMS. Discuss the reasons for your choice.
    a. Cost.
    b. Ability to handle different parts requirements.
    c. Advances in manufacturing technology.
    d. Reliability.

21. For each of the case situations described below, state which of the three types of manufacturing systems would be most appropriate: (1) stand-alone numerically controlled machines, (2) FMS, (3) fixed transfer line, or (4) other system. Explain your choice in each case.
    a. A local machine shop that accepts custom orders from a variety of clients in the electronics industry.
    b. An international supplier of standard-sized metal containers.
    c. The metal-working division of a large aircraft manufacturer, which must serve the needs of a variety of divisions in the firm.

22. For what reason might a traditional NPV (net present value) analysis not be an appropriate means for evaluating the desirability of purchasing an FMS?

23. With which decisions related to FMS design or control can the following methods assist? Also, briefly describe how each method would be used.
    a. Queuing theory.
    b. Simulation.
    c. Mathematical programming.

SNAPSHOT APPLICATION

Kraft Foods Uses Optimization and Simulation to Determine Best Layout

In the mid 1990s Kraft decided to renovate one of its major manufacturing facilities located in the Midwest. New higher capacity production lines were to replace the old lines. The new lines would have increased throughput and more mixing capabilities, and would be able to operate at a variety of speeds. One of the concerns in designing the new facility was determining the optimum number of new lines to install to maintain current capacity levels while also allowing for future growth. Management needed to decide the line configuration quickly because the lines required a six-month delivery lead time.

Kraft management asked the firm's operations research group to develop a detailed mathematical model of the system. A "seat of the pants" approach was simply unacceptable: a wrong decision would be costly. Too many lines would result in wasted space and money, and too few in a capacity bottleneck. A further difficulty was that none of Kraft's plants had used these lines before, so there was no prior experience from which to draw.

Staff members from the firm's operations research group were asked to consider both the problem of optimizing the number of new lines and the flow of material within the plant. To address the optimization problem, the group developed an integer program-ming formulation of the problem. The objective was to ensure that the makespan of any schedule Kraft was likely to encounter did not exceed one eight-hour shift. (Refer to Chapter 8 for a discussion of makespan.) They determined that six lines would easily take care of existing capacity requirements and include a substantial margin for future growth. The model was solved using the AMPL optimization package on a personal computer.

Once the optimal number of new lines was determined, the next step was to develop a detailed simulation of the factory floor to show graphically how the new design would work in practice. For this task, the group decided to use F&H Simulation's Taylor II software package, a graphical-based simulation package especially well-suited for considering the effects of different work schedules on a given layout. The simulation model consisted of 139 elements including lines, machines, work-in-process buffers, loading areas for raw materials, and shipping areas. By simulating the process with the heaviest schedules, the group demonstrated that the new layout would easily have enough capacity to handle the most severe demands the plant is likely to see. Furthermore, the simulation afforded factory personnel a means to see how the flow of materials would change prior to the time that the plant was renovated and the new layout implemented.

Source: Based on Sengupta and Combes (1995).

10.8 Locating New Facilities

Chapter 1 discussed several qualitative considerations when deciding on the best location for a new plant or other facility. The remainder of this chapter will consider quantitative techniques that can help with location decisions. There are many circumstances in which the objective can be quantified easily, and it is in these cases that analytical solution methods are most useful.

In some sense, the plant layout problem is a special case of the location problem. However, the problems considered in the remainder of this chapter differ from layout problems in the following way: we will assume that there are one or more existing facilities already in place and we wish to find the optimal location of a new facility. We will concentrate on the problem of locating a single new facility, although there are versions of the problem in which one must determine the locations of multiple facilities.

Examples of the type of one-facility location problems that can be solved by the methods discussed in this chapter include

1. Location of a new storage warehouse for a company with an existing network of production and distribution centers.
2. Determining the best location for a new machine in a job shop.
3. Locating the computer center in a university.
4. Finding the best location for a new hospital in a metropolitan area.
5. Finding the most suitable location for a power generating plant designed to serve a geographic region.
6. Determining the placement of an ATM in a neighborhood.
7. Locating a new police station in a community.

Undoubtedly, the reader can think of many other examples of location problems. Analytical methods assume that the objective is to locate the new facility to minimize some function of the distance of the new location from existing locations. For example, when locating a new warehouse, an appropriate objective would be to minimize the total distance traveled to the warehouse from production facilities and from the warehouse to retail outlets. A hospital would be located to be most easily accessible to the largest proportion of the population in the area it serves. A machine would be placed to minimize the weighted sum of materials handling trips to and from the machine. Clearly, choosing the location of a facility to minimize some function of the distance separating the new facility from existing facilities is appropriate for many real location problems.

Measures of Distance

Two measures of distance are most common: *Euclidean distance* and *rectilinear distance*. Euclidean distance is also known as straight-line distance. The Euclidean distance separating two points is simply the length of the straight line connecting the points. Suppose that an existing facility is located at the point (a, b) and let (x, y) be the location of the new facility. Then the Euclidean distance between (a, b) and (x, y) is

$$\sqrt{(x - a)^2 + (y - b)^2}.$$

The rectilinear distance (also known as metropolitan distance, recognizing the fact that streets usually run in a crisscross pattern) is given by the formula

$$|x - a| + |y - b|.$$

Figure 10–25 illustrates the difference between these two distance measures.

Rectilinear distance is appropriate for many location problems. Distances in metropolitan areas tend to be more closely approximated by rectilinear distances than by Euclidean distances even when the street pattern is not a perfect grid. In many manufacturing environments, material is transported across aisles arranged in regular patterns. It is a fortunate coincidence that rectilinear distance is more common than Euclidean distance, because the rectilinear distance problem is easier to solve.

FIGURE 10–25

*Euclidean and
rectilinear distances*

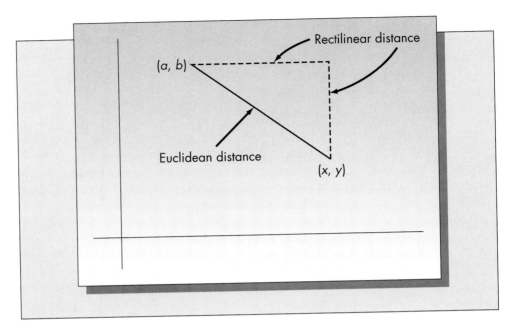

Problems for Section 8

24. Consider the problem of locating a new hospital in a metropolitan area. List the factors that can be quantified and those that cannot. Comment on the usefulness of quantitative methods in the decision-making process.

25. A coordinate system is superimposed on a map. Three existing facilities are located at (5, 15), (10, 20), and (6, 9). Compute both the rectilinear and the Euclidean distances separating each facility from a new facility located at $(x, y) = (8, 8)$.

26. For the situation described in Problem 25, suppose that there are only three feasible locations for the new facility: (8, 16), (6, 15), and (4, 18).
    a. What is the optimal location if the objective is to minimize the total rectilinear distance to the three existing facilities?
    b. What is the optimal location if the objective is to minimize the total Euclidean distance to the three existing facilities?

27. For each of the seven examples of location problems listed in this section, indicate which distance measure, Euclidean or rectilinear, would be more appropriate. (Discuss how, in some cases, one or the other objective could be appropriate for the same problem, depending upon the optimization criterion used.)

10.9 The Single-Facility Rectilinear Distance Location Problem

In this section we will present a solution to the general problem of locating a new facility among n existing facilities. The objective is to locate the new facility to minimize a weighted sum of the rectilinear distances from the new facility to existing facilities.

Assume that the existing facilities are located at points $(a_1, b_1), (a_2, b_2), \ldots, (a_n, b_n)$. Then the goal is to find values of x and y to minimize

$$f(x, y) = \sum_{i=1}^{n} w_i(|x - a_i| + |y - b_i|).$$

The weights are included to allow for different traffic rates between the new facility and the existing facilities. A simplifying property of the problem is that the optimal values of x and y may be determined separately, as

$$f(x, y) = g_1(x) + g_2(y).$$

where

$$g_1(x) = \sum_{i=1}^{n} w_i|x - a_i|$$

and

$$g_2(y) = \sum_{i=1}^{n} w_i|y - b_i|.$$

As we will see, there is always an optimal solution with x equal to some value of a_i and y equal to some value of b_i. (There may be other optimal solutions as well.) However, before presenting the general solution algorithm for finding the optimal location of the new facility, we consider a few simple examples in order to provide the reader with some intuition. Consider first the case in which there are exactly two existing facilities, located at (5, 10) and (20, 30) as pictured in Figure 10–26. Assume that the weight applied to each of these facilities is 1. If x assumes any value between 5 and 20, the value of $g_1(x)$ is equal to 15. (For example, if $x = 13$, then $g_1(x) = |5 - 13| + |13 - 20| = 8 + 7 = 15$.) Similarly, if y assumes any value between 10 and 30, then $g_2(y) = 20$. Any value of x outside the closed interval [5, 20] and any value of y outside

FIGURE 10–26

Optimal locations of the new facility for a rectilinear distance measure

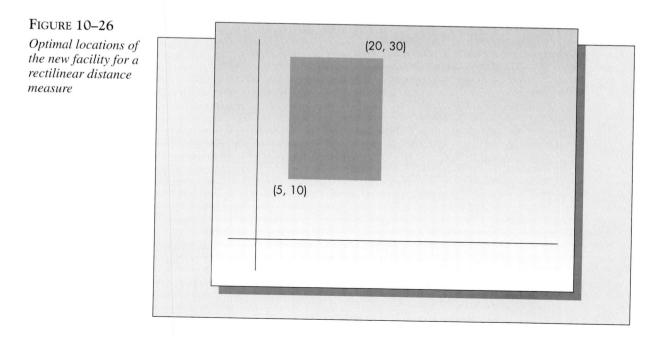

(20, 30)

(5, 10)

the closed interval [10, 30] results in larger values of $g_1(x)$ and $g_2(y)$. Hence, the optimal solution is (x, y) with $5 \leq x \leq 20$ and $10 \leq y \leq 30$. All locations in the shaded region pictured in Figure 10–26 are optimal.

As in the example, there always will be an optimal location of the new facility with coordinates coming from the set of coordinates of the existing facilities. Suppose that the existing facilities have locations (3, 3), (6, 9), (12, 8), and (12, 10). Again assume that the weight applied to these locations is 1. Ranking the x locations in increasing order gives 3, 6, 12, 12. A *median* value is such that half of the x values lie above it and half of the x values lie below it. Any value of x between 6 and 12 is a median location and is optimal for this problem. The optimal value of $g_1(x)$ is 15. (The reader should experiment with a number of different values of x between 6 and 12 to satisfy himself or herself that this is the case.) Ranking the y values in increasing order gives 3, 8, 9, 10. The median value of y is between 8 and 9, and the optimal value of $g_2(y) = 8$.

The optimal solution is to locate (x, y) at the median of the existing facilities. This result carries over to the case in which there are weights different from 1. Suppose in the previous example that we were given the locations of four machines in a job shop. The goal is to find the location of a fifth machine to minimize the total distance traveled to transport material between the new machine and the existing ones. Assume that on average there are respectively 2, 4, 3, and 1 materials handling trips per hour from the existing machines to the new machine. Summarizing the given information,

Location of Existing Machines	Weight
(3, 3)	2
(6, 9)	4
(12, 8)	3
(12, 10)	1

This problem is equivalent to one in which there are two machines at location (3, 3), four machines at location (6, 9), three machines at location (12, 8), and one machine at location (12, 10), with weights equal to 1. Hence, the x locations in increasing order are 3, 3, 6, 6, 6, 6, 12, 12, 12, 12. The median location is $x = 6$. The y locations in increasing order are 3, 3, 8, 8, 8, 9, 9, 9, 9, 10. The median location is any value of y on the interval [8, 9]. The reader should check that the value of the objective function at the optimal solution is $30 + 16 = 46$.

When the weights are large, this approach is inconvenient. A quicker method of finding the optimal location of the new facility is to compute the accumulated weights and determine the location or locations corresponding to half of the accumulated weight. The procedure is best illustrated by example.

Example 10.5

University of the Far West has purchased equipment that permits faculty to prepare videotapes of lectures. The equipment will be used by faculty from six schools on campus: business, education, engineering, humanities, law, and science. The locations of the buildings on the campus are pictured in Figure 10–27. The coordinates of the locations and the numbers of faculty that are anticipated to use the

FIGURE 10–27

Location of six campus buildings (refer to Example 10.5)

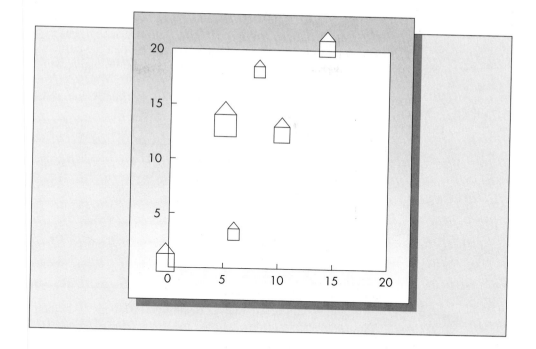

equipment are as follows:

School	Campus Location	Number of Faculty
Business	(5, 13)	31
Education	(8, 18)	28
Engineering	(0, 0)	19
Humanities	(6, 3)	53
Law	(14, 20)	32
Science	(10, 12)	41

The campus is laid out with large grassy areas separating the buildings, and walkways are mainly east–west or north–south, so that distances between buildings are rectilinear. The university planner would like to locate the new facility so as to minimize the total travel time of all faculty planning to use it.

We will find the optimal values of the x and y coordinates separately. Consider the optimal x coordinate value. We first rank x coordinates in increasing value and accumulate the weights.

School	x Coordinate	Weight	Cumulative Weight
Engineering	0	19	19
Business	5	31	50
Humanities	6	53	103
Education	8	28	131
Science	10	41	172
Law	14	32	204

The optimal value of the x coordinate is found by dividing the total cumulative weight by 2 and identifying the first location at which the cumulative weight exceeds this value. In the example this is the first time that the cumulative weight exceeds $204/2 = 102$. This occurs at $x = 6$, when the cumulative weight is 103. Hence, the optimal $x = 6$.

We use the same procedure to find the optimal value of the y coordinate. The rankings are given below.

School	y Coordinate	Weight	Cumulative Weight
Engineering	0	19	19
Humanities	3	53	72
Science	12	41	113
Business	13	31	144
Education	18	28	172
Law	20	32	204

In this case the cumulative weight first exceeds 102 when $y = 12$. Hence, $y = 12$ is optimal. The optimal location of the new facility is (6, 12). This solution is unique. Multiple optima will occur when a value of the cumulative weight exactly equals half of the total cumulative weight. For example, suppose that the weight for science in the example above were 19 instead of 41. Then the cumulative weight for science would be 91, exactly half of the total weight of 182, and all values of y on the closed interval [12, 13] would be optimal. ∎

Contour Lines

The previous example suggests an interesting question. Suppose that the location (6, 12) is infeasible. How can the university gauge the cost penalty when locating the new facility elsewhere? Contour lines, or isocost lines, can assist with determining the penalty of nonoptimal solutions. A contour line is a line of constant cost: locating a new facility anywhere along a contour line results in exactly the same cost.

We have pictured the contour lines for the simple example of Figure 10–26, in which there are only two existing facilities and the weights are equal, in Figure 10–28. The reader should convince him- or herself that the total rectilinear distance from any point along a contour line to the two points (5, 10) and (20, 30) is the same. Determining contour lines involves computing the appropriate slope for each of the regions obtained by drawing vertical and horizontal lines through each of the points (a_i, b_i). The procedure for determining contour lines is outlined in Appendix 10–B at the end of this chapter.

In Figure 10–29 we have pictured contour lines for the Example 10.5 problem in which the university must locate an audiovisual center. If the optimal location (6, 12) is infeasible, the university administration could use this map to see the penalties associated with alternative sites.

Minimax Problems

We have assumed thus far that the new facility should be placed so as to minimize the sum of the weighted distances to all existing facilities. There are circumstances in which this objective is inappropriate, however. Consider the following example. The

FIGURE 10–28

*Contour lines for
the two-facility
problem pictured in
Figure 10–26*

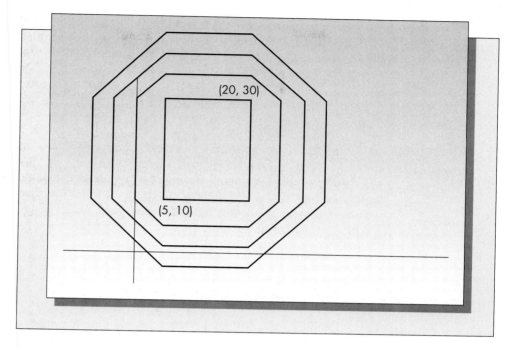

FIGURE 10–29

*Contour lines for
university location
Example 10.5*

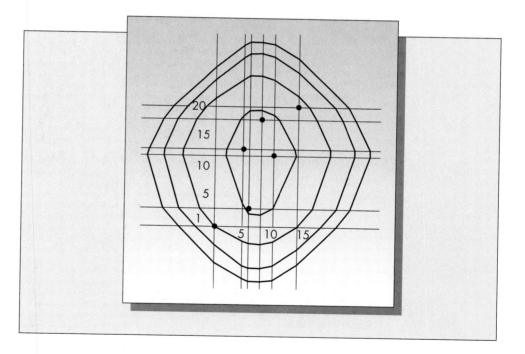

city is considering locations for a paramedic facility. The paramedics should be able to respond to emergency calls anywhere in the city. Certain conditions, such as a severe heart attack, must be treated quickly if the patient is to have any chance of surviving. Hence, the facility should be located so that *all* locations in the city can be reached in a given time.

In such a case the objective would be to determine the location of the new facility to minimize the maximum distance to the existing facilities rather than the total distance. Let $f(x, y)$ be the maximum distance from the new facility to the existing facilities. Then

$$f(x, y) = \max_{1 \le i \le n} (|x - a_i| + |y - b_i|).$$

The objective is to find (x^*, y^*) that satisfies

$$f(x^*, y^*) = \min_{x,y} f(x, y).$$

The procedure for finding the optimal minimax location is straightforward. Linear programming can be used to show that the procedure we outline below is optimal. (We will not present the details here. The interested reader should refer to Francis and White, 1974.) Define the numbers c_1, c_2, c_3, c_4, and c_5:

$$c_1 = \min_{1 \le i \le n} (a_i + b_i),$$
$$c_2 = \max_{1 \le i \le n} (a_i + b_i),$$
$$c_3 = \min_{1 \le i \le n} (-a_i + b_i),$$
$$c_4 = \max_{1 \le i \le n} (-a_i + b_i),$$
$$c_5 = \max(c_2 - c_1, c_4 - c_3).$$

Let

$$x_1 = (c_1 - c_3)/2,$$
$$y_1 = (c_1 + c_3 + c_5)/2,$$

and

$$x_2 = (c_2 - c_4)/2,$$
$$y_2 = (c_2 + c_4 - c_5)/2.$$

Then all points that lie along the line connecting (x_1, y_1) and (x_2, y_2) are optimal. That is, every optimal solution to the minimax problem, (x^*, y^*), can be expressed in the form

$$x^* = \lambda x_1 + (1 - \lambda)x_2,$$
$$y^* = \lambda y_1 + (1 - \lambda)y_2,$$

where λ is a constant satisfying $0 \le \lambda \le 1$. The optimal value of the objective function is $c_5/2$.

Example 10.6

Consider Example 10.5 of the University of the Far West above. As some faculty members have disabilities, the president has decided to locate the audiovisual facility to minimize the maximum distance from the facility to the six schools on campus. Recall that the locations of the schools are

(5, 13),　　(8, 18),
(0, 0),　　　(6, 3),
(14, 20),　　(10, 12).

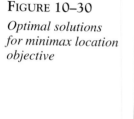

FIGURE 10–30

FIGURE 10–30

Optimal solutions for minimax location objective

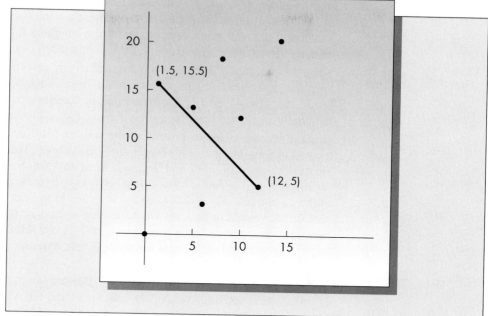

The values of the constants $c_1, \ldots, c_5$ are

$$c_1 = \min_{1 \leq i \leq n} (a_i + b_i) = \min(18, 26, 0, 9, 34, 22) = 0,$$

$$c_2 = \max_{1 \leq i \leq n} (a_i + b_i) = \max(18, 26, 0, 9, 34, 22) = 34,$$

$$c_3 = \min_{1 \leq i \leq n} (-a_i + b_i) = \min(8, 10, 0, -3, 6, 2) = -3,$$

$$c_4 = \max_{1 \leq i \leq n} (-a_i + b_i) = \max(8, 10, 0, -3, 6, 2) = 10,$$

$$c_5 = \max(c_2 - c_1, c_4 - c_3) = \max(34, 13) = 34.$$

Hence, it follows that

$$x_1 = (c_1 - c_3)/2 = (0 - (-3))/2 = 1.5,$$
$$y_1 = (c_1 + c_3 + c_5)/2 = (0 - 3 + 34)/2 = 15.5,$$

and

$$x_2 = (c_2 - c_4)/2 = (34 - 10)/2 = 12,$$
$$y_2 = (c_2 + c_4 - c_5)/2 = (34 + 10 - 34)/2 = 5.$$

All points on the line connecting (x_1, y_1) and (x_2, y_2) are optimal. The value of the objective function at the optimal solution(s) is $34/2 = 17$. The optimal locations for the minimax problem are pictured in Figure 10–30. Recall that the optimal solution when we used a weighted objective was (6, 12). It is interesting to note that one optimal solution to this problem, (6, 11), is quite close to that solution. ∎

Problems for Section 9

28. A machine shop has five machines, located at (3, 3), (3, 7), (8, 4), (12, 3), and (14, 6) respectively. A new machine is to be located in the shop with the following expected numbers of loads per hour transported to the existing

machines: $\frac{1}{8}, \frac{1}{8}, \frac{1}{4}, 1,$ and $\frac{1}{6}$. Material is transported along parallel aisles, so that a rectilinear distance measure is appropriate. Find the coordinates of the optimal location of the new machine to minimize the weighted sum of the rectilinear distances from the new machine to the existing machines.

29. Solve the problem of locating the new audiovisual center at the University of the Far West described in Example 10.5, assuming that the weights are respectively 80, 12, 56, 104, 42, and 17 for the schools of business, education, engineering, humanities, law, and science.

30. Armand Bender plans to visit six customers in Manhattan. Three are located in a building at 34th Street and 7th Avenue. The remaining customers are at 48th and 8th, 38th and 3rd, and 42nd and 5th. Streets are separated by 200 feet and avenues by 400 feet. He plans to park once and walk to all of the customers. Assume that he must return to his car after each visit to pick up different samples. At what location should he park in order to minimize the total distance traveled to the clients? (Hint: In designing your grid, be sure that you account for the fact that avenues are twice as far apart as streets.)

31. In the previous problem, suppose that Mr. Bender can park only in lots located at (1) 40th Street and 8th Avenue, (2) 46th Street and 6th Avenue, and (3) 33rd Street and 5th Avenue. Where should he plan to park?

32. An industrial park consists of 16 buildings. The corporations in the park are sharing the cost of construction and maintenance for a new first-aid center. Because of the park's layout, distances between buildings are most closely approximated by a rectilinear distance measure. Weights for the buildings are determined based on the frequency of accidents. Find the optimal location of the first-aid center to minimize the weighted sum of the rectilinear distances to the 16 buildings.

Building	a_i	b_i	w_i	Building	a_i	b_i	w_i
1	0	0	9	9	14	6	11
2	10	3	7	10	19	0	17
3	8	8	4	11	20	4	14
4	12	20	3	12	14	25	6
5	4	9	2	13	3	14	5
6	18	16	12	14	6	6	8
7	4	1	4	15	9	21	15
8	5	3	5	16	10	10	4

33. Draw contour lines for the problem of locating a machine shop described in Problem 28. (Refer to Appendix 10–B.)

34. Draw contour lines for the location problem described in Problem 30. (Refer to Appendix 10–B.)

35. Two facilities, located at (0, 0) and (0, 10), have respective weights 2 and 1. Draw contour lines for this problem. (Refer to Appendix 10–B.)

36. Solve Problem 28 assuming a minimax rectilinear objective.
37. Solve Problem 30 assuming a minimax rectilinear objective.
38. Solve Problem 32 assuming a minimax rectilinear objective.

10.10 Euclidean Distance Problems

Although the rectilinear distance measure is appropriate for many real problems, there are applications in which the appropriate measure of distance is the straight-line measure. An example is locating power-generating facilities in order to minimize the total amount of electrical cable that must be laid to connect the plant to the customers. This section will consider the Euclidean problem and a variant of it known as the gravity problem, which has a far simpler solution.

The Gravity Problem

The gravity problem corresponds to the case of an objective equal to the square of the Euclidean distance. Hence, the objective is to find values of (x, y) to minimize

$$f(x, y) = \sum_{i=1}^{n} w_i[(x - a_i)^2 + (y - b_i)^2].$$

This objective is appropriate when the cost of locating new facilities increases as a function of the square of the distance of the new facility to the existing facilities. Although such an objective is not common, the solution to this problem is straightforward and often has been used as an approximation to the more common straight-line distance problem.

The optimal values of (x, y) are easily determined by differentiation. The partial derivatives of the objective function with respect to x and y are

$$\frac{\partial f(x, y)}{\partial x} = 2 \sum_{i=1}^{n} w_i(x - a_i),$$

$$\frac{\partial f(x, y)}{\partial y} = 2 \sum_{i=1}^{n} w_i(y - b_i).$$

Setting these partial derivatives equal to zero and solving for x and y gives the optimal solution

$$x^* = \frac{\sum_{i=1}^{n} w_i a_i}{\sum_{i=1}^{n} w_i},$$

$$y^* = \frac{\sum_{i=1}^{n} w_i b_i}{\sum_{i=1}^{n} w_i}.$$

The term *gravity problem* arises for the following reason. Suppose that one places a map of the area in which the facility is to be located on a heavy piece of cardboard. Weights proportional to the numbers w_i are placed at the locations of the existing facilities. Then the gravity solution is the point on the map at which the entire thing would balance. (This particular description is by Keefer, 1934.) Although one could certainly solve the gravity problem this way, it is so easy to find (x^*, y^*) using the formulas above that there seems little reason to employ the physical model.

Example 10.7

We will find the solution to the problem of locating the audiovisual center for the University of the Far West assuming a squared Euclidean distance location measure. Substituting the values of the weights and the building locations into the formulas above, we obtain

$$x^* = 1{,}555/204 = 7.6,$$

$$y^* = 2{,}198/204 = 10.8,$$

which is somewhat different from the rectilinear solution $(6, 12)$. ∎

The Straight-Line Distance Problem

The straight-line distance measure arises much more frequently than does the squared-distance measure discussed in the previous section. The objective in this case is to find (x, y) to minimize

$$f(x, y) = \sum_{i=1}^{n} w_i \sqrt{(x - a_i)^2 + (y - b_i)^2}.$$

Unfortunately, it is not as easy to find the optimal solution mathematically when using a Euclidean distance measure as it is when using squared Euclidean distance. As with the gravity problem, there is also a physical model that one could construct to find the optimal solution. In this case one places a map of the area on a table top. Holes are punched in the table top at the locations of existing facilities and weights are suspended on strings through the holes. The size of the weights should be proportional to the relative weights of the locations. The strings are attached to a ring. If there is no friction, the ring will come to rest at the location of the optimal solution (see Figure 10–31). Although this method can be used to find a solution, it does have drawbacks. In particular, the friction between the string and the table must be negligible to be sure that the ring comes to rest in the correct position.

Determining the optimal solution mathematically is more difficult for Euclidean distance than for either rectilinear or squared Euclidean distance. There are no known simple algebraic solutions; all existing methods require an iterative procedure. We will describe a procedure that appears in Francis and White (1974) that will yield an optimal solution as long as the location of the new facility does not overlap with the location of an existing facility.

Define

$$g_i(x, y) = \frac{w_i}{\sqrt{(x - a_i)^2 + (y - b_i)^2}}.$$

FIGURE 10–31

Solution of the Euclidean distance problem using a physical model

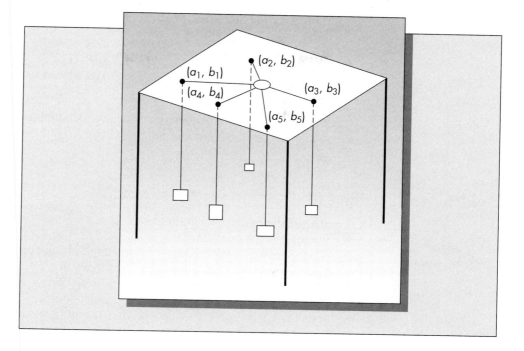

Let

$$x = \frac{\displaystyle\sum_{i=1}^{n} a_i g_i(x, y)}{\displaystyle\sum_{i=1}^{n} g_i(x, y)},$$

$$y = \frac{\displaystyle\sum_{i=1}^{n} b_i g_i(x, y)}{\displaystyle\sum_{i=1}^{n} g_i(x, y)}.$$

The procedure is as follows: Begin the process with an initial solution (x_0, y_0). The gravity solution is generally recommended to get the method started. Compute the values of $g_i(x, y)$ at this solution and determine new values of x and y from the formulas above. Then, recompute $g_i(x, y)$ using the new values of x and y, giving rise to yet another pair of (x, y) values. Continue to iterate in this fashion until the values of the coordinates converge. This procedure will give an optimal solution as long as the values of (x, y) at each iteration do not correspond to some existing location. The problem is that if we substitute $x = a_i$ and $y = b_i$, $g_i(x, y)$ is undefined, because the denominator is zero. It is unlikely that this will happen, but if it does, a modification of this method is required (see Francis and White, 1974).

Example 10.8

We will solve the university's location problem assuming a Euclidean distance measure. To get the procedure started we use the solution to the gravity problem, which, to three decimals, is $(x_0, y_0) = (7.622, 10.775)$. The sequence of (x, y) values obtained from iterating the equations above is

$$(x_1, y_1) = (7.895, 11.432), \quad (x_5, y_5) = (8.429, 11.880),$$
$$(x_2, y_2) = (8.102, 11.715), \quad (x_6, y_6) = (8.481, 11.889),$$
$$(x_3, y_3) = (8.251, 11.822), \quad (x_7, y_7) = (8.518, 11.894),$$
$$(x_4, y_4) = (8.356, 11.863), \quad (x_8, y_8) = (8.545, 11.899).$$

Clearly, these values are beginning to converge. Continuing with the iterations one eventually reaches the optimal solution, which is (8.621, 11.907). ■

Problems for Section 10

39. For the situation described in Problem 28,
    a. Find the gravity solution.
    b. Use the answer obtained in part (a) as an initial solution for the Euclidean distance problem. Determine (x_i, y_i) iteratively, for $1 \leq i \leq 5$ if you are solving the problem by hand, or for $1 \leq i \leq 20$ if you are solving with the aid of a computer. Based on your results, estimate the final solution.

40. Three existing facilities are located at (0, 0), (5, 5), and (10, 10). The weights applied to these facilities are 1, 2, and 3 respectively. Find the location of a new facility that minimizes the weighted Euclidean distance to the existing facilities.

41. A telecommunications system is going to be installed at a company site. The switching center that governs the system must be located in one of the five buildings to be served. The goal is to minimize the amount of underground cabling required. In which building should the switching center be located? Assume a straight-line distance measure.

Building	Location
1	(10, 10)
2	(0, 4)
3	(6, 15)
4	(8, 20)
5	(15, 0)

10.11 Other Location Models

The particular models discussed in Sections 9 and 10 have been referred to as "planar location models" by Francis, McGinnis, and White (1983). As they noted, there are seven assumptions associated with these models of location problems:

1. A plane is an adequate approximation of a sphere.
2. Any point on the plane is a valid location for a facility.
3. Facilities may be idealized as points.
4. Distances between facilities are adequately represented by planar distance.
5. Travel costs are proportional to distance.

6. Fixed costs are ignored.

7. Issues of distribution can be ignored.

Unless distances separating facilities are extremely large, assumption (1) should be a reasonable approximation of reality. Assumption (2) can be very restrictive. In some circumstances, only a small number of feasible locations exist, in which case the simplest approach is to evaluate the cost of locating the new facility at each of these locations and choose the one with the lowest cost. When the number of feasible alternatives is large, one could construct contour lines and consider only those locations along a given contour line with an acceptably low cost.

Depending upon the nature of the problem, assumption (3) may or may not be problematic. For example, if one uses location models to solve factory layout problems, then the size of the facilities becomes an issue. However, for a problem such as locating warehouses nationwide, idealizing facilities as points is reasonable. Assumption (4) requires that one use a particular measure of distance to compare various configurations. However, one distance measure may not be sufficient to explain all facility interactions. For example, rectilinear distances are typically used in problems in which facilities are to be located in cities. However, closed roads, rivers, or unusual street patterns could make this assumption inaccurate.

The final three assumptions deal with issues that typically arise in distribution problems. Transportation costs depend on the terrain: It is more expensive to transport goods over mountains than on flat interstate highways, for example. Fixed costs of transporting goods can be substantial depending on the mode of transportation. The fixed cost of transporting goods by air, for example, is very high.

More complex location models exist that deal with several of these shortcomings. We briefly review these in the following sections.

Locating Multiple Facilities

The previous section treated the problem of locating a single facility among n existing facilities. However, applications exist in which the goal is to locate multiple facilities among n existing facilities. For example, a nationwide consumer products producer might be considering where to locate five new regional warehouses.

In some circumstances, multifacility location problems can be solved as a sequence of single location problems. That is, the optimal locations of the new facilities can be determined one at a time. However, when there is any interaction among the new facilities this approach will not work.

This section will show how linear programming can be used to solve the multiple-facility rectilinear distance location problem. Assume that existing facilities have locations at points $(a_1, b_1), \ldots, (a_n, b_n)$ as above. Suppose that m new facilities are to be located at $(x_1, y_1), \ldots, (x_m, y_m)$. Then the objective function to be minimized may be written in the form

$$\text{Minimize } f_1(\mathbf{x}) + f_2(\mathbf{y}),$$

where

$$f_1(\mathbf{x}) = \sum_{1 \le j < k \le m} v_{jk} |x_j - x_k| + \sum_{j=1}^{m} \sum_{i=1}^{n} w_{ij} |x_j - a_i|$$

and

$$f_2(\mathbf{y}) = \sum_{1 \le j < k \le m} v_{jk} |y_j - y_k| + \sum_{j=1}^{m} \sum_{i=1}^{n} w_{ij} |y_j - b_i|.$$

The v_{jk} measure the interaction of new facilities j and k. It is the presence of these terms that prevents us from solving the multifacility problem as a sequence of one-facility problems. However, as with the single-facility problem, the optimal x and y locations may be determined independently. We present the linear programming formulation for finding the optimal x coordinates. The optimal y coordinates may be found in the same way.

The trick in transforming the problem of finding $\mathbf{x}$ to minimize $f_1(\mathbf{x})$ is to eliminate the absolute value function from the objective, as this is not a strictly linear function. The means for doing so is a standard trick in linear programming. (A similar technique was applied in the linear programming formulation of the aggregate planning problem in Chapter 3, Section 5.)

For any constants a and b write $|a - b| = c + d$, but require that $cd = 0$ (either c or d or both must be zero). If $a > b$ then $|a - b| = c$ and if $a < b$ then $|a - b| = d$. Hence, we may think of c as the positive part of $a - b$ and d as the negative part of $a - b$. Substituting $|x_j - x_k| = c_{jk} + d_{jk}$ and $|x_j - a_i| = e_{ij} + f_{ij}$, we obtain the linear programming formulation of the problem of determining $\mathbf{x}$:

$$\text{Minimize} \quad \sum_{1 \leq j < k \leq n} v_{jk}(c_{jk} + d_{jk}) + \sum_{j=1}^{m} \sum_{i=1}^{n} w_{ij}(e_{ij} + f_{ij})$$

subject to

$$
\begin{aligned}
x_j - x_k - c_{jk} + d_{jk} &= 0, & 1 \leq j < k \leq n; \\
x_j - a_i - e_{ij} + f_{ij} &= 0, & 1 \leq i \leq n, \quad 1 \leq j \leq n; \\
c_{jk} \geq 0, \qquad d_{jk} &\geq 0, & 1 \leq j < k \leq n; \\
e_{ij} \geq 0, \qquad f_{ij} &\geq 0, & 1 \leq i \leq n, \quad 1 \leq j \leq n;
\end{aligned}
$$

x_j unrestricted in sign.

We do not need to explicitly include the constraints $c_{jk}d_{jk} = 0$ and $e_{ij}f_{ij} = 0$. One can show that at the minimum cost solution these relationships always hold (which is certainly fortunate as these are not linear relationships). One additional substitution is necessary prior to solving the problem. Since linear programming codes require that all variables be nonnegative, we must substitute $x_j = x_j^+ - x_j^-$, where $x_j^+ \geq 0$ and $x_j^- \geq 0$. Commercial linear programming codes are based on the simplex method, which is extremely efficient. One can solve realistically sized problems easily even on a personal computer.

Multifacility gravity problems require the solution of a system of linear equations, so that gravity problems involving large numbers of facilities are easily solved as well. Multifacility Euclidean problems are solved by utilizing a multidimensional version of the iterative solution method described in the previous section. We will not review these methods here. The interested reader should refer to Francis and White (1974).

Further Extensions

Facilities Having Positive Areas. All previous models assumed that facilities are approximated by points in the plane. When the area of the facilities is small compared to the area covered by the available locations, this assumption is reasonable. However, in certain applications the areas of the facilities cannot be ignored. For example, when finding locations for machines in a job shop, the machines must be far enough apart for them to be able to operate efficiently. Tompkins and White (1984) present an approach that requires the assumptions that the facilities are rectangular in shape and that the weights are uniformly distributed over the areas. The method is based on developing an analogy between the location problem and the problem of locating forces on a beam, and is similar to the procedure for constructing contour lines discussed in Appendix 10–B.

Location-Allocation Problems. Often the decision of where to locate new facilities must be accompanied by the decision of which of the existing locations will be served by each new facility. For example, a firm may be considering where to locate several regional warehouses. In addition to determining the location and the number of these new warehouses, they also must decide which of the retail outlets will be serviced by which warehouses.

Location-allocation problems are difficult to solve owing to the large number of decision variables. The mathematical programming formulation of the problem, assuming that a rectangular distance measure is used, is

$$\text{Minimize} \sum_{j=1}^{m} \sum_{i=1}^{n} z_{ij} w_{ij} [|x_j - a_i| + |y_j - b_i|] + g(m)$$

subject to

$$\sum_{j=1}^{m} z_{ij} = 1 \qquad \text{for } 1 \leq i \leq n,$$

where

w_{ij} = Cost per unit time per unit distance if the existing facility i is serviced by new facility j;

$z_{ij} = \begin{cases} 1 & \text{if existing facility } i \text{ is serviced by new facility } j, \\ 0 & \text{otherwise}; \end{cases}$

m = Total number of new facilities, $1 \leq m \leq n$;

(x_j, y_j) = Coordinates of new facility j, $1 \leq j \leq m$;

(a_i, b_i) = Coordinates of existing facility i, $1 \leq i \leq n$;

$g(m)$ = Cost per unit time of providing m new facilities.

This problem formulation has decision variables m, the number of new facilities; z_{ij}, the specification of which of the existing locations i will be serviced by facility j; and (x_j, y_j), the location of the new facilities. The optimization is difficult due to the presence of the zero-one variables z_{ij} and the inclusion of m as a decision variable. The problem is typically solved by considering successive values of $m = 1, 2, \ldots,$ and enumerating all combinations of z_{ij} for each value of m. Given a fixed m and set of z_{ij} values, the solution can be obtained using the methods for locating multiple facilities discussed above. However, the number of different z_{ij} values grows quickly as a function of m, so that only moderately sized problems can be solved in this fashion.

Discrete Location Problems. The models considered in this chapter for location of new facilities assumed that the new facilities could be located anywhere in the plane. This is not the case for most applications. Contour lines assist with evaluating alternative locations but cannot be constructed for problems in which one must locate multiple facilities. An alternative approach is to restrict a priori the possible locations to some discrete set of possibilities. When there is only a single facility and the number of possible locations is small, the easiest approach is to evaluate the cost of each location and pick the smallest.

When there are multiple facilities, the assignment model discussed in Section 4 can be used to determine the optimal locations of the new facilities. In certain types of warehouse-layout problems, new facilities can take up more than one potential site. For example, suppose that we must determine in which locations in a warehouse to store k items. Suppose that the appropriate storage area in the warehouse is composed of n grid squares and each item stored takes up more than a single square. Each square would be numbered and the storage location of an item specified by the numbers of the grid squares covered by the item. The resulting model, discussed in Francis and White (1974), is a generalization of the simple assignment model appearing in Section 4. We will not present the details of the model here.

Network Location Models.   Planar location models assume that the goal is to locate one or more new facilities in order to minimize some function of the distance separating the new and the existing facilities. A rectilinear, Euclidean, or other distance measure is generally assumed. In certain applications, the distances should be measured over an existing network and are not accurately approximated by standard measures. Overland transport must follow road networks, water transport must follow shipping lanes and sea routes, and air transport is confined to predetermined air corridors. In other applications the network may correspond to a network of power cables or telephone wires. In many of these applications the new facility or facilities must be placed on or very near to a location on the network, and distances can be measured only in terms of the network. Network location models are beyond the scope of our coverage. The interested reader should refer to the review articles of Francis, McGinnis, and White (1983) and Tansel, Francis, and Lowe (1983).

International Issues.   The problem of locating facilities is a part of the larger issue of global supply chain management. It is truer and truer that businesses are evolving into global corporations. This is no longer just the case for the industry giants. Globalization now plays a greater role than ever before, and its importance will continue to grow. Arntzen et al. (1995) discuss the supply chain configuration for Digital Equipment Corporation. In one case, they show how various parts of computers are shipped from the United States to Europe, from Europe to Brazil and Taiwan, and from both Taiwan and China to Europe and back to the United States. Fabrication may be done in two or three different countries, and distribution networks may be equally complex.

Lower wage rates were traditionally the primary reason for firms based in developed countries to locate plants in less developed countries. This is certainly true today. General Motors does much of its auto assembly in Mexico. Virtually all of the large semiconductor manufacturers have fabrication facilities overseas, typically in places like Malaysia and the Philippines.

Cohen and Lee (1989) provide a good overview of some of the issues that one must take into account when locating facilities in other countries. They include

1. Duties and tariffs are based on material flows. Their impact must be incorporated into international shipping schedules of materials, intermediate product, and finished product.
2. Currency exchange rates fluctuate unpredictably and affect profit levels in each country.
3. Corporate tax rates vary considerably from country to country.
4. Global sourcing must take into account lead times, costs, new technologies, and dependence on particular countries.

5. Local content rules and quotas constrain material flow between countries.
6. Product designs may vary by national market.
7. Transfer price mechanisms must be put in place to take the place of centralized control.
8. Differences in language, cultural norms, education, and skills must be incorporated into location decisions.

Only recently have we begun to understand the complexity of the problem of locating new facilities and their effect on global supply chain operations. Well-constructed and well-thought-out mathematical models will continue to assist us in managing these increasingly complex networks. However, many of the issues alluded to above are difficult to quantify, thus making good judgment crucial.

Problems for Section 11

42. For each of the location problems described below, discuss which of the seven assumptions listed in this section are likely to be violated:
    a. Locating three new machines in a machine shop.
    b. Locating an international network of telecommunications facilities.
    c. Locating a hospital in a sparsely populated area.
    d. Locating spare parts depots to support a field repair organization.

43. Consider Problem 28 of this chapter. Suppose that two new machines, A and B, are to be located in the shop. Machine A has $\frac{1}{8}, \frac{1}{8}, \frac{1}{4}, 1$, and $\frac{1}{6}$ as the expected numbers of loads transported to the existing five machines respectively, and machine B has $\frac{1}{4}, \frac{1}{6}, 3, \frac{1}{5}$, and $\frac{1}{2}$ as the expected numbers of loads transported. Furthermore, suppose that there are two loads per hour on average transported between the new machines. Assume a rectilinear distance measure.
    a. Formulate the problem of determining the optimal locations of the new machines as a linear program.
    b. If you have access to a computerized linear programming code, solve the problem formulated in part (a).

44. Consider the University of the Far West described in Example 10.5 of Section 9. Suppose that the university administration has decided that two audiovisual centers are needed. Each center would have different facilities. The anticipated numbers of faculty using each center are

School	Faculty Using Center A	Faculty Using Center B
Business	13	18
Education	40	23
Engineering	24	17
Humanities	20	23
Law	30	9
Science	16	21

Furthermore, there will be a total of 16 staff persons at centers A and B. They will need to interact frequently.

    a. Formulate the problem of determining the optimal locations of the two audiovisual centers as a linear program. Assume that rectilinear distances are used throughout.

    b. If you have access to a computerized linear programming code, solve the problem formulated in part (*a*).

45. Describe the following location problems and how they differ from those previously treated in this chapter:

    a. Location-allocation problems.

    b. Discrete location problems.

    c. Network location problems.

10.12 Historical Notes

The problems discussed in this chapter have a long history. Determining suitable layouts for production facilities is a problem that dates back to the start of the industrial revolution, although it appears that the development of analytical techniques for finding layouts is recent. Little seems to have been published concerning analytical layout methods prior to 1950. Apple (1977) lists a number of texts dealing with the plant layout problem published in the early 1950s. The computerized layout techniques discussed in this chapter were developed in the 1960s and 1970s. CRAFT, one of the first computerized methods and most popular even today, is from Buffa, Armour, and Vollmann (1964). ALDEP is from Seehof and Evans (1967), CORELAP from Lee and Moore (1967), COFAD from Tompkins and Reed (1976), and PLANET from Deisenroth and Apple (1972). Both ALDEP and CRAFT are available from the IBM Corporation as part of its SHARE library.

Some of the location problems discussed in this chapter go back hundreds of years. The problem of finding the location of a single new facility to minimize the sum of the Euclidean distances to the existing facilities has been referred to as the Steiner-Weber problem or the general Fermat problem. Francis and White (1974) state that the problem with exactly three facilities was posed by Fermat and solved by the mathematician Torricelli prior to 1640. The work on the rectilinear distance problem is relatively recent and was sparked by a paper by Hakimi (1964). Research continues today on discovering efficient solution techniques for locating multiple facilities using various distance measures.

10.13 Summary

This chapter dealt with two important logistics problems: the most efficient layout of facilities and the best location of new facilities relative to the existing ones. In a sense, the layout problem is a special type of location problem, because the goal is to find the best location of facilities within a specified boundary.

The analytical methods for layout discussed in this chapter assume that the objective is to minimize some function of the distance separating facilities. This viewpoint is probably most appropriate for plant layout problems and less so for other problems in which qualitative factors play a greater role. Two charts are important for layout analysis: the *activity relationship chart* (or rel chart for short) and the *from-to chart*. In order to construct a rel chart, each pair of facilities is given a letter code *A* (absolutely necessary), *E* (especially important), *I* (important), *O* (ordinary importance), *U* (unimportant), or *X* (undesirable), representing the desirability of locating facilities near each other. A from-to chart may specify the distance between pairs of facilities, numbers of materials handling trips per unit time between facilities, or the cost of materials handling trips between facilities. Both rel charts and from-to charts are useful for evaluating the quality of a layout.

Section 3 discussed types of layouts. The two most common types are *product layouts* and *process layouts*. The product layout is usually a fixed transfer line arranged in the sequence of the manufacturing steps required. A process layout groups machines with similar functions. Part routings vary from product to product. The product layout is appropriate for high-volume production of a small number of products. The process layout is appropriate for a low-volume job-shop environment. *Fixed position layouts* are used for products that are too large to move. Recently, there has been considerable interest in layouts based on *group technology*. Parts are grouped into families, and machine cells are developed consistent with this grouping. Group technology layouts are appropriate for automated factories.

The *assignment model* can be used for solving relatively simple layout and location problems. In order to use the assignment algorithm, we assume that for each placement of a machine (say) in a location, we can evaluate the cost of that assignment. This assumes that there is no interaction between the machines. When interaction does occur, a *quadratic assignment* formulation exists, but quadratic assignment models are far more difficult to solve than simple assignment models.

We discussed five *computerized layout* techniques: CRAFT, COFAD, ALDEP, PLANET, and CORELAP. Both CRAFT and COFAD are improvement routines. That means that both require that the user specify an initial layout. The program proceeds to consider interchanging adjacent pairs of facilities in order to achieve an improvement. On the other hand, ALDEP, PLANET, and CORELAP are construction routines, which build the layout from scratch. Because construction routines often result in departments with odd shapes, improvement routines are generally preferred. We also noted that there are a host of new software products now available for the personal computer and for UNIX workstations. Many of these products are based on drafting programs with large libraries of graphical icons. These products use methods conceptually similar to the earlier programs mentioned above, but are far more user friendly.

The chapter included a discussion of *flexible manufacturing systems* (FMS). An FMS is a collection of machines linked by an automated materials handling system, and is generally controlled by a central computer. They are used primarily in the metalworking industries, and are an appropriate choice when there is medium to large volume and a moderate variety of part types required. The downside to these systems is cost, which can run as high as $10 million or more. Flexible manufacturing cells are a scaled-down lower-cost alternative.

The second part of the chapter was concerned with methods for locating new facilities. *Location models* are appropriate when locating one or more new facilities within a specified area already containing a finite number of existing facilities. The objective is to locate new facilities to minimize some function of the distance separating new and existing facilities. Three distance measures were considered: rectilinear, Euclidean, and squared Euclidean. The first two are the most common for describing real problems and depend on whether movement occurs according to a crisscross street pattern (rectilinear) or is measured by straight-line distances (Euclidean).

The optimal solution to the weighted rectilinear distance problem is to locate the (x, y) coordinates at the *median* of the existing coordinates. When using a squared Euclidean distance measure, the optimal location of the new facility is at the center of gravity of the existing coordinates. No simple algebraic solution for the Euclidean distance problem is known, but iterative solution techniques exist. The chapter also included a brief discussion of several more complex location problems, including location of multiple facilities, location of facilities having nonzero areas, location-allocation problems, discrete location problems, and network location models. Finally, the chapter concluded with a discussion of issues of concern when locating new facilities in other countries. As globalization of manufacturing and supply chain networks increases, these issues will play an even greater role in location planning.

Additional Problems on Layout and Location

46. A real estate firm wishes to open four new offices in the Boston area. There are six potential sites available. Based on the number of employees in each office and the location of the properties that each employee will manage, the firm estimated the total travel time in hours per day for each office and each location. Find the optimal assignment of offices to sites to minimize employee travel time.

		Offices			
		A	B	C	D
	1	10	3	3	8
	2	13	5	2	6
Sites	3	12	9	9	4
	4	14	2	7	7
	5	17	7	4	3
	6	12	8	5	5

47. A large supermarket chain in the Southeast requires five additional warehouses in the Atlanta area. It has identified five sites for these warehouses. The annual transportation costs (in $000) for each warehouse at each site are given below. Find the assignment of warehouses to sites to minimize the total annual transportation costs.

		Warehouses				
		A	B	C	D	E
	1	41	47	38	46	50
	2	39	37	42	36	45
Sites	3	43	46	45	42	46
	4	51	54	47	58	56
	5	44	40	42	41	45

48. A machine shop located on the outskirts of Los Angeles accepts custom orders from a number of hi-tech firms in southern California. The machine shop consists of four departments: A (lathes), B (drills), C (grinders), and D (sanders). The from-to chart showing distances in feet between department centers is given below.

		TO			
		A	B	C	D
F	A		45	63	32
R	B	29		27	46
O	C	63	75		68
M	D	40	30	68	

The shop has accepted orders for production of four products: P1, P2, P3, and P4. The routing for production of these products and the weekly production rates are

Product	Routing	Weekly Production
P1	A–B–C–D	200
P2	A–C–D	600
P3	B–D	400
P4	B–C–D	500

Assume that products are produced in batches of size 25.

a. Convert this information into a from-to chart giving numbers of materials handling trips per week between departments.

b. If the cost to transport one batch one foot is estimated to be $1.50, convert the from-to chart you found in part (a) to one giving the materials handling cost per week between departments.

c. Develop an activity relationship chart for these four departments based on the results of part (b). Assume that A is assigned to the highest cost and O to the least with the rankings E and I assigned to the costs falling between the extremes.

d. Suppose that the machine shop is located in a building that is 60 feet by 80 feet. Furthermore, suppose that departments are rectangularly shaped with the following dimensions:

Department	Dimensions
A	20 × 30
B	40 × 20
C	45 × 55
D	37 × 25

Sketch a layout consistent with the rel chart you obtained in part (c).

FIGURE 10–32

Shape of facility (for Problem 49)

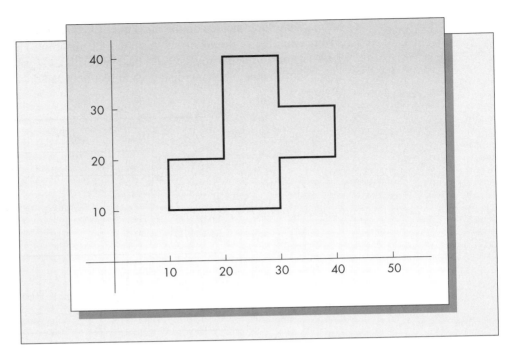

49. A facility has the shape shown in Figure 10–32 Using the methods described in Appendix 10–A, find the location of the centroid of this facility.

50. An initial layout for four departments and from-to charts giving distances separating departments and unit transportation costs appear in Figure 10–33. Using the CRAFT pairwise exchange technique, find the layout recommended by CRAFT to minimize total materials handling costs.

51. An initial layout for five departments and a from-to flow data chart are given in Figure 10–34. Assuming that departments A and D are in fixed locations and cannot be moved, find the layout recommended by CRAFT for departments B, C, and E. Assume that the objective is to minimize total distance traveled.

52. Consider the rel chart for the Meat Me fast-food restaurant, given in Figure 10–2. Assume that the areas required for each department are

Department	Area Required (square feet)
Cooking burgers	200
Cooking fries	200
Packing and storing	100
Drink dispensers	100
Counter servers	400
Drive-up server	100

Assume a sweep width of 2 squares and facility dimensions of 5 by 9 squares, where each square is 5 feet on a side. As a result, for example, the cooking-burgers department requires 8 squares. Use the ALDEP approach to

FIGURE 10–33

Layout and from-to charts (for Problem 50)

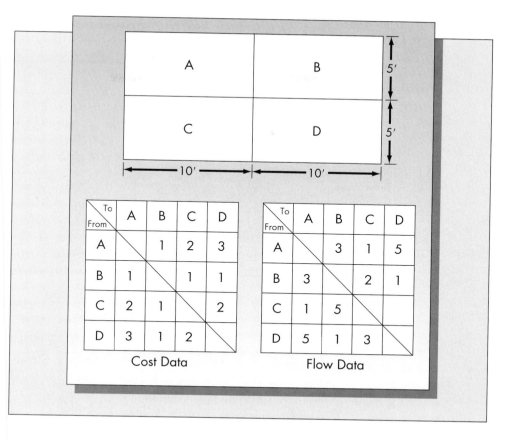

To From	A	B	C	D
A		1	2	3
B	1		1	1
C	2	1		2
D	3	1	2	

Cost Data

To From	A	B	C	D
A		3	1	5
B	3		2	1
C	1	5		
D	5	1	3	

Flow Data

FIGURE 10–34

Layout and from-to chart (for Problem 51)

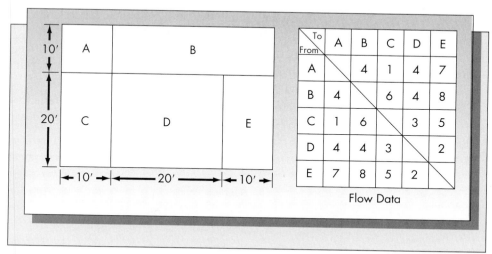

To From	A	B	C	D	E
A		4	1	4	7
B	4		6	4	8
C	1	6		3	5
D	4	4	3		2
E	7	8	5	2	

Flow Data

develop a layout for the restaurant. Comment on the practicality of your results.

53. Frank Green, an independent TV repairman, is considering purchasing a home in Ames, Iowa, that he will use as a base of operations for his repair business. Frank's primary sources of business are 10 industrial accounts located throughout the Ames area. He has overlaid a grid on a map of the city and determined the

following locations for these clients as well as the expected number of calls per month he receives:

Client	Grid Location	Expected Calls per Month
1	(5, 8)	2
2	(10, 3)	1
3	(14, 14)	1
4	(2, 2)	3
5	(1, 17)	1
6	(18, 25)	$\frac{1}{2}$
7	(14, 3)	$\frac{1}{4}$
8	(25, 4)	4
9	(35, 1)	3
10	(16, 21)	$\frac{1}{6}$

Find the optimal location of his house, assuming
  a. A weighted rectilinear distance measure.
  b. A squared Euclidean distance measure.
  c. The goal is to minimize the maximum rectilinear distance to any client.

54. An electronics firm located near Phoenix, Arizona, is considering where to locate a new phone switch that will link five buildings. The buildings are located at (0, 0), (2, 6), (10, 2), (3, 9), and (0, 4). The objective is to locate the switch to minimize the cabling required to those five buildings.
  a. Determine the gravity solution.
  b. Determine the optimal location assuming a straight-line distance measure. (If you are solving this problem by hand, iterate the appropriate equations at least five times and estimate the optimal solution.)

55. Consider three locations at (0, 0), (0, 6), and (3, 3) with equal weights. Using the methods described in Appendix 10–B, find contour lines for the rectilinear location problem.

56. A company is considering where to locate its cafeteria to service six buildings. The locations of the buildings and the fraction of the company's employees working at these locations are

Building	a_i	b_i	Fraction of Workforce
A	2	6	$\frac{1}{12}$
B	1	0	$\frac{1}{12}$
C	3	3	$\frac{1}{6}$
D	5	9	$\frac{1}{4}$
E	4	2	$\frac{1}{4}$
F	10	7	$\frac{1}{6}$

  a. Find the optimal location of the cafeteria to minimize the weighted rectilinear distance to all the buildings.
  b. Find the optimal location of the cafeteria to minimize the maximum rectilinear distance to all the buildings.

c. Find the gravity solution.

d. Suppose that the cafeteria must be located in one of the buildings. In which building should it be located if the goal is to minimize weighted rectilinear distance?

e. Solve part (d) assuming weighted Euclidean distance.

Spreadsheet Problems for Chapter 10

57. Design a spreadsheet to compute the total rectilinear distance from a set of up to 10 existing locations to any other location. Assume that existing locations are placed in Columns A and B and the new location in cells D1 and E1. Initialize column A with the value of cell D1 and column B with the value of cell E1, so that the total distance will be computed correctly when there are fewer than 10 locations.

a. Suppose that existing facilities are located at $(0, 0)$, $(5, 15)$, $(110, 120)$, $(35, 25)$, $(80, 10)$, $(75, 20)$, $(8, 38)$, $(50, 65)$, $(22, 95)$, and $(44, 70)$, and the new facility is to be located at $(50, 50)$. Determine the total rectilinear distance of the new facility to the existing facilities.

b. Suppose the new facility in part (a) can be located only at $x = 0, 5, 10, \ldots,$ 100 and $y = 0, 10, 20, \ldots, 100$. By systematically varying the x and y coordinates, find the optimal location of the new facility.

58. Solve Problem 57, assuming a Euclidean distance measure.

59. Solve Problem 32 using an electronic spreadsheet. To do so, enter the building numbers in column A, the x coordinates in column B, and the associated weights in column C. Sort columns A, B, and C in ascending order by using column B as the primary sort key. Now accumulate the weights (column C) in column D using the sum function. Divide the total accumulated weight by 2 and visually identify the optimal x coordinate value. It will be where the cumulative weight first exceeds half the total cumulative weight. Repeat the process for the y coordinates.

60. Design an electronic spreadsheet to compute the optimal solution to the gravity problem. Allow for up to 20 locations. Let column A be the location number, column B the x coordinates $(a_1, \ldots, a_n)$ of existing locations, and column C the y coordinates $(b_1, \ldots, b_n)$ of existing locations. Store the optimal solution in cell D1. Find the gravity solution to Problem 32 using your spreadsheet.

61. Extend the results of Problem 60 to find the optimal location of a new facility among a set of existing facilities assuming a straight-line Euclidean distance measure. Let column E correspond to $g_i(x, y)$, where the initial (x, y) values appear in F1 and F2. In locations G1 and G2, store

$$x = \frac{\displaystyle\sum_{i=1}^{n} a_i g_i(x, y)}{\displaystyle\sum_{i=1}^{n} g_i(x, y)},$$

$$y = \frac{\displaystyle\sum_{i=1}^{n} b_i g_i(x, y)}{\displaystyle\sum_{i=1}^{n} g_i(x, y)}.$$

Start with the gravity solution in cells F1 and F2. After calculation, replace the values in cells F1 and F2 with the values in cells G1 and G2. Continue in this manner until the solution converges. Using your spreadsheet,

a. Solve Problem 32.
b. Solve Problem 41.
c. Solve Problem 53 assuming a Euclidean distance measure.
d. Solve Problem 56 assuming a Euclidean distance measure.

APPENDIX 10–A   FINDING CENTROIDS

The centroid of any object is another term for the physical coordinates of the center of gravity. For a plate of uniform density, it would be the point at which the plate would balance exactly. Let R be any region in the plane. The centroid for R is defined by two points $\bar{x}$, $\bar{y}$. In order to find these two points, we first must obtain the moments of R, M_x, and M_y, which are given by the formulas

$$M_x = \int_R \int x \, dx \, dy,$$

$$M_y = \int_R \int y \, dx \, dy.$$

Let $A(R)$ be the area of R. Then the centroid of R is given by

$$\bar{x} = M_x/A(R),$$

$$\bar{y} = M_y/A(R).$$

We now obtain explicit expressions for the moments when R is a finite sum of rectangles. Suppose that R is a simple rectangle as pictured in Figure 10–35. Then

$$M_x = \int_{y_1}^{y_2} dy \int_{x_1}^{x_2} x \, dx = \int_{y_1}^{y_2} dy \frac{x^2}{2} \Big|_{x_1}^{x_2}$$

$$= \int_{y_1}^{y_2} dy \left(x_2^2 - x_1^2 \right)/2 = \frac{x_2^2 - x_1^2}{2}(y_2 - y_1),$$

$$M_y = \int_{x_1}^{x_2} dx \int_{y_1}^{y_2} y \, dy = \int_{x_1}^{x_2} dx \frac{y^2}{2} \Big|_{y_1}^{y_2}$$

$$= \int_{x_1}^{x_2} dx \left(y_2^2 - y_1^2 \right)/2 = \frac{y_2^2 - y_1^2}{2}(x_2 - x_1).$$

Note that because the area of the rectangle is $(x_2 - x_1)(y_2 - y_1)$, and $x_2^2 - x_1^2 = (x_2 - x_1)(x_1 + x_2)$ (and similarly for $y_1^2 - y_2^2$), we obtain

$$\bar{x} = (x_1 + x_2)/2,$$

$$\bar{y} = (y_1 + y_2)/2.$$

The formulas for the moments of a rectangle may be used to find the centroid when R consists of a collection of rectangles as well. Suppose that R can be subdivided into

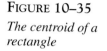

FIGURE 10–35

*The centroid of a
rectangle*

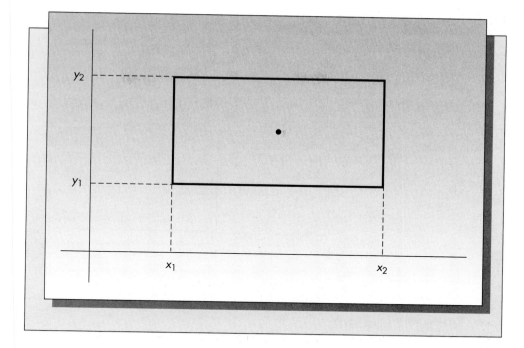

k rectangles labeled $R_1, R_2, \ldots, R_k$ with respective boundaries defined by $[(x_{1i}, x_{2i}),$ $(y_{1i}, y_{2i})]$ for $1 \le i \le k$. Since

$$M_x = \int_R \int x \, dx \, dy = \sum_{i=1}^{k} \int_{R_i} \int x \, dx \, dy,$$

it follows that

$$M_x = \sum_{i=1}^{k} \frac{x_{2i}^2 - x_{1i}^2}{2}(y_{2i} - y_{1i}).$$

Similarly,

$$M_y = \sum_{i=1}^{k} \frac{y_{2i}^2 - y_{1i}^2}{2}(x_{2i} - x_{1i}).$$

Example 10A.1

Consider the region R pictured in Figure 10–36. We will determine the centroid of the region using the formulas above. The region can be broken down into three rectangles in a number of different ways. For the one pictured in Figure 10–36, we have

$$
\begin{array}{ll}
x_{11} = 10, & x_{21} = 30, \\
y_{11} = 5, & y_{21} = 35, \\
x_{12} = 30, & x_{22} = 40, \\
y_{12} = 5, & y_{22} = 20, \\
x_{13} = 30, & x_{23} = 50, \\
y_{13} = 25, & y_{23} = 35.
\end{array}
$$

FIGURE 10–36

*Centroid of a figure
composed of
rectangles*

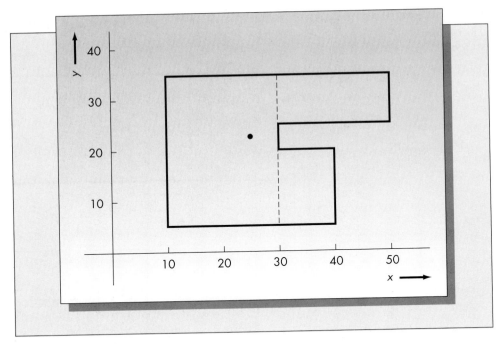

Substituting into the formulas above, we obtain

$$M_x = \frac{30^2 - 10^2}{2}(35 - 5) + \frac{40^2 - 30^2}{2}(20 - 5) + \frac{50^2 - 30^2}{2}(35 - 25)$$

$$= (400)(30) + (350)(15) + (800)(10) = 25{,}250,$$

$$M_y = \frac{35^2 - 5^2}{2}(30 - 10) + \frac{20^2 - 5^2}{2}(40 - 30) + \frac{35^2 - 25^2}{2}(50 - 30)$$

$$= (600)(20) + (187.5)(10) + (300)(20) = 19{,}875.$$

The total area of R is

$$A(R) = (20)(30) + (10)(15) + (20)(10) = 950.$$

It follows that the centroid is

$$\bar{x} = \frac{25{,}250}{950} = 26.579,$$

$$\bar{y} = \frac{19{,}875}{950} = 20.921.$$

The centroid is marked with a dot on Figure 10–36. ■

APPENDIX 10–B   COMPUTING CONTOUR LINES

This appendix outlines the procedure for computing contour lines, or isocost lines, such those pictured in Figures 10–28 and 10–29. The theoretical justification for this procedure appears in Francis and White (1974).

1. Plot the points (a_1, b_1), (a_2, b_2), . . . , (a_n, b_n) on graph paper. Draw a horizontal line (parallel to the x axis) and a vertical line (parallel to the y axis) through each point.

2. Number the horizontal and vertical lines in sequence from left to right and from top to bottom. (If none of the original points is collinear, there will be exactly n horizontal and n vertical lines.)

3. Let C_j be the sum of the weights associated with the points along vertical line j, and D_i the sum of the weights associated with the points along horizontal line i.

4. Compute the following numbers:

$$M_0 = -\sum_{i=1}^{n} w_i, \qquad N_0 = M_0 = -\sum_{i=1}^{n} w_i,$$
$$M_1 = M_0 + 2C_1, \qquad N_1 = N_0 + 2D_1,$$
$$M_2 = M_1 + 2C_2, \qquad N_2 = N_1 + 2D_2,$$

and so on.

(The final values of M_i and N_j will both be $+\sum_{i=1}^{n} w_i$.)

5. Define the region (i, j) as the region bounded by the ith and $(i + 1)$th vertical lines and the jth and $(j + 1)$th horizontal lines. The regions to the left of the first vertical line are labeled $(0, j)$, and those below the first horizontal line are labeled $(i, 0)$. The slope of any contour line passing through region (i, j) is given by

$$S_{i,j} = -M_i/N_j.$$

Once the slopes are determined, a contour line is constructed by starting at any point and moving through each region at the angle determined by the slope computed in step 5. We will present a simple example to illustrate the method.

Example 10B.1

Assume $(a_1, b_1) = (1, 1)$, $(a_2, b_2) = (5, 3)$, and $(a_3, b_3) = (3, 8)$, and $w_1 = 2$, $w_2 = 3$, and $w_3 = 6$. The first step is to plot these points on a grid as we have done in Figure 10–37. Drawing vertical and horizontal lines through each of the three points yields three vertical lines labeled 1, 2, and 3 and three horizontal lines labeled 1, 2, 3 as well. Regions are labeled from $(0, 0)$ to $(3, 3)$ as in the figure.

Next we compute $M_0, \ldots, M_3$ and $N_0, \ldots, N_3$.

$$
\begin{aligned}
M_0 &= -(2 + 3 + 6) = -11 = N_0,\\
M_1 &= -11 + (2)(2) = -7, & N_1 &= -11 + 2(2) = -7,\\
M_2 &= -7 + (2)(6) = +5, & N_2 &= -7 + (2)(3) = -1,\\
M_3 &= +5 + (2)(3) = +11, & N_3 &= -1 + (2)(6) = +11.
\end{aligned}
$$

Next the ratios are computed to find the slope for each region.

$$
\begin{aligned}
S_{0,0} &= -(-11)/(-11) = -1, & S_{0,2} &= -(-11)/(-1) = -11,\\
S_{1,0} &= -(-7)/(-11) = -0.64, & S_{1,2} &= -(-7)/(-1) = -7,\\
S_{2,0} &= -(5)/(-11) = +0.45, & S_{2,2} &= -(5)/(-1) = 5,\\
S_{3,0} &= -(11)/(-11) = +1, & S_{3,2} &= -(11)/(-1) = 11,\\
S_{0,1} &= -(-11)/(-7) = -1.57, & S_{0,3} &= -(-11)/(11) = 1,\\
S_{1,1} &= -(-7)/(-7) = -1, & S_{1,3} &= -(-7)/(11) = 0.64,\\
S_{2,1} &= -(5)/(-7) = +0.71, & S_{2,3} &= -(5)/(11) = -0.45,\\
S_{3,1} &= -(11)/(-7) = +1.57, & S_{3,3} &= -(11)/(11) = -1.
\end{aligned}
$$

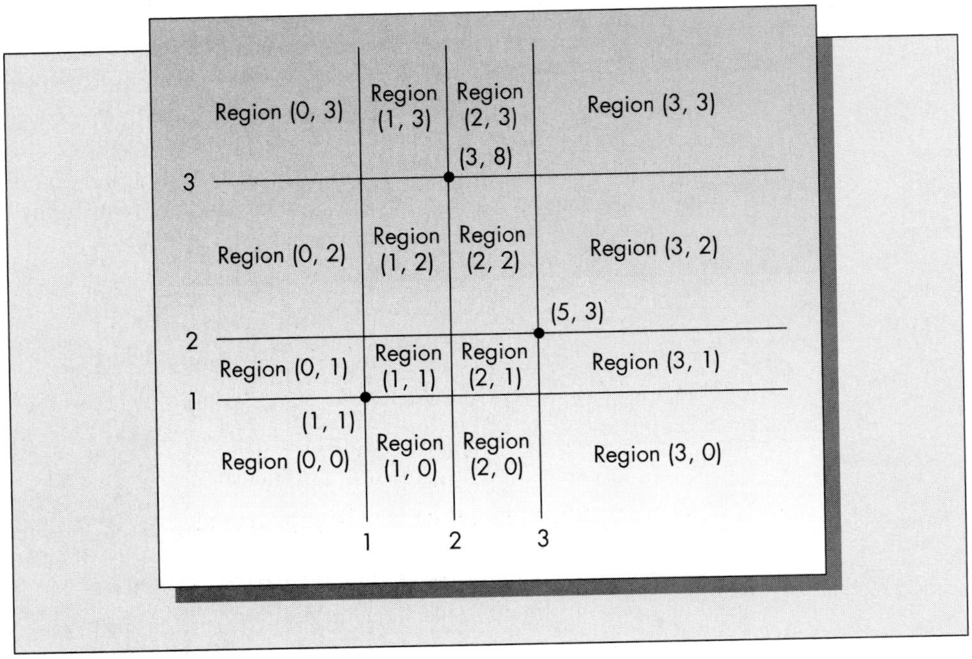

FIGURE 10–37

*Regions for
Example 10B.1*

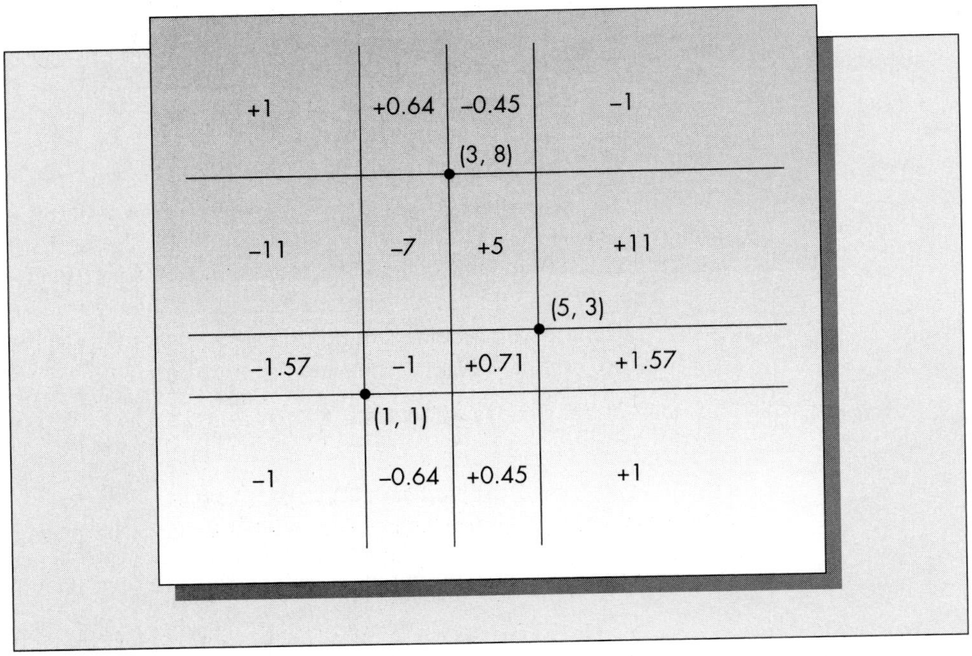

FIGURE 10–38

*Slopes for
Example 10B.1*

Before constructing the contour lines, it is convenient to place the slopes in the appropriate regions, as shown in Figure 10–38. A contour line may be started at any point on a region boundary. From the initial point one draws a line with the appropriate slope for that region to the boundary of the next region. At that point the slope changes to the value associated with the next region. One continues until the line segments return to the originating point. (If the slopes are correct and the

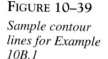

FIGURE 10–39

Sample contour lines for Example 10B.1

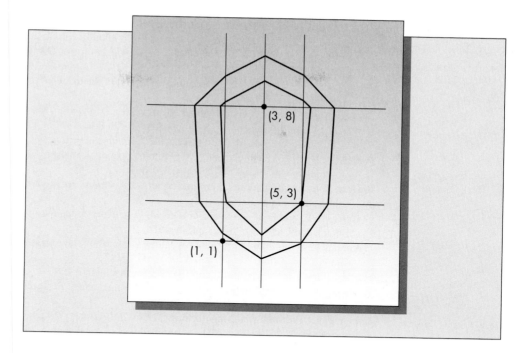

drawing accurate, one will always return to the point of origination.) Two typical contour lines for the example problem are shown in Figure 10–39. ∎

Bibliography

Apple, J. M. *Plant Layout and Material Handling.* 3rd ed. New York: John Wiley & Sons, 1977.

Armour, G. C., and E. S. Buffa. "A Heuristic Algorithm and Simulation Approach to Relative Allocation of Facilities." *Management Science* 9 (1963), pp. 294–309.

Arntzen, B. C.; G. G. Brown; T. P. Harrison; and L. L. Trafton. "Global Supply Chain Management at Digital Equipment Corporation." *Interfaces* 25, no.1 (1995), pp. 69–93.

Block, T. E. "A Note on 'Comparison of Computer Algorithms and Visual Based Methods for Plant Layout' by M. Scriabin and R. C. Vergin." *Management Science* 24 (1977), pp. 235–37.

Bruno, G., and P. Biglia. "Performance Evaluation and Validation of Tool Handling in Flexible Manufacturing Systems Using Petri Nets." In *Proceedings of the International Workshop on Timed Petri Nets,* pp. 64–71. Torino, Italy, 1985.

Buffa, E. S. "On a Paper by Scriabin and Vergin." *Management Science* 23 (1976), p. 104.

Buffa, E. S.; G. C. Armour; and T. E. Vollmann. "Allocating Facilities with CRAFT." *Harvard Business Review* 42 (1964), pp. 136–58.

Burbidge, J. L. *The Introduction of Group Technology.* London: Heinemann, 1975.

Buzacott, J. A., and D. D. Yao. "On Queuing Network Models of Flexible Manufacturing Systems." *Queuing Systems* 1 (1986), pp. 5–27.

Cohen, M., and H. L. Lee. "Resource Deployment Analysis of Global Manufacturing and Distribution Networks." *Journal of Manufacturing and Operations Management* 2 (1989), pp. 81–104.

Coleman, D. R. "Plant Layout: Computers versus Humans." *Management Science* 24 (1977), pp. 107–12.

Conway, R. W.; W. L. Maxwell; J. O. McClain; and S. L. Worona. *Users Guide to XCELL+Factory Modeling System.* Redwood City, CA: Scientific Press, 1987.

Cook, N. H. "Computer Managed Parts Manufacture." *Scientific American* 232 (1975), pp. 23–29.

Deisenroth, M. P., and J. M. Apple. "A Computerized Plant Layout Analysis and Evaluation Technique (PLANET)." In *Technical Papers 1962.* Norcross, GA: American Institute of Industrial Engineers, 1972.

Drezner, Z. "DISCON: A New Method for the Layout Problem." *Operations Research* 28 (1980), pp. 1375–84.

Eilon, S.; C. D. T. Watson-Gandy; and N. Christofides. *Distribution Management: Mathematical Modelling and Practical Analysis.* London: Griffin Press, 1971.

Foulds, L. R. "Techniques for Facilities Layout." *Management Science* 29 (1983), pp. 1414–26.

Francis, R. L.; L. F. McGinnis; and J. A. White. "Locational Analysis." *European Journal of Operational Research* 12 (1983), pp. 220–52.

Francis, R. L., and J. A. White. *Facility Layout and Location: An Analytical Approach.* Englewood Cliffs, NJ: Prentice Hall, 1974.

Groover, M. P., and E. W. Zimmers. *CAD/CAM: Computer Aided Design and Manufacturing.* Englewood Cliffs, NJ: Prentice Hall, 1984.

Hakimi, S. L. "Optimum Location of Switching Centers and the Absolute Centers and Medians of a Graph." *Operations Research* 12 (1964), pp. 450–59.

Hartley, J. *FMS at Work.* Amsterdam: IFS and North Holland, 1984.

Jaikumar, R. "Flexible Manufacturing Systems: A Management Perspective." Unpublished manuscript, Harvard School of Business, 1984.

Jaikumar, R. "Postindustrial Manufacturing." *Harvard Business Review* 64 (1986), pp. 69–76.

Johnson, Roger V. "SPACECRAFT for Multi-Floor Layout Planning." *Management Science* 28 (1982), pp. 407–17.

Kaplan, R. S. "Must CIM Be Justified by Faith Alone?" *Harvard Business Review* 64 (1986), pp. 69–76.

Keefer, K. B. "Easy Way to Determine the Center of Distribution." *Food Industries* 6 (1934), pp. 450–51.

Krouse, J. "Flexible Manufacturing Systems Begin to Take Hold." *High Technology* 6 (1986), p. 26.

Lee, R. C., and J. M. Moore. "CORELAP—Computerized Relationship Layout Planning." *Journal of Industrial Engineering* 18 (1967), pp. 194–200.

Muther, R. *Practical Plant Layout.* New York: McGraw-Hill, 1955.

Nicol, L. M., and R. H. Hollier. "Plant Layout in Practice." *Material Flow* 1, no. 3 (1983), pp. 177–88.

Nugent, C. E.; T. E. Vollmann; and J. Ruml. "An Experimental Comparison of Techniques for the Assignment of Facilities to Locations." *Operations Research* 16 (1968), pp. 150–73.

Pritsker, A. A. B. *Introduction to Simulation and SLAM II.* 3rd ed. New York: John Wiley Sons, 1986.

Rosenblatt, M. J. "The Dynamics of Plant Layout." *Management Science* 32, no. 1 (1986), pp. 76–86.

Sassani, F. "A Simulation Study on Performance Improvement of Group Technology Cells." *International Journal of Production Research* 28 (1990), pp. 293–300.

Schweitzer, P. J. "Maximum Throughput in Finite Capacity Open Queuing Networks with Product-Form Solutions." *Management Science* 24 (1977), pp. 217–23.

Scriabin, M., and R. C. Vergin. "Comparison of Computer Algorithms and Visual Based Methods for Plant Layout." *Management Science* 22 (1975), pp. 172–81.

Scriabin, M., and R. C. Vergin. "A Cluster Analytic Approach to Facility Layout." *Management Science* 31 (1985), pp. 33–49.

Seehof, J. M., and W. O. Evans. "Automated Layout Design Programs." *Journal of Industrial Engineering* 18 (1967), pp. 690–95.

Sengupta, S., and R. Combes. "Optimizing and General Food's Environment." *IIE Solutions,* August 1995, pp. 30–35.

Sly, D. "Computerized Facilities Design and Management: An Overview." *IIE Solutions,* August 1995, pp. 43–51.

Stecke, K. E. "Formulation and Solution of Nonlinear Integer Production Planning Problems for Flexible Manufacturing Systems." *Management Science* 29 (1983), pp. 273–88.

Suri, R., and R. R. Hildebrant. "Modeling Flexible Manufacturing Systems Using Mean Value Analysis." *SME Journal of Manufacturing Systems* 3 (1984), pp. 27–38.

Tansel, B. C.; R. L. Francis; and T. J. Lowe. "Location on Networks: A Survey (Parts 1 and 2)." *Management Science* 29 (1983), pp. 482–511.

Tompkins, J. A., and R. Reed Jr. "An Applied Model for the Facilities Design Problem." *International Journal of Production Research* 14 (1976), pp. 583–95.

Tompkins, J. A., and J. A. White. *Facilities Planning.* New York: John Wiley & Sons, 1984.

Vollman, T. E., and E. S. Buffa. "The Facilities Layout Problem in Perspective." *Management Science* 12 (1966), pp. 450–68.

Zygmont, J. "Flexible Manufacturing Systems: Curing the Cure-all." *High Technology* 6 (1986), pp. 22–27.

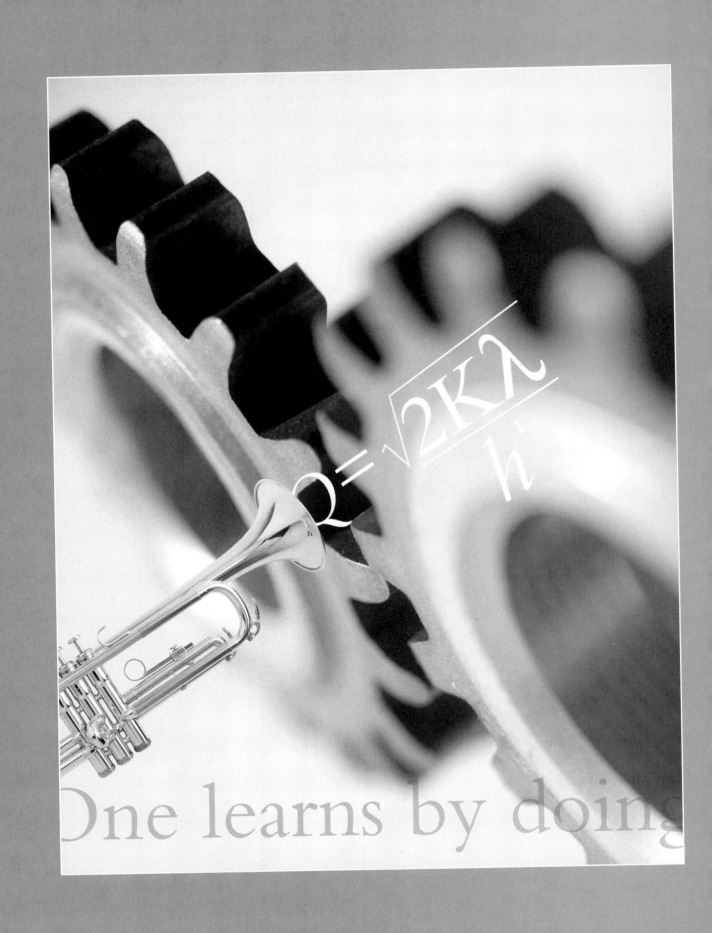

CHAPTER 11

QUALITY AND ASSURANCE

While the American economy was strong during the latter part of the 1990s, there are some disturbing trends. Our balance of trade in manufactured goods continues to be negative year after year. In particular, we have maintained a negative balance of trade with Japan for several decades, even during the recession that plagued that country in the 1990s. A negative balance of trade means that we import more than we export. In the case of Japan, this is due in large part to the fact that Americans are consuming Japanese-made products at an increasing rate. The success of Japan in consumer electronics and automobiles here are just two examples. But it is not only Japanese-made products whose consumption is increasing. Other examples are German- and Swedish-made automobiles, German-made kitchen appliances, bicycles made in Taiwan, and high-end watches made in Switzerland. In many cases, the imported products are considerably more expensive than their American counterparts, yet are still preferred. Why? The simple answer is that they are perceived to be of higher quality.

What is quality? Traditional thinking would say that quality is conformance to specifications; that is, does the product do what it was designed to do? Some feel that this definition is the only meaningful definition of quality, because conformance is something that can be measured. According to Philip Crosby (1979).

> That is precisely the reason we must define quality as "conformance to requirements" if we are to manage it. Thus, those who want to talk about quality of life must talk about that life in specific terms, such as desirable income, health, pollution control, political programs, and other items that can each be measured.

Crosby makes a good point. By defining quality in terms of conformance, we avoid making unreasonable comparisons. Is a Rolls-Royce a better-quality product than a Toyota Corolla? Not necessarily. The Toyota may be a higher-quality product relative to *what it was designed to do.*

This does not tell the entire story, however. Just as beauty is in the eye of the beholder, so is quality in the mind of the customer. If the customer is not happy with the product, it is not high quality. Viewed in this way, quality is a measure of the conformance of the product to the *customer's needs.*

Why is this different? Conformance to specifications assumes a given design and the specifications resulting from that design. Conformance to customer needs means that the design of the product is part of the evaluation. Given two washing machines with comparable repair records, what determines which one the consumer will buy? The answer is a combination of aesthetics, features, and design. Viewing quality in this broader way is both good and bad. It is good in that it gets at the heart of the issue:

Quality is what the customer thinks it is. It is bad in that it makes it difficult to measure quality and thus difficult to improve it.

There is little doubt that product reliability is an important part of the spectacular success of Japanese automobile manufacturers. Although several American automakers boast initial defect rates in their cars comparable to Japanese autos (as measured by the J. D. Power new car buyer survey), it is not the number of defects on new cars that is really important to most consumers. It is the car's reliability over its life. In this respect, few automakers can come close to the records posted by Honda and Toyota and other Japanese automakers. Each year the Consumer's Union conducts a survey of its readership to determine the readers' experiences that year with the products they own. Automobile reliability is rated on a five-point scale: much worse than average, worse than average, average, better than average, and much better than average. The ratings correspond to several factors, including mechanical failures, electrical failures, and body integrity. Japanese cars have consistently scored better than most of their American and European competitors.

Since the mid-1970s, Japan has experienced a steady growth in the American market share. Table 11–1 gives U.S. sales of automobiles by country of manufacture over the period 1974–1983.

The erosion of market share for U.S. automakers has continued to the present day. In 1991 the Japanese had increased their share to 34 percent of the total U.S. market and the U.S. automakers hit an all-time low of 60.1 percent (*Business Week,* 1991). General Motors, Chrysler, and Ford also posted their greatest losses, totaling about $7 billion.

If this trend continues, the big three American automakers will eventually go out of business. What are they doing to prevent this from happening? Serious quality control programs are under way in most automobile plants in the United States. Although new car defect rates do not tell the whole story, the fact that some American companies are achieving rates comparable to (and sometimes even better than) the Japanese is encouraging. GM is experimenting with Japanese management techniques in the Saturn facility in Tennessee. Findings from the Consumer's Union suggest that Saturn's reliability is close to that of Toyota and Honda, the most reliable Japanese cars.

Japanese automakers are not standing still, and neither are we. The success of Acura, Lexus, and Infiniti models attests to the Japanese ability to compete successfully

TABLE 11–1   **U.S. Passenger Car Sales by Country of Manufacture**

		Percentage of Sales by Country of Manufacture			
Year	*Total Sales (000)*	*U.S./Canada*	*W. Germany*	*Japan*	*Other*
1974	8,857	84.1	6.7	6.7	2.4
1975	8,633	81.7	5.7	9.5	3.1
1976	10,106	85.2	3.5	9.3	2.0
1977	11,179	81.4	4.1	12.4	2.0
1978	11,310	82.2	3.9	12.0	1.8
1979	10,660	78.1	3.3	16.6	2.0
1980	8,979	73.3	3.4	21.2	2.1
1981	8,533	72.7	2.9	21.8	2.6
1982	7,980	72.1	3.1	22.6	1.1
1983	9,181	74.0	3.0	20.9	2.1

Source: From Groocock, 1986.

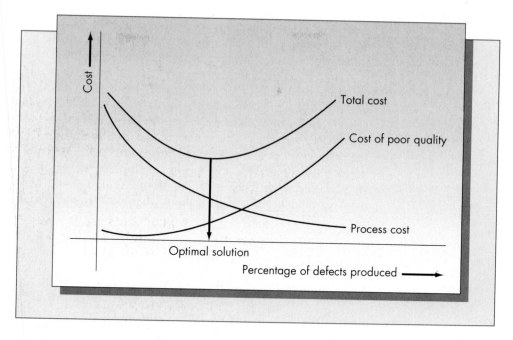

in the high end of the market. American automakers are trying to recapture some of this market. For example, the Cadillac Division of GM, a Malcolm Baldrige Award winner, employed the latest methods of design for manufacturability and concurrent engineering in developing the Seville STS.

The enormous strides made by the Japanese in consumer electronics and automobiles leaves the impression that American products and American industry are inferior. This simply is not true. American-based companies are leaders in many major industries. Airplane construction, mainframe computers, biotechnology, financial services, large appliances, chemicals, and telecommunications are a small sample of industries dominated by American firms. By transferring lessons learned in these industries, we will begin to regain our competitive edge in manufacturing in general.

Management must grapple with the difficult problems of knowing how much to invest in quality and determining the best way to go about making that investment. There is an optimal trade-off between the cost of poor quality and the investment required in the process to improve the quality, as represented in Figure 11–1. Although such curves can be drawn in principle, evaluating the costs of poor quality is difficult. Direct costs, such as those resulting from scrap, rework, and inspection, are relatively easy to determine. But how does one factor in the costs of lost consumer loyalty? No one would deny that marketing is essential, but has the emphasis on marketing in the United States been at the expense of manufacturing? We must acknowledge that the modern consumer is more educated and more discriminating than ever before. Clever advertisements will not sell second-rate products. It is time to put our investment where it belongs: in the design and manufacture of quality products.

Overview of This Chapter

Statistical quality control dates back to the 1930s. Its roots lie in the work of Walter Shewhart, a scientist employed by Bell Telephone Laboratories. W. Edwards Deming, the man credited with bringing the quality control message to the Japanese, was a student of Shewhart. Deming has stressed in his teachings that understanding the concept of statistical variation of processes is a key step in designing an effective quality control program. One needs to understand process variation in order to know how to produce products that conform to specifications. Deming has become a demigod in Japan, where his teachings ignited the Japanese quality revolution.

This chapter is aimed at providing the student with an understanding of the essentials underlying statistical quality control. We discuss two basic areas: *control charts* and *acceptance sampling*. Briefly, a control chart is a graphical means for determining if the underlying distribution of some measurable variable seems to have undergone a shift. Acceptance sampling is the set of procedures for inferring characteristics of a lot from the characteristics of a sample of items from that lot. Whereas "zero defects" thinking might suggest that these approaches are obsolete, we disagree. Process monitoring and statistical sampling continue to be used. It is important to understand the underlying theory behind these methods to know when and how they should be applied.

11.1 Statistical Basis of Control Charts

Control charts provide a simple graphical means of monitoring a process in real time. Although easy to construct and easy to use, control charts are based on rigorous statistical principles. They have gained wide acceptance in industry, and are preferred to more conventional statistical methods.

A control chart maps the output of a production process over time and signals when a change in the probability distribution generating observations seems to have occurred. To construct a control chart one uses information about the probability distribution of process variation and fundamental results from probability theory. A result that forms the basis for a class of control charts is known as the *central limit theorem*. Roughly, the central limit theorem says that the distribution of sums of independent and identically distributed random variables approaches the normal distribution as the number of terms in the sum increases.[1] Generally, the distribution of the sum converges very quickly to a normal distribution. In order to illustrate the central limit theorem, suppose that X is a random variable with the uniform distribution on the interval $(0, 1)$. The probability density function of X is pictured in Figure 11–2.

The density of X bears little resemblance to a normal density. Now let us assume that the three random variables X_1, X_2, and X_3 are independent random variables, each of which has the uniform distribution on the interval $(0, 1)$. Consider the random variable $W = X_1 + X_2 + X_3$. One can derive the distribution of W by convoluting the distributions of $X_1, X_2,$ and X_3. We will not present the details here. (The interested reader should refer to a graduate-level text in probability such as DeGroot, 1986.) The density function of W appears in Figure 11–3. The resemblance to a normal density is now quite striking. In Figure 11–4 we have graphed the probability density function of W and the associated normal approximation. Notice how closely the two curves agree. Were we to continue to add independent uniform random variables, the agreement would be even closer.

In quality control, the central limit theorem justifies the assumption that the distribution of $\overline{X}$, the sample mean, is approximately normally distributed. Recall the definition of the sample mean: If $(X_1, X_2, \ldots, X_n)$ is a random sample, then the sample mean $\overline{X}$ is defined as

$$\overline{X} = \frac{1}{n} \sum_{i=1}^{n} X_i.$$

---

[1]The central limit theorem was also used in Chapter 9 to justify the use of the normal distribution to describe project completion time in PERT networks.

FIGURE 11–2

Probability density of a uniform variate on (0, 1)

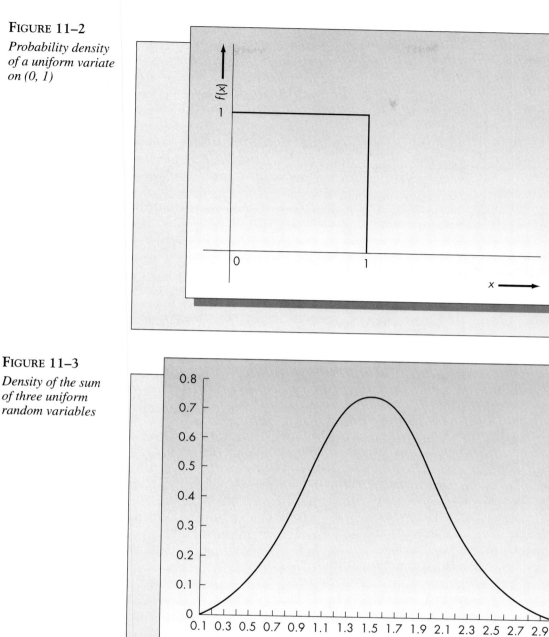

FIGURE 11–3

Density of the sum of three uniform random variables

Suppose that a variable Z has the standard normal distribution. Then, from Table A–1 at the back of this book,

$$P\{-3 \leq Z \leq 3\} = .9974.$$

In words, this means that the likelihood of obtaining a value of Z either larger than 3 or less than -3 is .0026, or roughly 3 chances in 1,000. This is the basis of the so-called three-sigma limits that have become the de facto standard in quality control. Now consider the sample mean $\overline{X}$, which the central limit theorem tells us is approximately normally distributed. Suppose that the mean of each sample value is μ and the standard

FIGURE 11–4

*Density of the sum
of three uniform
random variables
and the normal
approximation*

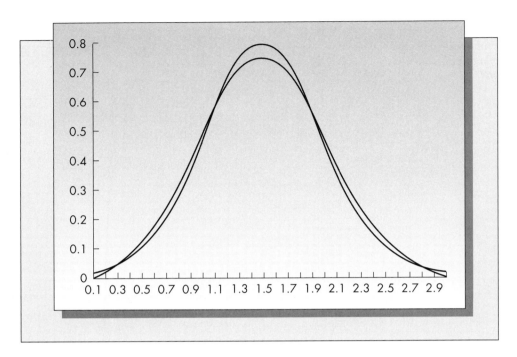

deviation of each sample value is σ. Then it is well known that the mean of $\overline{X}$ is also μ and the standard deviation of $\overline{X}$ is $\sigma/\sqrt{n}$ Therefore, the standardized variate

$$Z = \frac{\overline{X} - \mu}{\sigma/\sqrt{n}}$$

has (approximately) the normal distribution with zero mean and unit variance. It follows that

$$P\left\{-3 \le \frac{\overline{X} - \mu}{\sigma/\sqrt{n}} \le 3\right\} = .9974,$$

which is equivalent to

$$P\left\{\mu - \frac{3\sigma}{\sqrt{n}} \le \overline{X} \le \mu + \frac{3\sigma}{\sqrt{n}}\right\} = .9974.$$

That is, the likelihood of observing a value of $\overline{X}$ either larger than $\mu + 3\sigma/\sqrt{n}$ or less than $\mu - 3\sigma/\sqrt{n}$ is .0026. Such an event is sufficiently rare that if it were to occur, it is more likely to have been caused by a shift in the population mean, μ, than to have been the result of chance. This is the basis of the theory of control charts.

Problems for Section 1

1. Suppose that X_1 and X_2 are independent random variables having the uniform distribution on (0, 1, 2, 3, 4, 5). That is,

$$f(j) = P\{X_i = j\} = \tfrac{1}{6} \qquad \text{for } i = 1, 2 \text{ and } 0 \le j \le 5.$$

a. Compute $E(X_1)$ and $\text{Var}(X_1)$. (Hint: Use the formulas

$$E(X) = \sum jf(j),$$

$$\text{Var}(X) = \sum j^2 f(j) - (E(X))^2.)$$

b. Determine the probability distribution for $Y = X_1 + X_2$. (Hint: For each possible value of Y, determine all combinations of X_1 and X_2 that result in that value. For example, $Y = 3$ can be obtained by $(X_1, X_2) = (0, 3), (1, 2), (2, 1)$, and $(3, 0)$. Since each pair has probability $(\frac{1}{6})(\frac{1}{6}) = \frac{1}{36}$, we obtain $P\{Y = 3\} = \frac{4}{36} = \frac{1}{9}$. Repeat this process for all values of Y. As a check be sure that

$$\sum_y P\{Y = y\} = 1.$$

c. Using the results of part (b), find $P\{1.5 < Y < 6.5\}$.

d. Using the results $E(Y) = 2E(X_1)$ and $\text{Var}(Y) = 2\text{Var}(X_1)$, approximate the answer to part (c) using a normal distribution.

e. Suppose that $X_1, X_2, \ldots, X_{20}$ are independent identically distributed random variables having the uniform distribution on $(0, 1, 2, 3, 4, 5)$. Using a normal approximation, estimate

$$P\left\{\sum_{i=1}^{20} X_i \le 75\right\}.$$

f. Do you think that the approximation computed in part (d) or part (e) is more accurate? Why?

2. The following data represent the observed number of defective disks produced each hour based on observing the system for 30 successive hours.

0	3	5	2	6	8	3	5	4	6	6	9	5	1	2
1	2	5	3	3	0	1	0	7	1	7	5	4	4	3

a. Plot a frequency histogram for the numbers of failures per hour.

b. Compute the mean and the variance of the sample. Use the formulas

$$\overline{X} = \frac{1}{n}\sum_{i=1}^{n} X_i,$$

$$s^2 = \frac{1}{n-1}\left(\sum_{i=1}^{n} X_i^2 - n\overline{X}^2\right).$$

(Note: You can streamline the calculations by grouping the data first, as there are only 10 distinct values.)

c. If the number of defective disks produced hourly is normally distributed with mean and variance computed in part (b), determine the probability that fewer than five defectives are observed in any particular hour.

d. If each disk costs the company $5 to produce and defective disks are discarded, how long (in an expected-value sense) would it take to pay off a new piece of equipment costing $45,000 that would reduce defectives to half of the current level? Assume 40-hour production weeks and 48 weeks per year.

3. The tensile strength of a heavy-duty plastic bag used in trash compactors is normally distributed with mean 150 pounds per square inch and standard deviation 12 pounds per square inch. An independent landscape contractor uses them to haul refuse that requires 120-pounds-per-square-inch tensile strength. What proportion of the compactor bags will not meet his requirements?

4. A credit rating company recommends granting of credit cards based on several criteria. One is annual income. If the annual income of applicants is normally distributed with mean $22,000 and standard deviation $4,800 and the company recommends no applicant unless his or her income exceeds $15,000, what fraction of the applicants are denied on this basis?

5. *a.* What is the probability that a normal variable exceeds two-sigma limits? (That is, what is the probability of observing a value of the random variable larger than $\mu + 2\sigma$ or less than $\mu - 2\sigma$?)

   b. If "deciding that the process is out of control" means observing a realization of the sample mean exceeding k sigma limits, discuss the advantages and disadvantages of using $k = 2$ versus $k = 3$.

6. The members of a private golf club have handicaps that are normally distributed with mean 15 and standard deviation 3.5. In a particular event, foursomes are chosen by grouping four players chosen at random from the club. The handicap of the foursome is the arithmetic average of the handicaps of the four players comprising the foursome. In what proportion of the foursomes will the handicap of the foursome be less than 10 or more than 20? (Hint: The standard deviation of the average of four independent identically distributed random variables is exactly half the standard deviation of one of them.)

11.2 Control Charts for Variables: The $\overline{X}$ and R Charts

A process is in control if a stable system of chance causes is operating. That is, the underlying probability distribution generating observations is not changing with time. When the observed value of the sample mean of a group of observations falls outside the appropriate three-sigma limits, it is likely that there has been a change in the probability distribution generating observations. To illustrate how one develops and interprets control charts, consider the following case study.

Example 11.1

Wonderdisk produces a line of plug-compatible disks for IBM equipment. Building 35 is responsible for production of the read/write arms for the model A55C disk. The arms are approximately 2.875 inches in length. The design engineers have established a tolerance of ±0.025 inch for the arm lengths and advertise this figure in the published specifications.

    The company usually produces 40 arms per day. On 30 consecutive production days, five arms are sampled randomly from each day's production and measured. The resulting measurements appear in Table 11–2.

    These observations show that there is some variation in the length of the arms. However, there appears to be no discernible pattern to this variation. Define the random variable X to be the length of an arm selected at random. We may then interpret Table 11–2 as 150 independent observations on the random variable X.

    Howard Hamilton, an industrial engineer working for Wonderdisk, is given the job of analyzing these data. The first thing that Howard notices is that the established

TABLE 11–2　Tracking Arm Data

Sample Number	Measurements of the Length of a Tracking Arm					Average	Range
1	2.8971	2.8477	2.8624	2.8606	2.8971	2.8730	.0494
2	2.8863	2.8541	2.8677	2.8838	2.8854	2.8755	.0322
3	2.8772	2.8708	2.8920	2.8892	2.8840	2.8826	.0212
4	2.8808	2.8650	2.8686	2.8874	2.8804	2.8764	.0224
5	2.8633	2.8993	2.8650	2.8909	2.9131	2.8863	.0497
6	2.8743	2.8571	2.8863	2.8473	2.8739	2.8678	.0390
7	2.8820	2.8612	2.8805	2.8737	2.8933	2.8781	.0322
8	2.8847	2.8630	2.8846	2.8969	2.8916	2.8842	.0339
9	2.8569	2.8934	2.8926	2.8585	2.8721	2.8747	.0365
10	2.8784	2.8795	2.8794	2.8608	2.8672	2.8731	.0187
11	2.8821	2.8544	2.9053	2.8495	2.8670	2.8717	.0558
12	2.8643	2.8533	2.8718	2.8565	2.8724	2.8637	.0191
13	2.8675	2.8578	2.8971	2.8709	2.8908	2.8768	.0394
14	2.8495	2.8701	2.8741	2.8699	2.8766	2.8680	.0271
15	2.8822	2.8731	2.8551	2.8782	2.8687	2.8714	.0271
16	2.8731	2.8675	2.8743	2.8520	2.8900	2.8714	.0379
17	2.9054	2.9190	2.8752	2.8477	2.8639	2.8822	.0713
18	2.8759	2.8832	2.8660	2.8667	2.8674	2.8718	.0172
19	2.8676	2.8775	2.8793	2.8943	2.9048	2.8847	.0373
20	2.8765	2.8613	2.8737	2.8524	2.8767	2.8681	.0243
21	2.9052	2.8851	2.8895	2.8904	2.8723	2.8885	.0328
22	2.8606	2.8837	2.9017	2.8628	2.8455	2.8709	.0562
23	2.8752	2.8722	2.8618	2.8637	2.8725	2.8691	.0133
24	2.8566	2.8929	2.9035	2.9109	2.8594	2.8847	.0543
25	2.8495	2.8749	2.8873	2.8557	2.8673	2.8669	.0378
26	2.8736	2.8606	2.8797	2.8522	2.8802	2.8693	.0280
27	2.8449	2.8908	2.8851	2.8798	2.8610	2.8723	.0459
28	2.8589	2.8800	2.9025	2.8974	2.8606	2.8799	.0437
29	2.8910	2.8546	2.8744	2.8775	2.8634	2.8722	.0364
30	2.8607	2.8769	2.8771	2.8934	2.8706	2.8757	.0326

tolerances of ± 0.025 inch were often exceeded. This becomes most evident by computing the range of daily observations. The range of a sample is the maximum of the observations minus the minimum of the observations. For the 30 days of data, Howard observes that the range exceeds 0.05 in four cases. In order to obtain a clearer idea of what proportion of the population lies outside the specified tolerances, Howard develops a frequency histogram of the 150 measurements. This histogram appears in Figure 11–5. The histogram suggests that the measurements are normally distributed. Howard used a goodness-of-fit test to verify the normality of the observations. Figure 11–6 shows the theoretical normal curve.

One determines the theoretical normal curve in the following way. Because the normal distribution depends upon two parameters, μ and σ, we must estimate these values from the sample data. From the theory of statistics, we know that the "best" estimators of the population mean and variance are the sample mean, $\overline{X}$, and the

FIGURE 11–5

Frequency histogram of 150 measurements

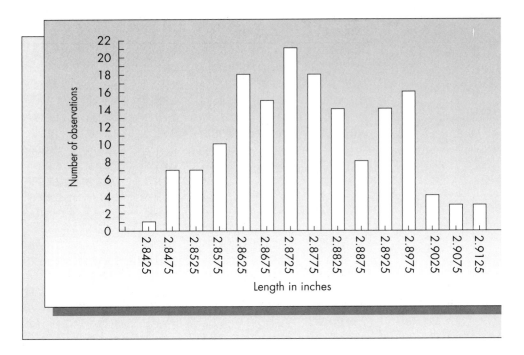

FIGURE 11–6

Theoretical normal curve of arm length

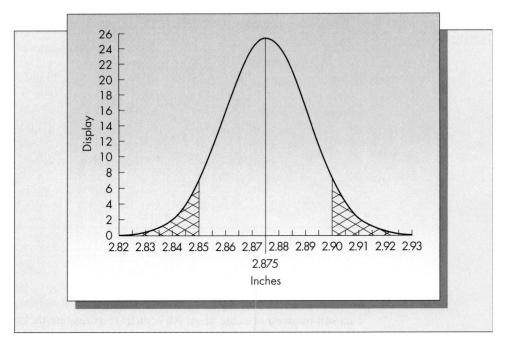

sample variance, s^2, given by

$$\overline{X} = \frac{1}{n} \sum_{i=1}^{n} X_i,$$

$$s^2 = \frac{1}{n-1} \sum_{i=1}^{n} (X_i - \overline{X})^2.$$

Interpret $X_1, X_2, \ldots, X_n$ as the sample values (the random sample). The sample mean based on all 150 observations is 2.875, and the sample variance is 0.0002434. The estimate of the population standard deviation is the square root of the sample variance, which is 0.0156.

Howard can now estimate the fraction of the arms produced that fall outside the advertised tolerances. An arm will exceed the tolerance if it is longer than 2.90 inches or shorter than 2.85 inches. The area of the crosshatched region in Figure 11–6 represents the probability that this will occur, which is evidently not negligible. In fact,

$$P\{X > 2.90 \text{ or } X < 2.85\} = P\{|X - \mu| > 0.025\}$$

$$= P\left\{\left|\frac{X - \mu}{\sigma}\right| > 1.602\right\}$$

$$= P\{|Z| > 1.602\} = .11.$$

We find the probability, .11, in a table of the normal distribution (Table A–1 at the back of this book). Hence, about 11 percent of the arms produced over the last 30 days do not meet the company's published specifications. However, failure rates for the disks due to incompatibility of the arms has been extremely low (less than 1 percent). Howard presented these results to the director of manufacturing, who discussed the problem with the company president and the director of engineering. After some additional investigation, Howard concluded that the original tolerances were not consistent with design requirements of the disk. It was found that a tolerance of ± 0.05 inch would be sufficient. The tighter tolerances were based on an earlier design, and the department simply forgot to revise its figures. Later testing showed that a tolerance of ± 0.05 inch was much more realistic and consistent with the operation of the disk. Howard satisfied himself that the revised tolerances would include more than 99 percent of the population. ■

This example raises an important point in the application of statistical principles to quality control. If control charts are to be used to compare the characteristics of manufactured items with preset design specifications, then the desired tolerances and the observed statistical variation in the sample must be consistent. If the tolerances are much tighter than the variation observed in the sample, as in the example above, then they often will be exceeded even when the process is in control. The opposite situation also can occur: the observed tolerances may be much *wider* than the observed variation in the population. In this case an observation may be out of control relative to other sample values, but may fall within desired tolerances. Whether the process is in control would yield little information about whether parts are meeting specifications.

$\overline{X}$ Charts

Consider Example 11.1. Let us say that Howard decides to construct an $\overline{X}$ chart for the data summarized in Table 11–1. An $\overline{X}$ chart requires that the data be broken down into subgroups of fixed size. The size of the subgroups for the example is $n = 5$. The subgroup size should be at least 4 for the central limit theorem to apply.

To construct an $\overline{X}$ chart, it is necessary to estimate the sample mean and the sample variance of the population. This can be done using the formulas above. However, it generally is not recommended that one use the sample standard deviation as an estimator of σ when constructing an $\overline{X}$ chart. For s to be an accurate estimator for σ, it is necessary that the underlying mean of the sample be constant. Because the purpose of an $\overline{X}$ chart is to determine whether a shift in the mean has occurred, we should not assume a priori that the mean is constant when estimating σ. An alternative method for estimating the sample variation that remains accurate when the population mean changes uses data ranges. Even if the process mean shifts, the ranges will be stable as long as the process variation is stable. There is a relationship between the standard deviation of the population and the range of the subgroups of a given size that depends on the subgroup size. That is, there exists a constant d_2 such that

$$\hat{\sigma} = \frac{\overline{R}}{d_2},$$

where $\overline{R}$ is the average of the observed ranges and $\hat{\sigma}$ is an estimate of the population standard deviation. The constants d_2 for various subgroup sizes appear in Table A–5 at the back of this book. For the data presented in Table 11–1, the average of the 30 ranges turns out to be 0.035756. The value of d_2 for subgroups of size 5 is 2.326. Hence the estimator for σ based on this data is

$$\hat{\sigma} = 0.035756/2.326 = 0.01537,$$

which is quite close to the estimate of the standard deviation using the sample standard deviation s.

Given estimators for the mean and the standard deviation of the group average, the control charts are constructed in the following way: Lines are drawn for the upper and the lower control limits at $\overline{X} \pm 3\sigma/\sqrt{n}$. The group averages are graphed on a daily basis. The process is said to be out of control if an observation falls outside of the control limits. The $\overline{X}$ chart for the sample of 30 days for the tracking arm appears in Figure 11–7. Notice that the process appears to be in control, as all observations fall within the 3σ limits.

Relationship to Classical Statistics

In this section we consider statistical control charts in the context of classical statistical hypothesis testing. The null hypothesis is that the underlying process is in control.

FIGURE 11–7

$\overline{X}$ *chart for tracking arm data*

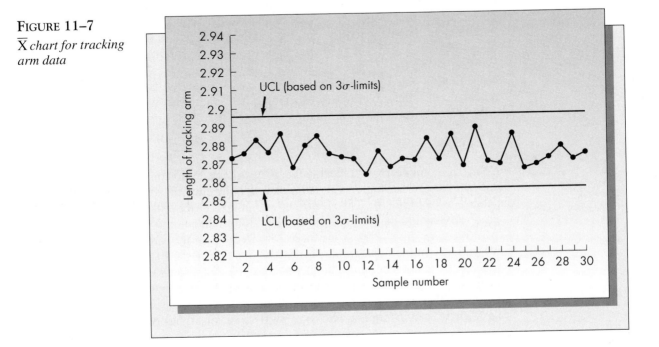

That is, we have the hypotheses

H_0: Process is in control.
H_1: Process is out of control.

We interpret the word *control* as meaning that the underlying chance mechanism generating observations over time is stable. For $\overline{X}$ charts, we test whether the process mean has undergone a shift. There are two ways that we can come to the wrong conclusion: reject the null hypothesis when it is true (conclude that the process is out of control when it is in control) and reject the alternative hypothesis when it is true (conclude the process is in control when it is out of control). These are called respectively the Type 1 and the Type 2 errors. We use the symbol α to represent the probability of a Type 1 error and β to represent the probability of a Type 2 error. A test is a rule that indicates when to reject H_0 based on the sample values. A test requires specification of an acceptable value of α. Conceptually, we are doing the same thing when we use control charts.

The hypothesis that the process is in control is rejected if an observed value of $\overline{X}$ falls outside the control limits. We can set the values of the upper control limit (UCL) and lower control limit (LCL) based on the specification of any value of α.

$$
\begin{aligned}
\alpha &= P\{\text{Type 1 error}\} \\
&= P\{\text{Out-of-control signal is observed} \mid \text{Process is in control}\} \\
&= P\{\overline{X} < \text{LCL or } \overline{X} > \text{UCL} \mid \text{True mean is } \mu\} \\
&= P\{\overline{X} < \text{LCL} \mid \mu\} + P\{\overline{X} > \text{UCL} \mid \mu\} \\
&= P\left\{\frac{\overline{X} - \mu}{\sigma/\sqrt{n}} < \frac{\text{LCL} - \mu}{\sigma/\sqrt{n}}\right\} + P\left\{\frac{\overline{X} - \mu}{\sigma/\sqrt{n}} > \frac{\text{UCL} - \mu}{\sigma/\sqrt{n}}\right\} \\
&= P\left\{Z < \frac{\text{LCL} - \mu}{\sigma/\sqrt{n}}\right\} + P\left\{Z > \frac{\text{UCL} - \mu}{\sigma/\sqrt{n}}\right\}.
\end{aligned}
$$

Because the normal distribution is symmetric, we set

$$
\frac{\text{LCL} - \mu}{\sigma/\sqrt{n}} = -z_{\alpha/2}, \qquad \frac{\text{UCL} - \mu}{\sigma/\sqrt{n}} = z_{\alpha/2},
$$

which gives

$$
\text{UCL} = \mu + \frac{\sigma z_{\alpha/2}}{\sqrt{n}},
$$

$$
\text{LCL} = \mu - \frac{\sigma z_{\alpha/2}}{\sqrt{n}}.
$$

Setting $z_{\alpha/2} = 3$, we obtain the popular three-sigma control limits. This is equivalent to choosing a value of $\alpha = .0026$. This particular value of α is the one that is traditionally used; it is not necessarily the only one that makes sense. In some applications, one might wish to increase the likelihood of recognizing when the process goes out of control. One would then use a larger value of α, which would result in tighter control limits. For example, a value of α of .05 would result in two-sigma rather than three-sigma limits.

R Charts

The $\overline{X}$ chart is used to test for a shift in the mean value of a process. In many instances we are also interested in testing for a shift in the variance of the process. Process variation can be monitored by examining the sample variances of the subgroup observations. However, the ranges of the subgroups give roughly the same information and are much easier to compute. The theory behind the R chart is that when the underlying population is normal, there is a relationship between the range of the sample and the standard deviation of the sample that depends on the sample size. If $\overline{R}$ is the average of the ranges of all of the subgroups of size n, then we have from the previous section

$$\hat{\sigma} = \overline{R}/d_2,$$

where d_2, which depends on n, appears in Table A–5 at the back of this book.

Normally, one would develop an R chart before an $\overline{X}$ chart in order to obtain a reliable estimator of the variance. The estimator $\hat{\sigma}$ is less sensitive to changes in the process mean than is the estimator s. The purpose of the R chart is to determine if the process variation is stable. The upper and the lower limits for this chart are given by the formulas

$$\text{LCL} = d_3\overline{R},$$

$$\text{UCL} = d_4\overline{R}.$$

The values of the constants d_3 and d_4 appear in Table A–6 at the back of this book. The values given for these constants assume three-sigma limits for the range process.

Example 11.1 (continued)

Again consider the data for the tracking arm in Table 11–1. The ranges of the samples of size $n = 5$ appear in the final column of the table. As stated earlier, the average of these 30 ranges is 0.035756. This is the value of $\overline{R}$ and becomes the center point for the R chart. The upper and the lower control limits for R are computed using the formulas above and Table A–6. For the case of $n = 5$, we have $d_3 = 0$ and $d_4 = 2.11$, thus resulting in the following control limits for Wonderdisk's R chart:

$$\text{LCL} = (0)(0.035756) = 0.$$
$$\text{UCL} = (2.11)(0.035756) = 0.07545.$$

These are three-sigma limits.

Wonderdisk's R chart appears in Figure 11–8. Because all observed values of R fall within the control limits, the process variation is in control. ■

R charts are used for testing whether there is a shift in the process variation. For the values of the estimators used to construct the $\overline{X}$ chart to be correct, the process variance should be constant. That is, it is recommended that the R chart be used *before* the $\overline{X}$ chart, since an $\overline{X}$ chart assumes that the process variation is stable.

R charts are not the only means for testing the stability of the process variation. One could also use a σ chart. One plots the sample standard deviations of subgroups over time to determine when and if a statistically significant shift in these values occurs. Sigma charts are rarely used in practice for two reasons:

1. It is more work to compute the sample standard deviations for each subgroup than it is to compute the ranges.
2. R charts and σ charts will almost always give the same results.

For these reasons we will not discuss σ charts in this text.

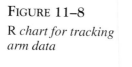

FIGURE 11–8

R *chart for tracking arm data*

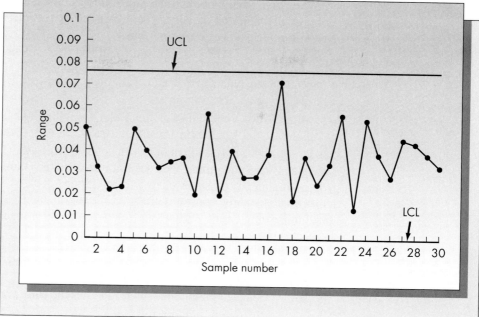

Problems for Section 2

7. The quality control group of a manufacturing company is planning to use control charts to monitor the production of a certain part. The specifications for the part require that each unit weigh between 13.0 and 15.5 ounces with a target value of 14.25. A sample of 75 observations results in the following:

$$\sum_{i=1}^{75} X_i = 1{,}065, \qquad \sum_{i=1}^{75} X_i^2 = 15{,}165.$$

   a. Are the specifications consistent with the statistical variation in the sample? (Hint: Use the computing formula for the sample variance:

$$s^2 = \frac{1}{n-1}\left[\sum_{i=1}^{n} X_i^2 - n\overline{X}^2\right].)$$

   b. What problems do you anticipate if the group attempts to use $\overline{X}$ and R charts for this process?
   c. If the 75 observations are statistically stable, what percentage of the manufactured items will fall outside the tolerances?

8. Control charts for $\overline{X}$ and R are maintained on the shear strength of spot welds. One hundred observations divided into subgroups of size five are used as a baseline to construct the charts, and estimates of μ and σ are computed from these observations. Assume that the 100 observations are $X_1, X_2, \ldots, X_{100}$ and the ranges of the 20 subgroups are $R_1, R_2, \ldots, R_{20}$. From these baseline data the following quantities are computed:

$$\sum_{i=1}^{100} X_i = 97{,}500, \qquad \sum_{j=1}^{20} R_j = 1{,}042.$$

Using this information, compute the values of three-sigma limits for both charts.

9. For Problem 8, suppose that the probability of concluding that the process is out of control when it is actually in control is set at .02. Find the upper and the lower control limits for the resulting $\overline{X}$ chart.

10. Suppose that the process mean shifts to 1,011 in Problem 8.
    a. What is the probability that the $\overline{X}$ chart will *not* indicate an out-of-control condition from a single subgroup sampling? (This is precisely the Type 2 error.)
    b. What is the probability that the $\overline{X}$ chart will not indicate an out-of-control condition after sampling 20 subgroups?

11. A film processing service monitors the quality of the developing process with light-sensitive equipment. The accuracy measure is a number with a target value of zero. Suppose that an $\overline{X}$ chart with subgroups of size five is used to monitor the process and the control limits are UCL = 1.5 and LCL = −1.5. Assume that the estimate for the process mean is zero and for the process standard deviation is 1.30.
    a. What is the value of α for this control chart?
    b. Find the UCL and the LCL based on three-sigma limits.
    c. Suppose that the process mean shifts to 1. What is the probability that the shift is detected on the first subgroup after the shift occurs?

12. An R chart is used to monitor the variation in the weights of packages of chocolate chip cookies produced by a large national producer of baked goods. An analyst has collected a baseline of 200 observations to construct the chart. Suppose the computed value of $\overline{R}$ is 3.825.
    a. If subgroups of size six are to be used, compute the value of three-sigma limits for this chart.
    b. If an $\overline{X}$ chart based on three-sigma limits is used, what is the difference between the UCL and the LCL?

13. A process is monitored using an $\overline{X}$ chart with UCL = 13.8 and LCL = 8.2. The process standard deviation is estimated to be 6.6. If the $\overline{X}$ chart is based on three-sigma limits,
    a. What is the estimate of the process mean?
    b. What is the size of each of the sampling subgroups?

11.3 Control Charts for Attributes: The *p* Chart

$\overline{X}$ and R charts are valuable tools for process control when the output of the process can be expressed as a single real variable. This is appropriate when there is a single quality dimension such as length, width, hardness, and so on. In two circumstances control charts for variables are not appropriate: (1) when one's concern is whether an item has a particular attribute (for example, the issue might be whether the item functions) and (2) when there are many different quality variables. In case (2) it is not practical or cost-effective to maintain separate control charts for each variable. Either the item has the desired attributes or it does not.

When using control charts for attributes, each sample value is either a 1 or a 0. A 1 means that the item is acceptable, and a 0 means that it is not. Let *n* be the size of the sampling subgroup and define the random variable *X* as the total number of defectives in the subgroup. We will assume that each subgroup represents a sampling from one day's production. The theory would be exactly the same whether the sampling interval is one hour, one day, or one month. Because X counts the number of defectives in a fixed sample size, the underlying distribution of X is binomial with parameters n and p. Interpret p as the proportion of defectives produced and n as the number of items sampled in each group (typically, n is the number of items sampled each day). A *p chart* would be used to determine if there is a significant shift in the true value of p.

Although one could construct p charts based on the exact binomial distribution, it is more common to use a normal approximation. Also, as our interest is in estimating the value of p, we track the random variable X/n, whose expectation is p, rather than X itself. It is easy to show that

$$E(X/n) = p,$$

$$\text{Var}(X/n) = p(1 - p)/n.$$

For large n, the central limit theorem tells us that X/n is approximately normally distributed with parameters $\mu = p$ and $\sigma = \sqrt{p(1 - p)/n}$. Using a normal approximation, the traditional three-sigma limits are

$$\text{UCL} = p + 3\sqrt{\frac{p(1 - p)}{n}},$$

$$\text{LCL} = p - 3\sqrt{\frac{p(1 - p)}{n}}.$$

The estimate for p, the true proportion of defectives in the population, is $\bar{p}$, the average fraction of defectives observed over some reasonable baseline period. The process is said to be in control as long as the observed fraction defective for each subgroup remains within the upper and the lower control limits.

Example 11.2

Xezet, a maker of floppy disks, inspects a sample of 50 disks from each day's output. Based on a variety of attributes, the quality inspector classifies each disk as acceptable or not. The experience over a typical 20-day period is summarized in Table 11–3.

To construct a control chart for the fraction defective, it is necessary to have an accurate estimate of the true fraction defective in the entire population, p. Based

TABLE 11–3 **Number of Rejected Disks**

Date	Number Rejected	Date	Number Rejected
3/18	3	4/1	0
3/19	10	4/2	4
3/20	13	4/3	9
3/21	4	4/4	22
3/22	12	4/5	7
3/25	14	4/8	6
3/26	8	4/9	18
3/27	7	4/10	3
3/28	19	4/11	9
3/29	1	4/12	7

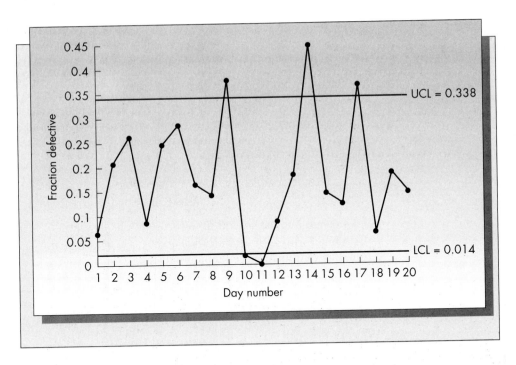

on the data in Table 11–3 we construct a preliminary control chart to determine if the baseline data are in control. The total number of defectives observed during the 20 days is 176. The total production over the same period of time is 1,000 disks. Hence, the current estimate of the proportion of defectives in the population is $176/1,000 = .176$. The current estimator for σ is

$$\hat{\sigma} = \sqrt{\frac{\bar{p}(1 - \bar{p})}{n}} = \sqrt{\frac{(.176)(.824)}{50}} = 0.054.$$

Based on three-sigma limits we obtain

$$\text{UCL} = 0.176 + (3)(0.054) = 0.338,$$
$$\text{LCL} = 0.176 - (3)(0.054) = 0.014.$$

Figure 11–9 is the preliminary control chart for the fraction defective. Notice that four points are out of control. These correspond to days 9 (3/28), 11 (4/1), 14 (4/4), and 17 (4/9). We need not worry about a point that falls below the lower control limit, because that shows a *better*-than-expected rate of defectives. The production manager considers the three remaining out-of-control points and realizes that they correspond to three days when a key employee was absent from work for personal reasons. The employee's job requires some very complex media-plating equipment. The high rate of defectives on these days was apparently the result of a temporary employee's lack of experience with media-plating equipment.

Because the out-of-control points were explained by assignable causes, we eliminate these points from our sample. The baseline data now consist of the data listed in Table 11–3 with the three days corresponding to the out-of-control points eliminated. That is, our database now consists of a total of 17 days. Our estimate of p is now recomputed based on the revised data. We obtain

$$\bar{p} = 117/(17)(50) = .138$$

FIGURE 11–10

Revised p *chart for
Xezet floppy disk
data (refer to
Example 11.2)*

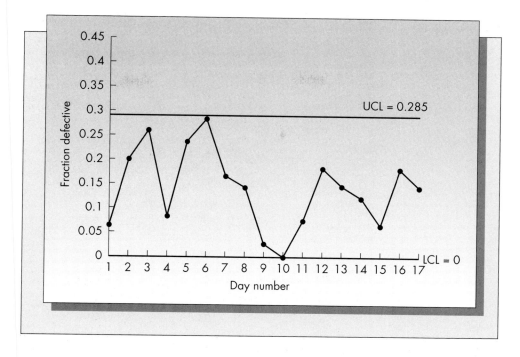

and

$$\hat{\sigma} = \sqrt{(.138)(.862)/50} = 0.049.$$

Based on these new estimators for p and σ we obtain

$$UCL = 0.138 + (3)(0.049) = 0.285,$$
$$LCL = 0.138 - (3)(0.049) = -0.009.$$

If LCL < 0, we set LCL $= 0$, because it is impossible to observe a value of p that is negative. In Figure 11–10 we graph the revised p chart for the floppy disk data. Notice that with the three out-of-control points eliminated, all remaining points now fall within the control limits. This will not always be the case, as the upper and the lower control limits are closer in Figure 11–10 than they were in Figure 11–9. The revised control limits would now be used to monitor future observations of the fraction defective for this process. ■

p *Charts for Varying Subgroup Sizes*

In the previous section we assumed that the number of items inspected in each subgroup was the same. This assumption is reasonable when the subgroups are sampled periodically from large lots. However, in many circumstances few items are produced each day and are consequently subject to 100 percent inspection. If daily production varies, the subgroup size also will vary. Here we base the analysis on the standardized variate Z:

$$Z = \frac{p - \bar{p}}{\sqrt{\dfrac{\bar{p}(1 - \bar{p})}{n}}},$$

which is approximately standard normal independent of n. The lower and the upper control limits would be set at -3 and $+3$ respectively to obtain three-sigma limits, and the control chart would monitor successive values of the standardized variate Z.

TABLE 11–4   **Observed Defectives for Industrial Lathes (Refer to Example 11.3)**

Day Number	Production	Number of Defectives	Standardized Z Value
1	82	8	−0.4451
2	76	12	1.2320
3	85	6	−1.2382
4	53	5	−0.4319
5	30	3	−0.2270
6	121	14	0.0893
7	63	11	1.5404
8	80	9	−0.0178
9	88	7	−0.9946
10	97	8	−0.9532
11	91	13	0.8953
12	77	14	1.9029
13	71	6	−0.7614
14	95	8	−0.8899
15	102	13	0.4566

Example 11.3

A producer of industrial lathes performs 100 percent inspection and testing of each day's production. The lathe production varies from day to day based on anticipated orders and the production schedule for the other machines the company produces. The observed number of defective lathes and the daily production over the past 15 days are given in Table 11–4.

One computes the standardized Z values in the last column in the following way. First, estimate p from the entire 15-day history by forming the ratio of the total number of defectives observed over the total number of items produced. For the data in Table 11–4 we obtain $\bar{p} = 137/1{,}211 = .1131$. The Z values are now computed by the formula above. For example, for day 1,

$$Z = \frac{8/82 - 137/1{,}211}{\sqrt{\dfrac{(137/1{,}211)(1 - 137/1{,}211)}{82}}} = -0.4451.$$

Because the three-sigma limits in Z are simply ± 3, it is clear from Table 11–4 that this process is in control. ∎

Problems for Section 3

14. A manufacturer of photographic film produces an average of 12,000 rolls of various types of film each day. An inspector samples 25 from each day's production and tests them for accuracy of color reproduction and overall film quality. During 23 consecutive production days, the inspector rejected the

following numbers of rolls of film:

Day	Number Rejected	Day	Number Rejected
1	2	13	1
2	3	14	1
3	2	15	0
4	1	16	0
5	3	17	0
6	4	18	2
7	7	19	3
8	1	20	1
9	3	21	2
10	3	22	2
11	0	23	1
12	0		

a. Based on these observations, compute three-sigma limits for a p chart.
b. Do any of the points fall outside the control limits? If so, recompute the three-sigma limits after eliminating the out-of-control points.

15. Applied Machines produces large test equipment for integrated circuits. The machines are made to order, so the production rate varies from month to month. Before shipping, each machine is subject to extensive testing. Based on the tests the machine is either passed or sent back for rework. During the past 20 months the firm has had to rework the following numbers of machines:

Month	Number Produced	Number Reworked	Month	Number Produced	Number Reworked
1	23	3	11	17	3
2	28	3	12	4	0
3	16	1	13	14	2
4	6	0	14	0	0
5	41	2	15	18	6
6	32	4	16	0	0
7	29	5	17	33	4
8	19	2	18	46	5
9	12	1	19	21	7
10	7	1	20	29	7

Determine if the process was in control for the 20-month period using a standardized version of the p chart. Assume three-sigma limits for the control chart. (Hint: Do you actually have 20 months of data?)

16. Consider the example of Applied Machines presented in Problem 15. Based on the estimate of the probability that a machine is sent back for rework computed from the 20 months of data, determine the following:
a. If the company produces 35 machines in one particular month, how many, on average, require rework?

b. Out of 100 machines produced, what is the probability that more than 20 percent of them require rework? (Use the normal approximation to the binomial for your calculations. It is discussed in Appendix 11–A.)

17. Over a period of 12 consecutive production hours, samples of size 50 resulted in the following proportions of defective items:

Sample Number	Proportion Defective	Sample Number	Proportion Defective
1	.04	7	.10
2	.02	8	.10
3	.06	9	.06
4	.08	10	.08
5	.08	11	.04
6	.04	12	.04

a. What are the three-sigma control limits for this process?
b. Do any of the sample points fall outside of the control limits?
c. The company claims a defect rate of 3 percent for these items. Are the observed proportions consistent with a target value of 3 percent defectives? What difficulty would arise if the control limits were based on a target value of 0.03? In view of the company's claims, what difficulty would arise if the control limits computed in part (a) were used?

11.4 The c Chart

There are control charts other than $\overline{X}$, R, and p charts. Although the form of the distribution of the appropriate random variable depends on the application, the basic approach is the same. In general, one must determine the probability distribution of the random variable of interest, and find upper and lower control limits that contain the universe of observations with a desired level of confidence. Usually one sets the probability of falling outside the control limits to be less than .01.

The p chart is appropriate when classifying an item as either good or bad. However, often we are concerned with the number of defects in an item or collection of items. An item is acceptable if the number of defects is not too large. For example, a refrigerator that has a few scratches might be considered acceptable, but one that has too many scratches might be considered unacceptable. As another example, for a textile mill that manufactures cloth, both the manufacturer and the consumer would be concerned with the number of defects per yard of cloth.

The *c chart* is based on the observation that if the defects are occurring completely at random, then the probability distribution of the number of defects per unit of production has the Poisson distribution. If c represents the true mean number of defects in a unit of production, then the likelihood that there are k defects in a unit is

$$P\{\text{Number of defects in one unit} = k\} = \frac{e^{-c}c^k}{k!} \qquad \text{for } k = 0, 1, 2, \ldots.$$

In using a control chart for number of defects, the sample size should be the same at each inspection. One estimates the value of c from baseline data by computing the sample mean of the observed number of defects per unit of production. When $c \geq 20$, the normal distribution provides a reasonable approximation to the Poisson. Because the mean and the variance of the Poisson are both equal to c, it follows that for large c,

$$Z = \frac{X - c}{\sqrt{c}}$$

is approximately standard normal. Using the traditional three-sigma limits, the upper and the lower control limits for the c chart are

$$LCL = c - 3\sqrt{c},$$
$$UCL = c + 3\sqrt{c}.$$

One develops and uses a c chart in the same way as $\overline{X}$, R, and p charts.

Example 11.4

Leatherworks produces various leather goods in its plant in Montpelier, Vermont. Inspections of the past 20 units of a leather portfolio revealed the following numbers of defects:

Unit Number	Number of Defects Observed	Unit Number	Number of Defects Observed
1	4	11	2
2	3	12	3
3	3	13	6
4	0	14	1
5	2	15	5
6	5	16	4
7	4	17	1
8	2	18	1
9	3	19	2
10	3	20	2

Most defects are the result of natural marks in the leather, but even so, the firm does not want to ship products out with too many defects in the leather. Using these 20 data points as a baseline, determine upper and lower control limits that include the universe of observations with probability .95. What control limits result from using a normal approximation of the Poisson?

Solution

To estimate c we compute the sample mean of the data, which is found by adding the total number of observed defects and dividing by the number of observations. This gives $c = 56/20 = 2.8$. To be certain that a c chart is appropriate here, we should do a goodness-of-fit test of these data for a Poisson distribution with parameter 2.8. We will leave it to the reader to check that the data do indeed fit a Poisson distribution. (Goodness-of-fit tests are described in almost every statistics text. The most common is the chi-square test.)

To determine exact control limits, we use Table A–3 in the back of the book. Because the Poisson is a discrete distribution, it is very unlikely that we will be able to find control limits that contain exactly 95 percent of the probability. From the table we see that the probability that the number of defects is less than or equal to zero is $1.0000 - .9392 = .0608$, which is too large. Hence, we will set the lower control limit to zero. By symmetry, the upper limit should correspond to a right tail of about .025, which occurs at $k = 7$. Hence we would recommend control limits of LCL $= 0$ and UCL $= 7$.

For a normal distribution, approximately 2 standard deviations from the mean include 95 percent of the probability. Hence, the control limits based on a normal approximation of the Poisson are

$$\text{LCL} = c - 2\sqrt{c} = 2.8 - (2)\sqrt{2.8} = -0.55 \qquad \text{(set to zero)},$$
$$\text{UCL} = c + 2\sqrt{c} = 2.8 + 2\sqrt{2.8} = 6.2.$$

The normal approximation is not very accurate in this case because c is too low. ■

Problems for Section 4

18. Amertron produces electrical wiring in 100-foot rolls. The quality inspection process involves selecting rolls of wire at random and counting the number of defects on each roll. The last 20 rolls examined revealed the following numbers of defects:

Roll	Number of Defects	Roll	Number of Defects
1	4	11	2
2	6	12	5
3	2	13	5
4	4	14	7
5	1	15	4
6	9	16	8
7	5	17	6
8	5	18	4
9	3	19	6
10	3	20	4

    a. If the number of defects per 100-foot roll of wire follows a Poisson distribution, what is the estimate of c obtained from these observations?

    b. Using a normal approximation to the Poisson, what are the three-sigma control limits that you would use to monitor this process?

    c. Are all 20 observations within the control limits?

19. Amertron, discussed in Problem 18, has established a policy of passing rolls of wire having five or fewer defects.

    a. Based on the exact Poisson distribution, what is the proportion of the rolls that pass inspection? (See Table A–3 in the back of the book.)

    b. Estimate the answer to part (a) using the normal approximation to the Poisson.

20. A large national producer of cookies and baked goods uses a c chart to monitor the number of chocolate chips in its chocolate chip cookies. The company would

like to have an average of six chips per cookie. One cookie is sampled each hour. The results of the last 12 hours were

Hour	Number of Chips per Cookie	Hour	Number of Chips per Cookie
1	7	7	3
2	4	8	6
3	3	9	3
4	3	10	2
5	5	11	4
6	4	12	4

    a. Assuming a target value of $c = 6$, what are the upper and the lower control limits for a c chart?

    b. Are the 12 observations above consistent with a target value of $c = 6$? If those 12 observations constitute a baseline, what upper and lower control limits result? (Use the normal approximation for your calculations.)

21. For the company mentioned in Problem 20 above, a purchaser of a bag of chocolate chip cookies discovers a cookie that has no chips in it and charges the company with fraudulent advertising. Suppose the company produces 300,000 cookies per year. If the expected number of chips per cookie is six, how many cookies baked each year would have no chips? See Table A–3 in the back of the book.

11.5 Classical Statistical Methods and Control Charts

Control charts signal nonrepresentative observations in a sample. The hypothesis that a shift in the process has occurred also can be tested by classical statistical methods. For example, consider the p chart. In constructing the p chart we are testing the hypothesis that a shift has occurred in the underlying value of p, the true proportion of defectives in the lot. A $2 \times n$ contingency table also can be used to test if p has changed. One variable is time, and the other variable is the proportion of defectives observed. The χ^2 test would be used to test whether or not there exists a relationship between the two variables; that is, whether the proportion of defectives changes with time.

    It is not necessarily true that the χ^2 test will give the same results as a p chart. In general, the χ^2 test will recommend rejection of the hypothesis that the data are homogeneous based on the average of the departures from the estimated mean, and the control chart will recommend rejection of the hypothesis that the process is in control based on a large deviation of a single observation. It is important to understand this difference to determine which would be a more appropriate procedure. It is probably true that in the context of manufacturing one is more concerned with extreme deviations of a few observations than the average of many deviations, thus providing one reason for the preference among practitioners for control chart methodology. Another reason for the preference for control charts is that they are easy to use and understand. Quality control managers are more familiar with control charts than they are with classical statistical methods.

Problem for Section 5

22. Consider the data presented in Problem 14. Problem 14 required testing whether the process was in control using a p chart. Test the hypothesis that the value of p is the same each day using classical statistical methods. That is, test the hypothesis

$$H_0: p_1 = p_2 = \cdots = p_k$$

versus

$$H_1: \text{Not all the } p_i \text{ are equal.}$$

where k is the number of days in the data set ($k = 23$ in this case) and p_i is the true proportion of defectives for day i. Define x_i as the number of rolls of film rejected in day i and p' as the estimate of p obtained from the data (p' was computed in Problem 14(a)). The test statistic is given by the formula

$$\chi^2 = \sum_{i=1}^{k} \frac{(x_i - np')^2}{np'(1 - p')},$$

where n is the number of items sampled each day ($n = 25$ in this case). The test is to reject H_0 if $\chi^2 > \chi^2_{\alpha,k-1}$, where $\chi^2_{\alpha,k-1}$ is a number obtained from a table of the χ^2 distribution. For $\alpha = .01$ (which is larger than the α value that yields 3σ limits on a p chart), $\chi^2_{.01,22} = 40.289$. Based on this value of α, does the χ^2 test indicate that this process is out of control? If the answer you obtained is different from that of Problem 14(b), how do you account for the discrepancy? Which method is probably better suited for this application?

*11.6 Economic Design of $\overline{X}$ Charts

The design of an $\overline{X}$ chart requires the determination of various parameters. These include the amount of time that elapses between sampling, the size of the sample drawn in each interval, and the upper and the lower control limits. The penalties associated with the upper and the lower control limits are reflected in the Type 1 and the Type 2 errors. This section will incorporate explicit costs of these errors into the analysis as well as costs of sampling, and considers the problem of designing an $\overline{X}$ chart based on cost minimization.

The model treated here does not include the sampling interval as a decision variable. In many circumstances the sampling interval is determined from considerations other than cost. There are convenient or natural times to sample based on the nature of the process, the items being produced, or personnel constraints.

We will consider the following three costs:

1. Sampling cost.
2. Search cost.
3. Cost of operating out of control.

1. *The sampling cost.* We assume that exactly n items are sampled each period. In most cases, sampling requires workers' time, so that personnel costs are incurred. There also may be costs associated with the equipment required for the sampling. Furthermore, sampling may require destructive testing, adding the cost of the item itself. We will assume that for each item sampled, there is a cost of a_1. It follows that the sampling cost incurred each period is $a_1 n$.

2. *The search cost.* When an out-of-control condition is signaled, the presumption is that there is an assignable cause for the condition. The search for the assignable cause

generally will require that the process be shut down. When an out-of-control signal occurs, there are two possibilities: either the process is truly out of control or the signal is a false alarm. In either case, we will assume that there is a cost a_2 incurred each time a search is required for an assignable cause of the out-of-control condition. The search cost could include the costs of shutting down the facility, labor time required to identify the cause of the signal, time required to determine if the out-of-control signal was a false alarm, and the cost of testing and possibly adjusting the equipment. Note that the search cost is probably a random variable: it might not be possible to predict the degree of effort required to search for an assignable cause of the out-of-control signal. When that is the case, interpret a_2 as the *expected* search cost.

3. *Operating out of control.* The third and final cost that we will consider is the cost of operating the process after it has gone out of control. There is a greater likelihood that defective items are produced if the process is out of control. If defectives are discovered during inspection, they would either be scrapped or repaired at a future time. An even more serious consequence is that a defective item becomes part of a larger subassembly, which must be disassembled or scrapped. Finally, defective items can make their way into the marketplace, resulting in possible costs of warranty claims, liability suits, and overall customer dissatisfaction. Assume that there is a cost of a_3 each period that the process is operated in an out-of-control condition.

We consider the economic design of $\overline{X}$ charts only. Assume that the process mean is μ and the process standard deviation is σ. A sufficient history of observations is assumed to exist so that μ and σ can be estimated accurately. We also assume that an out-of-control condition means that the underlying mean undergoes a shift from μ to $\mu + \delta\sigma$ or to $\mu - \delta\sigma$. Hence, *out of control* means that the mean shifts by δ standard deviations.

Define a cycle as the time interval from the start of production just after an adjustment to detection and elimination of the assignable cause of the next out-of-control condition. A cycle consists of two parts. Define T as the number of periods that the process remains in control directly following an adjustment and S as the number of periods that the process remains out of control until a detection is made. A cycle is the sum $T + S$. Successive cycles are pictured in Figure 11–11. Note that both T and S are random variables, so the length of each cycle is a random variable as well. The probability distribution of T is given below.

FIGURE 11–11

Successive cycles in process monitoring

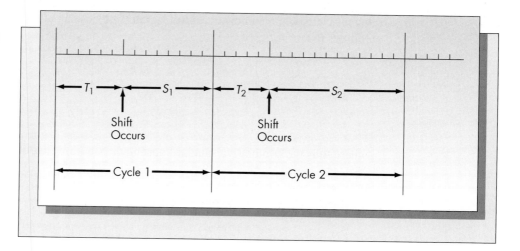

The $\overline{X}$ chart is assumed to be constructed using the following control limits:

$$\text{UCL} = \mu + \frac{k\sigma}{\sqrt{n}}.$$

$$\text{LCL} = \mu - \frac{k\sigma}{\sqrt{n}}.$$

Throughout this chapter we have assumed that $k = 3$, but this may not always be optimal. The goal of the analysis of this section is to determine the economically optimal values of both k and n. The method of analysis is to determine an expression for the total cost incurred in one cycle and an expression for the expected length of each cycle. In the spirit of regenerative process (see Ross, 1970, for example), we have the result that

$$E(\text{Cost per unit time}) = \frac{E(\text{Cost per cycle})}{E(\text{Length of cycle})}.$$

After determining an expression for the expected cost per unit time, we will find the optimal values of n and k that minimize this cost.[2]

Assume that T, the number of periods that the system remains in control following an adjustment, is a discrete random variable having the geometric distribution. That is,

$$P\{T = t\} = \pi(1 - \pi)^t \qquad \text{for } t = 0, 1, 2, 3, \dots.$$

The geometric model arises in the following fashion. Suppose that in any period the process is in control. Then π is the conditional probability that the process will shift out of control in the next period. The geometric distribution is the discrete analog of the exponential distribution. Like the exponential distribution, the geometric distribution also has the memoryless property.[3] In the present context, the memoryless property implies that there is no aging or decay in the production process. That is, the process is equally likely to shift out of control just after an assignable cause has been found and corrected as it is many periods later. This assumption will be accurate when process shifts are due to random causes, or when the process is recalibrated on an ongoing basis.

An out-of-control signal is indicated when

$$|\overline{X} - \mu| > \frac{k\sigma}{\sqrt{n}}.$$

As in earlier sections of this chapter, let α be the probability of Type 1 error. Type 1 error occurs when an out-of-control signal is observed but the process is in control. It follows that

$$\alpha = P\left\{ \left| \overline{X} - \mu \right| > \frac{k\sigma}{\sqrt{n}} \,\bigg|\, E(\overline{X}) = \mu \right\}$$

$$= P\left\{ \left| \frac{\overline{X} - \mu}{\sigma/\sqrt{n}} \right| > k \,\bigg|\, E(\overline{X}) = \mu \right\} = P\{|Z| > k\} = 2\Phi(-k),$$

where Φ is the cumulative standard normal distribution function.

---

[2]Similar ideas were used in inventory control models in Chapters 4 and 5 and will be used to analyze age replacement models in Chapter 12, Section 7.

[3]A detailed discussion of the memoryless property of the exponential distribution is given in Chapter 12, Section 2.

The Type 2 error probability, β, is the probability of not detecting an out-of-control condition. Here we assume that an out-of-control condition means that the process mean has shifted to $\mu + \delta\sigma$ or to $\mu - \delta\sigma$. Suppose that we condition on the event that the mean has shifted from μ to $\mu + \delta\sigma$. The probability that the shift is not detected after observing a sample of n observations is

$$\beta = P\left\{\left|\overline{X} - \mu\right| \le \frac{k\sigma}{\sqrt{n}} \;\middle|\; E(\overline{X}) = \mu + \delta\sigma\right\}$$

$$= P\left\{\frac{-k\sigma}{\sqrt{n}} \le \overline{X} - \mu \le \frac{k\sigma}{\sqrt{n}} \;\middle|\; E(\overline{X}) = \mu + \delta\sigma\right\}$$

$$= P\left\{-k - \delta\sqrt{n} \le \frac{\overline{X} - \mu - \delta\sigma}{\sigma/\sqrt{n}} \le k - \delta\sqrt{n} \;\middle|\; E(\overline{X}) = \mu + \delta\sigma\right\}$$

$$= P\{-k - \delta\sqrt{n} \le Z \le k - \delta\sqrt{n}\}$$

$$= \Phi(k - \delta\sqrt{n}) - \Phi(-k - \delta\sqrt{n}).$$

If we had conditioned on $E(\overline{X}) = \mu - \delta\sigma$, we would have obtained

$$\beta = \Phi(k + \delta\sqrt{n}) - \Phi(-k + \delta\sqrt{n}).$$

Using the symmetry of the normal distribution (specifically, that $\Phi(t) = 1 - \Phi(-t)$ for any t), it is easy to show that these two expressions for β are the same.

Consider the random variables T and S. We assume that T is a geometric random variable assuming values $0, 1, 2, \ldots$. One can show that

$$E(T) = \frac{1 - \pi}{\pi}$$

(see, for example, DeGroot, 1986). The random variable S is the number of periods that the process remains out of control after a shift occurs. The probability that a shift is not detected when the process is out of control is exactly β. It follows that S is also a geometric random variable except that it assumes only the values $1, 2, 3, \ldots$. That is,

$$P\{S = s\} = (1 - \beta)\beta^{s-1} \qquad \text{for } s = 1, 2, 3, \ldots.$$

Because S is defined on the set $1, 2, 3, \ldots$, the expected value of S is $E(S) = 1/(1 - \beta)$. It follows that the expected cycle length, say C, is given by

$$E(C) = E(T + S) = E(T) + E(S) = \frac{1 - \pi}{\pi} + \frac{1}{1 - \beta}.$$

Consider the expected sampling cost incurred in a cycle. Each period there are n items sampled. As there are, on the average, $E(C)$ periods per cycle, it follows that the sampling cost per cycle is $a_1 n E(C)$.

We now compute the expected search cost. The process is shut down each time an out-of-control signal is observed. One or more of these signals could be a false alarm. Suppose that there are exactly M false alarms in a cycle. The random variable M has the binomial distribution with probability of "success" (i.e., a false alarm) of α for a total of T trials. It follows that $E(M) = \alpha E(T)$. The expected number of searches per cycle is exactly $1 + E(M)$, as the final search is assumed to discover and correct the assignable cause. Hence, the total search cost in a cycle is

$$a_2[1 + \alpha E(T)] = a_2[1 + \alpha(1 - \pi)/\pi].$$

We also assume that there is a cost of a_3 for each period that the process is operated in an out-of-control condition. The process is out of control for exactly S periods. Hence, the expected out-of-control cost is $a_3 E(S) = a_3/(1 - \beta)$.

It follows that the expected cost per cycle is

$$a_1 n E(C) + a_2[1 + \alpha(1 - \pi)/\pi] + a_3/(1 - \beta).$$

Dividing by the expected length of a cycle, $E(C)$, gives the average cost per unit time as

$$a_1 n + \frac{a_2\left[1 + \alpha\dfrac{(1 - \pi)}{\pi}\right] + \dfrac{a_3}{1 - \beta}}{\dfrac{1 - \pi}{\pi} + \dfrac{1}{1 - \beta}}$$

$$= a_1 n + \frac{a_2\left[1 + \alpha\dfrac{(1 - \pi)}{\pi}\right] + \dfrac{a_3}{1 - \beta}}{\dfrac{1 - \beta(1 - \pi)}{(1 - \beta)\pi}}$$

$$= a_1 n + \frac{a_2(1 - \beta)[\pi + \alpha(1 - \pi)] + a_3\pi}{1 - \beta(1 - \pi)}.$$

We will write this as $G(n, k)$ to indicate that the optimization requires searching for the best n and k, where $n = 1, 2, 3, \ldots$ and $k > 0$. Note that α depends on k and β depends on both n and k. The goal is to find the values of n and k that minimize $G(n, k)$. This is a complex optimization problem because both α and β require evaluation of the cumulative normal distribution function.

Example 11.5

Consider Example 11.1 of Wonderdisk, introduced in Section 2. Howard Hamilton would like to design an $\overline{X}$ chart in an economically optimal fashion. Based on his experience with the process and an analysis of the past history of failures, he decides that the geometric distribution accurately describes changes in the process state.

In order to use the model described in this section, he must estimate various costs and system parameters. The first is the sampling cost. Here sampling requires measuring the length of a tracking arm. This requires moving the arm to a different location, mounting it on a special brace to protect it, and measuring the length with calipers designed for the purpose. The process requires about 12 minutes of a technician's time. The technician is paid $15 per hour, so that the sampling cost is $15/5 = $3 per item sampled.

The second cost to estimate is the search cost. The time spent searching for an assignable cause of an out-of-control signal is usually about 30 minutes. If a problem is not discovered within that time, it is generally assumed that the out-of-control signal was a false alarm. The arms generate a revenue for the company of about $1,200 daily. Assuming an eight-hour workday, the cost of shutting down production comes to about $1,200/8 = $150 per hour. Hence, the search cost is $75.

The third cost required by the model is the cost of operating the process in an out-of-control condition. If the process is out of control, the proportion of defective arms produced increases. Most of the defective arms show up in the final testing phase of the drives. If a drive has a defective arm, the drive is disassembled and the arm replaced. Some defective arms pass inspection and are shipped to the customer with the disk drive. Wonderdisk provides purchasers a 14-month warranty, and it is likely that a problem with the drive will develop during the warranty period if the arm is defective. Howard estimates that the cost of operating the process out of control is about $300 per hour, but he is not very confident about this estimate.

The model also requires estimates of π and δ. Recall that π represents the probability that the process will shift from an in-control state to an out-of-control state during one period. In the past, out-of-control signals have occurred at a rate of about one for every 10 hours of operation. As half of these have been false alarms, a reasonable estimate of the proportion of periods in which a shift has occurred is about one out of 20, or $\pi = .05$. The constant δ represents the degree of the shift as measured in multiples of the process standard deviation. In the past, the shifts have averaged about one standard deviation, so that the estimate of δ is 1.

In order to simplify the calculations, Howard decides to use an approximation to the standard normal cumulative distribution function. The one he uses is the following:

$$\Phi(z) = 0.500232 - 0.212159z^{2.08388} + 0.5170198z^{1.068529} + 0.041111z^{2.82894}.$$

This approximation, due to Herron (1985), is accurate to within 0.5 percent for $0 < z < 3$. Howard Hamilton decides that this is accurate enough for his purposes. The optimization scheme he adopts is the following. Because n is a discrete variable and represents the number of items sampled in each subgroup, it is unlikely that n would exceed 10. Furthermore, because k is the number of standard deviations of $\overline{X}$ used in the control chart, it is unlikely that k would exceed 3. Howard writes a computer program to evaluate $G(n, k)$ for $k = 0, 0.1, 0.2, \ldots, 2.8, 2.9, 3.0$ and $n = 1, 2, 3, \ldots, 10$. (These calculations were actually done using a popular spreadsheet program.) For each fixed value of n, the function $G(n, k)$ appears to be convex in the variable k (convex functions were discussed in Chapters 4 and 5). For the parameter values listed above and $n = 4$, Figure 11–12 shows the function $G(n, k)$ as a function of k. The graph shows that the minimum cost occurs at about $k = 1.7$ and equals about $45 per hour.

Table 11–5 gives the results of the calculations for these parameter settings. According to the table, the optimal subgroup size is four and the optimal k is 1.74. These results were reasonably close to the current policy and give Hamilton confidence that his estimates were at least in the right ballpark. When he presented the results to his boss, however, his boss expressed concern about the large value of β, the probability of Type 2 error:

> You mean to tell me that if we operate the system optimally, then there is almost a 40 percent chance that we won't be able to detect when the system has gone out of control? That sounds pretty high. Considering the push that is on all over the company for an improvement in quality, I find that figure very disturbing. How does that turn out to be optimal?

The reason that the Type 2 error was so large was the assumption that the cost to the company was $300 for each hour that the process was operated out of control. After giving the matter some thought, Howard's boss decided that a value of $1,000 was probably closer to the mark and was more consistent with the corporate goal of

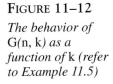

Figure 11–12

The behavior of G(n, k) as a function of k *(refer to Example 11.5)*

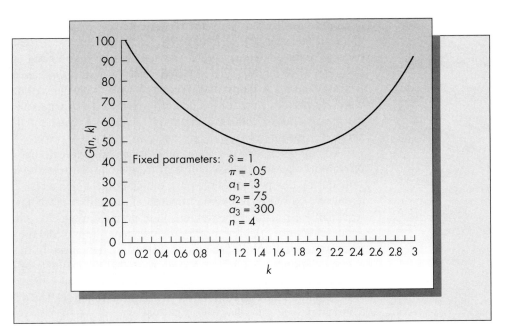

Fixed parameters: $\delta = 1$
$\pi = .05$
$a_1 = 3$
$a_2 = 75$
$a_3 = 300$
$n = 4$

Table 11–5   Optimal Values of k for Various n

		Fixed parameters are: $\delta = 1$,		
		$\pi = 0.05$,		
		$a_1 = 3$.		
		$a_2 = 75$,		
		$a_3 = 300$.		
n	*Optimal k*	α	β	*Cost*
1	1.13	.25	.54	$54.4
2	1.52	.14	.50	48.8
3	1.60	.11	.44	46.1
4	1.74	.08	.39	45.2
5	1.86	.07	.34	45.3
6	1.97	.05	.30	46.2
7	2.06	.04	.26	47.6
8	2.14	.03	.23	49.3
9	2.21	.02	.21	51.3
10	2.27	.02	.18	53.4

improving quality. Howard repeated his calculations substituting a value of $a_3 = 1,000$. The results are presented in Table 11–6.

With the revised value of $a_3 = 1,000$, the optimal value of the sample size n increased to 7 with a corresponding value of $k = 1.70$. The cost per hour at the optimal solution increased to $90.40. Although the limits on the control chart remained the same, the effect of increasing the sample size from 4 to 7 resulted in a

TABLE 11–6   **Optimal Values of k for Various n (Revised) Fixed Parameters: Same as in Table 11–4 except $a_3 = 1,000$**

n	Optimal k	α	β	Cost
1	0.67	.48	.30	$111.6
2	0.93	.33	.29	102.8
3	1.13	.25	.26	96.9
4	1.35	.18	.25	93.3
5	1.43	.15	.21	91.3
6	1.59	.11	.19	90.5
7	1.70	.09	.16	90.4
8	1.79	.07	.14	90.9
9	1.86	.06	.12	91.8
10	1.94	.05	.11	93.3

dramatic decrease in the value of β from .39 to .16. Howard's new boss was far happier with the values of α and β that resulted from the new design. ■

Problems for Section 6

23. A quality control engineer is considering the optimal design of an $\overline{X}$ chart. Based on his experience with the production process, there is a probability of .03 that the process shifts from an in-control to an out-of-control state in any period. When the process shifts out of control, it can be attributed to a single assignable cause; the magnitude of the shift is 2σ. Samples of n items are made hourly, and each sampling costs $0.50 per unit. The cost of searching for the assignable cause is $25 and the cost of operating the process in an out-of-control state is $300 per hour.
   a. Determine the hourly cost of operating the system when $n = 6$ and $k = 2.5$.
   b. Estimate the optimal value of k for the case $n = 6$. If you are doing the calculations by hand, use $k = 0.5, 1, 1.5, 2, 2.5$, and 3.0. If you are using a computer, use $k = 0.1, 0.2, \ldots, 2.9, 3.0$.
   c. Determine the optimal control chart design that minimizes average annual costs.

24. Consider the application of the economic design of $\overline{X}$ charts for Wonderdisk presented in this section. Without actually performing the calculations, discuss what the effect on the optimal values of n and k is likely to be if
   a. δ increased from 1 to 2.
   b. π increased from .05 to .10.
   c. a_1 decreased to 1.
   d. a_2 increased to 150.

25. Under what circumstances would the following assumptions not be accurate?
   a. The assumption that the probability law describing the number of periods until the process goes out of control follows the geometric distribution.
   b. The assumption that an out-of-control condition corresponds to a shift of the mean equal to $\delta\sigma$.
   c. The assumption that the search cost is a fixed constant, a_2.

26. Discuss the following pro and con positions on using optimization models to design control charts:

    Con: "These models are useless to me because I don't feel I can accurately estimate the values of the required inputs."

    Pro: "The choice of specific values of n and k in the construction of X bar charts means that you are assuming values for the various system costs and parameters. You might as well take the bull by the horns and obtain the best estimates you can and use those to design the X bar chart."

*27. Suppose that "the process going out of control" corresponds to a shift of the mean from μ to $\mu + \sigma$ with probability .25, $\mu + 2\sigma$ with probability .25, $\mu - \sigma$ with probability .25, and $\mu - 2\sigma$ with probability .25. What modifications in the model are required? In particular, if we write the shift in the form $\mu \pm \delta\sigma$, show how to compute β_1 and β_2 that would correspond to values of $\delta = 1$ and $\delta = 2$ respectively. If the out-of-control costs were now represented by a_3 when $\delta = 1$ and a_4 when $\delta = 2$, determine an expression for the average annual operating costs. (Assume all other costs and system parameters remain the same.)

28. A local contractor manufactures the bells used in telephones. The phone company requires the bells to ring at a specified noise level (in decibels). An $\overline{X}$ chart is being designed to monitor this variable. The process of sampling bells from the line requires hitting the bells with a fixed-force clapper and measuring the decibel level on a meter designed for that purpose. The cost of sampling is $1.25 per bell. When the process goes out of control, the thickness of the bells is incorrect. The cost of searching for an assignable cause is estimated to be $50. The cost of operating the process in an out-of-control state is estimated to be $180 per hour. Out of control corresponds to a shift of 2σ in the decibel level, and the probability that the process shifts out of control in any hour is .03.

    a. The company uses an $\overline{X}$ chart based on 3σ limits and subgroups of size 4. What is its hourly cost?

    b. What are the optimal values of n and k for this process and the associated optimal cost?

11.7 Overview of Acceptance Sampling

Control charts provide a convenient way to monitor a process in real time to determine if a shift in the process parameters appears to have occurred. Another important aspect of quality control is to determine the quality of manufactured goods *after* they have been produced. In most cases 100 percent inspection is either impossible or impractical. Hence, a sample of items is inspected and quality parameters of large lots of items are estimated based on the results of the sampling.

To be more specific, acceptance sampling addresses the following problem: If a sample is drawn from a large lot of items and the sample is subject to 100 percent inspection, what inferences can we draw about the quality of the lot based on the quality of the sample? Statistical analysis provides a means for extrapolating the characteristics of a sample to the characteristics of the lot, and a means for determining the probability of coming to the wrong conclusion.

Obviously, 100 percent inspection of all items in the lot will reduce the probability of an incorrect conclusion to zero. However, there are several reasons that 100 percent inspection is either not feasible or not desirable. Some of these include

1. In most cases 100 percent inspection is too costly. It is virtually impossible for high-volume transfer lines and continuous production processes.
2. In some cases 100 percent inspection may be impossible, such as when inspection involves destructive testing of the item. For example, determining the lifetime of a light bulb requires burning the bulb until it fails.
3. If the inspection is done by the consumer rather than the producer, 100 percent inspection by the consumer provides little incentive to the producer to improve quality. It is cheaper for the producer to repair or replace the items returned by the consumer than it is to improve the quality of the production process. However, if the consumer returns the entire lot based on the results of sampling, it provides a much greater motivation to the producer to improve the quality of outgoing lots.

In this chapter we treat the following three sampling plans.

1. *Single sampling plans.* Single sampling plans are by far the most popular and easiest to use of the plans we will discuss. Two numbers, n and c, determine a single sampling plan. If there are more than c defectives in a sample of size n, the lot is rejected; otherwise it is accepted.

2. *Double sampling plans.* In a double sampling plan, we first select a sample of size n_1. If the number of defectives in the sample is less than or equal to c_1, the lot is accepted. If the number of defectives is greater than c_2, then the lot is rejected. However, if the number of defectives is larger than c_1 and less than or equal to c_2, a second sample of size n_2 is drawn. The lot is now accepted if the cumulative number of defectives in both samples is less than or equal to a third number, c_3. (Often $c_3 = c_2$.)

3. *Sequential sampling.* A double sampling plan can obviously be extended to a triple sampling plan, which can be extended to a quadruple sampling plan, and so on. A sequential sampling plan is the logical conclusion of this process. Items are sampled one at a time and the cumulative number of defectives is recorded at each stage of the process. Based on the value of the cumulative number of defectives, there are three possible decisions at each stage:

a. Reject the lot.
b. Accept the lot.
c. Continue sampling.

A complex sampling plan may have desirable statistical properties but the acceptance and rejection regions could be difficult to calculate and the plan difficult to implement. The right sampling plan for a particular environment may not be the most mathematically sophisticated. As with any analytical tool, the potential benefits must be weighed against the potential costs.

Kolesar (1993) makes the point that with improving quality standards, the value of acceptance sampling may diminish in years to come. Motorola has become famous for instituting its "six-sigma" quality thrust (Motorola's quality initiatives are discussed in the Snapshot Application in this chapter). By this they mean that the defect rate should be no more than the area outside of $\pm 6\sigma$ under a normal curve. This translates to defect rates of less than 3.4 parts per million. (Few tables of the normal distribution go past 4σ, so you will have a difficult time verifying this probability.) When defect rates are so low, acceptance sampling becomes very inefficient. For example, suppose the defect rate increased by a factor of 10. In that case, the probability of finding a defect in a sample of, say, 1,000 units would be only .034. We wouldn't expect to see a single defect until we have sampled at least 29 lots on average! However, we should keep in mind that Motorola's six-sigma quality standard has not become an industry standard by a long shot. Acceptance sampling will remain a valuable tool for many years to come.

11.8 Notation

We will use the following notation throughout the remainder of this chapter.

N = Number of pieces in a given lot or batch.
n = Number of pieces in the sample ($n < N$).
M = Number of defectives in the lot.
β = Consumer's risk—the probability of accepting bad lots.
α = Producer's risk—the probability of rejecting good lots.
c = Rejection level.
X = Number of defectives in the sample.
p = Proportion of defectives in the lot.
p_0 = Acceptable quality level (abbreviated AQL).
p_1 = Lot tolerance percent defective (abbreviated LTPD).

Assume that N is a known constant. If N is very large relative to the sample size, n, it can be assumed to be infinite. In that case it will not enter into the calculations. Although M is a constant as well, its value is *not* known in advance. In fact, only 100 percent inspection will reveal the true value of M. Often we are interested in analyzing the behavior of the sampling plan for various values of M. The consumer's risk and the producer's risk depend upon the sampling plan. Finally, X, the number of defectives in the sample, is a random variable. This means that were we to repeat the sampling experiment with a different random sample of size n, we would not necessarily observe the same number of defectives. Based on statistical properties of the population as a whole, we can determine the form of the probability distribution of X.

The acceptable quality level, p_0, is the desired or target level of the proportion of defectives in the lot. If the true proportion of defectives in the lot is less than or equal to p_0, the lot is considered to be acceptable. The lot tolerance percent defective, p_1, is an unacceptable proportion of defectives in the lot. The lot is considered unacceptable if the proportion of defectives exceeds p_1. Because of the imprecision of statistical sampling, we allow a gray area between p_0 and p_1. When the AQL and the LTPD are equal, large sample sizes may be required to achieve acceptable values of α and β.

11.9 Single Sampling for Attributes

The goal of all sampling procedures is to estimate the properties of a population from the properties of the sample. In particular, we wish to test the hypotheses

H_0: Lot is of acceptable quality ($p \leq p_0$).
H_1: Lot is of unacceptable quality ($p \geq p_1$).

The test is of the form: Reject H_0 if $X > c$. The value of c depends on the choice of the Type 1 error probability α. The Type 1 error probability is the probability of rejecting H_0 when it is true. In the context of the quality control problem, this is the probability of rejecting the lot when it is acceptable. This is also known as the producer's risk. In equation form,

$$\alpha = P\{\text{Reject } H_0 \mid H_0 \text{ true}\} = P\{\text{Reject lot} \mid \text{Lot is good}\}$$
$$= P\{X > c \mid p = p_0\}.$$

The exact distribution of X is *hypergeometric* with parameters n, N, and M. That is,

$$P\{X = m\} = \frac{\binom{M}{m}\binom{N-M}{n-m}}{\binom{N}{n}} \quad \text{for } 0 \le m \le \min(M, n),$$

where

$$\binom{N}{n} = \frac{N!}{n!(N-n)!}.$$

In most applications, N is much larger than n, so the binomial approximation to the hypergeometric is satisfactory. In that case

$$P\{X = m\} = \binom{n}{m}p^m(1-p)^{n-m} \quad \text{for } 0 \le m \le n,$$

where $p = M/N$ is the true proportion of defectives in the lot.

Using the binomial approximation, the producer's risk and the consumer's risk are given by

$$\alpha = P\{X > c \mid p = p_0\} = \sum_{m=c+1}^{n} \binom{n}{m}p_0^m(1-p_0)^{n-m},$$

$$\beta = P\{X \le c \mid p = p_1\} = \sum_{m=0}^{c} \binom{n}{m}p_1^m(1-p_1)^{n-m}.$$

Most statistical tests require specification of the probability of Type 1 error, α. Values of α, n, and p_0 will determine a unique value of c, which can be found from tables of the cumulative binomial distribution. However, because the binomial is a discrete distribution, it may not be possible to find c to match exactly the desired value of α. When p is small and n is moderately large ($n > 25$ and $np < 5$), the Poisson distribution provides an adequate approximation to the binomial. For very large values of n such that $np(1 - p) > 5$, the normal distribution provides an adequate approximation to the binomial. Refer to Appendix 11–A for a detailed discussion of these approximations.

Example 11.6

Spire Records is a large West Coast retail chain of stores specializing in records and tapes. One of Spire's suppliers is B&G Records, which ships records to Spire in 100-record lots. After some negotiation, Spire and B&G have agreed that a 10 percent rate of defectives is acceptable and a 30 percent rate of defectives is unacceptable. From each lot of 100 records, Spire has established the following sampling plan: 10 records are sampled, and if more than 2 are found to be warped, scratched, or defective in some other way, the lot is rejected. Consider the consumer's and the producer's risk associated with this sampling plan.

From the given information, we have that $p_0 = .1, p_1 = .3, n = 10$, and $c = 2$. Hence,

$$\alpha = P\{X > c \mid p = p_0\} = P\{X > 2 \mid p = .1\} = 1 - P\{X \le 2 \mid p = .1\}$$

$$= 1 - \sum_{k=0}^{2} \binom{10}{k} (.1)^k (.9)^{10-k} = 1 - .9298 = .0702.$$

$$\beta = P\{X \le c \mid p = p_1\} = P\{X \le 2 \mid p = .3\}$$

$$= \sum_{k=0}^{2} \binom{10}{k} (.3)^k (.7)^{10-k} = .3828.$$

Note that the parameter values of $n = 10, p = .1$, and $n = 10, p = .3$ imply that neither the normal nor the Poisson approximation is accurate. (The reader may wish to check, using Table A–3 at the back of this book, that using the Poisson distribution with $\lambda = np$ gives α and β the approximate values of .0803 and .4216 respectively.) ∎

Derivation of the OC Curve

The operating characteristic (OC) curve measures the effectiveness of a test to screen lots of varying quality. The OC curve is a function of p, the true proportion of defectives in the lot, and is given by

$$OC(p) = P\{\text{Accepting the lot} \mid \text{True proportion of defectives} = p\}.$$

We will now derive the form of the OC curve for the particular case of a single sampling plan with sample size n and rejection level c. In that case,

$$OC(p) = P\{X \le c \mid \text{Proportion of defectives in lot} = p\}$$

$$= \sum_{k=0}^{c} \binom{n}{k} p^k (1 - p)^{n-k}.$$

Ideally, the sampling procedure would be able to distinguish perfectly between good and bad lots. Figure 11–13 shows the ideal OC curve.

FIGURE 11–13

The ideal OC curve

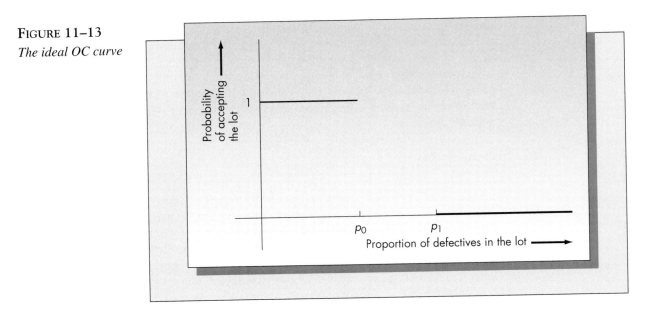

Example 11.6 (continued)

Consider again Example 11.6 of Spire Records. The OC curve for its single sampling plan is given by

$$OC(p) = \sum_{k=0}^{2} \binom{10}{k} p^k (1-p)^{10-k}.$$

The graph of Spire's OC curve appears in Figure 11–14. An examination of the figure shows that this particular sampling plan is more advantageous for the supplier, B&G, than it is for Spire. The value of $\beta = .3828$ means that Spire is passing almost 40 percent of the lots that contain 30 percent defectives. Furthermore, the probability of accepting lots with proportions of defectives as high as 40 percent and even 50 percent is not negligible. This accounts for Spire's experience that there seemed to be many customer returns of B&G label records.

Herman Sondle, an employee of Spire enrolled in a local Masters program, was asked to look into the problem with B&G records. He discovered the cause of the trouble by analysis of the OC curve pictured in Figure 11–14. In order to decrease the chances that Spire receives bad lots from B&G, he suggested that the sampling plan be modified by setting $c = 0$. The resulting consumer's risk is

$$\beta = P\{X \leq 0 \mid p = .3\} = (.3)^0 (.7)^{10} = .028,$$

or approximately 3 percent. This seemed to be an acceptable level of risk, so the firm instituted this policy. Unfortunately, the proportion of rejected batches *increased* dramatically. The resulting value of the producer's risk, α, is

$$\alpha = P\{X > 0 \mid p = .1\} = 1 - P\{X = 0 \mid p = .1\} = 1 - (.9)^{10} = .6513.$$

That is, about 65 percent of the good batches were being rejected by Spire under the new plan. B&G threatened to discontinue shipments to Spire unless it returned to its original sampling plan.

FIGURE 11–14

OC curve for Spire Records (n = 10)

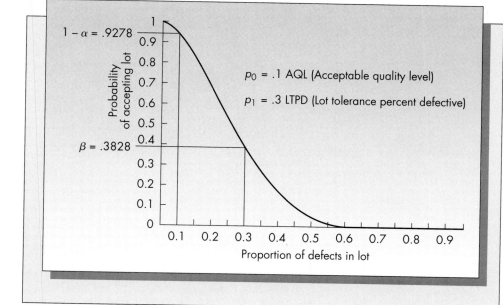

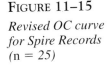

FIGURE 11–15

*Revised OC curve
for Spire Records
(n = 25)*

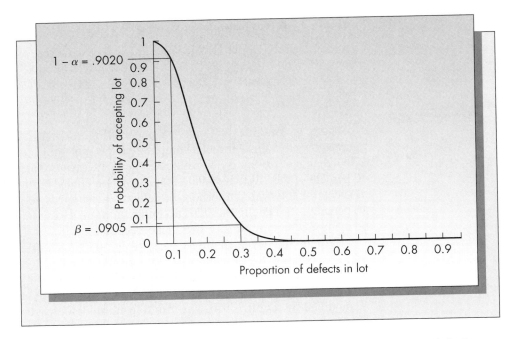

The Spire management didn't know what to do. If it returned to the original plan, it faced the risk of losing customers who would go elsewhere to purchase higher-quality records. If it continued with the current plan, it risked losing B&G as a supplier. Fortunately Sondle, who had been studying quality control methods, was able to propose a solution. If the sample size were increased, the power of the test would improve. Eventually, a test could be devised that would have acceptable levels of both the consumer's and the producer's risk. Because B&G insisted on no more than a 10 percent probability of rejecting good lots, Spire also wanted no more than a 10 percent probability of accepting bad lots.

After some experimentation, Herman found that a sample size of $n = 25$ with a rejection level of $c = 4$ seemed to meet the requirements of both B&G and Spire. The exact values of α and β for this test are

$$\alpha = P\{X > 4 \mid p = .1, n = 25\} = .0980.$$
$$\beta = P\{X \le 4 \mid p = .3, n = 25\} = .0905.$$

Of course, the improved efficiency of this plan did not come without some cost. The employee time required to inspect B&G records increased by two and a half times. B&G and Spire agreed to share the additional cost of the inspection. The OC curve for the sampling plan with $n = 25$ and $c = 4$ appears in Figure 11–15. Notice how much more closely this approximates the ideal curve than does the OC curve for the original plan pictured in Figure 11–14. ■

Problems for Section 9

29. Samples of size 20 are drawn from lots of 100 items and the lots are rejected if the number of defectives in the sample exceeds two. If the true proportion of defectives in the lot is 5 percent, determine the probability that a lot is accepted using
    a. The exact hypergeometric distribution.
    b. The binomial approximation to the hypergeometric.

c. The Poisson approximation to the binomial.

d. The normal approximation to the binomial.

30. A producer of pocket calculators purchases the main processor chips in lots of 1,000. They would like to have a 1 percent rate of defectives but will normally not refuse a lot unless it has 4 percent or more defectives. Samples of 50 are drawn from each lot, and the lot is rejected if more than two defectives are found.

a. What are p_0, p_1, n, and c for this problem?

b. Compute α and β. Use the Poisson approximation for your calculations.

31. A company employs the following sampling plan: It draws a sample of 10 percent of the lot being inspected. If 1 percent or less of the sample is defective, the lot is accepted. Otherwise the lot is rejected.

a. If a lot contains 500 items of which 10 are defective, what is the probability that the lot is accepted?

b. If a lot contains 1,000 items of which 20 are defective, what is the probability that the lot is accepted?

c. If a lot contains 10,000 items of which 200 are defective, what is the probability that the lot is accepted?

32. Hemispherical Conductor produces the 80J84 microprocessor, which the Sayle Company plans to use in a heart-lung machine. Because of the sensitivity of the application, Sayle has established a value of AQL of .001 and LTPD of .005. Sayle purchases the microprocessors in lots of 500 and tests 100 from each lot. The testing requires destruction of the microprocessor. The lot is rejected if any defectives are found.

a. What are the values of p_0, p_1, n, and c used?

b. Compute α and β.

c. In view of your answer to part (b), what problem could Sayle run into?

33. Determine a sampling plan for Spire Records that results in $\alpha = .05$, $\beta = .05$, AQL = .10, and LTPD = .30. Discuss the advantages and the disadvantages of the plan you obtain as compared to the current plan of $n = 25$ and $c = 4$. (Refer to Example 11.6.)

---

*11.10 Double Sampling Plans for Attributes

Five numbers define a double sampling plan: n_1, n_2, c_1, c_2, and c_3. The plan is implemented in the following way: One draws an initial sample of size n_1 and determines the number of defectives in the sample. If the number of defectives in the sample is less than or equal to c_1, the lot is accepted. If the number of defectives in the sample is larger than c_2, the lot is rejected. However, if the number of defectives is larger than c_1 but less than or equal to c_2, another sample of size n_2 is drawn. If the number of defectives in the combined samples is less than or equal to c_3, the lot is accepted. If not, the lot is rejected. Most double sampling plans assume that $c_3 = c_2$. We will make that assumption as well from this point on.

A double sampling plan is obviously more difficult to construct and more difficult to implement than a single sampling plan. However, it does have some advantages over single plans. First, a double sampling plan may give similar levels of the consumer's and the producer's risks but require less sampling in the long run than a single plan. Also, there is the psychological advantage in double sampling plans of providing a second chance before rejecting a lot.

Example 11.7

Consider again Example 11.6 concerning Spire Records. Herman Sondle decides to experiment with a few double sampling plans to see if he can achieve similar levels of efficiency with less sampling. Unfortunately, because the plan depends on four different numbers, considerable trial-and-error experimentation is necessary. Let us consider the computation of the consumer's and the producer's risks for the following sampling plan:

$$n_1 = 20, \quad c_1 = 3,$$
$$n_2 = 10, \quad c_2 = 5.$$

Define

X = Number of defectives observed in the first sample.
Y = Number of defectives observed in the second sample.
Z = Number of defectives observed in the combined samples ($Z = X + Y$).

The OC curve is

$$\text{OC}(p) = p\{\text{Lot is accepted} \mid p\} = P\{\text{Lot is accepted on first sample} \mid p\}$$
$$+ P\{\text{Lot is accepted on second sample} \mid p\}$$

where

$$P\{\text{Lot is accepted on the first sample} \mid p\} = P\{X \le 3 \mid p\}$$

and

$$P\{\text{Lot is accepted on the second sample} \mid p\}$$

$$= P\left\{\begin{array}{l} \text{Lot is neither accepted nor rejected on the} \\ \text{first sample and the lot is accepted on the} \\ \text{second sample} \end{array} \middle| p\right\}$$

$$= P\{3 < X \le 5, Z \le 5 \mid p\}.$$

Computation of this joint probability must be done carefully, as X and Z are *dependent* random variables.

Consider $p = \text{AQL} = .1$.

$$P\{\text{Lot is accepted on the first sample} \mid p = .1\}$$
$$= P\{X \le 3 \mid p = .1, n = 20\} = .8670,$$

$$P\{\text{Lot is accepted on the second sample} \mid p = .1\}$$
$$= P\{X = 4 \mid p = .1, n = 20\} P\{Y \le 1 \mid p = .1, n = 10\}$$
$$+ P\{X = 5 \mid p = .1, n = 20\} P\{Y \le 0 \mid p = .1, n = 10\}$$
$$= (.0898)(.7361) + (.0319)(.3487) = .0772.$$

Summing:

$$P\{\text{Lot is accepted} \mid p = .1\} = .8670 + .0772 = .9442.$$

Repeating similar calculations with $p = .3$ gives

$$P\{\text{Lot is accepted} \mid p = .3\}$$
$$= .1071 + (.1304)(.1493) + (.1789)(.0282)$$
$$= .1316.$$

Hence, it follows that for this case we obtain

$$\alpha = 1 - .9442 = .0558,$$
$$\beta = .1316. \quad \blacksquare$$

Experimentation with other values of n_1, n_2, c_1, and c_2 can lead to double sampling plans that more closely match the desired values of α and β. Tables are available for optimizing double sampling plans. (See, for example, Duncan, 1986, pp. 232–33.)

Problems for Section 10

34. Consider the double sampling plan for Spire Records presented in this section.
    a. Suppose that the true proportion of defectives in the lot is 10 percent. On average, how many items will have to be sampled before the lot is either accepted or rejected?
    b. Suppose that the true proportion of defectives in the lot is 30 percent. On average, how many items will have to be sampled before the lot is either accepted or rejected?

35. For the double sampling plan for Spire Records presented in this section, what is the probability that a lot is rejected on the first sample? Perform the computation for both $p = p_0$ and $p = p_1$.

36. Consider the double sampling plan for Spire Records described in this section. Over a period of one year, 3,860 boxes of records are subject to inspection using this plan. If 60 percent of these batches are "good" (that is, in 60 percent of the batches the proportion of defectives is exactly 10 percent) and 40 percent are "bad" (that is, in 40 percent of the batches the proportion of defectives is exactly 30 percent), then what is the expected number of batches
    a. Accepted?
    b. Rejected?
    c. Accepted on the first sample?
    d. Accepted on the second sample?
    e. Rejected on the first sample?
    f. Rejected on the second sample?

37. Graph the OC curve for the double sampling plan with $n_1 = 20$, $n_2 = 10$, $c_1 = 3$, and $c_2 = c_3 = 5$, as described in this section. If you are doing this by hand, evaluate the curve at $p = 0, .2, .4, .6, .8$, and 1 only. (Hint: The OC curve for this sampling plan has the form

$$\begin{aligned} \text{OC}(p) = \ &P\{X \le 3 \mid p, n = 20\} \\ &+ P\{X = 4 \mid p, n = 20\}\, P\{Y \le 1 \mid p, n = 10\} \\ &+ P\{X = 5 \mid p, n = 20\}\, P\{Y = 0 \mid p, n = 10\}.) \end{aligned}$$

38. By trial and error devise a double sampling plan for Spire Records that achieves $\alpha \approx .10$ and $\beta \approx .10$.

39. Consider the following double sampling plan. First select a sample of 5 from a lot of 100. If there are four or more defectives in the sample, reject the lot. If there is one or fewer defective, accept the lot. If there are two or three defectives, sample an additional five items and reject the lot if the combined number of defectives in both samples is five or more. If the lot has 10 defectives, what is the probability that a lot passes the inspection?

40. For the double sampling plan described in Problem 39, determine the following:
    a. The probability that the lot is rejected based on the first sample.
    b. The probability that the lot is rejected based on the second sample.
    c. The expected number of items sampled before the lot is accepted or rejected.

11.11 Sequential Sampling Plans

Double sampling plans may be extended to triple sampling plans, which also may be extended to higher-order plans. The logical conclusion of this process is the sequential sampling plan. In a sequential plan, items are sampled one at a time. After each sampling, two numbers are recorded: the number of items sampled and the cumulative number of defectives observed. Based on these numbers, one of three decisions is made: (1) accept the lot, (2) reject the lot, or (3) continue sampling. Unlike single and double sampling plans, there will always exist a sequential sampling plan that will give specific values of p_0, p_1, α, and β. Sequential sampling plans are defined by three regions: the acceptance region, the rejection region, and the sampling region. The three regions are separated by straight lines. The lines have the forms

$$L_1 = -h_1 + sn,$$
$$L_2 = h_2 + sn,$$

where n is the number of items sampled. Note that L_1 and L_2 are both linear functions of the variable n. The y intercepts are respectively $-h_1$ and h_2, and the slope of each line is s. As the lines have the same slope, they are parallel. The sequential sampling plan is implemented in the following manner: The cumulative number of defectives is graphed, together with the lines for L_1 and L_2. When the cumulative number of defectives exceeds L_2 the lot is rejected, and when the cumulative number of defectives falls below L_1 the lot is accepted. As long as the cumulative number of defectives lies between L_1 and L_2, sampling continues.

Figure 11–16 shows two examples of the results of sampling for the same sequential sampling plan. In Case A the sampling led to acceptance of the lot, and in Case B it led to rejection of the lot.

FIGURE 11–16

Two realizations of a sequential sampling plan

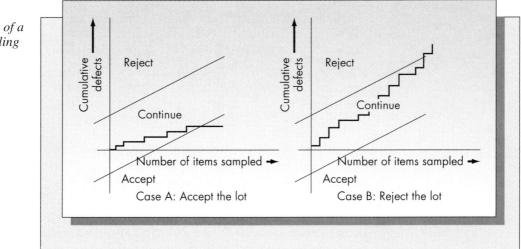

Case A: Accept the lot

Case B: Reject the lot

The equations for h_1, h_2, and s are

$$h_1 = \frac{\log \dfrac{1-\alpha}{\beta}}{\log \dfrac{p_1(1-p_0)}{p_0(1-p_1)}},$$

$$h_2 = \frac{\log \dfrac{1-\beta}{\alpha}}{\log \dfrac{p_1(1-p_0)}{p_0(1-p_1)}},$$

$$s = \frac{\log \dfrac{1-p_0}{1-p_1}}{\log \dfrac{p_1(1-p_0)}{p_0(1-p_1)}}.$$

Example 11.8

Consider again Example 11.6 of Spire Records. Herman Sondle has experimented with various sampling plans to achieve the desired levels of the consumer's and the producer's risks. He decides to construct a sequential sampling plan to see how it compares with the single and the double plans that are presented above. Spire and B&G agreed on the following values of the AQL, LTPD, consumer's risk, and producer's risk:

$$p_0 = .1, \quad \alpha = .1,$$
$$p_1 = .3, \quad \beta = .1.$$

Notice that the denominators in the expressions for h_1, h_2, and s are the same. We compute the denominator first. (We will use log to the base 10 in our calculations. Because all formulas involve the ratio of logarithms, the results will be the same whether one uses base 10 or base e.)

$$\log[(.3)(.9)/(.1)(.7)] = 0.58626.$$

Hence,

$$h_1 = \log(.9/.1)/0.58626 = 0.9542/0.58626 = 1.6277,$$
$$h_2 = h_1 = 1.6277 \quad \text{(since } \alpha = \beta \text{ for this case)},$$
$$s = \log(.9/.7)/0.58626 = 0.10914/0.58626 = 0.18617.$$

Figure 11–17 shows the decision regions for Spire Record's sequential sampling plan.

When Herman suggested that Spire Records implement the sequential sampling plan shown in Figure 11–17, he met with considerable resistance from some of his co-workers responsible for the inspection of incoming stock. With a single sampling plan with $n = 25$, they argued that at least they would know in advance how many records they would have to check. With the sequential plan, however, they argued that they might have to sample the entire lot without the plan recommending either acceptance or rejection. Although Herman had heard that

FIGURE 11–17

*Sequential sampling
plan for Spire
Records (refer to
Example 11.8)*

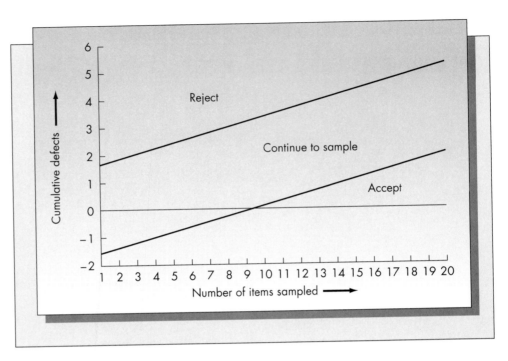

sequential plans were more efficient, he had difficulty convincing his co-workers to try the plan. ∎

The co-workers in the example were correct in seeing that the number of items sampled when using sequential sampling is a random variable. The *expected* sample size that results from a sequential sampling plan depends on the proportion of defectives in the lot, p. The ASN (average sample number) curve gives the expected sample size for a sequential sampling plan as a function of p. We will estimate the ASN curve by obtaining its value at five specific points: when $p = 0$, $p = p_0$, $p = s$, $p = p_1$, and $p = 1$. It is easy to find the ASN curve at these points. In most cases one can obtain an adequate approximation to the ASN curve knowing only its value at these five points. The five values are

$$\text{At } p = 0, \qquad \text{ASN} = \frac{h_1}{s}.$$

$$\text{At } p = p_0, \qquad \text{ASN} = \frac{(1 - \alpha)h_1 - \alpha h_2}{s - p_0}.$$

$$\text{At } p = s, \qquad \text{ASN} = \frac{h_1 h_2}{s(1 - s)}.$$

$$\text{At } p = p_1, \qquad \text{ASN} = \frac{(1 - \beta)h_2 - \beta h_1}{p_1 - s}.$$

$$\text{At } p = 1, \qquad \text{ASN} = \frac{h_2}{1 - s}.$$

Consider the application of these formulas to Spire Records:

$$\text{ASN}(0) = 1.6277/0.18617 = 8.74,$$

$$\text{ASN}(.1) = \frac{(.9)(1.6277) - (.1)(1.6277)}{0.18617 - .1} = 15.11,$$

$$\text{ASN}(.18617) = \frac{(1.6277)(1.6277)}{(0.18617)(1 - 0.18617)} = 17.49,$$

$$\text{ASN}(.3) = \frac{(.9)(1.6277) - (.1)(1.6277)}{(.3 - 0.18617)} = 11.44,$$

$$\text{ASN}(1) = \frac{1.6277}{(1 - 0.18617)} = 2.0.$$

The ASN curve is a unimodal curve whose maximum value lies between p_0 and p_1. Based on this and the five points computed above, we obtain the estimated ASN curve shown in Figure 11–18. We see from the figure that the expected sample size required for the sequential sampling plan for Spire Records will be at most 18 items. This is clearly an improvement over the single sampling plan with $n = 25$ and $c = 4$, which resulted in similar values of α and β. It is important to keep in mind, however, that the actual sample size required in the sequential plan for the inspection of any particular batch of records is a random variable. The ASN curve gives only the *expected* value of this random variable for any specified value of p. Thus, it is possible that in specific instances, the actual sample size could be larger than 18, or even larger than 25.

FIGURE 11–18

ASN curve for Spire Records (estimated)

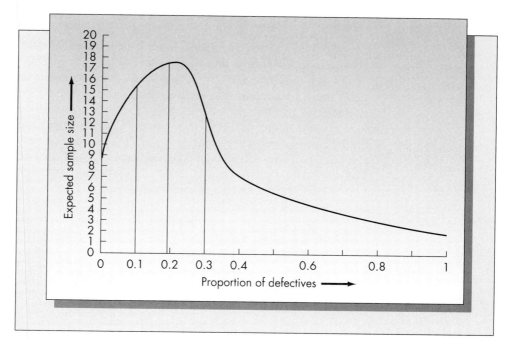

Problems for Section 11

41. A manufacturer of aircraft engines uses a sequential sampling plan to accept or reject incoming lots of microprocessors used in the engines. Assume an AQL of 1 percent and an LTPD of 5 percent. Determine a sequential sampling plan assuming $\alpha = .05, \beta = .10$. Graph the acceptance and rejection regions.

42. Consider the sequential sampling plan described in Problem 41. Suppose that a lot of 1,000 microprocessors is inspected. Suppose that the 31st, 89th, 121st, and 122nd chips tested are found defective. Assuming the sequential sampling plan derived in Problem 41, will the lot be accepted, rejected, or neither by the time the 122nd chip has been tested?

43. Estimate the ASN curve for the plan derived in Problem 41. According to your curve, what is the expected number of microprocessors that must be tested when the true proportion of defectives in the lot is
    a. 0.1 percent?
    b. 1.0 percent?
    c. 10 percent?

44. Consider the example of Hemispherical Conductor and the Sayle Company discussed in Problem 32. Devise a sequential sampling plan for Sayle that results in $\alpha = .05$ and $\beta = .20$. What are the advantages and disadvantages of this plan over the Sayle sampling plan derived in Problem 32?

45. Estimate the ASN curve for the sampling plan derived in Problem 44. On average, how many of the microprocessors would have to be tested if
    a. $p = .001$?
    b. $p = .005$?
    c. $p = .01$?

11.12 Average Outgoing Quality

The purpose of a sampling plan is to screen out lots of unacceptable quality. However, because sampling is a statistical process, it is possible that bad lots will be passed and good lots will be rejected. A fundamental issue related to the effectiveness of any sampling plan is to determine the quality of product that results *after* the inspection process is completed.

The calculation of the average outgoing quality of an inspection process depends on the assumption that one makes about lots that do not pass inspection and the assumption that one makes about defective items. Assume that rejected lots are subject to 100 percent inspection. We derive the average outgoing quality curve under two conditions: (1) defective items in samples and in rejected lots are not replaced and (2) defective items in samples and in rejected lots are replaced.

The average outgoing quality (AOQ) is the long-run ratio of the expected number of defectives and the expected number of items successfully passing inspection. That is,

$$\text{AOQ} = \frac{E\{\text{outgoing number of defectives}\}}{E\{\text{outgoing number of items}\}}.$$

The OC curve is the probability that a lot is accepted as a function of p. That is,

$$\text{OC}(p) = P\{\text{lot is accepted} \mid p\}.$$

For convenience we will refer to this term as P_a.

Case 1: Defective items are not replaced. Suppose that lots are of size N and samples of size n. Then the expected number of defectives and the expected number of items shipped are

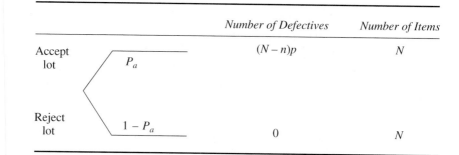

	Number of Defectives	Number of Items
Accept lot P_a	$(N-n)p$	$N-np$
Reject lot $1-P_a$	0	$N(1-p)$

From the tree diagram above we see that

$$E\{\text{outgoing number of defectives}\} = P_a(N-n)p + (1-P_a)(0)$$
$$= P_a(N-n)p$$

and

$$E\{\text{outgoing number of items}\} = P_a(N-np) + (1-P_a)N(1-p).$$

It follows that the ratio, AOQ, is given by

$$\text{AOQ} = \frac{P_a(N-n)p}{P_a(N-np) + (1-P_a)N(1-p)} = \frac{P_a(N-n)p}{N-np-p(1-P_a)(N-n)}.$$

When $N \gg n$ (N is much larger than n), which is a common assumption, this expression is approximately

$$\text{AOQ} \approx \frac{P_a p}{P_a + (1-P_a)(1-p)} = \frac{P_a p}{1 - p(1-P_a)}.$$

The formulas are somewhat simpler when defective items are replaced with good items.

Case 2: Defective items are replaced. In this case the tree diagram above becomes

	Number of Defectives	Number of Items
Accept lot P_a	$(N-n)p$	N
Reject lot $1-P_a$	0	N

The AOQ is given by

$$\text{AOQ} = \frac{P_a(N-n)p}{N},$$

which is approximately

$$\text{AOQ} \approx P_a p = \text{OC}(p)p$$

when $N \gg n$.

FIGURE 11–19
*AOQ curves for
Spire Records (refer
to Example 11.9)*

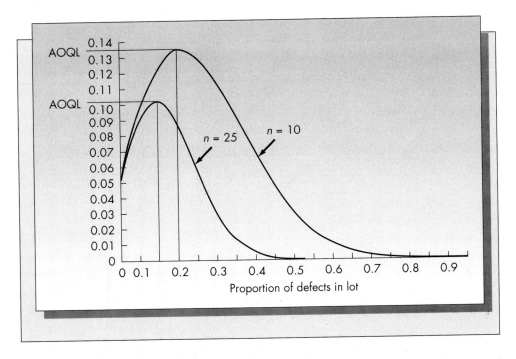

This last formula is the one most commonly used in practice, primarily because of its simplicity. An important measure of the effectiveness of a sampling plan is *the average outgoing quality limit* (AOQL), which is defined as the maximum value of the AOQ curve.

Example 11.9

Consider the case of Spire Records. In Spire's case we can assume that lots are large compared to samples and that all defectives are replaced. In that case AOQ $\approx$ OC(p)p. In Figure 11–19 we have generated the AOQ curves for Spire's single sampling plans with $n = 10$ and $n = 25$. Note how the larger sample size significantly improves the average outgoing quality. From Figure 11–19 we see that if Spire uses a single sampling plan with $n = 25$ and $c = 4$, the store can expect that the proportion of defective B&G records on its shelves will be no more than about 10.2 percent. ■

Problems for Section 12

46. If defective items are replaced and $N \gg n$, show by differential calculus that the value of p for which AOQ(p) achieves its maximum value satisfies

$$\frac{d\text{OC}(p)}{dp} = -\frac{\text{OC}(p)}{p}.$$

47. Consider the single sampling plan with $n = 10$ and $c = 0$.
    a. Derive an analytical expression for the OC curve as a function of p.
    b. Using the results of Problem 46 above, determine the value of p at which the AOQ curve is a maximum.

SNAPSHOT APPLICATION

Motorola Leads the Way with Six-Sigma Quality Programs

The Motorola Corporation has compiled an impressive record of defining and implementing new quality initiatives and translating those initiatives into profits. As we see in this chapter, traditional quality methods assume that an out-of-control condition corresponds to an observation falling outside of $\pm 2\sigma$ or $\pm 3\sigma$. Motorola decided that this standard was too loose, giving too many defects. In the 1980s, they moved toward a 6σ standard. To achieve this goal, Motorola established the practice of quality system reviews (QSRs) and placed a great deal of emphasis on classical statistical process control (SPC) techniques.

To achieve the 6σ standard, Motorola infused the total quality management philosophy into its entire organization. Quality programs were not assigned to and policed by a single group, but became part of everyone's job. Motorola's approach was based on the following key ingredients:

- Overriding objective of total customer satisfaction.
- Uniform quality metrics for all parts of the business.
- Consistent improvement expectations throughout the firm.
- Goal-directed incentive plans for management and employees.
- Coordinated training programs.

Motorola has carefully documented the road map to follow to its quality goals (Motorola, 1993). The first step in the process is a detailed audit of several key parts of the business. These include control of new product development, control of suppliers (both internal and external), monitoring of processes and equipment, human resource considerations, and assessing of customer satisfaction, among others. Even top management takes part in many of these audits by paying regular visits to customers, chairing meetings of the operating policy committees, and recognizing executives who have made outstanding contributions to the company's quality initiatives.

Kumar and Gupta (1993) report on their experience with a TQM program put in place in Motorola's Austin, Texas, assembly plant. In May of 1988, management began the process of implementing an SPC program at Austin. The process began by bringing in an outside consultant to design the program and assigning an internal coordinator to ultimately take over the duties of the consultant. To be sure that employees bought into the initiative, management organized participative problem-solving teams. Each team included a manufacturing manager, a group leader, operators from the two shifts, a representative from the QA Department, and an engineer. Austin had a total of six teams. To further ensure a buy-in from all employees, management initiated a plantwide training program in SPC. Training was tailored to job function.

The plant's QA Department instituted a certification program at Austin for vendors. As a result, about 60 percent of the vendors supplying the plant were certified. Within the plant, traditional SPC methods were employed: Attribute data were collected and charted and machines were shut down when out-of-control situations were detected. Members of the QA team employed design of experiments techniques to identify causes of problems.

What is the bottom line? Over the first two years of this initiative, the Austin plant reported a decrease in scrap rates of 56 percent. By developing a clear-cut and coordinated strategy, Motorola was able to achieve major improvements in traditional quality measures in this facility. Motorola's overall success is a testament to the fact that this was not an isolated example. It demonstrates that American companies can compete effectively with overseas competitors when the quality effort is a true companywide initiative.

    c. Using the results of parts (*a*) and (*b*), determine the maximum value of the average outgoing quality.

48. Consider the single sampling plan discussed in Problem 30. If defective items are replaced and $N \gg n$, graph the AOQ curve and determine the value of the AOQL.

11.13 Total Quality Management

This chapter reviewed the fundamentals of statistical quality control. Statistical quality control constitutes a set of techniques based on the theories of probability and statistical sampling for monitoring process variation and for determining if manufactured lots meet desired quality levels. However, delivering quality to the customer is a far broader problem than is addressed by statistical issues alone. This section considers quality from the management perspective.

Definitions

What is total quality management (TQM)? The term seems to have been first coined by Feigenbaum (1983) (in an earlier edition), who provided the following definition:

> Total quality control is an effective system for integrating the quality-development, quality-maintenance, and quality-improvement efforts of the various groups in an organization so as to enable marketing, engineering, production, and service at the most economical levels which allow for full customer satisfaction.

Feigenbaum's approach is to define quality in terms of the customer. As we noted in the introduction of this chapter, most definitions of quality concern either conformance to specifications or customer satisfaction. Garvin (1988) expands on these ideas and suggests that quality be considered along eight basic dimensions:

- Performance.
- Features.
- Reliability.
- Conformance.
- Durability.
- Serviceability.
- Aesthetics.
- Perceived quality.

We could lump the first five dimensions together under the general heading of conformance to requirements (the definition suggested by Crosby, 1979) and the last three under the heading of customer satisfaction (as suggested by Feigenbaum, 1983). However, by further breaking down these two categories, Garvin gives a better appreciation for the complexity of the quality issue.

Listening to the Customer

An important aspect of the process of designing quality products is giving people what they want. A perfectly designed and built coffee maker sold in a place where no one drinks coffee is, by definition, a failure. Hence, part of the process of delivering quality to the customer is knowing what the customer wants.

While listening to the customer is an important part of the manufacturing/design cycle link, it is generally more closely associated with marketing than with operations. Still, we are seeing the boundaries separating the functional areas of business becoming fuzzier. Manufacturing cannot operate in a vacuum. It must be part of the link with the customer.

Finding out what the customer wants and incorporating those wants into product design and manufacture is a multistep process. The steps of the process are

- Obtaining the data.
- Characterizing customer needs.
- Prioritizing customer needs.
- Linking needs to design.

There are several means for obtaining the raw data. Traditionally, customer opinion is solicited through interviews and surveys. There are many issues to be aware of when considering interviews with customers or potential customers. How many customer responses are enough? The right answer depends on several factors. How many market segments are there for the product? How many attributes are important? What methods will be used to interpret the results? Next, there is the question of how to solicit the information from the customer. Should one conduct interviews or surveys? The answer is unclear. Both have advantages. Interviews allow more open-ended responses, but the biases of the interviewer could slant the results. Both surveys and interviews depend on how questions are worded. For example, suppose Mr. Coffee is considering a new design for a coffee maker. A question like "What should the capacity of an automatic coffee maker be?" automatically assumes that the customer is concerned about capacity. The question "Do you prefer an 8- or a 12-cup coffee maker?" imposes even more assumptions (Dahan, 1995).

Focus groups are another popular technique for soliciting the voice of the customer. The focus group format has the advantage of being open-ended; the specific wording of questions is not as important as it is with surveys or interviews. However, focus groups have disadvantages. The moderator can affect the flow of the discussion. Also, participants with strong personalities are likely to dominate the group.

Once the database is developed, the customer needs and desires must be prioritized and grouped. Several methods are available for this. One that has received a great deal of attention in the marketing literature is conjoint analysis (due to Green and Rao, 1971). Conjoint analysis is a statistically based technique to estimate utilities for product attributes based on customer preference data.

Once attributes are determined and grouped, one needs to link those attributes to the design and manufacturing processes. This can be done with quality function deployment (QFD). With QFD, customer needs are related to the product attributes and/or aspects of the production process through a matrix. The user provides estimates of the correlation between attributes and needs in a "roof" portion of the matrix. The resulting figure looks a little like a house, hence the term *house of quality* to describe the resulting matrix (see Figure 11–20). QFD is used in conjunction with traditional methods such as surveys and focus groups. The strength of the correlations between customer needs and product attributes or design characteristics shows where the emphases should be placed when considering design of new products or design changes in existing products. The interested reader is referred to Cohen (1995) for an up-to-date comprehensive discussion of QFD methods.

This section only touched the surface of the issue of soliciting customer opinion and integrating that opinion into the process of product design and manufacture. However, one must be careful not to put *too* much emphasis on the voice of the customer. Real innovations do not come from the customer but from visionaries. By letting the customer be king, innovation could be stifled. According to Martin (1995):

> Still, more and more companies are learning that sometimes your customers can actually lead you astray. The danger lies in becoming a feedback fanatic, slavishly devoted to customers,

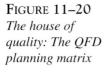

FIGURE 11–20
The house of quality: The QFD planning matrix

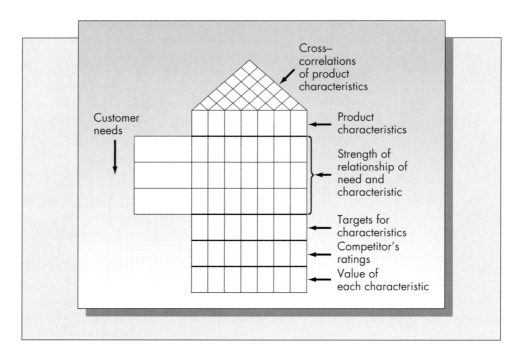

constantly trying to get in even better touch through more focus groups, more surveys. Not only can that distract you from the real work at hand, but it may also cause you to create new offerings that are safe and bland.

Home run new products like Post-it™ notes developed by 3M were not the result of marketing surveys. Many products that looked like winners from the market research flopped in the marketplace. One example is New Coke, which won hands down in taste tests but was never accepted by consumers. Of course, there are probably more examples of flops that resulted from not listening to the customer. The moral here is that while the customer is an important part of the equation, customer surveys and the like should not be used to the exclusion of ingenuity and vision.

Competition Based on Quality

Because quality can be defined in several ways, management must often choose along which dimensions of quality it will focus. The Cray Corporation developed a reputation for producing the fastest computers available. To do so, Cray put its energies into technology rather than reliability. The company's chairman was quoted as saying, "If a machine doesn't fail every month or so, it probably isn't properly optimized" (as quoted by T. Alexander, 1985). As Garvin (1988) notes, this approach allowed Cray to be a technological leader but left it vulnerable to competition with comparable performance and better reliability.

Other firms have chosen different dimensions of quality along which to compete. Tandem Computers' competitive edge is based on product reliability. With parallel processors, Tandem is able to guarantee essentially no downtime. Tandem's approach has been extremely successful because many customers, such as banks and utilities, are willing to pay a premium for improved reliability.

High reliability is also one of the primary factors for the success of the Japanese automakers. With a larger number of dual-career households, reliability is at a premium.

Based on the annual readers' survey of the Consumer's Union, the Japanese automakers consistently outperform their American and European competitors in this dimension of quality.

Banking is one industry that competes along the service dimension. The Wells Fargo Bank of California, for example, provides customers with a large network of branches, many automated tellers, and other services such as 24-hour account update information. Smaller banks attract customers with higher interest rates on savings, lower rates on credit card interest, and personalized service.

Some firms have based their strategies on providing the consumer with the highest-quality product, regardless of cost. Typical examples are Rolex watches, Leica cameras, Rolls-Royce automobiles, Cross pens, and Steinway pianos. Many American firms prefer to target their products to the mass market and leave the smaller high-end market for European and Japanese competitors.

Many successful companies rely on a business strategy based on being a market leader along one or two dimensions of quality. This may leave them vulnerable to competition, however.

Organizing for Quality

TQM requires total commitment on the part of management and workers. The organization must be structured to enhance product quality, not detract from it. Workers must be secure in their positions and have a stake in the success of the organization if quality programs are going to be successful. According to Feigenbaum (1983):

1. The quality function within a company should not be a single function housed within a single department. Quality should be recognized as a systematic group of disciplines to be applied on a coordinated basis by all functions throughout the company and the plant.

2. The quality function must have direct and ongoing contact with the buyers and the customers of the company's products and services.

3. The quality function must be organized to transcend individual functional organizational boundaries.

4. The overall quality function must be overseen from a high level of the firm as new products are developed, to be sure that quality issues are adequately dealt with, that "early warnings" of impending problems are easily recognized and corrected, and that management can properly manage quality.

What are the root causes of quality problems and what can be done to address those causes? Leonard and Sasser (1982) surveyed executives of 30 *Fortune* 500 firms to determine their opinions as to the causes of quality problems in the United States. The factor on the top of their list was workmanship and workforce. It is probably fair to assume that a similar survey of workers would point the finger at management. Management's policies determine workers' job security and working conditions. Reward structures determine incentives for behavior. For TQM to work, the entire organization must line up behind a quality imperative. To make this happen, management must be responsible for creating an organizational structure in which workers are empowered and in which workers have incentives to do their best, not one that stifles excellence.

A program that received a great deal of attention in the early 1980s was *quality circles*. Quality circles were an attempt to emulate the Japanese organizational structure. A quality circle is a small group of roughly 6 to 12 employees that meet on a regular basis with the purpose of identifying, analyzing, and suggesting solutions to problems affecting its work area. Typically, such a group would meet for about four hours per month on company time. The group might be given special training on statistical quality control methods, group dynamics, and general problem-solving techniques.

There was a dramatic increase in interest in quality circles in the United States starting around 1980. Lawler and Mohrman (1985) report that 44 percent of the companies in the United States with more than 500 employees had some type of quality circle program in place in 1982 and that 75 percent of these programs were started after 1980. There are several reasons for the sudden interest in quality circles in the United States:

1. *Cost.* Quality circles could be put into place relatively inexpensively. Consulting firms that specialize in the area would set up the program, train the appropriate individuals, and oversee the initial activities of the circle. From a cost perspective, an advantage of the quality circle is that not all company employees are necessarily involved.

2. *Control.* From a manager's point of view, a desirable feature of the quality circle is that it has no formal decision-making power, and thus is not perceived as a threat. The quality circle essentially serves the role of a formal "suggestion box" mechanism, and there is no requirement that the manager relinquish any management prerogatives in his or her area of responsibility.

3. *Fashion.* There is little doubt that quality circles have had an enormous fad appeal. The apparent success of programs of this type in Japanese industry has led to an early acceptance of QC in the United States.

With the interest in quality circles and the adoption of so many QC programs in the United States, what impact have these programs had? The experience, by and large, has been that few QC programs have evolved into other programs that could affect company procedures and practices. As a result, group interest in QC has tended to dwindle and the groups have met less frequently or not at all. Lawler and Mohrman (1985) suggest that the QC is inherently an unstable organizational structure. Because it has no decision-making power, its only role is to serve as a formal "suggestion box." It is natural that employees would eventually lose interest. Ultimately, quality circles could have served as a first step toward building a more participative approach to management, but that does not appear to have occurred.

Benchmarking Quality

Benchmarking means measuring one's performance against that of one's competitors. Competitive benchmarking is gaining importance in light of increasing global competition. Setting one's priorities and being certain that those priorities are consistent with the needs of the marketplace are essential. Based on a database developed at Boston University, Miller et al. (1992) report competitive priorities in Europe, Japan, and the United States. These are summarized in Table 11–7.

There are several interesting things to see from this table. First, consistent with our discussion above, the Japanese firms surveyed rank product reliability as their top priority. The Japanese understand that product reliability could be their greatest competitive asset and plan to continue to stress this important dimension of quality. It is also interesting to see that price is mentioned only by the U.S. companies.

TABLE 11–7   **Top Five Competitive Priorities**

Europe	Japan	United States
Conformance quality	Product reliability	Conformance quality
On-time delivery	On-time delivery	On-time delivery
Product reliability	Fast design change	Product reliability
Performance quality	Conformance quality	Performance quality
Delivery speed	Product customization	Price

Source: Miller et al., 1992.

The authors list four types of benchmarking:

1. Product benchmarking.
2. Functional or process benchmarking.
3. Best practices benchmarking.
4. Strategic benchmarking.

Product benchmarking refers to the practice of tearing down a competitor's product to see what can be learned from its design and construction. It is said that when Toyota initiated its program to produce the Lexus to compete with cars such as Mercedes and BMW, it carefully examined the competitor's products to determine how and where welds were placed, and how the cars were put together to achieve the look and feel of exceptional quality.

Functional benchmarking focuses on the process rather than on the product. Typical processes might be order entry, assembly, testing product development, and shipping. Functional benchmarking is possible only when companies are willing to cooperate and share information. It has the same goal as product benchmarking: to improve the process and ultimately the resultant product.

Best practices benchmarking is similar to functional benchmarking, except that it focuses on management practices rather than on specific processes. Best practices might consider factors such as the work environment and salary incentives for employees in firms with exceptional performance. General Electric is a strong advocate of best practices benchmarking (*Fortune,* 1991).

The goal of *strategic benchmarking* is to consider the results of other benchmarking comparisons in the light of the strategic focus of the firm. Specifically, what is the overall business strategy that has been articulated by the CEO, and are the results of other benchmarking studies consistent with this strategy?

Ultimately, what is the purpose of benchmarking? It is to ensure continuous improvement, and is only one of the means of achieving this. Continuous improvement in product and process is the ultimate goal of any quality program. Competitive benchmarking provides a means of learning from one's competitors. Although benchmarking can be a useful tool, it is not a substitute for a clearly articulated business strategy and a vision for the firm.

The Deming Prize and the Baldrige Award

The three leaders of the quality movement in the United States during the 1950s, W. Edwards Deming, Joseph M. Juran, and A. V. Feigenbaum, each contributed to the Japanese quality movement, although Deming is certainly the name that comes to mind first. Deming's success in Japan was the result of a seminar he presented on statistical quality control in 1950. He repeated that seminar several times, became active in the Japanese quality movement, and ultimately became a national hero in Japan. Deming recommended both the application of statistical methods and a systematic approach to solving quality problems. His approach later became known as the Plan, Do, Check, Act (PDCA) method.

Juran, another important leader in the quality movement, stressed the managerial aspects of quality rather than the statistical aspects. Juran also presented several seminars to the Japanese that were targeted at middle to upper management. The Juran Institute, located in Wilton, Connecticut, was founded in 1979 and continues to provide consulting and training in the quality area. [However, Philip Crosby Associates, founded the same year, generates almost 10 times the annual revenue (*Business Week*, 1991).]

Using the royalties from a book based on his 1950 lectures, Deming established a national quality prize in Japan. The funding for the prize continues to be primarily from Deming's royalties, but it is supplemented by a Japanese newspaper company and private donations. The Deming Prize is awarded to (1) those who have achieved excellence in research in the theory or application of statistical quality control, (2) those who have made remarkable contributions to the dissemination of statistical quality control methods, and (3) those corporations that have attained commendable results in the practice of statistical quality control.

The prize has been awarded almost every year in several categories since 1951. Recipients include such well-known firms as Toyota Motor Co., Hitachi, Fuji Film, Nippon Electric, NEC, and many others. Prizes also have been awarded to specific divisions of large firms and specific plants within a division. It is a highly sought-after national honor in Japan (Aguayo, 1990).

The Malcolm Baldrige National Quality Award was established by the U.S. Department of Commerce in 1987 largely as a response to success of the Deming Prize in Japan. The award is named for the late secretary of commerce, who died in an accident the same year that Congress and President Ronald Reagan enacted the Malcolm Baldrige National Quality Improvement Act.

The award is made each year in three categories: (1) manufacturing companies or subsidiaries, (2) service companies or subsidiaries, and (3) small businesses. Applying for the Baldrige Award is an involved process requiring a sincere commitment on the part of the firm seeking the award. A board of overseers appointed by the secretary of commerce has final say. They base their evaluation on the following nine-point value system:

1. Total quality management.
2. Human resource utilization.
3. Performance.
4. Measurables.
5. Customer satisfaction.
6. World-class quality.
7. Quality early in the process.
8. Innovation.
9. External leadership.

The evaluation process utilizes an elaborate scoring system and includes site visits from the evaluation committee. Funds to support the examining process are donated by individuals and firms. As of June 30, 1989, more than $10.4 million had been pledged to support the program (Pierce, 1991). The number of applicants grew each year for the first three years after the prize was announced. In 1989, 40 companies applied for awards; in 1990, 97 firms applied; in 1991, 106 applied; and in 1992, 90 firms applied. (There are several reasons for the drop in applications in 1992. One is that many firms are more concerned with ISO 9000 certification. Another is that the cost of applying is a problem for some companies. These costs are not inconsequential. For example, Xerox, a 1989 winner, spent $1 million preparing its application. Finally, being forced to lay off workers, many companies are not in a position to develop a credible application [Hillkirk, 1992].)

What has been the experience with the Baldrige Award now that more than 10 years have elapsed since its inception? One way of measuring the success of the program is to see how the firms that have won have fared. While not every Baldrige winner has been a winner in the marketplace, the evidence is that, on average, the firms seem to outperform the economy as a whole. According to an experiment conducted by the NIST (National Institute of Standards and Technology), five whole company winners—Eastman Chemical, Federal Express, Motorola, Solectron, and Zytec—have outperformed the S&P 500 by 6.5 to 1. The publicly traded parent companies of seven subsidiaries that won the award outperformed the S&P 500 by almost 3 to 1 (*Transportation and Distribution,* 1995). While not all Baldrige winners fared so well, these results, at least, are encouraging.

The Baldrige Award is not an end in itself but rather a means to an end. The process of preparing an application forces the firm to take a good hard look at its quality efforts and provides a blueprint for self-examination. The Baldrige application is an excellent means for ferreting out problems and suggesting where improvements need to be made.

ISO 9000

In 1992 the International Organization for Standards, based in Switzerland, established the guidelines for ISO 9000. This was the first attempt at developing a true uniform global standard for quality. For a firm to obtain ISO 9000 registration, they must carefully document their systems and procedures. Purchasing, materials handling, manufacturing, and distribution are all subject to ISO documentation and certification. ISO certification is very different from the Baldrige process. ISO certification is not an award, nor even necessarily a judgment about the quality of products. It is a certification of the manufacturing and business processes used by the firm.

With the establishment of the ISO standard came a host of consulting organizations specializing in taking a company through the certification process. Certification is not cheap. Aside from the direct cost of consultants, there are substantial indirect costs associated with developing the necessary documentation. According to John Rudin, president of Reynolds Aluminum Supply Company:

> From an out-of-pocket standpoint, just to get one facility registered gets to be somewhere in the $25,000 to $30,000 range. For RASCO as a whole, you're talking a million or so, if we were to do all the facilities. Right now we don't see the value in having a plaque to hang on the wall for a million dollars, but we do see the value in having a good, back-to-basics, well-documented process, which is what ISO is really all about anyway. (Kuster, 1995)

The firm seeking certification must document quality-related procedures, system and contract-related policies, and record keeping. According to Velury (1995), a firm should expect to document about 20 procedures for each section required for certification.

Given the expense in time and money, why should anyone bother? There are several advantages of achieving certification. One is from the process itself. Careful documentation of quality practices reveals where those practices fall short. The process of continuous improvement starts with knowing where one stands. Furthermore, many firms now require suppliers to obtain certification. It could be a prerequisite to doing business in many circumstances. According to Thomas Boldlund, president of Oregon Steel:

> ISO 9000 allows us to participate in markets that 10 years ago we could not serve even if we wanted to. When we meet a customer, the first thing they ask is "Can you meet the quality standards of your competition?" and second: "Can you produce the grades the Europeans and Japanese produce today?" With ISO 9000, the answer is yes. (Kuster, 1995)

The International Organization for Standards (ISO) continues to develop international standards. In 1996 the organization announced the ISO 14000 series of environmental standards (F. Alexander, 1996). ISO 14000 is a series of international standards for environmental management systems. The motivation for ISO in adopting a uniform environmental standard is the disparity of these standards in different countries. It is hoped that this new standard will not only spur international trade, but, more importantly, improve quality of life by improving the environment. Both ISO 9000 and ISO 14000 guidelines are important first steps in the movement towards a global uniform standard.

Quality: *The Bottom Line*

What ultimately determines whether a quality program is successful? It is the "bottom line." Whether the program is TQM, continuous improvement, or six sigma, the firm's management must believe that implementation will ultimately lead to higher profits. Traditional thinking is that quality initiatives are expensive. To achieve high reliability and conformance to specifications, substantial capital expenditures might be required (see Figure 11–1). Only if the expense translates to an improved bottom line will it be justified.

That quality costs more is not at all clear. In addition to erosion of the customer base, there are numerous direct costs of poor quality. These include costs of additional work-in-process inventories, scrap costs, rework costs, inspection costs, and so on. Firms face increased risks that design and/or reliability problems will lead to costly lawsuits and settlements.

Garvin's study of the room air conditioner industry from 1981 to 1982 (Garvin, 1988), summarized in Table 11–8, shows the dramatic increase in the costs of quality as product quality and reliability deteriorate.

Although not obvious from this table, the Japanese manufacturers as a group had much lower defect rates than the best U.S. manufacturers. (This is clearer in Garvin's 1983 study.) Before analyzing the results, we should point out some inconsistencies. First, the warranty period in the United States is five years and in Japan it is three years. This would tend to bias the results in favor of the Japanese, although the bias is small because most warranty costs are incurred in the first year. Second, the percentages in the third column include costs of prevention, inspection, rework, scrap, and warranties for the Japanese firms, and only rework, scrap, and warranties for the American firms. Hence, the comparable percentages in the third column for the American firms are even higher than those reported in the table.

What is clear from this study is that the costs of poor quality are substantial. In fact, they were observed to be as much as five times higher for poor U.S. manufacturers than for the Japanese manufacturers. This does not prove conclusively, however,

TABLE 11–8   **Quality and Quality Costs in the Room Air Conditioner Industry**

Grouping of Companies by Quality Performance	Average Warranty Costs as a Percentage of Sales	Total Cost of Quality (Japanese Companies), Total Failure Costs (U.S. Companies) as a Percentage of Sales
Japanese manufacturers	0.6%	1.3%
Best U.S. plants	1.8	2.8
Better U.S. plants	2.4	3.4
Fair U.S. plants	2.7	3.9
Poor U.S. plants	5.2	>5.8

Source: Garvin, 1988, p. 82.

that it is necessarily less expensive to produce higher-quality products. These data do not show the additional investments in capital equipment, processes, or statistical quality control made by the better manufacturers, and for that reason are not proof that "quality is free." However, other studies (see, in particular, Schoeffler et al., 1974) show that firms producing higher-quality products have higher return on investment, larger market share, and more profits than firms that produce lower-quality products.

11.14 Designing Quality into the Product

Traditional quality control methods focus on sampling and inspection. Sampling plans and inspection policies have the ultimate goal of producing an acceptable percentage of defects. Viewing quality in terms of the entire product cycle, including design, production, and consumption, shows that it is economical to design quality into the product. A recent development, applying statistical design of experiments to the problem of product design, has been developed by the Japanese statistician and consultant Genichi Taguchi. Taguchi's initial contributions were in *off-line* quality control methods. His later work incorporated economic issues as well.

Consider the following simple example from Taguchi et al. (1989). It is well known that, prior to World War II, the quality of manufactured goods in Japan was poor. At that time the Japanese sought to compete on price rather than on quality. Consider a Japanese product that sold for half the price of its American competitor. If one based a purchasing decision on price alone, one would have chosen the inferior Japanese product. However, factoring in consumption, we must account for the losses incurred to the customer in using that product. Suppose that the loss for using the American-made product was equal to the purchase, say P. Furthermore, assume that the loss for the Japanese-made product was nine times its purchase price, which we assume to be $0.5P$. Then the total cost to the customer for the American-made product would have been $P + P = 2P$ and the total cost to the customer for the Japanese-made product would have been $0.5P + (9)(0.5P) = 5P$. Hence, the Japanese-made product would have cost the consumer two and a half times as much as the American-made one. Given experience with both products, the consumer would eventually realize that his or her overall costs were greater with the less expensive product.

It is precisely this type of phenomenon that has led to maxims such as "you get what you pay for," and explains in simple economic terms why consumers are willing to pay more for quality.[4] The irony of the example above is how effectively the Japanese have managed to position themselves on the other side of the equation. It explains why Japanese cars continued to sell well in the United States even when the exchange rate between the dollar and the yen was so unfavorable to the Japanese.

Quality has significant economic value to the consumer and product design plays an important role in the product's quality. What does it mean to design for quality? It means that the number of parts that fail easily, or those that significantly complicate the manufacturing process, should be minimized. In particular, how can the design be simplified to eliminate small parts such as screws and latches that are difficult to assemble and may be likely trouble spots down the road? One example of a design in which the number of parts is reduced to a minimum is the IBM Proprinter. IBM developed an impressive video showing how easily the product can be assembled by hand. This simple design was an important factor in the product's reliability and success in the marketplace.

The design cycle is an important part of the quality chain. Taguchi et al. (1989) recommend the following three steps in the engineering design cycle:

1. *System design.* This is the basic prototype design that meets performance and tolerance specifications of the product. It includes selection of materials, parts, components, and system assembly.

2. *Parameter design.* After the system design is developed, the next step is optimization of the system parameters. Given a system design, there are generally several system parameters whose values need to be determined. A typical design parameter might be the gain for a transistor that is part of a circuit. One needs to find a functional relationship between the parameter and the measure of performance of the system to determine an optimal value of the parameter. In the example, the measure of performance might be the voltage output of the circuit. The goal is to find the parameter value that optimizes the performance measure. The Taguchi method considers this issue.

3. *Tolerance design.* The purpose of this step is to determine allowable ranges for the parameters whose values are optimized in step 2. Achieving the optimal value of a parameter may be very expensive, whereas a suboptimal value could give the desired quality at lower cost. The tolerance design step requires explicit evaluation of the costs associated with the system parameter values.

The same concepts can be applied to the design of the production process once the product design has been completed. The system design phase corresponds to the design of the actual manufacturing process. In the parameter design phase, one identifies parameters that affect the manufacturing process. Typical examples are temperature variation, raw material variation, and input voltage variation. In the tolerance design phase, one determines acceptable ranges for the parameters identified in phase 2.

The area of off-line quality control (as opposed to the subject of this chapter, which might be referred to as on-line quality control) involves techniques for achieving these

---

[4]However, the converse that more expensive products are necessarily superior is not always true. See Garvin (1988). p. 70, for a discussion of the correlation of quality and price.

three design objectives. Taguchi methods, based on the theory of design of experiments, give new approaches to solving these problems.

Optimizing a parameter value may not always be the overall optimal solution. In some cases, redesigning the product to be less sensitive to the parameter in question might be more economical. In this spirit, Kackar (1985) describes a Japanese tile manufacturer who solved the problem of sensitivity to temperature in this way. Quoting from the article.

> A Japanese ceramic tile manufacturer knew in 1953 that it is more costly to control causes of manufacturing variations than to make a process insensitive to these variations. The Ina Tile Company knew that an uneven temperature distribution in the kiln caused variation in the size of tiles. Since uneven temperature distribution was an assignable cause of variation, a process quality control approach would have been to devise methods for controlling the temperature distribution. This approach would have increased manufacturing cost. The company wanted to reduce the size variation without increasing cost. Therefore, instead of controlling temperature distribution they tried to find a tile formulation that reduced the effect of uneven temperature distribution on the uniformity of tiles. Through a designed experiment, the Ina Tile Company found a cost-effective method for reducing tile size variation caused by uneven temperature distribution in the kiln. The company found that increasing the content of lime in the tile formulation from 1 percent to 5 percent reduced the tile size variation by a factor of 10. This discovery was a breakthrough for the ceramic tile industry.

Taguchi's method is based on assuming a loss function, say $L(y)$, where y is the value of some functional characteristic and $L(y)$ is the quality loss measured in dollars. In keeping with much of the classical theory of statistics and control, Taguchi recommends a quadratic loss function. The quadratic form is the result of using the first two terms of a Taylor series expansion. Given an explicit form for the loss function, one can address such questions as the benefits of tightening tolerances and the value of various inspection policies. We refer the interested reader to Taguchi et al. (1989) and Logothetis and Wynn (1989) for a discussion of the general theory. Applications to specific industries are treated by Dehnad (1989).

Design, Manufacturing, and Quality

Quality starts with the product design and the way that design is integrated into the manufacturing process. The most creative design in the world is useless if it can't be manufactured economically into a reliable product. Successful linking of the design and the manufacturing processes is a hallmark of Japan's success in consumer products. The **design for manufacturability** (DFM) movement in the United States had its roots in Japanese manufacturing methods.

Boothroyd and Dewhurst (1989) and Boothroyd et al. (1994) were among the first to develop an effective scoring system for designs in terms of their ease of manufacturability. The two books summarize the methodology developed by these two engineers over a period of several years. The first book focuses on assembly efficiency. Assembly efficiency is the ratio of the theoretical minimum assembly time over an estimate of the actual assembly time based on the current product design. The later book deals with more general DFM issues including numbers of parts, types of parts, and types of fasteners. These rules, which are very detailed, recommend that simpler designs with fewer parts are preferred. Such designs lead to products that are easier and less expensive to manufacture, and are less likely to fail in use.

Ulrich and Eppinger (1995) recommend that designers keep track of design complexity via a scorecard approach. (See Example 11.10, below.) A scorecard provides a way to compare different designs objectively and a means of keeping track of the complexity of the manufacturing process for every product design.

Example 11.10

Scorecard of Manufacturing Complexity Example

Complexity Drivers	Revision 1	Revision 2
Number of new parts introduced	6	5
Number of new vendors introduced	3	2
Number of custom parts introduced	2	3
Number of new "major tools" introduced	2	2
Number of new production processes introduced	0	0
Total	13	12

Source: Ulrich and Eppinger, 1995.

This example is meant to be illustrative only. In practice, the team would have to decide on the relative importance of the drivers and apply suitable weights. Scorecards such as this force the design team to take a good hard look at the manufacturing consequences of their decisions. ■

An example of a successful DFM effort was the IBM Proprinter. The Proprinter was a dot matrix printer focused at the ever-expanding PC printer market dominated by the Japanese in the early 1980s. IBM developed a video showing someone assembling the printer by hand in a matter of minutes. In designing the Proprinter, IBM followed classic DFM methodology. The Proprinter had very few separate parts and virtually no screws and fasteners, without any compromise in functionality. The result was that IBM was able to assemble the Proprinter in the United States and remain cost competitive with Japanese rivals (Epson, in particular) that dominated the market at that time. The Proprinter was a very successful product for IBM.

The number of parts in a product is not the only measure of manufacturability. Exactly how parts are designed and put together also plays an important role. According to Boothroyd and Dewhurst (1989), the ideal characteristics of a part are

- Part should be inserted into the top of the assembly.
- Part is self-aligning.
- Part does not need to be oriented.
- Part requires only one hand for assembly.
- Part requires no tools.
- Part is assembled in a single, linear motion.
- Part is secured immediately upon insertion.

While there are some clear successes in applying DFM, the methodology has yet to gain universal acceptance. According to Boothroyd et al. (1994), the following reasons are the most common for not implementing DFM in the design phase:

1. *No time.* Designers are pushed to finish their designs quickly to minimize the design-to-manufacture time for a new product. The DFM approach is time intensive. Designing to reduce assembly costs and product complexity cannot be done haphazardly.

2. *Not invented here*. New ideas are always resisted. It would be better if the impetus for DFM came from the designers themselves, but more often it comes from management. Designers resent having a new approach thrust upon them by outsiders (as does anyone).

3. *Low assembly costs*. Since assembly costs often account for a small portion of total manufacturing costs, one might argue that there is little point to doing a DFA (design for assembly) analysis. However, savings often can be greater than one might think.

4. *Low volume*. One might argue that DFM analysis is not worthwhile for low-volume items. Boothroyd et al. (1994) argue that the opposite is true. When volumes are low, redesign is unlikely once production begins. This means that doing it right the first time is even more important.

5. *We already do it*. Many firms have used some simple rules of thumb for design (such as limiting the number of bends in a sheet metal part). While such rules make sense in isolation, they are unlikely to lead to the best overall design for the product.

6. *DFM leads to products that are difficult to service*. This is not likely to be true. Products that are easier to assemble are easier to disassemble, and thus easier to service.

Dvorak (1994) offers other reasons for the slow acceptance of DFM. Classical accounting systems may not be able to recognize the cost savings from new designs. A design that reduces fixed setup costs would not be viewed as cost effective since in many accounting systems fixed costs are considered part of overhead. An activity-based accounting system would not have this problem. However, most agree that the greatest obstacle to the acceptance of DFM is resistance to change.

Although the DFM movement may not be gaining acceptance at the rate that some would like, there is clearly a growing awareness and use of these powerful methods. Dvorak (1994) discusses several success stories of DFM implementation. One is at Coors, where production yield, quality, and delivery reliability were improved. According to Dvorak:

> Word is spreading throughout industry about how DFM can bring successes similar to those at Coors. Numerous organizations are taking up the DFM banner. The U.S. Department of Commerce, for one, has started DFM projects at six of their regional manufacturing centers. General Motors has relied on DFM to power its concurrent engineering efforts with startling response: It's saving 20% of the total car cost of the 1992 models on which it was applied. DFM and designing-for-assembly concepts have already worked wonders trimming material and manufacturing costs from a range of products. Now new ideas are infiltrating the discipline.

International competition continues to heat up. The Japanese, in particular, have demonstrated that one can be successful by careful analysis and thinking throughout the product design and development phase. They have been adept at concurrent product and process design. The result is products that can be manufactured more efficiently and work better than the competitors'. While DFM is only one piece of the pie, it provides a methodology for linking design and manufacturing.

Finally, we should not forget that product design extends far beyond issues of manufacturability only. Aesthetic issues are important as well, and may be the dominant factor for some products. Another important issue is the process of narrowing down the field of choices to a final design. Since the interest in this book is in manufacturing-related issues, we will not discuss these broader design issues but refer the interested reader to Pugh (1991).

11.15 Historical Notes

The desire to maintain quality of manufactured goods is far from new. The prehistoric man whose weapons and tools did not function did not survive. However, the statistical quality control methods discussed in this chapter were devised only in the last 70 years or so. Walter Shewhart, who was an employee of the Bell Telephone Laboratories, conceived the idea of the control chart. His 1931 monograph (Shewhart, 1931) summarized his contributions in this area. H. F. Dodge and H. G. Romig, also employees of Bell Labs, are generally credited with laying the foundations of the theory of acceptance sampling.

As with most new methodology, American industry was slow to adopt statistical quality control techniques. However, as part of the war effort, the U.S. government decided to adopt sampling inspection methods for army ordnance in 1942. As a result of this and other wartime governmental activities, knowledge and acceptance of the techniques discussed in this chapter became widespread after the war. In fact, sampling plans are often described in terms of the U.S. Army designation. Many of the acceptance sampling techniques discussed in this chapter are part of Military Standard 105. Military Standard 105 and its various revisions and additions form the basis for most of the acceptance sampling plans in use today.

Abraham Wald was responsible for developing the theory of sequential analysis, which forms the basis for the formulas that appear in Section 11. Wald's work proved to be a major milestone in the advance of the theory of acceptance sampling. Wald was part of a research team organized at Columbia University in 1942. The U.S. government considered his work to be so significant that they withheld publication until June of 1945.

W. Edwards Deming, who visited Japan in the early 1950s to deliver a series of lectures on quality control methods, is given much of the credit for transferring statistical quality control technology to Japan. Today, the highly prestigious Deming Prize in Quality Control, established by Deming in 1951, remains a symbol of the Japanese commitment to quality.

Most of the methods discussed in this chapter are in widespread use throughout industry in the United States and overseas. Optimization models for designing control charts have not enjoyed the same level of acceptance, however. The model outlined in Section 7 is from Baker (1971), although more complex and comprehensive economic models for design of $\overline{X}$ charts were developed earlier (see, for example, Duncan, 1956). For a comprehensive discussion of the history of the quality movement in the United States, see Kolesar (1993).

11.16 Summary

In this chapter we outlined the techniques for control chart design and acceptance sampling. The basis for the methodology of this chapter is classical probability and statistics. The underpinnings of control charts include fundamental results from probability theory such as the law of large numbers and the central limit theorem.

We discussed *control charts* in the first half of this chapter. A control chart is a graphical device that is used to determine when a shift in the value of a parameter of the underlying distribution of some measurable quantity has occurred. When this happens, the process is said to have gone out of control. The design of the control chart depends on the particular parameter that is being monitored.

The most common control chart is the $\overline{X}$ chart. An $\overline{X}$ chart is designed to monitor a single measurable variable, such as weight or length. Subgroups of size n are sampled on a regular basis and the sample mean of the subgroup is computed and placed on a graph. Because the sample mean is approximately normally distributed independently of the form of the distribution of the population, the likelihood that a single observation falls outside three-sigma limits is sufficiently small that when such an event occurs it is unlikely to be due to chance. Rather, it is more likely to be the result of a shift in the true mean of the process.

The second type of control chart treated in this chapter is the R chart. An R chart is designed to determine when a shift in the process variation has occurred. The symbol R stands for range. The range is the difference between the largest and the smallest observations in a subgroup. Because a close relationship exists between the value of the range of the sample and the underlying population variance, R charts are used to measure the stability of the variance of a process.

Often control charts for variables are inappropriate. The p chart is a control chart for attributes. When using p charts, items are classified as either acceptable or not. The p chart utilizes the normal approximation of the binomial distribution and may be used when subgroups are of equal or of varying sizes.

The last control chart presented was the c chart. The c chart is used to monitor the number of defectives in a unit of production. The parameter c is the average number or rate of defects per unit of production. The c chart is based on the Poisson distribution, and the control limits are derived using the normal approximation to the Poisson.

A model for the economic design of $\overline{X}$ charts was considered in Section 6. The decision variables for the model were the size of each sample subgroup, n, and the number of standard deviations used to signal an out-of-control condition, k. The model is based on the assumption that the process goes out of control randomly with a known probability π. Also assumed known are the cost of sampling, the cost of searching for an assignable cause when an out-of-control signal occurs, and the cost of operating the system in an out-of-control condition.

The chapter also considered *acceptance sampling*. The purpose of an acceptance sampling scheme is to determine if the proportion of defectives in a large lot of items is acceptable based on the results of sampling a relatively small number of items from the lot. The simplest sampling plan is single sampling. A *single sampling plan* is specified by two numbers: n and c. Interpret n as the size of the sample and c as the acceptance level. A *double sampling plan* requires specification of five numbers: n_1, n_2, c_1, c_2, and c_3, although often $c_2 = c_3$. Based on the results of an initial sample size of n_1, the lot is accepted or rejected, or an additional sample of size n_2 is drawn. The logical extension of double sampling is *sequential sampling,* in which items are sampled one at a time and a decision is made after each item is sampled about whether to accept the lot, reject the lot, or continue sampling.

Total quality management is a term that we hear more frequently as U.S. firms strive to compete with their European and Japanese competitors. We discussed Garvin's eight dimensions of quality: (1) performance, (2) features, (3) reliability, (4) conformance, (5) durability, (6) serviceability, (7) aesthetics, and (8) perceived quality. Methods for eliciting the voice of the customer such as *conjoint analysis* and *quality function deployment* (QFD) were discussed as well. In order for TQM to succeed, the quality activity must transcend functional and departmental boundaries. One approach that attempted to do this was *quality circles*. The program required minimal investment and restructuring on management's part, and as a result has not been very successful. Benchmarking provides a means for a firm to compare its performance with its competitors and learn the industry's "best practices." Two highly sought-after national prizes are the Deming Prize in Japan and the Baldrige Award in the United States. These awards recognize exceptional industry efforts in implementing quality.

Off-line quality methods are directed at the problem of designing quality into the product. Taguchi methods, largely based on the theory of design of experiments, are an important development in this area. The Taguchi methods identify important process and design parameters and attempt to find overall optimum values of these parameters, relative to some measure of performance of the system. The chapter concluded with a discussion of design for manufacturability and the contributions of Boothroyd and Dewhurst to this area.

Additional Problems on Quality and Assurance

49. In what ways could each of the factors listed below contribute to poor quality?
    a. Management.
    b. Labor.
    c. Equipment maintenance.
    d. Equipment design.
    e. Control and monitoring.
    f. Product design.

50. Figure 11–1 presents a conceptual picture of the trade-off between process cost and the costs of losses due to poor quality. What are the costs of poor quality and what difficulties might arise when attempting to measure these costs?

51. $\overline{X}$ and R charts are maintained on a single quality dimension. A sudden shift in the process occurs, causing the process mean to increase by 2σ, where σ is the true process standard deviation. No shift in the process variation occurs. Assuming that the $\overline{X}$ chart is based on 3σ limits and subgroups of size $n = 6$, what proportion of the points on the $\overline{X}$ chart would you expect to fall outside the limits after the shift occurs?

52. XYZ produces bearings for bicycle wheels and monitors the process with an $\overline{X}$ chart for the diameter of the bearings. The $\overline{X}$ chart is based on subgroups of size four. The target value is 0.37 inch, and the upper and the lower limits are 0.35 and 0.39 inch, respectively (assume that these are based on three-sigma limits). Wheeler, which purchases the bearings from XYZ to construct the wheels, requires tolerances of 0.39 ± 0.035 inch. Oversized bearings are ground down and undersized bearings are scrapped. What proportion of the bearings does Wheeler have to grind and what proportion must it scrap? Can you suggest what Wheeler should do to reduce the proportion of bearings that it must scrap?

53. A process that is in statistical control has an estimated mean value of 180 and an estimated standard deviation of 26.
    a. Based on subgroups of size four, what are the control limits for the $\overline{X}$ and R charts?

b. Suppose that a shift in the mean occurs so that the new value of the mean is 162. What is the probability that the shift is detected in the first subgroup after the shift occurs?

c. On average, how many subgroups would need to be sampled after the shift occurred before it was detected?

54. Consider the data presented in Table 11–2 for the tracking arm example. Suppose that an R chart is constructed based on sample numbers 1 to 15 only.
    a. What is the estimate of σ obtained from these 15 observations only?
    b. What are the values of the UCL and the LCL for an R chart based on these observations only?

55. Discuss the advantages and disadvantages of the following strategies in control chart design. In particular, what are the economic trade-offs attendant to each strategy?
    a. Choosing a very small value of α.
    b. Choosing a very small value of β.
    c. Choosing a large value of n.
    d. Choosing a small value for the sampling interval.

56. The construction of the p chart described in this chapter requires specification of both UCL and LCL levels. Is the LCL meaningful in this context? In particular, what does it mean when an observed value of p is less than the LCL? If only the UCL is used to signal an out-of-control condition, should the calculation of the UCL be modified in any way? (Hint: The definition of the Type 1 error probability, α, will be different. The hypothesis test for determining if the process is in control is one-sided rather than two-sided.)

57. A manufacturer of large appliances maintains a p chart for the production of washing machines. The machines may be rejected because of cosmetic or functional defectives. Based on sampling 30 machines each day for 50 consecutive days, the current estimate of p is .0855.
    a. What are the control limits for a p chart? (Assume 3σ limits.)
    b. Suppose that the percentage of defective washing machines increases to 20 percent. What is the probability that this shift is detected on the first day after it occurs?
    c. On average, how many days would be required to detect the shift?

58. A maker of personal computers, Noname, purchases 64K DRAM chips from two different manufacturers, A and B. Noname uses the following sampling plan: A sample of 10 percent of the chips is drawn and the lot is rejected if two or more defective chips are discovered. The two manufacturers supply the chips in lots of 100 and 1,000, respectively.
    a. For each manufacturer, determine the true proportion of defectives in the lot that would result in 90 percent of the lots being accepted. You may use the Poisson approximation for your calculations.
    b. Would you say that this plan is fair?

59. Graph the AOQ curves for manufacturers A and B mentioned in Problem 58. Estimate the values of the AOQL in each case.

60. Consider the sampling plan discussed in Problem 58 above. Would a fairer plan be to reject the lot if more than 10 percent of the chips in a sample are defective? Which of the two manufacturers mentioned above would be at an advantage if this plan were adopted?

61. Assuming AQL = 5 percent and LTPD = 10 percent, determine the values of α and β for the plan described in Problem 58 for manufacturers A and B, and for the plan described in Problem 60 for manufacturers A and B.

62. Graph the OC curves for the sampling plan described in Problem 58 for both manufacturers A and B.

63. $\overline{X}$ control charts are used to maintain control of the manufacture of the cases used to house a generic brand of personal computer. Separate charts are maintained for length, width, and height. The length chart has UCL = 20.5 inches, LCL = 19.5 inches, and a target value of 20 inches. This chart is based on using subgroups of size four and three-sigma limits. However, the customer's specifications require that the target length should be 19.75 inches with a tolerance of ±0.75 inch. What percentage of the cases shipped will fall outside the customer's specifications?

64. A p chart is used to monitor the fraction defective of an integrated circuit to be used in a commercial pacemaker. A sample of 15 circuits is taken from each day's production for 30 consecutive working days. A total of 17 defectives are discovered during this period.
   a. Determine the three-sigma control limits for this process.
   b. Suppose that α, the probability of Type 1 error (that is, the probability of drawing the conclusion that the process is out of control when it is in control), is set to be .05. What control limits do you now obtain? (Use a normal approximation for your calculations.)

65. A single sampling plan is used to determine the acceptability of shipments of a bearing assembly used in the manufacture of skateboards. For lots of 500 bearings, samples of $n = 20$ are taken. The lot is rejected if any defectives are found in the sample.
   a. Suppose that AQL = .01 and LTPD = .10. Find α and β.
   b. Is this plan more advantageous for the consumer or the producer?

66. A double sampling plan is constructed as follows. From a lot of 200 items, a sample of 10 items is drawn. If there are zero defectives, the lot is accepted. If there are two or more defectives, the lot is rejected. If there is exactly one defective, a second sample of 10 items is drawn. If the combined number of defectives in both samples is two or less, the lot is accepted; otherwise it is rejected. If the lot has 10 percent defectives, what is the probability that it is accepted?

67. Hammerhead produces heavy-duty nails, which are purchased by Modulo, a maker of prefabricated housing. Modulo buys the nails in lots of 10,000 and subjects a sample to destructive testing to determine the acceptability of the lot. Modulo has established an AQL of 1 percent and an LTPD of 10 percent.
   a. Assuming a single sampling plan with $n = 100$ and $c = 2$, find α and β.
   b. Derive the sequential sampling plan that achieves the same values of α and β as the single sampling plan derived in part (a).
   c. By estimating the ASN curve, find the maximum value of the expected sample size Modulo will require if it uses the sequential plan derived in part (b).
   d. Suppose that the sequential sampling plan derived in part (b) is used. One hundred nails are tested with the following result: The first 80 are acceptable,

the 81st is defective, and the remaining 19 are acceptable. By graphing the acceptance and the rejection regions, determine whether the sequential sampling plan derived in part (*b*) would recommend acceptance or rejection on or before testing the 100th nail.

68. For the single sampling plan derived in part (*a*) of Problem 67, suppose lots that are not passed are returned to Hammerhead.

    a. Estimate the graph of the AOQ curve by computing AOQ(*p*) for various values of *p*.

    b. Using the results of part (*a*), estimate the maximum proportion of defective nails that Modulo will be using in its construction.

69. Twenty sets of four measurements of the diameters in inches of Hot Shot golf balls were

Sample				
1	2.13	2.18	2.05	1.96
2	2.08	2.10	2.02	2.20
3	1.93	1.98	2.03	2.06
4	2.01	1.94	1.91	1.99
5	2.00	1.90	2.14	2.04
6	1.92	1.95	2.02	2.05
7	2.00	1.94	2.00	1.90
8	1.93	2.02	2.04	2.09
9	1.87	2.13	1.90	1.92
10	1.89	2.14	2.16	2.10
11	1.93	1.87	1.94	1.99
12	1.86	1.89	2.07	2.06
13	2.04	2.09	2.03	2.09
14	2.15	2.02	2.11	2.04
15	1.96	1.99	1.94	1.98
16	2.03	2.06	2.09	2.02
17	1.95	1.99	1.87	1.92
18	2.05	2.03	2.06	2.04
19	2.12	2.02	1.97	1.95
20	2.03	2.01	2.04	2.02

    a. Enter the data into a spreadsheet and compute the means and the ranges for each sample.

    b. Using the results of (*a*), develop $\overline{X}$ and R charts similar to Figures 11–7 and 11–8. Assume three-sigma limits for the $\overline{X}$ chart.

    c. Develop a histogram based on the 80 observations. Assume class intervals (1) 1.80–1.849, (2) 1.85–1.899, (3) 1.90–1.949, (4) 1.95–1.999, (5) 2.0–2.049, (6) 2.05–2.099, (7) 2.10–2.149, (8) 2.15–2.20. Based on your histogram, what distribution might accurately describe the diameter of a golf ball selected at random?

70. A *p* chart is used to monitor the number of riding lawn mowers produced. The numbers that are sent back for rework because they did not pass inspection are

Day	Number Produced	Number Rejected
1	400	23
2	480	18
3	475	24
4	525	34
5	455	17
6	385	17
7	372	12
8	358	19
9	395	24
10	405	29
11	385	16
12	376	19
13	395	23
14	405	14
15	415	25
16	440	34
17	380	26
18	318	19

Enter the data into a spreadsheet and compute standardized Z values for a p chart. Graph the Z values. Is this process in control?

71. Samples of size 50 are drawn from lots of 1,000 items. The lot is rejected if there are more than two defectives in the sample. Using a binomial approximation, graph the OC curve as a function of p, the proportion of defectives in the lot. For an AQL of .01 and an LTPD of .10, find α and β.

    a. Graph the OC curve and identify the Type 1 and the Type 2 error probabilities (that is, develop a graph similar to Figure 11–14).

    b. Graph the AOQ curve and identify the value of the AOQL.

72. *a.* Develop a spreadsheet from which one may obtain a graph such as Figure 11–17 for sequential sampling. Store the values of p_0, p_1, α, and β in cell locations so that these can be altered at will. Print a graph for $p_0 = .05$, $p_1 = .20$, $\alpha = .05$, and $\beta = .10$. Allow for $n \leq 100$.

    b. Sequential sampling resulted in the following: the first 40 items were good, item 41 was defective, item 68 was defective, and items 86 and 87 were defective. Place these results on the graph you obtained in part (*a*). Is the lot accepted, rejected, or neither on or before testing the 87th item?

APPENDIX 11–A   APPROXIMATING DISTRIBUTIONS

Several probability approximations were used in this chapter. This appendix will discuss the motivation and justification for these approximations.

The complexity of a probability distribution depends upon the number of parameters that are required to specify it. The distributions considered in this chapter in descending order of complexity are

Distribution	Parameters
1. Hypergeometric	n, N, M
2. Binomial	n, p
3. Poisson	λ
4. Normal	μ, σ

It is not clear from this chart why the normal distribution should be simpler than the Poisson, since the normal is a two-parameter distribution and the Poisson one. The reason is that all normal probabilities can be obtained from a single table of the standard normal distribution, and the Poisson distribution must be tabled separately for distinct values of λ.

The binomial approximation to the hypergeometric. The hypergeometric distribution (whose formula appears in Section 9) is the probability that if n items are drawn from a lot of N items of which M are defective, then there are exactly m defectives in the sample. The experiment that gives rise to the hypergeometric may be thought of as sampling the items one by one without replacement. If N is much larger than n, the probability that any item sampled is defective is very close to M/N. In that case the hypergeometric probability would be close to the binomial probability with $p = M/N$ and $n = n$. Note that the binomial distribution corresponds to sampling *with* replacement. If $N > 10n$, the binomial should provide an adequate approximation to the hypergeometric.

The Poisson approximation to the binomial. The Poisson distribution can be derived as the limit of the binomial as $n \to \infty$ and $p \to 0$, but with the product np remaining constant. Write $\lambda = np$. Then for large n and small p,

$$p\{X = m\} \approx \frac{e^{-\lambda}\lambda^m}{m!} \qquad \text{for } m = 0, 1, 2, \ldots.$$

It is not obvious under what circumstances this approximation is adequate. In general, $p < .1$ and $n > 25$ should hold, but if p is very small, then smaller values of n are acceptable, and if n is very large, then larger values of p are acceptable. For example, for $n = 10$ and $p = .01$, the binomial probability that $X = 1$ is .0914 and the Poisson probability is .0905. Values of p close to 1, such as $p = .99$, also would be acceptable because the binomial distribution with $p = .01$ is a mirror image of a binomial distribution with $p = .99$.

Normal approximations. The central limit theorem says (roughly) that the distribution of a sum of n independent identically distributed random variables approaches the normal distribution as n grows large. Because the binomial distribution is derived as the sum of n independent identically distributed Bernoulli random variables, when n is large the normal gives a good approximation to the binomial. As the normal approximation is more accurate when p is near .5, a good rule of thumb is that the approximation should be used only if $np(1 - p) > 5$.

Whenever the normal distribution is used to approximate any other distribution, it is necessary to express μ and σ in terms of the original parameters. In the binomial case, $\mu = np$ and $\sigma = \sqrt{np(1 - p)}$.

Because the normal random variable is continuous and the binomial random variable is discrete, the approximation can be improved by using the "continuity correction." In the binomial case, the events $\{X > 2\}$ and $\{X \geq 3\}$ are identical, but in the normal case they are not. The continuity correction would suggest approximating either of these cases by $\{X > 2.5\}$. The general rule is to express the original event in terms of both $>$ and $\geq$ (or $<$ and $\leq$) and to use a cutoff number halfway between the two.

For example, suppose that $n = 25$, $p = .40$, and we wish to determine $P\{X \leq 10\}$. Since $\{X \leq 10\} = \{X < 11\}$, the continuity correction cutoff is at 10.5. The exact binomial probability is .5858. The normal approximation at 10 gives

$$P\{X \leq 10\} \approx P\left\{Z < \frac{10 - (25)(.40)}{\sqrt{(25)(.40)(.60)}}\right\} = P\{Z < 0\} = 0.5,$$

and with the continuity correction

$$P\{X \leq 10\} \approx P\left\{Z < \frac{10.5 - (25)(.40)}{\sqrt{(5)(.40)(.60)}}\right\} = P\{Z < .2041\} = 0.5948.$$

The normal distribution also may be used to approximate the Poisson when λ is large ($\lambda > 10$). In that case, use $\mu = \lambda$, $\sigma = \sqrt{\lambda}$, and the continuity correction as described above. For example, suppose that we wish to use a normal approximation of the probability that a Poisson random variable with parameter $\lambda = 15$ exceeds 8. Since $\{X > 8\} = \{X \geq 9\}$, the continuity correction cutoff falls at 8.5. Hence,

$$P\{X > 8\} \approx P\left\{Z > \frac{8.5 - 15}{\sqrt{15}}\right\} = P\{Z > -1.68\} = .9535.$$

The exact Poisson probability is .9626.

APPENDIX 11–B GLOSSARY OF NOTATION FOR CHAPTER 11 ON QUALITY AND ASSURANCE

Note: This chapter uses accepted notation for control charts and acceptance sampling. As a result, the same symbol may have one meaning in the context of control charts and another in the context of acceptance sampling.

a_1 = Cost of sampling one item.
a_2 = Cost of searching for an assignable cause.
a_3 = Cost of operating the process in an out-of-control state.
AOQ = Average outgoing quality.
AOQL = Average outgoing quality limit. The maximum value of the AOQ curve.
AQL = Acceptable quality level.
α = $P\{$Type 1 error$\}$.
   Control chart usage: α represents the probability of obtaining an out-of-control signal when the process is in control.
   Acceptance sampling usage: α is the probability of rejecting good lots.
β = $P\{$Type 2 error$\}$.
   Control chart usage: β represents the probability of not obtaining an out-of-control signal when the process has gone out of control.
   Acceptance sampling usage: β is the probability of accepting bad lots.

$c = \begin{cases} \text{\textit{Control chart usage:} The expected number of defects per unit of} \\ \text{production.} \\ \text{\textit{Acceptance sampling usage:} The acceptance level for a single} \\ \text{sampling plan.} \end{cases}$

c_1 = Acceptance level for first sample for a double sampling plan.

c_2 = Rejection level for the first sample for a double sampling plan.

c_3 = Acceptance level for both the first and the second samples for a double sampling plan. (Often $c_2 = c_3$.)

d_2 = A constant depending on n that relates $\overline{R}$ and σ.

d_3 = A constant depending on n that when multiplied by $\overline{R}$ gives the lower control limit for an R chart.

d_4 = A constant depending on n that when multiplied by $\overline{R}$ gives the upper control limit for an R chart.

δ = Assumed magnitude of the shift in the mean as measured in standard deviations for the economic design of $\overline{X}$ charts.

LCL = Lower control limit for a control chart.

LTPD = Lot tolerance percent defective. Unacceptable quality level.

M = Number of defectives in a lot.

μ = Population mean.

N = Number of items in lot. Used in acceptance sampling.

$n = \begin{cases} \text{\textit{Control chart usage:} Size of each subgroup for an } \overline{X} \text{ chart.} \\ \text{\textit{Acceptance sampling usage:} Number of items sampled from} \\ \text{a lot for a single sampling plan.} \end{cases}$

n_1 = Size of the first sample for a double sampling plan.

n_2 = Size of the second sample in a double sampling plan.

OC(p) = Operating characteristic curve.

$p = \begin{cases} \text{\textit{Control chart usage:} True proportion of defective items produced.} \\ \text{\textit{Acceptance sampling usage:} True proportion of defectives in a lot.} \end{cases}$

p_0 = AQL.

p_1 = LTPD.

π = Probability that the process goes out of control in a single period.

R = Range of a sample. The difference between the largest and the smallest values in the sample.

s = Sample standard deviation of a random sample.

σ = Population standard deviation.

$\hat{\sigma}$ = Estimator for σ.

UCL = Upper control limit for a control chart.

$X = \begin{cases} \text{\textit{Control chart usage:} Value of a single measurement from} \\ \text{the population.} \\ \text{\textit{Acceptance sampling usage:} Number of defectives observed in a} \\ \text{sample of } n \text{ items.} \end{cases}$

$\overline{X}$ = Arithmetic average of a random sample of n independent measurements.

Z = Standard normal variate.

$z_{\alpha/2}$ = The number such that the probability of observing a value of Z that exceeds $z_{\alpha/2}$ is $\alpha/2$.

Bibliography

Aguayo, R. *Dr. Deming: The American Who Taught the Japanese about Quality.* New York: Lyle Stuart, 1990.

Alexander, F. "ISO 14001: What Does It Mean for IE's." *IIE Solutions* 2 (January 1996), pp. 15–18.

Alexander, T. "Cray's Way of Staying Super-Duper." *Fortune*, March 18, 1985, p. 76.

Baker, K. R. "Two Process Models in the Economic Design of an $\overline{X}$ Chart." *AIIE Transactions* 13 (1971), pp. 257–63.

Boothroyd, G., and P. Dewhurst. *Product Design for Assembly.* Wakefield, RI: Boothroyd Dewhurst, Inc., 1989.

Boothroyd, G.; P. Dewhurst; and W. A. Knight. *Product Design for Manufacturing.* New York: Marcel Dekker, 1994.

Business Week. "The Quality Imperative." Special issue devoted to quality. New York: McGraw-Hill, 1991.

Cohen, L. *Quality Function Deployment: How to Make QFD Work for You.* Reading, MA: Addison Wesley, 1995

Crosby, P. B. *Quality Is Free.* New York: McGraw-Hill, 1979.

Crosby, P. B. *Quality without Tears: The Art of Hassle-Free Management.* New York: McGraw-Hill, 1984.

Dahan, E. "Note on Listening to the Customer, Part I." Teaching note, Graduate School of Business, Stanford University, Stanford, CA, 1995.

DeGroot, M. H. *Probability and Statistics.* 2nd ed. Reading, MA: Addison Wesley, 1986.

Dehnad, K. *Quality Control, Robust Design, and the Taguchi Method.* Pacific Grove, CA: Wadsworth Cole, 1989.

Duncan, A. J. "The Economic Design of $\overline{X}$ Charts Used to Maintain Current Control of a Process." *Journal of the American Statistical Association* 51 (1956), pp. 228–42.

Duncan, A. J. *Quality Control and Industrial Statistics.* 5th ed. New York: McGraw-Hill/Irwin, 1986.

Dvorak, P. "Manufacturing Puts a New Spin on Design." *Machine Design* 67 (August 22, 1994), pp. 67–74.

Feigenbaum, A. V. *Total Quality Control.* 3rd ed. New York: McGraw-Hill, 1983.

Fortune. "How Jack Welch Keeps the Ideas Coming at GE." August 13, 1991.

Garvin, D. A. "Quality on the Line." *Harvard Business Review* 61 (1983), pp. 64–75.

Garvin, D. A. *Managing Quality.* New York: Free Press, 1988.

Grant, E. L., and R. S. Leavenworth. *Statistical Quality Control.* 6th ed. New York: McGraw-Hill, 1988.

Green, P., and V. R. Rao. "Conjoint Measurement for Quantifying Judgmental Data." *Journal of Marketing Research* 8 (1971), pp. 355–63.

Groocock, J. M. *The Chain of Quality: Market Dominance through Product Superiority.* New York: John Wiley & Sons, 1986.

Herron, D. A. Private communication, 1985.

Hillkirk, J. "Europe Upstages Quest for Baldrige Award." *USA Today,* April 22, 1992.

Ishikawa, K. *What Is Total Quality Control? The Japanese Way.* Translated from the Japanese by D. J. Lu. Englewood Cliffs, NJ: Prentice Hall, 1985.

Kackar, R. N. "Off-Line Quality Control, Parameter Design, and the Taguchi Method." *Journal of Quality Technology* 17 (1985), pp. 176–88.

Kolesar, P. "Scientific Quality Management and Management Science." In *Handbooks in Operations Research and Management Science,* vol. 4, *Logistics of Production and Inventory,* ed. S. Graves, A. H. G. Rinnooy Kan; and P. Zipkin. Chapter 13. Amsterdam: North Holland, 1993.

Kumar, S., and Y. Gupta. "Statistical Process Control at Motorola's Austin Assembly Plant." *Interfaces* 23, no. 2 (March–April 1993), pp. 84–92.

Kuster, T. "ISO 9000: A 500-lb Gorilla?" *Metal Center News* 35, no. 10 (September 1995), pp. 5–6.

Lawler, E. E., and S. A. Mohrman. "Quality Circles after the Fad." *Harvard Business Review* 63 (1985), pp. 65–71.

Leonard, H., and E. Sasser. "The Incline of Quality." *Harvard Business Review* 60 (1982), pp. 163–71.

Logothetis, N., and H. P. Wynn. *Quality through Design.* Oxford: Clarendon Press, 1989.

Martin, J. "Ignore Your Customer." *Fortune,* May 1, 1995, pp. 121–26.

Miller, J. G.; A. D. Meyer; and J. Nakane. *Benchmarking Global Manufacturing.* New York: McGraw-Hill/Irwin, 1992.

Montgomery, D. C. *Introduction to Statistical Quality Control.* New York: John Wiley & Sons, 1985.

Motorola, Inc. "Motorola Corporate Quality System Review Guidelines." March 1991 edition. Referenced in S. Kumar and Y. Gupta, "Statistical Process Control at Motorola's Austin Assembly Plant," *Interfaces* 23, no. 2 (March–April 1993), pp. 84–92.

Pierce, R. J. *Leadership Perspective, and Restructuring for Total Quality.* Milwaukee: ASQC Quality Press, 1991.

Pugh, S. *Total Design.* Workingham, England: Addison Wesley, 1991.

Ross, S. M. *Applied Probability Models with Optimization Applications.* San Francisco: Holden Day, 1970.

Schoeffler, S.; R. D. Buzzell; and D. F. Heang. "Impact of Strategic Planning on Profit Performance." *Harvard Business Review* 52 (March–April 1974).

Shewhart, W. A. *Economic Control of the Quality of Manufactured Product.* New York: D. Van Nostrand, 1931.

Taguchi, G.; A. E. Elsayed; and T. Hsiang. *Quality Engineering in Production Systems.* New York: McGraw-Hill, 1989.

Transportation and Distribution 36, no. 5 (May 1995), pp. 26–28.

Ulrich, K. T., and S. D. Eppinger. *Product Design and Development.* New York: McGraw-Hill, 1995.

Velury, J. "Integrating ISO 9000 into the Big Picture." *IIE Solutions* 1 (October 1995), pp. 26–29.

CHAPTER 12

RELIABILITY AND MAINTAINABILITY

In operations management quality has been a key issue in recent years. The dramatic success of the Japanese has been attributed to a large extent to the quality of their manufactured goods. Quality is multidimensional (see Chapter 11, Section 13), but reliability is certainly a key component. In Table 11–8 of Chapter 11, we reported on competitive priorities in Europe, Japan, and the United States in the next five years. Product reliability ranked number one for the group of Japanese firms surveyed.

When we think of Japan's economic success, it is the automobile industry that many of us think of first. Japanese automakers have had a steadily growing market share in the United States over the past two decades, even during periods with import quotas or unfavorable exchange rates. Why have the Japanese been so successful in the United States? Perceived product quality is probably the key reason that so many Americans choose to purchase Japanese automobiles. But what dimension of quality is most important? A likely answer is product reliability. Annual surveys conducted by the Consumer's Union attest to the exceptional reliability of Japanese-made automobiles.

Reliability as a field separated from the mainstream of statistical quality control in the 1950s with the postwar growth of the aerospace and the electronics industries in the United States. The Department of Defense took a keen interest in reliability studies when it became painfully apparent that there was a serious problem with the reliability of military components and systems. Garvin (1988) reports that in 1950 only one-third of the Navy's electronic devices were working properly at any given time, and that for every vacuum tube the military had in an operational state there were nine tubes in warehouses or on order. According to Amstadter (1971), the yearly cost of maintaining some military systems in an operable state has been as high as 10 times the original cost of the equipment.

What is the difference between statistical quality control and reliability? Statistical quality control is concerned with monitoring processes to ensure that the manufactured product conforms to specifications. The random variables of interest are numbers of defects and degree of conformance variation. Reliability considers the performance of a product over *time*. The random variables of interest concern the amount of elapsed time between failures after the product is placed into service. A definition of reliability that emphasizes its close association with quality has been suggested by O'Connor (1985): reliability is a time-based concept of quality. Alternatively, reliability is the probability that a product will operate adequately for a given period in its intended application (Amstadter, 1971).

A reviewer of the first edition said that from the student's point of view, reliability is "a narrow engineering issue which almost never makes it to the front page of

The Wall Street Journal." This couldn't be further from the truth. Three of the most significant disasters of recent times were the result of reliability failures: The accidents at the nuclear plants at Three Mile Island in Pennsylvania and at Chernobyl in the former Soviet Union, and the dramatic loss of the *Challenger,* the worst disaster suffered in the U.S. space program. Reliability concerns each of us every day. Our lives depend on the reliability of cars, commuter trains, and airplanes. Our livelihoods depend on the reliability of power generation, telephones, and computers. Our health depends on the reliability of pollution control systems, heating and air-conditioning systems, and emergency medical care systems.

Reliability and risk are closely related. Risks of poor reliability are of concern to both the producer and the consumer. Some aspects of risk from the producer's point of view include

1. *Competition.* Product reliability is an important component of perceived quality by the consumer. Highly unreliable products do not gain customer loyalty and do eventually disappear.

2. *Customer requirements.* The U.S. government required weapons systems with clearly specified reliability levels when it found that maintenance costs for these systems were becoming prohibitive. Today, reliability requirements established by the buyer are common.

3. *Warranty and service costs.* Warranties, which will be treated in the second part of this chapter, are a significant financial burden to the manufacturer when products are unreliable. American automobile firms provided extended warranties as an incentive to the consumer several years ago to boost sales (the most notable was Chrysler's 7-year, 70,000-mile warranty). There is no evidence that these warranties were accompanied by improved reliability. As a result, these programs proved to be very costly to the automobile firms.

4. *Liability costs.* Largely as an outgrowth of the efforts of Ralph Nader, the U.S. Congress has enacted legislation that makes a manufacturer liable for the consequences of failures in product performance resulting from faulty design or manufacture. Liability losses have had the effect of shifting some of the costs of poor reliability from the consumer to the manufacturer.

Some of the risks of poor reliability borne by the consumer include

1. *Safety.* There is no doubt that equipment failure results in human death. Approximately 50,000 Americans die in automobile accidents on the nation's roads each year. Undoubtedly, some portion of these are attributable to mechanical failure. Travelers die in airplane accidents, many of which are the result of equipment failure. The failure of the nuclear plants at Three Mile Island and Chernobyl resulted in human death and injury from radiation exposure. Safety and reliability are closely linked.

2. *Inconvenience.* Even though many failures do not result in death, they can be a source of frustration and delay. Delays at airports are common because some piece of equipment on board the plane fails to operate properly. Automobile breakdowns may leave motorists stranded for hours. Failure of communication equipment, computer equipment, or power generation plants can cripple businesses.

3. *Cost.* Poor reliability costs everyone in the end. For this reason, consumers are willing to pay a premium for products with higher reliability. The

Japanese have learned this lesson well, and every indication is that they will continue the strategy of increasing market share by producing more reliable products than their competitors.

Why study reliability? We need to understand the probability laws governing failure patterns in order to better design processes to build reliable systems. Incorrect analysis can lead to disastrous consequences. For example, the so-called Rasmussen report (U.S. Nuclear Regulatory Commission, *Reactor Safety Study,* 1975) predicted that it would be hundreds of years before we could expect a major accident in a nuclear plant. Given that we have since observed two major accidents, the analysis in this report is clearly flawed.

Reliability is an issue of concern for operations management from two perspectives. First, to implement total quality management, we must understand how and why products fail. Reliability will continue to be a key component of quality. Designing an effective quality delivery system will require understanding the randomness of failure patterns. Second, we need to understand the failure patterns of the equipment used in the manufacturing process. In this way, we can develop effective maintenance policies for that equipment.

Overview of This Chapter

Reliability theory is based on sophisticated mathematics. For that reason this is the most mathematically rigorous chapter in this book. The student studying this material should have a solid grounding in probability. Understanding random variables and distribution functions is crucial.

This chapter considers three broad areas: (1) component and system reliability, (2) age replacement strategies, and (3) warranties. Much of classical reliability theory concerns failure patterns for individual components and for systems of components. This is the subject of the first part of the chapter. Age replacement strategies concern the economics of replacing operating equipment before failure to decrease the likelihood that a failure will occur during operation. Warranty analysis, a subject that has received little attention in texts, is the final major topic of this chapter. Warranties will continue to play an important part in the value of a product to the consumer.

There is some overlap of the basic probability theory required in this chapter and that treated in Supplement 2 on queuing. In particular, the material on the exponential distribution and on the Poisson process is relevant to both areas.

12.1 Reliability of a Single Component

Introduction to Reliability Concepts

In order for readers to better understand and appreciate some of the definitions and concepts introduced in this chapter, this section starts with an example.

Example 12.1

In 1970 the U.S. Army purchased 1,000 identical capacitors for use in short-distance radio transmitters. The army maintained detailed records on the failure pattern of the capacitors with the following results:

Number of years of operation	1	2	3	4	5	6	7	8	9	10	>10
Number of failures	220	158	121	96	80	68	47	40	35	25	110

Based on these data, the army wanted to estimate the probability distribution associated with the failure of a capacitor chosen at random.

Define the random variable T as the time that the capacitor will operate before failure. We can estimate the cumulative distribution function of T, $F(t)$, using the data above. In symbols, $F(t) = P\{T \leq t\}$, and in words, $F(t)$ is the probability that a component chosen at random fails at or before time t.

In order to estimate $F(t)$ from the given data, we find the cumulative number of failures and the proportion of the total this number represents each year. We have

Number of years of service	1	2	3	4	5	6	7	8	9	10	>10
Cumulative failures	220	378	499	595	675	743	790	830	865	890	1,000
Proportion of total	.220	.378	.499	.595	.675	.743	.790	.830	.865	.890	1.0

The proportions are estimates of $F(t)$ for $t = 1, 2, \ldots$. These probabilities may be used directly to compute various quantities of interest by treating T as a discrete random variable, or they may be used to estimate the parameters of a continuous distribution. We will use the discrete version to answer several questions about the lifetime of the capacitors. For example, suppose that we wish to determine

    a. The probability that a capacitor chosen at random lasts more than five years.

    b. The proportion of the original 1,000 capacitors put into operation that fail in year 6.

    c. The proportion of components that survive for at least five years that fail in year 6.

    d. The proportion of components that survive at least eight years that fail in year 9.

Solution

    a. $P\{T > 5\} = 1 - P\{T \leq 5\} = 1 - .675 = .325$.

    b. $P\{T = 6\} = P\{T \leq 6\} - P\{T \leq 5\} = .743 - .675 = .068$

c. At first glance, this question might appear to be the same as part (2). However, there is an important difference. Part (2) asks for the proportion of the original set of capacitors failing in year 6, whereas part (3) asks for the proportion of components lasting *at least five years* that fail in year 6. This is a *conditional* probability.

$$P\{T = 6 \mid T > 5\} = \frac{P\{T = 6, T > 5\}}{P\{T > 5\}} = \frac{P\{T = 6\}}{P\{T > 5\}} = \frac{.068}{.325} = 209.$$

Notice that the events $\{T = 6, T > 5\}$ and $\{T = 6\}$ are equivalent since $\{T = 6\} \subset \{T > 5\}$.

d.

$$P\{T = 9 \mid T > 8\} = \frac{P\{T = 9\}}{P\{T > 8\}} = \frac{.035}{.170} = 206,$$

which is almost precisely the same as the proportion of components surviving more than five years that fail in year 6. In fact, the proportion of components that survive n years and fail in year $n + 1$ is very close to .20 for $n = 0, 1, 2, \ldots, 9$. As we will later see, this results in a failure distribution with some unusual properties. ∎

Preliminary Notation and Definitions

Many new concepts were introduced in Example 12.1. This section formalizes the definitions to be used throughout this chapter.

As above, define the random variable T as the lifetime of the component. We assume that T has cumulative distribution function $F(t)$ given by

$$F(t) = P\{T \le t\}.$$

In what follows we will treat $F(t)$ as a differentiable function of t, so that the probability density function $f(t)$, given by the equation

$$f(t) = \frac{dF(t)}{dt},$$

will exist.

In addition to the distribution and density functions of the random variable T, we will be interested in related functions. One is the reliability function (also known as the survival function). The reliability function of the component, which we call $R(t)$, is given by

$$R(t) = P\{T > t\} = 1 - F(t).$$

In words, $R(t)$ is the probability that a new component will survive past time t. Notice that this implies that $F(t)$ is the probability that a new component will *not* survive past time t.

Consider the following conditional probability:

$$P\{t < T \le t + s \mid T > t\}.$$

This is the conditional probability that a new component will fail between t and $t + s$ given that it lasts beyond t. We may think of this conditional probability in the following way: Interpret t as now and s as an increment of time into the future. The event $\{T > t\}$ means that the component has survived until the present, or in other words, that it is still working. The conditional event $\{t < T \le t + s \mid T > t\}$ means that the component is working now but will fail before an additional s units of time have passed.

Recall from elementary probability theory that for any events A and B

$$P\{A \mid B\} = \frac{P\{A \cap B\}}{P\{B\}}.$$

In the special case where $A \subset B$, $A \cap B = A$, so that

$$P\{A \mid B\} = \frac{P\{A\}}{P\{B\}} \qquad \text{when } A \subset B.$$

Identify the event $A = \{t < T \le t + s\}$ and $B = \{T > t\}$. A little reflection shows that $A \subset B$ in this particular case, so that

$$P\{t < T \le t + s \mid T > t\} = \frac{P\{t < T \le t + s\}}{P\{T > t\}} = \frac{F(t + s) - F(t)}{R(t)}.$$

We will divide by s and let s approach zero.

$$\lim_{s \to 0} \frac{1}{s} \frac{F(t + s) - F(t)}{R(t)} = \frac{f(t)}{R(t)}.$$

This ratio turns out to be a fundamental quantity in reliability theory.

Define

$$r(t) = \frac{f(t)}{R(t)}.$$

We call $r(t)$ the *failure rate function*. The derivation above is the best way to understand what the failure rate function means. For positive s, the conditional probability used to derive $r(t)$ is the probability that a component that has survived up until time t fails between times t and $t + s$. Dividing by s and letting s go to zero is the same way one derives a first derivative. Hence, the failure rate function is the rate of change of the conditional probability of failure at time t. It can be considered a measure of the likelihood that a component that has survived up until time t fails in the next instant of time.

The failure rate function is a fundamental quantity in reliability theory but, like the probability density function, does not have a direct physical interpretation. However, for values of Δt sufficiently small, the term $r(t)\Delta t$ is the probability that an item that survives to time t fails between t and $t + \Delta t$. How does one determine if Δt is sufficiently small? In general, Δt should be small relative to the lifetime of a typical component. However, the only way to be certain as to whether $r(t)\Delta t$ is a good approximation to this conditional probability is to compute $P\{t < T \le t + \Delta t \mid T > t\}$ directly.

Example 12.2

The length of time that a particular piece of equipment operates before failure is a random variable with cumulative distribution function

$$F(t) = 1 - e - 0.043t^{2.6}.$$

Consider the following:

a. The failure rate function.

b. The probability that the equipment operates for more than five years without experiencing failure.

c. Suppose that 100 pieces of the equipment are placed into service in year 0. What fraction of the units surviving four years fail in year 5? Can one accurately estimate this proportion using only the failure rate function?

d. What fraction of units surviving four years fail in the first month of year 5? Can this be accurately estimated using the failure rate function?

Solution

a.

$$f(t) = \frac{dF(t)}{dt} = -e^{-0.043t^{2.6}} \frac{d}{dt}(-0.043t^{2.6})$$

$$= (0.043)(2.6)t^{1.6} e^{-0.043t^{2.6}}$$

$$= 0.1118t^{1.6} e^{-0.043t^{2.6}}$$

Since $R(t) = 1 - F(t) = e^{-0.043t^{2.6}}$, it follows that

$$r(t) = \frac{f(t)}{R(t)} = 0.1118t^{1.6}.$$

b.

$$P\{T > 5\} = R(5) = e^{-0.043(5)^{2.6}} = e^{-2.8235} = 0.0594.$$

c. We will compute this directly and compare the result with $r(t)\Delta t$. The proportion of units surviving four years that fail in year 5 is

$$P\{4 < T \le 5 \mid T > 4\} = \frac{F(5) - F(4)}{R(4)} = \frac{R(4) - R(5)}{R(4)}$$

$$= \frac{e^{-0.043(4)^{2.6}} - e^{-0.043(5)^{2.6}}}{e^{-0.043(4)^{2.6}}}$$

$$= \frac{0.2059 - 0.0594}{0.2059}$$

$$= 0.7115.$$

This means that about 71 percent of the machines surviving four years will fail in the fifth year. As about 94 percent of the units fail in the first five years of operation ($F(5) = .9406$), a value of $\Delta t = 1$ year is probably too large for $r(t)\Delta t$ to give a good approximation. In fact, we see that $r(4)(1) = (0.1118)(4)^{1.6} = 1.0274$.

d. Because one month corresponds to $\frac{1}{12} = 0.0833$ year, we wish to compute

$$P\{4 < T \le 4.0833 \mid T > 4\} = \frac{F(4.0833) - F(4)}{R(4)} = \frac{0.2059 - 0.1887}{0.2059}$$

$$= 0.0836.$$

Here $\Delta t = \frac{1}{12}$ should be sufficiently small to use the failure rate function to estimate this probability. We obtain $r(4) \frac{1}{12} = 0.0856$. ∎

The Exponential Failure Law

The exponential distribution plays a fundamental role in reliability theory and practice because it accurately describes the failure characteristics of many types of operating equipment. The exponential law can be derived in several ways. We will consider a derivation that utilizes the failure rate function.

We know that the failure rate function $r(t)$ is given by the formula

$$r(t) = f(t)/R(t)$$

and is a measure of the likelihood that a unit that has been operating for t units of time fails in the next instant. Consider the following case: $r(t) = \lambda$, for some constant $\lambda > 0$. This means that the likelihood that a working unit fails in the next instant of time is independent of how long it has been operating. This implies that the unit does not exhibit any signs of aging. It is equally likely to fail in the next instant whether it is new or old. We will derive the probability distribution of the lifetime T that corresponds to a constant failure rate function.

We can determine a solution to the equation $r(t) = \lambda$ by noting that since $R(t) = 1 - F(t)$,

$$f(t) = \frac{dR(t)}{dt} = -R'(t).$$

Hence, the equation $r(t) = \lambda$ may be written in the form

$$\frac{-R'(t)}{R(t)} = \lambda$$

or

$$R'(t) = -\lambda R(t).$$

This is the simplest first-order linear differential equation. Its solution is

$$R(t) = e^{-\lambda t}.$$

It follows that the distribution function $F(t)$ is given by

$$F(t) = 1 - e^{-\lambda t}$$

and the density function $f(t)$ is given by

$$f(t) = \lambda e^{-\lambda t}.$$

This is known as the exponential distribution. It depends on the single parameter λ, which represents a rate of occurrence. If T has the exponential distribution with parameter λ, then T corresponds to the lifetime of a component that exhibits no aging over time; that is, a component that has survived up until time t_1 is equally likely to fail in the next instant of time as one that has survived up until time t_2 for any times t_1 and t_2. The expected failure time is $1/\lambda$. The standard deviation of the failure time is also $1/\lambda$. The exponential density and distribution functions appear in Figure 12–1.

FIGURE 12–1

The exponential density and distribution functions

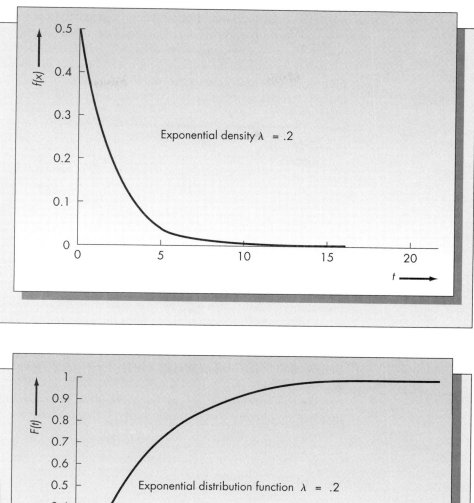

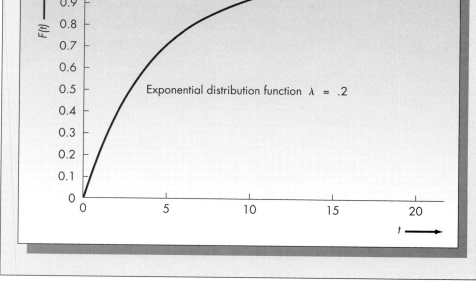

Example 12.3

Because the exponential distribution has a constant failure rate function, it is likely that the failure law for the capacitor described in the example presented at the start of the chapter is exponential. This follows because we observed that the proportion of the capacitors having survived for n years that failed in year $n + 1$ was the same for

$n = 0, 1, \ldots$, which was approximately 20 percent. In order to estimate the value of λ, note that the proportion failing in the first year is also 20 percent, which gives $F(1) = 0.20 = 1 - e^{-\lambda}$. Solving for λ results in $e^{-\lambda} = 0.8$, or $\lambda = -\ln(0.8) = 0.223$.

We can compare the expected number of capacitors that would fail if the true lifetime distribution were exponential versus the actual number that failed, to see if the exponential distribution provides a reasonable fit of the data.

Number of years of operation	1	2	3	4	5	6	7	8	9	10	>10
Number of failures	220	158	121	96	80	68	47	40	35	25	110
Expected number of failures under exponential law with $\lambda = 0.223$	200	160	128	102	82	66	52	42	34	27	107

The expected number of failures for each year is obtained by multiplying the probability of failure for a given year, assuming the exponential law, by the total number of units. For example, one finds the expected number of failures for year 3 by multiplying 1,000 by $F(3) - F(2) = e^{-2\lambda} - e^{-3\lambda} = 0.1280$. Clearly there is a close agreement between the actual number of failures and the expected number, indicating that the exponential distribution provides a good fit of the observed historical data.[1]

We can now answer several questions about the capacitors using the exponential distribution directly. For example, suppose that we wish to find

    a. The probability that a capacitor chosen at random lasts more than eight years.

    b. The proportion of capacitors that survive three years that also survive at least three additional years.

Solution

    a. $P\{T > 8\} = e^{-\lambda t} = e^{-(.223)(8)} = e^{-1.784} = .1680.$

    b. We wish to compare $P\{T > 6 \mid T > 3\}$. Using the laws of conditional probability we have

$$P\{T > 6 \mid T > 3\} = \frac{P\{T > 6, T > 3\}}{P\{T > 3\}} = \frac{P\{T > 6\}}{P\{T > 3\}} = \frac{.2624}{.5122} = .5122.$$

It is not a coincidence that this is the same as the unconditional probability that a new capacitor lasts more than three years. This property is known as the memoryless property of the exponential distribution. ∎

The memoryless property of the exponential distribution relates to the following conditional probability:

$$P\{T > t + s \mid T > t\}.$$

---

[1]It is easy to verify by a formal goodness-of-fit test that the exponential distribution fits these data very closely.

This is the probability that the component survives past time $t + s$ given that it has survived until time t. If we think of time t as now, then this is the probability that a component that is currently functioning continues to function for at least another s units of time. If T follows an exponential failure law, then

$$P\{T > t + s \mid T > t\} = \frac{P\{T > t + s, T > t\}}{P\{T > t\}}$$

$$= \frac{P\{T > t + s\}}{P\{T > t\}}$$

$$= \frac{e^{-\lambda(t+s)}}{e^{-\lambda t}}$$

$$= e^{-\lambda s}$$

$$= P\{T > s\}.$$

Note that as $\{T > t + s\} \subset \{T > t\}$, the events $\{T > t + s, T > t\}$ and $\{T > t + s\}$ are equivalent. The last expression, $P\{T > s\}$, is the *unconditional* probability that a new component will last at least s units of time. That is, we have demonstrated that if the component has been operating for t units of time without failure, then the probability that it continues to operate for at least another s units of time is the same as the probability that a new component operates for at least s units of time. This means that there is no aging. The likelihood of failure is independent of how long the component has been operating. However, we required that the lifetime distribution be exponential. In fact, the exponential distribution is the *only* continuous distribution possessing the memoryless property; that is, it is the only one for which $P\{T > t + s \mid T > t\} = P\{T > s\}$.

When we say that an item fails completely at random, we mean that the failure law for the item is exponential. Events that occur completely at random over time follow a *Poisson process*. The Poisson process is discussed in a later section of this chapter.

Problems for Section 1

1. Three hundred identical cathode ray tubes (CRTs) placed into service simultaneously on January 1, 1976, experienced the following numbers of failures through December 31, 1988:

Year	1983	1984	1985	1986	1987	1988
Number of failures	13	19	16	34	21	38

Assume that there were no failures before 1983.
   a. Based on these data, estimate the cumulative distribution function (CDF) of a CRT chosen at random.

   Using the results of part (*a*), estimate the probability that a CRT chosen at random
   b. Lasts more than 5 years.
   c. Lasts more than 10 years.
   d. Lasts more than 12 years.
   e. That has survived for 10 years fails in the 11th year of operation.

2. Suppose that the cumulative distribution function of the lifetime of a piece of operating equipment is given by

$$F(t) = 1 - e^{-0.6t} - 0.6te^{-0.6t},$$

where t is measured in years of continuous operation.

a. Determine the reliability function.
b. Determine the failure rate function.
c. What is the probability that this piece of equipment fails in the first year of operation?
d. What is the probability that this piece of equipment fails in the fifth year of operation?
e. What proportion of the equipment surviving four years fails in the fifth year? (Calculate without using the failure rate function.)
f. Does $r(4)$ closely approximate the answer to part (e)? Why or why not?
g. What proportion of the equipment surviving four years fails in the first month of the fifth year? (Calculate using the failure rate function.)

3. A large number of identical items are placed into service at time 0. The items have a failure rate function given by

$$r(t) = 1.105 + 0.30t,$$

where t is measured in years of operation.

a. Derive $R(t)$ and $F(t)$.
b. If 300 items are still operating at time $t = 1$ year, approximately how many items would you expect to fail between year 1 and year 2?
c. Does the value of $r(1)$ yield a good approximation to the conditional probability computed in part (b)? Why or why not?
d. Repeat the calculation of part (b), but determine the expected number of items that fail between $t = 1$ year and $t = 1$ year plus 1 week. Does $r(t)\Delta t$ provide a reasonable approximation to the conditional probability in this case? Why or why not?

4. A microprocessor that controls the tuner in color TVs fails completely at random (that is, according to the exponential distribution). Suppose that the likelihood that a microprocessor that has survived for k years fails in year $k + 1$ is .0036. What is the cumulative distribution function of the time until failure of the microprocessor?

5. A pressure transducer regulates a climate control system in a factory. The transducer fails according to an exponential distribution with rate one failure every five years on average.

a. What is the cumulative distribution function of the time until failure?
b. What is the probability that a transducer chosen at random functions for eight years without failure?
c. What is the probability that a transducer that has functioned for eight years continues to function for another eight years?

6. For the pressure transducer mentioned in the previous problem, use the failure rate function to estimate the likelihood that a transducer that has been operating for six years fails in the seventh year. How close is the approximation to the exact answer?

12.2 Increasing and Decreasing Failure Rates

Although the constant failure rate function that leads to the exponential law is significant in reliability theory, there are other important failure laws as well. Most of us are more familiar with items that possess increasing failure rate functions. That is, they are more likely to fail as they get older. Decreasing failure rate functions also occur frequently. New products often have a high failure rate because of the "burn-in" phase, in which the defective items in the population are weeded out.

An important class of failure rate functions that includes both increasing and decreasing failure rate functions is of the form

$$r(t) = \alpha\beta t^{\beta-1} \qquad \text{where } \alpha \text{ and } \beta > 0.$$

Here $r(t)$ is a polynomial function in the variable t that depends on the two parameters α and β. When $\beta > 1$, $r(t)$ is increasing, and when $0 < \beta < 1$, $r(t)$ is decreasing. Typical failure rate functions for these cases are pictured in Figure 12–2.

This form of $r(t)$ will yield another differential equation in $R(t)$. It can be shown that the solution in this case will be

$$R(t) = e^{-\alpha t^{\beta}} \qquad \text{for all } t \geq 0$$

or

$$F(t) = 1 - e^{-\alpha t^{\beta}} \qquad \text{for all } t \geq 0.$$

This distribution is known as the Weibull distribution. It depends on the two parameters α and β, and as we saw above, when $0 < \beta < 1$, it corresponds to the lifetime of an item with a decreasing failure rate, and when $\beta > 1$, it corresponds to the lifetime of an item with an increasing failure rate. Because it is often true that empirical failure rate functions (i.e., those that are observed from test data) are closely approximated by polynomials, the Weibull distribution is an accurate description of the failure law of many types of operating equipment. Note that when $\beta = 1$ the Weibull reduces to the exponential. Various Weibull densities appear in Figure 12–3.

FIGURE 12–2

Failure rate functions for the Weibull lifetime distribution

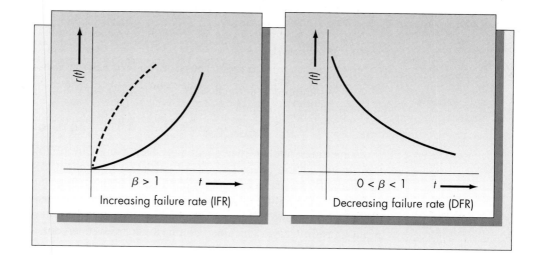

Increasing failure rate (IFR)   Decreasing failure rate (DFR)

FIGURE 12–3

*Weibull densities for
various values of
$\beta(\alpha = 0.5)$*

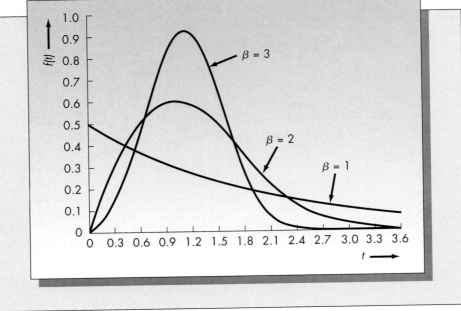

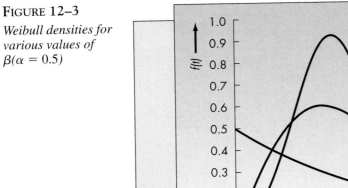

Example 12.4

A local manufacturer of copying equipment includes a repair warranty with each copier. Virtually all of his equipment exhibit an increasing failure rate. Based on historical repair data, the failure rate for Model 25cc7 is accurately described by the function $r(t) = 2.7786t^{1.3}$, where t is measured in months of continuous operation. What is the probability that the time between two successive failures of this equipment exceeds two months of operation?

Solution

Because $r(t)$ is a polynomial in t, the distribution of the time until failure is the Weibull distribution. It is necessary to identify the values of α and β. We have that

$$2.7786t^{1.3} = \alpha\beta t^{\beta-1}.$$

It follows that $\beta - 1 = 1.3$, or $\beta = 2.3$. Since $\alpha\beta = 2.7786$, we obtain

$$\alpha = 2.7786/\beta = 2.7786/2.3 = 1.208.$$

We are required to compute $P\{T > 2\} = R(2)$. Substituting $t = 2$ into the equation for $R(t)$ we obtain

$$R(2) = e^{-1.208 \times (2)^{2.3}} = 4.977 \times 10^{-4}. \quad \blacksquare$$

Sometimes neither increasing nor decreasing failure rate functions accurately describe the failure characteristics of particular equipment. A typical case in point is the "bathtub" failure rate function pictured in Figure 12–4. In the early phases of the product life, the failure rate is decreasing. This follows because defective components fail quickly, causing failure rates to be high initially. This is commonly known as the infant mortality stage. Once bad components are weeded out, the failure rate remains constant until aging begins. At that time we enter the wear-out phase and the failure rate starts to increase.

FIGURE 12–4

*"Bathtub" failure
rate function*

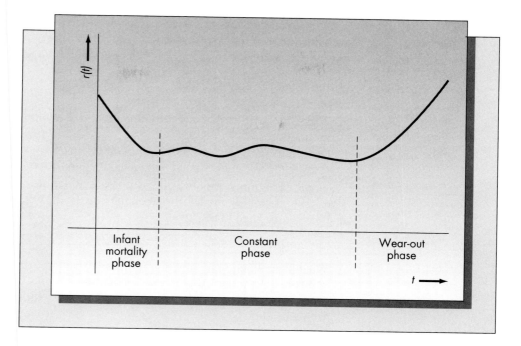

If $r(t)$ is an arbitrary failure rate function, then it can be shown that the reliability function $R(t)$ is given by

$$R(t) = \exp\left(-\int_0^t r(u)\, du\right).$$

It is easy to verify that both the exponential and the Weibull cases satisfy this relationship. Sometimes, such as with the bathtub failure rate function, finding an explicit representation for the function $r(t)$ is difficult. In such cases it is possible to approximate $r(t)$ by step functions. We will not illustrate the procedure here.

Problems for Section 2

7. A piece of equipment has a lifetime T (measured in years) that is a continuous random variable with cumulative distribution function

$$F(t) = 1 - e^{-t/10} - (t/10)\, e^{-t/10} \qquad \text{for all } t \geq 0.$$

   a. What is the probability density function of T?
   b. What is the probability that a piece of equipment survives more than 20 years?
   c. What is the probability that a piece of equipment survives more than 10 years but fewer than 20 years?
   d. What is the probability that a piece of equipment survives more than 20 years given that it has survived for 10 years?

8. For the equipment mentioned in the previous problem,
   a. Derive the failure rate function $r(t)$, and draw a graph of the function.
   b. Without using the failure rate function, determine the probability that a piece of equipment that has survived 20 years of operation fails in the 21st year.
   c. Does $r(20)$ accurately estimate your answer to part (b)? Why or why not?

9. The Air Force maintains enormous amounts of data on engine failure times. A particular engine has experienced a failure pattern whose failure rate function is closely approximated by

$$r(t) = 0.000355e^{2.2t},$$

where t is in flying hours.

   a. What are the reliability and the cumulative distribution functions of the time until failure?

   b. Determine the value of t such that the likelihood that an engine fails before t is the same as the likelihood that an engine fails after t.

10. A sample of high-capacity resistors is tested until failure and the results fitted to a Weibull probability model. Based on these tests, the reliability function of a resistor is estimated to be

$$R(t) = e^{-0.0013\,t^{1.83}}.$$

   a. What is the failure rate function for these resistors?

   b. Is this resistor more likely to fail as it ages?

   c. What is the probability that a resistor will function for more than 30 hours without failure?

   d. Suppose that a resistor has been operating for 50 hours. What is the probability that it fails in the 51st hour? (Use the results of part (*a*) for your calculations.)

12.3 The Poisson Process in Reliability Modeling

Consider a single piece of operating equipment that fails completely at random. As we saw in the previous section, that means that the time until failure follows the exponential distribution. Suppose that when the item fails, it is immediately repaired, or that the repair time is sufficiently small compared with the interfailure time that it can be ignored. Thus we have a process in which events (failures) occur over time and the times between successive failures are independent identically distributed exponential random variables. Let $T_1, T_2, \ldots$ be random variables corresponding to the times between successive failures. Each of the random variables has distribution function $F(t)$ and reliability function $R(t)$ given by

$$F(t) = 1 - e^{-\lambda t},$$

$$R(t) = e^{-\lambda t},$$

where λ is the rate at which failures occur. We also define a related sequence of random variables $W_1, W_2, \ldots$, which are defined by the equations

$$W_1 = T_1,$$
$$W_2 = T_1 + T_2,$$
$$W_3 = T_1 + T_2 + T_3,$$

and so on.

Interpret W_n as the time of the nth failure. Finally, we introduce the process $N(t)$. Define $N(t)$ as the number of failures that occur up until time t. $N(t)$ is a *stochastic* process because for each fixed value of t, $N(t)$ is a random variable. Clearly, $N(t)$ is closely related to both the times between failures, $T_1, T_2, \ldots$, and the times of failures

FIGURE 12–5

*Realization of a
Poisson process*

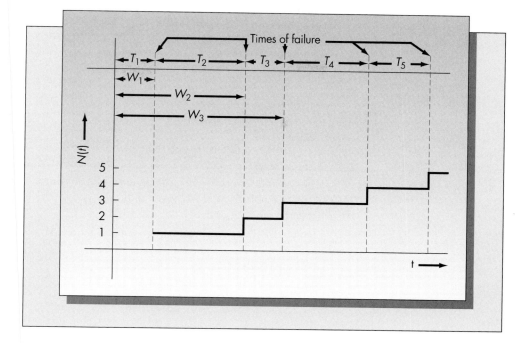

W_1, W_2, ... When T_1, T_2, . . . are independent exponentially distributed random variables, $N(t)$ is a *Poisson process*. A realization of a Poisson process is pictured in Figure 12–5.

We will proceed with the analysis of the Poisson process in the following way: We start with the knowledge of the distribution of the interfailure times, T_1, T_2, ... From that we can derive the distribution of the times until failure W_1, W_2, ... We obtain the distribution of $N(t)$ by using the following equivalence of events:

$$\{N(t) < n\} = \{W_n > t\}.$$

A little reflection should convince you of the truth of this identity. In order for the left-hand side to be true, the number of failures up until time t must be fewer than n. If that is the case, then the time of the nth failure must be after t, which gives the right-hand side. Similarly, if the nth failure occurs after t, then the number of failures up until time t must be fewer than n.

The analysis requires obtaining the cumulative distribution function of the times until failure. Since

$$W_n = T_1 + T_2 + \cdots + T_n,$$

the distribution of W_n can be obtained by forming the n-fold convolution of the exponential distribution, which leads to the *Erlang* distribution:

$$P\{W_n > t\} = \sum_{k=0}^{n-1} \frac{e^{-\lambda t}(\lambda t)^k}{k!}.$$

The Erlang distribution is named for A. K. Erlang in recognition of his pioneering work in the area of queuing. (That W_n has an Erlang distribution is shown in Hillier and Lieberman, 1986, pp. 565–66.) We can now derive the distribution of the number of failures up until time t, $N(t)$.

TABLE 12–1   **Summary Results for Poisson Process**

Random Variable	Distribution	Parameter(s)	Mean	Variance
Time between failure, T_n	Exponential	λ	$1/\lambda$	$1/\lambda^2$
Time of nth failure, W_n	Erlang	λ, n	n/λ	n/λ^2
Number of failures until time t, $N(t)$	Poisson	λt	λt	λt

We have that

$$P\{N(t) = n\} = P\{N(t) < n + 1\} - P\{N(t) < n\}$$

$$= \sum_{k=0}^{n} \frac{e^{-\lambda t}(\lambda t)^k}{k!} - \sum_{k=0}^{n-1} \frac{e^{-\lambda t}(\lambda t)^k}{k!}$$

$$= \frac{e^{-\lambda t}(\lambda t)^n}{n!}.$$

This is exactly the Poisson distribution with parameter λt. The Poisson distribution is a discrete distribution that assumes values only on the nonnegative integers 0, 1, 2, ... It has mean and variance given by λt. The process by which (1) events occur completely at random over time, (2) the interevent times are exponential random variables, (3) the times of events are Erlang random variables, and (4) the number of events up until any time t is a Poisson random variable is called the Poisson process. It is a fundamental stochastic process that crops up in many fields besides reliability. The single parameter λ, the rate at which events occur, defines the process. Table 12–1 summarizes the important points about the Poisson process.

As W_n is the sum of independent identically distributed random variables, when n is reasonably large, the central limit theorem implies that the normal distribution may be used to approximate the Erlang. Also note that the reliability function of the Erlang is of precisely the same form as the cumulative Poisson distribution. Hence, a table of the Poisson distribution (Table A–3 at the back of this book) may be used to obtain exact Erlang probabilities. Furthermore, note that for large values of λt, the normal distribution is an adequate approximation of the Poisson distribution. (However, the normal should not be used to approximate the exponential.)

Example 12.5

A local military base maintains a variety of different equipment. One of these is a sensitive radar device that signals incursion of enemy planes into American airspace. Breakdowns of the device occur completely at random at an average rate of three per year. The equipment is generally repaired the same day that it fails. Determine the following:

    a. The probability that the time between two successive failures is less than one month.

    b. The probability that there are exactly five breakdowns in any given year.

    c. The probability that there are more than 15 failures in a four-year period.

d. The average time for 100 failures to occur.

e. The probability that the 25th failure occurs after 10 years of operation.

Solution

As with all probability problems, it is a good idea to have a ballpark estimate of the answer before beginning formal calculations. This will serve as a verification of your calculations.

a. Because there are three failures per year on average, there will be on average four months between failures. Hence, the probability that the time between two successive failures is less than a month should be less than .5.

Let T be the time between any two successive failures. Then we know that T has the exponential distribution with parameter $\lambda = 3$ per year. We must be sure to express all units of time in terms of years as we solve this problem. We compute

$$P\{T < 1/12\} = 1 - \exp(-\lambda t) = 1 - \exp(-3/12) = .22.$$

b. Here we count the number of breakdowns in a given year, so that the appropriate distribution is Poisson with parameter $\lambda t = 3 \times 1 = 3$.

$$P\{N(1) = 5\} = \frac{e^{-3}3^5}{5!} = .1008.$$

c. We wish to compute $P\{N(4) > 15\}$, where $N(4)$ has the Poisson distribution with parameter $\lambda t = 3 \times 4 = 12$. From Table A–3, we obtain

$$P\{N(4) > 15\} = P\{N(4) \geq 16\} = .1556.$$

This probability also can be approximated using the normal distribution. In order to use a normal approximation, the standardized variate Z is constructed by subtracting the mean and dividing by the standard deviation. Hence,

$$P\{N(4) > 15\} \approx P\{Z > (15 - 12)/\sqrt{12}\} = P\{Z > 0.8660\} = .1922.$$

This approximation can be improved by using the continuity correction. The continuity correction is appropriate when approximating a discrete random variable with a continuous random variable. Since $\{N(4) > 15\} = \{N(4) \geq 16\}$, we go halfway between and use 15.5. (The continuity correction is discussed in detail in Appendix 11–A of Chapter 11.) Hence, $P\{N(4) > 15\} \approx P\{Z > (15.5 - 12)/\sqrt{12}\} = P\{Z > 1.01\} = .1562.$ This is very close to the exact answer, .1556.

d. Here we are interested in $E(W_{100})$. From Table 12–1, $E(W_{100}) = 100/\lambda = 100/3 = 33.33$ years.

e. The time of the 25th failure is the random variable W_{25}. Again we will use the normal approximation, but we do not require the continuity correction because both the Erlang and the normal distributions are continuous. From the table we have that $E(W_{25}) = 25/\lambda = 25/3 = 8.33$, and $\text{Var}(W_{25}) = 25/\lambda^2 = 25/9$. Hence, $\sigma = 5/3 = 1.67$. It follows that

$$P\{W_{25} > 10\} \approx P\left\{Z > \frac{10 - 8.33}{1.67}\right\} = P\{Z > 1\} = .1587.$$

(The continuity correction is not required because W_{25} is continuous.) ∎

　　Many applications involve monitoring many pieces rather than a single piece of equipment. For example, a repairman is responsible for maintaining all equipment in his or her location, and the airlines must maintain an entire fleet of planes. Failures of collections of equipment are treated below.

Series Systems Subject to Purely Random Failures

Consider a bank of items labeled $1, 2, \ldots, N$ and assume that each of the items fails completely at random, that is, according to an exponential failure law. Furthermore, we assume that the items fail independently. A series system implies that the bank fails when the first item in the bank fails. Let $T_1, T_2, \ldots, T_N$ be the failure times associated with each piece of equipment. Then

$$P\{T_i > t\} = \exp(-\lambda_i t) \qquad \text{for } 1 \le i \le N.$$

　　Define the random variable $T = \min(T_1, T_2, \ldots, T_N)$. Then T represents the time that the next component fails. It is also the time the bank fails.

$$P\{T > t\} = P\{\min(T_1, T_2, \ldots, T_N) > t\}$$
$$= P\{T_1 > t, T_2 > t, \ldots, T_N > t\}$$

(this follows because if the minimum of a group of numbers exceeds a fixed number, then it must be true that all members of the group exceed it as well)

$$= P\{T_1 > t\} \times P\{T_2 > t\} \times \cdots \times P\{T_N > t\}$$

(this follows from the independence of the individual failure times)

$$= e^{-\lambda_1 t} e^{-\lambda_2 t} \cdots e^{-\lambda_N t}$$
$$= \exp\left(-\sum_{i=1}^{N} \lambda_i t\right),$$

which is exactly the exponential failure law with $\lambda = \Sigma \lambda_i$. If we assume that units that fail are repaired quickly, then the number of failures of the bank up until any time t, say $N(t)$, will be a Poisson process with rate λ.

Problems for Section 3

11. Automobiles arrive at a tollbooth on a highway completely at random according to a Poisson process with rate $\lambda = 4$ cars per hour. Determine the following:
    a. The probability that the time between any two successive arrivals exceeds 20 minutes.
    b. The probability that exactly four cars arrive in any given hour.
    c. The probability that more than five cars arrive in any given hour.
    d. The probability that more than 10 cars arrive in any given two-hour period.
    e. A person working for the transportation department begins counting arrivals at the tollbooth at 8 A.M. What is the probability that he counts 20 arrivals before 12 noon? (Use a normal approximation for your calculations.)

12. Herman's Hardware uses a neon light in its store window that is left burning continuously. The light has an average lifetime of 1,250 hours and fails completely at random. Lights that burn out are replaced instantly.
    a. On average, how many neon lights does Herman's use in one year?

b. Suppose that the lights cost $37.50 each and Herman, the store owner, has budgeted $300 annually for them. What is the probability that Herman exceeds his annual budget in any given year?

c. What is the probability that two bulbs will be used within the same month? (Assume that one month equals 30 days for your calculation.)

13. An electronic module used by the Navy in a sonar device requires replacement on the average once every 16 months and fails according to a Poisson process. Suppose that the Navy places these sonar devices into service on the same date in eight different aircraft carriers. If the modules are replaced immediately after failure and the budget allows for exactly 40 spares over five years, what is the probability that the budget is exceeded? (Hint: Use a normal approximation to the Poisson.)

14. For the previous problem determine the following:

a. The probability that a single carrier sent on a six-month mission will not require replacement of the module during that time.

b. The probability that the time of the fifth failure is more than one year after the devices are placed into service.

c. The expected time to use all 40 spares.

12.4 Failures of Complex Equipment

Many applications of reliability theory involve predicting the failure patterns of equipment from knowledge of the failure patterns of the components comprising that equipment. For example, a well-known study published by the U.S. Nuclear Regulatory Commission (1975) claimed that nuclear plants are safe. In the study, reliability theory was used to predict the likelihood of a major problem occurring in a nuclear plant by analyzing the failure rate of the various components comprising the plant. Unfortunately, incorrect assumptions about the independence of these components led to the conclusion that major nuclear accidents were virtually impossible. That is obviously not the case.

The previous section showed that a bank of items connected in series, each of which has the exponential failure law, will also have the exponential failure law. That result is a special case of one that will be derived in this section.

Components in Series

A series system will function only if every component functions. A schematic diagram of a series system appears in Figure 12–6a.

Define T_i as the time until failure of the ith component and let T_S be the time of failure of the entire series system. As above, we have that $T_S = \min(T_1, T_2, \ldots, T_N)$. Recall the definition of the reliability function, $R(t) = P\{T > t\}$. We will derive $R_S(t)$, the reliability function of the system in terms of the reliability functions of each of the components, $R_i(t)$. Using essentially the same arguments as in the previous section, we have

$$\begin{aligned}
R_S(t) = P\{T_S > t\} &= P\{\min(T_1, T_2, \ldots, T_N) > t\} \\
&= P\{T_1 > t, T_2 > t, \ldots, T_N > t\} \\
&= P\{T_1 > t\} \times P\{T_2 > t\} \times \cdots \times P\{T_N > t\} \\
&= R_1(t) \times R_2(t) \times \cdots \times R_N(t).
\end{aligned}$$

FIGURE 12–6

*Systems of
components in both
series and parallel*

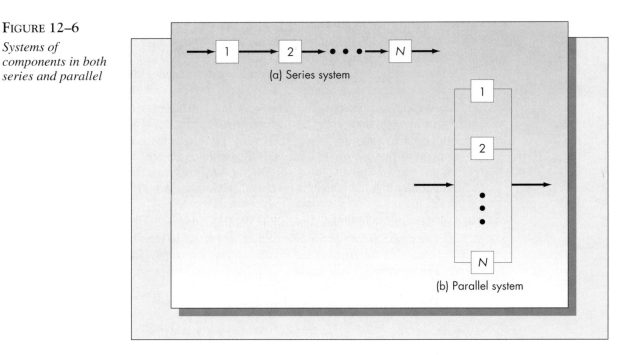

For N identical components each having reliability function $R(t)$, this becomes simply $[R(t)]^N$.

To express the cumulative distribution function of the series system in terms of the distribution function of each of the individual components, substitute $F(t) = 1 - R(t)$. For N identical components in series we obtain

$$F_S(t) = 1 - [1 - F(t)]^N.$$

Components in Parallel

Figure 12–6*b* shows a parallel system of components. A parallel system functions if any one of the components functions. Parallel systems occur when redundancy is included to increase reliability. Define T_P as the time of failure of a parallel system of identical components. It is clear that $T_P = \max(T_1, T_2, \ldots, T_N)$. For a parallel system, it is more convenient to determine the distribution function of the time until failure rather than that of the reliability function. We have that

$$F_P(t) = P\{\max(T_1, T_2, \ldots, T_N) \le t\}$$
$$= P\{T_1 \le t, T_2 \le t, \ldots, T_N \le t\}$$

(because if the largest member of a group is less than a given number, then all of the members of the group are also less than that number)

$$= F_1(t) \times F_2(t) \times \cdots \times F_N(t),$$

which reduces to $[F(t)]^N$ in the case of N identical components.

The reliability function of a parallel system with N identical components is $R_P(t) = 1 - [1 - R(t)]^N$.

Expected Value Calculations

A useful result from probability theory that may simplify calculating expected values is the following:

Theorem

If T is a nonnegative random variable with cumulative distribution function $F(t)$ and probability density function $f(t)$, then one can compute the expected value two ways:

$$E(T) = \int_0^\infty tf(t)\,dt = \int_0^\infty [1 - F(t)]\,dt.$$

The second equation can streamline calculations of the expected time until failure. We will use it to compute the expected time until failure of a parallel system of N identical components, each of which has the exponential failure law with parameter λ.

We have that

$$E(T_P) = \int_0^\infty [1 - (1 - e^{-\lambda t})^N]\,dt.$$

In order to perform the integration, we make the change of variable $v = 1 - e^{-\lambda t}$, which gives $dv = \lambda e^{-\lambda t}\,dt$, or

$$dt = \frac{1}{\lambda}\frac{1}{e^{-\lambda t}}\,dv = \frac{1}{\lambda}\frac{1}{1-v}\,dv.$$

Hence,

$$E(T_P) = \frac{1}{\lambda}\int_0^1 \frac{1 - v^N}{1 - v}\,dv.$$

The expression in the integrand is just the finite geometric series $1 + v + v^2 + \cdots + v^{N-1}$. Hence, it follows that

$$E(T_P) = \frac{1}{\lambda}\int_0^1 (1 + v + v^2 + \cdots + v^{N-1})\,dv$$

$$= \frac{1}{\lambda}(1 + \tfrac{1}{2} + \tfrac{1}{3} + \cdots + 1/n).$$

This implies that for a system with k components in parallel, the expected lifetime of the system is increased by $1/(k + 1)\lambda$ by adding one additional component.

Example 12.6

Wizard, a popular brand of electric garage door opener, includes two 40-watt bulbs that go on when the garage door is opened. A bulb will generally last about one year in normal operation. Three neighbors, James, Smith, and Walker, each has a Wizard opener in their respective garages. Each time a bulb burns out, James replaces both bulbs. Smith, on the other hand, replaces only the bulb that has burned out, and Walker replaces both bulbs only after both have burned out. Assume that light bulbs fail according to an exponential law.

  a. Over a 10-year period, how many bulbs, on average, will each neighbor require?

b. What percentage of the time will Walker have only one bulb burning?

c. Is there any advantage of James's strategy over Smith's?

Solution

a. Both James and Smith treat the system as a pure series system of two components: the system fails when one of the bulbs fails. If each bulb has failure rate $\lambda_i = 1$, then the system has failure rate $\lambda = \lambda_1 + \lambda_2 = 2$. Because James replaces two bulbs at each occurrence of a failure, he uses an average of 4 bulbs per year, or 40 bulbs in 10 years. Smith, on the other hand, only requires one bulb each time the system fails, so that he uses an average of 2 bulbs per year, or 20 bulbs during 10 years.

Walker's policy of replacing both bulbs only after both have failed is equivalent to a pure parallel system of two components. The expected lifetime of a parallel system of two components with failure rate 1 is $1 + \frac{1}{2} = 1.5$ years. Hence, every 1.5 years he requires two bulbs, thus resulting in an average of 6.67 replacements over the 10 years, which amounts to a total of 13.33 bulbs.

b. One might think that half the time he would have one bulb operating and half the time he would have two bulbs operating. However, this turns out not to be the case. Because the failure rate of the series system is $\lambda = 2$, the first bulb will fail on average after six months. As the parallel system has, on average, a 1.5-year lifetime, the remaining bulb will last an average of one year. Hence, he will be operating his garage door an average of 66.67 percent of the time with only one bulb.

c. Because the failure law is exponential, there is absolutely no advantage of James's strategy over Smith's. They will both have to make replacements equally often (twice a year) but James will use twice as many light bulbs. However, if the failure law is *not* exponential, it is possible that James's method will result in fewer occasions in which a replacement must be made. ∎

This problem raises an interesting point. In installations in which many lights are used, such as a Las Vegas hotel sign, it is a common strategy to periodically replace all bulbs at once. However, as we saw in the problem, if the bulbs follow an exponential failure law, this strategy will result in no fewer unplanned replacements, but will lead to using far more bulbs.

K *Out of* N *Systems*

Assume that a system consists of N components. A K out of N system is one in which the system functions only if at least K of the components function, where $1 \le K \le N$. A typical example of a K out of N system is a four-engine airplane that can fly as long as at least two of its engines are operating.

To analyze a K out of N system, we use a binomial framework. Think of each component as a separate Bernoulli trial: Identify a success with a functioning component and a failure with a nonfunctioning component. We assume that all components are identical, so that each has the same reliability function $R(t)$ and the same distribution function $F(t)$. Fix a point in time, t. Let $p = P\{$a component functions at time $t\} = R(t)$. The probability that the system functions at time t, $R_k(t)$, is the probability that there are at least K successes in N trials of a binomial experiment with $p = P\{$success$\}$ at each trial.

FIGURE 12–7

Reliability of 2 out of 4 and 3 out of 4 systems

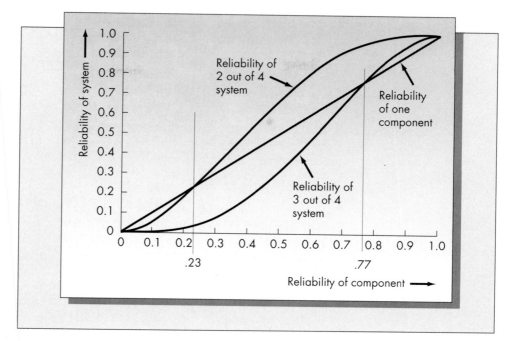

FIGURE 12–7

Reliability of 2 out of 4 and 3 out of 4 systems

Hence,

$$R_K(t) = \sum_{j=K}^{N} \binom{N}{j} p^j (1 - p)^{N-j}$$

$$= \sum_{j=K}^{N} \binom{N}{j} R(t)^j F(t)^{N-j}.$$

Note that both series and parallel systems are special cases of K out of N systems. A series system is an N out of N system, and a parallel system is a 1 out of N system. A series system of N identical components will always have lower reliability than a single component, whereas a parallel system of N identical components will always have higher reliability than a single component.

However, whether a K out of N system is more reliable than a single component depends upon the reliability of the individual components. Figure 12–7 graphs the reliability of 2 out of 4 and 3 out of 4 systems as a function of the reliability of each component. The 45-degree line represents the reliability of a single component. The reliability curve for the 2 out of 4 system crosses the 45 degree line at about $p = .23$ and for the 3 out of 4 system at about $p = .77$. This means that a 2 out of 4 system is preferred to a single component only if the reliability of the components comprising the system exceeds .23 and a 3 out of 4 system is preferred to a single component only if the component reliability exceeds .77.

Problems for Section 4

15. A subassembly in an industrial robot consists of 12 components in series, each of which fails completely at random at a rate of once every 50 years.
   a. What is the mean time between failures for this part of the robot?
   b. What is the probability that this subassembly does not fail within eight years of operation?

FIGURE 12–8

*System of compo-
nents (for
Problem 19)*

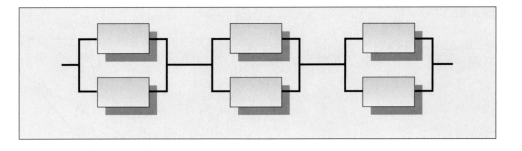

16. Show that the time until failure of a series system of n identical components, each of which has the Weibull lifetime distribution, also has a distribution of the Weibull type.

17. Consider the following three systems: (*a*) a single component with failure rate one per year, (*b*) two components in series, each of which has failure rate one every two years, and (*c*) two components in parallel, each of which has failure rate two per year. Compare the reliability of these three systems assuming an exponential failure law.

18. A design engineer is considering the number of levels of redundancy to build into a particular circuit. The circuit will be part of a sensitive piece of equipment with cost estimated at \$500 per failure. Each additional level of redundancy costs \$100. If each component fails at random at a rate of one failure every five years, what level of redundancy most closely equates the cost of the design with the expected failure cost of the equipment over its 10-year life cycle?

19. Consider the system of six identical components pictured in Figure 12–8. If each component has constant failure rate λ, derive the distribution function of the time until failure of the system.

20. An aircraft engine fails with probability p. Assume that for an aircraft to successfully complete a flight, at least half the engines must operate. Show that for $0 < p < \frac{1}{3}$, a four-engine plane is preferred to a two-engine plane and for $\frac{1}{3} < p < 1$, a two-engine plane is preferred to a four-engine plane.

12.5 Introduction to Maintenance Models

The maintenance of complex equipment often can account for a large portion of the costs associated with that equipment. It has been estimated, for example, that the maintenance costs in the military comprise almost one-third of all of the operating costs incurred. Clearly, the issues of reliability and maintenance are closely connected.

This section will introduce some standard maintenance terminology:

1. MTBF = Mean time between failures. This corresponds to the expected time between failures in our previous notation and equals $1/\lambda$.

2. MTTR = Mean time to repair. This is the expected value of the repair time R.

FIGURE 12–9

Realization of failure and repair times

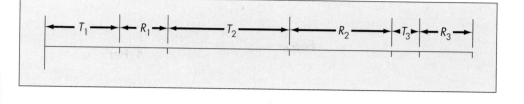

3. Availability = Average fraction of time the equipment operates. It is given by the formula

$$\text{Availability} = \frac{E(T_i)}{E(T_i) + E(R_i)} = \frac{\text{MTBF}}{\text{MTBF} + \text{MTTR}}.$$

We may think of a single piece of equipment that has successive failure times T_1, $T_2, \ldots$ and successive repair times $R_1, R_2, \ldots$ A schematic diagram of a such a system appears in Figure 12–9.

Example 12.7

A copier machine has a mean time between failures of 400 operating hours. Repairs typically require an average of 10 hours from the time that the repair call is received until service is completed. Determine the availability of this copier.

Solution

The availability is $400/(400 + 10) = 400/410 = .9756.$ ■

Define a repair cycle as the time between two successive repairs. Often we must determine the distribution of a repair cycle rather than just its expectation. A single repair cycle is the sum of the failure time T_i and the repair time R_i. The exact distribution of the sum of two random variables is the *convolution* of the individual distributions. (See DeGroot, 1986, for example.)

If the interfailure and the repair times can be reasonably approximated by the normal distribution, then the sum of the two also will be approximately normal.

Example 12.8

Suppose that the time between failures, T_i, is approximately normally distributed with mean 400 hours and variance 10,000. The repair time of the equipment is also approximately normally distributed with mean 10 hours and variance 11.6. Find the probability that there are more than six repair cycles within a one-year period. Assume that one year corresponds to 2,000 hours of operation.

Solution

We have that

$$E(T_i + R_i) = 400 + 10 = 410,$$
$$\text{Var}(T_i + R_i) = 10,000 + 11.6 = 10,011.6.$$

It follows that

$$E\left[\sum_{i=1}^{6} (T_i + R_i)\right] = 6 \times 410 = 2{,}460,$$

$$\text{Var}\left[\sum_{i=1}^{6} (T_i + R_i)\right] = 6 \times 10{,}011.6 = 60{,}069.6,$$

$$P\left\{\sum_{i=1}^{6} (T_i + R_i) \le 2{,}000\right\} = P\left\{Z \le \frac{2{,}000 - 2{,}460}{\sqrt{60{,}069.6}}\right\}$$

$$= P\{Z \le -1.88\} = .03. \quad \blacksquare$$

12.6 Deterministic Age Replacement Strategies

For operating equipment that does not exhibit an exponential failure law, there are often advantages to replacing a piece of equipment *before* it fails. This is true when the cost of repair is much higher if the equipment fails while it is operating. In some cases, such as in military operations, an equipment failure might be impossible to correct and could result in the loss of life.

This section will consider age replacement models that do not explicitly account for the uncertainty of the failure process. Rather, the aging mechanism is subsumed in the cost structure; in particular, it is assumed that the cost of maintaining the equipment increases as the equipment ages. The next section will consider models of planned replacement that explicitly include the uncertainty of the failure process.

The models we consider are appropriate for both continuously operating equipment, such as radar or power generating units, and intermittently operating equipment, such as automobiles. In the latter case, we would keep track of operating time rather than clock time.

Let us assume that the replacement cost of the item is K. Also assume that the instantaneous cost rate of operating an item of age u is $C(u)$. We will consider various forms for $C(u)$ but assume initially that $C(u) = au$.

Based on the values of the various costs, there will be some optimal point at which to replace the item in order to minimize the total cost per unit time. The total cost function may be thought of as the sum of two components: maintenance and replacement. The marginal cost of maintenance increases over time and the marginal cost of replacement decreases over time. The optimal replacement age minimizes the average cost function. In the cases we consider, the average cost function is convex, thus making the optimal solution easy to find.

The Optimal Policy in the Basic Case

We make the following assumptions:

1. The equipment used is operating continuously.
2. We ignore downtime for repair and maintenance.
3. The planning horizon is infinite.
4. Every new piece of equipment has identical characteristics.
5. Only maintenance and replacement costs are considered.

FIGURE 12–10

*Optimal age
replacement strategy*

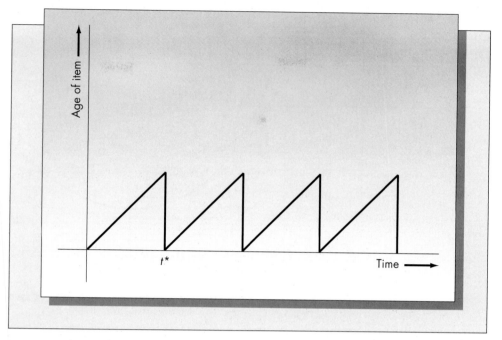

6. The objective is to minimize the long-run costs of replacement and maintenance.

7. The cost rate of maintaining an item of age u is au, and the replacement cost of the item is K. There is no salvage value.

The decision variable is the amount of time that elapses from the point that a piece of equipment is purchased until it is replaced with a new item. Figure 12–10 shows successive replacement cycles.

The object of the analysis is to determine the value of t that minimizes the total cost of maintenance and replacement over an infinite horizon. A replacement cycle is the time between successive replacements. Because all replacement cycles are identical, we may restrict attention only to the costs incurred in a single cycle.

$$\text{Total replacement cost per cycle} = K,$$

$$\text{Total maintenance costs per cycle} = \int_0^t C(u)\,du = \int_0^t au\,du = \frac{at^2}{2}.$$

The average cost per unit time is just the total cycle cost divided by the length of the cycle. Let $G(t)$ be the average cost per unit time if the replacement time is t. Then

$$G(t) = \frac{1}{t}\left(K + \frac{at^2}{2}\right) = \frac{K}{t} + \frac{at}{2}.$$

As $G''(t) = K/t^3 > 0$, it follows that $G(t)$ is a convex function of the single variable t. The function $G(t)$ is pictured in Figure 12–11. The goal is to find the optimal value of t, say t^*, that minimizes $G(t)$. Since $G(t)$ is convex, it follows that the optimal solution satisfies

$$G'(t) = \frac{-K}{t^2} + \frac{a}{2} = 0,$$

FIGURE 12–11

Computation of the optimal replacement age

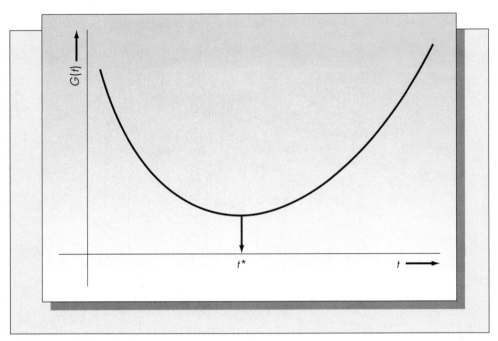

which results in

$$t^* = \sqrt{\frac{2k}{a}}.$$

Example 12.9

We will use the simple version of the age replacement model to estimate the number of years that one should keep a car. Although the model does not exactly describe the car replacement problem, we can use it as an approximation. Let us assume that the maintenance cost rate of a car u years old is $400u$ dollars. This means that the maintenance cost during the first year is $200, during the second year is $600, during the third year is $1,000, and so on. (These numbers are obtained by computing $at^2/2$ for $t = 1$, 2, and 3 and subtracting the costs of the previous years.) This is probably rising a bit faster than our actual maintenance costs would. Assume that a new car costs $10,000. According to the formula, the optimal number of years that we should hold the car is

$$t^* = \sqrt{\frac{2K}{a}} = \sqrt{\frac{(2)(10,000)}{400}} = 7.07 \text{ years.} \quad \blacksquare$$

A General Age Replacement Model

When we include a salvage value and allow for more general maintenance cost functions, the model becomes considerably more complex. As above, suppose that $C(u)$ is an arbitrary function representing the maintenance cost rate of an item of age u. Define $S(u)$ as the salvage value of an item of age u. Then the total cost incurred in the cycle is

$$K + \int_0^t C(u) \, du - S(t).$$

The average cost per unit time is

$$G(t) = \frac{K}{t} + \frac{1}{t} \int_0^t C(u) \, du - \frac{S(t)}{t}.$$

The optimal value of t, t^*, is the solution to

$$G'(t) = \frac{-K}{t^2} + \frac{H(t)}{t^2} + \frac{C(t)}{t} + \frac{S(t)}{t^2} - \frac{S'(t)}{t} = 0,$$

or

$$tC(t) + S(t) = K + H(t) + tS'(t),$$

where for convenience we let

$$H(t) = \int_0^t C(u) \, du.$$

Finding t^* can be very difficult. (For example, try to obtain a solution assuming $C(u) = au$ and $S(u) = K - bu$.) For many real problems the exponential distribution provides an accurate description of the increase in maintenance costs and the decrease in resale value of operating equipment. If we let

$$C(u) = ae^{bu}, \qquad \text{where } a, b > 0,$$

and

$$S(u) = ce^{-du}, \qquad \text{where } c, d > 0,$$

then the optimal value of t satisfies

$$tae^{bt} + ce^{-dt} = K + \int_0^t ae^{bu} \, du + t\frac{d}{dt}(ce^{-dt}).$$

It is easy to show that

$$H(t) = \int_0^t ae^{bu} \, du = \frac{a}{b}(e^{bt} - 1)$$

and

$$\frac{d}{dt}(ce^{-dt}) = -cde^{-dt},$$

so that the equation defining an optimal solution is

$$tae^{bt} + ce^{-dt} = K + \frac{a}{b}(e^{bt} - 1) - tcde^{-dt}.$$

Rearranging terms gives

$$ae^{bt}\left(t - \frac{1}{b}\right) + ce^{-dt}(1 + dt) + \frac{a}{b} = K.$$

The goal is to find the value of t that makes the left-hand side of the equation as close to K, the replacement cost, as possible. This is a difficult equation to solve for t because it involves both exponentials and constants.[2] Spreadsheets provide a convenient method of obtaining a solution. One simply computes the left-hand side of the

---

[2]It is called a transcendental equation.

equation for various values of t and graphically determines the point at which this function crosses the value of K.

Example 12.10

Consider again finding the optimal time to replace an automobile. Exponential functions are more realistic than linear functions and should give an accurate estimate of the true optimal time for replacement. As in Example 12.9, assume that the replacement cost of the automobile is $10,000. Furthermore, assume that the car loses 15 percent of its value each year. This is probably a reasonable estimate of the decline in the resale value of most new cars. This means that the car is worth $(0.85)(10,000) = \$8,500$ after one year, $(0.85)(0.85)(10,000) = \$7,225$ after two years, and so on.

We wish to determine c and d so that $S(t) = ce^{-dt}$ agrees with these values. Because the salvage value at time $t = 0$ is exactly the replacement cost, we have $S(0) = 10,000$. Substituting, we obtain

$$S(0) = ce^{-d(0)} = c = 10,000.$$

The value of the car after one year is $(0.85)(10,000)$, which corresponds to $S(1)$. Hence,

$$S(1) = (0.85)(10,000) = ce^{-d(1)} = ce^{-d}.$$

Since $c = 10,000$, we obtain

$$e^{-d} = 0.85,$$
$$d = -\ln(0.85) = 0.1625.$$

Hence, it follows that

$$S(t) = 10,000e^{-0.1625t}.$$

Now consider the maintenance costs. Assume, as in Example 12.9, that maintenance costs for the first year of operation amount to $200. This is equivalent to

$$H(1) = 200$$

or

$$(a/b)(e^b - 1) = 200.$$

Furthermore, suppose that the maintenance costs increase at a rate of 40 percent per year. This means that

$$\frac{C(t)}{C(t-1)} = 1.4.$$

Substituting for $C(t)$ gives

$$\frac{ae^{bt}}{ae^{b(t-1)}} = e^b = 1.40,$$

or $b = \ln(1.4) = 0.3365$. It follows that

$$a = \frac{(200)(b)}{e^b - 1} = \frac{(200)(0.3365)}{0.4} = 168.25.$$

FIGURE 12–12

*Optimal number of
years to replace
auto (M = .4)*

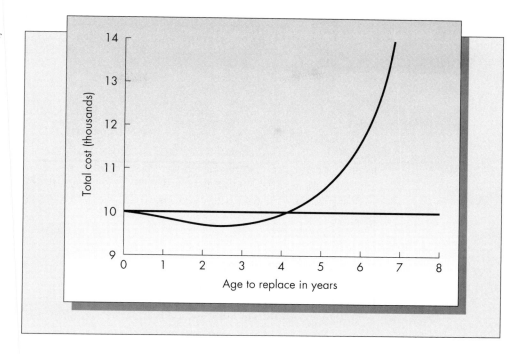

FIGURE 12–12

*Optimal number of
years to replace
auto (M = .4)*

Combining these results, it follows that the optimal time to replace the automobile, t^*, is the value of t that solves

$$168.25e^{0.3365t}(t - 2.972) + 10{,}000e^{-0.1625t}(1 + 0.1625t) + 500 = 10{,}000.$$

One method of estimating the solution to this equation is to graph the function represented by the left-hand side of the equation. We have done so in Figure 12–12. The minimum-cost replacement age is about 4.5 years. That means that for an automobile costing \$10,000 that declines in value at the rate of 15 percent per year, and for which the maintenance cost is \$200 the first year and increases at the rate of 40 percent per year, the optimal strategy, which minimizes average costs of replenishment less salvage plus maintenance, is to replace the car about once every four and a half years. ∎

To see the effect of the maintenance cost, we have re-solved the example with a value of 20 percent rather than 40 percent for the rate of increase of maintenance costs per year. This is probably more accurate for most cars. The solution is represented graphically in Figure 12–13. Here, the optimal replacement time is approximately nine years.

Let M represent the rate at which maintenance costs increase each year expressed as a fraction ($M = .20$ in Figure 12–13). Let I_0 be the maintenance cost in the first year, and D the yearly rate of depreciation, also expressed as a fraction ($D = 0.15$) in the example). Then it can be shown that

$b = \ln(1 + M),$
$a = I_0 b/M,$
$c = K,$
$d = \ln(1 - D).$

FIGURE 12–13

Optimal number of years to replace auto (M = .2)

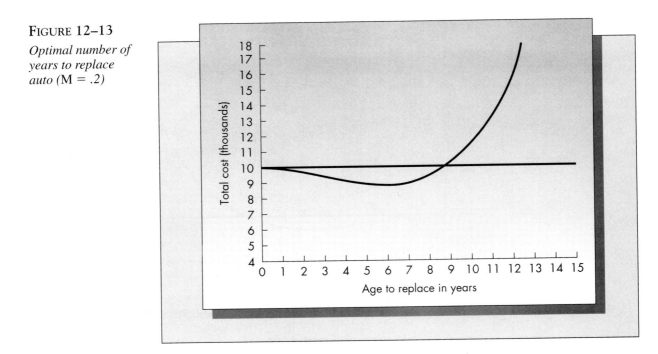

The models presented in this section do not consider the effects of inflation. If inflation effects were included, the resulting solutions would not change appreciably, because the inflation is applied to both the replacement and the maintenance costs. The method is to compute the present value of all future discounted costs and determine the replacement strategy that minimizes the resulting function. We will not consider discounted cash flows in this section.

Problems for Section 6

21. For the basic age replacement model, consider a piece of equipment that costs $18,000 to replace. The *total* maintenance costs for five years of operation are estimated to be $2,400. Assuming a linear maintenance cost rate, find the value of a and the optimal age at which the equipment should be replaced.

22. For the basic age replacement model, derive the optimal replacement age when the maintenance cost rate $C(u)$ has the form $C(u) = a\sqrt{u}$ for some constant $a > 0$.

23. Suppose for the simple age replacement model that the maintenance cost is the same every year (that is, $C(u) = a$ for all $u \geq 0$). What is the optimal replacement age? Why is this so?

24. The army is attempting to determine the optimal replacement age for a piece of field equipment. The equipment costs $280,000 to replace. The manufacturer will supply a rebate toward the next purchase that declines at a rate of 20 percent per year. Maintenance costs for the first year are estimated to be $1,000, and they increase roughly at the rate of 18 percent per year. Estimate the number of years that the army should hold the equipment before making a replacement.

25. Try to determine the optimal replacement age when $C(u) = au$ and $S(u) = K - bu$. What difficulty do you encounter?

12.7 Planned Replacement under Uncertainty

The purpose of preventive maintenance is to decrease the likelihood that an item will require replacement because of failure. At the heart of such a policy is the assumption that it costs more to make a repair or replacement at the time of failure than at some predetermined time. For example, if failure means that a production line must be stopped to determine the cause of the failure and repair the problem, whereas preventive maintenance can be accomplished at a convenient time when the system is not operating, then the cost of planned replacements is less than the cost of unplanned replacements.

Because of the memoryless property of the exponential distribution, if an item or group of items obeys an exponential failure law, then there is no advantage to replacing prior to failure. In the exponential case, the likelihood that failure will occur in a time Δt is the same just after a planned replacement as it is for an item that has been operating for an arbitrary amount of time. Hence, planned replacement strategies can have value only if the items exhibit aging, that is, have an increasing failure rate function.

Planned Replacement for a Single Item

Consider a single piece of continuously operating equipment whose lifetime is a random variable T with known cumulative distribution function $F(t)$. We assume that T is a continuous random variable. Suppose that it costs c_1 to replace the item when it fails and $c_2 < c_1$ to replace the item prior to failure. We assume that planned replacements are made exactly t units of time after the last replacement. The goal is to find the optimal value of t to minimize the average cost per unit time of both planned and unplanned replacements.

A cycle is the time between successive replacements. Because the process "restarts" itself after each replacement, irrespective of whether the replacement was planned or unplanned, we may use the renewal method to obtain an expression for the expected cost per unit time. That is,

$$E \text{ (cost per unit time)} = \frac{E \text{ (cost per cycle)}}{E \text{ (length of a cycle)}}.$$

Renewal arguments have been used before in the text (Chapter 11, Section 6). This approach also was used in a variety of other places, including the previous section on age replacement and in much of Chapters 4 and 5 on inventory modeling. Successive replacement cycles are pictured in Figure 12–14.

We have that

$$E \text{ (cost per cycle)} = c_1 P\{\text{replacement is result of failure}\}$$
$$+ c_2 P\{\text{replacement is planned}\}.$$

Notice that $P\{\text{replacement is result of failure}\} = P\{T \leq t\} = F(t)$, and $P\{\text{replacement is planned}\} = P\{T > t\} = 1 - F(t)$, where T is the lifetime of the item placed into

FIGURE 12–14

Successive cycles for planned replacement of a single item

FIGURE 12–14

Successive cycles for planned replacement of a single item

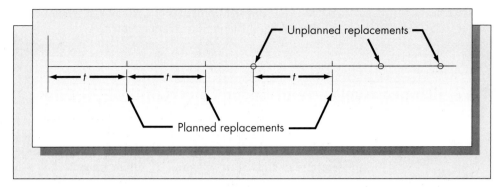

service at the end of the previous cycle. It follows that

$$E \text{ (cost per cycle)} = c_1 F(t) + c_2(1 - F(t)).$$

Let T be the time of failure of the item placed into service at the end of the previous cycle. Then, clearly, the next replacement will occur at $\min(T, t.)$ Hence,

$$E \text{ (length of cycle)} = E[\min(T, t)] = \int_0^\infty \min(x, t) f(x)\, dx$$

$$= \int_0^t x f(x)\, dx + t \int_0^\infty f(x)\, dx$$

$$= \int_0^t x f(x)\, dx + t[1 - F(t)].$$

It follows that the expected cost per unit time, say $G(t)$, is given by

$$G(t) = \frac{c_1 F(t) + c_2[1 - F(t)]}{\int_0^t x f(x)\, dx + t[1 - F(t)]}.$$

The goal is to find t to minimize $G(t)$. The optimization may be cumbersome depending upon the form of the lifetime distribution $F(t)$.

We will now show that there is no advantage to planned replacement when the lifetime distribution is exponential. Suppose that $F(t) = 1 - e^{-\lambda t}$. Then the expected length of each cycle is

$$\int_0^t x e^{-\lambda x}\, dx + t e^{-\lambda t} = \frac{1}{\lambda}[1 - e^{-\lambda t}(1 + \lambda t)] + t e^{-\lambda t}$$

$$= \frac{1}{\lambda}(-e^{-\lambda t}).$$

(The expression for $\int_0^t x e^{-\lambda x}\, dx$ can be obtained by integration by parts or can be found in a table of integrals.) It follows that

$$G(t) = \frac{c_1(1 - e^{-\lambda t}) + c_2 e^{-\lambda t}}{\frac{1}{\lambda}(1 - e^{-\lambda t})} = \frac{c_1 - (c_1 - c_2)\, e^{-\lambda t}}{\frac{1}{\lambda} - \frac{1}{\lambda} e^{-\lambda t}}.$$

As $t \to \infty$, the term $e^{-\lambda t} \to 0$, so that $G(\infty) = \lambda c_1$. Furthermore, substituting $t = 0$ results in $G(0) = \infty$. It can be shown by either calculus or direct computation that the

FIGURE 12–15

The function G(t)

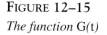

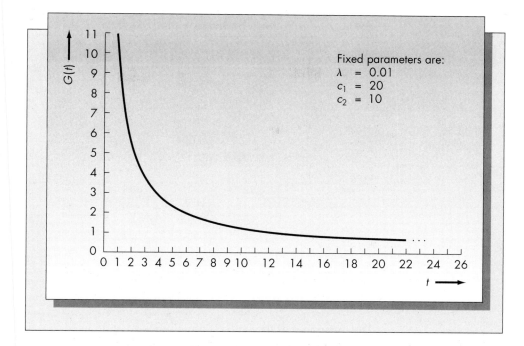

Fixed parameters are:
$\lambda = 0.01$
$c_1 = 20$
$c_2 = 10$

function $G(t)$ is monotonically decreasing; a typical case is pictured in Figure 12–15. Hence, the optimal solution is $t = \infty$, which means that a planned replacement should never be made.

We have shown that if the lifetime distribution is exponential (constant failure rate), then there is no economy in replacing an item prior to the time it fails. This also holds if the failure rate is decreasing.

Example 12.11

A large trucking company, Harley Brown Inc., maintains detailed records on the mortality of the tires used on company-owned trucks. A statistical analysis of the data on tire failure shows that the lifetime of a tire as measured in thousands of miles of use is closely approximated by the Weibull probability law with parameters $a = 0.00235$ and $\beta = 2.3$. The company estimates a cost of $450 if a tire fails during use. This is a result of the lost time and the potential liability of an accident. Tires replaced before failure cost $220 each. The company would like to find the optimal timing of tire replacement.

The difficulty with calculating the optimal solution for this problem is determining an expression for the term $\int_0^t x f(x)\, dx$ that appears in the denominator of $G(t)$.

To circumvent the problem of finding an analytical expression for this integral, we will obtain a discrete approximation to the failure law and perform the calculation as if the lifetime distribution were discrete rather than continuous. The probability of failure within the first 1,000 miles is

$$P\{T \le 1\} = 1 - F(1) = .0023.$$

The probability of failure after 1,000 miles but before 2,000 miles of wear is

$$P\{T \le 2\} - P\{T \le 1\} = R(1) - R(2) = .0092.$$

TABLE 12–2   **Failure Probabilities for Example 12.11**

Lifetime (thousands of miles)	Probability of Failure	Lifetime (thousands of miles)	Probability of Failure
1	.0023	14	.0625
2	.0092	15	.0580
3	.0175	16	.0527
4	.0264	17	.0469
5	.0354	18	.0408
6	.0440	19	.0348
7	.0517	20	.0291
8	.0582	21	.0238
9	.0632	22	.0191
10	.0664	23	.0150
11	.0679	24	.0116
12	.0677	25	.0087
13	.0658		

The remainder of the failure probabilities are computed in a similar fashion and appear in Table 12–2.

Using these discrete probabilities we may now compute $G(t)$ directly. The partial expectation term

$$\int_0^t xf(x)\,dx \approx \sum_{k=1}^t kp_k,$$

where the probabilities p_k are as given in Table 12–2. Note that this approximation assumes that all failures occur at multiples of 1,000 miles.

The remainder of the terms comprising $G(t)$ can be obtained directly from the Weibull reliability function

$$R(t) = e^{-\alpha t^\beta}.$$

Hence, the approximate form of $G(t)$ may be written

$$G(t) = \frac{c_1 - (c_1 - c_2)R(t)}{\sum_{k=1}^t kp_k + tR(t)}.$$

$G(t)$ appears in Table 12–3 for the parameter values $\alpha = 0.00235$, $\beta = 2.3$, $c_1 = 450$, and $c_2 = \$220$. The function appears to be convex and is minimized at $t = 13$. Hence, the optimal policy calls for replacing the tires after about 13,000 miles of wear. The value of the objective function at the optimal solution is $G(13) = 33.21$. This means that at the optimal solution, the replacement cost is $33.21 per thousand miles of use per tire. ∎

Block Replacement for a Group of Items

In certain circumstances it is more economical to replace groups of items at the same time rather than one by one. The example above showed that the optimal policy was to replace a truck tire after about 13,000 miles of use. Depending upon the time and the expense involved in changing truck tires, it could be more economical to replace all the tires on a

TABLE 12–3   The Function $G(t)$ for Example 12.11

t	$G(t)$	t	$G(t)$
1	220.54	14	33.24
2	111.45	15	33.35
3	75.91	16	33.53
4	58.82	17	33.72
5	49.14	18	33.93
6	43.20	19	34.13
7	39.38	20	34.32
8	36.89	21	34.48
9	35.27	22	34.62
10	34.25	23	34.74
11	33.64	24	34.84
12	33.33	25	34.91
13	33.21		

truck when a replacement is made. The costs of transporting the truck to a service area, placing the truck on a lift, and paying a technician to mount and balance the tires could be comparable to the cost of the tire itself. If all the tires were replaced simultaneously, this cost would be incurred less often than if the tires were individually replaced.

This section will consider a model to determine the optimal time to replace an entire group of items. In order to avoid intricate mathematics that are beyond the scope of this book, we will assume that the lifetime of each operating unit is a discrete random variable with a known distribution. That is, suppose that p_k is the probability that an item fails in period k assuming the item was placed into service at period 0. These probabilities may be estimated directly from historical data or computed from a continuous distribution as in the example above.

Assume that n_0 items are placed into service at time 0. Suppose there is no block replacement and all items that fail in a period are replaced at the end of that period. We also will assume for simplicity that p_k is the actual proportion of units k periods old that fail. Then the number of failures occurring in period 1 is $n_1 = n_0 p_1$.

In period 2 the proportion of the original group of items that fail is $n_0 p_2$, and the proportion of the items placed into service in period 1 that fail is $n_1 p_1$. Hence, the expected number of failures in period 2 is $n_2 = n_0 p_2 + n_1 p_1$. Continuing with this argument, we obtain

$$n_k = n_0 p_k + n_1 p_{k-1} + \cdots + n_{k-1} p_1.$$

Now suppose that individual replacements cost a_1 each and the entire block of n_0 can be replaced for a_2. If all n_0 items were replaced at the end of each period, the cost each period would be $a_2 + a_1 n_1$. If all n_0 items were replaced at the end of every other period, the cost incurred every two periods would be $a_2 + a_1(n_1 + n_2)$ or an average per period cost of $[a_2 + a_1(n_1 + n_2)]/2$. Similarly, the average per period cost of replacing all n_0 items after k periods is

$$G(k) = \frac{a_2 + a_1 \sum_{j=1}^{k} n_j}{k}.$$

The optimal number of periods to replace all n_0 items is the value of k that minimizes $G(k)$. The minimum value of $G(k)$ should be compared to the expected cost per period assuming that items are replaced as they fail. Let

$$E(T) = \sum_{k=1}^{\infty} kp_k$$

represent the expected lifetime of a single item. Then $\lambda = 1/T$ is the failure rate of a single item. It follows that the cost of making replacements to items on a one-at-a-time basis is $a_1\lambda$ per item or $n_0 a_1 \lambda$ for the entire block of items. This should be compared to the optimal value of $G(k)$ to determine if a block replacement strategy is economical.

Example 12.12

A large sign is lit by 8,000 bulbs. The bulbs cost $2 each to replace as they fail but can be replaced for 30 cents each when they are replaced all at once. Based on past experience, bulbs fail according to the following probability law:

Months of Service	Probability of Failure
1	.02
2	.03
3	.03
4	.05
5	.08
6	.09
7	.07
8	.10
9	.11
10	.13
11	.15
12	.14

The first step is to compute n_k. We have

$n_0 = 8,000,$
$n_1 = n_0 p_1 = (8,000)(.02) = 160,$
$n_2 = n_0 p_2 + n_1 p_1 = (8,000)(.03) = (160)(.02) = 243,$
and so on.

These values are used to compute $G(k)$. The results of the calculation appear in Table 12–4. Using values of $a_2 = (0.30)(8,000) = \$2,400$ and $a_1 = 2$, we see that the optimal time to replace the block of bulbs is after four months with an expected monthly cost of $1,135.

It is interesting to compare block replacement with a policy of replacing the bulbs only if they fail. Each bulb has an expected lifetime of

$$E(T) = \sum_{k=1}^{12} kp_k = 8.22 \text{ months.}$$

TABLE 12–4 $G(k)$ for Example 12.12

k	p_k	n_k	G_k
1	.02	160.0	2,720.0
2	.03	243.2	1,603.2
3	.03	249.2	1,235.2
4	.05	417.1	1,135.0
5	.08	671.1	1,176.4
6	.09	778.4	1,239.8
7	.07	654.6	1,249.7
8	.10	930.5	1,326.1
9	.11	1,064.0	1,415.2
10	.13	1,298.4	1,533.4
11	.15	1,542.9	1,674.5
12	.14	1,562.4	1,795.4

Hence, the failure rate is $1/8.22 = 0.12165$ failure per month per bulb. For 8,000 bulbs this amounts to an average number of failures of 973.24 per month. The resulting replacement cost is $1,946.47 monthly, which is considerably more than the cost of replacing bulbs as a block every four months, which has an expected monthly cost of $1,135. ∎

Problems for Section 7

26. Consider Example 12.11 of Harley Brown Inc. Without performing the calculations, discuss what the effect on the optimal replacement policy would likely be if the parameter values were $c_1 = \$800$, $c_2 = \$300$, and $\beta = 1$.

27. Repeat the calculations for the Harley Brown trucking company using the following parameter values: $\alpha = 0.0156$, $\beta = 1.8$, $c_1 = 1,000$, $c_2 = 600$.

28. An expensive piece of equipment is used in the masking operation for semiconductor manufacture. A capacitor in the equipment fails randomly. The capacitor costs $7.50, but if it burns out while the machine is in use, the production process must be halted. Here the replacement cost is estimated to be $150. Based on past experience, the lifetime distribution of the capacitor is estimated to be

Number of Months of Service	Probability of Failure
1	.08
2	.12
3	.16
4	.26
5	.22
6	.16

How often should the capacitors be replaced in order to minimize the expected monthly cost of planned and unplanned replacement?

29. A large electronic pipe organ contains 100 fuses. Because of the power demands of the organ, the fuses burn out at a fairly regular rate, but newer fuses last longer than older ones. The probability distribution of the lifetime of a fuse is closely approximated by the Weibull law with $\alpha = 0.0204$ and $\beta = 1.8$. Assume that t is hours of playing time. The fuses cost $1.35 each when replaced as a block but $12 each when replaced just after a failure.
   a. Express the lifetime distribution as a discrete distribution assuming t is measured in hours. (Follow the procedure used in Example 12.11 for the Harley Brown trucking company.)
   b. Determine the optimal time to replace all 100 fuses and the average hourly cost of that policy.
   c. Compare the answer you obtained in part (b) with the cost of replacing the fuses as they fail. Which policy would you recommend?

30. Tires that fail in service result in significantly higher replacement costs than those replaced before failure. For an 18-wheeler (that is, a truck with 18 tires), failure on the road costs $300, whereas all 18 tires can be replaced prior to failure at a cost of $75 per tire. The probability of failure is given below.

Number of Miles	Probability of Failure
0–5,000	.05
5,001–10,000	.15
10,001–15,000	.20
15,001–20,000	.40
20,001–25,000	.20

If tires are replaced at multiples of 5,000 miles only, what is the optimal age replacement policy?

31. The navy uses a certain type of vacuum tube in a sonar scanning device. Based on past experience, the vacuum tube exhibits the following failure pattern:

Number of Months of Operation	Probability of Failure
1	.1
2	.1
3	.2
4	.1
5	.3
6	.2

Failures during operation cost $200 each, but the tube can be replaced before failure for $50. Find the optimal replacement strategy.

32. A local newsletter is printed on a printer with a ribbon that may break or run out of ink during operation. The ribbons can be replaced for $7.50, but if they fail

when the newsletter is being printed, the cost is estimated to be $25 because of the delay in publication. The failure distribution of the ribbons is

Weeks of Use	Probability of Failure
1	.1
2	.2
3	.3
4	.4

Determine the optimal time to replace the ribbons.

33. Mactronics produces industrial robots. Each robot contains a part with a lifetime of five years at most, and at least one year. Lifetimes between one and five years are equally likely. If the part fails during operation, replacement costs are estimated to be $400, whereas the part can be replaced before failure for $50. When should the part be replaced? (Hint: The lifetime distribution is uniform on $\{1, 2, 3, 4, 5\}$. Assuming discrete variables, this means that $f(x) = \frac{1}{5}$ for $x = 1, 2, \ldots, 5$.)

34. A firm has purchased 30 Mactronics robots, described in the previous problem. If replaced as a block, the parts cost $20 each, but cost $400 each when replaced after failure.
    a. Find the optimal block replacement strategy.
    b. Compare this to the cost of replacing the items as they fail.
    c. Compare the cost of the block replacement policy you obtained in part (a) with the solution you obtained in part (b). Is the block replacement strategy preferred to an individual replacement strategy?

*12.8 Analysis of Warranty Policies

An important issue related to the reliability of operating equipment is the protection afforded to the consumer who experiences failure of the equipment prior to its intended lifetime. Buyers and sellers perceive warranties differently. From the seller's point of view, the warranty is a means of limiting liability by specifying consumer responsibilities. These responsibilities include proper use of the product and following the warnings. From a marketing perspective, the warranty also can serve as an inducement to purchase the product. From the buyer's point of view, the warranty is a means of reducing or eliminating the economic penalty if the product fails to operate properly for a reasonable period of time. Warranties are particularly important to the consumer for products that are likely to experience high failure rates early in the product lifetime.

This section will present mathematical models for determining the economic value of a warranty. Such models could be used to find the portion of the cost of an item that could reasonably be attributed to the costs of satisfying a warranty commitment. In the models, we consider both the structure of the warranty and the reliability of the product to find the value of a warranty.

We must distinguish between repairable and nonrepairable items. Nonrepairable items include most electronic components, items in which failure corresponds to destruction of the item (burning out of a bulb or blowout of a tire, for example), or items typically not repaired but replaced (such as batteries that fail to hold a charge). Most major appliances, such as washing machines, televisions, and so on, fall into the category of repairable items. Repairable items that fail during the warranty period are typically repaired rather than replaced. The mathematical models presented in this section assume nonrepairable items.

Warranties for nonrepairable consumer goods generally take one of two forms. One is the free replacement warranty: if a failure occurs during the warranty period, a new item is supplied without charge. The second type of warranty is the pro rata warranty. Here, the consumer is given a rebate proportional to the amount of time remaining in the warranty period. The rebate is used to reduce the cost of a replacement item.

The Free Replacement Warranty

Assume that a single piece of operating equipment is placed into service and fails completely at random (that is, according to an exponential failure law) with known failure rate λ. The item is assumed to operate continuously. For intermittently operating equipment, clock time could be measured in operating hours rather than elapsed hours.

We will use the following notation:

T = Lifetime of an item chosen at random.

λ = Failure rate of an item chosen at random.

$F(t)$ = Cumulative distribution function of the random variable t.

C_1 = Cost of purchasing a new item with free replacement warranty.

K = Cost of purchasing a new item without any warranty.

W_1 = Time that the free replacement warranty is in effect after purchase.

If a failure occurs during the warranty period, the item is replaced free of charge. Assume that the consumer purchases a new item when a failure occurs after the expiration of the warranty. The new item has an identical free replacement warranty. Let Y be a random variable representing the time between successive purchases by the consumer. From Figure 12–16, we see that

$$Y = W_1 + \text{Time until the first failure after the warranty expires.}$$

It can be shown that

$$E(Y) = W_1 + 1/\lambda.$$

FIGURE 12–16

Replacement cycles for free replacement warranty

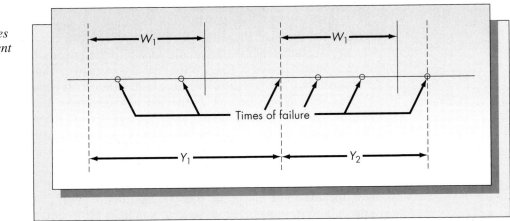

This expression for $E(Y)$ is valid only when the lifetime distribution is exponential. It results from the property of the exponential failure law that the time of the first failure after a fixed time (known as the forward recurrence time in probability theory) has the same distribution as the time between two successive failures. This result is true *only* if the failure law is exponential.

We will now compute the cost per unit time for an item that is replaced infinitely many times and has a free replacement warranty.

Each time that an item is purchased after the warranty expires constitutes the start of a new cycle. As the cost per cycle is C_1, it follows that the average cost per unit time under the free replacement warranty is

$$\frac{C_1}{E(Y)} = \frac{C_1}{W_1 + 1/\lambda} = \frac{\lambda C_1}{\lambda W_1 + 1}.$$

Without the warranty, the cost of replacement per unit time is simply λK. Let C_1^* be the cost of an item with a free replacement warranty that is indifferent to the cost of an item without the warranty, K. Then C_1^* solves

$$\lambda K = \frac{\lambda C_1^*}{\lambda W_1 + 1}$$

or

$$C_1^* = (\lambda W_1 + 1)K.$$

By definition $C_1^* - K$ is the economic value of the warranty. If $C_1 > C_1^*$, then the warranty should be purchased.

The Pro Rata Warranty

Assume all notation of the previous section. We also define

C_2 = Cost of a new item with a pro rata warranty.
W_2 = Effective warranty period with a pro rata warranty.

Consider an item purchased under a pro rata warranty that fails at a random time T. There are two cases:

Case 1: $T < W_2$. In this case the fraction of the warranty period that has expired is T/W_2. The cost of the replacement item is then $C_2(T/W_2)$.

Case 2: $T \geq W_2$. Here, the warranty has expired and the replacement cost is C_2.

Both of these cases can be represented mathematically by the expression

$$\frac{C_2}{W_2} \min(W_2, T).$$

To determine the expected life-cycle cost, we need to find an expression for $E[\min(W_2, T)]$. We have that

$$E[\min(W_2, T)] = \int_0^\infty \min(W_2, t)\lambda e^{-\lambda t}\, dt$$

$$= \int_0^{W_2} t\lambda e^{-\lambda t}\, dt + W_2 \int_{W_2}^\infty \lambda e^{-\lambda t}\, dt,$$

where $\lambda e^{-\lambda t}$ is the probability density of the time until failure, T.

The first integral requires integration by parts and the second is the reliability function evaluated at W_2. It is easy to show that

$$\int_0^{W_2} t\lambda e^{-\lambda t}\, dt = \frac{1}{\lambda}[1 - e^{-\lambda W_2}(1 + \lambda W_2)]$$

and

$$W_2 \int_{W_2}^{\infty} \lambda e^{-\lambda t}\, dt = W_2 e^{-\lambda W_2}.$$

Combining terms, it follows that

$$E[\min(W_2, T)] = \frac{1}{\lambda}(1 - e^{-\lambda W_2}).$$

The pro rata warranty starts anew with each purchase. Hence, each purchase begins a new cycle. It follows that the expected cost per unit time following a pro rata warranty is

$$\frac{C_2}{W_2}\frac{E[\min(W_2, T)]}{E(T)} = \frac{C_2}{W_2}\frac{(1/\lambda)(1 - e^{-\lambda W_2})}{1/\lambda} = \frac{C_2(1 - e^{-\lambda W_2})}{W_2}.$$

The cost of the pro rata warranty that is indifferent to the cost of the item without the warranty, C_2^*, solves

$$\frac{C_2^*(1 - e^{-\lambda W_2})}{W_2} = \lambda K,$$

or

$$C_2^* = \frac{\lambda K W_2}{1 - e^{-\lambda W_2}}.$$

The value of the pro rata warranty is $C_2^* - K$. If $C_2 > C_2^*$, then the pro rata warranty should not be purchased, and if $C_2 < C_2^*$, the pro rata warranty should be purchased. This assumes that the consumer is willing to base his decision on expected values. If the consumer is risk averse, the indifference value will be slightly higher than C_2^*.

Example 12.13

You are considering purchasing a battery for your automobile. You find the same battery offered at three different stores. Store A sells the battery for $21 and offers no warranty or guarantee. Store B sells the battery for $40 and offers a free replacement if the battery fails to hold a charge for the first two years of operation. Store C sells the battery for $40 as well but offers a pro rata warranty for the anticipated lifetime of the battery, which is advertised to be five years. The failure rate of the battery depends on the usage and the conditions, but from past experience you estimate that the time between failure is about once every three years.

We will determine the values of C_1^* and C_2^* for both warranties. For the full replacement warranty, we have that

$$K = 21,$$
$$C_1 = 40,$$
$$\lambda = \tfrac{1}{3},$$
$$W_1 = 2.$$

It follows that

$$C_1^* = (\lambda W_1 + 1)K = [(\tfrac{1}{3})(2) + 1](21) = \$35.$$

This means that the value of the free replacement warranty is $\$35 - \$21 = \$14$, which is less than the $\$19$ difference in the prices. On the basis of expected costs, the battery with no warranty (store A) is preferred to the one with the free replacement warranty (store B).

For the case of the pro rata warranty offered by store C, $W_2 = 5$ and the remaining parameters are as defined above. In that case we obtain

$$C_2^* = \frac{\lambda K W_2}{1 - e^{-\lambda W_2}} = \frac{\tfrac{1}{3}(21)(5)}{1 - e^{-5/3}} = \$43.15.$$

The value of the pro rata warranty is $\$43.15 - \$21.00 = \$22.15$, which exceeds the difference between the price of the battery with the warranty and without, and exceeds the value of the free replacement warranty. Hence, on the basis of this analysis, the pro rata warranty is the preferred choice. ∎

Extensions and Criticisms

A criticism of Example 12.13 is the assumption that the failure law for the batteries is exponential. This means that a new battery has the same probability of failing in its first year of operation as a four-year-old battery has of failing in its fifth year of operation. In the case of batteries, it would seem that a failure law incorporating aging, such as the Weibull, would be more accurate. The models discussed above can be extended to include more general types of failure patterns, but the calculations required to determine an optimal policy are very complex. In particular, one must determine the renewal function, which is generally a difficult computation. In Problem 41 an example is considered for the case in which the failure law follows an Erlang distribution.

The two types of warranties discussed in this section, the free replacement warranty and the pro rata warranty, account for the majority of consumer warranties for nonrepairable items. In the military context, another type of warranty is very common. This is known as the *reliability improvement warranty* (RIW). In this case, the supplier agrees to repair or replace items that fail within a specified warranty period and provides a pool or pools of spares and perhaps one or more repair facilities. This type of warranty is intended to provide an incentive to the supplier to make initial reliability high and to provide improvements in the reliability of existing units if possible.

For repairable items, warranties usually cover all or some of the cost of effecting repairs during the warranty period. Analysis of warranties for repairable items is more complex than that for nonrepairable items. Different levels of repair are possible and the likelihood of failure during the warranty period may depend upon usage, which can vary significantly from one user to another.

For repairable items, an issue closely related to warranties is service contracts. The primary difference between a warranty and a service contract is that the cost of the warranty is usually assumed to be included in the purchase price of the item, whereas a service contract is an additional item whose purchase is at the option of the buyer. A service contract can be thought of as an extended warranty beyond the normal warranty period. The analytical approach discussed here also can be applied to determining the economic value of a service contract. However, complex failure laws and different costs associated with various levels of repair should be allowed. In addition, the pricing of consumer service contracts relies heavily on the assumption that the consumer is risk averse. That is, a typical consumer would prefer to pay more than a service contract is worth in an expected value sense to reduce the risk of incurring an expensive repair. In that sense, the service contract serves the role of an insurance policy. Perhaps it is the tendency of consumers to be risk averse that justifies the rather high prices that are charged for service contracts for some consumer products such as videocassette recorders. (See Lund, 1978, pp. 4–63.)

Problems for Section 8

35. For Example 12.13, what value of the warranty period equates the full replacement warranty with no warranty? (That is, for what value of W_1 is the consumer indifferent to purchasing from store A or store B?)

36. For Example 12.13, what value of the warranty period for the full replacement warranty equates the full replacement and the pro rata warranties? (For what value of W_1 is the consumer indifferent to purchasing from store B or store C, assuming $W_2 = 5$?)

37. A producer of pocket calculators estimates that the calculators fail at a rate of one every five years. The calculators are sold for $25 each with a one-year free replacement warranty but can be purchased from an unregistered mail-order source for $18.50 without the warranty. Is it worth purchasing the calculator with the warranty?

38. For the previous problem, what length of period of the warranty equates the replacement costs of the calculator with and without the warranty?

39. Zemansky's sells tires with a pro rata warranty. The tires are warranted to deliver 50,000 miles with the rebate based on the remaining tread on the tire. The tires fail on the average after 35,000 miles of wear. Suppose the tires sell for $50 each with the warranty. If failures occur completely at random, what would be a consistent price for the tires if no warranty were offered?

40. Habard's, a chain of hardware stores, sells a variety of tools and home repair items. One of their best wrenches sells for $5.50. Habard's will include a three-year free replacement warranty for an additional $1.50. The wrench is expected to be subject to heavy use and, based on past experience, will fail randomly at a rate of one every eight years. Is it worth purchasing the warranty?

41. Consider the case in which the failure mechanism for the product does not obey the exponential law. In that case, the cost under the free replacement warranty that is indifferent to the cost of buying the item without a warranty is given by

$$C_1^* = K[M(W_1) + 1],$$

where $M(t)$ is known as the renewal function.

    If the time between failures, T, follows an Erlang law with parameters λ and 2, then

$$M(t) = \frac{\lambda t}{2} - 0.25 + 0.25e^{-2\lambda t} \qquad \text{for all } t \geq 0.$$

(See, for example, Barlow and Proschan, 1965, p. 57.)

a. For Example 12.13, presented in this section, determine the indifference value of the item with a free replacement warranty when the failure law follows an Erlang distribution. Assume that $\lambda = \frac{2}{3}$ to give the same value of $E(T)$ as in the example.

b. Is the value of the warranty larger or smaller than in the corresponding exponential case? Explain the result intuitively.

12.9 Software Reliability

Software reliability is a problem with characteristics different from hardware reliability problems. Typically, new software possesses a few "bugs," or errors. Ideally, one would like to remove all the bugs from the software before its release, but that may be impossible. It is more reasonable to release the software when the number of bugs has been reduced to an acceptable level. Predicting the number of remaining bugs is, however, a difficult problem.

The importance of software reliability cannot be overemphasized. Quoting from *The Wall Street Journal* (Davis, 1987):

> The tiniest software bug can fell the mightiest machine—often with disastrous consequences. During the past five years, software defects have killed sailors, maimed patients, wounded corporations and threatened to cause the government-securities market to collapse. Such problems are likely to grow as industry and the military increasingly rely on software to run systems of phenomenal complexity, including President Reagan's proposed "Star Wars" anti-missile defense system.

Several models have been proposed for estimating software reliability. However, we will not present these models in detail because their utility has yet to be determined. Jelinski and Moranda (1972) have suggested the following approach. Let N be the total initial error content (i.e., the number of bugs) in the software. As the software undergoes testing, the number of bugs is reduced. They assume that the failure rate (that is, the likelihood of detecting a bug) is proportional to the number of bugs remaining in the program, where ϕ is the proportionality constant. That is, the time until detection of the first bug has the exponential distribution with parameter $N\phi$; the time between detection of the first and the second bugs has exponential distribution with parameter $(N - 1)\phi$; and so on.

Hence, as bugs are removed from the program, the amount of time required to detect the next bug increases. After n bugs have been removed, one will have observed the values of $T_1, T_2, \ldots, T_n$ representing the time between successive detections. These observations are used to estimate ϕ and N using the maximum likelihood principle. Based on these estimates, one could predict exactly how much testing would be required in order to achieve a certain level of reliability in the software.

Shooman (1972) suggests using a normalized error rate to measure the error content in the program. He defines

$p(t) =$ Errors per total number of instructions per month of debugging time

and develops a reliability model based on first principles. He demonstrates how this model can be used to build a functional relationship between the amount of time devoted to debugging and the reliability of the program.

The works of Jelinski and Moranda and of Shooman represent the foundation of the theory of software reliability. Extensions of their methods have been considered. It remains to be seen, however, if these methods provide accurate descriptions of the

SNAPSHOT APPLICATION

Reliability-Centered Maintenance Improves Operations at Three Mile Island Nuclear Plant

The Three Mile Island nuclear facility located on the Susquehanna River about 10 miles from Harrisburg, Pennsylvania, is notorious in one respect. It was the site of the worst nuclear power generating plant accident in the United States. In March of 1979, Unit 2 underwent a core meltdown as safety systems failed to lift nuclear fuel rods from the core. The facility was shut down as a result of the accident for the next six and one-half years, finally reopening in October 1985. The plant, operated by GPU Nuclear Corporation, has compiled one of the most impressive records in the industry since it has reopened. According to Fox et al. (1994), the plant was ranked top in the world in 1989 on the basis of its capacity factor (proportion of up-time).

In 1987, GPU began to consider the benefits of a reliability-centered maintenance (RCM) approach to preventive maintenance. They identified 28 out of a total of 134 systems as viable candidates for RCM. These 28 systems included the main turbine, the cooling water system, the main generator, circulating water, and so on. The RCM process relied on the following four basic principles:

- Preserve system functions.
- Identify equipment failures that defeat those functions.
- Prioritize failure modes.
- Define preventive maintenance tasks for high-priority failure modes.

The RCM project spanned the period of September 1988 to June 1994. A total of 3,778 components in the 28 subsystems came under consideration. By the end of the program, preventive maintenance policies included more than 5,400 tasks for these components. The cost of implementing RCM was substantial: about $30,000 per system. However, these costs were more than offset by the benefits. Over the period 1990 to 1994, records show a significant decline in plant equipment failures. In addition, a reliability-based maintenance program can have other benefits, including

- Increased plant availability.
- Optimized spare parts inventories.
- Identification of component failure modes.
- Discovery of new plant failure scenarios.
- Training for engineering personnel.
- Identification of components that benefit from revised preventive maintenance strategies.
- Identification of potential design improvements.
- Improved documentation.

Fox et al. (1994) report several lessons learned from this experience. One is that it is better for the internal maintenance organization, rather than an outside agency, to direct the process. This avoids the "we versus they" syndrome. Successful implementation is also more likely in this case. A cost analysis checklist was developed to screen failure modes. Finally, the team evolved an efficient multiuser relational database software system to facilitate RCM evaluations. This system reduced the time required to perform the necessary analyses by 50 percent.

The lesson learned from this case is that a carefully designed and implemented reliability-based preventive maintenance program can have big payoffs for high stakes systems.

problem and whether they ultimately will assist in predicting the time required to achieve an acceptable level of reliability.

12.10 Historical Notes

Much of the theory of reliability, life testing, and maintenance strategies has its roots in actuarial theory developed by the insurance industry. Sophisticated mathematical models for predicting survival probabilities date back to the turn of the century. Lotka

(1939) discusses some of the connections between equipment replacement models and actuarial studies. The work of Weibull (1939 and 1951) laid the foundations for the subject of fatigue life in materials.

Interest in reliability problems became considerably more widespread during World War II when attempts were made to understand the failure laws governing complex military systems. During the 1950s, problems concerning life testing and missile reliability began to receive serious attention. In 1952 the Department of Defense established the Advisory Group on Reliability of Electronic Equipment, which published its first report on reliability in June of 1957.

The origins of the specific age replacement models presented in this chapter are unclear. However, sophisticated age replacement models date back as far as the early 1920s (see Taylor, 1923, and Hotelling, 1925). The stochastic planned replacement models presented in Section 7 form the basis for much of the research in replacement theory, but the origins of these models are unclear as well.

The section on warranties is based on the paper by Blischke and Scheuer (1975). Extensions and corrections of their work can be found in Mamer (1982). Readers interested in pursuing further reading should refer to the excellent texts by Barlow and Proschan (1965 and 1975) on reliability models, and by Gertsbakh (1977) on maintenance strategies. Issues concerning the application of maintenance models are discussed by Turban (1967) and Mann (1976).

12.11 Summary

The purpose of this chapter was to review the terminology and the methodology of the theory and application of reliability and maintenance models. *Reliability theory* is an area of study that has received considerable attention from mathematicians. However, the mathematics is of interest not only for its own sake. These models are extremely useful in an operational setting in considering such issues as failure characteristics of operating equipment, economically sound maintenance strategies, and the value of product warranties and service contracts.

The complexity of the analysis depends upon the assumptions made about the random variable T, which represents the lifetime of a single item or piece of operating equipment. The *distribution function* of T, $F(t)$, is the probability that the item fails at or before time t ($P\{T \leq t\}$), whereas the *reliability function* of T, $R(t)$, is the probability that the item fails after time t ($P\{T > t\}$). An important quantity related to these functions is the *failure rate function* $r(t)$, which is the ratio $f(t)/R(t)$ of the probability density function and the reliability function. If Δt is sufficiently small, the term $r(t)\Delta t$ can be interpreted as the conditional probability that the item will fail in the next Δt units of time given that it has survived up until time t.

The failure rate function provides considerable information about the aging characteristics of operating equipment. In a manufacturing environment, we would expect that most operating equipment would have an increasing failure rate function. That means it would be more likely to fail as it ages. A decreasing failure rate function can arise when the likelihood of early failure is high due to defectives in the population. The *Weibull* probability law can be used to describe the failure characteristics of equipment having either an increasing or a decreasing failure rate function.

Of interest is the case in which the failure rate function is constant. This case gives rise to the *exponential distribution* for the lifetime of a single component. The exponential distribution is the only continuous distribution possessing the *memoryless property*. This means that the conditional probability that an item that has been operating up until time t fails in the next s units of time is independent of t.

The *Poisson process* describes the situation in which a single piece of operating equipment fails according to the exponential distribution and is replaced immediately upon failure. When this occurs, the number of failures in a given time has the Poisson distribution, the time between successive failures has the exponential distribution, and the time for n failures to occur has the Erlang distribution.

The chapter considered the reliability functions of complex systems of components. It showed how to obtain the reliability functions for *components in series and parallel* from the reliability functions of the individual components. The chapter also considered K out of N systems, which function only if at least K components function.

Reliability issues form the basis of the *maintenance models* discussed in the latter half of the chapter. An important measure of a system's performance is the *availability*, which is the proportion of the time that the equipment operates. We treated both deterministic age replacement models, which do not explicitly include the likelihood of equipment failure, and stochastic age replacement models, which do. The stochastic models allow for replacing the equipment before failure. This is of interest when items have an increasing failure rate function and unplanned failures are more costly than planned failures.

Finally, we concluded the chapter with a discussion of the economic value of *warranties*. A warranty is a promise supplied by the seller to the buyer to either replace the item with a new one if it fails during the warranty period (*free replacement warranty*) or provide a discount on the purchase of a new item proportional to the remaining amount of time (or wear) in the warranty period (*pro rata warranty*). The issues surrounding warranties and service contracts are similar, but service contract models are considerably more complex, owing to the need to include multiple levels of repair.

Additional Problems on Reliability and Maintainability

42. A large national producer of appliances has traced customer experience with a popular toaster oven. A survey of 5,000 customers who purchased the oven early in 1975 has revealed the following:

Year	Number of Breakdowns
1975	188
1976	58
1977	63
1978	72
1979	54
1980	71

a. Using these data, estimate p_k = the probability that a toaster oven fails in its kth year of operation, for $k = 1, \ldots, 6$.

b. What is the likelihood that a toaster oven will last at least six years without failure based on these data?

c. The discrete failure rate function has the form $r_k = p_k/R_{k-1}$, where R_k is the probability that a unit survives through period k. Determine the failure rate function for the first five years of operation from the given data.

d. Suppose that you purchased a toaster oven at the beginning of 1984 and it is still operating at the end of 1987. If the reliability has not changed appreciably from 1975 to 1984, use the results of part (c) to obtain the probability that it will fail during the first two months of calendar year 1988.

43. Six thousand light bulbs light a large hotel and casino marquee. Each bulb fails completely at random, and each has an average lifetime of 3,280 hours. Assuming that the marquee stays lit continuously and bulbs that burn out are replaced immediately, how many replacements must be made each year on the average?

44. The owner of the hotel mentioned in Problem 43 has decided that in order to decrease the number of burned-out bulbs, he will replace all six thousand bulbs at the start of each year in addition to replacing the bulbs as they burn out. Comment on the effectiveness of this strategy.

45. The owner of the hotel mentioned in Problem 43 falls on hard times and dispenses with replacement of the bulbs. He notices that more than half of the bulbs have burned out before the advertised average lifetime of 3,280 hours and decides to sue the light bulb manufacturer for false advertising. Do you think he has a case? (Hint: What fraction of the bulbs would be expected to fail prior to the mean lifetime?)

46. Continuing with the example of Problem 43, determine the following:

a. The proportion of bulbs lasting more than two years.

b. The probability that a bulb chosen at random fails in the first three months of operation.

c. The probability that a bulb that has lasted for 10 years fails in the next three months of operation.

47. Assume that the bulbs in Problem 43 are not replaced as they fail.

a. What fraction of the 6,000 bulbs are expected to fail in the first year?

b. What fraction of the bulbs surviving the first year are expected to fail in the second year?

c. What fraction of the bulbs surviving the nth year are expected to fail in year $n + 1$ for any value of $n = 1, 2, \ldots$?

d. Using the results of part (c), of the original 6,000 bulbs, how many would be expected to fail in the fourth year of operation?

48. The mean value of a Weibull random variable is given by the formula

$$\mu = a^{-1/\beta}\,\Gamma(1 + 1/\beta),$$

where Γ represents the gamma function. The gamma function has the property that $\Gamma(k) = (k - 1)\Gamma(k - 1)$ for any value of $k > 1$ and $\Gamma(1) = 1$. Notice that if k is an integer, this results in $\Gamma(k) = (k - 1)!$. If k is not an integer, one must use the recursive definition for $\Gamma(k)$ coupled with the table below. For values of

$1 \leq k \leq 2$, $\Gamma(k)$ is given by

k	$\Gamma(k)$	k	$\Gamma(k)$
1.00	1.0000	1.55	.8889
1.05	.9735	1.60	.8935
1.10	.9514	1.65	.9001
1.15	.9330	1.70	.9086
1.20	.9182	1.75	.9191
1.25	.9064	1.80	.9314
1.30	.8975	1.85	.9456
1.35	.8912	1.90	.9612
1.40	.8873	1.95	.9799
1.45	.8857	2.00	1.0000
1.50	.8862		

For example, this table would be used as follows: $\Gamma(3.6) = (2.6)\Gamma(2.6) = (2.6)(1.6)\Gamma(1.6) = (2.6)(1.6)(.8935) = 3.717$.

  a. Compute the expected failure time for Example 12.4 regarding copier equipment.
  b. Compute the expected failure time for a piece of operating equipment whose failure law is given in Example 12.2.
  c. Determine the mean failure time for $\alpha = 1.35$ and $\beta = 0.20$.
  d. Determine the mean failure time for $\alpha = 0.90$ and $\beta = 0.45$.

49. Suppose that a particular light bulb is advertised as having an average lifetime of 2,000 hours and is known to satisfy an exponential failure law. Suppose for simplicity that the bulb is used continuously. Find the probability that the bulb lasts
  a. More than 3,000 hours.
  b. Less than 1,500 hours.
  c. Between 2,000 and 2,500 hours.

50. Applicational Materials sells several pieces of equipment used in the manufacture of silicon-based microprocessors. In 1983 the company filled 130 orders for model a55212. Suppose that the machines fail according to a Weibull law. In particular, the cumulative distribution function $F(t)$ of the time until failure of any machine is given by

$$F(t) = 1 - e^{-0.0475t^{1.2}} \qquad \text{for all } t \geq 0,$$

where t is in years.
  a. What is the failure rate function for this piece of equipment?
  b. Of the original 130 sold in 1983, how many machines would one expect would not experience a breakdown before January 1987? Assume for the sake of simplicity that all of the machines were sold on January 1, 1983.
  c. Using the results of part (a), estimate the fraction of machines that have survived 10 years of use that will break down during the 11th year of operation [or you may compute this directly if you did not get the answer to part (a)].

51. A local cab company maintains a fleet of 10 cabs. Each time a cab breaks down, it is repaired the same day. Assume that breakdowns of individual cabs occur completely at random at a rate of two per year.

    a. What is the probability that any particular cab will run for a full year without suffering a breakdown?

    b. What is the probability that the entire fleet will run for one month without a breakdown?

    c. On the average, how many breakdowns would the fleet expect in a typical three-month period?

    d. What is the probability that there are more than five breakdowns between Thanksgiving Day (November 28) and New Year's Day (January 1)?

    e. For what reason might your answer in part (*d*) be too low?

52. A collection of 30 Christmas tree lights are arranged in a pure series circuit; that is, if one of the lights burns out, then the entire string goes out. Suppose that each light fails completely at random at a rate of one failure every year. What is the probability that the lights will burn from the beginning of Christmas Eve (December 24) to the end of New Year's Day (January 1) without failure?

53. A piece of industrial machinery costs $48,000 to replace and has essentially no salvage value. Over the first five years of operation, maintenance costs amounted to $8,000. If the maintenance cost rate is a linear function of time, what is the optimal age at which to replace the machinery?

54. For an automobile that you own or would like to own, estimate the correct values of the replacement cost, the rate of depreciation, the initial maintenance cost, and the rate at which the maintenance cost increases. Based on these estimates, determine the optimal number of years that you should wait before replacing your car.

APPENDIX 12–A   GLOSSARY OF NOTATION ON RELIABILITY AND MAINTAINABILITY

a = Maintenance cost rate per unit time for simple age replacement model. Also used as a parameter of the exponential maintenance cost function for the exponential age replacement model.

a_1 = Cost of replacing an item when it fails. Used in the model of block replacement.

a_2 = Cost of replacing the entire block of n_0 items. Used in the model of block replacement.

α = A parameter of the Weibull probability function.

b = A parameter of the exponential maintenance cost function for the exponential age replacement model.

β = A parameter of the Weibull probability function.

c = A parameter of the exponential salvage value function for the exponential age replacement model.

c_1 = Cost of replacing an item when it fails. Used in the model of planned replacement under uncertainty for a single item.

c_2 = Cost of replacing an item before it fails. Used in the model of planned replacement under uncertainty for a single item.

C_1 = Cost of purchasing a new item with a free replacement warranty.

C_2 = Cost of purchasing a new item with a pro rata warranty.

d = A parameter of the exponential salvage value function for the exponential age replacement model.

$f(t)$ = Density function of the random variable T.

$F(t) = P\{T \le t\}$ = Cumulative distribution function of T.

$F_K(t)$ = Cumulative distribution function of the random variable T_K, the lifetime of a K out of N system.

$F_P(t)$ = Cumulative distribution function of the random variable T_P, the lifetime of a parallel system.

$F_S(t)$ = Cumulative distribution function of the random variable T_S, the lifetime of a series system.

$G(t)$ = Average cost per unit time for age replacement and warranty models.

K = Fixed replacement cost for age replacement models. Also used to denote the replacement cost of a new item without a warranty in warranty models.

λ = The parameter of the exponential failure law; the expected number of failures per unit time when failures occur at random.

MTBF = Mean time between failures.

MTTR = Mean time to repair.

n_0 = Total number of items in a block. Used in the block replacement model.

$N(t)$ = Number of failures occurring in the interval $(0, t)$ when failures follow a Poisson process. $N(t)$ has the Poisson distribution with parameter λt.

$r(t)$ = Failure rate function of T.

$R(t) = P\{T > t\}$ = Reliability function of T.

$R_K(t)$ = Reliability function of a K out of N system of components.

$R_P(t)$ = Reliability function of a parallel system of components.

$R_S(t)$ = Reliability function of a series system of components.

T = Random variable corresponding to the lifetime of an operating unit.

T_K = Lifetime of a K out of N system of components.

T_P = Lifetime of a parallel system of components.

T_S = Lifetime of a series system of components.

W_1 = Warranty period for a free replacement warranty.

W_2 = Warranty period for a pro rata warranty.

W_n = Time for n failures to occur when failures follow a Poisson process. W_n has the Erlang distribution with parameters λ and n.

Bibliography

Amstadter, B. L. *Reliability Mathematics.* New York: McGraw-Hill, 1971.

Barlow, R. E., and F. Proschan. *Mathematical Theory of Reliability.* New York: John Wiley & Sons, 1965.

Barlow, R. E., and F. Proschan. *Statistical Theory of Reliability and Life Testing.* New York: Holt, Rinehart & Winston, 1975.

Blischke, W. R., and E. M. Scheuer. "Calculation of the Cost of Warranty Policies as a Function of Estimated Life Distributions." *Naval Research Logistics Quarterly* 22 (1975), pp. 681–96.

Davis, R. "As Complexity Rises, Tiny Flaws in Software Pose a Growing Threat." *The Wall Street Journal,* January 28, 1987, p. 1.

DeGroot, M. H. *Probability and Statistics.* 2nd ed. Reading, MA: Addison-Wesley Publishing, 1986.

Fox, B. H.; M. G. Snyder; and A. M. Smith. "Reliability-Centered Maintenance Improves Operations at TMI Nuclear Plant." *Power Engineering,* November 1994, pp. 75–78.

Garvin, D. A. *Managing Quality.* New York: The Free Press, 1988.

Gertsbakh, I. B. *Models of Preventive Maintenance.* Amsterdam: North Holland, 1977.

Hillier, F. S., and G. J. Lieberman. *Introduction to Operations Research.* Oakland, CA: Holden Day, 1986.

Hotelling, H. "A General Mathematical Theory of Depreciation." *Journal of the American Statistical Association* 20 (September 1925), pp. 340–53.

Jelinski, Z., and P. Moranda. "Software Reliability Research." In *Statistical Computer Performance Evaluation,* ed. W. Freiberger. New York: Academic Press, 1972.

Lotka, A. J. "A Contribution to the Theory of Self-Renewing Aggregates with Special Reference to Industrial Replacement." *Annals of Mathematical Statistics* 10 (1939), pp. 1–25.

Lund, R. T. *Consumer Durables: Warranties, Service Contracts, and Alternatives,* vol. 4, *Analyses of Consumer Product and Warranty Relationships.* Center for Policy Alternatives, Massachusetts Institute of Technology, 1978.

Mamer, J. "Cost Analysis of Pro Rata and Free Replacement Warranties." *Naval Research Logistics Quarterly* 29 (1982), pp. 345–56.

McCall, J. J. "Maintenance Policies for Stochastically Failing Equipment: A Survey." *Management Science* 11 (1965), pp. 493–524.

Mann, L. L. *Maintenance Management.* Lexington, MA: Lexington Books, 1976.

O'Connor, P. D. T. *Practical Reliability Engineering.* 2nd ed. New York: John Wiley & Sons, 1985.

Shooman, M. L. "Probabilistic Models for Software Reliability Prediction." In *Statistical Computer Performance Evaluation,* ed. W. Freiberger, pp. 485–502. New York: Academic Press, 1972.

Sivazlian, B. D., and L. E. Stanfel. *Analysis of Systems in Operations Research.* Englewood Cliffs, NJ: Prentice Hall, 1975.

Taylor, J. S. "A Statistical Theory of Depreciation." *Journal of the American Statistical Association,* December 1923, pp. 1010–23.

Terborgh, B. *Dynamic Equipment Policy.* New York: McGraw-Hill, 1949.

Turban, E. "The Use of Mathematical Models in Plant Maintenance Decision Making." *Management Science* 13 (1967), pp. 20–27.

U.S. Nuclear Regulatory Commission. *Reactor Safety Study.* WASH–1400, NU REG 75–01. 1975.

Weibull, W. "A Statistical Theory of the Strength of Materials." *Ing. Vetenskaps Akad Handl.,* no. 151 (1939).

Weibull, W. "A Statistical Distribution Function of Wide Applicability." *Journal of Applied Mechanics* 18 (1951), pp. 293–97.

Wilkerson, J. J. "How to Manage Maintenance." *Harvard Business Review* 46 (1968), pp. 100–111.

Appendix
Tables

TABLE A–1   Areas under the Normal Curve

Z	.00	.01	.02	.03	.04	.05	.06	.07	.08	.09
.00	.0000	.0040	.0080	.0120	.0160	.0199	.0239	.0279	.0319	.0359
.10	.0398	.0438	.0478	.0517	.0557	.0596	.0636	.0675	.0714	.0753
.20	.0793	.0832	.0871	.0910	.0948	.0987	.1026	.1064	.1103	.1141
.30	.1179	.1217	.1255	.1293	.1331	.1368	.1406	.1443	.1480	.1517
.40	.1554	.1591	.1628	.1664	.1700	.1736	.1772	.1808	.1844	.1879
.50	.1915	.1950	.1985	.2019	.2054	.2088	.2123	.2157	.2190	.2224
.60	.2257	.2291	.2324	.2357	.2389	.2422	.2454	.2486	.2517	.2549
.70	.2580	.2611	.2642	.2673	.2703	.2734	.2764	.2793	.2823	.2852
.80	.2881	.2910	.2939	.2967	.2995	.3023	.3051	.3078	.3106	.3133
.90	.3159	.3186	.3212	.3238	.3264	.3289	.3315	.3340	.3365	.3389
1.00	.3413	.3438	.3461	.3485	.3508	.3531	.3554	.3577	.3599	.3621
1.10	.3643	.3665	.3686	.3708	.3729	.3749	.3770	.3790	.3810	.3830
1.20	.3849	.3869	.3888	.3907	.3925	.3944	.3962	.3980	.3997	.4015
1.30	.4032	.4049	.4066	.4082	.4099	.4115	.4131	.4147	.4162	.4177
1.40	.4192	.4207	.4222	.4236	.4251	.4265	.4279	.4292	.4306	.4319
1.50	.4332	.4345	.4357	.4370	.4382	.4394	.4406	.4418	.4429	.4441
1.60	.4452	.4463	.4474	.4484	.4495	.4505	.4515	.4525	.4535	.4545
1.70	.4554	.4564	.4573	.4582	.4591	.4599	.4608	.4616	.4625	.4633
1.80	.4641	.4649	.4656	.4664	.4671	.4678	.4686	.4693	.4699	.4706
1.90	.4713	.4719	.4726	.4732	.4738	.4744	.4750	.4756	.4761	.4767
2.00	.4772	.4778	.4783	.4788	.4793	.4798	.4803	.4808	.4812	.4817
2.10	.4821	.4826	.4830	.4834	.4838	.4842	.4846	.4850	.4854	.4857
2.20	.4861	.4864	.4868	.4871	.4875	.4878	.4881	.4884	.4887	.4890
2.30	.4893	.4896	.4898	.4901	.4904	.4906	.4909	.4911	.4913	.4916
2.40	.4918	.4920	.4922	.4925	.4927	.4929	.4931	.4932	.4934	.4936
2.50	.4938	.4940	.4941	.4943	.4945	.4946	.4948	.4949	.4951	.4952
2.60	.4953	.4955	.4956	.4957	.4959	.4960	.4961	.4962	.4963	.4964
2.70	.4965	.4966	.4967	.4968	.4969	.4970	.4971	.4972	.4973	.4974
2.80	.4974	.4975	.4976	.4977	.4977	.4978	.4979	.4979	.4980	.4981
2.90	.4981	.4982	.4982	.4983	.4984	.4984	.4985	.4985	.4986	.4986
3.00	.4987	.4987	.4987	.4988	.4988	.4989	.4989	.4989	.4990	.4990
3.10	.4990	.4991	.4991	.4991	.4992	.4992	.4992	.4992	.4993	.4993
3.20	.4993	.4993	.4994	.4994	.4994	.4994	.4994	.4995	.4995	.4995
3.30	.4995	.4995	.4995	.4996	.4996	.4996	.4996	.4996	.4996	.4997
3.40	.4997	.4997	.4997	.4997	.4997	.4997	.4997	.4997	.4997	.4998
3.50	.4998	.4998	.4998	.4998	.4998	.4998	.4998	.4998	.4998	.4998
3.60	.4998	.4998	.4999	.4999	.4999	.4999	.4999	.4999	.4999	.4999
3.70	.4999	.4999	.4999	.4999	.4999	.4999	.4999	.4999	.4999	.4999
3.80	.4999	.4999	.4999	.4999	.4999	.4999	.4999	.4999	.4999	.4999

The values in the body of the table are the areas between the mean and the value of Z. Example: If we want to find the area under the standard normal curve between $Z = 0$ and $Z = 1.96$, we find the $Z = 1.90$ row and .06 column (for $Z = 1.90 + .06 = 1.96$) and read .4750 at the intersection. For convenience, this information also appears in Table A–4.

Source: P. Billingsley, D. J. Croft, D. V. Huntsberger, and C. J. Watson, *Statistical Inference for Management and Economics* (Boston: Allyn & Bacon, 1986). Reprinted with permission.

TABLE A–2 Cumulative Binomial Probabilities

$$P(X \leq r) = \sum_{k=0}^{r} \binom{n}{k} p^k (1-p)^{n-k}$$

where X is the number of successes in n trials

n = 5

r	.01	.05	.10	.20	.30	.40	p .50	.60	.70	.80	.90	.95	.99	r
0	0.9510	0.7738	0.5905	0.3277	0.1681	0.0778	0.0313	0.0102	0.0024	0.0003	0.0000	0.0000	0.0000	0
1	0.9990	0.9774	0.9185	0.7373	0.5282	0.3370	0.1875	0.0870	0.0308	0.0067	0.0005	0.0000	0.0000	1
2	1.0000	0.9988	0.9914	0.9421	0.8369	0.6826	0.5000	0.3174	0.1631	0.0579	0.0086	0.0012	0.0000	2
3	1.0000	1.0000	0.9995	0.9933	0.9692	0.9130	0.8125	0.6630	0.4718	0.2627	0.0815	0.0226	0.0010	3
4	1.0000	1.0000	1.0000	0.9997	0.9976	0.9898	0.9688	0.9222	0.8319	0.6723	0.4095	0.2262	0.0490	4

n = 10

r	.01	.05	.10	.20	.30	.40	p .50	.60	.70	.80	.90	.95	.99	r
0	0.9044	0.5987	0.3487	0.1074	0.0282	0.0060	0.0010	0.0001	0.0000	0.0000	0.0000	0.0000	0.0000	0
1	0.9957	0.9139	0.7361	0.3758	0.1493	0.0464	0.0107	0.0017	0.0001	0.0000	0.0000	0.0000	0.0000	1
2	0.9999	0.9885	0.9298	0.6778	0.3828	0.1673	0.0547	0.0123	0.0016	0.0001	0.0000	0.0000	0.0000	2
3	1.0000	0.9990	0.9872	0.8791	0.6496	0.3823	0.1719	0.0548	0.0106	0.0009	0.0000	0.0000	0.0000	3
4	1.0000	0.9999	0.9984	0.9672	0.8497	0.6331	0.3770	0.1662	0.0473	0.0064	0.0001	0.0000	0.0000	4
5	1.0000	1.0000	0.9999	0.9936	0.9527	0.8338	0.6230	0.3669	0.1503	0.0328	0.0016	0.0001	0.0000	5
6	1.0000	1.0000	1.0000	0.9991	0.9894	0.9452	0.8281	0.6177	0.3504	0.1209	0.0128	0.0010	0.0000	6
7	1.0000	1.0000	1.0000	0.9999	0.9984	0.9877	0.9453	0.8327	0.6172	0.3222	0.0702	0.0115	0.0001	7
8	1.0000	1.0000	1.0000	1.0000	0.9999	0.9983	0.9893	0.9536	0.8507	0.6242	0.2639	0.0861	0.0043	8
9	1.0000	1.0000	1.0000	1.0000	1.0000	0.9999	0.9990	0.9940	0.9718	0.8926	0.6513	0.4013	0.0956	9

n = 15

r	.01	.05	.10	.20	.30	.40	p .50	.60	.70	.80	.90	.95	.99	r
0	0.8601	0.4633	0.2059	0.0352	0.0047	0.0005	0.0000	0.0000	0.0000	0.0000	0.0000	0.0000	0.0000	0
1	0.9904	0.8290	0.5490	0.1671	0.0353	0.0052	0.0005	0.0000	0.0000	0.0000	0.0000	0.0000	0.0000	1
2	0.9996	0.9638	0.8159	0.3980	0.1268	0.0271	0.0037	0.0003	0.0000	0.0000	0.0000	0.0000	0.0000	2
3	1.0000	0.9945	0.9444	0.6482	0.2969	0.0905	0.0176	0.0019	0.0001	0.0000	0.0000	0.0000	0.0000	3
4	1.0000	0.9994	0.9873	0.8358	0.5155	0.2173	0.0592	0.0093	0.0007	0.0000	0.0000	0.0000	0.0000	4
5	1.0000	0.9999	0.9978	0.9389	0.7216	0.4032	0.1509	0.0338	0.0037	0.0001	0.0000	0.0000	0.0000	5
6	1.0000	1.0000	0.9997	0.9819	0.8689	0.6098	0.3036	0.0950	0.0152	0.0008	0.0000	0.0000	0.0000	6
7	1.0000	1.0000	1.0000	0.9958	0.9500	0.7869	0.5000	0.2131	0.0500	0.0042	0.0000	0.0000	0.0000	7
8	1.0000	1.0000	1.0000	0.9992	0.9848	0.9050	0.6964	0.3902	0.1311	0.0181	0.0003	0.0000	0.0000	8
9	1.0000	1.0000	1.0000	0.9999	0.9963	0.9662	0.8491	0.5968	0.2784	0.0611	0.0022	0.0001	0.0000	9
10	1.0000	1.0000	1.0000	1.0000	0.9993	0.9907	0.9408	0.7827	0.4845	0.1642	0.0127	0.0006	0.0000	10
11	1.0000	1.0000	1.0000	1.0000	0.9999	0.9981	0.9824	0.9095	0.7031	0.3518	0.0556	0.0055	0.0000	11
12	1.0000	1.0000	1.0000	1.0000	1.0000	0.9997	0.9963	0.9729	0.8732	0.6020	0.1841	0.0362	0.0004	12
13	1.0000	1.0000	1.0000	1.0000	1.0000	1.0000	0.9995	0.9948	0.9647	0.8329	0.4510	0.1710	0.0096	13
14	1.0000	1.0000	1.0000	1.0000	1.0000	1.0000	1.0000	0.9995	0.9953	0.9648	0.7941	0.5367	0.1399	14

(continued)

TABLE A–2 (*continued*)

n = 20

r	.01	.05	.10	.20	.30	.40	p .50	.60	.70	.80	.90	.95	.99	r
0	0.8179	0.3585	0.1216	0.0115	0.0008	0.0000	0.0000	0.0000	0.0000	0.0000	0.0000	0.0000	0.0000	0
1	0.9831	0.7358	0.3917	0.0692	0.0076	0.0005	0.0000	0.0000	0.0000	0.0000	0.0000	0.0000	0.0000	1
2	0.9990	0.9245	0.6769	0.2061	0.0355	0.0036	0.0002	0.0000	0.0000	0.0000	0.0000	0.0000	0.0000	2
3	1.0000	0.9841	0.8670	0.4114	0.1071	0.0160	0.0013	0.0000	0.0000	0.0000	0.0000	0.0000	0.0000	3
4	1.0000	0.9974	0.9568	0.6296	0.2375	0.0510	0.0059	0.0003	0.0000	0.0000	0.0000	0.0000	0.0000	4
5	1.0000	0.9997	0.9887	0.8042	0.4164	0.1256	0.0207	0.0016	0.0000	0.0000	0.0000	0.0000	0.0000	5
6	1.0000	1.0000	0.9976	0.9133	0.6080	0.2500	0.0577	0.0065	0.0003	0.0000	0.0000	0.0000	0.0000	6
7	1.0000	1.0000	0.9996	0.9679	0.7723	0.4159	0.1316	0.0210	0.0013	0.0000	0.0000	0.0000	0.0000	7
8	1.0000	1.0000	0.9999	0.9900	0.8867	0.5956	0.2517	0.0565	0.0051	0.0001	0.0000	0.0000	0.0000	8
9	1.0000	1.0000	1.0000	0.9974	0.9520	0.7553	0.4119	0.1275	0.0171	0.0006	0.0000	0.0000	0.0000	9
10	1.0000	1.0000	1.0000	0.9994	0.9829	0.8725	0.5881	0.2447	0.0480	0.0026	0.0000	0.0000	0.0000	10
11	1.0000	1.0000	1.0000	0.9999	0.9949	0.9435	0.7483	0.4044	0.1133	0.0100	0.0001	0.0000	0.0000	11
12	1.0000	1.0000	1.0000	1.0000	0.9987	0.9790	0.8684	0.5841	0.2277	0.0321	0.0004	0.0000	0.0000	12
13	1.0000	1.0000	1.0000	1.0000	0.9997	0.9935	0.9423	0.7500	0.3920	0.0867	0.0024	0.0000	0.0000	13
14	1.0000	1.0000	1.0000	1.0000	1.0000	0.9984	0.9793	0.8744	0.5836	0.1958	0.0113	0.0003	0.0000	14
15	1.0000	1.0000	1.0000	1.0000	1.0000	0.9997	0.9941	0.9490	0.7625	0.3704	0.0432	0.0026	0.0000	15
16	1.0000	1.0000	1.0000	1.0000	1.0000	1.0000	0.9987	0.9840	0.8929	0.5886	0.1330	0.0159	0.0000	16
17	1.0000	1.0000	1.0000	1.0000	1.0000	1.0000	0.9998	0.9964	0.9645	0.7939	0.3231	0.0755	0.0010	17
18	1.0000	1.0000	1.0000	1.0000	1.0000	1.0000	1.0000	0.9995	0.9924	0.9308	0.6083	0.2642	0.0169	18
19	1.0000	1.0000	1.0000	1.0000	1.0000	1.0000	1.0000	1.0000	0.9992	0.9885	0.8784	0.6415	0.1821	19

n = 25

r	.01	.05	.10	.20	.30	.40	p .50	.60	.70	.80	.90	.95	.99	r
0	0.7778	0.2774	0.0718	0.0038	0.0001	0.0000	0.0000	0.0000	0.0000	0.0000	0.0000	0.0000	0.0000	0
1	0.9742	0.6424	0.2712	0.0274	0.0016	0.0001	0.0000	0.0000	0.0000	0.0000	0.0000	0.0000	0.0000	1
2	0.9980	0.8729	0.5371	0.0982	0.0090	0.0004	0.0000	0.0000	0.0000	0.0000	0.0000	0.0000	0.0000	2
3	0.9999	0.9659	0.7636	0.2340	0.0332	0.0024	0.0001	0.0000	0.0000	0.0000	0.0000	0.0000	0.0000	3
4	1.0000	0.9928	0.9020	0.4207	0.0905	0.0095	0.0005	0.0000	0.0000	0.0000	0.0000	0.0000	0.0000	4
5	1.0000	0.9988	0.9666	0.6167	0.1935	0.0294	0.0020	0.0001	0.0000	0.0000	0.0000	0.0000	0.0000	5
6	1.0000	0.9998	0.9905	0.7800	0.3407	0.0736	0.0073	0.0003	0.0000	0.0000	0.0000	0.0000	0.0000	6
7	1.0000	1.0000	0.9977	0.8909	0.5118	0.1536	0.0216	0.0012	0.0000	0.0000	0.0000	0.0000	0.0000	7
8	1.0000	1.0000	0.9995	0.9532	0.6769	0.2735	0.0539	0.0043	0.0001	0.0000	0.0000	0.0000	0.0000	8
9	1.0000	1.0000	0.9999	0.9827	0.8106	0.4246	0.1148	0.0132	0.0005	0.0000	0.0000	0.0000	0.0000	9
10	1.0000	1.0000	1.0000	0.9944	0.9022	0.5858	0.2122	0.0344	0.0018	0.0000	0.0000	0.0000	0.0000	10
11	1.0000	1.0000	1.0000	0.9985	0.9558	0.7323	0.3450	0.0778	0.0060	0.0001	0.0000	0.0000	0.0000	11
12	1.0000	1.0000	1.0000	0.9996	0.9825	0.8462	0.5000	0.1538	0.0175	0.0004	0.0000	0.0000	0.0000	12
13	1.0000	1.0000	1.0000	0.9999	0.9940	0.9222	0.6550	0.2677	0.0442	0.0015	0.0000	0.0000	0.0000	13
14	1.0000	1.0000	1.0000	1.0000	0.9982	0.9656	0.7878	0.4142	0.0978	0.0056	0.0000	0.0000	0.0000	14
15	1.0000	1.0000	1.0000	1.0000	0.9995	0.9868	0.8852	0.5754	0.1894	0.0173	0.0001	0.0000	0.0000	15
16	1.0000	1.0000	1.0000	1.0000	0.9999	0.9957	0.9461	0.7265	0.3231	0.0468	0.0005	0.0000	0.0000	16
17	1.0000	1.0000	1.0000	1.0000	1.0000	0.9988	0.9784	0.8464	0.4882	0.1091	0.0023	0.0000	0.0000	17
18	1.0000	1.0000	1.0000	1.0000	1.0000	0.9997	0.9927	0.9264	0.6593	0.2200	0.0095	0.0002	0.0000	18
19	1.0000	1.0000	1.0000	1.0000	1.0000	0.9999	0.9980	0.9706	0.8065	0.3833	0.0334	0.0012	0.0000	19

TABLE A–2 (*concluded*)

							n = 25							

							p							
r	.01	.05	.10	.20	.30	.40	.50	.60	.70	.80	.90	.95	.99	*r*
20	1.0000	1.0000	1.0000	1.0000	1.0000	1.0000	0.9995	0.9905	0.9095	0.5793	0.0980	0.0072	0.0000	20
21	1.0000	1.0000	1.0000	1.0000	1.0000	1.0000	0.9999	0.9976	0.9668	0.7660	0.2364	0.0341	0.0001	21
22	1.0000	1.0000	1.0000	1.0000	1.0000	1.0000	1.0000	0.9996	0.9910	0.9018	0.4629	0.1271	0.0020	22
23	1.0000	1.0000	1.0000	1.0000	1.0000	1.0000	1.0000	0.9999	0.9984	0.9726	0.7288	0.3576	0.0258	23
24	1.0000	1.0000	1.0000	1.0000	1.0000	1.0000	1.0000	1.0000	0.9999	0.9962	0.9282	0.7226	0.2222	24

Source: P. Billingsley, D. J. Croft, D. V. Huntsberger, and C. J. Watson, *Statistical Inference for Management and Economics* (Boston: Allyn & Bacon, 1986). Reprinted with permission.

TABLE A–3 Cumulative Poisson Probabilities

$$\sum_{x=x'}^{\infty} \frac{e^{-m}m^x}{x!}$$

x'	0.1	0.2	0.3	0.4	*m* 0.5	0.6	0.7	0.8	0.9	1.0
0	1.0000	1.0000	1.0000	1.0000	1.0000	1.0000	1.0000	1.0000	1.0000	1.0000
1	.0952	.1813	.2592	.3297	.3935	.4512	.5034	.5507	.5934	.6321
2	.0047	.0175	.0369	.0616	.0902	.1219	.1558	.1912	.2275	.2642
3	.0002	.0011	.0036	.0079	.0144	.0231	.0341	.0474	.0629	.0803
4	.0000	.0001	.0003	.0008	.0018	.0034	.0058	.0091	.0135	.0190
5	.0000	.0000	.0000	.0001	.0002	.0004	.0008	.0014	.0023	.0037
6	.0000	.0000	.0000	.0000	.0000	.0000	.0001	.0002	.0003	.0006
7	.0000	.0000	.0000	.0000	.0000	.0000	.0000	.0000	.0000	.0001

x'	1.1	1.2	1.3	1.4	*m* 1.5	1.6	1.7	1.8	1.9	2.0
0	1.0000	1.0000	1.0000	1.0000	1.0000	1.0000	1.0000	1.0000	1.0000	1.0000
1	.6671	.6988	.7275	.7534	.7769	.7981	.8173	.8347	.8504	.8647
2	.3010	.3374	.3732	.4082	.4422	.4751	.5068	.5372	.5663	.5940
3	.0996	.1205	.1429	.1665	.1912	.2166	.2428	.2694	.2963	.3233
4	.0257	.0338	.0431	.0537	.0656	.0788	.0932	.1087	.1253	.1429
5	.0054	.0077	.0107	.0143	.0186	.0237	.0296	.0364	.0441	.0527
6	.0010	.0015	.0022	.0032	.0045	.0060	.0080	.0104	.0132	.0166
7	.0001	.0003	.0004	.0006	.0009	.0013	.0019	.0026	.0034	.0045
8	.0000	.0000	.0001	.0001	.0002	.0003	.0004	.0006	.0008	.0011
9	.0000	.0000	.0000	.0000	.0000	.0000	.0001	.0001	.0002	.0002

x'	2.1	2.2	2.3	2.4	*m* 2.5	2.6	2.7	2.8	2.9	3.0
0	1.0000	1.0000	1.0000	1.0000	1.0000	1.0000	1.0000	1.0000	1.0000	1.0000
1	.8775	.8892	.8997	.9093	.9179	.9257	.9328	.9392	.9450	.9502
2	.6204	.6454	.6691	.6916	.7127	.7326	.7513	.7689	.7854	.8009
3	.3504	.3773	.4040	.4303	.4562	.4816	.5064	.5305	.5540	.5768
4	.1614	.1806	.2007	.2213	.2424	.2640	.2859	.3081	.3304	.3528

(*continued*)

TABLE A–3 *(continued)*

x'	2.1	2.2	2.3	2.4	2.5	2.6	2.7	2.8	2.9	3.0
					m					
5	.0621	.0725	.0838	.0959	.1088	.1226	.1371	.1523	.1682	.1847
6	.0204	.0249	.0300	.0357	.0420	.0490	.0567	.0651	.0742	.0839
7	.0059	.0075	.0094	.0116	.0142	.0172	.0206	.0244	.0287	.0335
8	.0015	.0020	.0026	.0033	.0042	.0053	.0066	.0081	.0099	.0119
9	.0003	.0005	.0006	.0009	.0011	.0015	.0019	.0024	.0031	.0038
10	.0001	.0001	.0001	.0002	.0003	.0004	.0005	.0007	.0009	.0011
11	.0000	.0000	.0000	.0000	.0001	.0001	.0001	.0002	.0002	.0003
12	.0000	.0000	.0000	.0000	.0000	.0000	.0000	.0000	.0001	.0001

x'	3.1	3.2	3.3	3.4	3.5	3.6	3.7	3.8	3.9	4.0
					m					
0	1.0000	1.0000	1.0000	1.0000	1.0000	1.0000	1.0000	1.0000	1.0000	1.0000
1	.9550	.9592	.9631	.9666	.9698	.9727	.9753	.9776	.9798	.9817
2	.8153	.8288	.8414	.8532	.8641	.8743	.8838	.8926	.9008	.9084
3	.5988	.6201	.6406	.6603	.6792	.6973	.7146	.7311	.7469	.7619
4	.3752	.3975	.4197	.4416	.4634	.4848	.5058	.5265	.5468	.5665
5	.2018	.2194	.2374	.2558	.2746	.2936	.3128	.3322	.3516	.3712
6	.0943	.1054	.1171	.1295	.1424	.1559	.1699	.1844	.1994	.2149
7	.0388	.0446	.0510	.0579	.0653	.0733	.0818	.0909	.1005	.1107
8	.0142	.0168	.0198	.0231	.0267	.0308	.0352	.0401	.0454	.0511
9	.0047	.0057	.0069	.0083	.0099	.0117	.0137	.0160	.0185	.0214
10	.0014	.0018	.0022	.0027	.0033	.0040	.0048	.0058	.0069	.0081
11	.0004	.0005	.0006	.0008	.0010	.0013	.0016	.0019	.0023	.0028
12	.0001	.0001	.0002	.0002	.0003	.0004	.0005	.0006	.0007	.0009
13	.0000	.0000	.0000	.0001	.0001	.0001	.0001	.0002	.0002	.0003
14	.0000	.0000	.0000	.0000	.0000	.0000	.0000	.0000	.0001	.0001

x'	4.1	4.2	4.3	4.4	4.5	4.6	4.7	4.8	4.9	5.0
					m					
0	1.0000	1.0000	1.0000	1.0000	1.0000	1.0000	1.0000	1.0000	1.0000	1.0000
1	.9834	.9850	.9864	.9877	.9889	.9899	.9909	.9918	.9926	.9933
2	.9155	.9220	.9281	.9337	.9389	.9437	.9482	.9523	.9561	.9596
3	.7762	.7898	.8026	.8149	.8264	.8374	.8477	.8575	.8667	.8753
4	.5858	.6046	.6228	.6406	.6577	.6743	.6903	.7058	.7207	.7350
5	.3907	.4102	.4296	.4488	.4679	.4868	.5054	.5237	.5418	.5595
6	.2307	.2469	.2633	.2801	.2971	.3142	.3316	.3490	.3665	.3840
7	.1214	.1325	.1442	.1564	.1689	.1820	.1954	.2092	.2233	.2378
8	.0573	.0639	.0710	.0786	.0866	.0951	.1040	.1133	.1231	.1334
9	.0245	.0279	.0317	.0358	.0403	.0451	.0503	.0558	.0618	.0681
10	.0095	.0111	.0129	.0149	.0171	.0195	.0222	.0251	.0283	.0318
11	.0034	.0041	.0048	.0057	.0067	.0078	.0090	.0104	.0120	.0137
12	.0011	.0014	.0017	.0020	.0024	.0029	.0034	.0040	.0047	.0055
13	.0003	.0004	.0005	.0007	.0008	.0010	.0012	.0014	.0017	.0020
14	.0001	.0001	.0002	.0002	.0003	.0003	.0004	.0005	.0006	.0007
15	.0000	.0000	.0000	.0001	.0001	.0001	.0001	.0001	.0002	.0002
16	.0000	.0000	.0000	.0000	.0000	.0000	.0000	.0000	.0001	.0001

TABLE A–3   (*continued*)

x'	5.1	5.2	5.3	5.4	5.5	5.6	5.7	5.8	5.9	6.0
					m					
0	1.0000	1.0000	1.0000	1.0000	1.0000	1.0000	1.0000	1.0000	1.0000	1.0000
1	.9939	.9945	.9950	.9955	.9959	.9963	.9967	.9970	.9973	.9975
2	.9628	.9658	.9686	.9711	.9734	.9756	.9776	.9794	.9811	.9826
3	.8835	.8912	.8984	.9052	.9116	.9176	.9232	.9285	.9334	.9380
4	.7487	.7619	.7746	.7867	.7983	.8094	.8200	.8300	.8396	.8488
5	.5769	.5939	.6105	.6267	.6425	.6579	.6728	.6873	.7013	.7149
6	.4016	.4191	.4365	.4539	.4711	.4881	.5050	.5217	.5381	.5543
7	.2526	.2676	.2829	.2983	.3140	.3297	.3456	.3616	.3776	.3937
8	.1440	.1551	.1665	.1783	.1905	.2030	.2159	.2290	.2424	.2560
9	.0748	.0819	.0894	.0974	.1056	.1143	.1234	.1328	.1426	.1528
10	.0356	.0397	.0441	.0488	.0538	.0591	.0648	.0708	.0772	.0839
11	.0156	.0177	.0200	.0225	.0253	.0282	.0314	.0349	.0386	.0426
12	.0063	.0073	.0084	.0096	.0110	.0125	.0141	.0160	.0179	.0201
13	.0024	.0028	.0033	.0038	.0045	.0051	.0059	.0068	.0078	.0088
14	.0008	.0010	.0012	.0014	.0017	.0020	.0023	.0027	.0031	.0036
15	.0003	.0003	.0004	.0005	.0006	.0007	.0009	.0010	.0012	.0014
16	.0001	.0001	.0001	.0002	.0002	.0002	.0003	.0004	.0004	.0005
17	.0000	.0000	.0000	.0001	.0001	.0001	.0001	.0001	.0001	.0002
18	.0000	.0000	.0000	.0000	.0000	.0000	.0000	.0000	.0000	.0001

x'	6.1	6.2	6.3	6.4	6.5	6.6	6.7	6.8	6.9	7.0
					m					
0	1.0000	1.0000	1.0000	1.0000	1.0000	1.0000	1.0000	1.0000	1.0000	1.0000
1	.9978	.9980	.9982	.9983	.9985	.9986	.9988	.9989	.9990	.9991
2	.9841	.9854	.9866	.9877	.9887	.9897	.9905	.9913	.9920	.9927
3	.9423	.9464	.9502	.9537	.9570	.9600	.9629	.9656	.9680	.9704
4	.8575	.8658	.8736	.8811	.8882	.8948	.9012	.9072	.9129	.9182
5	.7281	.7408	.7531	.7649	.7763	.7873	.7978	.8080	.8177	.8270
6	.5702	.5859	.6012	.6163	.6310	.6453	.6594	.6730	.6863	.6993
7	.4098	.4258	.4418	.4577	.4735	.4892	.5047	.5201	.5353	.5503
8	.2699	.2840	.2983	.3127	.3272	.3419	.3567	.3715	.3864	.4013
9	.1633	.1741	.1852	.1967	.2084	.2204	.2327	.2452	.2580	.2709
10	.0910	.0984	.1061	.1142	.1226	.1314	.1404	.1498	.1505	.1695
11	.0469	.0514	.0563	.0614	.0668	.0726	.0786	.0849	.0916	.0985
12	.0224	.0250	.0277	.0307	.0339	.0373	.0409	.0448	.0490	.0534
13	.0100	.0113	.0127	.0143	.0160	.0179	.0199	.0221	.0245	.0270
14	.0042	.0048	.0055	.0063	.0071	.0080	.0091	.0102	.0115	.0128
15	.0016	.0019	.0022	.0026	.0030	.0034	.0039	.0044	.0050	.0057
16	.0006	.0007	.0008	.0010	.0012	.0014	.0016	.0018	.0021	.0024
17	.0002	.0003	.0003	.0004	.0004	.0005	.0006	.0007	.0008	.0010
18	.0001	.0001	.0001	.0001	.0002	.0002	.0002	.0003	.0003	.0004
19	.0000	.0000	.0000	.0000	.0001	.0001	.0001	.0001	.0001	.0001

x'	7.1	7.2	7.3	7.4	7.5	7.6	7.7	7.8	7.9	8.0
					m					
0	1.0000	1.0000	1.0000	1.0000	1.0000	1.0000	1.0000	1.0000	1.0000	1.0000
1	.9992	.9993	.9993	.9994	.9994	.9995	.9995	.9996	.9996	.9997

(*continued*)

TABLE A–3   (*continued*)

					m					
x'	7.1	7.2	7.3	7.4	7.5	7.6	7.7	7.8	7.9	8.0
2	.9933	.9939	.9944	.9949	.9953	.9957	.9961	.9964	.9967	.9970
3	.9725	.9745	.9764	.0781	.9797	.9812	.9826	.9839	.9851	.9862
4	.9233	.9281	.9326	.9368	.9409	.9446	.9482	.9515	.9547	.9576
5	.8359	.8445	.8527	.8605	.8679	.8751	.8819	.8883	.8945	.9004
6	.7119	.7241	.7360	.7474	.7586	.7693	.7797	.7897	.7994	.8088
7	.5651	.5796	.5940	.6080	.6218	.6354	.6486	.6616	.6743	.6866
8	.4162	.4311	.4459	.4607	.4754	.4900	.5044	.5188	.5330	.5470
9	.2840	.2973	.3108	.3243	.3380	.3518	.3657	.3796	.3935	.4075
10	.1798	.1904	.2012	.2123	.2236	.2351	.2469	.2589	.2710	.2834
11	.1058	.1133	.1212	.1293	.1378	.1465	.1555	.1648	.1743	.1841
12	.0580	.0629	.0681	.0735	.0792	.0852	.0915	.0980	.1048	.1119
13	.0297	.0327	.0358	.0391	.0427	.0464	.0504	.0546	.0591	.0638
14	.0143	.0159	.0176	.0195	.0216	.0238	.0261	.0286	.0313	.0342
15	.0065	.0073	.0082	.0092	.0103	.0114	.0127	.0141	.0156	.0173
16	.0028	.0031	.0036	.0041	.0046	.0052	.0059	.0066	.0074	.0082
17	.0011	.0013	.0015	.0017	.0020	.0022	.0026	.0029	.0033	.0037
18	.0004	.0005	.0006	.0007	.0008	.0009	.0011	.0012	.0014	.0016
19	.0002	.0002	.0002	.0003	.0003	.0004	.0004	.0005	.0006	.0006
20	.0001	.0001	.0001	.0001	.0001	.0001	.0002	.0002	.0002	.0003
21	.0000	.0000	.0000	.0000	.0000	.0000	.0001	.0001	.0001	.0001

					m					
x'	8.1	8.2	8.3	8.4	8.5	8.6	8.7	8.8	8.9	9.0
0	1.0000	1.0000	1.0000	1.0000	1.0000	1.0000	1.0000	1.0000	1.0000	1.0000
1	.9997	.9997	.9998	.9998	.9998	.9998	.9998	.9998	.9999	.9999
2	.9972	.9975	.9977	.9979	.9981	.9982	.9984	.9985	.9987	.9988
3	.9873	.9882	.9891	.9900	.9907	.9914	.9921	.9927	.9932	.9938
4	.9604	.9630	.9654	.9677	.9699	.9719	.9738	.9756	.9772	.9788
5	.9060	.9113	.9163	.9211	.9256	.9299	.9340	.9379	.9416	.9450
6	.8178	.8264	.8347	.8427	.8504	.8578	.8648	.8716	.8781	.8843
7	.6987	.7104	.7219	.7330	.7438	.7543	.7645	.7744	.7840	.7932
8	.5609	.5746	.5881	.6013	.6144	.6272	.6398	.6522	.6643	.6761
9	.4214	.4353	.4493	.4631	.4769	.4906	.5042	.5177	.5311	.5443
10	.2959	.3085	.3212	.3341	.3470	.3600	.3731	.3863	.3994	.4126
11	.1942	.2045	.2150	.2257	.2366	.2478	.2591	.2706	.2822	.2940
12	.1193	.1269	.1348	.1429	.1513	.1600	.1689	.1780	.1874	.1970
13	.0687	.0739	.0793	.0850	.0909	.0971	.1035	.1102	.1171	.1242
14	.0372	.0405	.0439	.0476	.0514	.0555	.0597	.0642	.0689	.0739
15	.0190	.0209	.0229	.0251	.0274	.0299	.0325	.0353	.0383	.0415
16	.0092	.0102	.0113	.0125	.0138	.0152	.0168	.0184	.0202	.0220
17	.0042	.0047	.0053	.0059	.0066	.0074	.0082	.0091	.0101	.0111
18	.0018	.0021	.0023	.0027	.0030	.0034	.0038	.0043	.0048	.0053
19	.0008	.0009	.0010	.0011	.0013	.0015	.0017	.0019	.0022	.0024
20	.0003	.0003	.0004	.0005	.0005	.0006	.0007	.0008	.0009	.0011
21	.0001	.0001	.0002	.0002	.0002	.0002	.0003	.0003	.0004	.0004
22	.0000	.0000	.0001	.0001	.0001	.0001	.0001	.0001	.0002	.0002
23	.0000	.0000	.0000	.0000	.0000	.0000	.0000	.0000	.0001	.0001

TABLE A–3 *(continued)*

					m					
x'	9.1	9.2	9.3	9.4	9.5	9.6	9.7	9.8	9.9	10
0	1.0000	1.0000	1.0000	1.0000	1.0000	1.0000	1.0000	1.0000	1.0000	1.0000
1	.9999	.9999	.9999	.9999	.9999	.9999	.9999	.9999	1.0000	1.0000
2	.9989	.9990	.9991	.9991	.9992	.9993	.9993	.9994	.9995	.9995
3	.9942	.9947	.9951	.9955	.9958	.9962	.9965	.9967	.9970	.9972
4	.9802	.9816	.9828	.9840	.9851	.9862	.9871	.9880	.9889	.9897
5	.9483	.9514	.9544	.9571	.9597	.9622	.9645	.9667	.9688	.9707
6	.8902	.8959	.9014	.9065	.9115	.9162	.9207	.9250	.9290	.9329
7	.8022	.8108	.8192	.8273	.8351	.8426	.8498	.8567	.8634	.8699
8	.6877	.6990	.7101	.7208	.7313	.7416	.7515	.7612	.7706	.7798
9	.5574	.5704	.5832	.5958	.6082	.6204	.6324	.6442	.6558	.6672
10	.4258	.4389	.4521	.4651	.4782	.4911	.5040	.5168	.5295	.5421
11	.3059	.3180	.3301	.3424	.3547	.3671	.3795	.3920	.4045	.4170
12	.2068	.2168	.2270	.2374	.2480	.2588	.2697	.2807	.2919	.3032
13	.1316	.1393	.1471	.1552	.1636	.1721	.1809	.1899	.1991	.2084
14	.0790	.0844	.0900	.0958	.1019	.1081	.1147	.1214	.1284	.1355
15	.0448	.0483	.0520	.0559	.0600	.0643	.0688	.0735	.0784	.0835
16	.0240	.0262	.0285	.0309	.0335	.0362	.0391	.0421	.0454	.0487
17	.0122	.0135	.0148	.0162	.0177	.0194	.0211	.0230	.0249	.0270
18	.0059	.0066	.0073	.0081	.0089	.0098	.0108	.0119	.0130	.0143
19	.0027	.0031	.0034	.0038	.0043	.0048	.0053	.0059	.0065	.0072
20	.0012	.0014	.0015	.0017	.0020	.0022	.0025	.0028	.0031	.0035
21	.0005	.0006	.0007	.0008	.0009	.0010	.0011	.0013	.0014	.0016
22	.0002	.0002	.0003	.0003	.0004	.0004	.0005	.0005	.0006	.0007
23	.0001	.0001	.0001	.0001	.0001	.0002	.0002	.0002	.0003	.0003
24	.0000	.0000	.0000	.0000	.0001	.0001	.0001	.0001	.0001	.0001

					m					
x'	11	12	13	14	15	16	17	18	19	20
0	1.0000	1.0000	1.0000	1.0000	1.0000	1.0000	1.0000	1.0000	1.0000	1.0000
1	1.0000	1.0000	1.0000	1.0000	1.0000	1.0000	1.0000	1.0000	1.0000	1.0000
2	.9998	.9999	1.0000	1.0000	1.0000	1.0000	1.0000	1.0000	1.0000	1.0000
3	.9988	.9995	.9998	.9999	1.0000	1.0000	1.0000	1.0000	1.0000	1.0000
4	.9951	.9977	.9990	.9995	.9998	.9999	1.0000	1.0000	1.0000	1.0000
5	.9849	.9924	.9963	.9982	.9991	.9996	.9998	.9999	1.0000	1.0000
6	.9625	.9797	.9893	.9945	.9972	.9986	.9993	.9997	.9998	.9999
7	.9214	.9542	.9741	.9858	.9924	.9960	.9979	.9990	.9995	.9997
8	.8568	.9105	.9460	.9684	.9820	.9900	.9946	.9971	.9985	.9992
9	.7680	.8450	.9002	.9379	.9626	.9780	.9874	.9929	.9961	.9979
10	.6595	.7576	.8342	.8906	.9301	.9567	.9739	.9846	.9911	.9950
11	.5401	.6528	.7483	.8243	.8815	.9226	.9509	.9696	.9817	.9892
12	.4207	.5384	.6468	.7400	.8152	.8730	.9153	.9451	.9653	.9786
13	.3113	.4240	.5369	.6415	.7324	.8069	.8650	.9083	.9394	.9610
14	.2187	.3185	.4270	.5356	.6368	.7255	.7991	.8574	.9016	.9339
15	.1460	.2280	.3249	.4296	.5343	.6325	.7192	.7919	.8503	.8951
16	.0926	.1556	.2364	.3306	.4319	.5333	.6285	.7133	.7852	.8435
17	.0559	.1013	.1645	.2441	.3359	.4340	.5323	.6250	.7080	.7789
18	.0322	.0630	.1095	.1728	.2511	.3407	.4360	.5314	.6216	.7030
19	.0177	.0374	.0698	.1174	.1805	.2577	.3450	.4378	.5305	.6186

(continued)

TABLE A-3 (concluded)

x'	11	12	13	14	15	16	17	18	19	20
20	.0093	.0213	.0427	.0765	.1248	.1878	.2637	.3491	.4394	.5297
21	.0047	.0116	.0250	.0479	.0830	.1318	.1945	.2693	.3528	.4409
22	.0023	.0061	.0141	.0288	.0531	.0892	.1385	.2009	.2745	.3563
23	.0010	.0030	.0076	.0167	.0327	.0582	.0953	.1449	.2069	.2794
24	.0005	.0015	.0040	.0093	.0195	.0367	.0633	.1011	.1510	.2125
25	.0002	.0007	.0020	.0050	.0112	.0223	.0406	.0683	.1067	.1568
26	.0001	.0003	.0010	.0026	.0062	.0131	.0252	.0446	.0731	.1122
27	.0000	.0001	.0005	.0013	.0033	.0075	.0152	.0282	.0486	.0779
28	.0000	.0001	.0002	.0006	.0017	.0041	.0088	.0173	.0313	.0525
29	.0000	.0000	.0001	.0003	.0009	.0022	.0050	.0103	.0195	.0343
30	.0000	.0000	.0000	.0001	.0004	.0011	.0027	.0059	.0118	.0218
31	.0000	.0000	.0000	.0001	.0002	.0006	.0014	.0033	.0070	.0135
32	.0000	.0000	.0000	.0000	.0001	.0003	.0007	.0018	.0040	.0081
33	.0000	.0000	.0000	.0000	.0000	.0001	.0004	.0010	.0022	.0047
34	.0000	.0000	.0000	.0000	.0000	.0001	.0002	.0005	.0012	.0027
35	.0000	.0000	.0000	.0000	.0000	.0000	.0001	.0002	.0006	.0015
36	.0000	.0000	.0000	.0000	.0000	.0000	.0000	.0001	.0003	.0008
37	.0000	.0000	.0000	.0000	.0000	.0000	.0000	.0001	.0002	.0004
38	.0000	.0000	.0000	.0000	.0000	.0000	.0000	.0000	.0001	.0002
39	.0000	.0000	.0000	.0000	.0000	.0000	.0000	.0000	.0000	.0001
40	.0000	.0000	.0000	.0000	.0000	.0000	.0000	.0000	.0000	.0001

Source: *CRC Standard Mathematical Tables,* 16th edition, The Chemical Rubber Company, 1968.

TABLE A–4 Normal Probability Distribution and Partial Expectations

Standardized Variate z	Probabilities		Partial Expectations	
	$F(z)$	$1 - F(z)$	$L(z)$	$L(-z)$
.00	.5000	.5000	.3989	.3989
.01	.5040	.4960	.3940	.4040
.02	.5080	.4920	.3890	.4090
.03	.5120	.4880	.3841	.4141
.04	.5160	.4840	.3793	.4193
.05	.5200	.4800	.3744	.4244
.06	.5239	.4761	.3697	.4297
.07	.5279	.4721	.3649	.4349
.08	.5319	.4681	.3602	.4402
.09	.5359	.4641	.3556	.4456
.10	.5398	.4602	.3509	.4509
.11	.5438	.4562	.3464	.4564
.12	.5478	.4522	.3418	.4618
.13	.5517	.4483	.3373	.4673
.14	.5557	.4443	.3328	.4728
.15	.5596	.4404	.3284	.4784
.16	.5636	.4364	.3240	.4840
.17	.5685	.4325	.3197	.4897
.18	.5714	.4286	.3154	.4954
.19	.5753	.4247	.3111	.5011
.20	.5793	.4207	.3069	.5069
.21	.5832	.4168	.3027	.5127
.22	.5871	.4129	.3027	.5186
.23	.5910	.4090	.2944	.5244
.24	.5948	.4052	.2904	.5304
.25	.5987	.4013	.2863	.5363
.26	.6026	.3974	.2824	.5424
.27	.6064	.3936	.2784	.5484
.28	.6103	.3897	.2745	.5545
.29	.6141	.3859	.2706	.5606
.30	.6179	.3821	.2668	.5668
.31	.6217	.3783	.2630	.5730
.32	.6255	.3745	.2592	.5792
.33	.6293	.3707	.2555	.5855
.34	.6331	.3669	.2518	.5918
.35	.6368	.3632	.2481	.5981
.36	.6406	.3594	.2445	.6045
.37	.6443	.3557	.2409	.6109
.38	.6480	.3520	.2374	.6174
.39	.6517	.3483	.2339	.6239
.40	.6554	.3446	.2304	.6304
.41	.6591	.3409	.2270	.6370
.42	.6628	.3372	.2236	.6436
.43	.6664	.3336	.2203	.6503
.44	.6700	.3300	.2169	.6569
.45	.6736	.3264	.2137	.6637
.46	.6772	.3228	.2104	.6704
.47	.6808	.3192	.2072	.6772
.48	.6844	.3156	.2040	.6840
.49	.6879	.3121	.2009	.6909
.50	.6915	.3085	.1978	.6978

(continued)

TABLE A–4 (*continued*)

Standardized Variate z	Probabilities		Partial Expectations	
	F(z)	1 − F(z)	L(z)	L(−z)
.51	.6950	.3050	.1947	.7047
.52	.6985	.3015	.1917	.7117
.53	.7019	.2981	.1887	.7187
.54	.7054	.2946	.1857	.7287
.55	.7088	.2912	.1828	.7328
.56	.7123	.2877	.1799	.7399
.57	.7157	.2843	.1771	.7471
.58	.7190	.2810	.1742	.7542
.59	.7224	.2776	.1714	.7614
.60	.7257	.2743	.1687	.7687
.61	.7291	.2709	.1659	.7759
.62	.7324	.2676	.1633	.7833
.63	.7357	.2643	.1606	.7906
.64	.7389	.2611	.1580	.7980
.65	.7422	.2578	.1554	.8054
.66	.7454	.2546	.1528	.8128
.67	.7486	.2514	.1503	.8203
.68	.7517	.2483	.1478	.8278
.69	.7549	.2451	.1453	.8353
.70	.7580	.2420	.1429	.8429
.71	.7611	.2389	.1405	.8505
.72	.7642	.2358	.1381	.8581
.73	.7673	.2327	.1358	.8658
.74	.7703	.2297	.1334	.8734
.75	.7733	.2267	.1312	.8812
.76	.7764	.2236	.1289	.8889
.77	.7793	.2207	.1267	.8967
.78	.7823	.2177	.1245	.9045
.79	.7852	.2148	.1223	.9123
.80	.7881	.2119	.1202	.9202
.81	.7910	.2090	.1181	.9281
.82	.7939	.2061	.1160	.9360
.83	.7967	.2033	.1140	.9440
.84	.7996	.2004	.1120	.9520
.85	.8023	.1977	.1100	.9600
.86	.8051	.1949	.1080	.9680
.87	.8067	.1922	.1061	.9761
.88	.8106	.1894	.1042	.9842
.89	.8133	.1867	.1023	.9923
.90	.8159	.1841	.1004	1.0004
.91	.8186	.1814	.0986	1.0086
.92	.8212	.1788	.0968	1.0168
.93	.8238	.1762	.0955	1.0250
.94	.8264	.1736	.0953	1.0330
.95	.8289	.1711	.0916	1.0416
.96	.8315	.1685	.0899	1.0499
.97	.8340	.1660	.0882	1.0582
.98	.8365	.1635	.0865	1.0665
.99	.8389	.1611	.0849	1.0749
1.00	.8413	.1587	.0833	1.0833
1.01	.8438	.1562	.0817	1.0917
1.02	.8461	.1539	.0802	1.1002

Table A–4 (*continued*)

Standardized Variate z	Probabilities		Partial Expectations	
	F(z)	1 − F(z)	L(z)	L(−z)
1.03	.8485	.1515	.0787	1.1087
1.04	.8508	.1492	.0772	1.1172
1.05	.8531	.1469	.0757	1.1257
1.06	.8554	.1446	.0742	1.1342
1.07	.8577	.1423	.0728	1.1428
1.08	.8599	.1401	.0714	1.1514
1.09	.8621	.1379	.0700	1.1600
1.10	.8643	.1357	.0686	1.1686
1.11	.8665	.1335	.0673	1.1773
1.12	.8686	.1314	.0659	1.1859
1.13	.8708	.1292	.0646	1.1946
1.14	.8729	.1271	.0634	1.2034
1.15	.8749	.1251	.0621	1.2121
1.16	.8770	.1230	.0609	1.2209
1.17	.8790	.1210	.0596	1.2296
1.18	.8810	.1190	.0584	1.2384
1.19	.8830	.1170	.0573	1.2473
1.20	.8849	.1151	.0561	1.2561
1.21	.8869	.1131	.0550	1.2650
1.22	.8888	.1112	.0538	1.2738
1.23	.8907	.1093	.0527	1.2827
1.24	.8925	.1075	.0517	1.2917
1.25	.8943	.1057	.0506	1.3006
1.26	.8962	.1038	.0495	1.3095
1.27	.8980	.1020	.0485	1.3185
1.28	.8997	.1003	.0475	1.3275
1.29	.9015	.0985	.0465	1.3365
1.30	.9032	.0968	.0455	1.3455
1.31	.9049	.0951	.0446	1.3446
1.32	.9066	.0934	.0436	1.3636
1.33	.9082	.0918	.0427	1.3727
1.34	.9099	.0901	.0418	1.3818
1.35	.9115	.0885	.0409	1.3909
1.36	.9131	.0869	.0400	1.4000
1.37	.9147	.0853	.0392	1.4092
1.38	.9162	.0838	.0383	1.4183
1.39	.9177	.0823	.0375	1.4275
1.40	.9192	.0808	.0367	1.4367
1.41	.9207	.0793	.0359	1.4459
1.42	.9222	.0778	.0351	1.4551
1.43	.9236	.0764	.0343	1.4643
1.44	.9251	.0749	.0336	1.4736
1.45	.9265	.0735	.0328	1.4828
1.46	.9279	.0721	.0321	1.4921
1.47	.9292	.0708	.0314	1.5014
1.48	.9306	.0694	.0307	1.5107
1.49	.9319	.0681	.0300	1.5200
1.50	.9332	.0668	.0293	1.5293
1.51	.9345	.0655	.0286	1.5386
1.52	.9357	.0643	.0280	1.5480
1.53	.9370	.0630	.0274	1.5574

(*continued*)

TABLE A–4 (*continued*)

Standardized Variate z	Probabilities		Partial Expectations	
	F(z)	1 − F(z)	L(z)	L(−z)
1.54	.9382	.0618	.0267	1.5667
1.55	.9394	.0606	.0261	1.5761
1.56	.9406	.0594	.0255	1.5855
1.57	.9418	.0582	.0249	1.5949
1.58	.9429	.0571	.0244	1.6044
1.59	.9441	.0559	.0238	1.6138
1.60	.9460	.0540	.0232	1.6232
1.61	.9463	.0537	.0227	1.6327
1.62	.9474	.0526	.0222	1.6422
1.63	.9484	.0516	.0216	1.6516
1.64	.9495	.0505	.0211	1.6611
1.65	.9505	.0495	.0206	1.6706
1.66	.9515	.0485	.0201	1.6801
1.67	.9525	.0475	.0197	1.6897
1.68	.9535	.0465	.0192	1.6992
1.69	.9545	.0455	.0187	1.7087
1.70	.9554	.0446	.0183	1.7183
1.71	.9564	.0436	.0178	1.7278
1.72	.9573	.0427	.0174	1.7374
1.73	.9582	.0418	.0170	1.7470
1.74	.9591	.0409	.0166	1.7566
1.75	.9599	.0401	.0162	1.7662
1.76	.9608	.0392	.0158	1.7558
1.77	.9616	.0384	.0154	1.7854
1.78	.9625	.0375	.0150	1.7950
1.79	.9633	.0367	.0146	1.8046
1.80	.9641	.0359	.0143	1.8143
1.81	.9649	.0351	.0139	1.8239
1.82	.9656	.0344	.0136	1.8436
1.83	.9664	.0336	.0132	1.8432
1.84	.9671	.0329	.0129	1.8529
1.85	.9678	.0322	.0126	1.8626
1.86	.9685	.0314	.0123	1.8723
1.87	.9693	.0307	.0119	1.8819
1.88	.9699	.0301	.0116	1.8916
1.89	.9706	.0294	.0113	1.9013
1.90	.9713	.0287	.0111	1.9111
1.91	.9719	.0281	.0108	1.9208
1.92	.9726	.0274	.0105	1.9305
1.93	.9732	.0268	.0102	1.9402
1.94	.9738	.0262	.0100	1.9500
1.95	.9744	.0256	.0097	1.9597
1.96	.9750	.0250	.0094	1.9694
1.97	.9756	.0244	.0092	1.9792
1.98	.9761	.0239	.0090	1.9890
1.99	.9767	.0233	.0087	1.9987
2.00	.9772	.0228	.0085	2.0085
2.01	.9778	.0222	.0083	2.0183
2.02	.9783	.0217	.0080	2.0280
2.03	.9788	.0212	.0078	2.0378
2.04	.9793	.0207	.0076	2.0476
2.05	.9798	.0202	.0074	2.0574
2.06	.9803	.0197	.0072	2.0672

Table A–4 (*continued*)

| Standardized Variate | Probabilities | | Partial Expectations | |
z	$F(z)$	$1 - F(z)$	$L(z)$	$L(-z)$
2.07	.9808	.0192	.0072	2.0770
2.08	.9812	.0188	.0068	2.0868
2.09	.9817	.0183	.0066	2.0966
2.10	.9821	.0179	.0065	2.1065
2.11	.9826	.0174	.0063	2.1163
2.12	.9830	.0170	.0061	2.1261
2.13	.9834	.0166	.0060	2.1360
2.14	.9838	.0162	.0058	2.1458
2.15	.9842	.0158	.0056	2.1556
2.16	.9846	.0154	.0055	2.1655
2.17	.9850	.0150	.0053	2.1753
2.18	.9854	.0146	.0052	2.1852
2.19	.9857	.0143	.0050	2.1950
2.20	.9861	.0139	.0049	2.2049
2.21	.9864	.0136	.0048	2.2148
2.22	.9868	.0132	.0046	2.2246
2.23	.9871	.0129	.0045	2.2345
2.24	.9875	.0125	.0044	2.2444
2.25	.9878	.0122	.0042	2.2542
2.26	.9881	.0119	.0041	2.2641
2.27	.9884	.0116	.0040	2.2740
2.28	.9887	.0113	.0039	2.2839
2.29	.9890	.0110	.0038	2.2938
2.30	.9893	.0107	.0037	2.3037
2.31	.9896	.0104	.0036	2.3136
2.32	.9898	.0102	.0035	2.3235
2.33	.9901	.0099	.0034	2.3334
2.34	.9904	.0096	.0033	2.3433
2.35	.9906	.0094	.0032	2.3532
2.36	.9909	.0091	.0031	2.3631
2.37	.9911	.0089	.0030	2.3730
2.38	.9913	.0087	.0029	2.3829
2.39	.9916	.0084	.0028	2.3928
2.40	.9918	.0082	.0027	2.4027
2.41	.9920	.0080	.0026	2.4126
2.42	.9922	.0078	.0026	2.4226
2.43	.9925	.0075	.0025	2.4325
2.44	.9927	.0073	.0024	2.4424
2.45	.9929	.0071	.0023	2.4523
2.46	.9931	.0069	.0023	2.4623
2.47	.9932	.0068	.0022	2.4722
2.48	.9934	.0066	.0021	2.4821
2.49	.9936	.0064	.0021	2.4921
2.50	.9938	.0062	.0020	2.5020
2.51	.9940	.0060	.0019	2.5119
2.52	.9941	.0059	.0019	2.5219
2.53	.9943	.0057	.0018	2.5318
2.54	.9945	.0055	.0018	2.5418
2.55	.9946	.0054	.0017	2.5517
2.56	.9948	.0052	.0017	2.5617
2.57	.9949	.0051	.0016	2.5716
2.58	.9951	.0049	.0016	2.5816

(*continued*)

TABLE A–4 *(concluded)*

Standardized Variate	Probabilities		Partial Expectations	
z	F(z)	1 − F(z)	L(z)	L(−z)
2.59	.9952	.0048	.0015	2.5915
2.60	.9953	.0047	.0015	2.6015
2.61	.9955	.0045	.0014	2.6114
2.62	.9956	.0044	.0014	2.6214
2.63	.9957	.0043	.0013	2.6313
2.64	.9959	.0041	.0013	2.6413
2.65	.9960	.0040	.0012	2.6512
2.66	.9961	.0039	.0012	2.6612
2.67	.9962	.0038	.0012	2.6712
2.68	.9963	.0037	.0011	2.6811
2.69	.9964	.0036	.0011	2.6911
2.70	.9965	.0035	.0011	2.7011
2.71	.9966	.0034	.0010	2.7110
2.72	.9967	.0033	.0010	2.7210
2.73	.9968	.0032	.0010	2.7310
2.74	.9969	.0031	.0009	2.7409
2.75	.9970	.0030	.0009	2.7509
2.76	.9971	.0029	.0009	2.7609
2.77	.9972	.0028	.0008	2.7708
2.78	.9973	.0027	.0008	2.7808
2.79	.9974	.0026	.0008	2.7908
2.80	.9974	.0026	.0008	2.8008
2.81	.9975	.0025	.0007	2.8107
2.82	.9976	.0024	.0007	2.8207
2.83	.9977	.0023	.0007	2.8307
2.84	.9977	.0023	.0007	2.8407
2.85	.9978	.0022	.0006	2.8506
2.86	.9979	.0021	.0006	2.8606
2.87	.9979	.0021	.0006	2.8706
2.88	.9980	.0020	.0006	2.8806
2.89	.9981	.0019	.0006	2.8906
2.90	.9981	.0019	.0005	2.9005
2.91	.9982	.0018	.0005	2.9105
2.92	.9982	.0018	.0005	2.9205
2.93	.9983	.0017	.0005	2.9305
2.94	.9984	.0016	.0005	2.9405
2.95	.9984	.0016	.0005	2.9505
2.96	.9985	.0015	.0004	2.9604
2.97	.9985	.0015	.0004	2.9704
2.98	.9986	.0014	.0004	2.9804
2.99	.9986	.0014	.0004	2.9904
3.00	.9986	.0014	.0004	3.0004

Source: R. G. Brown, *Decision Rules for Inventory Management* (Hinsdale, IL.: Dryden, 1967). Adapted from Table VI, pp. 95–103.

TABLE A–5 Factors d_2 for R Charts

Number of Observations in Subgroup, n	Factor d_2, $d_2 = \dfrac{\overline{R}}{\sigma}$
2	1.128
3	1.693
4	2.059
5	2.326
6	2.534
7	2.704
8	2.847
9	2.970
10	3.078
11	3.173
12	3.258
13	3.336
14	3.407
15	3.472
16	3.532
17	3.588
18	3.640
19	3.689
20	3.735
21	3.778
22	3.819
23	3.858
24	3.895
25	3.931
30	4.086
35	4.213
40	4.322
45	4.415
50	4.498
55	4.572
60	4.639
65	4.699
70	4.755
75	4.806
80	4.854
85	4.898
90	4.939
95	4.978
100	5.015

Note: These factors assume sampling from a normal population.
Source: E. L. Grant and R. S. Leavenworth, *Statistical Quality Control,* 6th ed. (New York: McGraw-Hill, 1988).

TABLE A–6 Factors d_3 and d_4 for R Charts

Number of Observations in Subgroup, n	Factors for R Chart	
	Lower Control Limit d_3	Upper Control Limit d_4
2	0	3.27
3	0	2.57
4	0	2.28
5	0	2.11
6	0	2.00
7	.08	1.92
8	.14	1.86
9	.18	1.82
10	.22	1.78
11	.26	1.74
12	.28	1.72
13	.31	1.69
14	.33	1.67
15	.35	1.65
16	.36	1.64
17	.38	1.62
18	.39	1.61
19	.40	1.60
20	.41	1.59

Note: These factors assume sampling from a normal population. They give the upper and the lower control limits for an R chart as follows:

$$\text{LCL} = d_3 R$$
$$\text{UCL} = d_4 R$$

Source: E. L. Grant and R. S. Leavenworth, *Statistical Quality Control,* 6th ed. (New York: McGraw-Hill, 1988).

Index

M